Digital Design Using VHDL

A Systems Approach

This introductory textbook provides students with a system-level perspective and the tools they need to understand, analyze, and design digital systems. It goes beyond the design of simple combinational and sequential modules to show how such modules are used to build complete systems.

- All the essential topics needed to understand modern design practice are covered, including:
  - Design and analysis of combinational and sequential modules
  - Composition of combinational and sequential modules
  - Data and control partitioning
  - Factoring and composition of finite-state machines
  - Interface specification
  - System timing
  - Synchronization
- Teaches how to write VHDL-2008 HDL in a productive and maintainable style that enables CAD tools to do much of the tedious work.
- Covers the fundamentals of logic design, describing an efficient method to design combinational logic and state machines both manually and using modern CAD tools.

A complete introduction to digital design is given through clear explanations, extensive examples, and online VHDL files. The teaching package is completed with lecture slides, labs, and a solutions manual for instructors (available via www.cambridge.org/dallyvhdl). Assuming no previous digital knowledge, this textbook is ideal for undergraduate digital design courses that will prepare students for modern digital practice.

William J. Dally is the Willard R. and Inez Kerr Bell Professor of Engineering at Stanford University and Chief Scientist at NVIDIA Corporation. He and his group have developed system architecture, network architecture, signaling, routing, and synchronization technology that can be found in most large parallel computers today. He is a Member of the National Academy of Engineering, a Fellow of the IEEE, a Fellow of the ACM, and a Fellow of the American Academy of Arts and Sciences. He has received numerous honors, including the ACM Eckert-Mauchly Award, the IEEE Seymour Cray Award, and the ACM Maurice Wilkes Award.

R. Curtis Harting is a Software Engineer at Google and holds a Ph.D. from Stanford University. He graduated with honors in 2007 from Duke University with a B.S.E., majoring in Electrical & Computer Engineering and Computer Science. He received his M.S. in 2009 from Stanford University.

Tor M. Aamodt is an Associate Professor in the Department of Electrical and Computer Engineering at the University of British Columbia. Alongside his graduate students, he developed the GPGPU-Sim simulator. Three of his papers related to the architecture of general purpose GPUs have been selected as "Top Picks" by *IEEE Micro Magazine* and one as a "Research Highlight" by *Communications of the ACM magazine*. He was a Visiting Associate Professor in the Computer Science Department at Stanford University during his 2012–2013 sabbatical, and from 2004 to 2006 he worked at NVIDIA on the memory system architecture ("framebuffer") of the GeForce 8 Series GPU.

"Dally and Harting blend circuit and architecture design in a clear and constructive manner on the basis of their exceptional experience in digital design."

"Students will discover a modern and effective way to understand the fundamental underpinning of digital design, by being exposed to the different abstraction levels and views of computing systems."

Giovanni De Micheli, *EPFL Switzerland*

"Bill and Curt have combined decades of academic and industry experience to produce a textbook that teaches digital system design from a very practical perspective without sacrificing the theoretical understanding needed to train tomorrow's engineers. Their approach pushes students to understand not just what they are designing, but also what they are building. By presenting key advanced topics, such as synthesis, delay and logical effort, and synchronization, at the introductory level, this book is in the rare position of providing both practical advice and deep understanding. In doing so, this book will prepare students well even as technology, tools, and techniques change in the future."

David Black-Schaffer, *Uppsala University*

"Everything you would expect from a book on digital design from Professor Dally. Decades of practical experience are distilled to provide the tools necessary to design and compose complete digital systems. A clear and well-written text that covers the basics and system-level issues equally well. An ideal starting point for the microprocessor and SoC designers of the future!"

Robert Mullins, *University of Cambridge and the Raspberry Pi Foundation*

"This textbook sets a new standard for how digital system design is taught to undergraduates. The practical approach and concrete examples provide a solid foundation for anyone who wants to understand or design modern complex digital systems."

Steve Keckler, *The University of Texas at Austin*

"This book not only teaches how to do digital design, but more importantly shows how to do *good* design. It stresses the importance of modularization with clean interfaces, and the importance of producing digital artifacts that not only meet their specifications, but which can also be easily understood by others. It uses an aptly chosen set of examples and the Verilog code used to implement them."

"It includes a section on the design of asynchronous logic, a topic that is likely to become increasingly important as energy consumption becomes a primary concern in digital systems."

"The final appendix on Verilog coding style is particularly useful. This book will be valuable not only to students, but also to practitioners in the area. I recommend it highly."

Chuck Thacker, *Microsoft*

"A terrific book with a terrific point-of-view of systems. Everything interesting – and awful – that happens in digital design happens because engineers must integrate ideas from bits to blocks, from signals to CPUs. The book does a great job of focusing on the important stuff, moving from foundations to systems, with the right amount of HDL (Verilog) focus to make everything practical and relevant."

Rob A. Rutenbar, *University of Illinois at Urbana-Champaign*

Digital Design Using VHDL

A Systems Approach

WILLIAM J. DALLY
Stanford University

R. CURTIS HARTING
Google, Inc.

TOR M. AAMODT
The University of British Columbia

CAMBRIDGE
UNIVERSITY PRESS

CAMBRIDGE
UNIVERSITY PRESS

University Printing House, Cambridge CB2 8BS, United Kingdom

One Liberty Plaza, 20th Floor, New York, NY 10006, USA

477 Williamstown Road, Port Melbourne, VIC 3207, Australia

314-321, 3rd Floor, Plot 3, Splendor Forum, Jasola District Centre, New Delhi - 110025, India

79 Anson Road, #06-04/06, Singapore 079906

Cambridge University Press is part of the University of Cambridge.

It furthers the University's mission by disseminating knowledge in the pursuit of education, learning and research at the highest international levels of excellence.

www.cambridge.org
Information on this title: www.cambridge.org/9781107098862

First published 2016
Reprinted 2018

A catalogue record for this publication is available from the British Library

Library of Congress Cataloging in Publication data
Dally, William J., author.
Digital design using VHDL : a systems approach / William J. Dally, Stanford University, California, R. Curtis Harting, Google, Inc., New York, Tor M. Aamodt, The University of British Columbia.
pages cm
Includes bibliographical references and index.
ISBN 978-1-107-09886-2 (Hardback : alk. paper)
1. Digital integrated circuits–Computer-aided design. 2. Electronic digital computers–Computer-aided design. 3. Digital electronics–Data processing. 4. VHDL (Computer hardware description language) I. Harting, R. Curtis, author. II. Aamodt, Tor M., author. III. Title.
TK7868.D5D3285 2015
621.38150285′5133–dc23 2015021269

ISBN 978-1-107-09886-2 Hardback

Additional resources for this publication at www.cambridge.org/9781107098862

CONTENTS

Part III Arithmetic circuits

Part IV Synchronous sequential logic

Part VI System design

PREFACE

This book is intended to teach an undergraduate student to understand and design digital *systems*. It teaches the skills needed for current industrial digital system design using a hardware description language (VHDL) and modern CAD tools. Particular attention is paid to system-level issues, including factoring and partitioning digital systems, interface design, and interface timing. Topics needed for a deep understanding of digital circuits, such as timing analysis, metastability, and synchronization, are also covered. Of course, we cover the manual design of combinational and sequential logic circuits. However, we do not dwell on these topics because there is far more to digital system design than designing such simple modules.

Upon completion of a course using this book, students should be prepared to practice digital design in industry. They will lack experience, but they will have all of the tools they need for contemporary practice of this noble art. The experience will come with time.

This book has grown out of more than 25 years of teaching digital design to undergraduates (CS181 at Caltech, 6.004 at MIT, EE121 and EE108A at Stanford). It is also motivated by 35 years of experience designing digital systems in industry (Bell Labs, Digital Equipment, Cray, Avici, Velio Communications, Stream Processors, and NVIDIA). It combines these two experiences to teach what students need to know to function in industry in a manner that has been proven to work on generations of students. The VHDL guide in Appendix B is informed by nearly a decade of teaching VHDL to undergraduates at UBC (EECE 353 and EECE 259).

We wrote this book because we were unable to find a book that covered the system-level aspects of digital design. The vast majority of textbooks on this topic teach the manual design of combinational and sequential logic circuits and stop. While most texts today use a hardware description language, the vast majority teach a TTL-esque design style that, while appropriate in the era of 7400 quad NAND gate parts (the 1970s), does not prepare a student to work on the design of a three-billion-transistor GPU. Today's students need to understand how to factor a state machine, partition a design, and construct an interface with correct timing. We cover these topics in a simple way that conveys insight without getting bogged down in details.

Outline of the book

A flow chart showing the organization of the book and the dependences between chapters is shown in Figure 1. The book is divided into an introduction, five main sections, and chapters about style and verification. Appendix B provides a summary of VHDL-2008 syntax.

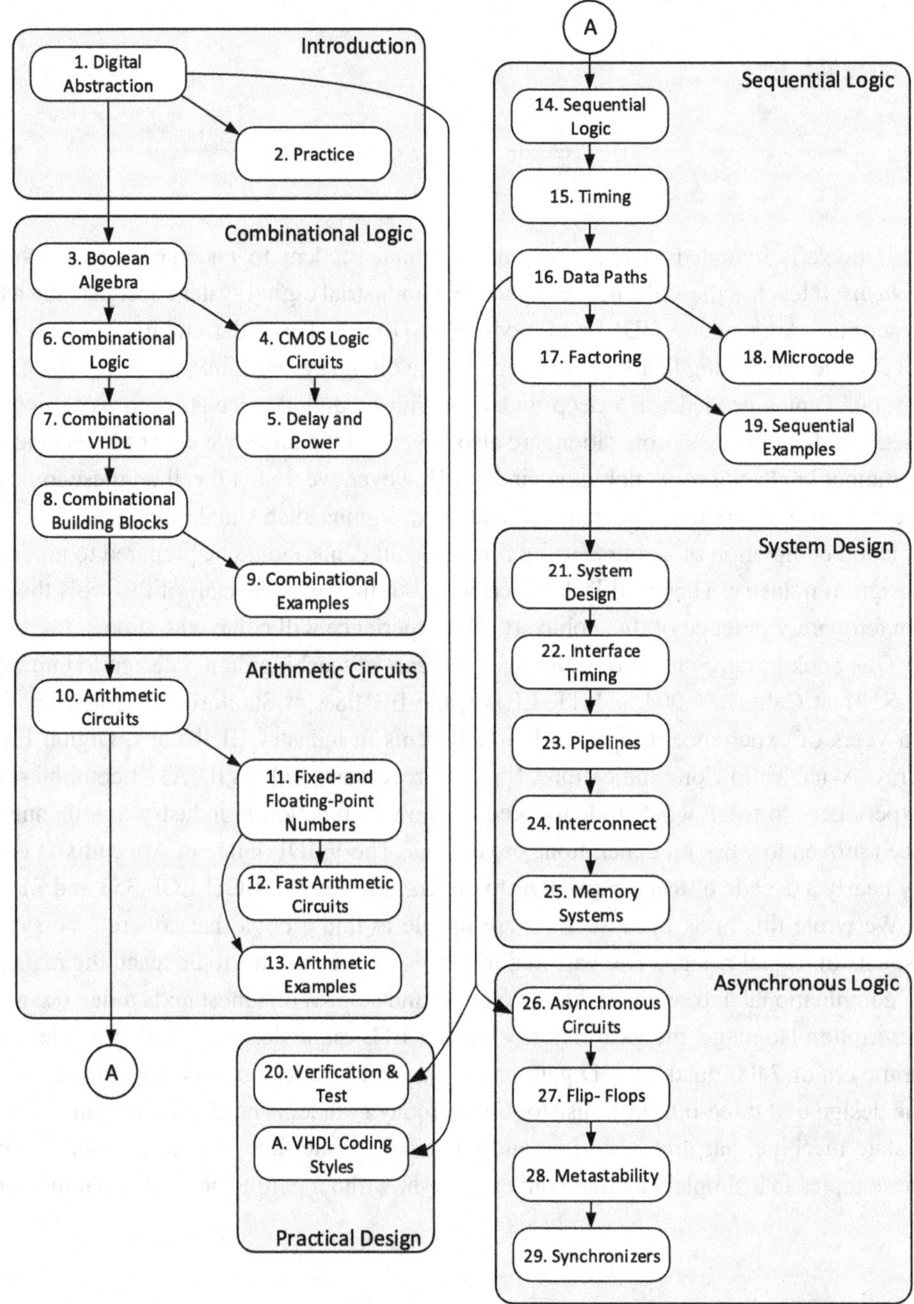

Figure 1. Organization of the book and dependences between chapters.

Part I Introduction

Chapter 1 introduces digital systems. It covers the representation of information as digital signals, noise margins, and the role of digital logic in the modern world. The practice of digital design in industry is described in Chapter 2. This includes the design process, modern implementation technologies, computer-aided design tools, and Moore's law.

Part II Combinational logic

Chapters 3–9 deal with combinational logic circuits – digital circuits whose outputs depend only on the current values of their inputs. Boolean algebra, the theoretical underpinning of logic design, is discussed in Chapter 3. Switching logic and CMOS gate circuits are introduced in Chapter 4. Chapter 5 introduces simple models for calculating the delay and power of CMOS circuits. Manual methods for designing combinational circuits from basic gates are described in Chapter 6. Chapter 7 discusses how to automate the design process by coding behavioral descriptions of combinational logic in the VHDL hardware description language. Building blocks for combinational logic, decoders, multiplexers, etc. are described in Chapter 8, and several examples of combinational design are given in Chapter 9.

Part III Arithmetic circuits

Chapters 10–13 describe number systems and arithmetic circuits. Chapter 10 describes the basics of number representation and arithmetic circuits that perform the *four functions* $+$, $-$, $\times$, and $\div$ on integers. Fixed-point and floating-point number representations and their accuracy are presented in Chapter 11. This chapter includes a discussion of floating-point unit design. Techniques for building fast arithmetic circuits, including carry look-ahead, Wallace trees, and Booth recoding, are described in Chapter 12. Finally, examples of arithmetic circuits and systems are presented in Chapter 13.

Part IV Synchronous sequential logic

Chapters 14–19 describe synchronous sequential logic circuits – sequential circuits whose state changes only on clock edges – and the process of designing finite-state machines. After describing the basics in Chapter 14, timing constraints are covered in Chapter 15. The design of *datapath* sequential circuits – whose behavior is described by an equation rather than a state table – is the topic of Chapter 16. Chapter 17 describes how to factor complex state machines into several smaller, simpler state machines. The concept of stored program control, and how to build finite-state machines using microcoded engines, is described in Chapter 18. This section closes with a number of examples in Chapter 19.

Part V Practical design

Chapter 20 and the Appendix discuss two important aspects of working on digital design projects. The process of verifying the correctness of logic and testing that it works after manufacturing are the topics of Chapter 20. The Appendix teaches the student proper VHDL coding style. It is a style that is readable, maintainable, and enables CAD tools to produce optimized hardware. Students should read this chapter before, during, and after writing their own VHDL.

Part VI System design

Chapters 21–25 discuss system design and introduce a systematic method for the design and analysis of digital systems. A six-step process for system design is introduced in Chapter 21. System-level timing and conventions for the timing of interfaces are discussed in Chapter 22. Chapter 23 describes pipelining of modules and systems, and includes several example pipelines. System interconnects including buses, crossbar switches, and networks are described in Chapter 24. A discussion of memory systems is given in Chapter 25.

Part VII Asynchronous logic

Chapters 26–29 discuss asynchronous sequential circuits – circuits whose state changes in response to any input change, without waiting for a clock edge. The basics of asynchronous design including flow-table analysis and synthesis and the problem of races are introduced in Chapter 26. Chapter 27 gives an example of these techniques, analyzing flip-flops and latches as asynchronous circuits. The problem of *metastability* and synchronization failure is described in Chapter 28. This section, and the book, closes with a discussion of synchronizer design – how to design circuits that safely move signals across asynchronous boundaries – in Chapter 29.

Teaching using this book

This book is suitable for use in a one-quarter (10-week) or one-semester (13-week) introductory course on digital systems design. It can also be used as the primary text of a second, advanced, course on digital systems.

There need not be any formal prerequisites for a course using this book. A good understanding of high-school-level mathematics is the only required preparation. Except for Chapters 5 and 28, the only place derivatives are used, the material does not require a knowledge of calculus. At Stanford, E40 (Introduction to Electrical Engineering) is a prerequisite for EE108A (Digital Systems I), but students often take EE108A without the prerequisite with no problems.

A one-quarter introductory course on digital systems design covers the material in Chapters 1, 3, 6, 7, 8, 10, (11), 14, 15, 16, (17), 21, 22, (23), 26, 28, and 29. For the one-quarter course we omit the details of CMOS circuits (Chapters 4 and 5), microcode (Chapter 18), and the more advanced systems topics (Chapters 24 and 25). The three chapters in parentheses are optional and can be skipped to give a slightly slower-paced course. In offerings of this course at Stanford, we typically administer two midterm examinations: one after covering Chapter 11, and the second after covering Chapter 22.

A one-semester introductory course on digital systems can use the three additional weeks to include the material on CMOS circuits and a few of the more advanced systems topics. A typical semester-long course covers the material in Chapters 1, 2, 3,

4, (5), 6, 7, 8, 9, 10, (11), 13, 14, 15, 16, (17), (18), (19), 21, 22, (23), (24), (25), 26, (27), 28, and 29.

This book can be used for an advanced course on digital systems design. Such a course covers the material from the introductory courses in more depth and includes advanced topics that were omitted from the introductory courses. Such a course usually includes a significant student project.

Materials

To support teaching with this book, the course website includes teaching materials: lecture slides, a series of laboratories, and solutions to selected exercises. The laboratories are intended to reinforce the material in the course and can be performed via simulation or a combination of simulation and implementation on FPGAs.

ACKNOWLEDGMENTS

We are deeply indebted to the many people who have made contributions to the creation of this book. This book has evolved over many years of teaching digital design at MIT (6.004) and Stanford (EE108A). We thank the many generations of students who took early versions of this class and provided feedback that led to constant refinement of our approach. Professors Subhasish Mitra, Phil Levis, and My Le have taught at Stanford using early versions of this material, and have provided valuable comments that led to many improvements. The course and book benefited from contributions by many great teaching assistants over the years. Paul Hartke, David Black-Shaffer, Frank Nothaft, and David Schneider deserve special thanks. Frank also deserves thanks for contributing to the exercise solutions. Teaching 6.004 at MIT with Gill Pratt, Greg Papadopolous, Steve Ward, Bert Halstead, and Anant Agarwal helped develop the approach to teaching digital design that is captured in this book. An early draft of the VHDL edition of this book was used for EECE 259 at the University of British Columbia (UBC). We thank the students who provided feedback that led to refinements in this version. The treatment of VHDL-2008 in Appendix B has been informed by several years of experience teaching earlier VHDL versions in EECE 353 at UBC using a set of slides originally developed by Professor Steve Wilton. Steve also provided helpful feedback on an early draft of the VHDL edition.

Julie Lancashire and Kerry Cahill at Cambridge University Press helped throughout the original Verilog edition, and Julie Lancashire, Karyn Bailey, and Jessica Murphy at Cambridge Press helped with the current VHDL edition. We thank Irene Pizzie for careful copy editing of the original Verilog edition and Abigail Jones for shepherding the original Verilog edition through the sometimes difficult passage from manuscript to finished project. We thank Steven Holt for careful copy editing of the present VHDL edition you see before you.

Finally, our families: Sharon, Jenny, Katie, and Liza Dally, and Jacki Armiak, Eric Harting, and Susanna Temkin, and Dayna and Ethan Aamodt have offered tremendous support and made significant sacrifices so we could have time to devote to writing.

Part I

Introduction

1 The digital abstraction

Digital systems are pervasive in modern society. Some uses of digital technology are obvious – such as a personal computer or a network switch. However, there are also many other applications of digital technology. When you speak on the phone, in almost all cases your voice is being digitized and transmitted via digital communications equipment. When you listen to an audio file, the music, recorded in digital form, is processed by digital logic to correct errors and improve the audio quality. When you watch TV, the image is transmitted in a digital format and processed by digital electronics. If you have a DVR (digital video recorder) you are recording video in digital form. DVDs are compressed digital video recordings. When you play a DVD or stream a movie, you are digitally decompressing and processing the video. Most communication radios, such as cell phones and wireless networks, use digital signal processing to implement their modems. The list goes on.

Most modern electronics uses analog circuitry only at the edge – to interface to a physical sensor or actuator. As quickly as possible, signals from a sensor (e.g., a microphone) are converted into digital form. All real processing, storage, and transmission of information is done digitally. The signals are converted back to analog form only at the output – to drive an actuator (e.g., a speaker) or control other analog systems.

Not so long ago, the world was not as digital. In the 1960s digital logic was found only in expensive computer systems and a few other niche applications. All TVs, radios, music recordings, and telephones were analog.

The shift to digital was enabled by the scaling of integrated circuits. As integrated circuits became more complex, more sophisticated signal processing became possible. Complex techniques such as modulation, error correction, and compression were not feasible in analog technology. Only digital logic, with its ability to perform computations without accumulating noise and its ability to represent signals with arbitrary precision, could implement these signal processing algorithms.

In this book we will look at how the digital systems that form such a large part of our lives function and how they are designed.

1.1 DIGITAL SIGNALS

Digital systems store, process, and transport information in digital form. Digital information is represented as discrete symbols that are encoded into ranges of a physical quantity. Most often we represent information with just two symbols, "0" and "1," and encode these symbols into voltage ranges as shown in Figure 1.1. Any voltage in the ranges labeled "0" and "1" represents a "0" or "1" symbol, respectively. Voltages between these two ranges, in the region labeled "?,"

Table 1.1. Encoding of binary signals for 2.5 V LVCMOS logic

Signals with voltage in $[-0.3, 0.7]$ are considered to be a 0. Signals with voltage in $[1.7, 2.8]$ are considered to be a 1. Voltages in $[0.7, 1.7]$ are undefined. Voltages outside $[-0.3, 2.8]$ may cause permanent damage.

Parameter	Value	Description
V_{min}	−0.3 V	absolute minimum voltage below which damage occurs
V_0	0.0 V	nominal voltage representing logic "0"
V_{OL}	0.2 V	maximum output voltage representing logic "0"
V_{IL}	0.7 V	maximum voltage considered to be a logic "0" by a module input
V_{IH}	1.7 V	minimum voltage considered to be a logic "1" by a module input
V_{OH}	2.1 V	minimum output voltage representing logic "1"
V_1	2.5 V	nominal voltage representing logic "1"
V_{max}	2.8 V	absolute maximum voltage above which damage occurs

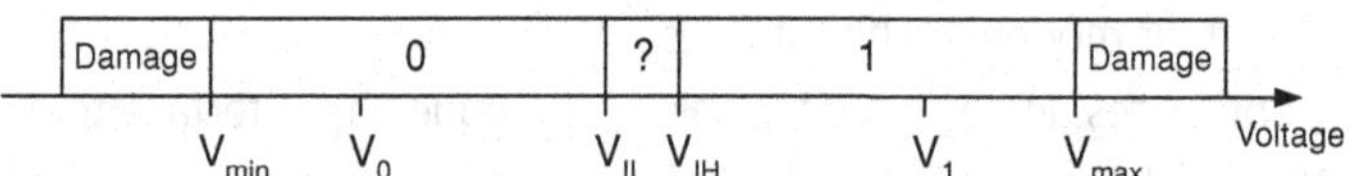

Figure 1.1. Encoding of two symbols, 0 and 1, into voltage ranges. Any voltage in the range labeled 0 is considered to be a 0 symbol. Any voltage in the range labeled 1 is considered to be a 1 symbol. Voltages between the 0 and 1 ranges (the ? range) are undefined and represent neither symbol. Voltages outside the 0 and 1 ranges may cause permanent damage to the equipment receiving the signals.

are undefined and represent neither symbol. Voltages outside the ranges, below the "0" range, or above the "1" range are not allowed and may permanently damage the system if they occur. We call a signal encoded in the manner shown in Figure 1.1 a *binary* signal because it has two valid states.

Table 1.1 shows the JEDEC JESD8-5 standard [62] for encoding a binary digital signal in a system with a 2.5 V power supply. Using this standard, any signal with a voltage between −0.3 V and 0.7 V is considered to be a "0," and a signal with a voltage between 1.7 V and 2.8 V is considered to be a "1." Signals that do not fall into these two ranges are undefined. If a signal is below −0.3 V or above 2.8 V, it may cause damage.[1]

Digital systems are not restricted to binary signals. One can generate a digital signal that can take on three, four, or any finite number of discrete values. However, there are few advantages to using more than two values, and the circuits that store and operate on binary signals are simpler and more robust than their multi-valued counterparts. Thus, except for a few niche applications, binary signals are universal in digital systems today.

Digital signals can also be encoded using physical quantities other than voltage. Almost any physical quantity that can be easily manipulated and sensed can be used to represent a digital

[1] The actual specification for V_{max} is $V_{DD} + 0.3$, where V_{DD}, the power supply, is allowed to vary between 2.3 and 2.7 V.

signal. Systems have been built using electrical current, air or fluid pressure, and physical position to represent digital signals. However, the tremendous capability of manufacturing complex systems as low-cost CMOS integrated circuits has made voltage signals universal.

1.2 DIGITAL SIGNALS TOLERATE NOISE

The main reason why digital systems have become so pervasive, and what distinguishes them from *analog* systems, is that they can process, transport, and store information without it being distorted by noise. This is possible because of the discrete nature of digital information. A binary signal represents either a 0 or a 1. If you take the voltage that represents a 1, V_1, and disturb it with a small amount of noise, ϵ, it still represents a 1. There is no loss of information with the addition of noise, until the noise becomes large enough to push the signal out of the 1 range. In most systems it is easy to bound the noise to be less than this value.

Figure 1.2 compares the effect of noise on an analog system (Figure 1.2(a)) and on a digital system (Figure 1.2(b)). In an analog system, information is represented by an analog voltage, V. For example, we might represent temperature (in degrees Fahrenheit) with voltage according to the relation $V = 0.2(T-68)$. So a temperature of 72.5 °F is represented by a voltage of 900 mV. This representation is continuous; every voltage corresponds to a different temperature. Thus, if we disturb the signal V with a noise voltage ϵ, the resulting signal $V + \epsilon$ corresponds to a different temperature. If $\epsilon = 100$ mV, for example, the new signal $V + \epsilon = 1$ V corresponds to a temperature of 73 °F ($T = 5V + 68$), which is different from the original temperature of 72.5 °F.

In a digital system, each bit of the signal is represented by a voltage, V_1 or V_0 depending on whether the bit is 1 or 0. If a noise source perturbs a digital 1 signal V_1, for example, as shown in Figure 1.2(b), the resulting voltage $V_1 + \epsilon$ still represents a 1 and applying a function to this noisy signal gives the same result as applying a function to the original signal. Moreover, if a temperature of 72 °F is represented by a three-bit digital signal with value 010 (see Figure 1.7(c)), the signal still represents a temperature of 72 °F even after all three bits of the signal have been disturbed by noise – as long as the noise is not so great as to push any bit of the signal out of the valid range.

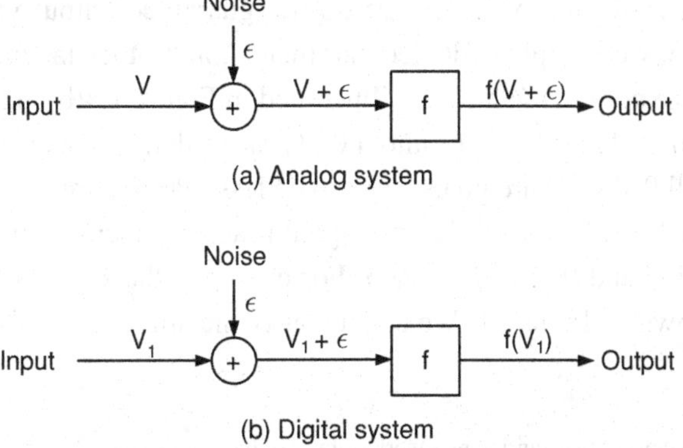

Figure 1.2. Effects of noise in analog and digital systems. (a) In an analog system, perturbing a signal V by noise ϵ results in a degraded signal $V + \epsilon$. Operating on this degraded signal with a function f gives a result $f(V + \epsilon)$ that is different from the result of operating on the signal without noise. (b) In a digital system, adding noise ϵ to a signal V_1 representing a symbol, 1, gives a signal $V_1 + \epsilon$ that still represents the symbol 1. Operating on this signal with a function f gives the same result $f(V_1)$ as operating on the signal without the noise.

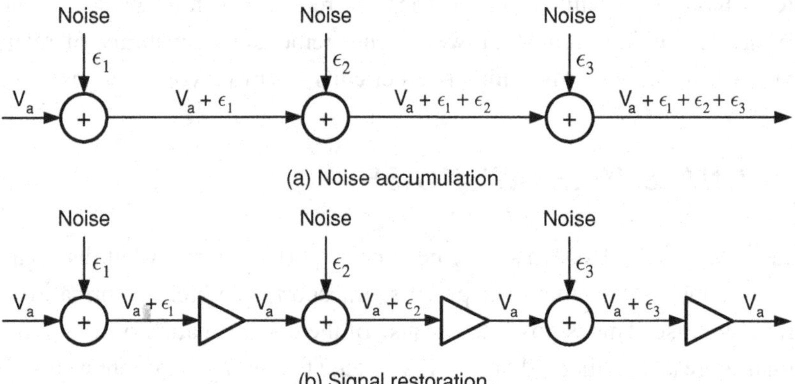

Figure 1.3. Restoration of digital signals. (a) Without restoration, signals accumulate noise and will eventually accumulate enough noise to cause an error. (b) By restoring the signal to its proper value after each operation, noise is prevented from accumulating.

To prevent noise from accumulating to the point where it pushes a digital signal out of the valid 1 or 0 range, we periodically restore digital signals as illustrated in Figure 1.3. After transmitting, storing, and retrieving, or operating on a digital signal, it may be disturbed from its nominal value V_a (where a is 0 or 1) by some noise ϵ_i. Without restoration (Figure 1.3(a)), the noise accumulates after each operation and eventually will overwhelm the signal. To prevent accumulation, we restore the signal after each operation. The restoring device, which we call a *buffer*, outputs V_0 if its input lies in the 0 range and V_1 if its output lies in the 1 range. The buffer, in effect, restores the signal to be a pristine 0 or 1, removing any additive noise.

This capability to restore a signal to its noiseless state after each operation enables digital systems to carry out complex high-precision processing. Analog systems are limited to performing a small number of operations on relatively low-precision signals because noise is accumulated during each operation. After a large number of operations, the signal is swamped by noise. Since all voltages are valid analog signals, there is no way to restore the signal to a noiseless state between operations. Analog systems are also limited in precision. They cannot represent a signal with an accuracy finer than the background noise level. Digital systems can perform an indefinite number of operations, and, as long as the signal is restored after each operation, no noise is accumulated. Digital systems can also represent signals of arbitrary precision without corruption by noise.[2]

In practice, buffers and other restoring logic devices do not guarantee output voltages of exactly V_0 or V_1. Variations in power supplies, device parameters, and other factors lead the outputs to vary slightly from these nominal values. As illustrated in Figure 1.4(b), all restoring logic devices guarantee that their 0 (1) outputs fall into a 0 (1) range that is narrower than the input 0 (1) range. Specifically, all 0 signals are guaranteed to be less than V_{OL} and all 1 signals are guaranteed to be greater than V_{OH}. To ensure that the signal is able to tolerate some amount of noise, we insist that $V_{OL} < V_{IL}$ and that $V_{IH} < V_{OH}$. For example, the values of V_{OL} and V_{OH} for 2.5 V LVCMOS are shown in Table 1.1. We can quantify the amount of noise that can

[2] Of course, one is limited by analog input devices in acquiring real-world signals of high precision.

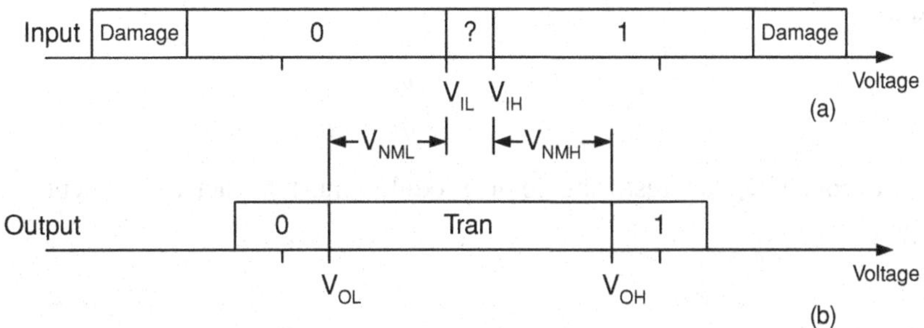

Figure 1.4. Input and output voltage ranges. (a) Inputs of logic modules interpret signals as shown in Figure 1.1. (b) Outputs of logic modules restore signals to narrower ranges of valid voltages.

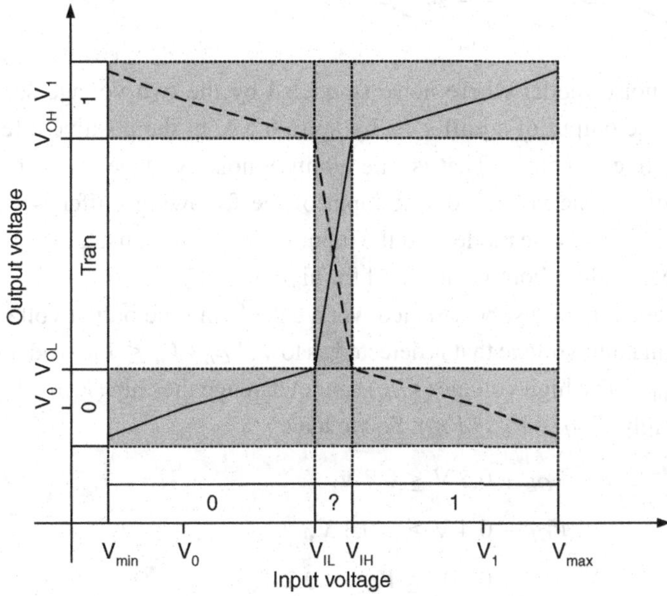

Figure 1.5. DC transfer curve for a logic module. For an input in the valid ranges, $V_{min} \leq V_{in} \leq V_{IL}$ or $V_{IH} \leq V_{in} \leq V_{max}$, the output must be in the valid output ranges $V_{out} \leq V_{OL}$ or $V_{OH} \leq V_{out}$. Thus, all valid curves must stay in the shaded region. This requires that the module have gain > 1 in the invalid input region. The solid curve shows a typical transfer function for a non-inverting module. The dashed curve shows a typical transfer function for an inverting module.

be tolerated as the *noise margins* of the signal:

$$\begin{aligned} V_{NMH} &= V_{OH} - V_{IH}, \\ V_{NML} &= V_{IL} - V_{OL}. \end{aligned} \tag{1.1}$$

While one might assume that a bigger noise margin would be better, this is not necessarily the case. Most noise in digital systems is induced by signal transitions, and hence tends to be proportional to the signal swing. Thus, what is really important is the *ratio* of the noise margin to the signal swing, $V_{NM}/(V_1 - V_0)$, rather than the absolute magnitude of the noise margin.

Figure 1.5 shows the relationship between DC input voltage and output voltage for a logic module. The horizontal axis shows the module input voltage and the vertical axis shows the module output voltage. To conform to our definition of *restoring*, the transfer curve for all modules must lie entirely within the shaded region of the figure so that an input signal in the valid 0 or 1 range will result in an output signal in the narrower output 0 or 1 range. Non-inverting modules, like the buffer of Figure 1.3, have transfer curves similar to the solid line. Inverting modules have transfer curves similar to the dashed line. In either case, gain is required to implement a restoring logic module. The absolute value of the maximum slope of the signal is bounded by

$$\max \left| \frac{dV_{\text{out}}}{dV_{\text{in}}} \right| \geq \frac{V_{OH} - V_{OL}}{V_{IH} - V_{IL}}. \tag{1.2}$$

From this we conclude that restoring logic modules must be active elements capable of providing gain.

EXAMPLE 1.1 Noise toleration

Figure 1.6. Noise model for Example 1.1.

Figure 1.6 illustrates a noise model where noise (modeled by the two voltage sources) can displace the voltage on the output of a buffer V_a by up to 0.5 V in the positive direction and up to 0.4 V in the negative direction. That is, the additive noise voltage $V_n \in [-0.4, 0.5]$. For an output voltage of V_a, the voltage on the input of the following buffer is $V_a + V_n \in [V_a - 0.4, V_a + 0.5]$. Using this noise model and the input and damage constraints of Table 1.1, calculate the range of legal values both for low and for high outputs.

There are four constraints that must be satisfied when calculating the output voltage. A low output (V_{OL}) must give an input voltage that is detected as low ($V_{OL}+V_n \leq V_{IL}$) and not damage the chip ($V_{OL}-V_n \geq V_{\min}$). The high voltage (V_{OH}) cannot damage the chip ($V_{OH}+V_n \leq V_{\max}$) and must be sensed as high ($V_{OH} - V_n \geq V_{IH}$). So we have

$$V_{OL} + 0.5\text{ V} \leq 0.7\text{ V},$$
$$V_{OL} - 0.4\text{ V} \geq -0.3\text{ V},$$
$$0.1\text{ V} \leq V_{OL} \leq 0.2\text{ V},$$

$$V_{OH} + 0.5\text{ V} \leq 2.8\text{ V},$$
$$V_{OH} - 0.4\text{ V} \geq 1.7\text{ V},$$
$$2.1\text{ V} \leq V_{OH} \leq 2.3\text{ V}.$$

The output voltage from each buffer must be between 0.1 V and 0.2 V to represent “0” and between 2.1 V and 2.3 V to represent “1.” Although not acceptable with this noise model, almost all circuits will run with nominal output voltages of 0 V and V_{DD} (the supply) to represent 0 and 1.

1.3 DIGITAL SIGNALS REPRESENT COMPLEX DATA

Some information is naturally binary in nature and can be represented with a single binary digital signal (Figure 1.7(a)). Truth propositions or predicates fall into this category. For example, a single signal can indicate that a door is open, a light is on, a seatbelt is buckled, or a button is pressed.

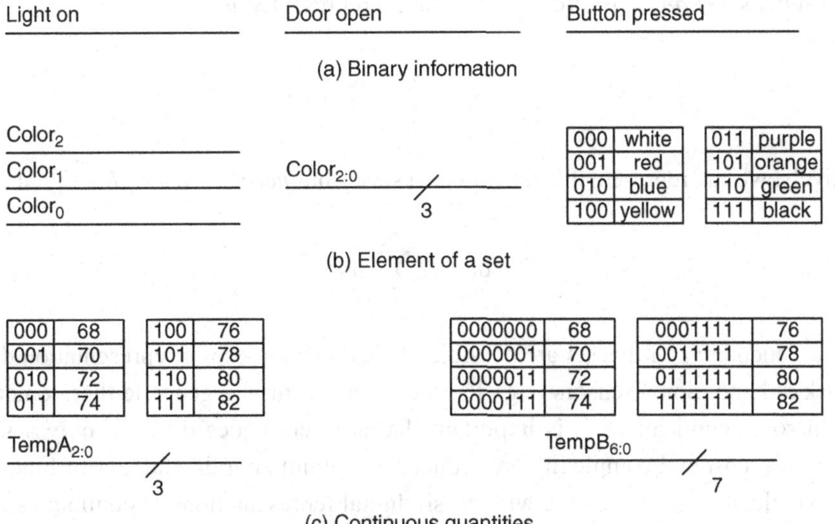

Figure 1.7. Representing information with digital signals. (a) Binary-valued predicates are represented by a single-bit signal. (b) Elements of sets with more than two elements are represented by a group of signals. In this case one of eight colors is denoted by a three-bit signal $Color_{2:0}$. (c) A continuous quantity, like temperature, is *quantized*, and the resulting set of values is encoded by a group of signals. Here one of eight temperatures can be encoded as a three-bit signal $TempA_{2:0}$ or as a seven-bit *thermometer-coded* signal $TempB_{6:0}$ with at most one transition from 0 to 1.

By convention, we often consider a signal to be "true" when the voltage is high. This need not be the case, as nothing precludes using low voltages to represent the above conditions. Throughout the book we will try to make it clear when we are using this low–true convention. The signal description in such cases will often be changed instead, e.g., "a seatbelt is unbuckled."

Often we need to represent information that is not binary in nature: a day of the year, the value and suit of a playing card, the temperature in a room, or a color. We encode information with more than two natural states using a group of binary signals (Figure 1.7(b)). The elements of a set with N elements can be represented by a signal with $n = \lceil \log_2 N \rceil$ bits. For example, the eight colors shown in Figure 1.7(b) can be represented by three one-bit signals, $Color_0$, $Color_1$, and $Color_2$. For convenience we refer to this group of three signals as a single multi-bit signal $Color_{2:0}$. In a circuit or schematic diagram, rather than drawing three lines for these three signals, we draw a single line with a slash indicating that it is a multi-bit signal and the number "3" near the slash to indicate that it is composed of three bits.[3]

Continuous quantities, such as voltage, temperature, and pressure, are encoded as digital signals by *quantizing* them. This reduces the problem to one of representing elements of a set. Suppose, for example, that we need to represent temperatures between 68 °F and 82 °F and that it suffices to resolve the temperature to an accuracy of 2 °F. We quantize this temperature range into eight discrete values as shown in Figure 1.7(c). We can represent this range with binary

[3] This notation for multi-bit signals is discussed in more detail in Section 8.1.

weighted signals $TempA_{2:0}$, where the temperature represented is

$$T = 68 + 2\sum_{i=0}^{2} 2^i TempA_i. \tag{1.3}$$

Alternatively, we can represent this range with a seven-bit *thermometer-coded* signal $TempB_{6:0}$:

$$T = 68 + 2\sum_{i=0}^{6} TempB_i. \tag{1.4}$$

Many other encodings of this set are possible. A designer chooses a representation depending on the task at hand. Some sensors (e.g., thermometers) naturally generate thermometer-coded signals. In some applications it is important that adjacent codes differ in only a single bit. At other times, cost and complexity are reduced by minimizing the number of bits needed to represent an element of the set. We will revisit digital representations of continuous quantities when we discuss numbers and arithmetic in Chapter 10.

1.3.1 Representing the day of the year

Suppose we wish to represent the day of the year with a digital signal. (We will ignore for now the problem of leap years.) The signal is to be used for operations that include determining the next day (i.e., given the representation of today, compute the representation of tomorrow), testing whether two days are in the same month, determining whether one day comes before another, and whether a day is a particular day of the week.

One approach is to use a $\lceil \log_2 365 \rceil = 9$-bit signal that represents the integers from 0 to 364, where 0 represents January 1 and 364 represents December 31. This representation is compact (you cannot do it in fewer than nine bits), and it makes it easy to determine whether one day comes before another. However, it does not facilitate the other two operations we need to perform. To determine the month a day corresponds to requires comparing the signal with ranges for each month (January is 0–30, February is 31–58, etc.). Determining the day of the week requires taking the date integer modulo 7.

A better approach, for our purposes, is to represent the signal as a four-bit month field (January $= 1$, December $= 12$) and a five-bit day field (1–31, leaving 0 unused). With this representation, for example, July 4 (US Independence Day) is $0111\ 00100_2$. The $0111_2 = 7$ represents July and $00100_2 = 4$ represents the day. With this representation we can still directly compare whether one day comes before another and also easily test whether two days are in the same month by comparing the upper four bits. However, it is even more difficult with this representation to determine the day of the week.

To solve the problem of the day of the week, we use a redundant representation that consists of a four-bit month field (1–12), a five-bit day of the month field (1–31), and a three-bit day of the week field (Sunday $= 1, \ldots,$ Saturday $= 7$). With this representation, July 4 (which is a Monday in 2016) would be represented as the 12-bit binary number 0111 00100 100. The 0111 means month 7, or July, 00100 means day 4 of the month, and 100 means day 4 of the week, or Wednesday.

Table 1.2. **Three-bit representation of colors**

This can be derived by filtering white light with zero or more primary colors; the representation is chosen so that mixing two colors is the equivalent of OR-ing the representations together

Color	Code
White	000
Red	001
Yellow	010
Blue	100
Orange	011
Purple	101
Green	110
Black	111

1.3.2 Representing subtractive colors

We often pick a representation to simplify carrying out operations. For example, suppose we wish to represent colors using a *subtractive* system. In a subtractive system we start with white (all colors) and filter this with one or more primary color (red, blue, or yellow) transparent filters. For example, if we start with white then use a red filter we get red. If we then add a blue filter we get purple, and so on. By filtering white with the primary colors we can generate the derived colors purple, orange, green, and black.

One possible representation for colors is shown in Table 1.2. In this representation we use one bit to denote each of the primary colors. If this bit is set, a filter of that primary color is in place. We start with white represented as all zeros – no filters in place. Each primary color has exactly one bit set – only the filter of that primary color in place. The derived colors orange, purple, and green each have two bits set since they are generated by two primary color filters. Finally, black is generated by using all three filters, and hence has all three bits set.

It is easy to see that, using this representation, the operation of mixing two colors together (adding two filters) is equivalent to the operation of taking the logical OR of the two representations. For example, if we mix red 001 with blue 100 we get purple 101, and $001 \vee 100 = 101$.[4]

1.4 DIGITAL LOGIC FUNCTIONS

Once we have represented information as digital signals, we use digital logic circuits to compute logical functions of our signals. That is, the logic computes an output digital signal that is a function of the input digital signal(s).

[4] The symbol $\vee$ denotes the logical OR of two binary numbers; see Chapter 3.

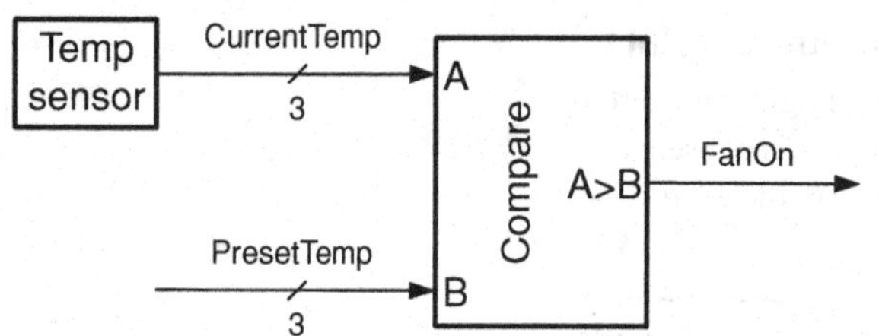

Figure 1.8. A digital thermostat is realized with a comparator. The comparator turns a fan on when the current temperature is larger than a preset temperature.

Suppose we wish to build a thermostat that turns on a fan if the temperature is higher than a preset limit. Figure 1.8 shows how this can be accomplished with a single *comparator*, a digital logic block that compares two numbers and outputs a binary signal that indicates whether one is greater than the other. (We will examine how to build comparators in Section 8.6.) The comparator takes two temperatures as input: the current temperature from a temperature sensor and the preset limit temperature. If the current temperature is greater than the limit temperature, the output of the comparator goes high, turning the fan on. This digital thermostat is an example of a *combinational logic circuit*, a logic circuit whose output depends only on the current state of its inputs. We will study combinational logic in Chapters 6–13.

EXAMPLE 1.2 Current day circuit

Suppose we wish to build a calendar circuit that always outputs the current day in the month, day of month, day of week representation described in Section 1.3.1. This circuit, shown in Figure 1.9, requires storage. A *register* stores the current day (current month, day of month, and day of week), making it available on its output and ignoring its input until the clock rises. When the clock signal rises, the register updates its contents with the value on its input and then resumes its storage function.[5] A logic circuit computes the value of tomorrow from the value of today. This circuit increments the two day fields and takes appropriate action if they overflow. We present the implementation of this logic circuit in Section 9.2. Once a day (at midnight) the clock signal rises, causing the register to update its contents with tomorrow's value. Our digital calendar is an example of a *sequential* logic circuit. Its output depends not only on current inputs (the clock), but also on the internal state (today), which reflects the value of past inputs. We will study sequential logic in Chapters 14–19.

We often build digital systems by composing subsystems. Or, from a different perspective, we design a digital system by partitioning it into combinational and sequential subsystems and then designing each subsystem. As a very simple example, suppose we want to modify our thermostat so that the fan does not run on Sundays. We can do this by combining our thermostat circuit with our calendar circuit, as shown in Figure 1.10. The calendar circuit is used only for its day of week (DoW) output. This output is compared with the constant Sunday = 1. The output of the comparator is true if today is Sunday (ItsSunday). An *inverter*, also called a NOT gate, complements this value. Its output (ItsNotSunday) is true if it is not Sunday. Finally, an AND gate combines the inverter output with the output of the thermostat. The output of

[5] We leave unanswered for now how the register is initially set with the correct date.

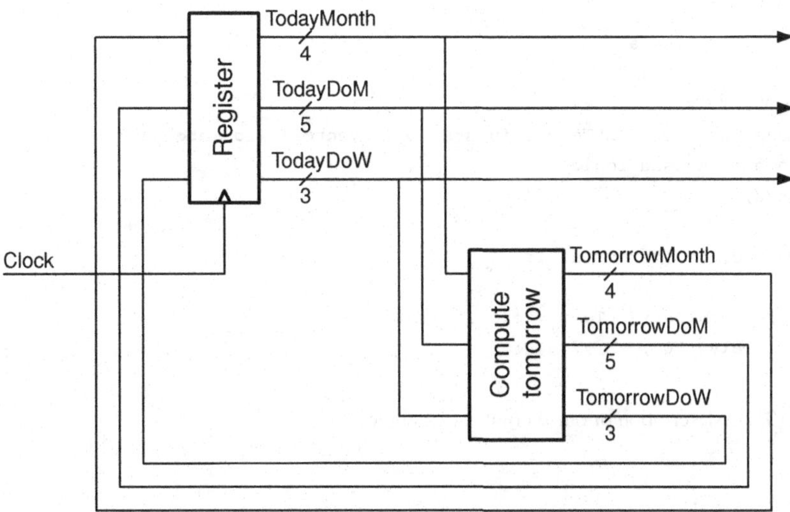

Figure 1.9. A digital calendar outputs the current day in the month, day of month, day of week format. A register stores the value of the current day (today). A logic circuit computes the value of the next day (tomorrow).

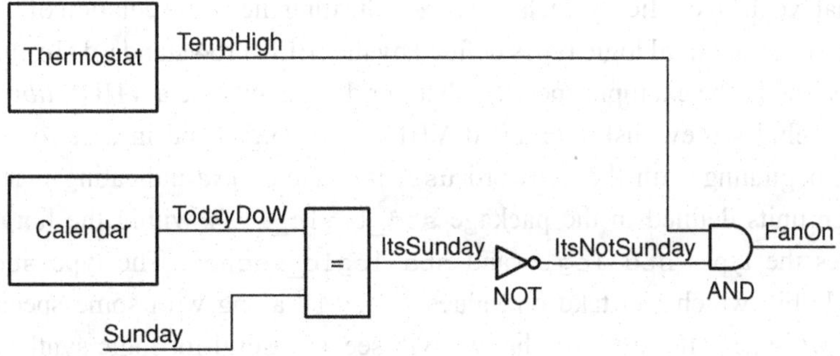

Figure 1.10. By combining our thermostat and calendar circuits, we realize a circuit that turns the fan on when the temperature is high, except on Sundays, when the fan remains off.

the AND gate is true only when the temperature is high AND it's not Sunday. System-level design – at a somewhat higher level than this simple example – is the topic of Chapters 21–25.

1.5 VHDL DESCRIPTION OF DIGITAL CIRCUITS AND SYSTEMS

VHDL is a *hardware description language* (HDL) that is used to describe digital circuits and systems. Once a system has been described in VHDL, we can simulate its operation using a VHDL simulator. We can also *synthesize* the circuit using a synthesis program (similar to a compiler) to convert the VHDL description to a *gate*-level description to be mapped to standard cells or an FPGA. VHDL and Verilog are the two HDLs in wide use today. Most chips and systems in industry are designed by writing descriptions in one of these two languages.

We will use VHDL throughout this book both to illustrate principles and also to teach VHDL coding style. By the end of a course using this book, the reader should be proficient in both reading and writing VHDL. Appendix A gives a style guide for writing VHDL and Appendix B provides an overview of VHDL syntax.

```
library ieee;
use ieee.std_logic_1164.all;

entity THERMOSTAT is
  port ( presetTemp, currentTemp : in std_logic_vector( 2 downto 0 );
         fanOn : out std_logic );
end thermostat;

architecture impl of Thermostat is
begin
  fanOn <= '1' when currentTemp > presetTemp
               else '0';
end impl;
```

Figure 1.11. VHDL description of our thermostat example.

A VHDL description of our thermostat example is shown in Figure 1.11. Note that this VHDL describes the *behavior* of the circuit but not how the circuit is implemented. A synthesis program can transform this VHDL into hardware, but neither the synthesis program nor that hardware can be said to "execute" the VHDL. VHDL is a strongly typed language. This means that for each signal we must explicitly declare a *type* indicating the representation of that signal. VHDL includes a set of standard logic types defined by the IEEE 1164 standard. To enable us to use them, the first line in the example specifies that our design will use a VHDL *library* called IEEE. A library includes previously compiled VHDL code organized in a set of *packages*. The second line, beginning with the keyword **`use`**, is a *use clause* indicating that we wish to use **`all`** design units defined in the package `std_logic_1164` within the library `IEEE`. Our example uses the types `std_logic` and `std_logic_vector`. The type `std_logic` represents a single bit, which can take on values `'0'`, `'1'` along with some special values `'-'`, `'Z'`, `'X'`, `'W'`, `'L'`, `'H'` and `'U'` that we will see are useful for logic synthesis and/or simulation. The type `std_logic_vector` represents a multi-bit signal, where each bit has type `std_logic`. Our thermostat will need to compare the value of the current temperature against the preset temperature (e.g., using >).

In the example the thermostat is described using a VHDL *design entity* consisting of two parts, an *entity declaration* and an *architecture body*. The entity declaration defines the interface between the design entity and the outside world, whereas the architecture body defines the internal operation of the design entity.

The entity declaration begins with the keyword **`entity`**. In our example the entity declaration indicates that the design entity's name is `THERMOSTAT`. Note that VHDL is case insensitive. For example, this means `THERMOSTAT` is considered by the tools as identical if written all in lower case (`thermostat`) or using mixed case (e.g., `Thermostat`). This case insensitivity applies both to identifiers and to keywords. In Figure 1.11 we have used different capitalization to emphasize VHDL's case insensitivity. The inputs and outputs to the design entity are specified in a port list that begins with the keyword **`port`**. The port list contains three signals: `presetTemp`, `currentTemp`, and `fanOn`. The keyword **`in`** indicates that the two temperature signals `presentTemp` and `currentTemp` are inputs. These three-bit signals are declared to have type `std_logic_vector` and can take on values between

"000" and "111", representing the unsigned integers between 0 and 7. The "(2 downto 0)" indicates that presentTemp has subsignals presentTemp(2), presentTemp(1), and presentTemp(0). Here the most significant bit is presentTemp(2). Note that we could have declared presentTemp with "(0 to 2)," in which case the most significant bit would be presentTemp(0). The keyword **out** indicates that the signal fanOn is an output of the design entity. Since it represents a single bit, fanOn is declared with type std_logic.

The architecture body begins with the keyword **architecture** followed by an identifier for the architecture body, impl (short for "implementation"), then the keyword **of** and then the name of the entity declaration this architecture body corresponds to (Thermostat). Although most design entities have a single architecture body, the VHDL language allows multiple architecture bodies, each of which should have a unique identifier. Next are the keywords **is** and **begin** followed by the code that defines the behavior of the architecture body. In this example the architecture body consists of a single *concurrent* assignment statement. The signal assignment compound delimiter, <=, indicates that the expression on the right is assigned to the signal on the left. The keywords **when** and **else** indicate that this is a *conditional signal assignment* statement. The conditional signal assignment sets fanOn to logic value '1' whenever currentTemp is greater than presetTemp and to '0' otherwise. The conditional signal assignment statement ends with a semicolon (;).

The result of simulating this design entity with presetTemp "011" (i.e., 3) and currentTemp sweeping from "000" to "111" is shown in Figure 1.12.

Despite the unfamiliar syntax, at first glance VHDL code may look somewhat like a conventional programming language such as C or Java. However, VHDL, or any other HDL, is fundamentally different than a programming language. In a programming language like C, only one statement is *active* at a time. Statements are executed one at a time in sequence. In VHDL, on the other hand, all design entities and all concurrent assignment statements in each component are active all of the time. That is, all of the statements are executed all of the time.

It is very important in coding VHDL to keep in mind that the code is ultimately being compiled into hardware. Each component instantiated adds a hardware component to the design. Each assignment statement in each design entity adds gates to each instance of that design entity. VHDL can be a tremendous productivity multiplier – allowing the designer to work at a much higher level than if the gates had to be manually synthesized. At the same time, VHDL can be an impediment if its abstraction causes the designer to lose touch with the end product and write an inefficient design.

```
# 011 000 -> 0
# 011 001 -> 0
# 011 010 -> 0
# 011 011 -> 0
# 011 100 -> 1
# 011 101 -> 1
# 011 110 -> 1
# 011 111 -> 1
```

Figure 1.12. Result of simulating the VHDL of Figure 1.11 with presetTemp = "011" (i.e., 3) and currentTemp sweeping from "000" to "111" (i.e., 7).

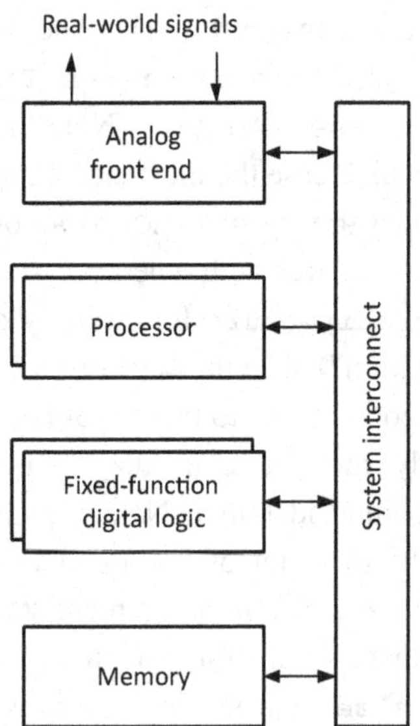

Figure 1.13. A modern digital electronic system consists of an analog front end, one or more processors, which perform the complex, but less demanding, parts of an application, and digital logic blocks that perform the computationally intensive parts of the application.

1.6 DIGITAL LOGIC IN SYSTEMS

Most modern electronic systems, from cell phones to televisions to the embedded engine controller in your car, have the form shown in Figure 1.13. These systems are divided into an analog front end, one or more processors, and fixed-function digital logic blocks. These components represent the *hardware* of a system. Because analog circuitry accumulates noise and has limited levels of integration, we use it as little as possible. In a modern system, analog circuitry is limited to the periphery of the system and to signal conditioning and conversion from analog to digital and vice versa.

Programmable processors, which are built using digital logic, perform the system functions that are complex but not computationally intense. The processor itself can be fairly simple; the complexity is in the *software* running on the processor, which can involve millions of lines of code. Complex user interfaces, system logic, undemanding applications, and other such functions are implemented in software running on the processor. However, because the area (and energy) required to implement a function on a processor is from 10 to 1000 times higher than that required to implement the function in a fixed-function logic block, the most computationally demanding functions are implemented directly in logic, not in software running on the processor.

The fixed-function logic blocks in a typical system perform most of the system's computation, as measured by the number of operations, but account for only a small fraction of the system's complexity, as measured by lines of code. Fixed-function logic is used to implement functions such as radio modems, video encoders and decoders, encryption and decryption functions, and other such blocks. It is not unusual for a modem or codec to perform 10^{10} to 10^{11} operations per second while consuming the same amount of power as a processor that is performing 10^8 to 10^9 operations per second. If you watch a streaming video on your cell phone, most of the operations being performed are taking place in the radio modem, which decodes the RF waveform into symbols, and the video codec, which decodes those symbols into pixels. Very little computation, relatively speaking, is taking place on the ARM processor that controls the process.

This book deals with the design and analysis of digital logic. In the system of Figure 1.13, digital logic can be found in all of the blocks, including in the glue that connects them together. Digital logic controls, sequences, calibrates, and corrects the A/D and D/A converters in the analog front end. The processor, while its function is defined by software, is built from digital logic. Digital logic implements the fixed-function blocks. Finally, digital logic implements the buses and networks that allow these blocks to communicate with one another.

Digital logic is the foundation on which modern electronic systems are built. When you finish this book, you will have a basic understanding of this foundational technology.

Summary

This chapter has given you an overview of digital design. We have seen that representing signals digitally makes them tolerant of noise. Unless a *noise margin* is exceeded, a digital signal corrupted by noise can be restored to its original state. This enables us to perform complex calculations without accumulating noise.

Information is represented as digital signals. Truth propositions, e.g., "the door is open," have just two values and thus can be directly represented with a digital signal. We represent elements of a set, e.g., the day of the week, with a *multi-bit* digital signal by assigning a *binary code* to each element of the set. Continuous values, such as the voltage on a wire or the temperature in a room, are represented by first *quantizing* the value – mapping a continuous range onto a finite number of discrete steps – and then representing these steps as elements of a set.

Digital logic circuits perform logical functions – computing digital output signals whose values are functions of other digital signals. A *combinational* logic circuit computes an output that is only a function of its present input. A *sequential* logic circuit includes *state feedback*, and computes an output that is a function not only of its current input, but also of its state and hence, indirectly, of past inputs.

We describe digital logic functions in the VHDL hardware description language for purposes of simulation and synthesis. We can *simulate* a circuit described in VHDL, determining its response to a test input, to verify its behavior. Once verified, we can *synthesize* the VHDL description into actual circuit for implementation. While VHDL at first glance looks a lot like a conventional programming language, it is quite different. A conventional program describes a step-by-step process where one line of the program is executed at a time; however, a VHDL description describes *hardware*, where all modules are executed simultaneously all of the time. A style guide for VHDL is given in the Appendix.

Digital logic design is important because it forms the foundation of modern electronic systems. These systems consist of analog front ends, one or more processors, and digital fixed-function logic. Digital logic is used both to implement the processors – which perform the functions that are complex but not computationally intense – and the fixed-function logic – which is where most of the computation occurs.

BIBLIOGRAPHIC NOTES

Charles Babbage is credited with designing the first calculating machines: his difference engines [5]. Although he designed his Difference Engine No. 2 in the 1840s, it was complex and not completely built until 2002 [105]. He also planned an even more complex Analytical Engine with many similarities to modern computers [18]. Other mechanical calculators existed by the early 1900s. John Atanasoff was ruled by the US court system[6] to have invented the first digital computer at Iowa State in 1940–41. Atanasoff writes about his invention 40 years later in ref. [4].

Robert Noyce (biography [11]) and Jack Kilby were co-inventors of the integrated circuit [89]. After leaving Fairchild Semiconductor, Noyce founded Intel with Gordon Moore. While at Intel, Moore made an observation about the growth of the density of transistors on an integrated circuit. This observation, Moore's law [84], of exponential growth has held for over 40 years.

One of the first examples of the high-level synthesis tools was Berkeley's BDSyn [99]. For further information about the VHDL language, please refer to later chapters of this book, or the numerous online resources about the language.

Exercises

1.1 *Noise margins, I.* Suppose you have a module that uses the encoding described in Table 1.1 but you have freedom to choose either $(V_{OL}, V_{OH}) = (0.3, 2.2)$ or $(V_{OL}, V_{OH}) = (0.1, 2.1)$. Which of these output ranges would you choose and why?

1.2 *Noise margins, II.* Two wires have been placed close together on a chip. They are so close, in fact, that the larger wire (the aggressor) couples to the smaller wire (the victim) and causes the voltage on the victim wire to change. Using the data from Table 1.1, determine the following.

(a) If the victim wire is at V_{OL}, what is the most the aggressor can push it down without causing a problem?

(b) If the victim wire is at 0 V, what is the most the aggressor can push it down without causing a problem?

(c) If the victim wire is at V_{OH}, what is the most the aggressor can push it down without causing a problem?

(d) If the victim wire is at V_1, what is the most the aggressor can push it up without causing a problem?

1.3 *Power-supply noise.* Two systems, A and B, use the encoding of Table 1.1 to send logic signals to one another. Suppose there is a voltage shift between the two systems' power supplies so that all voltages in A are V_N higher than in B. A voltage of V_x in system A appears as a voltage of $V_x + V_N$ in system B. A voltage of V_x in system B appears as a voltage of $V_x - V_N$ in system A. Assuming that there are no other noise sources, over what range of V_N will the system operate properly?

[6] Honeywell v. Sperry Rand.

1.4 *Ground noise.* A logic family has signal levels as shown in Table 1.3. We connect device A's output to B's inputs using this logic family. All signal levels are relative to the local ground. How much can the ground voltages of the two devices (*GNDA* and *GNDB*) differ before an error occurs? Compute the difference both for how much higher *GNDA* can be relative to *GNDB* and for how much lower *GNDA* can be relative to *GNDB*.

Table 1.3. Voltage levels for Exercise 1.4

Parameter	Value
V_{OL}	0.1 V
V_{IL}	0.4 V
V_{IH}	0.6 V
V_{OH}	0.8 V

1.5 *Proportional signal levels.* A logic device encodes signals with levels proportional to its power supply voltage (V_{DD}) according to Table 1.4.

Table 1.4. Voltage levels for Exercise 1.5

Parameter	Value
V_{OL}	$0.1V_{DD}$
V_{IL}	$0.4V_{DD}$
V_{IH}	$0.6V_{DD}$
V_{OH}	$0.9V_{DD}$

Suppose two such logic devices A and B send signals to one another and the supply of device A is $V_{DDA} = 1.0$ V. Assuming that there are no other noise sources and that the two devices have a common ground (i.e., 0 V is the same level in both devices), what is the range of supply voltages for device B, V_{DDB}, over which the system will operate properly?

1.6 *Noise margins, III.* Using the proportional scheme of Exercise 1.5, what is the lowest supply voltage (V_{DD}) that can tolerate noise up to 100 mV?

1.7 *Noise margins, IV.* A logic family uses signal levels relative to V_{SS} and proportional to V_{DD} as shown in Table 1.5. We connect two logic devices A and B using this logic family with signals traveling in both directions between the two devices. Both systems have $V_{DD} = 1$ V, and system A has $V_{SSA} = 0$ V. Over what range of values for V_{SSB} will the system operate properly?

Table 1.5. Voltage levels for Exercise 1.7

Parameter	Value
V_{OL}	$0.9V_{SS} + 0.1V_{DD}$
V_{IL}	$0.7V_{SS} + 0.3V_{DD}$
V_{IH}	$0.3V_{SS} + 0.7V_{DD}$
V_{OH}	$0.1V_{SS} + 0.9V_{DD}$

1.8 *Gain of restoring devices.* What is the minimum absolute value of gain for a circuit that restores signals according to the values in Table 1.1?

1.9 *Gray codes.* A continuous value that has been quantized into N states can be encoded into an $n = \lceil \log_2 N \rceil$ bit signal in which adjacent states differ in at most one bit position. Show how the eight temperatures of Figure 1.7(c) can be encoded into three bits in this manner. Make your encoding such that the encodings of 82 °F and 68 °F also differ in just one bit position.

1.10 *Encoding rules.* Equations (1.3) and (1.4) are examples of *decoding* rules that return the value represented by a multi-bit digital signal. Write down the corresponding *encoding* rules. These rules give the value of each bit of the digital signal as a function of the value being encoded.

1.11 *Encoding playing cards.* Suggest a binary representation for playing cards – a set of binary signals that uniquely identifies one of the 52 cards in a standard deck. What different representations might be used to (i) optimize the density (the minimum number of bits per card) or (ii) simplify operations such as determining whether two cards are of the same suit or rank? Explain how you can check to see whether two cards are adjacent (rank differing by one) using a specific representation.

1.12 *Day of the week.* Explain how to derive the day of the week from the month/day representation we discussed in Section 1.3.1.

1.13 *Colors, I.* Derive a representation for colors that supports the operation of additive composition of primary colors. You start with black and add colored light that is red, green, or blue.

1.14 *Colors, II.* Extend the representation of Exercise 1.13 to support three levels of intensity for each of the primary colored lights; that is, each color can be off, weakly on, medium on, or strongly on.

1.15 *Encoding and decoding, I.* A four-core chip is arranged as a 4×1 array of processors, where each processor is connected to its east and west neighbors. There are no connections on the ends of the array. The processors' addresses start at 0 on the east-most processor and go up by 1 to address 3 at the west-most processor. Given the current processor's address and the address of a destination processor, how do you determine whether to go east or west to reach the destination processor?

1.16 *Encoding and decoding, II.* A 16-core chip is arranged as a 4×4 array of processors where each processor is connected to its north, south, east, and west neighbors. There are no connections on the edges. Pick an encoding for the address of each processor (0–15) such that when a datum is moving through the processors it is easy (i.e., similar to Exercise 1.15) to determine whether it should move north, south, east, or west at each processor on the basis of the destination address and the address of the current processor.

(a) Draw the array of processors, labeling each core with its address according to your encoding.

(b) Describe how to determine the direction the data should move on the basis of current and destination addresses.

(c) How does this encoding or its interpretation differ from simply labeling the processors 0–15 starting at the north-west corner?

1.17 *Circular Gray code.* Come up with a way of encoding the numbers 0–5 onto a four-bit binary signal so that adjacent numbers differ in only one bit and also so that the representations of 0 and 5 differ in only one bit.

2 The practice of digital system design

Before we dive into the technical details of digital system design, it is useful to take a high-level look at the way systems are designed in industry today. This will allow us to put the design techniques we learn in subsequent chapters into the proper context. This chapter examines four aspects of contemporary digital system design practice: the design process, implementation technology, computer-aided design tools, and technology scaling.

We start in Section 2.1 by describing the design process – how a design starts with a specification and proceeds through the phases of concept development, feasibility studies, detailed design, and verification. Except for the last few steps, most of the design work is done using English-language[1] documents. A key aspect of any design process is a systematic – and usually quantitative – process of managing technical risk.

Digital designs are implemented on very-large-scale integrated (VLSI) circuits (often called *chips*) and packaged on printed-circuit boards (PCBs). Section 2.2 discusses the capabilities of contemporary implementation technology.

The design of highly complex VLSI chips and boards is made possible by sophisticated computer-aided design (CAD) tools. These tools, described in Section 2.3, amplify the capability of the designer by performing much of the work associated with capturing a design, synthesizing the logic and physical layout, and verifying that the design is both functionally correct and meets timing.

Approximately every two years, the number of transistors that can be economically fabricated on an integrated-circuit chip doubles. We discuss this growth rate, known as *Moore's law*, and its implications for digital systems design in Section 2.4.

2.1 THE DESIGN PROCESS

As in other fields of engineering, the digital design process begins with a specification. The design then proceeds through phases of concept development, feasibility, partitioning, and detailed design. Most texts, like this one, deal with only the last two steps of this process. To put the design and analysis techniques we will learn into perspective, we will briefly examine the other steps here. Figure 2.1 gives an overview of the design process.

2.1.1 Specification

All designs start with a specification that describes the item to be designed. Depending on the novelty of the object, developing the specification may be a straightforward or elaborate process

[1] Or some other natural (human) language.

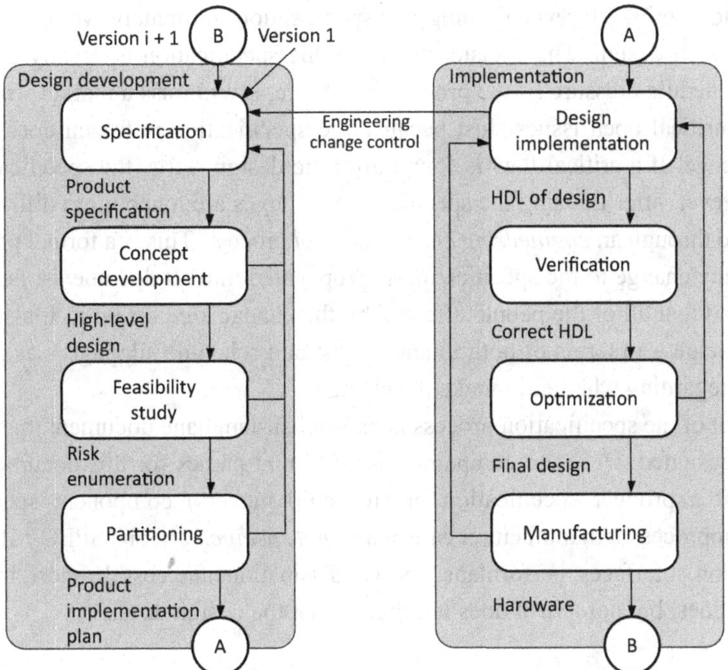

Figure 2.1. The design process. Initial effort is focused on design specification and risk analysis. Only after the creation of a *product implementation plan* does implementation begin. The implementation process itself has many iterations of verification and optimization. Many products also have multiple internal hardware revisions before being shipped.

in itself. The vast majority of designs are evolutionary – the design of a new version of an existing product. For such evolutionary designs, the specification process is one of determining how much better (faster, smaller, cheaper, more reliable, etc.) the new product should be. At the same time, new designs are often constrained by the previous design. For example, a new processor must usually execute the same *instruction set* as the model it is replacing, and a new I/O device must usually support the same standard I/O interface (e.g., a PCI bus) as the previous generation.

On rare occasions, the object being specified is the first of its kind. For such revolutionary developments, the specification process is quite different. There are no constraints of backward compatibility, although the new object may need to be compatible with one or more *standards*. This gives the designer more freedom, but also less guidance, in determining the function, features, and performance of the object.

Whether revolutionary or evolutionary, the specification process is an iterative process – like most engineering processes. We start by writing a *straw man* specification for the object – and in doing so we identify a number of questions or open issues. We then iteratively refine this initial specification by gathering information to answer the questions or resolve the open issues. We meet with customers or end users of the product to determine the features they want, how much they value each feature, and how they react to our proposed specification. We commission engineering studies to determine the cost of certain features. Examples of cost metrics include how much die area it will take to reach a certain level of performance, or how much power will be dissipated by adding a branch predictor to a processor. Each time a new piece of information comes in, we revise our specification to account for the new information. A history of this revision process is also kept to give a rationale for the decisions made.

While we could continue forever refining our specification, ultimately we must *freeze* the specification and start design. The decision to freeze the specification is usually driven by a combination of schedule pressure (if the product is too late, it will miss a market *window*) and resolution of all critical open issues. Just because the specification is frozen does not mean that it cannot change. If a critical flaw is found after the design starts, the specification must be changed. However, after freezing the specification, changes are much more difficult in that they must proceed through an *engineering change control* process. This is a formal process that makes sure that any change to the specification is propagated into all documents, designs, test programs, etc., and that all of the people affected by the change *sign off* on it. It also assesses the cost of the change – in terms of both financial cost and schedule slippage – as part of the decision process regarding whether to make the change.

The end product of the specification process is an English-language document that describes the object to be designed. Different companies use different names for this document. Many companies call it a product specification or (for chip makers) component specification. A prominent microprocessor manufacturer calls it a *target specification* or TSPEC.[2] It describes the object's function, interfaces, performance, power dissipation, and cost. In short, it describes *what* the product does, but not *how* it does it – that's what the design does.

2.1.2 Concept development and feasibility

During the concept development phase the high-level design of the system is performed. Block diagrams are drawn, major subsystems are defined, and the rough outline of system operation is specified. More importantly, key engineering decisions are made at this stage. This phase is driven by the specification. The concept developed must meet the specification, or, if a requirement is too difficult to meet, the specification must be changed.

In the partitioning, as well as in the specification of each subsystem, different approaches to the design are developed and evaluated. For example, to build a large communication switch we could use a large crossbar, or we could use a multi-stage network. During the concept development phase, we would evaluate both approaches and select the one that best meets our needs. Similarly, we may need to develop a processor that operates at $1.5\times$ the speed of the previous model. During the concept development phase we would consider increasing the clock rate, using a more accurate branch predictor, increasing the cache size, and/or increasing the issue width. We would evaluate the costs and benefits of these approaches in isolation and in combination.

Technology selection and vendor qualification is also a part of concept development. During these processes, we select what components and processes we are going to use to build our product and determine who is going to supply them. In a typical digital design project, this involves selecting suppliers of standard chips (like memory chips and FPGAs), suppliers of custom chips (either an ASIC vendor or a foundry), suppliers of packages, suppliers of circuit boards, and suppliers of connectors. Particular attention is usually paid to components, processes, or suppliers that are new, since they represent an element of risk. For example, if

[2] Often the product specification is accompanied by a business plan for the new product that includes sales forecasts and computes the return on investment for the new product development. However, that is a separate document.

we consider using a new optical transceiver or optical switch that has never been built or used before, we need to evaluate the probability that it may not work, may not meet specifications, or may not be available when we need it.

A key part of technology selection is making *make vs. buy* decisions about different pieces of the design. For example, you may need to choose between designing your own Ethernet interface and buying the VHDL for the interface from a vendor. The two (or more) alternatives are evaluated in terms of cost, schedule, performance, and risk. A decision is then made on the basis of the merits of each. Often information needs to be gathered (from design studies, reference checks on vendors, etc.) before making the decision. Too often, engineers favor building things themselves when it is often much cheaper and faster to buy a working design from a vendor. On the other hand, "caveat emptor"[3] applies to digital design. Just because someone is selling a product does not mean that it works or meets specification. You may find that the Ethernet interface you purchased does not work on certain packet lengths. Each piece of technology acquired from an outside supplier represents a risk and needs to be carefully verified before it is used. This verification can often take up a large fraction of the effort that would have been required to do the design yourself.

A large part of engineering is the art of managing technical risk – of setting the level of ambition high enough to give a winning product, but not so high that the product can't be built in time. A good designer takes a few calculated risks in selected areas that give big returns, and manages them carefully. Being too conservative (taking no risks, or too few risks) usually results in a non-competitive product. On the other hand, being too aggressive (taking too many risks – particularly in areas that give little return) results in a product that is too late to be relevant. Far more products fail for being too aggressive (often in areas that do not matter) than fail on account of being too conservative.

To manage technical risks effectively, risks must be identified, evaluated, and mitigated. Identifying risks calls attention to them so they can be monitored. Once we have identified a risk, we evaluate it along two axes – importance and danger. For importance, we ask "what do we gain by taking this risk?" If it doubles our system performance or halves its power requirement, it might be worth taking. However, if the gain (compared with a more conservative alternative) is negligible, there is no point taking the risk. For danger, we quantify or classify risks according to how likely they are to succeed. One approach is to assign two numbers between 1 and 5 to each risk, one for importance and one for danger. Risks that are (1,5) – low importance and high danger – are abandoned. Risks that are (5,1) – nearly sure bets with big returns – are kept and managed. Risks that rank (5,5) – very important and very dangerous – are the trickiest. We can't afford to take too many risks, so some of these have to go. Our approach is to reduce the danger of these risks through mitigation: turning a (5,5) into a (5,4) and eventually into a (5,1).

Many designers manage risks informally – mentally following a process similar to the one described here and then making instinctive decisions as to which risks to take and which to avoid. This is a bad design practice for several reasons. It does not work with a large design team (written documents are needed for communication) or for large designs (there are too

[3] Let the buyer beware.

many risks to keep in one head). Designers often make poor and uninformed decisions when using an informal, non-quantitative risk management scheme. These schemes also leave no written rationale behind risk management decisions.

We often mitigate risks by gathering information. For example, suppose our new processor design calls for a single pipeline stage to check dependences, rename registers, and issue instructions to eight ALUs (a complex logical function). We have identified this as both important (it buys us lots of performance) and dangerous (we are not sure it can be done at our target clock frequency). We can reduce the danger to level 1 by carrying out the design early and establishing that it can be done. This is often called a *feasibility study*, establishing that a proposed design approach is, in fact, feasible. We can often establish feasibility (to a high degree of probability) with much less effort than completing a detailed design.

Risks can also be mitigated by developing a backup plan. For example, suppose that one of the (5,5) risks in our conceptual design is the use of a new SRAM part made by a small manufacturer, which is not going to be available until just before we need it. We can reduce this risk by finding an alternative component, with better availability but worse performance, and designing our system so it can use either part. If the high-risk part is not available in time, rather than not having a system at all, we have a system that has just a little less performance, and can be upgraded when the new component is out.

Risks cannot be mitigated by ignoring them and hoping that they go away. This is called *denial*, and is a sure-fire way to make a project fail.

With a formal risk management process, identified risks are typically reviewed on a periodic basis (e.g., once every week or two). At each review the importance and danger of the risk are updated on the basis of new information. This review process makes risk mitigation visible to the engineering team. Risks that are successfully being mitigated, whether through information gathering or backup plans, will have their danger drop steadily over time. Risks that are not being properly managed will have their danger level remain high – drawing attention to them so that they can be more successfully managed.

The result of the concept development phase is a second English-language document that describes in detail *how* the object is to be designed. It describes the key aspects of the design approach taken, giving a rationale for each. It identifies all of the outside players: chip suppliers, package suppliers, connector suppliers, circuit-board supplier, CAD tool providers, design service providers, etc. This document also identifies all risks and gives a rationale for why they are worth taking, and describes completed and ongoing actions to mitigate them. Different companies use different names for this *how* document. It has been called an implementation specification and a product implementation plan.

2.1.3 Partitioning and detailed design

Once the concept phase is complete and design decisions have been made, what remains is to partition the design into modules and then perform the detailed design of each module. The high-level system partitioning is usually done as part of the conceptual design process. A specification is written for each of these high-level modules, with particular attention to interfaces. These specifications enable the modules to be designed independently and, if they all conform to the specification, work when plugged together during system integration.

In a complex system the top-level modules will themselves be partitioned into submodules, and so on. The partitioning of modules into submodules is often referred to as block-level design since it is carried out by drawing *block diagrams* of the system, where each block represents a module or submodule and the lines between the modules represent the interfaces over which the modules interact.

Ultimately, we subdivide a module to the level where each of its submodules can be directly realized using a synthesis procedure. These bottom-level modules may be combinational logic blocks to compute a logical function on its inputs, arithmetic modules to manipulate numbers, and finite-state machines that sequence the operation of the system. Much of this book focuses on the design and analysis of these bottom-level modules. It is important to maintain perspective of where they fit in a larger system.

2.1.4 Verification

In a typical design project, more than half of the effort goes not into design, but into verifying that the design is correct. Verification takes place at all levels: from the conceptual design down to individual modules. At the highest level, architectural verification is performed on the conceptual design. In this process, the conceptual design is checked against the specification to ensure that every requirement of the specification is satisfied by the implementation.

Unit tests are written to verify the functionality of each individual module. Typically, there are far more lines of test code than there are lines of VHDL implementing the modules. After the individual modules have been verified, they are integrated into the enclosing subsystem and the process is repeated at the next level of the module hierarchy. Ultimately the entire system is integrated, and a complete suite of tests is run to validate that the system implements all aspects of the specification.

The verification effort is usually performed according to yet another written document, called a *test plan*.[4] In the test plan, every feature of the device under test (DUT) is identified, and tests are specified to *cover* all of the identified features. A large fraction of tests will deal with error conditions (how the system responds to inputs that are outside its normal operating modes) and boundary cases (inputs that are just inside or just outside the normal operating mode).

A large subset of all tests will typically be grouped into a *regression test suite* and run on the design periodically (often every night) and any time a change is *checked in* to the revision control system. The purpose of the regression suite is to make sure that the design does not regress – i.e., to make sure that, in fixing one bug, a designer has not caused other tests to fail.

When time and resources are short, engineers are sometimes tempted to take shortcuts and skip some verification. This is hardly ever a good idea. A healthy philosophy toward verification is the following: *If it hasn't been tested, it doesn't work.* Every feature, mode, and boundary condition needs to be tested. In the long run, the design will get into production more quickly if you complete each step of the verification and resist the temptation to take shortcuts.

Bugs occur in all designs, even those by top engineers. The earlier a bug can be detected, the less time and money it takes to fix. A rule of thumb is that the cost of fixing a bug increases by

[4] As you can see, most engineers spend more time writing English-language documents than writing VHDL or C code.

a factor of ten each time the design proceeds forward by one major step. Bugs are cheapest to fix if they are detected during unit test. A bug that slips through unit test and is caught during integration test costs 10× more to fix.[5] One that makes it through integration test and is caught at full system test costs 10× more again (100× more than fixing the bug at unit test). If the bug makes it through system test and is not caught until silicon debug, after the chip has been taped out and fabricated, it costs 10× more again. A bug that is not caught until the chip is in production (and must be recalled) costs another 10× – 10 000× more than fixing the bug in unit test. You get the point – do not skimp on testing. The earlier you find a bug, the easier it is to fix.

2.2 DIGITAL SYSTEMS ARE BUILT FROM CHIPS AND BOARDS

Modern digital systems are implemented using a combination of standard integrated circuits and custom integrated circuits interconnected by circuit boards that, in turn, are interconnected by connectors and cables.

Standard integrated circuits are parts that can be ordered from a catalog and include memories of all types (SRAM, DRAM, ROM, EPROM, EEPROM, etc.), programmable logic (FPGAs, see below), microprocessors, and standard peripheral interfaces. Designers make every effort possible to use a standard integrated circuit to realize a function; since these components can simply be purchased, there is no development cost or effort and usually little risk associated with these parts. However, in some cases, a performance, power, or cost specification cannot be realized using a standard component, and a custom integrated circuit must be designed.

Custom integrated circuits (sometimes called ASICs, for application-specific integrated circuits) are chips built for a specific function. Or, put differently, they are chips you design yourself because you can't find what you need in a catalog. Most ASICs are built using a *standard-cell* design method in which standard modules (cells) are selected from a library and instantiated and interconnected on a silicon chip. Typical standard cells include simple gate circuits, SRAM and ROM memories, and I/O circuits. Some vendors also offer higher-level modules such as arithmetic units, microprocessors, and standard peripherals – either as cells, or written in synthesizable HDL (e.g., VHDL). Thus, designing an ASIC from standard cells is similar to designing a circuit board from standard parts. In both cases, the designer (or CAD tool) selects cells from a catalog and specifies how they are connected. Using standard cells to build an ASIC has the same advantages as using standard parts on a board: reduced development cost and reduced risk. In rare cases, a designer will design their own non-standard cell at the transistor level. Such custom cells can provide significant performance, area, and power advantages over standard-cell logic, but should be used sparingly because they involve significant design effort and are major risk items.

As an example of a custom integrated circuit, Figure 2.2 shows an NVIDIA "Fermi" GPU [88, 114]. This chip is fabricated in 40 nm CMOS technology and has over 3×10^9 transistors. It consists of 16 streaming multiprocessors (in four rows at the top and bottom of the die) with

[5] There may be one or more subsystem test levels, each of which multiplies the cost by ten.

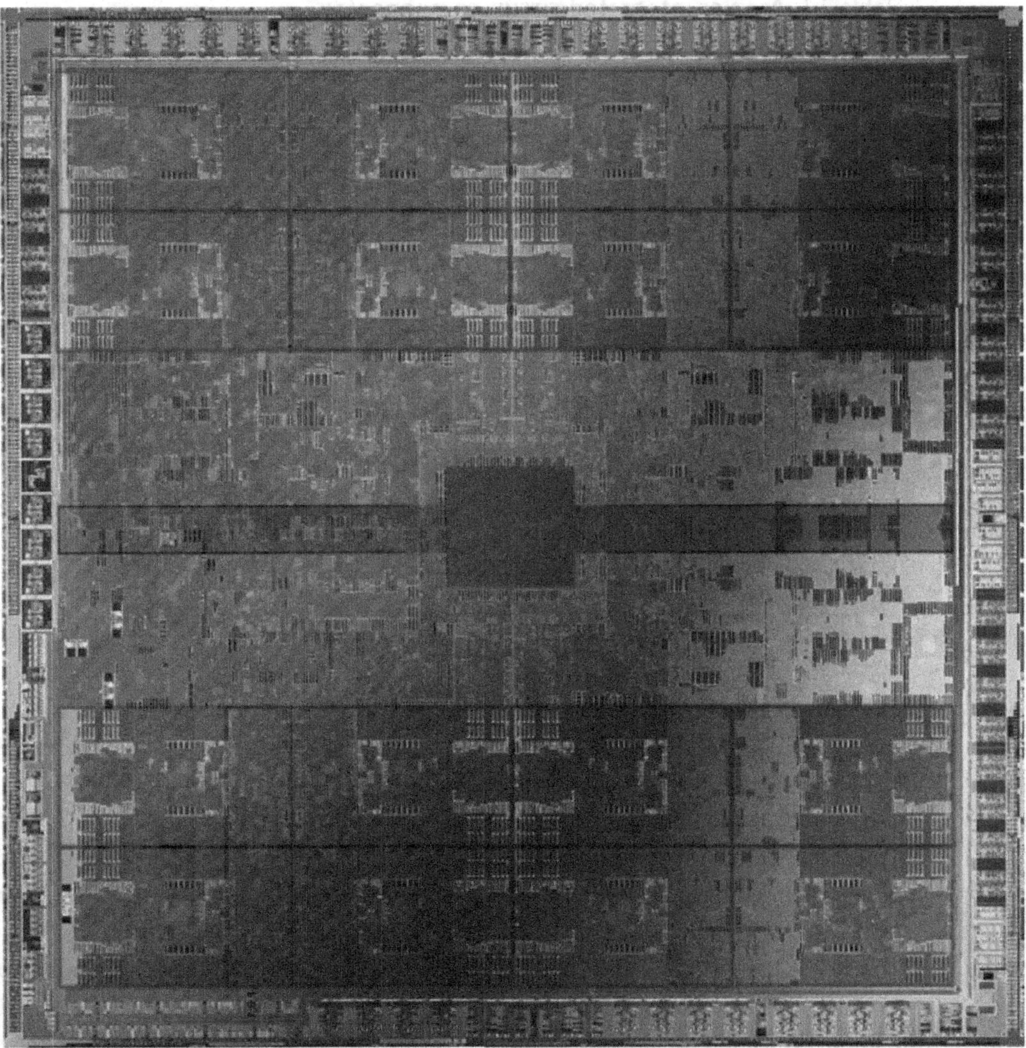

Figure 2.2. Die photo of an NVIDIA Fermi GPU. This chip is fabricated in 40 nm CMOS technology and contains over three billion transistors.

32 CUDA cores each, for a total of 512 cores. Each core contains an integer unit and a double-precision floating-point unit. A crossbar switch (see Section 24.3) can be seen in the center of the die. Connections to six 64-bit partitions (384 bits total) of GDDR5 DRAM (see Section 25.1.2) consume most of the periphery of the chip.

Field-programmable gate arrays (FPGAs) are an intermediate point between standard parts and ASICs. They are standard parts that can be programmed to realize an arbitrary function. Although they are significantly less efficient than ASICs, they are ideally suited to realizing custom logic in less demanding, low-volume applications. Large FPGAs, like the Xilinx Virtex 7, contain up to two million logic cells, over 10 MB of SRAM, several microprocessors, and hundreds of arithmetic building blocks. The programmable logic is significantly (over an order of magnitude) less dense, less energy-efficient, and slower than fixed standard-cell logic. This makes it prohibitively costly in high-volume applications. However, in low-volume

Table 2.1. **Area of integrated circuit components in grids**

Module	Area (grids)
One bit of DRAM	2
One bit of ROM	2
One bit of SRAM	24
Two-input NAND gate	40
Static latch	100
Flip-flop	300
One bit of a ripple-carry adder	500
32-bit carry-look-ahead adder	30 000
32-bit multiplier	300 000
32-bit RISC microprocessor (w/o caches)	500 000

applications, the high per-unit cost of an FPGA is attractive compared with the tooling costs for an ASIC. Manufacturing a 28 nm ASIC can cost up to $3 million, not including design costs. The total non-recurring cost[6] of an ASIC is about $20 million to $40 million.

To give you an idea what will fit on a typical ASIC, Table 2.1 lists the area of a number of typical digital building blocks in units of grids (χ^2). A *grid* is the area between the centerlines of adjacent minimum spaced wires in the *x*- and *y*-directions. In a contemporary 28 nm process, the minimum wire pitch is $\chi = 90$ nm and one grid is of area $\chi^2 = 8100\ \text{nm}^2$. In such a process, there are 1.2×10^8 grids/mm^2 and 1.2×10^9 grids on a relatively small 10 mm^2 die – enough room for 30 million NAND gates. A simple 32-bit RISC processor, which used to fill a chip in the mid 1980s, now fits into less than 0.01 mm^2. As described in Section 2.4, the number of grids per chip doubles every 18 months, so the number of components that can be packed on a chip is constantly increasing.

EXAMPLE 2.1 Estimating chip area

Estimate the total amount of chip area occupied by an eight-tap FIR filter. All inputs are 32 bits wide (i_i), and the filter stores eight 32-bit weights (w_i) in flip-flops. The output (X) is calculated as follows:

$$X = \sum_{i=0}^{7} i_i \times w_i.$$

The area for storing the weights, the multipliers, and the adders can be computed as follows:

$$A_w = 8 \times 32 \times A_{\text{ff}} = 7.68 \times 10^4 \text{ grids},$$

[6] The cost that is paid once, regardless of the number of chips manufactured.

$$A_m = 8 \times A_{\text{mul}} = 2.4 \times 10^6 \text{ grids},$$
$$A_a = 7 \times A_{\text{add}} = 2.1 \times 10^5 \text{ grids}.$$

We need only seven adders since we use a tree structure to do a pairwise reduction of addends (see Section 12.3). To get the total area in 28 nm technology, we sum the area of each of the components; the area is dominated by the multipliers:

$$A_{FIR} = A_w + A_m + A_a = 2.69 \times 10^6 \text{ grids} = 0.022 \text{ mm}^2.$$

Unfortunately, chip I/O bandwidth does not increase as fast as the number of grids per chip. Modern chips are limited to about 1000 signal pins by a number of factors, and such high pin counts come at a significant cost. One of the major factors limiting pin count and driving cost is the achievable density of printed-circuit boards. Routing all of the signals from a high-pin-count integrated circuit out from under the chip's package, the *escape pattern*, stresses the density of a printed-circuit board and often requires additional layers (and hence cost).

Modern circuit boards are laminated from copper-clad glass–epoxy boards interleaved with *pre-preg* glass–epoxy sheets.[7] The copper-clad boards are patterned using photolithography to define wires and are then laminated together. Connections between layers are made by drilling the boards and electroplating the holes. Boards can be made with a large number of layers – 20 or more is not unusual, but is costly. More economical boards have ten or fewer layers. Layers typically alternate between an x signal layer (carrying signals in the x-direction), a y signal layer, and a power plane. The power planes distribute power supplies to the chips, isolate the signal layers from one another, and provide a return path for the transmission lines of the signal layers. The signal layers can be defined with minimum wire width and spacing of 3 mils (0.003 inches, about 75 μm). Less expensive boards use 5 mil width and spacing rules.

Holes to connect between layers are the primary factor limiting board density. Because of electroplating limits, the holes must have an aspect ratio (ratio of the board thickness to the hole diameter) no greater than 10:1. A board with a thickness of 0.1 inch requires a minimum hole diameter of 0.01 inch. The minimum hole-to-hole centerline spacing is 25 mils (40 holes per inch). Consider, for example, the escape pattern under a chip in a 1 mm ball-grid-array (BGA) package. With 5 mil lines and spacing, there is room to escape just one signal conductor between the through holes (with 3 mil width and spacing, two conductors fit between holes), requiring a different signal layer for each row of signal balls after the first around the periphery of the chip.

Figure 2.3 shows a board from a Cray XT6 supercomputer. The board measures 22.58×14.44 inches and contains a number of integrated circuits and modules. On the left side of the board are two *Gemini* router chips – Cray ASICs that form a system-wide interconnection network (see Section 24.4). All that you see of the Gemini chips (and most of the other chips) is the metal heat sinks that draw heat out of the chips into the forced-air cooling. Next to the Gemini chips are 16 DRAM DIMM modules that provide the main memory for the supercomputer node. Next in line are four AMD Opteron eight-core CPU chips. On the right side of the board,

[7] A fiberglass cloth that is impregnated with an epoxy resin that has not been cured.

Figure 2.3. A node board from a Cray XT6 supercomputer. The board contains (left to right) two router chips, 16 DRAM DIMM modules, four AMD Opteron eight-core CPUs, and four NVIDIA Fermi GPU modules (see Figure 2.2). All that is visible of most of the chips and modules is their heat sinks.

under large copper heat sinks, are four NVIDIA Fermi C2090X GPU modules. Each of these modules is itself a small printed-circuit board that contains a Fermi GPU chip (shown in Figure 2.2), 24 GDDR5 DRAM chips, and voltage regulators for the GPU and memories.

Connectors carry signals from one board to another. Right-angle connectors connect cards to a *backplane* or *midplane* that carries signals between the cards. The Cray module of Figure 2.3 has such a connector at its far left side. This connector inserts into a backplane that is itself a printed-circuit board. The backplane contains signal layers that provide connections between modules in the backplane. The backplane also houses cable connectors that connect the modules via electrical or optical cables to modules on other backplanes.

Co-planar connectors connect daughter cards to a mother card. The four NVIDIA GPU modules at the right side of Figure 2.3 are connected to the Cray module via such co-planar connectors.

Packaging of electronic systems is described in more detail in ref. [33].

2.3 COMPUTER-AIDED DESIGN TOOLS

The modern digital designer is assisted by a number of computer-aided design (CAD) tools. CAD tools are computer programs that help manage one or more aspects of the design process. They fall into three major categories: capture, synthesis, and verification. CAD tools exist for doing logical, electrical, and physical design. We show an example design flow in Figure 2.4.

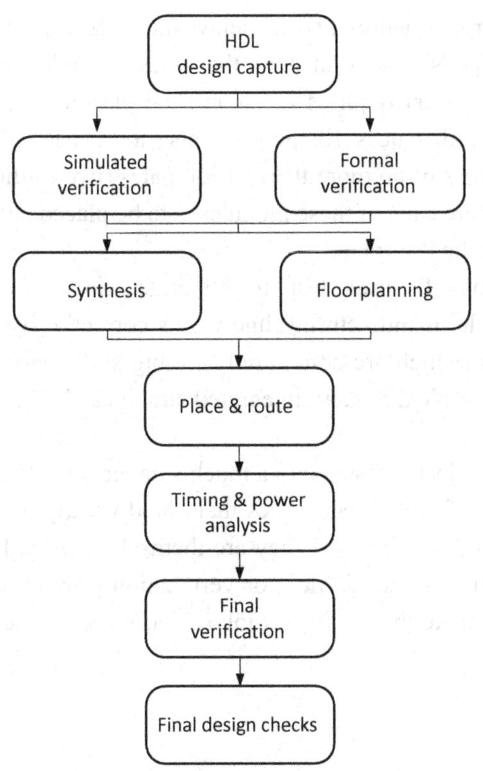

Figure 2.4. A typical CAD toolflow. The HDL design is first captured and verified. Engineers then floorplan and synthesize the design into logic gates, which are then placed and routed. Final tests on the design include timing analysis, final verification of the placed and routed netlist, and final physical design checks. Not pictured are the back-edges from all boxes to HDL design as the implementation is constantly refined and optimized.

As the name implies, capture tools help *capture* the design. The most common capture tool is a schematic editor. A designer uses the tool to enter the design as a hierarchical drawing showing the connections between all modules and submodules. For many designs a textual hardware description language (HDL) such as VHDL is used instead of a schematic, and a text editor is used to capture the design. Textual design capture is, with few exceptions, far more productive than schematic capture.

Once a design has been captured, verification tools are used to ensure that it is correct. Before a design can be shipped, it must be functionally correct, meet timing constraints, and have no electrical rule violations. A simulator is used to test the functionality of a schematic or HDL design. Test scripts are written to drive the inputs and observe the outputs of the design, and an error is flagged if the outputs are not as expected. Simulators, however, are only as good as their test cases. Without a test that exposes an error, it won't be found by a simulator. *Formal* verification tools use mathematical proof techniques to prove that a design meets a specification independently of test vectors. A static timing analyzer verifies that a design meets its timing constraints (also independently of test cases). We detail verification further in Chapter 20.

A synthesis tool reduces a design from one level of abstraction to a lower level of abstraction. For example, a logic synthesis tool takes a high-level description of a design in an HDL like VHDL, and reduces it to a gate-level netlist. Logic synthesis tools have largely eliminated manual combinational logic design, making designers significantly more productive. A place-and-route tool takes a gate-level netlist and reduces it to a physical design by placing the individual gates and routing the wires between them.

In modern ASICs and FPGAs, a large fraction of the delay and power is due to the wires interconnecting gates and other cells, rather than to the gates or cells themselves. Achieving high performance (and low power) requires managing the placement process to ensure that critical signals travel only short distances. Keeping signals short is best achieved by manually partitioning a design into modules of no more than 50 000 gates (two million grids), constructing a *floorplan* that indicates where each of these modules is to be placed, and placing each module separately into its *region* of the floorplan.

CAD tools are also used to generate manufacturing tests for integrated circuits. These tests prove that a particular chip coming off the manufacturing line works correctly. By scanning test patterns into the flip-flops of a chip (which are configured as a big shift register for the purpose), every transistor and wire of a complex, modern integrated circuit can be verified with a relatively small number of test patterns.

CAD tools, however, often limit a large design space into a much smaller set of designs that can be easily made with a particular set of tools. Techniques that could greatly improve the efficiency of a design are often disallowed, not because they are themselves risky, but rather because they do not work with a particular synthesis flow or verification procedure. This is unfortunate; CAD tools are supposed to make the designer's job easier, not limit the scope of what they can design.

2.4 MOORE'S LAW AND DIGITAL SYSTEM EVOLUTION

In 1965, Gordon Moore predicted that the number of transistors on an integrated circuit would double every year. This prediction, that circuit density increases exponentially, has held for 40 years so far, and has come to be known as Moore's law. Over time, the doubling every year has been revised to doubling every 18–20 months, but, even so, the rate of increase is very rapid. The number of components (or grids) on an integrated circuit is increasing with a compound annual growth rate of over 50%, growing by nearly an order of magnitude roughly every five years. This is plotted in Figure 2.5.

For many years (from the 1960s until about 2005), voltage scaled linearly with gate length. With this *constant-field* or Dennard scaling [37], as the number of devices increased, the devices also got faster and dissipated less energy. To a first approximation, when the linear dimension L of a semiconductor technology is halved, the area required by a device, which scales as L^2, is quartered, hence we can get four times as many devices in the same area. With constant-field scaling, the delay of the device was also proportional to L, and hence was also halved – so each of these devices runs twice as fast. The energy consumed by switching a single device $E_{sw} = CV^2$ scaled as L^3, since both C and V were proportional to L. Thus, back in the good old days of constant-field scaling, each time L was halved, our circuits could do eight times as much work (four times the number of devices running twice as fast) for the same power.

Unfortunately, the linear scaling of both device speed and supply voltage ended in 2005. Since then, supply voltages have remained roughly constant at about 1 V while L continues to reduce. In this new regime of *constant-voltage scaling*, each time L is halved we still get four

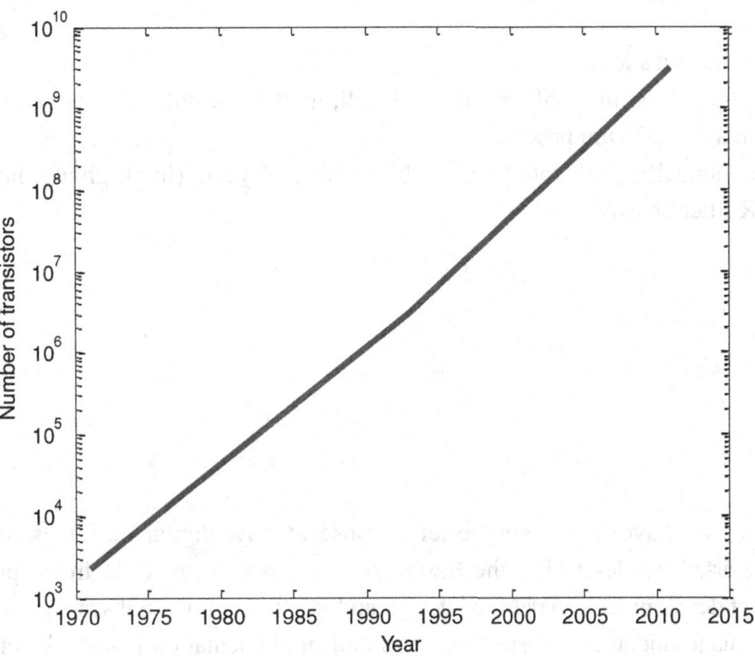

Figure 2.5. The number of transistors on commercial processors over several decades, illustrating Moore's law.

times as many devices per unit area. However, now they run only slightly faster (about 25% faster), and, most importantly, E_{sw} decreases only linearly, because V remains constant. Thus, in this new era, when we halve L we can do about five times as much work per unit area (four times as many devices running 1.25× faster) for 2.5× the power. It is easy to see that, in the regime of constant-voltage scaling, chips quickly get to the point where they are limited not by the number of devices that can integrated on the chip, but rather by power.

Moore's law makes the world an interesting place for a digital system designer. Each time the density of integrated circuits increases by an order of magnitude or two (every five to ten years), there is a qualitative change both in the type of systems being designed and in the methods used to design them. In contrast, most engineering disciplines are relatively stable – with slow, incremental improvements. You don't see cars getting a factor of eight more energy efficient every three years. Each time such a qualitative change occurs, a generation of designers gets a clean sheet of paper to work on, because much of the previous wisdom about how best to build a system is no longer valid. Fortunately, the basic principles of digital design remain invariant as technology scales; however, design practices change considerably with each technology generation.

The rapid pace of change in digital design means that digital designers must be students throughout their professional careers, constantly learning to keep pace with new technologies, techniques, and design methods. This continuing education typically involves reading the trade press (*EE Times* is a good place to start), keeping up with new-product announcements from fabrication houses, chip vendors, and CAD tool vendors, and occasionally taking a formal course to learn a new set of skills or update an old set.

EXAMPLE 2.2 Moore's law

Estimate how many FIR filters of Example 2.1 will, in 2017, fit into an area of a single filter implemented in a 2012 28 nm process.

We raise the annual growth rate (1.5) to the number of years (five), giving the following increase in FIR filter density:

$$N = 1.5^{2017-2012} = 7.6.$$

Summary

In this chapter you have had a very brief glimpse at how digital design is practiced in industry. We started by describing the *design process*, which proceeds from specification, through concept development, to detailed design and verification. Digital systems are designed in English-language documents – specifications and implementation plans – before they are captured in VHDL. A large part of the design process is managing risk. By quantifying risk and developing techniques to mitigate risk (e.g., backup plans), one can get the benefits of being technically aggressive without putting the overall project in jeopardy. Verification is a large part of the development process. The best rule of verification is that if it hasn't been tested, it doesn't work. The cost of fixing a bug increases by an order of magnitude at each stage of the design.

Modern digital systems are implemented with integrated circuits (chips) and printed-circuit boards. We presented a simple model for estimating the chip area required by a particular digital logic function and looked at an example chip and board.

Modern digital design practice makes heavy use of, and is constrained by, *computer-aided design* (CAD) tools. These tools are used for design capture, simulation, synthesis, and verification. Tools operate both at the logical level – manipulating VHDL designs and gate-level netlists – and at the physical level – manipulating the geometry of layers on integrated circuits.

The number of transistors that can be economically implemented on an integrated circuit increases exponentially with time – doubling every 18 months. This phenomenon, known as *Moore's law*, makes digital design a very dynamic field, since this rapidly evolving technology constantly presents new design challenges and makes possible new products that were not previously feasible.

BIBLIOGRAPHIC NOTES

Moore's original article in *Electronics* [84] not only predicts that the number of devices increases exponentially over time, but also explains why, with constant-field scaling, this results in constant chip power. Brunvand [21] gives an overview of the modern design process from hardware description language to manufacturing, showing how to use state-of-the-art CAD tools and providing many examples. Two books that discuss the engineering process as a whole are Brooks's

The Mythical Man-Month [19] and Kidder's *The Soul of a New Machine* [64]. Brooks's text provides several essays about software engineering, though many of its lessons also apply to hardware engineering. Kidder's book recounts the building of a minicomputer over the course of a year. Both are worth reading.

Exercises

2.1 *Specification, I.* You have decided to build an economical (cheap) video game system. Provide a specification of your design. Include components, inputs and outputs, and device media.

2.2 *Specification, II.* Congratulations, your video game console from Exercise 2.1 has become a great success! Provide a specification for version 2 of the console. Focus specifically on the changes from version 1, such as whether you will support backwards compatibility.

2.3 *Specification, III.* Give a specification for a traffic-light system to be placed at a busy intersection. Points to consider include how many lights, turn lanes, pedestrians, and light duration. You can assume that traffic is equally heavy in all directions.

2.4 *Buying vs. building, I.* Provide a buy vs. build decision and rationale for three of the components from the video game system of Exercise 2.1. Include at least one "buy" item and one "build" item.

2.5 *Buying vs. building, II.* If tasked with creating an air-bag deployment system for cars, what components will you need? What will you buy off the shelf and what will you design? Why? At minimum, you will need accelerometers, an actuator, and a centralized controller.

2.6 *Buy vs. building, III.* As part of your design, you have been tasked with purchasing a USB controller. Find a vendor (online) that sells USB controllers and download two different data sheets and pricings. What are the key differences?

2.7 *Risk management.* You are in charge of building the next-generation electronic car. Assign (and explain) a reward, risk ordered pair (on a scale of 1–5) to each of the following.

(a) Using an experimental new battery that can hold 5× the energy of your current batteries.

(b) Installing seat belts.

(c) Adding a cup holder.

(d) Installing sensors and a control system to alert the driver of cars in their blind spot.

(e) Adding a full satellite and navigation system. What are the risks?

(f) For your car, give another example of a (1,5) feature.

(g) For your car, give another example of a (5,5) feature.

2.8 *Feasibility and mitigation.* Design a feasibility study or a way of mitigating the risk for each of the following high-risk tasks:

(a) verifying a critical component with a new verification methodology;

(b) adding a new instruction to your next-generation processor;

(c) putting 16 cores onto a chip for a next-generation processor;

(d) porting your design from a 0.13 μm process to 28 nm;

(e) your answer to Exercise 2.7(g).

2.9 *Chip area, I.* Estimate how much area you will need, using Table 2.1, to implement a module that outputs the average of the last four 32-bit input values. You will need flip-flops to store the last three inputs. What about the weighted average with 32-bit arbitrary weights in ROM? How much area does it cost to store the weights in SRAM?

2.10 *Chip area, II.* A basic matrix multiply operation of two $n \times n$ matrices requires $3n^2$ units of storage and n^3 fused multiplication-additions. For this problem, assume that each element of a matrix is 32 bits and use the component sizes of Table 2.1. Assume that doing a fused multiply-add takes a combined 5 ns; that is, each 5 ns, one of the n^3 operations completes. The functional unit has an area equal to the sum of a 32-bit multiplier and a 32-bit adder.

(a) How much SRAM area is required to do an $n = 500$ matrix multiply? Give your answer in both grids and mm^2 for a 28 nm process.

(b) How long will the matrix multiply take if you have just one fused multiplier-adder?

(c) Assume you have an area budget of 10 mm^2. If you fill all the die area that is not SRAM with multiply-adders, how long does the operation take? Assume that *each* functional unit performs one multiply-add operation every 5 ns.

2.11 *Chip area, III.* Using the assumptions from Exercise 2.10, what is the largest matrix multiplication you can do in under 1 ms? Feel free to modify the proportion of die area devoted to storage and computation.

2.12 *Chip area, IV.* Using the assumptions from Exercise 2.10, what is the largest matrix multiplication you can do in under 1 μs?

2.13 *Chips and boards.* Find an image of a computer motherboard online or examine your own computer. Identify and explain the functions of at least three different chips found on the motherboard. You are not allowed to select the CPU, graphics processor, or DRAM as one of your chips.

2.14 *BGA escape pattern.* Sketch the escape pattern of 32 wires (eight on each side) from a chip to a connector at a different part of a board. Assume all wires must be routed on the surface of the board and cannot cross over each other.

2.15 *CAD tools, I.* Pick two functions from Figure 2.4, and find and describe three different computer programs that perform each function. Are any of them free? To get you started, leading design vendors include Synopsys and Cadence.

2.16 *CAD tools, II.* Why is the final verification stage needed at the end of Figure 2.4?

2.17 *Moore's law, I.* In 2015, a 133 mm^2 chip could have 1.9 billion transistors. Using Moore's law, how many transistors will be on a chip in (a) 2020 and (b) 2025?

2.18 *Moore's law, II.* In 2015, some manufacturers introduced 14 nm processors. If we assume that gate lengths scale with the square root of the number of transistors as given by Moore's law, in what year will gate lengths be five silicon atoms across?

2.19 *Moore's law, III.* Table 2.1 shows the area of a RISC processor and SRAM in *grids*. In a 28 nm process there are about 1.2×10^8 grids mm^2. How many RISC processors with 64 K Bytes (or 64 KB) of SRAM each will fit on a 20 mm $\times$ 20 mm chip in a 2011 28 nm process? How many will fit on a chip in 2020?

Part II

Combinational logic

3 Boolean algebra

We use Boolean algebra to describe the logic functions from which we build digital systems. Boolean algebra is an algebra over two elements, 0 and 1, with three operators: AND, which we denote as $\wedge$, OR, which we denote as $\vee$, and NOT, which we denote with a prime or overbar, e.g., NOT(x) is x' or $\bar{x}$. These operators have their natural meanings: $a \wedge b$ is 1 only if both a and b are 1, $a \vee b$ is 1 if either a or b is 1, and $\bar{a}$ is true only if a is 0.

We write logical expressions using these operators and binary variables. For example, $a \wedge \bar{b}$ is a logic expression that is true when binary variable a is true and binary variable b is false. An instantiation of a binary variable or its complement in an expression is called a *literal*. For example, the expression above has two literals, a and $\bar{b}$. Boolean algebra gives us a set of rules for manipulating such expressions so we can simplify them, put them in *normal form*, and check two expressions for equivalence.

We use the $\wedge$ and $\vee$ notation for AND and OR to make it clear that Boolean AND and OR are not multiplication and addition over the real numbers. Many sources, including many textbooks, unfortunately use $\times$ or $\cdot$ to denote AND and $+$ to denote OR. We avoid this practice because it can lead students to simplify Boolean expressions as if they were conventional algebraic expressions, that is expressions in the algebra of $+$ and $\times$ over the integers or real numbers. This can lead to confusion since the properties of Boolean algebra, while similar to conventional algebra, differ in some crucial ways.[1] In particular, Boolean algebra has the property of duality – which we shall discuss below – while conventional algebra does not. One manifestation of this is that in Boolean algebra $a \vee (b \wedge c) = (a \vee b) \wedge (a \vee c)$, whereas in conventional algebra $a + (b \times c) \neq (a + b) \times (a + c)$.

We will use Boolean algebra in our study of CMOS logic circuits (Chapter 4) and combinational logic design (Chapter 6).

3.1 AXIOMS

All of Boolean algebra can be derived from the definitions of the AND, OR, and NOT functions. These are most easily described as truth tables, shown in Tables 3.1 and 3.2. Mathematicians like to express these definitions in the form of *axioms*, a set of mathematical statements that we assert to be true. All of Boolean algebra derives from the following axioms:

$$\text{identity} \quad 1 \wedge x = x, \qquad 0 \vee x = x; \tag{3.1}$$

$$\text{annihilation} \quad 0 \wedge x = 0, \qquad 1 \vee x = 1; \tag{3.2}$$

$$\text{negation} \quad \bar{0} = 1, \qquad \bar{1} = 0. \tag{3.3}$$

[1] For historical reasons, we still refer to a set of signals ANDed (ORed) together as a product (sum).

Table 3.1. **Truth tables for AND and OR operations**

a	b	$a \wedge b$	$a \vee b$
0	0	0	0
0	1	0	1
1	0	0	1
1	1	1	1

Table 3.2. **Truth table for NOT operation**

a	$\bar{a}$
0	1
1	0

The *duality* of Boolean algebra is evident in these axioms. The principle of duality states that if a Boolean equation is true, then replacing the expressions on both sides with one where (a) all $\vee$s are replaced by $\wedge$s and vice versa and (b) all 0s are replaced by 1s and vice versa also gives an equation that is true. Since this duality holds in the axioms, and all of Boolean algebra is derived from these axioms, duality holds for all of Boolean algebra.

3.2 PROPERTIES

From our axioms we can derive a number of useful properties about Boolean expressions:

$$
\begin{array}{rll}
\text{commutative} & x \wedge y = y \wedge x, & x \vee y = y \vee x; \\
\text{associative} & x \wedge (y \wedge z) = (x \wedge y) \wedge z, & x \vee (y \vee z) = (x \vee y) \vee z; \\
\text{distributive} & x \wedge (y \vee z) = (x \wedge y) \vee (x \wedge z), & x \vee (y \wedge z) = (x \vee y) \wedge (x \vee z); \\
\text{idempotence} & x \wedge x = x, & x \vee x = x; \\
\text{complementation} & x \wedge \bar{x} = 0, & x \vee \bar{x} = 1; \\
\text{absorption} & x \wedge (x \vee y) = x, & x \vee (x \wedge y) = x; \\
\text{combining} & (x \wedge y) \vee (x \wedge \bar{y}) = x, & (x \vee y) \wedge (x \vee \bar{y}) = x; \\
\text{DeMorgan's} & \overline{(x \wedge y)} = \bar{x} \vee \bar{y}, & \overline{(x \vee y)} = \bar{x} \wedge \bar{y}; \\
\text{consensus} & (x \wedge y) \vee (\bar{x} \wedge z) \vee (y \wedge z) = (x \wedge y) \vee (\bar{x} \wedge z), & \\
 & (x \vee y) \wedge (\bar{x} \vee z) \wedge (y \vee z) = (x \vee y) \wedge (\bar{x} \vee z). &
\end{array}
$$

These properties can all be proved by checking their validity for all four possible combinations of x and y or for all eight possible combinations of x, y, and z. For example, we can prove De Morgan's theorem as shown in Table 3.3. Mathematicians call this proof technique *perfect induction*.

Table 3.3. **Proof of De Morgan's theorem by perfect induction**

x	y	$\overline{(x \wedge y)}$	$\bar{x} \vee \bar{y}$
0	0	1	1
0	1	1	1
1	0	1	1
1	1	0	0

This list of properties is by no means exhaustive. We can write down other logic equations that are always true. This set is chosen because it has proven to be useful in simplifying logic equations.

The commutative and associative properties are identical to the properties you are already familiar with from conventional algebra. We can reorder the arguments of an AND or OR operation, and an AND or OR with more than two inputs can be grouped in an arbitrary manner. For example, we can rewrite $a \wedge b \wedge c \wedge d$ as $(a \wedge b) \wedge (c \wedge d)$ or $(d \wedge (c \wedge (b \wedge a)))$. Depending on delay constraints and the library of available logic circuits, there are times when we would use both forms.

The distributive property is also similar to the corresponding property from conventional algebra. It differs, however, in that it applies both ways. We can distribute OR over AND as well as AND over OR. In conventional algebra we cannot distribute $+$ over $\times$.

The next four properties (idempotence, complementation, absorption, and combining) have no equivalent in conventional algebra. These properties are very useful in simplifying equations. For example, consider the following logic function:

$$f(a, b, c) = (a \wedge c) \vee (a \wedge b \wedge c) \vee (\bar{a} \wedge b \wedge c) \vee (a \wedge b \wedge \bar{c}). \tag{3.4}$$

First, we apply idempotence twice to triplicate the second term and apply the commutative property to regroup the terms:

$$f(a, b, c) = (a \wedge c) \vee (a \wedge b \wedge c) \vee (\bar{a} \wedge b \wedge c) \vee (a \wedge b \wedge c) \vee (a \wedge b \wedge \bar{c}) \vee (a \wedge b \wedge c). \tag{3.5}$$

Now we can apply the absorption property to the first two terms[2] and the combining property twice (to terms 3 and 4 and to terms 5 and 6), giving

$$f(a, b, c) = (a \wedge c) \vee (b \wedge c) \vee (a \wedge b). \tag{3.6}$$

In this simplified form it is easy to see that this is the famous *majority function* that is true whenever two or three of its input variables are true.

[2] The astute reader will notice that this gets us back to where we started before making a copy of the second term. However, it is useful to demonstrate the absorption property.

EXAMPLE 3.1 Proof of combining property

Prove, using perfect induction, the combining property of Boolean expressions.

We show the proof by enumerating all possible inputs and calculating the output of each function, as shown in Table 3.4.

Table 3.4. **Proof of combining property using perfect induction**

x	y	$(x \wedge y) \vee (x \wedge \bar{y})$	x	$(x \vee y) \wedge (x \vee \bar{y})$	x
0	0	0	0	0	0
0	1	0	0	0	0
1	0	1	1	1	1
1	1	1	1	1	1

EXAMPLE 3.2 Function simplification

Simplify, using the properties listed previously, the Boolean expression $f(x,y) = (x \wedge (y \vee \bar{x})) \vee (\overline{\bar{x} \vee \bar{y}})$.

The solution is as follows:

$$
\begin{aligned}
f(x,y) &= (x \wedge (y \vee \bar{x})) \vee (\overline{\bar{x} \vee \bar{y}}) & \\
&= ((x \wedge y) \vee (x \wedge \bar{x})) \vee (\overline{\bar{x} \vee \bar{y}}) & \text{distributive} \\
&= ((x \wedge y) \vee 0) \vee (\overline{\bar{x} \vee \bar{y}}) & \text{complementation} \\
&= (x \wedge y) \vee (\overline{\bar{x} \vee \bar{y}}) & \text{identity} \\
&= (x \wedge y) \vee (x \wedge y) & \text{DeMorgan} \\
&= (x \wedge y) & \text{idempotence.}
\end{aligned}
$$

3.3 DUAL FUNCTIONS

The dual of a logic function, f, is the function f^D derived from f by substituting a $\wedge$ for each $\vee$, a $\vee$ for each $\wedge$, a 1 for each 0, and a 0 for each 1.

For example, if

$$f(a,b) = (a \wedge b) \vee (b \wedge c), \tag{3.7}$$

then

$$f^D(a,b) = (a \vee b) \wedge (b \vee c). \tag{3.8}$$

A very useful property of duals is that the dual of a function applied to the complement of the input variables equals the complement of the function. That is,

$$f^D(\bar{a}, \bar{b}, \ldots) = \overline{f(a, b, \ldots)}. \tag{3.9}$$

This is a generalized form of De Morgan's theorem, which states the same result for simple AND and OR functions. We will use this property in Section 4.3 to use dual switch networks to construct the pull-up networks for CMOS gates.

EXAMPLE 3.3 Finding the dual of a function

Give the dual of the following unsimplified function:

$$f(x,y) = (1 \wedge x) \vee (0 \vee \bar{y}).$$

The solution is given by

$$f^D(x,y) = (0 \vee x) \wedge (1 \wedge \bar{y}).$$

Table 3.5. **Evaluation of Equation (3.9) using our example function**

x	y	$f^D(\bar{x},\bar{y})$	$\overline{f(x,y)}$
0	0	0	0
0	1	1	1
1	0	0	0
1	1	0	0

Table 3.5 shows that Equation (3.9) is valid for this particular function. You are asked to provide a more general proof in Exercise 3.6.

3.4 NORMAL FORM

Often we would like to compare two logical expressions to see whether they represent the same function. We could verify equivalence by testing them on every possible input combination – essentially filling out the truth tables and comparing them. However, an easier approach is to put both expressions into *normal form*, as a sum of product terms.[3]

For example, the normal form for the three-input majority function of Equations (3.4)–(3.6) is given by

$$f(a,b,c) = (a \wedge b \wedge \bar{c}) \vee (a \wedge \bar{b} \wedge c) \vee (\bar{a} \wedge b \wedge c) \vee (a \wedge b \wedge c). \tag{3.10}$$

Each product term of a logic expression in normal form corresponds to one row of the truth table with a true output for that function. These ANDed expressions are called a *minterm*. In normal form, each of the products must correspond to only one truth table row (minterm) and no more.

[3] This sum-of-products normal form is often called the conjunctive normal form. Because of duality, it is equally valid to use a product-of-sums normal form – often called the disjunctive normal form.

We can transform any logic expression into normal form by *factoring* it about each input variable using the identity:

$$f(x_1,\ldots,x_i,\ldots,x_n) = (x_i \wedge f(x_1,\ldots,1,\ldots,x_n)) \vee (\overline{x_i} \wedge f(x_1,\ldots,0,\ldots,x_n)). \tag{3.11}$$

For example, we can apply this method to factor the variable a out from the majority function of Equation (3.6) as follows:

$$f(a,b,c) = (a \wedge f(1,b,c)) \vee (\bar{a} \wedge f(0,b,c)) \tag{3.12}$$

$$= (a \wedge (b \vee c \vee (b \wedge c))) \vee (\bar{a} \wedge (b \wedge c)) \tag{3.13}$$

$$= (a \wedge b) \vee (a \wedge c) \vee (a \wedge b \wedge c) \vee (\bar{a} \wedge b \wedge c). \tag{3.14}$$

Repeating the expansion about b and c gives the majority function in normal form, Equation (3.10).

EXAMPLE 3.4 Normal form

Write the following equation in normal form:

$$f(a,b,c) = a \vee (b \wedge c).$$

Table 3.6. **Truth table for Example 3.48**

a	b	c	$f(a,b,c)$
0	0	0	0
0	0	1	0
0	1	0	0
0	1	1	1
1	0	0	1
1	0	1	1
1	1	0	1
1	1	1	1

We can find the normal form for this expression by writing the truth table, shown in Table 3.6, and selecting all of the minterms for which the output is one. The answer is as follows:

$$f(a,b,c) = (\bar{a} \wedge b \wedge c) \vee (a \wedge \bar{b} \wedge \bar{c}) \vee (a \wedge \bar{b} \wedge c) \vee (a \wedge b \wedge \bar{c}) \vee (a \wedge b \wedge c).$$

3.5 FROM EQUATIONS TO GATES

We often represent logical functions using a *logic diagram*, i.e., a schematic drawing of gate symbols connected by lines. Three basic gate symbols are shown in Figure 3.1. Each gate takes one or more binary inputs on its left side and generates a binary output on its right side. The

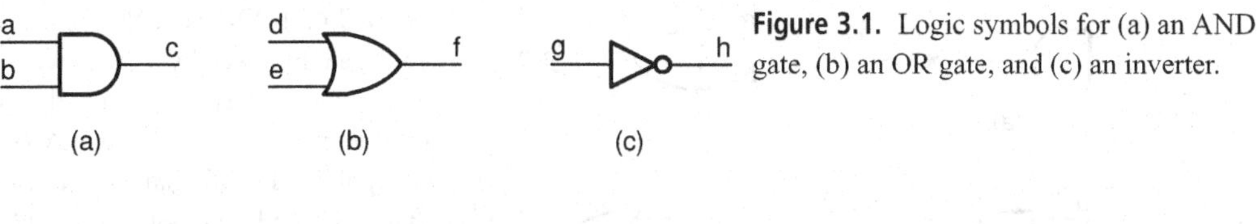

Figure 3.1. Logic symbols for (a) an AND gate, (b) an OR gate, and (c) an inverter.

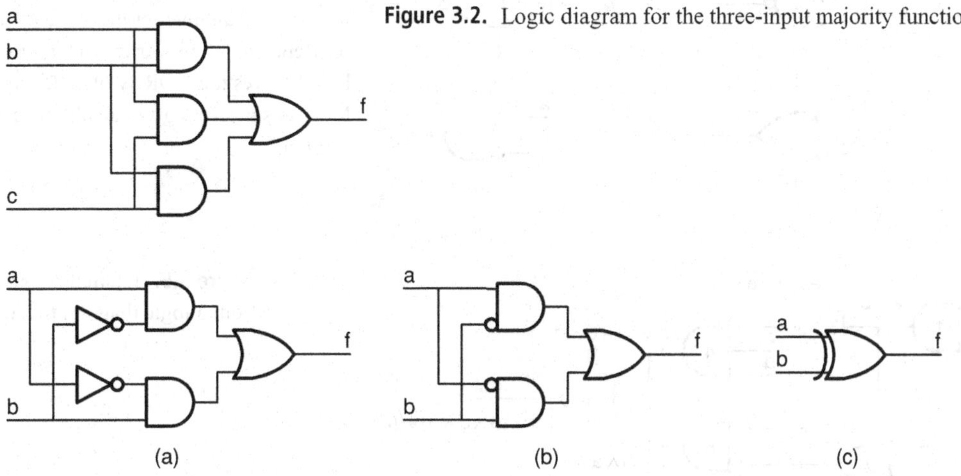

Figure 3.2. Logic diagram for the three-input majority function.

Figure 3.3. The exclusive-or function: (a) logic diagram with inverters, (b) logic diagram with inversion bubbles, and (c) gate symbol.

AND gate (Figure 3.1(a)) outputs a binary signal that is the AND of its inputs, $c = a \wedge b$. The OR gate of Figure 3.1(b) computes the OR of its inputs, $f = d \vee e$. The inverter (Figure 3.1(c)) generates a signal that is the complement of its single input, $h = \overline{g}$. AND gates and OR gates may have more than two inputs. Inverters always have a single input.

Using these three gate symbols, we can draw a logic diagram for any Boolean expression. To convert from an expression to a logic diagram, pick an operator ($\vee$ or $\wedge$) at the top level of the expression and draw a gate of the corresponding type. Label the inputs to the gate with the argument subexpressions. Repeat this process on the subexpressions.

For example, a logic diagram for the majority function of Equation (3.6) is shown in Figure 3.2. We start by converting the two $\vee$s at the top level into a three-input OR gate at the output. The inputs to this OR gate are the products $a \wedge b$, $a \wedge c$, and $b \wedge c$. We then use three AND gates to generate these three products. The net result is a logic circuit that computes the expression $f = (a \wedge b) \vee (a \wedge c) \vee (b \wedge c)$.

Figure 3.3(a) shows a logic diagram for the *exclusive-or* or XOR function, a logic function where the output is high only if exactly one of its inputs is high (i.e., if one input is exclusively high): $f = (a \wedge \overline{b}) \vee (\overline{a} \wedge b)$. The two inverters generate $\overline{b}$ and $\overline{a}$, respectively. The AND gates then form the two products $a \wedge \overline{b}$ and $\overline{a} \wedge b$. Finally, the OR gate forms the final sum. The exclusive-or function is used frequently enough that we give it its own gate symbol, shown in Figure 3.3(c). It also has its own symbol, $\oplus$, for use in logic expressions: $a \oplus b = (a \wedge \overline{b}) \vee (\overline{a} \wedge b)$.

Because we are frequently complementing signals in logic diagrams, we often drop the inverters and replace them with *inversion bubbles*, as shown in Figure 3.3(b). This diagram represents the same function as Figure 3.3(a); we have simply used a more compact notation

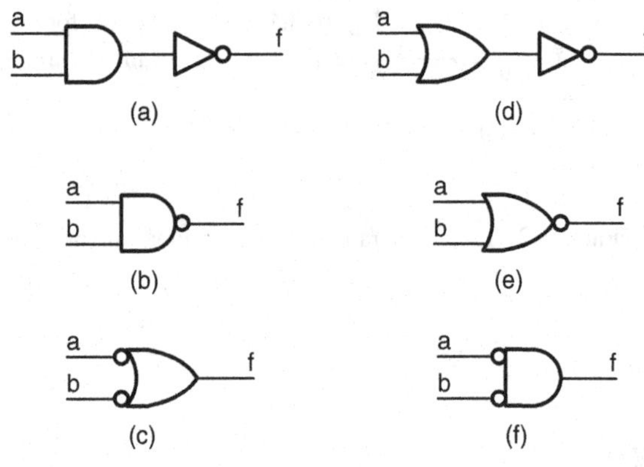

Figure 3.4. NAND and NOR gates. (a) An AND gate followed by an inverter realizes the NAND function. (b) Replacing the inverter with an inversion bubble gives the NAND symbol. (c) Applying De Morgan's theorem gives an alternate NAND symbol. (d) An OR gate followed by an inverter gives the NOR function. (e) Replacing the inverter with an inversion bubble gives the NOR symbol. (f) Applying De Morgan's theorem gives an alternate NOR symbol.

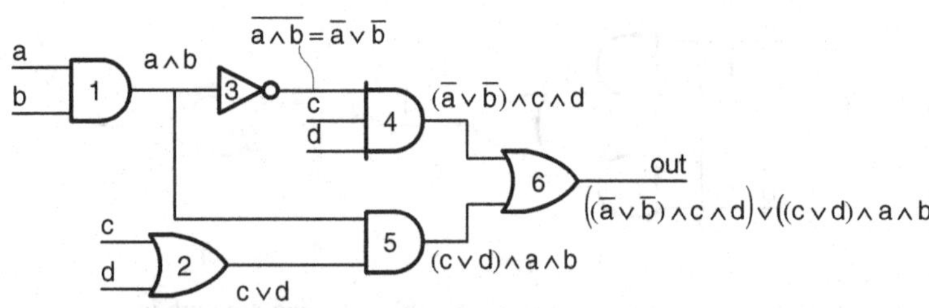

Figure 3.5. Example of converting from a logic diagram to an equation.

for the inversion of a and b. An inversion bubble may be placed on the input or the output of a gate. In either location it inverts the sense of the signal. Putting an inversion bubble on the input of a gate is equivalent to passing the input signal through an inverter and then connecting the output of the inverter to the gate input.

An inversion bubble can be used on the output of a gate as well as the input. Figure 3.4 shows this notation. An AND gate followed by an inverter (Figure 3.4(a)) is equivalent to an AND gate with an inversion bubble on its output (Figure 3.4(b)). By De Morgan's theorem, this is also equivalent to an OR gate with inversion bubbles on its input (Figure 3.4(c)). We refer to this gate, which performs the function $f = \overline{(a \wedge b)}$, as a NAND gate (for NOT-AND).

We can apply the same transformation to an OR gate followed by an inverter (Figure 3.4(d)). We replace the inverter with an inversion bubble to yield the NOR gate symbol of Figure 3.4(e), and by applying De Morgan's theorem we get the alternative NOR gate symbol of Figure 3.4(f). Because common logic families, such as CMOS, provide only inverting gates, we often use NAND and NOR gates as our primitive building blocks rather than AND and OR.

Figure 3.5 shows how we convert from logic diagrams to equations. Starting at the input, label the output of each gate with an equation. For example, the AND gate labeled 1 (AND-1) computes $a \wedge b$ and OR-2 computes $c \vee d$ directly from the inputs. Inverter-3 inverts $a \wedge b$, giving $\overline{(a \wedge b)} = \bar{a} \vee \bar{b}$. Note that this inverter could be replaced with an inversion bubble on the input of AND-4. AND-4 combines the output of the inverter with inputs c and d to generate $(\bar{a} \vee \bar{b}) \wedge c \wedge d$. AND-5 combines the outputs of gates 1 and 2 to give $(c \vee d) \wedge a \wedge b$. Finally, OR-6 combines the outputs of AND-4 and AND-5 to give the final result: $((\bar{a} \vee \bar{b}) \wedge c \wedge d) \vee ((c \vee d) \wedge a \wedge b)$.

EXAMPLE 3.5 From equation to schematic

Draw the schematic for a three-input majority function that uses only NAND gates.

We begin with the initial function and apply De Morgan's theorem to convert our function to use NAND gates:

$$\begin{aligned} f(a,b,c) &= (a \wedge b) \vee (a \wedge c) \vee (b \wedge c) \\ &= \overline{\overline{(a \wedge b) \vee (a \wedge c) \vee (b \wedge c)}} \\ &= \overline{\overline{(a \wedge b)} \wedge \overline{(a \wedge c)} \wedge \overline{(b \wedge c)}}. \end{aligned}$$

The resulting schematic, using the alternate form of a three-input NAND gate, is shown in Figure 3.6.

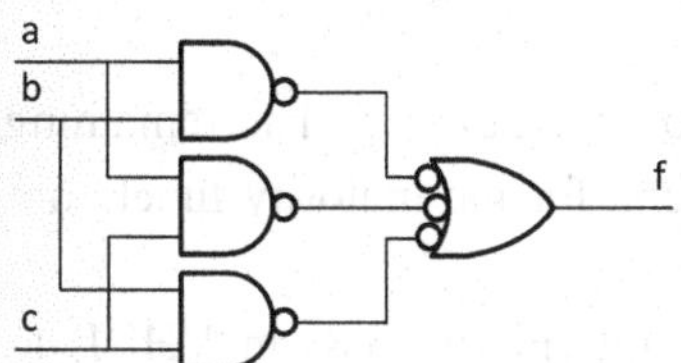

Figure 3.6. A NAND-only implementation of the majority function, as derived in Example 3.5.

To convert back to an equation to check the answer, we first write the values of the intermediate nodes. From top to bottom, they are $\overline{a \wedge b}$, $\overline{a \wedge c}$, and $\overline{b \wedge c}$. The final three-input NAND gate returns us to the original function:

$$f(a,b,c) = \overline{\overline{(a \wedge b)} \wedge \overline{(a \wedge c)} \wedge \overline{(b \wedge c)}}.$$

3.6 BOOLEAN EXPRESSIONS IN VHDL

In this book we will be implementing digital systems by describing them in a hardware description language named *VHDL*. VHDL code can be simulated, compiled to a field-programmable gate array (FPGA), or synthesized into a manufacturable chip. In this section we will introduce VHDL by showing how it can be used to describe logic expressions.

VHDL uses the keywords **`and`**, **`or`**, **`xor`** and **`not`** to represent bitwise AND, OR, XOR, and NOT, respectively. Using these symbols, we can write a VHDL expression for our majority function Equation (3.6) as follows:

```
output <= (a and b) or (a and c) or (b and c) ;
```

The signal assignment compound delimiter "`<=`" indicates that this statement assigns a value to signal `output`. The statement is terminated by a semicolon (`;`). Note that **`and`**, **`or`** and **`xor`** have identical operator precedence in VHDL and are evaluated from left to right. Hence, we must group subexpressions using parentheses.

We can declare the majority gate to be a VHDL *design entity*, as shown in Figure 3.7. The first two lines include definitions for the type `std_logic` (described in Section 1.5). The next four lines are an *entity declaration*, which specifies the design entity name, `majority`, and

```
library ieee;
use ieee.std_logic_1164.all;

entity majority is
  port( a, b, c : in std_logic;
        output : out std_logic );
end majority;

architecture impl of majority is
begin
  output <= (a and b) or (a and c) or (b and c);
end impl;
```

Figure 3.7. VHDL description of a majority gate.

declares the design entity's inputs `a`, `b`, and `c`, and output `output`. The remaining lines are an *architecture body* that contain our VHDL expression for the majority function as a single *concurrent assignment statement.*

To test our majority gate, we can write a test script (Figure 3.8) in VHDL to simulate the gate for all eight possible combinations of input variables. At the outset it is important to understand that this VHDL is not meant to be synthesized into hardware (i.e., gates). To ensure that synthesis tools do not try to compile this VHDL into gates we can surround the testbench VHDL with a *`--pragma translate_off`* and *`--pragma translate_on`* pair. Alternatively, we can place the testbench in a separate file from the design entity and ensure that the testbench VHDL is not used as input for synthesis.

The test script declares a top-level design entity called `test_maj` with no inputs or outputs. The architecture body begins with a declarative section after the keyword **`is`** and before the keyword **`begin`** that includes a *component* declaration followed by two signal declarations.

A *component declaration* starts with the keyword **`component`** and follows the same syntax as the entity declaration except that it ends with **`end component`**. It is required so that the testbench can later *instantiate* the `majority` design entity from Figure 3.7. It may seem redundant to specify the interface of `majority` twice – once in the `entity` declaration part of Figure 3.7 and again as a component declaration inside the design entity in 3.8. VHDL employs component declarations because they enable the hardware designer to specify multiple architecture bodies for a given component. Selecting the architecture body to use for a given component can be done with an optional VHDL *configuration declaration*, which we will describe in Section 7.1.8. In the absence of a configuration declaration, the last architecture body for the design entity is used.

A *signal declaration* starts with the keyword **`signal`** followed by an identifier, then a colon (:) delimiter, and then the type of the signal. A signal, such as the three-bit signals `count` and `output` in Figure 3.8, can be accessed only from within the architecture body where the signal is declared.

The rest of the architecture body after the keyword **`begin`** includes a component instantiation statement, which creates an instance of the majority design entity along with a *process* statement used to generate inputs to the majority unit and for displaying its output. Note

```
-- pragma translate_off
library ieee;
use ieee.std_logic_1164.all;
use ieee.std_logic_unsigned.all;

entity test_maj is
end test_maj;

architecture test of test_maj is
  component majority is
    port( a, b, c : in std_logic;
          output : out std_logic );
  end component;
  signal count: std_logic_vector(2 downto 0); -- input (three-bit counter)
  signal output: std_logic; -- output of majority
begin
  -- instantiate the gate
  DUT: majority port map(count(0), count(1), count(2), output);

  -- generate all eight input patterns
  process begin
    count <= "000";
    for i in 0 to 7 loop
      wait for 10 ns;
      report "count = " & to_string(count) & ", output = " & to_string(output);
      count <= count + 1;
    end loop;
    std.env.stop(0);
  end process;
end test;
-- pragma translate_on
```

Figure 3.8. Test script to instantiate and exercise a majority gate.

that unlike in a conventional software language, these two statements are active at exactly the same time.

The component instantiation statement creates an instance of the majority gate and gives it a *label*, `DUT` (device under test), of our choice. The component instantiation statement connects the majority gate to the signals `count` and `output` using a *port map aspect* identified by the keywords **`port map`**.

The process statement begins with the keyword **`process`**. We will see later that the process statement can be used both for specifying scripts and in synthesizable logic. In this example we use a process statement to specify a script and so we employ VHDL syntax including **`wait`** and **`loop`** that will simplify the job of testing but should *never* be used for VHDL meant to describe synthesizable logic. The body of the process statement includes a set of *sequential* statements, the first of which sets the values of `count` to `"000"`. This is followed by a **`for`** loop that iterates eight times. The *wait* statement inside the body of the for loop inserts a delay of 10 nanoseconds to allow the output of the majority gate to stabilize. The next line, beginning with the keyword **`report`**, prints out the values of `count` and `output`. The VHDL-2008 function `to_string()`, defined in `ieee.std_logic_1164`, is called twice and it converts

```
count = 000, output = 0
count = 001, output = 0
count = 010, output = 0
count = 011, output = 1
count = 100, output = 0
count = 101, output = 1
count = 110, output = 1
count = 111, output = 1
```

Figure 3.9. Output from test script of Figure 3.8.

its input argument from `std_logic_vector` to a string that can be printed by **`report`**. The next line increments the value of `count`. To use the addition operator, +, we need to include the *use clause* "**`use`** `ieee.std_logic_unsigned.`**`all`**`;`". The call to the VHDL-2008 function `std.env.stop()` after the loop stops simulation.

The result of running this test script is shown in Figure 3.9.

EXAMPLE 3.6 VHDL design entity

Write the VHDL that directly corresponds to the form of the majority function

$$f(a,b,c) = \overline{\overline{(a \wedge b)} \wedge \overline{(a \wedge c)} \wedge \overline{(b \wedge c)}}.$$

The VHDL design entity declaration is equivalent to Figure 3.7. We replace the signal assignment statement with the following:

```
output <= not (not (a and b) and not (a and c) and not (b and c));
```

Note that in VHDL **`not`** has higher operator precedence than **`and`**, **`or`**, and **`xor`**. This form of the majority function is less readable than the one shown in Figure 3.7; needlessly so, since synthesis tools will transform the first form into the second automatically when this is warranted.

Summary

In this chapter you have learned the basics of Boolean algebra, the algebra with elements 0 and 1 and operators ∧, ∨, and NOT. Boolean algebra is used to analyze digital logic, which operates on 0s and 1s using these operators.

Boolean algebra is completely defined by the three *axioms* of *identity*, *annihilation*, and *negation*. From the three axioms, we can infer a number of useful *properties*, including the *commutative*, *associative*, and *distributive* properties and *De Morgan's theorem*. Using these and other properties, we can manipulate equations in Boolean algebra. To compare Boolean functions, we can express them in *normal form*, as a sum of *minterms*, product terms containing all input variables.

Boolean algebra has the property of *duality*. We find the dual of a function by substituting $\wedge$ for $\vee$, $\vee$ for $\wedge$, 1 for 0, and 0 for 1. Dual functions have the useful property that $f^D(\bar{a}, \bar{b}, \ldots) = \overline{f(a, b, \ldots)}$. For example, the dual of $a \wedge b$ is $a \vee b$ and $\overline{a \wedge b} = \bar{a} \vee \bar{b}$.

A Boolean function can be expressed as a gate-level circuit by substituting the gate symbols of Figure 3.1 for the $\wedge$, $\vee$, and NOT operations in the function. We use an inversion *bubble* as a shorthand for NOT. We can also convert gate-level schematics to equations by propagating partial expressions from the inputs to the output of each gate.

We can express Boolean functions in VHDL using a signal assignment statement (<=) and substituting AND for $\wedge$, OR for $\vee$, and NOT for NOT. Once expressed in VHDL, we can simulate a Boolean function, or encapsulate it in a *design entity* that can be used to build more complex functions.

BIBLIOGRAPHIC NOTES

George Boole formulated Boolean logic in the mid 1800s. He presented his results in two texts that are now freely available online [13, 14]. De Morgan described much of his work on logic formulations in an 1860 text [36].

Exercises

3.1 *Prove absorption.* Prove that the absorption property is true by using perfect induction (i.e., enumerate all the possibilities).

3.2 *Prove idempotence.* Prove that the idempotence property is true by using perfect induction.

3.3 *Prove the associative property.* Prove that the associative property is true.

3.4 *Prove the distributive property.* Prove that the distributive property – both $\wedge$ over $\vee$ and $\vee$ over $\wedge$ – is true.

3.5 *It's not $+$ and $\times$.* Prove that the distributive property does not work for distributing $+$ over $\times$ with integers.

3.6 *De Morgan's theorem, I.* Using perfect induction, prove De Morgan's theorem with four variables, specifically

$$\overline{w \wedge x \wedge y \wedge z} = \bar{w} \vee \bar{x} \vee \bar{y} \vee \bar{z}$$

and

$$\overline{w \vee x \vee y \vee z} = \bar{w} \wedge \bar{x} \wedge \bar{y} \wedge \bar{z}.$$

3.7 *De Morgan's theorem, II.* Show that Equation (3.9) is indeed true by successively applying De Morgan's theorem to a logic function in normal form.

3.8 *Simplifying Boolean equations, I.* Reduce the following Boolean expression to a minimum number of literals: $(x \vee y) \wedge (x \vee \bar{y})$.

3.9 *Simplifying Boolean equations, II.* Reduce the following Boolean expression to a minimum number of literals: $(x \wedge y \wedge z) \vee (\bar{x} \wedge y) \vee (x \wedge y \wedge \bar{z})$.

3.10 *Simplifying Boolean equations, III.* Reduce the following Boolean expression to a minimum number of literals: $((y \wedge \bar{z}) \vee (\bar{x} \wedge w)) \wedge ((x \wedge \bar{y}) \vee (z \wedge \bar{w}))$.

3.11 *Simplifying Boolean equations, IV.* Reduce the following Boolean expression to a minimum number of literals: $(x \wedge y) \vee (x \wedge ((w \wedge z) \vee (w \wedge \bar{z})))$.

3.12 *Simplifying Boolean equations, V.* Reduce the following Boolean expression to a minimum number of literals: $(w \wedge \bar{x} \wedge \bar{y}) \vee (w \wedge \bar{x} \wedge \bar{y} \wedge z) \vee (w \wedge x \wedge \bar{y} \wedge z)$.

3.13 *Dual functions, I.* Find the dual function of the following function and write it in normal form: $f(x, y) = (x \wedge \bar{y}) \vee (\bar{x} \wedge y)$.

3.14 *Dual functions, II.* Find the dual function of the following function and write it in normal form: $f(x, y, z) = (x \wedge y) \vee (x \wedge z) \vee (y \wedge z)$.

3.15 *Dual functions, III.* Find the dual function of the following function and write it in normal form: $f(x, y, z) = (x \wedge ((y \wedge z) \vee (\bar{y} \wedge \bar{z}))) \vee (\bar{x} \wedge ((y \wedge \bar{z}) \vee (\bar{y} \wedge z)))$.

3.16 *Normal form, I.* Rewrite the following Boolean expression in normal form: $f(x, y, z) = (x \wedge \bar{y}) \vee (\bar{x} \wedge z)$.

3.17 *Normal form, II.* Rewrite the following Boolean expression in normal form: $f(x, y, z) = x$.

3.18 *Normal form, III.* Rewrite the following Boolean expression in normal form: $f(x, y, z) = (x \wedge ((y \wedge z) \vee (\bar{y} \wedge \bar{z}))) \vee (\bar{x} \wedge ((y \wedge \bar{z}) \vee (\bar{y} \wedge z)))$.

3.19 *Normal form, IV.* Rewrite the following Boolean expression in normal form: $f(x, y, z) = 1$ if exactly 0 or 2 inputs are 1.

3.20 *Equation from schematic, I.* Write down a simplified Boolean expression for the functions computed by the logic circuit of Figure 3.10(a).

3.21 *Equation from schematic, II.* Write down a simplified Boolean expression for the functions computed by the logic circuit of Figure 3.10(b).

3.22 *Equation from schematic, III.* Write down a simplified Boolean expression for the functions computed by the logic circuit of Figure 3.10(c).

3.23 *Schematic from equation, I.* Draw a schematic for the following unsimplified logic equation: $f(x, y, z) = (\bar{x} \wedge y \wedge \bar{z}) \vee (\bar{x} \wedge \bar{y} \wedge \bar{z}) \vee (x \wedge \bar{y} \wedge \bar{z})$.

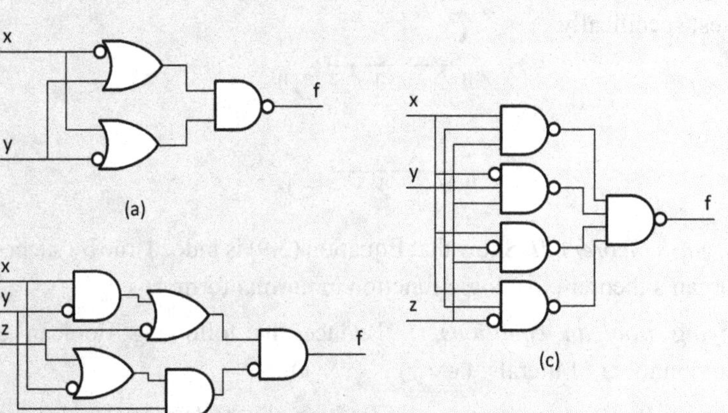

Figure 3.10. Logic circuits for Exercises 3.20, 3.21, and 3.22.

3.24 *Schematic from equation, II.* Draw a schematic for the following unsimplified logic equation: $f(x,y,z) = ((x \wedge y) \vee z) \wedge (x \wedge \bar{z})$.

3.25 *Schematic from equation, III.* Draw a schematic for the following unsimplified logic equation:

$$f(x,y,z) = \overline{\overline{(x \wedge y)} \vee z}.$$

3.26 *Schematic from equation, IV.* Draw a schematic for the following unsimplified logic equation: $f(x,y,z) = 1$ if 1 or 2 inputs are 1.

3.27 *VHDL* Write a VHDL design entity that implements the logic function

$$f(x,y,z) = (x \wedge y) \vee (\bar{x} \wedge z).$$

Write a test script to verify the operation of your design entity on all eight combinations of x, y, and z. What function does this circuit realize?

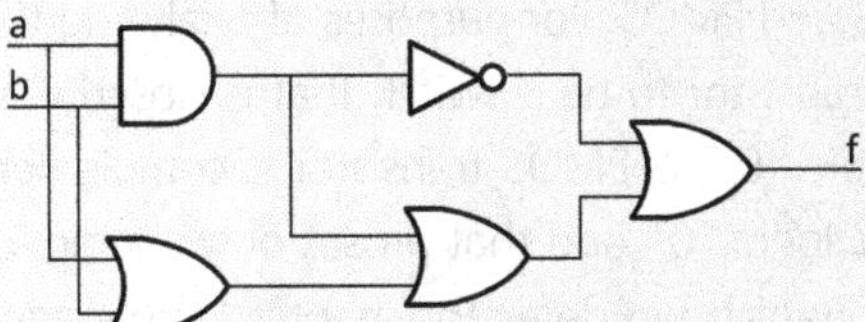

Figure 3.11. Logic circuit for Exercise 3.28.

3.28 *Logic equations.*

(a) Write out the unsimplified logic equation for the circuit of Figure 3.11.

(b) Write the dual form with no simplification.

(c) Draw the circuit for the unsimplified dual form.

(d) Simplify the original equation.

(e) Explain how the inverter and the last OR gate in the original circuit work together to allow this simplification.

3.29 *Choosing a representation.* Which representation, (a), (b), or (c), for a playing card requires the fewest gate inputs to check whether three playing cards are all of the same suit (i.e., all hearts, all spades, all diamonds, or all clubs). For the purposes of this answer, assume that an XOR gate costs the same as three normal gates, e.g., a two-input XOR costs as much as six gate inputs.

(a) Representing the suit as a four-bit one-hot number. A one-hot number has exactly one bit set to 1. For example, clubs would be represented by 0001, spades by 0010, diamonds by 0100, and hearts by 1000.

(b) Representing the suit as a two-bit Gray-coded number. Gray coding is explained in Exercise 1.9, an example of which is 00 for clubs, 01 for spades, etc.

(c) Representing the suit with three bits that can be either one-hot or zero-hot. This encoding can have zero (000) or one bit set (001, 010, 100).

4 CMOS logic circuits

In this chapter we will see how to build logic circuits (*gates*) using complementary metal–oxide–semiconductor (CMOS) transistors. We start in Section 4.1 by examining how logic functions can be realized using switches. A series combination of switches performs an AND function while a parallel combination of switches performs an OR function. We can build up more complex *switch-logic* functions by building more complex series–parallel switch networks.

In Section 4.2 we present a very simple switch-level model of an MOS transistor. CMOS transistors come in two flavors: NMOS and PMOS. For purposes of analyzing the function of logic circuits, we consider an NMOS transistor to be a switch that is closed when its gate is a logic "1" and that passes only a logic "0." A PMOS transistor is complementary – it is a switch that is closed when its gate is a logic "0" and that passes only a logic "1." To model the delay and power of logic circuits (which we defer to Chapter 5), we add a resistance and a capacitance to our basic switch. This switch-level model is *much* simpler than the models used for MOS circuit design, but is perfectly adequate to analyze the functionality and performance of digital logic circuits.

Using our switch-level model, we see how to build *gate circuits* in Section 4.3 by building a pull-down network of NMOS transistors and a complementary pull-up network of PMOS transistors. A NAND gate, for example, is realized with a series pull-down network of NMOS transistors and a parallel pull-up network of PMOS transistors.

4.1 SWITCH LOGIC

In digital systems we use binary variables to represent information and switches controlled by these variables to process information. Figure 4.1 shows a simple switch circuit. When binary variable a is false (0), Figure 4.1(a), the switch is open and the light is off. When a is true (1), the switch is closed, current flows in the circuit, and the light is on (Figure 4.1(b)).

We can perform simple logic with networks of switches as illustrated in Figure 4.2. Here, we omit the voltage source and light bulb for clarity, but we still think of the switching network as being *true* when its two terminals are connected; i.e., the light bulb, if connected, would be on.

Suppose we want to build a switch network that will launch a missile only if two switches (activated by responsible individuals) are closed. We can do this as illustrated in Figure 4.2(a) by placing two switches in series controlled by logic variables a and b, respectively. For clarity we usually omit the switch symbols and denote a switch as a break in the wire labeled by the variable controlling the switch as shown in the middle of the figure. Only when both a and b are true are the two terminals connected. Thus, we are assured that the missile will be launched

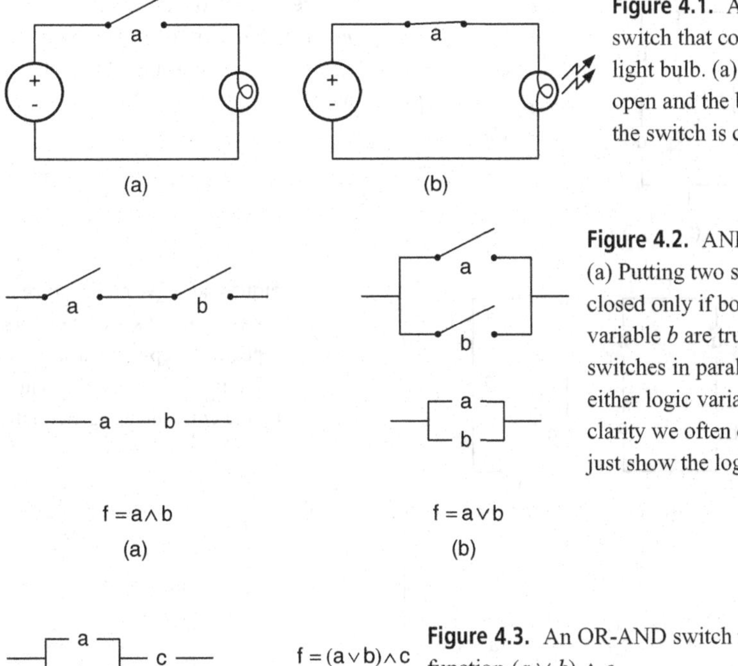

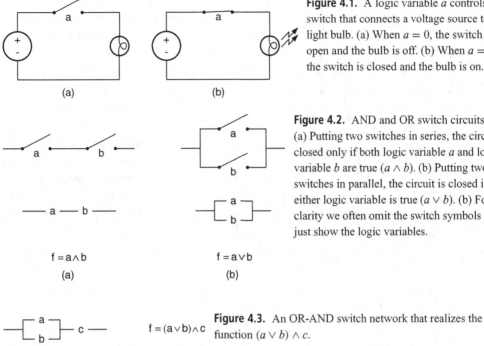

Figure 4.1. A logic variable a controls a switch that connects a voltage source to a light bulb. (a) When $a = 0$, the switch is open and the bulb is off. (b) When $a = 1$, the switch is closed and the bulb is on.

Figure 4.2. AND and OR switch circuits. (a) Putting two switches in series, the circuit is closed only if both logic variable a and logic variable b are true ($a \wedge b$). (b) Putting two switches in parallel, the circuit is closed if either logic variable is true ($a \vee b$). (b) For clarity we often omit the switch symbols and just show the logic variables.

Figure 4.3. An OR-AND switch network that realizes the function $(a \vee b) \wedge c$.

only if both a and b agree that it should be launched. Either a or b can stop the launch by not closing its switch. The logic function realized by this switch network is $f = a \wedge b$.[1]

When launching missiles we want to make sure that everyone agrees to launch before going forward. Hence we use an AND function. When stopping a train, on the other hand, we would like to apply the brakes if anyone sees a problem. In that case, we use an OR function, as shown in Figure 4.2(b), placing two switches in parallel controlled by binary variables a and b, respectively. In this case, the two terminals of the switch network are connected if either a, b, or both a and b are true. The function realized by the network is $f = a \vee b$.

We can combine series and parallel networks to realize arbitrary logic functions. For example, the network of Figure 4.3 realizes the function $f = (a \vee b) \wedge c$. To connect the two terminals of the network, c must be true, and either a or b must be true. For example, you might use a circuit like this to engage the starter on a car if the key is turned c and either the clutch is depressed a or the transmission is in neutral b.

More than one switch network can realize the same logical function. For example, Figure 4.4 shows two different networks that both realize the three-input majority function. Recall that a majority function returns true if the majority of its inputs are true; in the case of a three-input function, this occurs if at least two inputs are true. The logic function realized by both of these networks is $f = (a \wedge b) \vee (a \wedge c) \vee (b \wedge c)$.

There are several ways to analyze a switch network to determine the function it implements. One can enumerate all 2^n combinations of the n inputs to determine the combinations for which the network is connected. Alternatively, one can trace all paths between the two terminals

[1] Recall from Chapter 3 that $\wedge$ denotes the logical AND of two variables and $\vee$ denotes the logical OR of two variables.

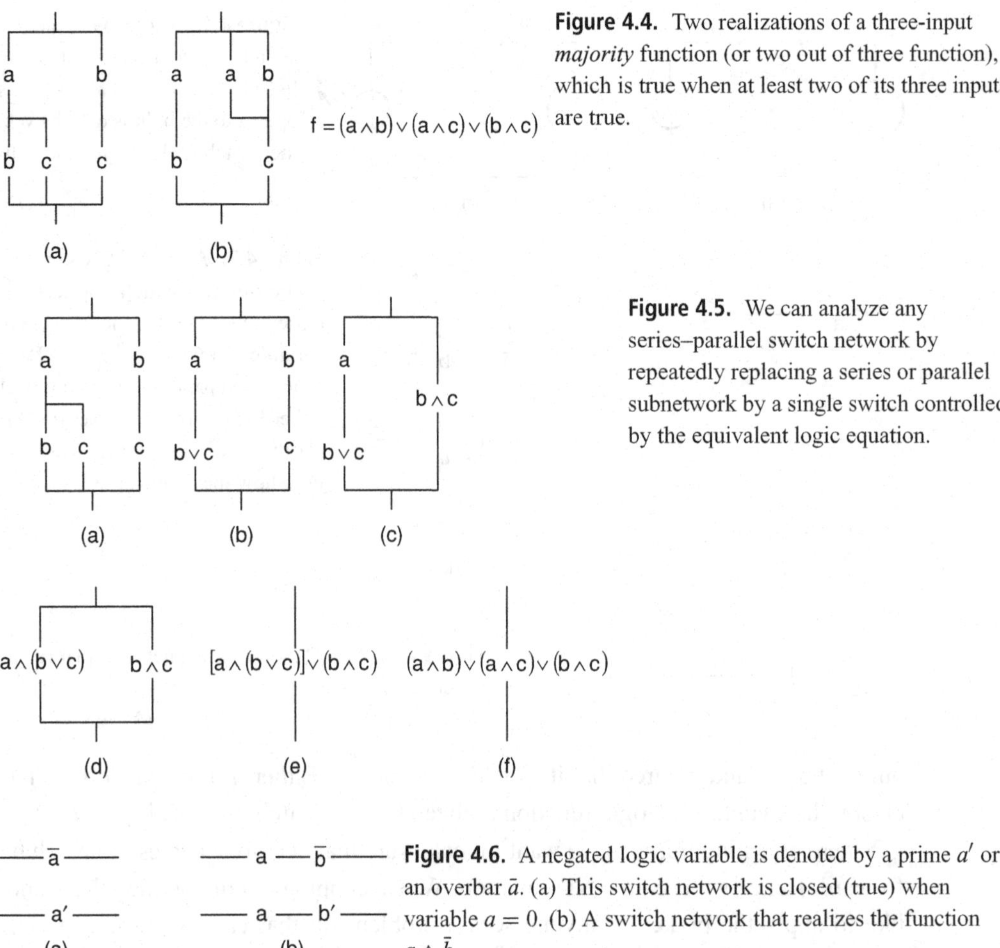

Figure 4.4. Two realizations of a three-input *majority* function (or two out of three function), which is true when at least two of its three inputs are true.

Figure 4.5. We can analyze any series–parallel switch network by repeatedly replacing a series or parallel subnetwork by a single switch controlled by the equivalent logic equation.

Figure 4.6. A negated logic variable is denoted by a prime a' or an overbar $\bar{a}$. (a) This switch network is closed (true) when variable $a = 0$. (b) A switch network that realizes the function $a \wedge \bar{b}$.

to determine the sets of variables that, if true, make the function true. For a series–parallel network, one can also reduce the network one step at a time by replacing a series or parallel combination of switches with a single switch controlled by an AND or OR of the previous switches' expressions.

Figure 4.5 shows how the network of Figure 4.4(a) is analyzed by replacement. The original network is shown in Figure 4.5(a). We first combine the parallel branches labeled b and c into a single switch labeled $b \vee c$ (Figure 4.5(b)). The series combination of b and c is then replaced by $b \wedge c$ (Figure 4.5(c)). In Figure 4.5(d) the switches labeled a and $b \vee c$ are replaced by $a \wedge (b \vee c)$. The two parallel branches are then combined into $[a \wedge (b \vee c)] \vee (b \wedge c)$ (Figure 4.5(e)). If we distribute the AND of a over $(b \vee c)$ we get the final expression in Figure 4.5(f).

So far we have used only positive switches in our network – that is, switches that are closed when their associated logic variable or expression is true (1). The set of logic functions we can implement with only positive switches is limited to monotonically increasing functions. To allow us to implement all possible functions we need to introduce negative switches – switches that are closed when their controlling logic variable is false (0). As shown in Figure 4.6(a) we denote a negative switch by labeling its controlling variable with either a prime a' or an overbar $\bar{a}$. Both of these indicate that the switch is closed when a is false (0). We can build

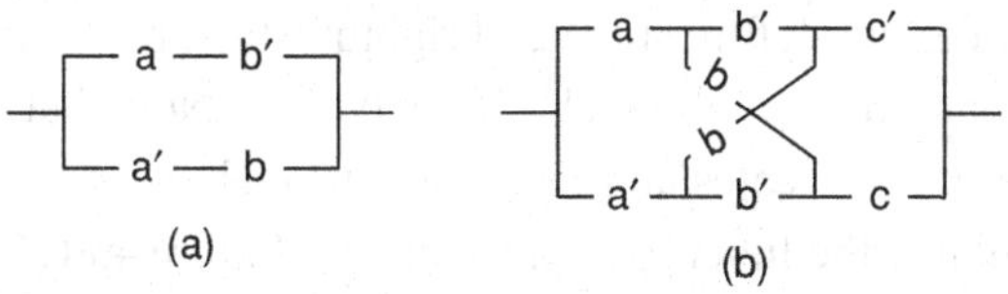

Figure 4.7. Exclusive-or (XOR) switch networks are true (closed) when an odd number of their inputs are true. (a) Two-input XOR network. (b) Three-input XOR network.

logic networks that combine positive and negative switches. For example, Figure 4.6(b) shows a network that realizes the function $f = a \wedge \bar{b}$.

Often we will control both positive and negative switches with the same logic variable. For example, Figure 4.7(a) shows a switch network that realizes the two-input exclusive-or (XOR) function. The upper branch of the circuit is connected if a is true and b is false, while the lower branch is connected if a is false and b is true. Thus this network is connected (true) if exactly one of a or b is true. It is open (false) if both a and b are true or false.

This circuit should be familiar to anyone who has ever used a light in a hallway or stairway controlled by two switches, one at either end of the hall or stairs. Changing the state of either switch changes the state of the light. Each switch is actually two switches – one positive and one negative – controlled by the same variable: the position of the switch control.[2] They are wired exactly as shown in the figure – with switches $a, \bar{a}$ at one end of the hall, and $b, \bar{b}$ at the other end.

In a long hallway, we sometimes would like to be able to control the light from the middle of the hall as well as from the ends. This can be accomplished with the three-input XOR network shown in Figure 4.7(b). An n-input XOR function is true when an odd number of the inputs are true. This three-input XOR network is connected if exactly one of the inputs a, b, or c is true or if all three of them are true. To see this, you can enumerate all eight combinations of a, b, and c or you can trace paths. You cannot, however, analyze this network by replacement as with Figure 4.5 because it is not a series–parallel network. If you want to have more fun analyzing non-series–parallel networks, see Exercises 4.1 and 4.2.

In the hallway application, the switches associated with a and c are placed at either end of the hallway, and the switches associated with b are placed in the center of the hall. As you have probably observed, if we want to add more switches controlling the same light, we can repeat the four-switch pattern of the b switches as many times as necessary, each time controlled by a different variable.[3]

EXAMPLE 4.1 Series–parallel networks

Draw and simplify a series–parallel circuit that realizes the function $f(d, c, b, a) = 1$ if $dcba$ is a legal thermometer encoding (0000, 0001, 0011, 0111, or 1111).

Figure 4.8 shows the steps we take to find a solution. First, we begin in Figure 4.8(a) by drawing the networks represented by the five minterms in parallel. Next, in Figure 4.8(b), we

[2] Electricians call these three-terminal, two-switch units *three-way switches*.

[3] Electricians call this four-terminal, four-switch unit, where the connections are straight through when the variable is false (switch handle down) and crossed when the variable is true (switch handle up), a *four-way switch*. To control one light with $n \geq 2$ switches requires two three-way switches and $n - 2$ four-way switches. Of course, one can always use a four-way switch as a three-way switch by leaving one terminal unconnected.

eliminate unnecessary variables from the two leftmost (d) and rightmost (a) paths. Finally, we collapse the $a \wedge b$ terms into a single path in parallel with both c and $\bar{d} \wedge \bar{c}$. Note that we cannot combine the two $\bar{d} \wedge \bar{c}$ terms, since doing so creates an illegal path through $c \wedge \bar{b}$.

To check our solution, we can formulate the function represented by Figure 4.8(c). We show our work in Figures 4.8(d)–(f). We arrive at the following equation:

$$f(d, c, b, a) = ((c \vee (\bar{d} \wedge \bar{c})) \wedge b \wedge a) \vee (\bar{d} \wedge \bar{c} \wedge \bar{b}).$$

This is another form of our original function, and verifies our answer.

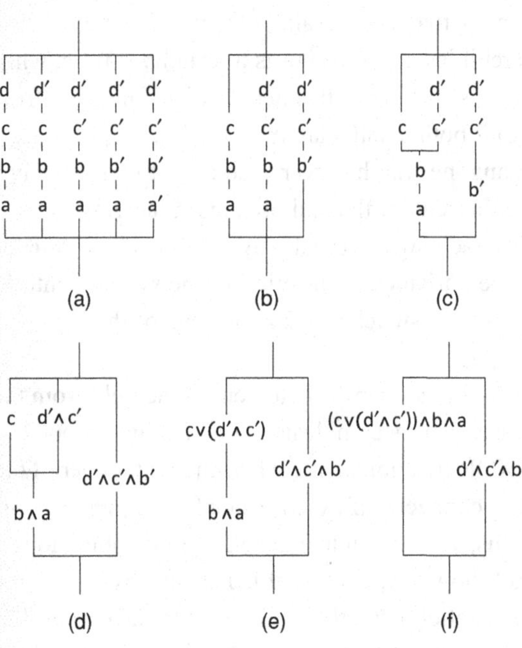

Figure 4.8. The solution for Example 4.1. In parts (a)–(c) we iteratively reduce our initial series–parallel network. In parts (d)–(f) we derive the Boolean expression shown by (c).

4.2 SWITCH MODEL OF MOS TRANSISTORS

Most modern digital systems are built using CMOS (complementary metal–oxide–semiconductor) field-effect transistors as switches. Figure 4.9 shows the physical structure and schematic symbol for a MOS transistor. A MOS transistor is formed on a semiconductor substrate and has three terminals:[4] the gate, source, and drain. The source and drain are identical terminals formed by diffusing an impurity into the substrate. The gate terminal is formed from polycrystalline silicon (called *polysilicon* or just *poly* for short) and is insulated from the substrate by a thin layer of oxide. The name MOS, a holdover from the days when the gate terminals were metal (aluminum), refers to the layering of the gate (metal), gate oxide (oxide), and substrate (semiconductor).[5]

[4] The substrate is a fourth terminal that we will ignore at present.

[5] Some very advanced processes are moving back to metal gates.

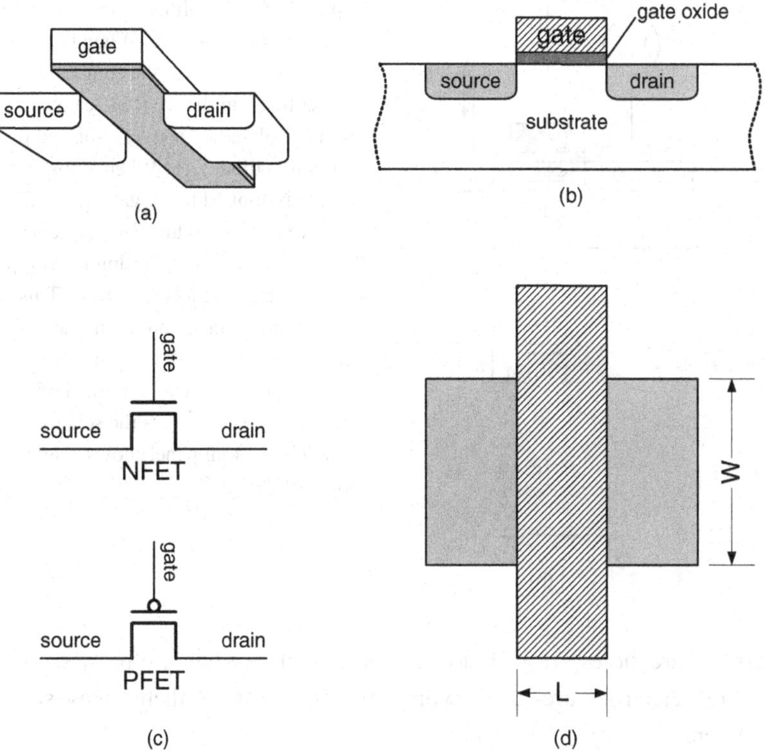

Figure 4.9. A MOS field-effect transistor (FET) has three terminals. Current passes between the source and drain (identical terminals) when the device is on. The voltage on the gate controls whether the device is on or off. (a) The structure of a MOSFET with the substrate removed. (b) A side view of a MOSFET. (c) Schematic symbols for an n-channel FET (NFET) and a p-channel FET (PFET). (d) Top view of a MOSFET, showing its width W and length L.

Figure 4.9(d), a top view of the MOSFET, shows the two dimensions that can be varied by the circuit or logic designer to determine transistor performance:[6] the device *width* W and the device *length* L. The gate length L is the distance that charge carriers (electrons or holes) must travel to get from the source to the drain, and thus is directly related to the speed of the device. Gate length is so important that we typically refer to a semiconductor process by its gate length. For example, most new designs today (2015) are implemented in 20 nm CMOS processes (i.e., CMOS processes with a minimum gate length of 20 nm). Almost all logic circuits use the minimum gate length supported by the process. This gives the fastest devices with the least power dissipation.

The channel width W controls the strength of the device. The wider the device, the more charge carriers can traverse the device in parallel. Thus the larger W, the lower the on resistance of the transistor and the higher the current the device can carry. A large W makes the device faster by allowing it to discharge a load capacitance more quickly. Alas, reduced resistance comes at a cost – the gate capacitance of the device also increases with W. Thus as W increases it takes longer to charge or discharge the gate of a device.

Figure 4.9(c) shows the schematic symbols for an n-channel MOSFET (NFET) and a p-channel MOSFET (PFET). In an NFET, the source and drain are n-type semiconductor in a p-type substrate and the charge carriers are electrons. In a PFET the types are reversed – the source and drain are p-type in an n-type substrate (usually an n-well diffused in a p-type

[6] The gate oxide thickness is also a critical dimension, but it is set by the process and cannot be varied by the designer. In contrast, W and L are determined by the mask set, and hence can be adjusted by the designer.

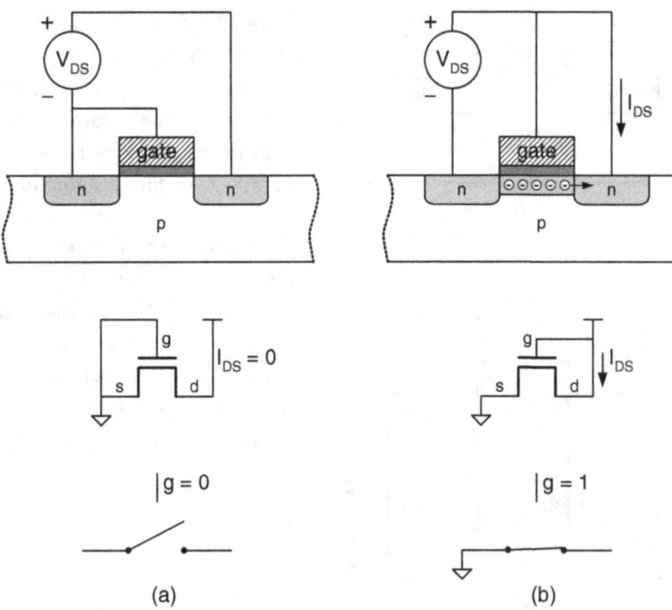

Figure 4.10. Simplified operation of an n-channel MOSFET. (a) When the gate is at the same voltage as the source, no current flows in the device because the drain is isolated by a reverse-biased p–n junction (a diode). (b) When a positive voltage is applied to the gate, it induces negative carriers in the *channel* beneath the gate, effectively inverting the p-type silicon to become n-type silicon. This connects the source and drain, allowing a current I_{DS} to flow. The top panel shows what happens physically in the device. The middle panel shows the schematic view. The bottom panel shows a switch model of the device.

substrate) and the carriers are holes. If you do not have a clue what n-type and p-type semiconductors, holes, and electrons are, don't worry; we will abstract them away shortly. Bear with us for the moment.

Figure 4.10 illustrates a simple digital model of operation for an n-channel FET.[7] As shown in Figure 4.10(a), when the gate of the NFET is a logic 0, the source and drain are isolated from one another by a pair of p–n junctions (back-to-back diodes), and hence no current flows from drain to source, $I_{DS} = 0$. This is reflected in the schematic symbol in the middle panel. We model the NFET in this state with an open switch, as shown in the bottom panel.

When the gate terminal is a logic 1 *and* the source terminal is a logic 0, as shown in Figure 4.10(b), the NFET is turned on. The positive voltage between the gate and source induces a negative charge in the *channel* beneath the gate. The presence of these negative charge carriers (electrons) makes the channel effectively n-type and forms a conductive region between the source and drain. The voltage between the drain and the source accelerates the carriers in the channel, resulting in a current flow from drain to source, I_{DS}. The middle panel shows the schematic view of the on NFET. The bottom panel shows a switch model of the on NFET. When the gate is 1 and the source is 0, the switch is closed.

It is important to note that if the source[8] is 1, the switch will *not* be closed, even if the gate is 1, because there is no net voltage between the gate and source to induce the channel charge. The switch is not open in this state either, because it will turn on if either terminal drops a threshold voltage below the 1 voltage. With source = 1 and gate = 1, the NFET is in an undefined state (from a digital perspective). The net result is that an NFET can reliably pass only a logic 0 signal. To pass a logic 1 requires a PFET.

[7] A detailed discussion of MOSFET operation is far beyond the scope of this book. Consult a textbook on semiconductor devices for more details.

[8] Physically the source and drain are identical, and the distinction is a matter of voltage. The source of an NFET (PFET) is the most negative (positive) of the two non-gate terminals.

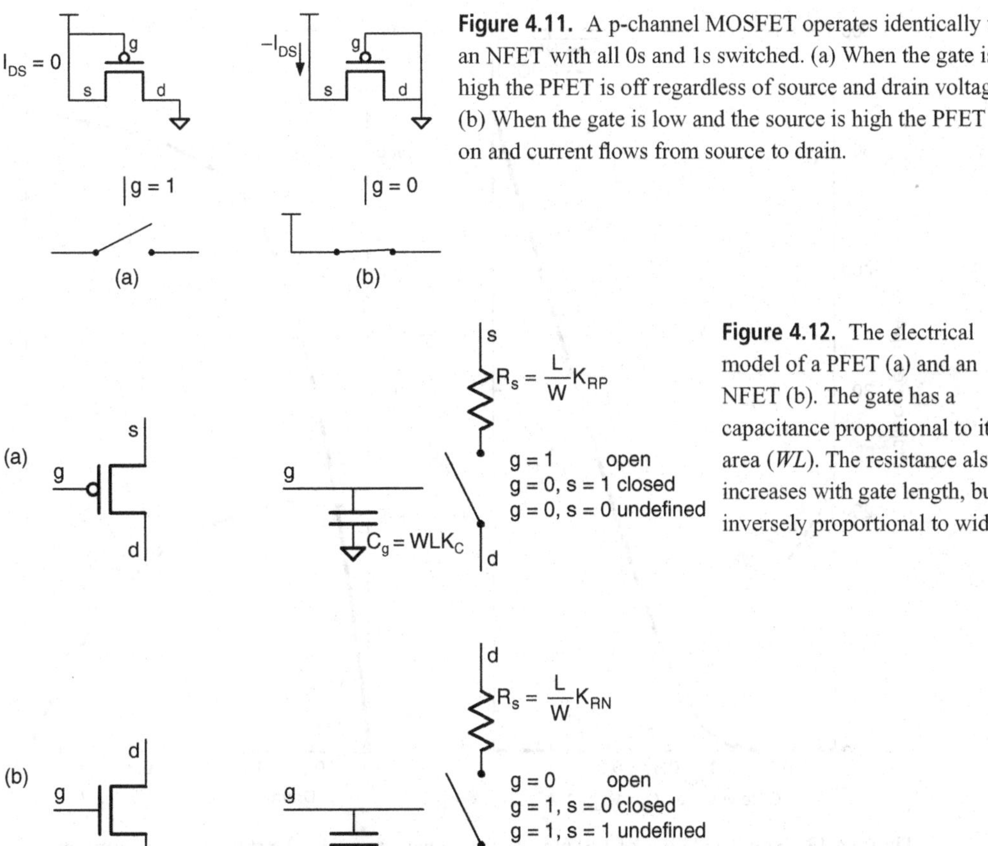

Figure 4.11. A p-channel MOSFET operates identically to an NFET with all 0s and 1s switched. (a) When the gate is high the PFET is off regardless of source and drain voltages. (b) When the gate is low and the source is high the PFET is on and current flows from source to drain.

Figure 4.12. The electrical model of a PFET (a) and an NFET (b). The gate has a capacitance proportional to its area (*WL*). The resistance also increases with gate length, but is inversely proportional to width.

Operation of a PFET, illustrated in Figure 4.11, is identical to the NFET with the 1s and 0s reversed. When the gate is 0 and the source is 1, the device is on. When the gate is 1, the device is off. When the gate is 0 and the source is 0, the device is in an undefined state. Because the source must be 1 for the device to be reliably on, the PFET can reliably pass only a logic 1. This nicely complements the NFET, which can only pass a 0.

The NFET and PFET models of Figures 4.10 and 4.11 accurately model the function of most digital logic circuits. However, to model the delay and power of logic circuits we must complicate our model slightly by adding a resistance in series with the source and drain and a capacitance from the gate to ground, as shown in Figure 4.12.[9] The capacitance on the gate node is proportional to the area of the device, *WL*. The resistance, on the other hand, is proportional to the aspect ratio of the device, L/W. We write the capacitance and resistance equations as follows:

$$C_g = WLK_C, \tag{4.1}$$

$$R_s = \frac{L}{W}K_R, \tag{4.2}$$

[9] In reality, there is capacitance on the source and drain nodes as well – usually each has a capacitance equal to about half of the gate capacitance (depending on device size and geometry). For the purposes of this book, however, we'll lump all of the capacitance on the gate node.

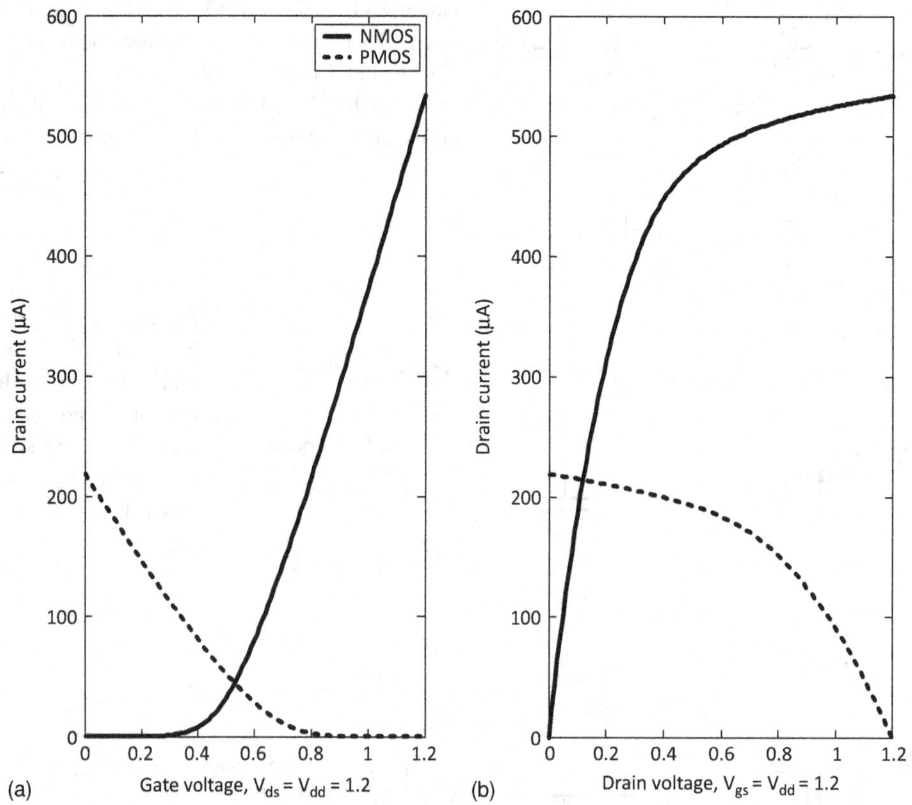

Figure 4.13. Simulated plots of 130 nm transistor characteristics. We show the drain current as a function of gate voltage (a) and drain voltage (b) for both an NFET (solid line) and a PFET (dashed line).

where K_C is a constant for capacitance measured in farads per unit area, and K_R is a resistance constant measured in ohms per square. For NFETs and PFETs, K_C is the same, but K_R is different by the ratio K_P. PFETs have a higher resistance than NFETs.

Figure 4.13(a) shows the current of a transistor (both an NFET and a PFET) as a function of the input voltage. In steady-state digital systems, we operate only at the two edges of the plot, where the difference in current is orders of magnitude. This validates our switch model. Figure 4.13 also shows that the ratio of NFET drive current to PFET drive current (K_P) is approximately 2.5. Figure 4.13(b) indicates that as the source to drain voltage reaches 0, transistors stop conducting current (and dissipating power). Figure 4.14 shows the steady-state output of an inverter as a function of the input. This graph shows the noise immunity of a CMOS circuit at both low and high inputs; small changes, or noise, in the input voltage do not affect the output voltage.

For convenience, and to make our discussion independent of a particular process generation, we will express W and L in units of $L_{\min}$, the minimum gate length of a technology. For example, in a 28 nm technology, we will refer to a device with $L = 28$ nm and $W = 224$ nm as an $L = 1$, $W = 8$ device, or just as an $W = 8$ device since $L = 1$ is the default. In most cases we will scale W by $W_{\min} = 8L_{\min}$ and refer to a minimum-sized $W/L = 8$ device as a unit-sized device. We then size other devices in relation to this unit-sized device.

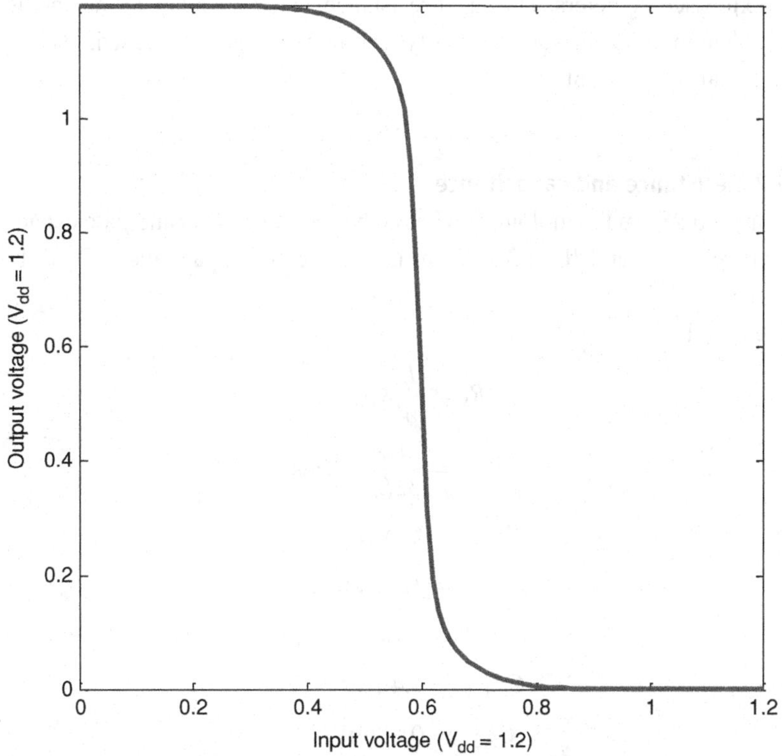

Figure 4.14. Output voltage of an inverter as a function of input voltage.

Table 4.1. **Device parameters for typical 130 nm and 28 nm CMOS processes**

All numbers assume a minimum gate length and variable gate width

Parameter	Value (130 nm)	Value (28 nm)	Units
K_C	2.0×10^{-16}	2.8×10^{-17}	farads/L^2_{min}
K_{RN}	2×10^4	4.2×10^4	ohms/square
K_P	2.5	1.3	
$K_{RP} = K_P K_{RN}$	5×10^4	5.5×10^4	ohms/square
$\tau_N = K_C K_{RN}$	3.9×10^{-12}	1.2×10^{-12}	seconds
$\tau_P = K_C K_{RP}$	9.8×10^{-12}	1.5×10^{-12}	seconds

Table 4.1 gives typical values of K_C, K_{RN}, and K_{RP} for 130 nm and 28 nm technologies. The key parameters here are τ_N and τ_P, the basic time constants of the technology. As technology scales, K_C (expressed as farads/L^2_{min}) remains roughly proportional to gate length and can be approximated as

$$K_C \approx 1.25 \times 10^{-9} L_{\text{min}}, \tag{4.3}$$

where $L_{\min}$ is expressed in meters. The sheet resistance, and thus delay, of the gate is no longer proportional to length. The resistance of n-type transistors has increased, while the PFET resistance has remained constant.[10]

EXAMPLE 4.2 Resistance and capacitance

For both 130 nm and 28 nm technologies, what is the resistance (R_s) and gate capacitance (C_g) of an NFET where $L = 1$ and $W = 32$? What is the size and capacitance of a PFET with the same R_s?

First, for the NFET:

$$
\begin{aligned}
R_s &= \frac{L}{W} K_{RN} \\
&= \frac{L_{\min}}{32L_{\min}} K_{RN}, \\
R_{s,130} &= 625\ \Omega, \\
R_{s,28} &= 1313\ \Omega, \\
C_g &= WLK_C, \\
C_{g,130} &= 6.4\ \text{fF}, \\
C_{g,28} &= 0.896\ \text{fF}.
\end{aligned}
$$

For the PFET, we know the resistance and solve for width and capacitance:

$$
\begin{aligned}
W_P &= \frac{LK_{RP}}{R_s}, \\
W_P &= \frac{LK_{RN}K_P}{R_s}, \\
W_P &= K_P W_n, \\
W_{P,130} &= 80L_{\min}, \\
W_{P,28} &= 41.6L_{\min}, \\
C_{g,130} &= 16\ \text{fF}, \\
C_{g,28} &= 1.16\ \text{fF}.
\end{aligned}
$$

4.3 CMOS GATE CIRCUITS

In Section 4.1 we learned how to do logic with switches, and in Section 4.2 we saw that MOS transistors can, for most digital purposes, be modeled as switches. Putting this information together, we can see how to make logic circuits with transistors.

[10] For device scaling projections and information, consult the International Technology Roadmap for Semiconductors.

A well-formed logic circuit should support the digital abstraction by generating an output that can be applied to the input of another, similar logic circuit. Thus, we need a circuit that generates a voltage on its output – not just one that connects two terminals together. The circuit must also be restoring, so that degraded input levels will result in restored output levels. To achieve this, the voltage on the output must be derived from a supply voltage, not from one of the inputs.

4.3.1 Basic CMOS gate circuit

A *static CMOS gate* circuit realizes a logic function *f* while generating a restoring output that is compatible with its input as shown in Figure 4.15. When function *f* is true, a PFET switch network connects output terminal x to the positive supply (V_{DD}). When function f is false, output x is connected to the negative supply by an NFET switch network. This obeys our constraints of passing only logic 1 (high) signals through PFET switch networks and logic 0 (low) signals through NFET networks. It is important that the functions realized by the PFET network and the NFET network be complements. If the functions should overlap (both be true at the same time), a short circuit from the power supply to ground results, drawing a large amount of current, and possibly causing permanent damage to the circuit. If the two functions do not cover all input states (there are some input states where neither is true), then the output is undefined in these states.

Because NFETs turn on with a high input and generate a low output and PFETs are the opposite, we can only generate *inverting* logic functions with static CMOS gates. A positive (negative) transition on the input of a single CMOS gate circuit can either cause a negative (positive) transition on the output or no change at all. Such a logic function, where transitions in one direction on the inputs cause transitions in just a single direction on the output, is called a *monotonic* logic function. If the transitions on the outputs are in the opposite direction to the transitions on the inputs, it is a monotonic decreasing or inverting logic function. If the transitions are in the same direction, it is a monotonic increasing function. To realize a non-inverting or non-monotonic logic function requires multiple stages of CMOS gates.

We can use the principle of duality, Equation (3.9), to simplify the design of gate circuits. If we have an NFET pull-down network that realizes a function $f_n(x_1,\ldots,x_n)$, we know that our gate will realize function $f_p = \overline{f_n(x_1,\ldots,x_n)}$. By duality, we know that

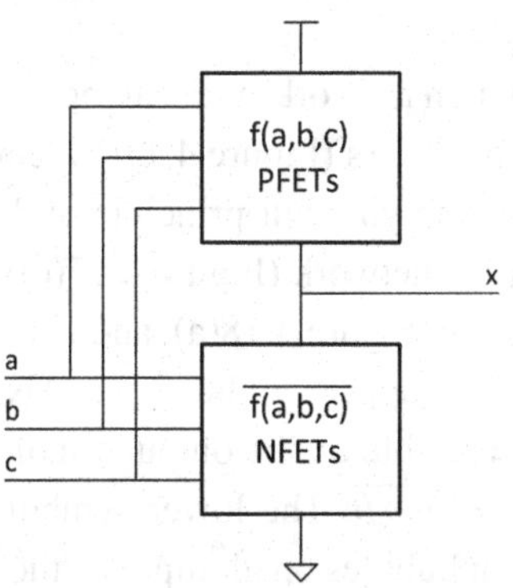

Figure 4.15. A CMOS gate circuit consists of a PFET switch network that pulls the output high when function *f* is true and an NFET switch network that pulls the output low when *f* is false.

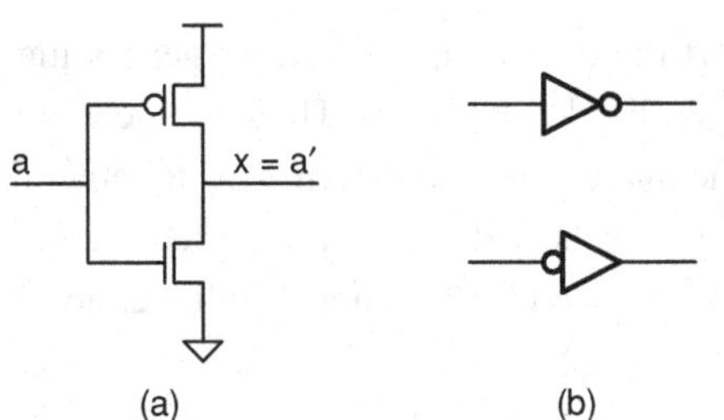

Figure 4.16. A CMOS inverter circuit. (a) A PFET connects x to 1 when $a = 0$ and an NFET connects x to 0 when $a = 1$. (b) Logic symbols for an inverter. The bubble on the input or output denotes the NOT operation.

$f_p = \overline{f_n(x_1, \ldots, x_n)} = f_n^D(\overline{x_1}, \ldots, \overline{x_n})$. So, for the PFET pull-up network, we want the dual function with inverted inputs. The PFETs give us the inverted inputs, since they are "on" when the input is low. To get the dual function, we take the pull-down network and replace ANDs with ORs and vice versa. In a switch network, this means that a series connection in the pull-down network becomes a parallel connection in the pull-up network and vice versa.

4.3.2 Inverters, NANDs, and NORs

The simplest CMOS gate circuit is the inverter, shown in Figure 4.16(a). Here the PFET network is a single transistor that connects output x to the positive supply whenever input a is low: $x = \overline{a}$. Similarly, the NFET network is a single transistor that pulls output x low whenever the input is high.

Figure 4.16(b) shows the schematic symbols for an inverter. The symbol is a rightward-facing triangle with a *bubble* on its input or output. The triangle represents an amplifier – indicating that the signal is restored. The bubble (sometimes called an *inversion bubble*) implies negation. The bubble on the input is considered to apply a NOT operation to the signal before it is input to the amplifier. Similarly, a bubble on the output is considered to apply a NOT operation to the output signal after it is amplified. Logically, the two symbols are equivalent. It doesn't matter whether we consider the signal to be inverted before or after amplification. We choose one of the two symbols to obey the *bubble rule*, which is as follows:

> **Bubble rule** Where possible, signals that are output from a gate with an inversion bubble on its output shall be input to a gate with an inversion bubble on its input.

Schematics drawn using the bubble rule are easier to read than schematics where the polarity of logic signals changes from one end of the wire to the other. We shall see many examples of this in Chapter 6.

Figure 4.17 shows some example NFET and PFET switch networks that can be used to build NAND and NOR gate circuits. A parallel combination of PFETs (Figure 4.17(b)) connects the output high if either input is low, so $f = \overline{a} \vee \overline{b} = \overline{a \wedge b}$. Applying our principle of duality, this switch network is used in combination with a series NFET network (Figure 4.17(c)) to realize a NAND gate. The complete NAND gate circuit is shown in Figure 4.18(a), and two schematic symbols for the NAND are shown in Figure 4.18(b). The upper symbol is an AND symbol (square left side, half-circle right side) with an inversion bubble on the output – indicating that we AND inputs a and b and then invert the output, $f = \overline{a \wedge b}$. The lower symbol is an OR symbol (curved left side, pointy right side) with inversion bubbles on all inputs – the inputs are

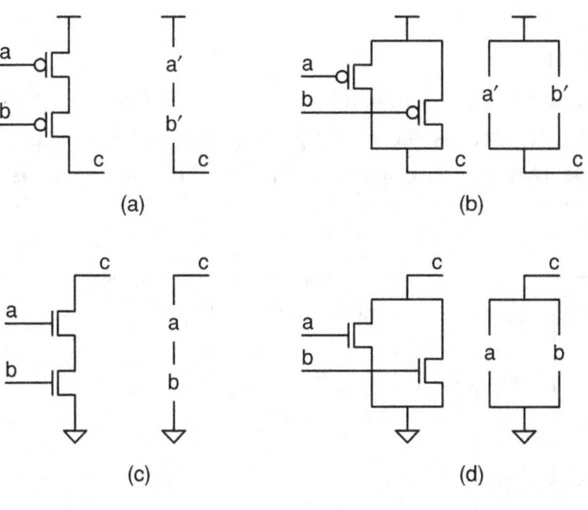

Figure 4.17. Switch networks used to realize NAND and NOR gates. (a) Series PFETs connect the output c high when all inputs are low, $f = \overline{a} \wedge \overline{b} = \overline{a \vee b}$. (b) Parallel PFETs connect the output if either input is low, $f = \overline{a} \vee \overline{b} = \overline{a \wedge b}$. (c) Series NFETs pull the output low when both inputs are high, $f = \overline{a \wedge b}$. (d) Parallel NFETs pull the output low when either input is true, $f = \overline{a \vee b}$. NAND gates are composed of networks (b) and (c), whereas NOR gates connect (a) and (d).

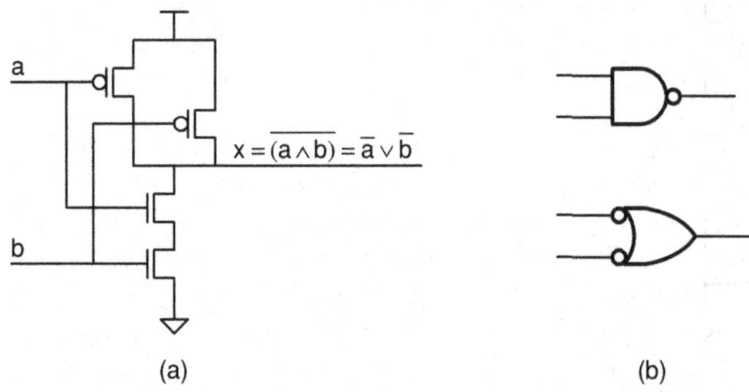

Figure 4.18. A CMOS NAND gate. (a) Circuit diagram: the NAND has a parallel PFET pull-up network and a series NFET pull-down network. (b) Schematic symbols: the NAND function can be thought of as an AND with an inverted output (top) or an OR with inverted inputs (bottom).

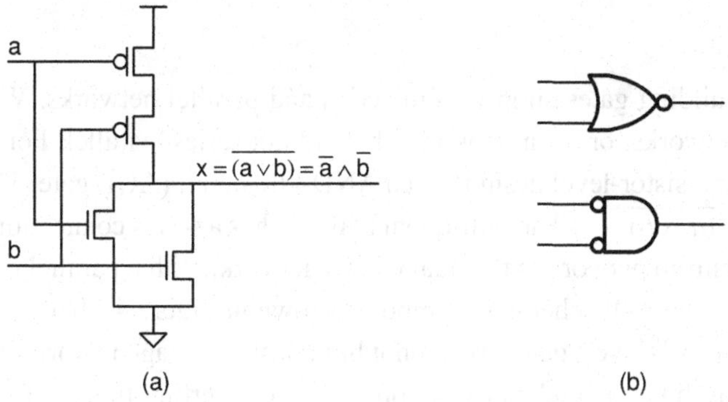

Figure 4.19. A CMOS NOR gate. (a) Circuit diagram: the NOR has a series PFET pull-up network and a parallel NFET pull-down network. (b) Schematic symbols: the NOR can be thought of as an OR with an inverted output or an AND with inverted inputs.

inverted and then the inverted inputs are ORed, $f = \overline{a} \vee \overline{b}$. By De Morgan's law (and duality), these two functions are equivalent. As with the inverter, we select between these two symbols to observe the bubble rule.

A NOR gate is constructed with a series network of PFETs and a parallel network of NFETs, as shown in Figure 4.19(a). A series combination of PFETs (Figure 4.17(a)) connects the output to 1 when a and b are both low, $f = \overline{a} \wedge \overline{b} = \overline{a \vee b}$. Applying duality, this circuit is used in combination with a parallel NFET pull-down network (Figure 4.17(d)). The schematic symbols for the NOR gate are shown in Figure 4.19(b). As with the inverter and the NAND, we can choose between inverted inputs and inverted outputs depending on the bubble rule.

EXAMPLE 4.3 Four-input NAND

Draw the transistor-level implementation of a four-input NAND gate, $f = \overline{a \wedge b \wedge c \wedge d}$.

The implementation of this gate extends the two-input NAND gate by adding two more NFETs in series with the first two and two parallel PFETs. The final gate is shown in Figure 4.20.

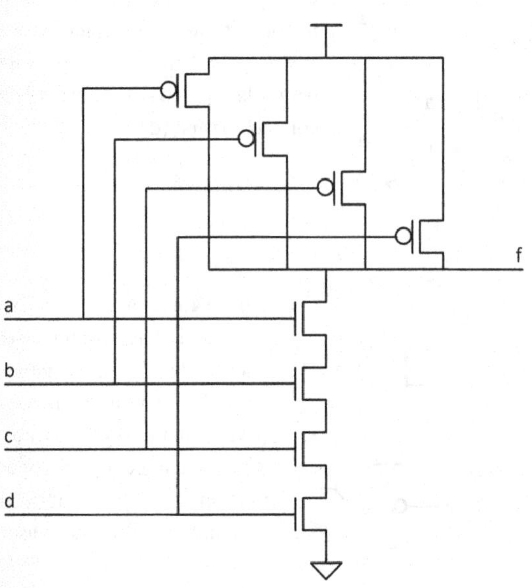

Figure 4.20. The transistor-level implementation of a four-input NAND gate.

4.3.3 Complex gates

We are not restricted to building gates simply from series and parallel networks. We can use arbitrary series–parallel networks, or even networks that are not series–parallel. For example, Figure 4.21(a) shows the transistor-level design for an AND-OR-invert (AOI) gate. This circuit computes the function $f = \overline{(a \wedge b) \vee c}$. The pull-down network has a series connection of a and b in parallel with c. The pull-up network is the dual of this network, with a parallel connection of a and b in series with c. The AOI schematic symbol is shown in Figure 4.21(b).

Figure 4.22 shows a majority-invert gate. We cannot build a single-stage majority gate since it is a monotonic increasing function and gates can only realize inverting functions. However, we can build the complement of the majority function as shown. The majority is an interesting function in that it is its own dual. That is, $\text{maj}(\bar{a}, \bar{b}, \bar{c}) = \overline{\text{maj}(a, b, c)}$. Because of this we can implement the majority gate with a pull-up network that is identical to the pull-down network, as shown in Figure 4.22(a). The majority function is also a *symmetric* logic function in that the inputs are all equivalent. Thus we can permute the inputs to the PFET and NFET networks without changing the function.

A more conventional implementation of the majority-invert gate is shown in Figure 4.22(b). The NFET pull-down network here is the same as for Figure 4.22(a), but the PFET pull-up network has been replaced by a dual network – one that replaces each series element with a parallel element and vice versa. The parallel combination of b and c in series with a in the

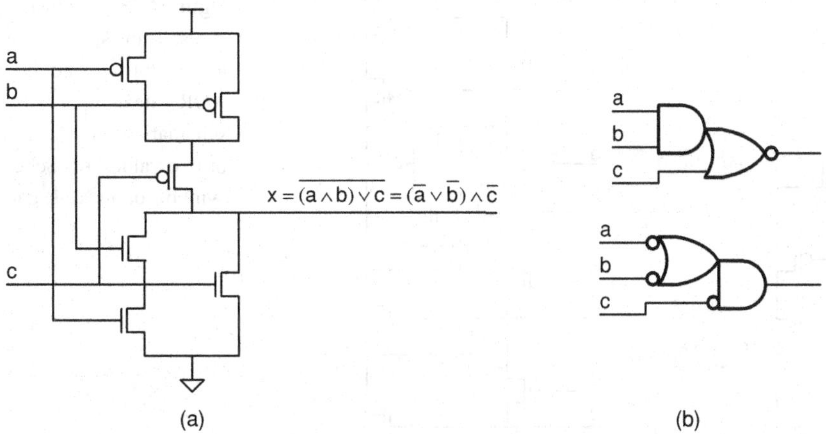

Figure 4.21. An AND-OR-invert (AOI) gate. (a) Transistor-level implementation uses a parallel–series NFET pull-down network and its dual series–parallel PFET pull-up network. (b) Two schematic symbols for the AOI gate.

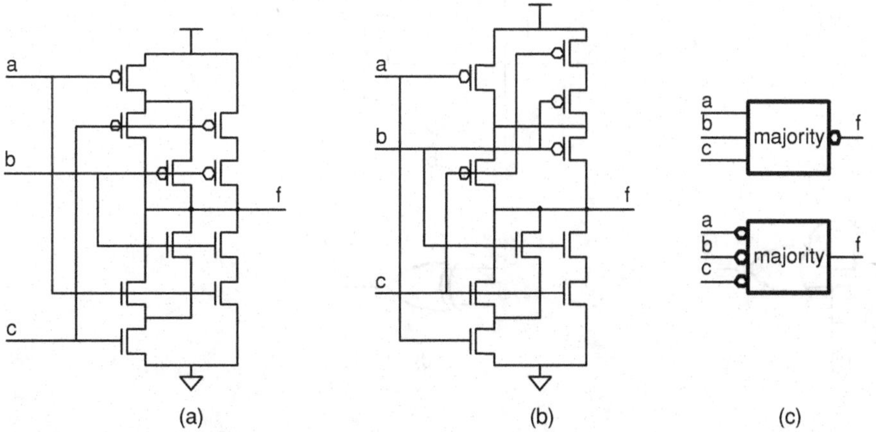

Figure 4.22. A majority-invert gate. The output is false if two or three of the inputs are true. (a) Implementation with symmetric pull-up and pull-down networks. (b) Implementation with a pull-up network that is a dual of the pull-down network. (c) Schematic symbols: the function is still majority irrespective of whether the inversion is on the input or the output.

pull-down network, for example, translates to a series combination of *b* and *c* in parallel with *a* in the pull-up network. A PFET pull-up network that is the dual of the NFET pull-down network will always give a switching function that is the complement of the pull-down network because of Equation (3.9).

Figure 4.22(c) shows two possible schematic symbols for the majority-invert gate. Because the majority function is self-dual, it does not matter whether we put the inversion bubbles on the inputs or the output. The function is a majority either way. If at least two out of the three inputs are high, the output will be low – a majority with a low-true output. It is also the case that if at least two of the three inputs are low, the output will be high – a majority with low-true inputs.

Strictly speaking, we cannot make a single-stage CMOS exclusive-or (XOR) gate because XOR is a non-monotonic function. A positive transition on an input may cause either a positive

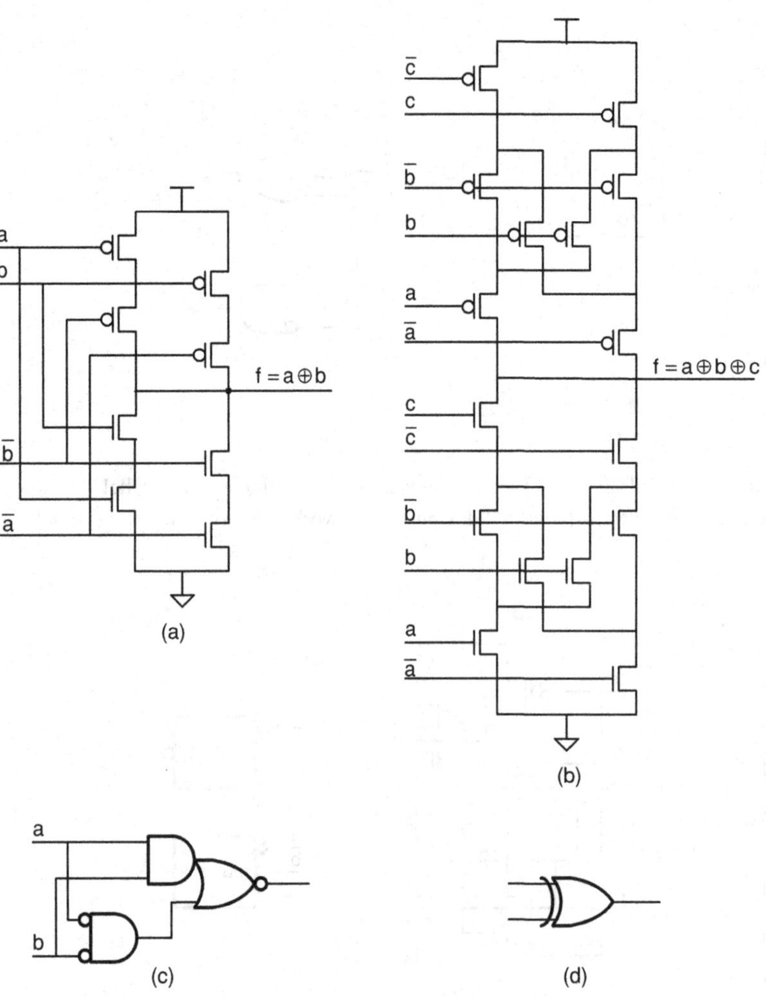

Figure 4.23. Exclusive-or (XOR) gates. Transistor-level diagrams for a two-input (a) and three-input (b) XOR gate at top. (c) Two-level schematic for computing the XOR of two values. (d) Schematic symbol for an XOR gate.

or a negative transition on an output, depending on the state of the other inputs. However, if we have inverted versions of the inputs, we can realize a two-input XOR function as shown in Figure 4.23(a), taking advantage of the switch network of Figure 4.7(a). A three-input XOR function can be realized as shown in Figure 4.23(b). The switch networks here are not series–parallel networks. If inverted inputs are not available, it is more efficient to realize a two-input XOR gate using two CMOS gates in series, as shown in Figure 4.23(c). We leave the transistor-level design of this circuit as an exercise. An XOR symbol is shown in Figure 4.23(d).

EXAMPLE 4.4 CMOS gate synthesis

Draw a complex gate that outputs a 0 if and only if the input, *cba*, is a legal thermometer-coded signal.

First, we write the Boolean function and reduce it as follows:

$$\begin{aligned} f(c,b,a) &= \overline{(\bar{c} \wedge \bar{b} \wedge \bar{a}) \vee (\bar{c} \wedge \bar{b} \wedge a) \vee (\bar{c} \wedge b \wedge a) \vee (c \wedge b \wedge a)} \\ &= \overline{(\bar{c} \wedge \bar{b}) \vee (b \wedge a)}. \end{aligned}$$

Thus, our NFET switch-network function is given by

$$f_n = (\bar{c} \wedge \bar{b}) \vee (b \wedge a).$$

Next, in order to draw the PFET array correctly, we find the dual of our NFET function:

$$f_p(c, b, a) = (\bar{c} \vee \bar{b}) \wedge (b \vee a).$$

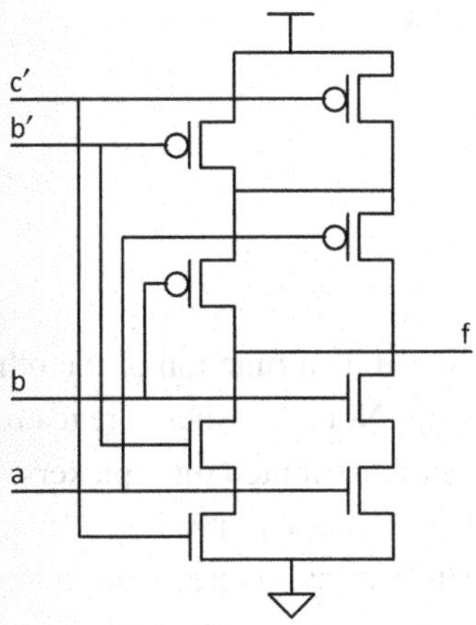

Figure 4.24. Schematic for the CMOS circuit to detect whether inputs *abc* represent a legal thermometer encoding.

Figure 4.24 shows our final answer. In order to implement this circuit in a single stage, the complements to inputs c and b are required.

4.3.4 Tri-state circuits

On occasion we want to build a distributed multiplexer function where we drive a value onto a signal node from point A when logic signal a is true and from point B when logic signal b is true. We can realize this functionality with a *tri-state* inverter. This circuit, shown in Figures 4.25(a) and (b), drives $\bar{a}$ onto x when e is true and presents a high impedance (drives neither 1 nor 0) to the output when e is false. When the enable e is high ($\bar{e}$ is low), the middle two transistors are on. The outer transistors, controlled by a, then function as a normal inverter. With e low ($\bar{e}$ high), the enable transistors are off and no value is driven to the output. In this state, the value at the output is initially equal to the previous output, but will eventually leak into an unknown state. In VHDL, a wire that is not being driven is denoted with a `'Z'` symbol.

The tri-state inverter is *not* a *gate* as defined in Section 4.3.1 because its output is *not* a restored logic function of its input. Its pull-down function $f_n(a, e) = a \wedge e$ is not equal to the complement of its pull-up function $f_p = \bar{a} \wedge e$. The *no-pull* function $f_z = \overline{f_n} \wedge \overline{f_p} = \bar{e}$ is true when f_n and f_p are both false, and indicates when x goes to the high impedance or z state. Because this circuit produces three output states, 0, 1, and z, we refer to it as a *tri-state* circuit. This is not to be confused with *ternary* logic circuits that use signals that can take on three values.

Tri-state circuits are not restricted to inverters. We can modify any CMOS gate circuit to gate the output off when an enable signal e is low. We are also not restricted to a single enable signal. An arbitrary logic function can determine when the circuit is enabled. This function can

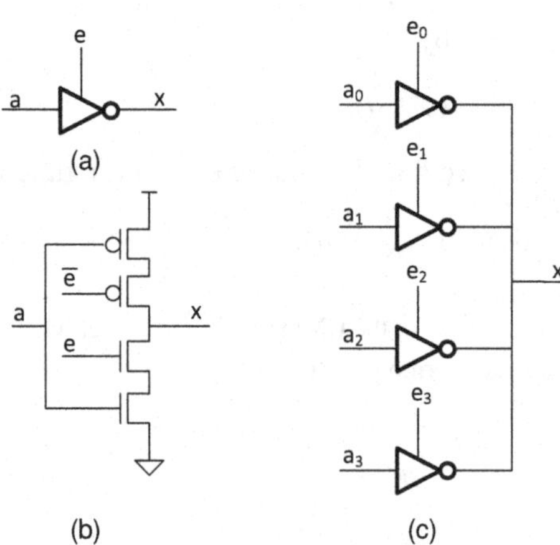

Figure 4.25. A tri-state inverter has two inputs and a single output, (a). Transistor-level schematic, (b), when enable e is low, there is no supply connected to the output. When e is high, the output is driven to $\overline{a}$. Four tri-state inverters implement a bus or distributed multiplexer, (c). When exactly one enable e_i is high, the output is $\overline{a_i}$. When all enables are low, the output is indeterminate. If multiple enables are high, permanent damage may occur.

differ for pull-up and pull-down, and the enable function can be a function of the other circuit inputs. Any CMOS circuit where $f_z \neq 0$ is a tri-state circuit. Not all of these are useful.

Tri-state inverters can be used to realize multiplexers and distributed multiplexers, as shown in Figure 4.25(c). Here, four tri-state inverters all drive the same output wire, x. If exactly one enable signal e_i is asserted, $x = \overline{a_i}$. If all enable signals are low, there is no drive to x – it is in the high-impedance or z state. If more than one enable is asserted simultaneously (even for a very short period of time), a short circuit may form, causing static current to flow from the power supply of one gate to the ground of another. At the very least, this increases power dissipation and gives an invalid output. At worst, the high currents involved can vaporize metal traces, irreversibly damaging the chip. Even if only one enable signal is asserted each clock cycle, skew (see Chapter 15) between the clocks generating the different enables can lead to a fatal overlap. Distributed multiplexers, which often motivate the use of tri-state circuits, are particularly vulnerable because a large clock skew can develop over the distance between point A and point B.

Because of the potential for a short circuit, tri-state circuits should be avoided. If they are used, strong measures should be taken to ensure that there is no possibility of overlap between enable signals. One approach is to leave an idle cycle – with all enable signals low – between the assertion of one enable signal and another. Alternatively, one can ensure a *break before make* action by designing the circuits driving the enable signals to have a long delay on the rising edge and a short delay on the falling edge – with the difference in delays being large enough to compensate for clock skew and other factors. In most cases, the best approach is to build the required function – even a distributed multiplexer or bus (see Section 24.2) – with static CMOS gates (NANDs and NORs).

4.3.5 Circuits to avoid

Before closing this chapter it is worth examining a few more circuits that are not gates and do not work but represent common errors in CMOS circuit design. Figure 4.26 shows four representative mistakes. The would-be buffer in Figure 4.26(a) does not work because it attempts to pass a 1 through a PFET and a 0 through an NFET. The transistors cannot reliably pass those values and so the output signal is undefined – attenuated from the input swing at best.

These are NOT static gates: do not use these circuits

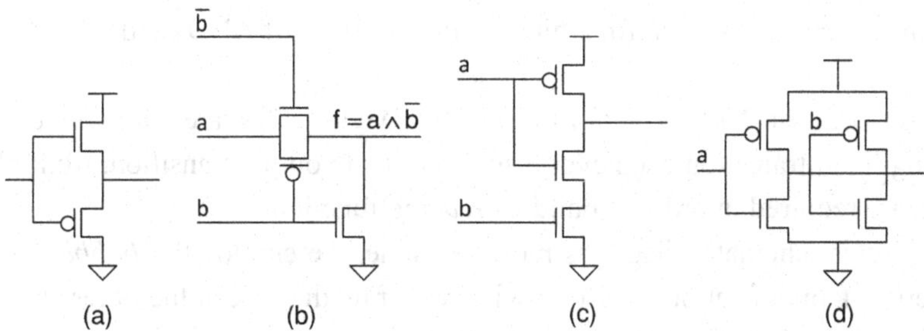

Figure 4.26. Four circuits that are *not* gates and should not be used: circuit (a) attempts to pass a 1 through an NFET and a 0 through a PFET; circuit (b) does not restore high output values; circuit (c) does not drive the output when $a = 1$ and $b = 0$; circuit (d) conducts static current when $a \neq b$.

The AND-NOT circuit of Figure 4.26(b) does in fact implement the logical function $f = a \wedge \bar{b}$. However, it violates the digital abstraction in that it does not *restore* its output. If $b = 0$, a noise on input a is passed directly to the output.[11] Any noise on the output is also passed onto the input signal and all of the input's driven gates.

The circuit of Figure 4.26(c) leaves its output disconnected when $a = 1$ and $b = 0$. Owing to parasitic capacitance, the previous output value will be *stored* for a short period of time on the output node. However, after a period, the stored charge will leak off and the output node becomes an undefined value. The final circuit, Figure 4.26(d), will have both the pull-up and pull-down networks conducting when a and b are not equal. This static current from power to ground outputs an undefined logic value, wastes power, and potentially damages the chip.

Summary

CMOS *gate* circuits are the basic building blocks of modern digital systems. We derive gate circuits from *switch logic* using complementary MOS (CMOS) transistors.

A switch connects its terminals when it is closed. A series connection of switches connects its terminals only when all switches in the circuit are closed, an AND function. Similarly, a parallel connection of switches performs an OR function. Series–parallel networks of switches can perform arbitrary combinations of AND and OR functions.

NMOS and PMOS (n-type and p-type MOS) transistors act as switches with restrictions. An NMOS transistor, or NFET, is *on* (switch closed) when its gate terminal is high (logic 1) but it passes only low signals (logic 0s). A PMOS transistor, or PFET, the complementary device, is *on* when its gate is low and passes only high signals.

We construct a CMOS gate circuit that implements function f by constructing a pull-down network of NFETs that pulls the output low when $\bar{f}$ is true and a pull-up network of PFETs that pulls the output high when f is true. Because the inputs of PFETs are inverting, using a pull-up network that is the *dual* of the pull-down network gives the required function. For example, to make a NAND gate, we use a series connection of NFETs for the pull-down network and a parallel connection of PFETs for the pull-up network. In addition to NAND gates (series

[11] Such circuits can be used with care in isolated areas, but must be followed by a restoring stage before a long wire or another non-restoring gate. In most cases it is better to steer clear of such *shortcut* gates.

NFETs, parallel PFETs) and NOR gates (parallel NFETs, series PFETs), we can build complex gates, like AND-OR-invert gates, that perform arbitrary combinations of ANDs and ORs, with a final inversion.

Because of the properties of NFETs and PFETs, all CMOS gates are *monotonically decreasing*; i.e., a rising input transition can never result in a rising output transition. Multiple stages of CMOS gates are required in order to build *increasing* functions.

To make our logic circuit schematic diagrams more readable, we employ the *bubble rule*. Where possible, if there is a bubble at one end of a wire, we draw the gate at the other end of the wire so there is a bubble at that end of the wire as well.

A *tri-state* circuit has an output that can be put in a *high-impedance* or unconnected state allowing several tri-state circuits to drive a single signal node. Because of the potential for short circuits between tri-state circuits driving the same node, these circuits should be avoided when possible.

BIBLIOGRAPHIC NOTES

Two of the first studies into using switch networks came from Shannon [100] and Montgomerie [82]. One of the first to simulate MOSFET circuits using the switch model was Bryant in the early 1980s [22].

A digital circuit design text such as Rabaey *et al.* [94] is a good source of more detailed information on digital logic circuits.

Weste and Harris's textbook [112] provides further information about gate layouts and VLSI.

Finally, readers interested in the physics of MOSFETs should refer to Muller, Kamins, and Chan's text [85].

Exercises

4.1 *Non-series–parallel analysis I.* Write down the logic function that describes the conditions under which the switch network of Figure 4.27(a) connects its two terminals. Note that this is not a series–parallel network.

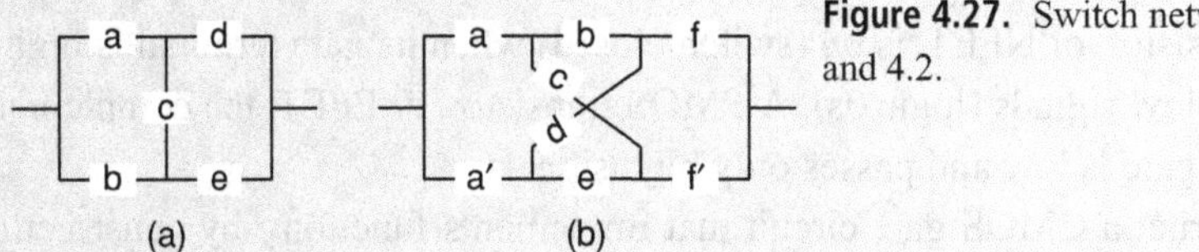

Figure 4.27. Switch network for Exercises 4.1 and 4.2.

4.2 *Non-series–parallel analysis II.* Write down the logic function that describes the conditions under which the switch network of Figure 4.27(b) connects its two terminals.

4.3 *Series–parallel synthesis I.* Draw a series–parallel switch circuit that implements the function $f(x, y, z) = ((x \vee \bar{y}) \wedge ((y \wedge z) \vee \bar{x})) \vee (x \wedge y \wedge z)$.

4.4 *Series–parallel synthesis II.* Draw a series–parallel switch circuit that implements the function $f(w, x, y, z) = 1$ if at least one input is true.

4.5 *Series–parallel synthesis III.* Draw a series–parallel switch circuit that implements the function $f(w, x, y, z) = 1$ if at least two inputs are true.

4.6 *Series–parallel synthesis IV.* Draw a series–parallel switch circuit that implements the function $f(w, x, y, z) = 1$ if at least three inputs are true.

4.7 *Series–parallel synthesis V.* Draw a series–parallel switch circuit that implements the function $f(w, x, y, z) = 1$ if all four inputs are true.

4.8 *Series–parallel synthesis VI.* Draw a series–parallel switch circuit that implements the function $f(w, x, y, z) = 1$ if exactly one or three inputs are true.

4.9 *Series–parallel synthesis VII.* Draw a series–parallel switch circuit that implements the function $f(x, y, z) = 1$ if inputs xyz represent either 1 or a prime number in binary ($xyz =$ 001, 010, 011, 101, 111).

4.10 *Series–parallel synthesis VIII.* Draw a series–parallel switch circuit that implements the function $f(w, x, y, z) = 1$ if inputs $wxyz$ represent 1 or a prime number in binary ($wxyz =$ 0001, 0010, 0011, 0101, 0111, 1011, 1101).

4.11 *CMOS switch models, I.* Draw the switch models (using open and closed switches in place of the transistors) for Figure 4.22(a) and inputs $a = 1, b = 0, c = 0$.

4.12 *CMOS switch models, II.* Draw the switch models (using open and closed switches in place of the transistors) for Figure 4.22(a) and inputs $a = 1, b = 1, c = 1$.

4.13 *CMOS switch models, III.* Draw the switch models (using open and closed switches in place of the transistors) for the following diagram and input: Figure 4.23(a), $a = 1, b = 0$.

4.14 *CMOS switch models, IV.* Draw the switch models (using open and closed switches in place of the transistors) for the following diagram and input: Figure 4.23(b), $a = 1, b = 0$, $c = 1$.

4.15 *Resistance and capacitance, I.* For both 130 nm and 28 nm technologies, calculate the resistance and gate capacitance of an NFET with $W = 20L_{\text{min}}$. What are the width and gate capacitance of a PFET with half the resistance of the $W = 20L_{\text{min}}$ NFET? Assume all transistors have $L = L_{\text{min}}$.

4.16 *Resistance and capacitance, II.* For both 130 nm and 28 nm technologies, calculate the resistance and gate capacitance of a PFET with $W = 40L_{\text{min}}$. What are the width and gate capacitance of an NFET with twice the resistance of the $W = 40L_{\text{min}}$ PFET? Assume all transistors have $L = L_{\text{min}}$.

4.17 *Scaling throughput.* You are designing a logic module that has an area of $1 \times 10^6 L_{\text{min}}^2$. It does one unit of work every $400(1 + K_P)\tau_n$ seconds. Use Table 4.1 for all scaling parameters.

(a) How many modules can you place on a 1 mm^2 chip in both 130 nm and 28 nm technologies?

(b) What is the time taken for one module to do one unit of work (latency) in both 130 nm and 28 nm technologies?

(c) Using both 130 nm and 28 nm technologies, what is the total amount of work we can do on one chip in one second (throughput)?

4.18 *Simple gates, I.* Draw the transistor implementation for a three-input NAND gate.

4.19 *Simple gates, II.* Draw the transistor implementation for a four-input NOR gate.

4.20 *CMOS schematics, I.* Draw a schematic using NFETs and PFETs for a restoring logic gate that implements the function $f = \overline{a \wedge (b \vee c)}$.

4.21 *CMOS schematics, II.* Draw a schematic using NFETs and PFETs for a restoring logic gate that implements the function $f = \overline{((a \wedge b) \vee c) \vee (d \wedge e)}$.

4.22 *CMOS schematics, III.* Draw a schematic using NFETs and PFETs for a restoring logic gate that implements the function $f = \overline{(\bar{a} \wedge \bar{b} \wedge \bar{c}) \vee (a \wedge b \wedge c)}$. Assume that all inputs and their complements are available.

4.23 *CMOS schematics, IV.* Draw a schematic using NFETs and PFETs for a restoring logic gate that implements the function $f = 0$ if and only if $cba = 010, 011, 101$, or 111. Assume that all inputs and their complements are available.

4.24 *CMOS schematics, V.* Draw a schematic using NFETs and PFETs for a restoring logic gate that implements the XOR gate of Figure 4.23(c). Assume that all inputs and their complements are available.

4.25 *CMOS schematics, VI.* Draw a schematic using NFETs and PFETs for a restoring logic gate that implements a five-input majority-invert function. Assume that all inputs and their complements are available.

4.26 *CMOS schematics, VII.* Draw a schematic using NFETs and PFETs for a restoring logic gate that implements the function $f = 1$ if $cba = 001, 010, 011$, or 101 (Fibonacci numbers). Assume that all inputs and their complements are available.

4.27 *CMOS schematics, VIII.* Draw a schematic using NFETs and PFETs for a restoring logic gate that implements the function $f = 0$ if zero or two of inputs cba are true. Assume that all inputs and their complements are available.

4.28 *CMOS schematics, IX.* Draw a schematic using NFETs and PFETs for a restoring logic gate that implements the function $f = 1$ if one or two of inputs cba are true. Assume that all inputs and their complements are available.

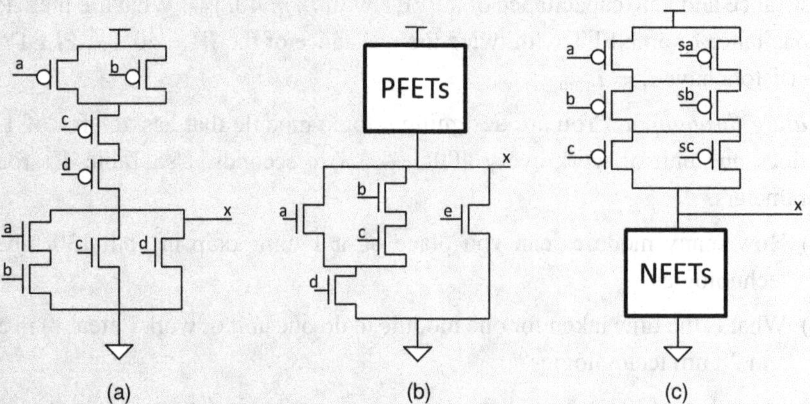

Figure 4.28. CMOS circuits for Exercises 4.30, 4.31, and 4.32. You may assume that the NFET and PFET blocks are implemented correctly.

4.29 *CMOS schematics, X.* Draw a schematic using NFETs and PFETs for a restoring logic gate that implements the function $f = 0$ if $dcba = 0010, 0011, 0101, 0111, 1011, 1101$ (four-bit prime numbers). Assume that all inputs and their complements are available.

4.30 *CMOS to logical equation, I.* Write down the logic function implemented by the CMOS circuit of Figure 4.28(a).

4.31 *CMOS to logical equation, II.* Write down the logic function implemented by the CMOS circuit of Figure 4.28(b).

4.32 *CMOS to logical equation, III.* Write down the logic function implemented by the CMOS circuit of Figure 4.28(c).

4.33 *Tri-state buffers.* Describe two static implementations of the logic performed by four connected tri-state inverters, as in Figure 4.25(c). Your first description should be for a single gate, while the second includes multiple gates.

5 Delay and power of CMOS circuits

The specification for a digital system typically includes not only its function, but also the delay and power (or energy) of the system. For example, a specification for an adder describes (i) the function, that the output is to be the sum of the two inputs; (ii) the delay, that the output must be valid within 1 ns after the inputs are stable; and (iii) its energy, that each add consumes no more than 2 pJ. In this chapter we shall derive simple methods to estimate the delay and power of CMOS logic circuits.

5.1 DELAY OF STATIC CMOS GATES

As illustrated in Figure 5.1, the delay of a logic gate, t_p, is the time from when the input of the gate crosses the 50% point between V_0 and V_1 to when the output of the gate crosses the same point. Specifying delay in this manner allows us to compute the delay of a chain of logic gates by simply summing the delays of the individual gates. For example, in Figure 5.1 the delay from a to c is the sum of the delay of the two gates. The 50% point on the output of the first inverter is also the 50% point on the input of the second inverter.

Because the resistance of the PFET pull-up network may be different than that of the NFET pull-down network, a CMOS gate may have a rising delay that is different from its falling delay. When the two delays differ, we denote the rising delay, the delay from a falling input to a rising output, as t_{pr} and the falling delay as t_{pf}, as shown in Figure 5.1.

We can use the simple switch model derived in Section 4.2 to estimate t_{pr} and t_{pf} by calculating the RC time constant of the circuit formed by the output resistance of the driving gate and the input capacitance of its load(s).[1] Because this time constant depends in equal parts on the driving and receiving gates, we cannot specify the delay of a gate by itself, but only as a function of output load.

Consider, for example, a CMOS inverter with a pull-up of width W_P and a pulldown of width W_N driving an identical inverter, as shown in Figures 5.2(a) and (b).[2] For both rising and falling edges, the input capacitance of the second inverter is the sum of the capacitance of the PFET and NFET: $C_{\text{inv}} = (W_P + W_N)C_G$. When the output of the first inverter rises, the output resistance is that of the PFET with width W_P, as shown in Figure 5.2(c): $R_P = K_{RP}/W_P = K_P K_{RN}/W_P$. Thus for a rising edge we have

$$t_{pr} = R_P C_{\text{inv}} = \frac{K_P K_{RN}(W_P + W_N)C_G}{W_P}. \tag{5.1}$$

[1] In reality, the driving gate has output capacitance roughly equal to its input capacitance. We ignore that capacitance here to simplify the model.

[2] W_P and W_N are in units of $W_{\text{min}} = 8L_{\text{min}}$; C_G is the capacitance of a gate with width $8L_{\text{min}}$, so $C_G = 0.22$ fF.

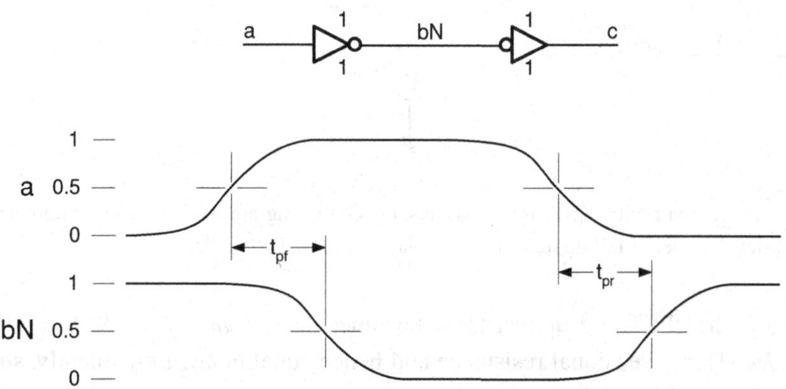

Figure 5.1. We measure delay from the 50% point of an input transition to the 50% point of an output transition. This figure shows the waveforms on input a and output bN, with the falling and rising propagation delays, t_{pf} and t_{pr}, labeled.

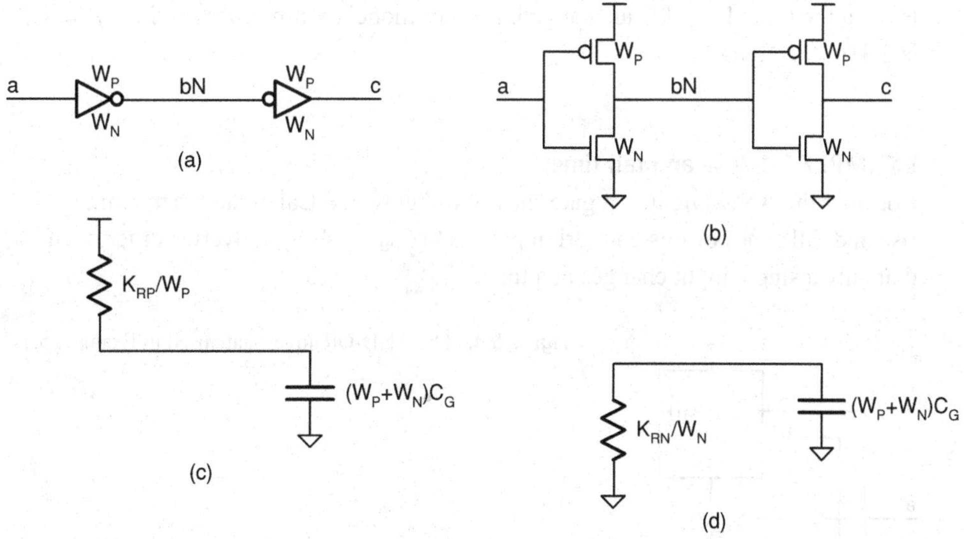

Figure 5.2. Delay of an inverter driving an identical inverter. (a) Logic diagram (all numbers are device widths). (b) Transistor-level circuit. (c) Switch-level model to compute rising delay. (d) Switch-level model for falling delay.

Similarly, for a falling edge, the output resistance is the resistance of the NFET pull-down, as shown in Figure 5.2(d): $R_N = K_{RN}/W_N$. This gives a falling delay of

$$t_{pf} = R_N C_{\text{inv}} = \frac{K_{RN}(W_P + W_N)C_G}{W_N}. \tag{5.2}$$

Most of the time we wish to size CMOS gates so that the rise and fall delays are equal; that is, so $t_{pr} = t_{pf}$. For an inverter, this implies that $W_P = K_P W_N$, as shown in Figure 5.3. We make the PFET K_P times wider than the NFET to account for the fact that its resistivity (per square) is

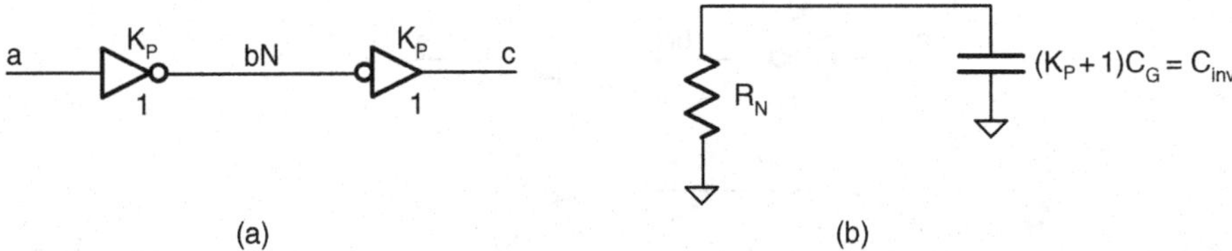

Figure 5.3. An inverter pair with equal rise/fall delays. (a) Logic diagram (sizings reflect parameters of Table 4.1). (b) Switch-level model of falling delay (rising delay is identical).

K_P times larger. The PFET pull-up resistance becomes $R_P = K_{RP}/W_P = (K_P K_{RN})/(K_P W_N) = K_R N/W_N = R_N$. This gives equal resistance and hence equal delay. Equivalently, substituting for W_P in the formula above gives

$$t_{\text{inv}} = \frac{K_{RN}}{W_N}(K_P + 1)W_N C_G = (K_P + 1)K_{RN} C_G = (K_P + 1)\tau_N. \tag{5.3}$$

Note that the W_N term cancels out. The delay of an inverter driving an identical inverter, t_{inv}, is independent of device width. As the devices are made wider, R decreases and C increases, leaving the total delay RC unchanged. For our model 28 nm process with $K_P = 1.3$, this delay is $2.3\tau_N = 2.70$ ps.[3]

EXAMPLE 5.1 Rise and fall times

Consider the AND-OR-invert gate shown in Figure 5.4. Calculate the maximum and minimum rise and fall times of this gate driving a FO4 ($C_{\text{out}} = 4C_{\text{inv}}$) inverter in terms of t_{inv}. Assume that only a single input changes at a time.

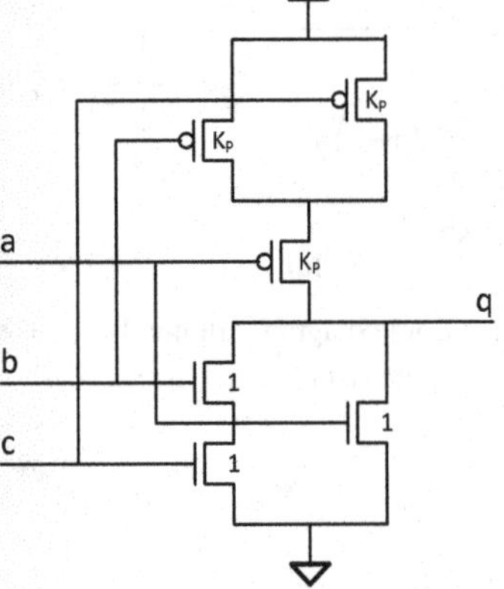

Figure 5.4. The AND-OR-invert gate used in Example 5.1.

The minimum fall time occurs on the transition from $abc = 000$ to 100 because this switches on the pull-down path with only a single transistor. Into a load of $4C_{\text{inv}}$, this gives $t_{\text{fmin}} = 4t_{\text{inv}}$. The maximum fall time occurs on the transition from 010 or 001 to 011. This turns on the pull-down path with two NFETs. The delay is twice that of the path with a single NFET, so $t_{\text{fmax}} = 8t_{\text{inv}}$.

[3] For a minimum-sized $W_N = 8L_{\text{min}}$ inverter, with equal rise/fall delay, $C_{\text{inv}} = 0.5$ fF in our model process.

The minimum rise time occurs on the transition from 100 to 000; this turns on the PFET controlled by a when both of the other PFETs are on. Hence the equivalent series resistance is $R_P = 1.5K_{RN}$ and into a load of $4C_{\text{inv}}$ we have $t_{\text{rmin}} = 6t_{\text{inv}}$. The maximum rise time occurs on the transition from 101 to 001 (or 110 to 010). In this case, only one of the two parallel PFETs is on and we have $R_P = 2K_{RN}$, giving $t_{\text{rmax}} = 8t_{\text{inv}}$.

5.2 FAN-OUT AND DRIVING LARGE LOADS

Consider the case where a single inverter of size 1 ($W_N = W_{\min}$) sized for equal rise/fall delay ($W_P = K_P W_N$) drives four identical inverters as shown in Figure 5.5(a). The equivalent circuit for calculating the RC time constant is shown in Figure 5.5(c). Compared with the situation with identical inverters (fan-out of 1), this fan-out-of-4 situation has the same driving resistance, R_N, but four times the load capacitance, $4C_{\text{inv}}$. The result is that the delay for a fan-out of four is four times the delay of the fan-out of one circuit. In general, the delay for a fan-out of F is F times the delay of a fan-out of one circuit:

$$t_F = Ft_{\text{inv}}. \tag{5.4}$$

For the case where $F = 4$ we have $t_4 = 4t_{\text{inv}} = 10.8$ ps. This fan-out-of-4 (FO4) number is often used to compare processes, and designers often refer to their cycle time and the depth of their logic in terms of FO4 delays (t_4).

The delay is the same if the unit-sized inverter drives a single inverter that is sized four times larger, as shown in Figure 5.5(b). The load capacitance on the first inverter is still four times its input capacitance.

When we have a very large fan-out, it is advantageous to increase the drive of a signal in stages rather than all at once. This gives a delay that is logarithmic, rather than linear in the size of the fan-out. Consider the situation shown in Figure 5.6(a). Signal bN, generated by a unit-sized inverter,[4] must drive a load that is 1024 times larger than a unit-sized inverter (a fan-out

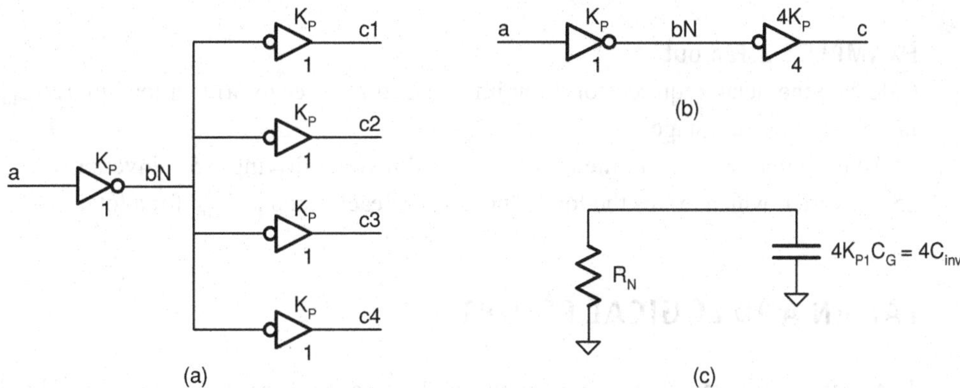

Figure 5.5. An inverter driving four times its own load: (a) driving four other inverters; (b) driving one large (4×) inverter. (c) Switch-level model of falling delay.

[4] From now on, we may drop W_P from our diagrams whenever gates are sized for equal rise and fall.

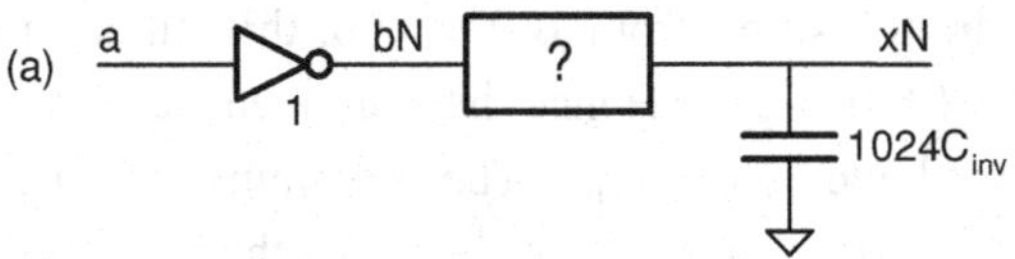

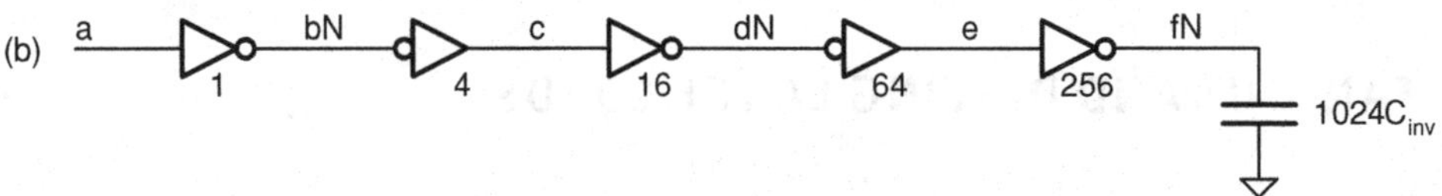

Figure 5.6. Driving a large capacitive load. (a) The output of a unit-sized inverter needs to drive a fan-out of 1024. We need a circuit to buffer up the signal *bN* to drive this large capacitance. (b) Minimum delay is achieved by using a chain of inverters that increases the drive by the same factor (in this case 4) at each stage.

of $F = 1024$). If we simply connect *bN* to *xN* with a wire, the delay will be $1024t_{\text{inv}}$. If we increase the drive in stages, as shown in Figure 5.6(b), however, we have a circuit with five stages. Each stage has a fan-out of 4, giving a much smaller total delay of $20t_{\text{inv}}$.

In general, if we divide a fan-out of F into n fan-outs of $\alpha = F^{1/n}$ stages, our delay will be given by

$$t_{Fn} = nF^{1/n}t_{\text{inv}} = \alpha t_{\text{inv}} \log_\alpha F. \tag{5.5}$$

We can solve for the minimum delay by taking the derivative of Equation (5.5) with respect to n (or α) and setting this derivative to zero. Solving shows that the minimum delay occurs for a fan-out per stage of $\alpha = e$. In practice, fan-outs between 3 and 6 give good results. Fan-outs much smaller than 3 result in too many stages, while fan-outs larger than 6 give too much delay per stage. A fan-out of 4 is often used in practice. Overall, driving a large fan-out, F, using multiple stages with a fan-out of α reduces the delay from one that increases linearly with F to one that increases logarithmically with F – as $\log_\alpha F$.

EXAMPLE 5.2 Fan-out

Calculate the delay required for a minimum-sized inverter to drive a load of $125C_{\text{inv}}$ using a fan-out of 5 at each stage.

We use three stages with the minimum-sized inverter driving a 5× inverter, which drives a 25× inverter, which drives the load. The delay of each stage is $5t_{\text{inv}}$ for a total of $15t_{\text{inv}}$.

5.3 FAN-IN AND LOGICAL EFFORT

Just as fan-out increases delay by increasing load capacitance, fan-in increases the delay of a gate by increasing output resistance – or equivalently input capacitance. To keep the output drive constant, we size the transistors of a multi-input gate so that the pull-up series resistance and the pull-down series resistance are equal to the resistance of an equal rise/fall inverter with the same relative size.

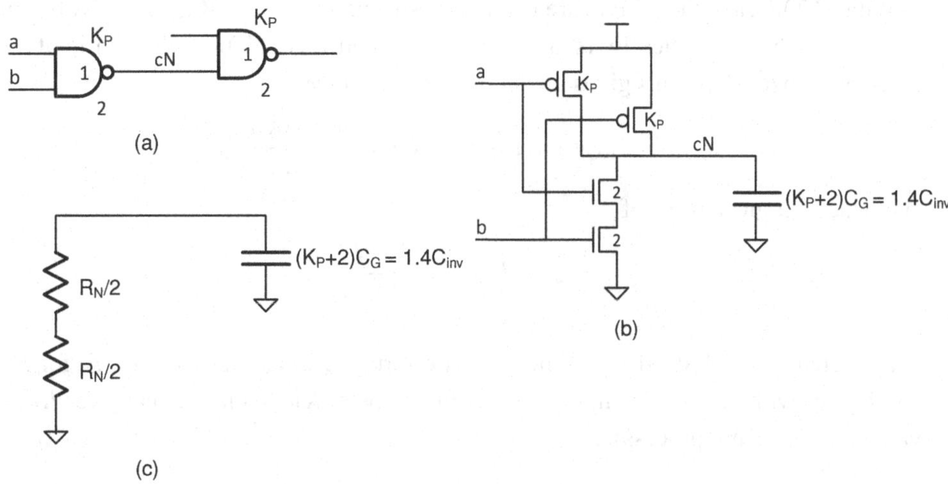

Figure 5.7. (a) A NAND gate driving an identical NAND gate. Both are sized for equal rise and fall delays. (b) Transistor-level schematic. (c) Switch-level model.

For example, consider a two-input NAND gate driving an identical NAND gate as shown in Figure 5.7(a). We size the devices of each NAND gate so the pull-up and pull-down networks have the same output resistance as a unit-drive (equal rise/fall) inverter, as shown in Figure 5.7(b). Since in the worst case only a single pull-up PFET is on, we size these PFETS $W_P = K_P$, just as in the inverter. We get no credit for the parallel combination of PFETs since both are on in only one of the three input states where the output is high (both inputs zero). To give a pull-down resistance equal to R_N, each NFET in the series chain is sized at twice the minimum width. As shown in Figure 5.7(c) putting these two $R_N/2$ devices in series gives a total pull-down resistance of R_N. The capacitance of each input of this unit-drive NAND gate is the sum of the PFET and NFET capacitance:

$$(2 + K_P)C_G = \frac{2 + K_P}{1 + K_P}C_{\text{inv}}.$$

We refer to this increase in input capacitance for the same output drive as the *logical effort* of the two-input NAND gate. It represents the effort (in additional charge that must be moved compared with an inverter) to perform the two-input NAND logic function. The delay of a gate driving an identical gate (as in Figure 5.7(a)) is the product of its logical effort and t_{inv}.

In general, for a NAND gate with fan-in F, we size the PFETs K_P and the NFETs F giving an input capacitance of

$$C_{\text{NAND}} = (F + K_P)C_G = \frac{F + K_P}{1 + K_P}C_{\text{inv}}, \tag{5.6}$$

and hence a logical effort of

$$LE_{\text{NAND}} = \frac{F + K_P}{1 + K_P}, \tag{5.7}$$

and a delay, driving an identical NAND gate, of

$$t_{\text{NAND}} = LE_{\text{NAND}}t_{\text{inv}} = \frac{F + K_P}{1 + K_P}t_{\text{inv}}. \tag{5.8}$$

With a NOR gate, the NFETs are in parallel, so a unit-drive NOR gate has NFET pull-downs of size 1. In the NOR, the PFETs are in series, so a unit-drive NOR with a fan-in of F has PFET pull-ups of size FW_P. This gives a total input capacitance of

$$C_{\mathrm{NOR}} = (1 + FK_P)C_G = \frac{1 + FK_P}{1 + K_P} C_{\mathrm{inv}}, \tag{5.9}$$

and hence a logical effort of

$$LE_{\mathrm{NOR}} = \frac{1 + FK_P}{1 + K_P}. \tag{5.10}$$

For reference, Table 5.1 gives the logical effort as a function of fan-in, F, for NAND and NOR gates with one to five inputs both as functions of K_P and numerically for $K_P = 1.3$ (the value for our 28 nm process).

Table 5.1. **Logical effort as a function of fan-in for NAND and NOR gates (ignoring source/drain capacitance)**

Fan-in (F)	Logical effort			
	$f(K_P)$		$K_P = 1.3$	
	NAND	NOR	NAND	NOR
1	1	1	1.00	1.00
2	$\frac{2+K_P}{1+K_P}$	$\frac{1+2K_P}{1+K_P}$	1.43	1.56
3	$\frac{3+K_P}{1+K_P}$	$\frac{1+3K_P}{1+K_P}$	1.87	2.13
4	$\frac{4+K_P}{1+K_P}$	$\frac{1+4K_P}{1+K_P}$	2.30	2.70
5	$\frac{5+K_P}{1+K_P}$	$\frac{1+5K_P}{1+K_P}$	2.74	3.26

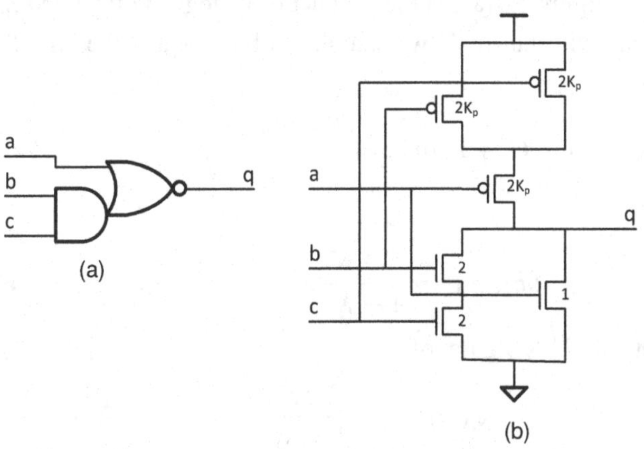

Figure 5.8. Logical effort of an AND-OR-invert (AOI) gate. (a) Gate symbol. (b) Transistor-level schematic showing devices sized for equal rise/fall delays with unit drive.

We follow a similar process to compute the logical effort of a complex gate. Figure 5.8 shows a three-input AND-OR-invert (AOI) gate sized to match the drive strength of a minimum inverter. The input capacitance is not the same for all three inputs, and the logical effort must be calculated individually. First, for a:

$$C_{\mathrm{AOI},a} = (1 + 2K_P)C_G = \frac{1 + 2K_P}{1 + K_P}C_{\mathrm{inv}}, \tag{5.11}$$

$$LE_{\mathrm{AOI},a} = \frac{1 + 2K_P}{1 + K_P}; \tag{5.12}$$

and next for the b and c inputs:

$$C_{\mathrm{AOI},b,c} = (2 + 2K_P)C_G = \frac{2 + 2K_P}{1 + K_P}C_{\mathrm{inv}}, \tag{5.13}$$

$$LE_{\mathrm{AOI},b,c} = \frac{2 + 2K_P}{1 + K_P} = 2. \tag{5.14}$$

EXAMPLE 5.3 Logical effort

Calculate the logical effort of the 2–2 OR-AND-Invert gate shown in Figure 5.9.

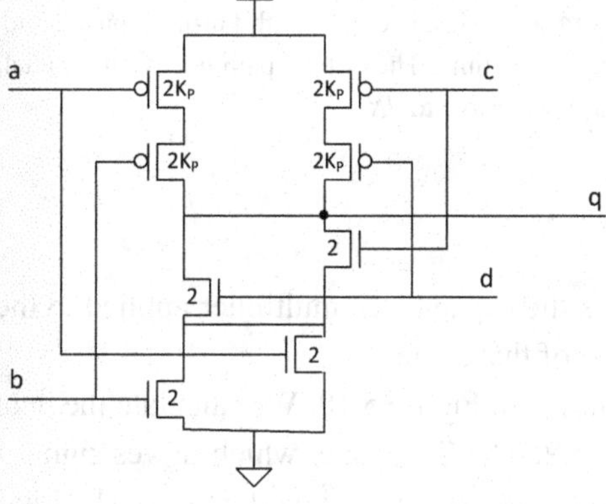

Figure 5.9. A minimum-sized AND-OR-invert gate.

The input capacitance of this gate on each input is given by

$$C_{\mathrm{OAI}} = 2 + 2K_P.$$

Thus the logical effort is

$$LE_{\mathrm{OAI}} = \frac{2 + 2K_P}{1 + K_P} = 2.$$

5.4 DELAY CALCULATION

The delay of each stage i of a logic circuit is the product of its fan-out or *electrical effort* from stage i to stage $i + 1$ and the logical effort of stage $i + 1$. The fan-out is the ratio of the drive

Table 5.2. Computing the delay of a logic circuit
For each stage along the path, we compute the fan-out of the signal and the logical effort of the gate receiving the signal. Multiplying the fan-out by the logical effort gives the delay per stage. Summing over the stages gives the total delay, normalized with respect to t_{inv}.

Driver i	Signal i to $i+1$	Fan-out i to $i+1$	Logical effort $i+1$	Delay i to $i+1$
1	bN	4.00	1.00	4.00
2	c	1.00	2.13	2.13
3	dN	2.00	1.43	2.86
4	e	4.00	1	4.00
Total				13.0

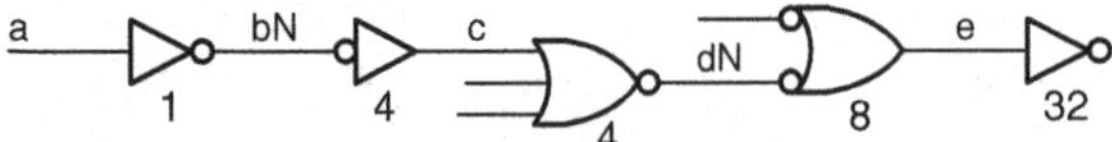

Figure 5.10. Logic circuit for an example delay calculation. The number under each gate is its output drive (conductance) relative to a minimum-sized inverter with equal rise/fall delays.

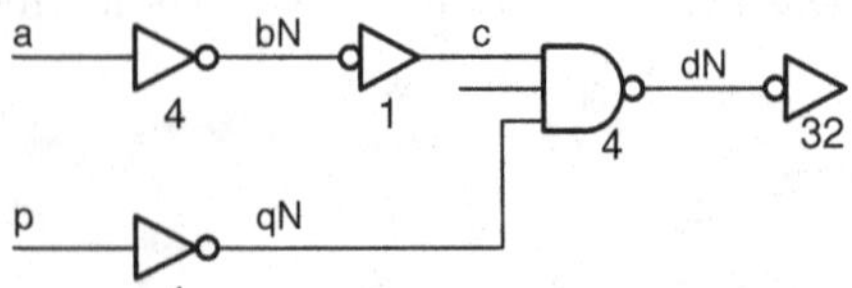

Figure 5.11. Logic circuit with fan-in. Inputs a and p change at the same time. The critical path for maximum delay is the path from a to c to dN.

of stage i to stage $i+1$. The logical effort is the capacitance multiplier applied to the input of stage $i+1$ to implement the logical function of that stage.

For example, consider the logic circuit shown in Figure 5.10. We calculate the delay from a to e one stage at a time, as shown in Table 5.2. The first stage, which drives signal bN, has a fan-out of 4, and the logical effort of the following stage (an inverter) is 1. The total delay of this stage is $4t_{\text{inv}}$. The second stage, driving signal c, has a fan-out of 1, since both this stage and the next have a drive of 4. Signal c drives a three-input NOR gate which has a logical effort of 2.13, so the total delay of this stage is 2.13. The third stage, driving signal dN, has both fan-out and logical effort. The fan-out of this stage is 2 (four driving eight), and the logical effort is that of the two-input NAND, 1.43, for a total delay of $2 \times 1.43 = 2.86$. Finally the fourth stage, driving signal e, has a fan-out of 4 and logical effort of 1. We do not compute the delay of the final inverter (with drive 32). It is shown simply to provide the load on signal e. The total delay is determined by summing the delays of the four stages $t_{pae} = (4 + 2.13 + 2.86 + 4)t_{\text{inv}} = 13.0t_{\text{inv}} = 35$ ps.

When we are computing the maximum delay of a circuit with fan-in, we need to determine the longest (or critical) path. We do so by calculating the delay along each path (as shown in Table 5.2), and taking the maximum value. For example, in Figure 5.11 suppose input signals a and p change at the same time, $t = 0$. The delay calculations are shown in Table 5.3. The

Table 5.3. **Delay calculation for both paths of Figure 5.11**

Signal i **to** $i+1$	**Fan-out** i **to** $i+1$	**Logical effort** $i+1$	**Delay** i **to** $i+1$
bN	0.25	1	0.25
c	4	1.87	7.48
Subtotal a to c			7.73
qN	1	1.87	1.87
Subtotal p to qN			1.87
dN	8	1	8
Total a to dN			15.73
Total p to dN			9.87

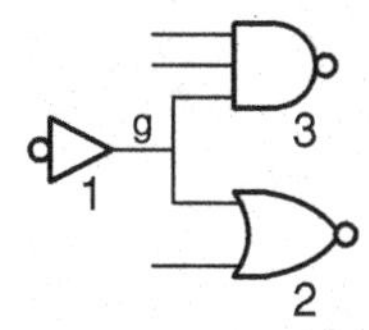

Figure 5.12. Logic circuit with fan-out to different gate types. The total effort of signal g is calculated by summing the product of fan-out and logic effort across all receiving gates.

delay from a to c is $7.73t_{\text{inv}}$, while the delay from p to qN is $1.87t_{\text{inv}}$. Thus, when calculating maximum delay, the critical path is from a to c to dN – a total delay of $15.73t_{\text{inv}}$. If we are concerned with the minimum delay of the circuit, then we use the path from p to qN to dN – with total delay $9.87t_{\text{inv}}$.

Some logic circuits include fan-out to different gate types, as shown for signal g in Figure 5.12. In this case, we compute the electrical and logical effort for each fan-out for each driven gate. Summing these products gives the total delay of signal g. The upper NAND gate has a fan-out of 3 with a logical effort of 1.87 for a total effort of 5.61. The lower NOR gate has a fan-out of 2 and a logical effort of 1.56 for a total effort of 3.12. Thus, the total delay (or effort) of signal g is $8.73t_{\text{inv}}$.

EXAMPLE 5.4 Delay calculation

Calculate the delay of the logic circuit shown in Figure 5.13, which is one slice of the 6:64 decoder of Figure 8.5. The circuit drives a load of $256C_{\text{inv}}$. The fan-out of each stage is shown. Signal b drives both the 2× inverter and two copies of the two-input NOR gate P. Signal c drives a total of two copies of P; d drives 16 copies of Q. The logical effort of the two-input NOR P and the three-input NAND Q are as given in Table 5.4.

The delay calculation is shown in the table below. For each stage we calculate the load by summing the product of the fan-out, logical effort, and size of each type of gate in the next stage. The delay, in units of t_{inv} is then the load divided by the drive.

Table 5.4. **Delay calculation for the circuit shown in Figure 5.13**

Signal	Drive	Load (C_{inv})	Delay (t_{inv})
b	2×	$2+2\times1.56\times8=27.0$	13.5
c	2×	$2\times1.56\times8=25.0$	12.5
d	8×	$16\times1.87\times4=120$	15.0
e	4×	32	8
f	32×	256	8
Total			57.0

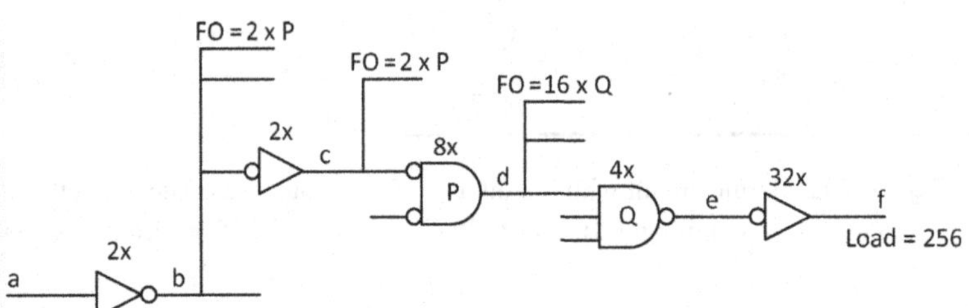

Figure 5.13. Circuit used for calculating delay in Example 5.4.

5.5 OPTIMIZING DELAY

To minimize the delay of a logic circuit we size the stages so that there is an equal amount of delay per stage. For a single n-stage path, a simple way to perform this optimization is to compute the total effort along the path, TE, and then divide this effort evenly across the stages by sizing each stage to have a total effort (product of fan-out and logical effort) of $TE^{1/n}$.

Consider, for example, the circuit of Figure 5.14. The delay calculation for this circuit is shown in Table 5.5. The ratio of the first and last gates specifies the total amount of fan-out required, 96. We multiply this electrical effort by the logical effort of stages 3 and 4, 1.43 and 1.87, respectively, to give the total effort of 257. We then take $257^{1/4} \approx 4$ as the total effort (or delay) per stage. The electrical effort of the gate is found by dividing this stage effort of 4 by the logical effort. Thus, the gate sizes are $x = 4$, $y = 4 \times 4/1.43 = 11.2$, and $z = 24.0$. This gives a total delay of just over $16t_{inv}$.

Suppose the final inverter in Figure 5.14 was sized with a drive of 2048 rather than 96. In that case the total effort is $TE = 2048 \times 1.43 \times 1.87 \approx 5477$. If we attempt to divide this into four stages, we would get a delay of $5477^{1/4} = 8.6t_{inv}$ per stage, which is a bit high, giving a total delay of about $34.4t_{inv}$. In this case, we can reduce the delay by adding an even number of inverter stages, as in the example of Figure 5.6. The optimum number of stages is $\ln 5477 \approx 8$. With eight stages, each stage must have an effort of 2.93, giving a total delay of $23.4t_{inv}$. A compromise circuit is to aim for a delay of four per stage, which requires $\log_4 5477 \approx 6$ stages for a total delay of $25.2t_{inv}$.

Table 5.5. **Optimizing gate sizes to minimize delay, the total effort is determined and divided evenly across the stages**

Driver	Signal	Fan-out	Logical effort	Size	Delay
i	i to $i+1$	i to $i+1$	$i+1$	i	i to $i+1$
1	bN	4.00	1.00	1	$x=4$
2	c	2.80	1.43	4	$1.43y/x=4$
3	dN	2.14	1.87	11.2	$1.87z/y=4$
4	e	4	1	24.0	$96/z=4$
Total					16

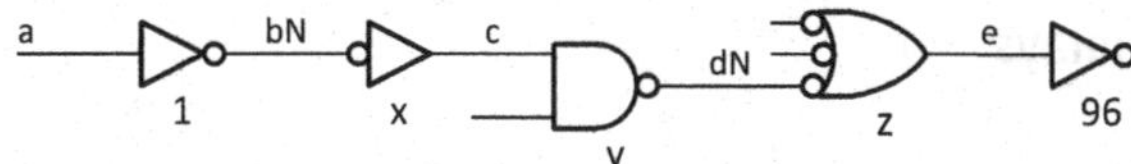

Figure 5.14. Unsized logic circuit. The sizes x, y, and z of the three middle stages must be chosen to minimize delay by equalizing the delay of each stage and adding stages, if needed.

If we are to add either two or four inverters to the circuit of Figure 5.14 we must decide where to add them. We could insert a pair of inverters at any stage of the circuit without changing its function or delay. We could even insert individual inverters at arbitrary points if we are willing to convert the NANDs to NORs (which is generally a bad idea because it increases total effort.) However, it is usually best to place the extra stages *last* to avoid the extra power that would otherwise be consumed if the high logical effort stages were sized larger. However, if one of the signals has a large wire load, it may be advantageous to insert one or more of the extra stages before that point to ensure adequate drive for the wire.

In our discussion of delay optimization, we have omitted the self (or parasitic) capacitance of logic gates. Adding it to our model would have two impacts, aside from increasing the delay of each gate. First, the optimal stage effort becomes 3–4 instead of e as adding stages becomes more expensive. Second, large fan-in gates lead to unfeasibly large gate delays. For example, a 64-input NAND gate would have 65 transistors attached to the output node. Do not build high fan-in gates.

EXAMPLE 5.5 Optimizing delay

Resize the gates of Example 5.4 to give minimum delay. The input inverters, driving signals b and c, must remain $2\times$.

Starting at point c, we calculate the total effort assuming unit-sized gates by multiplying the gate fan-out (the number of driven gates) of each stage by the logical effort of the following stage.

From Table 5.6, we see that the total effort due to fan-out and logical effort is $3.12 \times 29.9 = 93.4$. Multipying this by the increase in load of 128 gives a total effort of 11.9×10^3. Dividing this effort evenly across the four stages gives an effort per stage of $(11.9 \times 10^3)^{1/4} = 10.5$.

Table 5.6. **The effort calculation for each stage of the circuit shown in Figure 5.13**

Signal	Gate fan-out	Logical effort	Effort
c	2	1.56	3.12
d	16	1.87	29.9
e	1	1	1
f	1	1	1

Thus, we resize the gates to give a total effort per stage of 10.5. The resulting sizes and delays are shown in Table 5.7. We reduce the delay from that calculated in Example 5.4 by $3.7t_{\text{inv}}$.

Table 5.7. **Optimized size and delays for each gate in Figure 5.13**

Signal	Size	Effort	Delay
b	2		11.5
c	2	3.12	10.5
d	6.73	29.9	10.5
e	2.36	1	10.5
f	24.8	1	10.3
Total			53.3

5.6 WIRE DELAY

On modern integrated circuits a large fraction of delay and power is due to driving the wires that connect gates. An on-chip wire has both resistance and capacitance, with typical values for 130 nm and 28 nm processes shown in Table 5.8. Our examples below assume minimum-sized wires, and we ask you to explore the impacts of larger, less resistant wires, in Exercise 5.20c.

Wires that are short enough that their total resistance is small compared with the output resistance of the driving gate can be modeled as a lumped capacitance. For example, a minimum-sized ($W_N = 8L_{\text{min}}$) inverter has an output resistance of 5.25 kΩ. A wire of less than 105 μm in length has a total resistance of less than one-fifth this amount and can be considered a lumped capacitance. A wire of exactly 105 μm, for example, can be modeled as a capacitance of 19 fF, the equivalent of a fan-out of 36 compared with the 0.52 fF input capacitance of the minimum-sized inverter.

For larger drivers, shorter wires have a resistance that is comparable to the driver output resistance. For a 16× minimum-sized inverter with an output resistance of 328 Ω, for example, a wire of length 33 μm has a resistance equal to the output resistance of the driver and one must get down to a length of 6.1 μm for the resistance to be less than one-fifth of the driver resistance.

Table 5.8. **Resistance and capacitance of minimum-sized wires in 130 nm and 28 nm processes**

Parameter	130 nm value	28 nm value	Units	Description
R_w	0.25	0.45	Ω/square	resistance per square
w_w	0.25	0.045	μm	wire width
R_w	1	10	Ω/μm	resistance per μm
C_w	0.2	0.18	fF/μm	capacitance per μm
τ_w	0.2	1.8	fs/μm^2	RC time constant

(a)

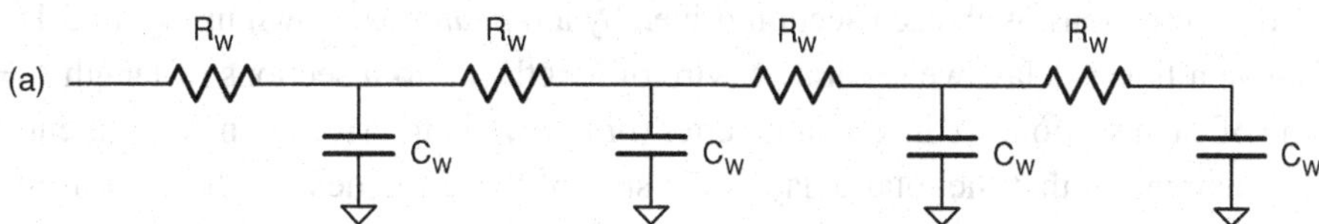

(b)

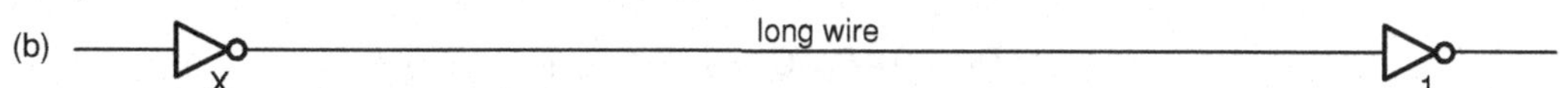

(c)

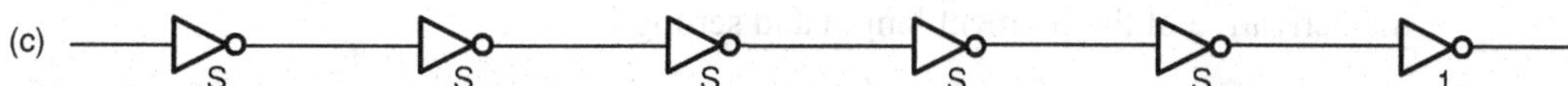

Figure 5.15. (a) A long on-chip wire has significant series resistance R_w and parallel capacitance C_w, giving it a delay that grows quadratically with length. (b) Driving a long wire often gives unacceptable delay and rise time. Increasing the size X of the driver does not help due to the resistivity of the line. (c) The delay of the line can be made linear, rather than quadratic, with length by inserting repeaters of size S at a fixed interval in the line.

For wires that are long enough for their resistance to be significant compared with the resistance of their driver, the delay of the wire increases quadratically with wire length. As illustrated in Figure 5.15(a), as the wire gets longer both the resistance and the capacitance of the wire increase linearly, causing the RC time constant to increase quadratically. Increasing the size of the driver as shown in Figure 5.15(b) does not improve the situation because the total resistance is dominated by the wire resistance.

We can calculate the resistance, capacitance, and intrinsic delay of a 1 mm minimum-sized wire as follows. When modeling distributed capacitances and resistances, like wires, the delay is not RC, but rather $0.4RC$. (We ignore output capacitance in our calculations.) So

$$R_{w,1\text{mm}} = R_w L = (10)(10^3) = 10\,\text{k}\Omega, \tag{5.15}$$

$$C_{w,1\text{mm}} = C_w L = (0.18 \times 10^{-15})(10^3) = 0.18\,\text{pF}, \tag{5.16}$$

$$D = 0.4RC = 0.4(R_{w,1\text{mm}})(C_{w,1\text{mm}}) = 720\,\text{ps}. \tag{5.17}$$

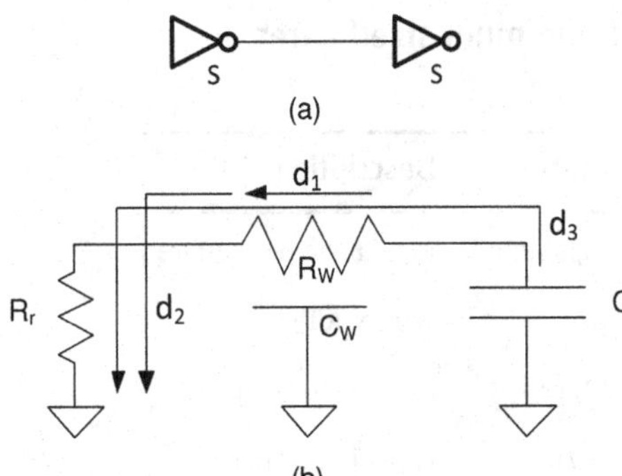

Figure 5.16. (a) Schematic of one wire segment being driven and terminated by repeaters of size S. (b) The delay modeling we use is the sum of three delays: the intrinsic wire delay (d_1), the discharge of the wire capacitance through the driving resistor (d_2), and the discharge of subsequent driver's gate through both the driver and the wire (d_3).

To make the delay of a long wire linear (rather than quadratic) with length, the wire can be divided into sections, with each section driven by a *repeater* as shown in Figure 5.15(c).

To give a linear delay, we separate a wire of length L into n sections of length $l = L/n$. At the end of each section, we insert an inverter (or *repeater*) to drive the next segment. We make the approximation that the total delay is the sum of three RC delays: the wire itself, the wire capacitance passing through the driver resistance, and the next driver's capacitance through both the wire and driver resistance.[5] We show this model in Figure 5.16.

$$D_l = 0.4R_{w,l}C_{w,l} + R_rC_{w,l} + C_r(R_{w,l} + R_r), \tag{5.18}$$

$$D_L = \frac{L}{l}(0.4l^2R_wC_w + lR_rC_w + C_r(lR_w + R_r)). \tag{5.19}$$

We can use Equation (5.19) to derive the repeater spacing that gives minimal delay by taking the derivative of the segment length and setting it to 0:

$$\frac{d}{dl}D_L = 0.4R_wC_w - \frac{R_rC_r}{l^2} = 0, \tag{5.20}$$

$$l = \sqrt{\frac{t_{\mathrm{inv}}}{0.4R_wC_w}} = 61\ \mu\mathrm{m}. \tag{5.21}$$

In Figure 5.17(a), we show how the delay of a line varies with the repeater spacing. At very short distances, the delay of the repeaters dominate. Once past about 60 μm, the delay begins to increase. An 8× increase in repeater spacing, for example, causes a 2× increase in delay. We can also follow a similar procedure to find the optimal repeater size, S. So

$$D_L = \frac{L}{l}\left(0.4l^2R_wC_w + \frac{lR_{\mathrm{inv}}C_w}{S} + SC_{\mathrm{inv}}\left(lR_w + \frac{R_{\mathrm{inv}}}{S}\right)\right), \tag{5.22}$$

$$\frac{d}{dS}D_L = C_{\mathrm{inv}}R_w - \frac{R_{\mathrm{inv}}C_w}{S^2} = 0, \tag{5.23}$$

$$S = \sqrt{\frac{R_{\mathrm{inv}}C_w}{C_{\mathrm{inv}}R_w}} = 13.5, \tag{5.24}$$

$$D_{L,1\mathrm{mm}} = 228\ \mathrm{ps}. \tag{5.25}$$

[5] This is a very simple model that underestimates the total delay, but it gives a reasonable estimate of the optimal repeater spacing. For more accurate and complex models, refer to refs. [8] and [29].

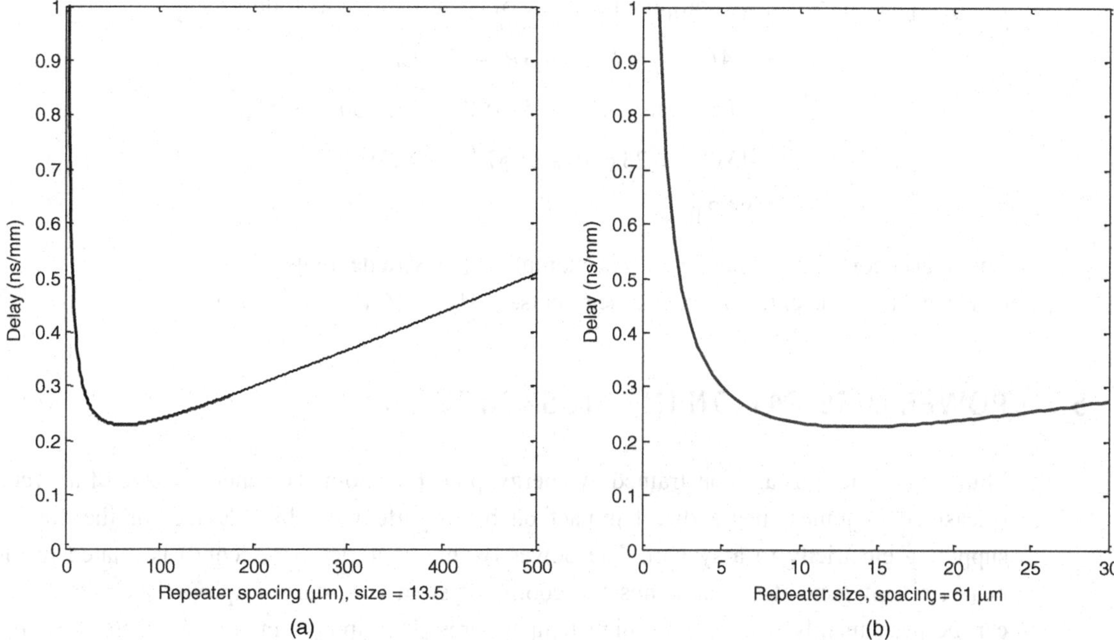

Figure 5.17. (a) The delay of a 1 mm wire as a function of the distance between two repeaters. (b) The delay of a 1 mm wire as a function of the size of the optimally spaced repeaters. The non-repeated delay is 720 ps for 1 mm.

We see that, independently of the segment length, the optimal driver size is 13.5× that of a minimum inverter. The relationship between driver size and delay, Figure 5.17(b), is nearly flat. Once drivers pass a certain size threshold, they have minimal impact on delay. Large drivers, however, do increase energy consumption.

EXAMPLE 5.6 Wire delay

Compute the delay of a 1 mm minimum-width wire in 28 nm technology that is divided into four segments, each driven by a 10× minimum-size inverter. Use the values from Tables 5.8 and 4.1.

We calculate the resistance and capacitance of a wire segment and a repeater as follows. A minimum-sized inverter has $W_N = 8L_{\min}$, so our 10× inverter has $W_N = 80L_{\min}$. Then

$$R_w = 10\,\frac{\Omega}{\mu\text{m}} \times 250\,\mu\text{m} = 2500\,\Omega,$$

$$C_w = 0.18\,\frac{\text{fF}}{\mu\text{m}} \times 250\,\mu\text{m} = 45\,\text{fF},$$

$$R_r = \frac{K_{RN}}{W_N} = \frac{4.2 \times 10^4}{80} = 525\,\Omega,$$

$$C_r = W_N(1 + K_P)K_C = 80(1 + 1.3)2.8 \times 10^{-17} = 5.15\,\text{fF}.$$

Using Equation (5.18), we compute the delay of each segment as follows:

$$\begin{aligned} D_l &= 0.4R_wC_w + R_rC_w + (R_r + R_w)C_r \\ &= (0.4)(2500)(45) + (525)(45) + (2500 + 525)(5.15) \\ &= 45\,000 + 23\,600 + 15\,579 = 84\,179\text{ fs} \\ &\approx 84.2\text{ ps}. \end{aligned}$$

The largest contributer to delay in this example is the wire delay itself.

The delay of the entire wire, four segments, is $4D_l = (4)(84.2) = 337$ ps.

5.7 POWER DISSIPATION IN CMOS CIRCUITS

Many digital designs are constrained by energy, power, or both. The energy usage of a circuit (measured in joules) has a direct impact on battery life in mobile devices or the cost of supplying electricity to a system. The power (watts, or joules per second) is related to the heat a circuit generates – and thus the cooling system necessary. Improperly cooled chips can be permanently damaged by high temperatures. It is important for the digital designer to understand how to calculate and optimize both energy and power.

5.7.1 Dynamic power

In a CMOS chip, most of the energy dissipation is due to charging and discharging the capacitance of gates and wires. The energy consumed in charging a capacitor from V_0 to V_1 and then discharging it again to V_0 through a resistor is given by

$$E = CV^2. \tag{5.26}$$

For our 28 nm process, with $C_{\text{inv}} = 0.6$ fF and $V = V_1 - V_0 = 0.9$ V, $E_{\text{inv}} = 0.49$ fJ. That is, charging and discharging the gate of a minimum-sized inverter consumes half a femtojoule.

As an example, we will compute the energy dissipated by the circuit of Figure 5.10 when input a cycles up and down. This input cycles all internal nodes up to and including e. Recall that the input capacitance of each gate is the product of its size (s), logical effort (LE), and C_{inv}. Thus

$$\begin{aligned} E &= CV^2 \\ &= V^2 \sum_i C_i \\ &= V^2 C_{\text{inv}} \sum_i s_i LE_i \\ &= E_{\text{inv}} \sum_i s_i LE_i \\ &= E_{\text{inv}}(1 + 4 + 4LE_{\text{NOR3}} + 8LE_{\text{NAND2}} + 32) \\ &= 27.9\text{ fJ}. \end{aligned}$$

Circuits are typically designed so that each stage of logic has equal delay. As the preceding worked example showed, however, energy consumption is dominated by the largest gates in a chain.

The power ($P = E/T = Ef$) consumed in charging and discharging capacitance depends on the signal transition frequency. For a circuit with capacitance C that operates at a frequency f and has α transitions each cycle,[6] the power consumed is given by

$$P = 0.5CV^2 f\alpha. \quad (5.27)$$

The factor of 0.5 is due to the fact that half of the energy is consumed on the charging transition and the other half on the discharge. For an inverter with activity factor $\alpha = 0.33$ and a clock rate of $f = 2$ GHz, $P = 162$ nW.

To reduce the power dissipated by a circuit we can reduce any of the terms of Equation (5.27). If we reduce voltage, power reduces quadratically. However, the circuit also operates slower at a lower voltage. For this reason we often reduce V and f together, getting a factor of 8 reduction in power each time we halve V and f. Reducing capacitance is typically accomplished by making our circuit as physically small as possible – minimizing wire length, and hence wire capacitance.

The activity factor, α, can be reduced through a number of measures. First, it is important that the circuit does not make unnecessary transitions. For a combinational circuit, each transition of the inputs should result in at most one transition of each output. Glitches or hazards (see Section 6.10) should be eliminated because they result in unnecessary power dissipation. The activity factor can also be reduced by gating (stopping) the clock to unused portions of the circuit, so that these unused portions have no activity at all. For example, if an adder is not being used on a particular cycle, stopping the clock to the adder prevents the adder's inputs and combinational logic from toggling, saving considerable power.

5.7.2 Static power

Up to now we have focused on dynamic power – the power due to charging and discharging capacitors. As gate lengths and supply voltages have shrunk, however, static *leakage* power has become an increasingly important factor. Leakage current is the current that flows through a MOSFET when it is in the off state ($V_{GS} = 0$, $V_{DS} = V_{DD}$) and is proportional to e^{-V_T}. Thus, as threshold voltage decreases, leakage current increases exponentially. The slope of this curve is often referred to as the *subthreshold slope* and is typically about 70 mV/dec. That is, for each 70 mV decrease in threshold voltage, leakage current increases by 10×. The scaling of supply voltage has largely stopped because lower supply voltages require lower threshold voltages, resulting in higher leakage current.

Static power can be reduced by using transistors with a higher V_T. These transistors come at the expense of lower speed or a higher supply voltage (dynamic energy) or both. Many digital designs use low-V_T transistors on critical paths and high-V_T transistors everywhere else. Another way of eliminating leakage current is to turn off or *power-gate* circuits. Power-gating

[6] We count a signal that transitions from 0 to 1 *or* 1 to 0 each cycle as $\alpha = 1$. A clock signal has $\alpha = 2$. Some references use an activity factor that counts full cycles rather than transitions and hence is given by half of this amount.

is done at a coarse granularity in both space and time because it takes significant time and control logic to cycle between on and off states.

Consider a process that is tailored for high-speed applications, with a supply voltage of 0.9 V and a leakage current of 100 nA/μm. This means that every minimum-sized transistor ($W =$ 224 nm) burns about 20 nW of power ($P = IV$). Remember that commercial chips can have the equivalent of one to two billion minimum inverters, making chip-wide leakage power 20–40 W. This is a significant portion of these chips' 60–120 W power budgets.

5.7.3 Power scaling

The capacitance of a CMOS transistor scales with L. This is because all three dimensions of the parallel-plate capacitor scale linearly with L and $C = LW/H$. For a constant supply voltage, the energy of any given logic module also scales with L. The energy density of the chip thus increases as $1/L$; i.e., if we halve L, we double the energy consumed per unit area. Even at constant frequency, the power density increases as $1/L$, quickly outstripping our ability to cool the chip. The problem gets worse when we attempt to increase performance by increasing the frequency, causing power density to grow as $1/L^2$. Many designers now turn to parallelism, running more modules slower, to increase performance. Parallelism is a more energy-efficient approach to performance.

EXAMPLE 5.7 Energy calculation

Compute the dynamic energy dissipated by the circuit of Figure 5.14 with $x = 4$, $y = 12$, and $z = 24$ when signal a goes through a full cycle from 0 to 1 and back to 0. Assume that the transitions on a propagate all the way to e. You may ignore wire capacitance and assume the supply voltage to be $V_{DD} = 1$ V.

Table 5.9. **Energy calculation for Example 5.7**

Signal	Capacitance (C_{inv})
a	1
bN	4
c	$12 \times 1.43 = 17.2$
dN	$24 \times 1.87 = 44.9$
e	96
Total	163

We compute the capacitance of each stage and sum. The dynamic energy is $E = CV^2$, and, since $V = 1$, $E = C$. We first compute the capacitance in terms of C_{inv} and then convert to farads after summing (see Table 5.9). Thus, the total capacitance is $163 \times C_{\text{inv}} =$ (163)(0.515 fF) $= 84$ fF. Thus the dynamic energy from cycling a is 84 fJ.

Summary

In this chapter you have learned a simple method to estimate the delay and power of CMOS logic circuits. While this is no substitute for detailed circuit simulations, this method will allow you to estimate the delay and dynamic energy of typical CMOS logic circuits with about 20% accuracy. Just as importantly, it provides you with a means of comparing different circuit designs and selecting the right solution.

Using the simple switch-level model of a MOSFET that we developed in Chapter 4, we estimated the delay of a CMOS gate using a simple *RC* model. The delay is the output resistance of the driving gate multiplied by the total capacitance being driven. We see that if we drive a load directly, delay increases linearly with the *fan-out* of our driver, the ratio of load capacitance to driver input capacitance. To drive large loads more quickly, we build a multi-stage driver increasing our driver size by a fixed multiple (typically $4\times$) each stage.

Increasing the complexity of a CMOS gate while holding output resistance constant increases the input capacitance of the gate. We refer to this increased input capacitance as the *logical effort* of the gate. The product of a driving gate's fan-out (or electrical effort) and the driven gate's logical effort is the *total effort* of a stage of logic. The delay of the stage is proportional to this total effort. Delay is minimized when the total effort of each stage is balanced and near an optimal total effort – approximately 4.

The resistance and capacitance of an on-chip wire both increase linearly with length. This gives a wire delay, *RC*, that increases quadratically with wire length. To make wire delay linear with length, we break the wire into fixed-length segments and drive each segment with a *repeater*.

CMOS chips dissipate dynamic power, due to charging and discharging capacitance as signals switch, and static power, due to the leakage of transistors. The dynamic energy associated with one gate switching is $E = CV^2$. This energy is about 0.5 fJ for a typical inverter in a 28 nm process. The dynamic power is $P = 0.5Ef\alpha$ for frequency f and *activity factor* α.

Static power is largely due to subthreshold leakage current, which varies exponentially with threshold voltage. For every 70 mV decrease in threshold voltage, leakage current increases by $10\times$. By adjusting the threshold voltage of a device, we can set the leakage current wherever we want – at a cost in performance. Typical high-speed processes have a worst-case leakage current that accounts for about 30% of total power, while low-leakage processes have negligible leakage current but much slower gates.

BIBLIOGRAPHIC NOTES

Mead and Rem first described the exponential horn for driving large capacitive loads in ref. [77]. Further work on CMOS delay models can be found in ref. [50]. Both of these references use an Elmore delay model to find the *RC* delay [41].

Sutherland and Sproull introduced the notion of logical effort in 1991 [104], and Sutherland, Sproull, and Harris have written a monograph describing this concept and its application in detail [102].

Scaling, before leakage power became dominant, followed the rules laid out by Dennard *et al.* [37]. Dennard's scaling estimates that power densities do not increase with smaller gate lengths are no longer true.

Exercises

5.1 *Rise and fall times, I.* Compute the maximum and minimum rise/fall times of the gates shown in Figure 5.18(a). Assume that only one input toggles at a time and that the gates drive an output of $4C_{inv}$. You may leave your answer in terms of t_{inv}.

5.2 *Rise and fall times, II.* Compute the maximum and minimum rise/fall times of the gates shown in Figure 5.18(b). Assume that only one input toggles at a time and that the gates drive an output of $4C_{inv}$. You may leave your answer in terms of t_{inv}.

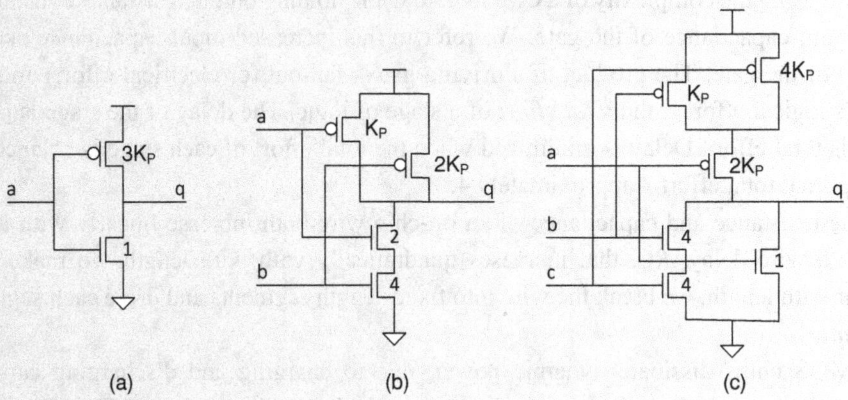

Figure 5.18. Circuits for Exercises 5.1–5.3.

5.3 *Rise and fall times, III.* Compute the maximum and minimum rise/fall times of the gates shown in Figure 5.18(c). Assume that only one input toggles at a time and that the gates drive an output of $4C_{inv}$. You may leave your answer in terms of t_{inv}.

5.4 *Inverter chain delay and energy, I.* Compute the delay and energy of driving an inverter of size 256 from a minimum-size inverter (size 1) with a series of FO2 inverters. Express your answer in terms of t_{inv} and E_{inv}.

5.5 *Inverter chain delay and energy, II.* Compute the delay and energy of driving an inverter of size 256 from a minimum-size inverter (size 1) with a series of FO4 inverters. Express your answer in terms of t_{inv} and E_{inv}.

5.6 *Inverter chain delay and energy, III.* Compute the delay and energy of driving an inverter of size 256 from a minimum-size inverter (size 1) with a series of FO8 inverters. Express your answer in terms of t_{inv} and E_{inv}.

5.7 *Inverter chain delay and energy, IV.* Compute the delay and energy of driving an inverter of size 256 from a minimum-size inverter (size 1) with a series of FO16 inverters. Express your answer in terms of t_{inv} and E_{inv}.

5.8 *Sizing of CMOS gates, I.* Consider a four-input static CMOS gate that implements the function $f = \overline{a \wedge (b \vee (c \wedge d))}$.

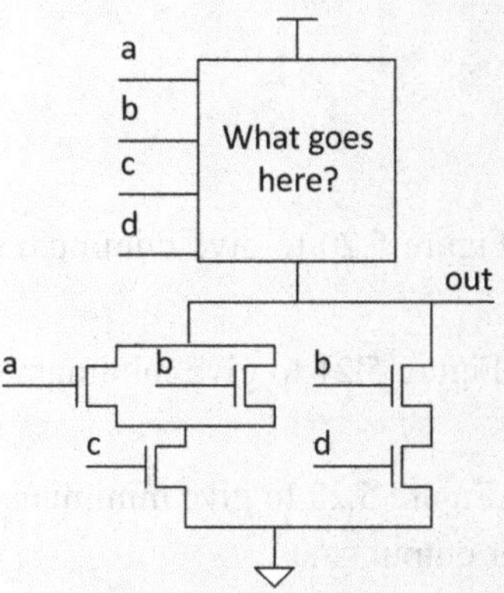

Figure 5.19. Circuit for Exercise 5.10.

(a) Draw a schematic symbol for this gate with the bubble on the output.

(b) Draw a transistor schematic for this gate and size the transistors for rise and fall delay equal to a minimum-sized inverter with equal rise/fall.

(c) Compute the logical effort of this gate for each of its inputs.

5.9 *Sizing of CMOS gates, II.* Repeat Exercise 5.8 for a gate that implements the function $f = \overline{(a \wedge b) \vee (c \wedge d)}$.

5.10 *Sizing of CMOS gates, III.* For Figure 5.19, do the following:

(a) Draw the correct PFET network.

(b) Size each of the transistors to provide the same rise and fall resistance as a minimum-sized inverter.

(c) Compute the logical effort for each input.

5.11 *Delay calculation I.* Calculate the delay of the circuit in Figure 5.20 in terms of t_{inv}.

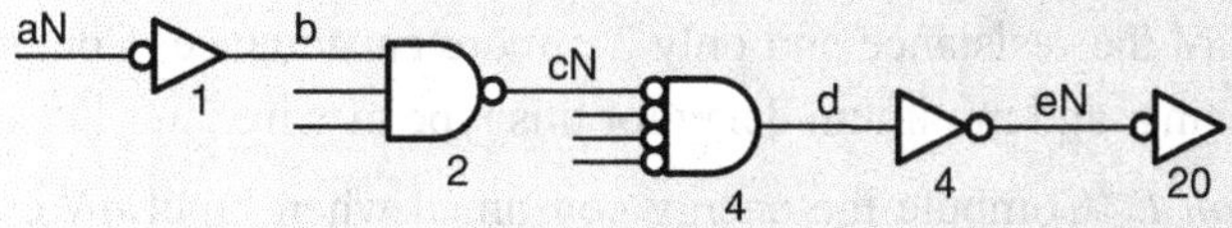

Figure 5.20. Circuit for Exercises 5.11, 5.14, and 5.22. Gate sizes are given by the numbers under each gate.

5.12 *Delay calculation II.* Calculate the delay of the circuit in Figure 5.21 in terms of t_{inv}.

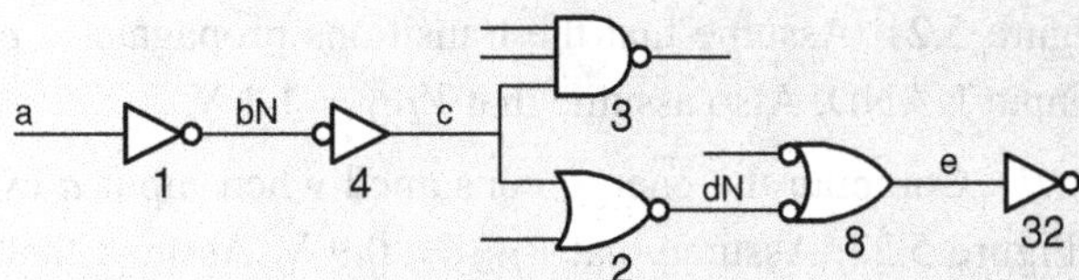

Figure 5.21. Circuit for Exercises 5.12, 5.15, and 5.23. Gate sizes are given by the numbers under each gate.

5.13 *Delay calculation II.* Calculate the delay of the circuit in Figure 5.22 from inputs *a* and *p* to signal *dN* in terms of t_{inv}.

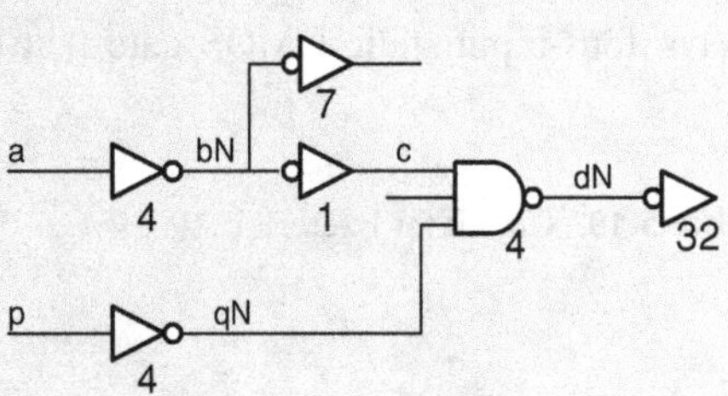

Figure 5.22. Circuit for Exercises 5.13, 5.16, and 5.22. Gate sizes are given by the numbers under each gate.

5.14 *Delay optimization I.* Resize the gates in Figure 5.20 to give minimum delay. You may not change the size of input or output gates.

5.15 *Delay optimization II.* Resize the gates in Figure 5.21 to give minimum delay. You may not change the size of input or output gates.

5.16 *Delay optimization III.* Resize the gates in Figure 5.22 to give minimum delay from *a* to *dN*. You may not change the size of input or output gates.

5.17 *Wire delay I.* Calculate the delay of a 10 mm wire in 28 nm technology that is divided into 20 0.5 mm segments with a 20× minimum-size inverter driving each segment.

5.18 *Wire delay II.* Calculate the delay of a 1 mm wire in 28 nm technology that is divided into five 200 μm segments with a 10× minimum-size inverter driving each segment.

5.19 *Wire delay III.* Calculate the delay of a 1 mm wire in 28 nm technology that is divided into ten 100 μm segments with a 10× minimum-size inverter driving each segment.

5.20 *Wire delay and energy, I.* Use a repeater that is 13.5 times larger than minimum in the following.

(a) Calculate the minimum time to transmit a bit across a 5 mm wire. What is the total energy required to transmit a bit using this circuit?

(b) If we doubled the spacing between repeaters, what are the new delay and energy?

(c) Plot the following graphs: delay vs. segment length, energy vs. segment length, and a scatter plot of energy vs. delay.

5.21 *Wire delay and energy, II.* Intermediate wires that are larger than minimum-sized wires provide a lower resistance and higher capacitance. For example, a wire with a 3× width may have one-third the resistance and only 2× more capacitance. Compute the optimal repeater size, spacing, and minimum delay for this type of wire.

5.22 *Energy calculation I.* Compute the energy consumed when input *aN* cycles from 0 to 1 and back to 0 in Figure 5.20. Assume that the transitions propagate to *eN* and that $V_{DD} = 1$ V.

5.23 *Energy calculation II.* Compute the energy consumed when input *a* cycles from 0 to 1 and back to 0 in Figure 5.21. Assume that the transitions propagate to *e* but not to the output of the three-input NAND. Also assume that $V_{DD} = 1.1$ V.

5.24 *Energy calculation III.* Compute the energy consumed when input *a* cycles from 0 to 1 and back to 0 in Figure 5.22. Assume that $V_{DD} = 0.9$ V. Assume that the transitions propagate to *dN* but that *p* remains at 0. You may ignore the output load on the 7× inverter.

5.25 *Design for power.* How does designing for power differ between a cellular phone radio chip and a high-utilization server processor? What power reduction mechanisms would you use for each? What are the differing constraints?

6 Combinational logic design

Combinational logic circuits implement logical functions on a set of inputs. Used for control, arithmetic, and data steering, combinational circuits are the heart of digital systems. Sequential logic circuits (see Chapter 14) use combinational circuits to generate their next state functions.

In this chapter we introduce combinational logic circuits and describe a procedure to design these circuits given a specification. At one time, before the mid 1980s, such manual synthesis of combinational circuits was a major part of digital design practice. Today, however, designers write the specification of logic circuits in a hardware description language (like VHDL) and the synthesis is performed automatically by a computer-aided design (CAD) program.

We describe the manual synthesis process here because every digital designer should understand how to generate a logic circuit from a specification. Understanding this process allows the designer to better use the CAD tools that perform this function in practice, and, on rare occasions, to generate critical pieces of logic manually.

6.1 COMBINATIONAL LOGIC

As illustrated in Figure 6.1, a combinational logic circuit generates a set of outputs whose state depends only on the *current* state of the inputs. Of course, when an input changes state, some time is required for an output to reflect this change. However, except for this *delay* the outputs do not reflect the *history* of the circuit. With a combinational circuit, a given input state will always produce the same output state regardless of the sequence of previous input states. A circuit where the output depends on previous input states is called a *sequential* circuit (see Chapter 14).

For example, a majority circuit, a logic circuit that accepts n inputs and outputs a 1 if at least $\lfloor n/2+1 \rfloor$ of the inputs are 1, is a combinational circuit. The output depends only on the number of 1s in the present input state. Previous input states do not affect the output.

On the other hand, a circuit that outputs a 1 if the number of 1s in the n inputs is greater than the previous input state is sequential (not combinational). A given input state, e.g., $i_k = 011$, can result in $o = 1$ if the previous input was $i_{k-1} = 010$, or it can result in $o = 0$ if the previous input was $i_{k-1} = 111$. Thus, the output depends not just on the present input, but also on the history (in this case, very recent history) of previous inputs.

Combinational logic circuits are important because their static nature makes them easy to design and analyze. As we shall see, general sequential circuits are quite complex in comparison. In fact, to make sequential circuits tractable we usually restrict ourselves to

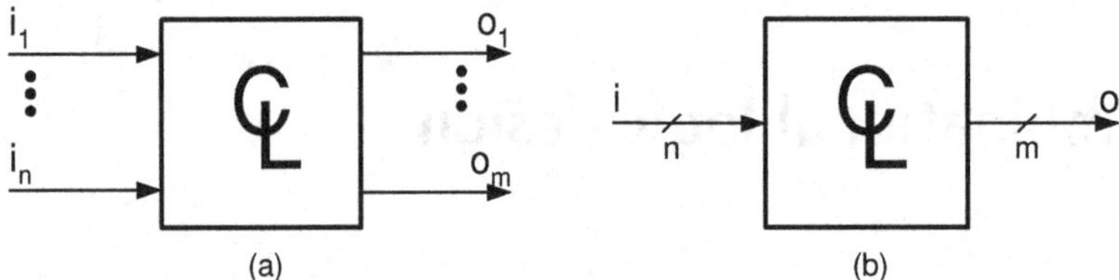

Figure 6.1. A combinational logic circuit produces a set of outputs $\{o_1, \ldots, o_m\}$ that depend only on the *current* state of a set of inputs $\{i_1, \ldots, i_n\}$. (a) Block CL is shown with n inputs and m outputs. (b) Equivalent block with n inputs and m outputs shown as buses.

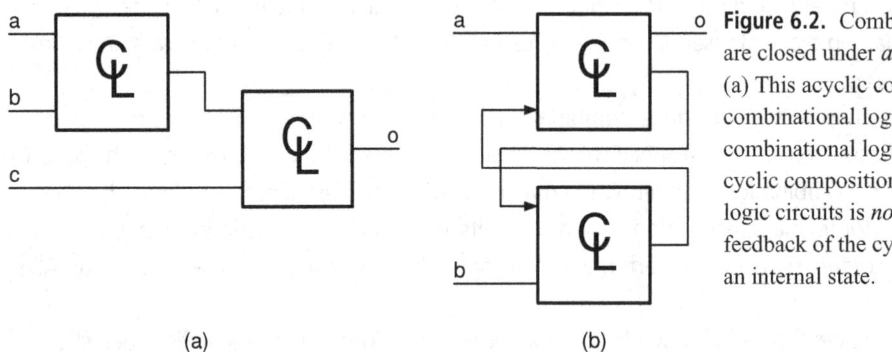

Figure 6.2. Combinational logic circuits are closed under *acyclic* composition. (a) This acyclic composition of two combinational logic circuits is itself a combinational logic circuit. (b) This cyclic composition of two combinational logic circuits is *not* combinational. The feedback of the cyclic composition creates an internal state.

synchronous sequential circuits, which use combinational logic to generate a next-state function (see Chapter 14).

Please note that logic circuits that depend only on their inputs are *combinational* and **not** *combinatorial*. While these two words sound similar, they mean different things. The word *combinatorial* refers to the mathematics of counting, not to logic circuits. To keep them straight, remember that combinational logic circuits *combine* their inputs to generate an output.

6.2 CLOSURE

A valuable property of combinational logic circuits is that they are closed under *acyclic* composition. That is, if we connect together a number of combinational logic circuits – connecting the outputs of one to the inputs of another – and avoid creating any loops (that would be cyclic), the result will be a combinational logic circuit. Thus we can create large combinational logic circuits by connecting together small combinational logic circuits.

An example of acyclic, and of cyclic, composition is shown in Figure 6.2. A combinational circuit realized by acyclically composing two smaller combinational circuits is shown in Figure 6.2(a). The circuit in Figure 6.2(b), on the other hand, is not combinational. The cycle created by feeding the output of the upper block into the input of the lower block creates a state. The value of this feedback variable can *remember* the history of the circuit. Hence the output of this circuit is not just a function of its inputs. In fact, we shall see that *flip-flops*, the building blocks of most sequential logic circuits, are built using exactly the type of feedback shown in Figure 6.2(b).

It is easy to prove that acyclic compositions of combinational circuits are themselves combinational by induction, starting at the input and working toward the output. Let a combinational block whose inputs are connected only to primary inputs (i.e., not to the outputs

Table 6.1. **Truth table for a four-bit prime number, or 1, circuit**

The column "Out" shows the output of the circuit for each of the 16 input combinations.

No.	In	Out
0	0000	0
1	0001	1
2	0010	1
3	0011	1
4	0100	0
5	0101	1
6	0110	0
7	0111	1
8	1000	0
9	1001	0
10	1010	0
11	1011	1
12	1100	0
13	1101	1
14	1110	0
15	1111	0

of other blocks) be a rank 1 block. Similarly, let a block whose inputs are connected only to primary inputs and/or to the outputs of blocks of ranks 1 through k be a rank $k+1$ block. By definition, all rank 1 blocks are combinational. Then, if we assume that all blocks of ranks 1 to k are combinational, a rank $k+1$ block is also combinational. Since its outputs depend only on the current state of its inputs, and since all of its inputs depend only on the current state of the primary inputs, its outputs also depend only on the current state of the primary inputs.

6.3 TRUTH TABLES, MINTERMS, AND NORMAL FORM

Suppose we want to build a combinational logic circuit that outputs a 1 when its four-bit input represents a prime number in binary. One way to represent the logic function realized by this circuit is with an English-language description – as we have just specified it. However, we generally prefer a more precise definition.

Often we start with a *truth table* that shows the output value for each input combination. Table 6.1 shows a truth table for the four-bit prime number function.[1] For an n-input function, a truth table has 2^n rows (16 in this case), one for each input combination. Each row lists the output of the circuit for that input combination (0 or 1 for a one-bit output).

[1] Note that this is really a "prime-or-one" function since it is true when the input is "1" and "1" is not a prime number [45]. We leave it as an exercise (Exercise 6.5) to design a prime number function that does not include "1."

Table 6.2. **Abbreviated truth table for a four-bit prime number circuit**

Only inputs for which the output is 1 are listed explicitly.

No.	In	Out
1	0001	1
2	0010	1
3	0011	1
5	0101	1
7	0111	1
11	1011	1
13	1101	1
Otherwise		0

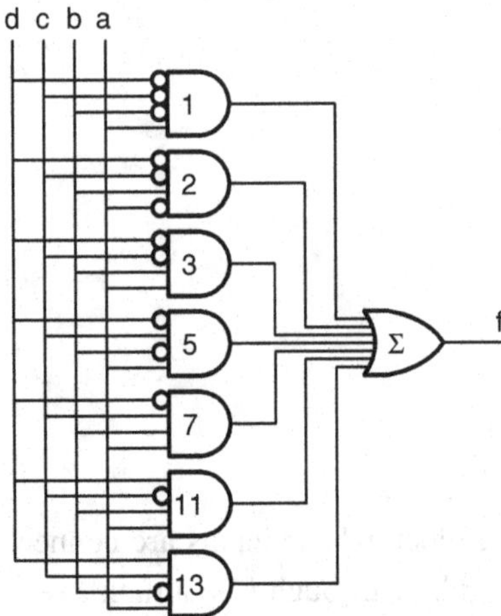

Figure 6.3. A four-bit prime number circuit in conjunctive (sum-of-products) normal form. An AND gate generates the minterm associated with each row of the truth table that gives a true output. An OR gate combines the minterms, giving an output that is true when the input matches any of these rows.

Of course, it is a bit redundant to show both the 0 and 1 outputs in the table. It suffices to show just those input combinations for which the output is 1. Such an abbreviated table for our prime number function is shown in Table 6.2.

The reduced table (Table 6.2) suggests one way to implement a logic circuit that realizes the prime function. For each row of the table, an AND gate is connected so that the output of the AND is true only for the input combination shown in that row. For example, for the first row of the table, we use an AND gate connected to realize the function $f_1 = \bar{d} \wedge \bar{c} \wedge \bar{b} \wedge a$ (where *d*, *c*, *b*, and *a* are the four bits of *in*). If we repeat this process for each row of the table, we get the complete function:

$$f = (\bar{d} \wedge \bar{c} \wedge \bar{b} \wedge a) \vee (\bar{d} \wedge \bar{c} \wedge b \wedge \bar{a}) \vee (\bar{d} \wedge \bar{c} \wedge b \wedge a) \vee (\bar{d} \wedge c \wedge \bar{b} \wedge a)$$
$$\vee\ (\bar{d} \wedge c \wedge b \wedge a) \vee (d \wedge \bar{c} \wedge b \wedge a) \vee (d \wedge c \wedge \bar{b} \wedge a). \tag{6.1}$$

Figure 6.3 shows a schematic logic diagram corresponding to Equation (6.1). The seven AND gates correspond to the seven product terms of Equation (6.1), which in turn correspond to the seven rows of Table 6.2. The output of each AND gate goes high when the inputs match the input values listed in the corresponding row of the truth table. For example, the output of the AND gate labeled 5 goes high when the inputs are 0101 (binary 5). The AND gates feed a seven-input OR gate, which outputs high if any of the AND gates have a high output, that is, if the input matches 1, 2, 3, 5, 7, 11, or 13 – which is the desired function.

Each product term in Equation (6.1) is called a *minterm*. A minterm is a product term that includes each input of a circuit or its complement. Each of the terms of Equation (6.1) includes all four inputs (or their complements). Thus they are minterms. The name minterm derives from the fact that these four-input product terms represent a minimal (single) number of input states, or rows of the truth table. As we shall see in Section 6.4, we can write product terms that represent multiple input states – in effect combining minterms.

We can write Equation (6.1) in shorthand as follows:

$$f = \sum_{\text{in}} m(1, 2, 3, 5, 7, 11, 13), \tag{6.2}$$

to indicate that the output is the sum (OR) of the minterms listed in the parentheses. Because each minterm corresponds to a row of a truth table, a list of minterms, as in Equation (6.2), is a list of the rows of the truth table for which the function is true.

You will recall from Section 3.4 that expressing a logic function as a sum of minterms is a *normal form* that is unique for each logic function. While this form is unique, it is not particularly efficient. We can do much better by combining minterms into simpler product terms that each represent multiple lines of our truth table.

EXAMPLE 6.1 Truth tables

Draw an abbreviated truth table for a four-bit *multiple of 3* function. The output of the function should be true if the input is a multiple of 3: 3, 6, 9, 12, or 15. Also express this function as a sum of minterms.

The abbreviated truth table, Table 6.3, simply lists the input combinations for which the output is true. Expressing this as a sum of minterms, we write

$$f = \sum_{\text{in}} m(3, 6, 9, 12, 15).$$

Table 6.3. **Abbreviated truth table for a multiple-of-3 function**

No.	In	Out
3	0011	1
6	0110	1
9	1001	1
12	1100	1
15	1111	1
Otherwise		0

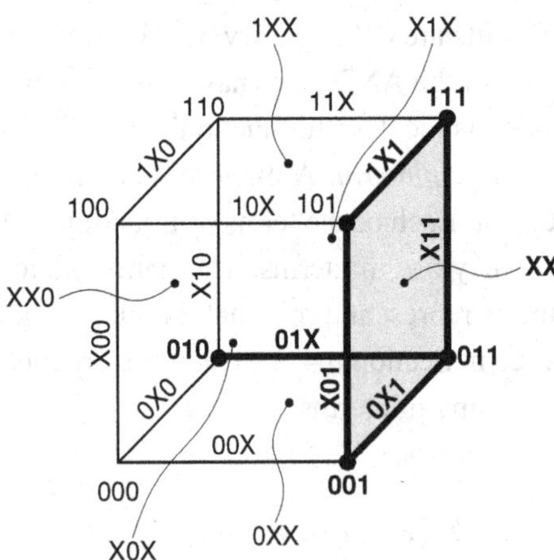

Figure 6.4. Cube visualization of the three-bit prime number function. Each vertex corresponds to a minterm, each edge to a product of two variables, and each face to a single variable. The bold vertices, the bold edges, and the shaded face show implicants of the three-bit prime number function.

6.4 IMPLICANTS AND CUBES

An examination of Table 6.2 reveals several rows that differ in only one position. For example, the rows 0010 and 0011 differ only in the rightmost (least significant) position. Thus, if we allow bits of *in* to be X (matches either 0 or 1), we can replace the two rows 0010 and 0011 by the single row 001X. This new row 001X corresponds to a product term that includes just three of the four inputs (or their complements):

$$f_{001\mathrm{X}} = \bar{d} \wedge \bar{c} \wedge b = (\bar{d} \wedge \bar{c} \wedge b \wedge \bar{a}) \vee (\bar{d} \wedge \bar{c} \wedge b \wedge a). \tag{6.3}$$

The 001X product term subsumes the two minterms corresponding to 0010 and 0011 because it is true only when at least one of them is true. Thus, in a logic function we can replace the two minterms for 0010 and 0011 with the simpler product term for 001X without changing the function.

A product term like 001X ($\bar{d} \wedge \bar{c} \wedge b$) that is true only when a function is true is called an *implicant* of the function. This is just a way of saying that the product term *implies* the function. A minterm may or may not be an implicant of a function. The minterm 0010 ($\bar{d} \wedge \bar{c} \wedge \bar{b} \wedge a$) is an implicant of the prime function because it implies the function – when 0010 is true, the function is true. Note that 0100 ($\bar{d} \wedge c \wedge \bar{b} \wedge \bar{a}$) is also a minterm (it is a product that includes each input or its complement), but it is not an implicant of the prime function. When 0100 is true, the prime function is false because 4 is not a prime. If we say that a product is a *minterm of a function* we are saying that it is both a minterm and an implicant of the function.

It is often useful to visualize implicants on a *cube*, as shown in Figure 6.4. This figure shows a three-bit prime number function mapped onto a three-dimensional cube. Each vertex of the cube represents a minterm. The cube makes it easy to see which minterms and implicants can be combined into larger implicants.[2] Minterms that differ in just one variable (e.g., 001 and 011) are adjacent to each other, and the edge between two vertices (e.g., 01X) represents the product that includes the two minterms (the OR of the two adjacent minterms). Edges that differ in just one variable (e.g., 0X1 and 1X1) are adjacent on the cube, and the face between

[2] One implicant is larger than another if it contains more minterms. For example, implicant 001 has size 1 because it contains just one minterm. Implicant 01X has size 2 because it contains two minterms (001 and 011) and hence is larger.

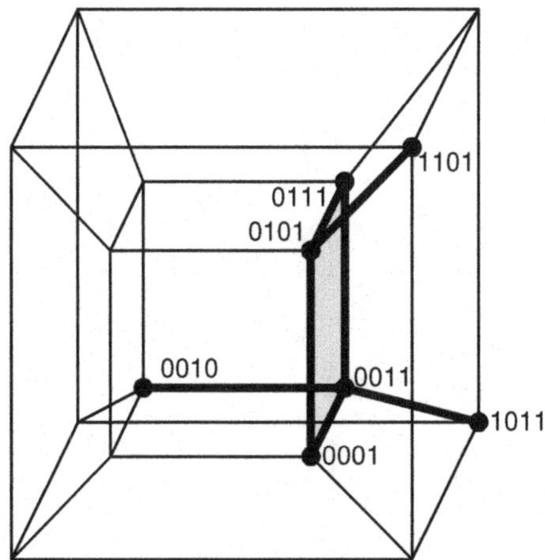

Figure 6.5. Cube visualization of the four-bit prime number function.

the edges represents the product that includes the two edge products (e.g., XX1). In this figure, the three-bit prime number function is shown as five bold vertices (001, 010, 011, 101, and 111). Five bold edges connecting these vertices represent the five two-variable implicants of the function (X01, 0X1, 0X1, X11, and 1X1). Finally, the shaded face (XX1) represents the single one-variable implicant of the function.

A cube representation of the full four-bit prime number function is shown in Figure 6.5. Only the minterms of the function are labeled. To represent four variables, we draw a four-dimensional cube as two three-dimensional cubes, one within the other. As before, vertices represent minterms, edges represent products with one X, and faces represent products with two Xs. In four dimensions, however, we also have eight volumes that represent products with three Xs. For example, the outside cube represents 1XXX – all minterms where the leftmost (most significant) bit *d* is true. The four-bit prime number function has seven vertices (minterms). Connecting adjacent vertices gives seven edges (implicants with a single X). Finally, connecting adjacent edges gives a single face (implicant with two Xs). All of these implicants of the four-bit prime number function are shown in Table 6.4.

Computer programs that synthesize and optimize logic functions use an internal representation of logic functions as a set of implicants, where each implicant is represented as a vector with elements 0, 1, or X. To simplify a function, the first step is to generate all of the implicants of the function, such as that shown in Table 6.4. A systematic procedure to achieve this is to start with all minterms of the function (the "4" column of Table 6.4). For each minterm, attempt to insert an X into each variable position. If the result is an implicant of the function, insert it in a list of single-X implicants (the "3" column of Table 6.4). Then, for each implicant with one X, attempt to insert an X into each of the remaining non-X positions, and if the result is an implicant, insert it in a list of two-X implicants. The process is repeated for two-X implicants and so on, until no further implicants are generated. Such a procedure will, given a list of minterms, generate a list of implicants.

If an implicant *x* has the property that replacing any 0 or 1 digit of *x* with an X results in a product that is not an implicant, then we call *x* a *prime implicant*.[3] A prime implicant is an

[3] The use of the word "prime" here has nothing to do with the prime number function.

Table 6.4. **All implicants of the four-bit prime number function; prime implicants are shown in bold**

Number of variables			
4	**3**	**2**	**1**
0001	**001X**	**0XX1**	
0010	00X1		
0011	0X01		
0101	0X11		
0111	01X1		
1011	**X011**		
1101	**X101**		

implicant that cannot be made any larger and still be an implicant. The prime implicants of the prime number function are shown in bold in Table 6.4.

If a prime implicant of a function, x, is the only prime implicant that contains a particular minterm of the function y, we say that x is an *essential prime implicant*; x is essential because no other prime implicant includes y. Without x a collection of prime implicants will not include minterm y. All four of the prime implicants of the four-bit prime number function are essential. Implicant 0XX1 is the only prime implicant that includes 0001 and 0111. Minterm 0010 is included only in prime implicant 001X, X101 is the only prime implicant that includes 1101, and 1011 is only included in prime implicant X011.

EXAMPLE 6.2 Implicants

Write down all of the implicants of the following function and indicate which implicants are prime implicants:

$$f = \sum_{\text{in}} m(0, 1, 4, 5, 7, 10).$$

Table 6.5 lists the implicants. We start by writing down the six minterms that are implicants in the far left column. We then check whether changing one bit of each of these minterms gives another minterm that is an implicant. If it does, we enter an implicant with an X in that position in the next column. For example, changing the LSB of 0000 gives 0001, which is also an implicant, so we put 000X in the three-variable column.

We repeat the process by seeing whether complementing any non-X bit in each of the three-variable implicants gives another three-variable implicant. If so, we enter the two-variable implicant that covers both in the next column. For example, flipping the second bit of 000X gives 010X, which is also an implicant, so we add 0X0X to the list of two-variable implicants. This is the only two-variable implicant of this function.

Table 6.5. **The implicants and prime implicants of the function given in Example 6.2**

Number of variables			
4	3	2	1
0000	000X	**0X0X**	
0001	0X00		
0100	0X01		
0101	010X		
0111	**01X1**		
1010			

The three prime implicants of the function are shown in bold; 1010 is a prime implicant because changing any bit gives a minterm that is not an implicant. Thus, 1010 is not covered by any of the three-variable implicants. Similarly, 01X1 is prime because it is not covered by 0X0X. The largest implicant 0X0X is prime because making it any larger gives an implicant that includes minterms not in the function.

6.5 KARNAUGH MAPS

Because it is inconvenient to draw cubes (especially in four or more dimensions), we often use a version of a cube flattened into two dimensions called a *Karnaugh map* (or K-map for short). Figure 6.6(a) shows how four-variable minterms are arranged in a four-variable K-map. Each square of a K-map corresponds to a minterm, and the squares of the K-map in Figure 6.6(a) are labeled with their minterm numbers. A pair of variables is assigned to each dimension and sequenced using a Gray code so that only one variable changes as we move from one square to another across a dimension – including the wrap-around from the end back to the beginning. In Figure 6.6(a), for example, we assign the rightmost two bits *ba* of the input *dcba* to the horizontal axis. As we move along this axis, these two bits (*ba*) take on the values 00, 01, 11, and 10 in turn. We map the leftmost bits *dc* to the vertical axis in a similar manner. Because only one variable changes from column to column and from row to row (including wrap-arounds), two minterms that differ in only one variable are adjacent in the K-map, just as they are adjacent in the cube representation.

Figure 6.6(b) shows a K-map for the four-bit prime number function. The content of each square is either a 1, which indicates that this minterm is an implicant of the function, or a 0, to indicate that it is not. Later we will allow squares to contain an X to indicate that the minterm may or may not be an implicant – i.e., it is a *don't care*.

Figure 6.6(c) shows how the adjacency property of a K-map, just like the adjacency property of a cube, makes it easy to find larger implicants. The figure shows the prime implicants of the

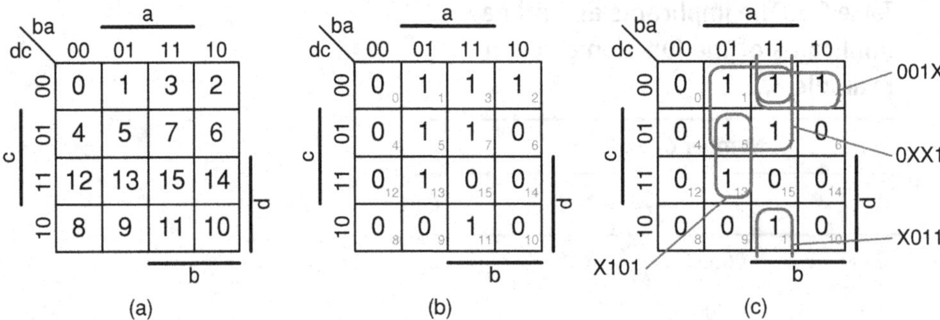

Figure 6.6. A Karnaugh map (K-map) for the four-bit prime number function. Inputs a and b change along the horizontal axis, while inputs c and d change along the vertical axis. The map is arranged so that each square is adjacent (including wrap-around) to all squares that correspond to changing exactly one input variable. (a) The arrangement of minterms in a four-variable K-map. (b) The K-map for the four-bit prime number function. (c) The same K-map with the four prime implicants of the function identified. Note that implicant X011 wraps around from top to bottom.

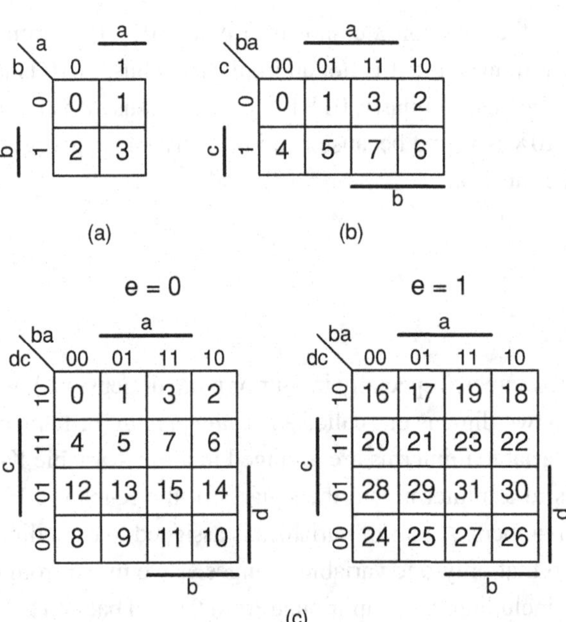

Figure 6.7. Position of minterms in K-maps of different sizes. (a) Two-variable K-map; (b) three-variable K-map; (c) five-variable K-map.

prime number function identified on the K-map. The three implicants of size 2 (single X) are pairs of adjacent 1s in the map. For example, implicant X011 is the pair of 1s in the $ab = 11$ column that wraps from top to bottom ($c = 0$). An implicant of size 4 contains four 1s and may be either a square, as is the case for 0XX1, or a full row or column, none in this function. For example, the product XX00 corresponds to the leftmost column of the K-map.

Figure 6.7 shows the arrangement of minterms for K-maps with two, three, and five variables. The five-variable K-map consists of two four-variable K-maps side by side. Corresponding squares of the two K-maps are considered to be *adjacent* in that their minterms differ only in the value of variable e. K-maps with up to eight variables can be handled by creating a 4×4 array of four-variable K-maps.

EXAMPLE 6.3 Karnaugh map

Draw a Karnaugh map for the function of Example 6.2 and circle the prime implicants.

The Karnaugh map for this function is shown in Figure 6.8, with the prime implicants identified. We draw the Karnaugh map by putting 1s in the squares corresponding to the minterms that are implicants of the function. We identify larger implicants by combining adjacent 1s.

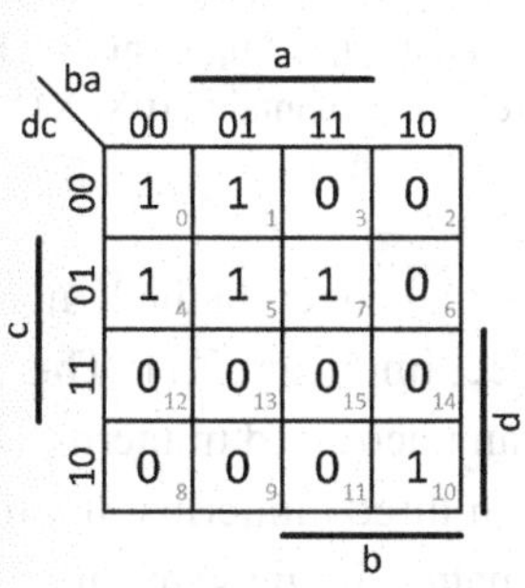
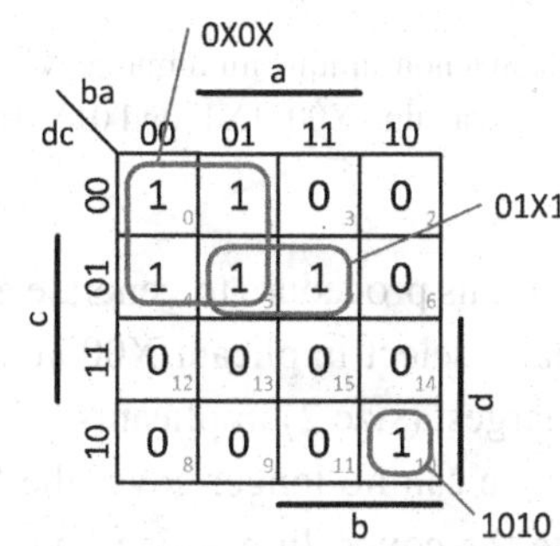

Figure 6.8. Karnaugh map and cover for Example 6.3, a function with minterms 0, 1, 4, 5, 7, and 10.

6.6 COVERING A FUNCTION

Once we have a list of implicants for a function, the problem remains how to select the least expensive set of implicants that *cover* the function. A set of implicants is a cover of a function if each minterm of the function is included in at least one implicant of the cover. We define the cost of an implicant as the number of variables in the product. Thus, with a four-variable function, a minterm like 0011 has cost 4, a one-X implicant like 001X has cost 3, a two-X implicant like 0XX1 has cost 2, and so on.

A procedure to select an inexpensive set of implicants is as follows:

(1) start with an empty cover;
(2) add all essential prime implicants to the cover;
(3) for each remaining uncovered minterm, add the largest implicant that covers that minterm to the cover.

This procedure will always result in a *good* cover. However, there is no guarantee that it will give the lowest-cost cover. Depending on the order in which minterms are covered in step (3), and the method used to select between equal-cost implicants to cover each minterm, different covers of possibly different cost may result.

For the four-bit prime number function, the function is completely covered by the four essential prime implicants. Thus, the synthesis process is done after step (2), and the cover is both minimum and unique.

Consider, however, the logic function shown in Figure 6.9(a). This function has no essential prime implicants so our process moves to step (3) with an empty cover. At step (3), suppose we select uncovered minterms in numerical order. Hence we start with minterm 000. We can cover 000 with either X00 or 0X0. Both are implicants of the function. If we choose X00, the cover shown in Figure 6.9(b) will result. If instead we choose 0X0, we get the cover shown in Figure 6.9(c). Both of these covers are minimal – even if they are not unique.

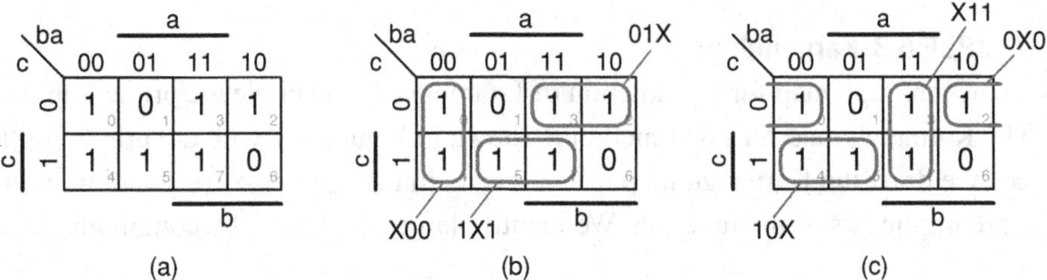

Figure 6.9. A function with a non-unique minimum cover and no essential prime implicants. (a) K-map of the function. (b) One cover contains X00, 1X1, and 01X. (c) A different cover contains 10X, X11, and 0X0.

It is also possible for this procedure to generate a non-minimal cover. In the K-map of Figure 6.9, suppose we initially select implicant X00 and then select implicant X11. This is possible since it is one of the largest (size 2) implicants that covers an uncovered minterm. However, if we make this choice, we can no longer cover the function in three minterms; it will take four minterms to complete the cover. In practice, this doesn't matter. Logic gates are inexpensive and, except in rare cases, no one cares whether your cover is minimal or not.

EXAMPLE 6.4 Covering a function

Derive the minimal cover of the three-variable function

$$f = \sum_{\text{in}} m(1, 3, 4, 5).$$

As shown on the Karnaugh map in Figure 6.10, there are three two-variable implicants 0X1, X01, and 10X. Of these, however, only 0X1 and 10X are *essential*. The function is entirely covered by these two prime implicants. Thus, we can write

$$f = 0X1 \vee 10X$$

or

$$f(c, b, a) = (\bar{c} \wedge a) \vee (c \wedge \bar{b}).$$

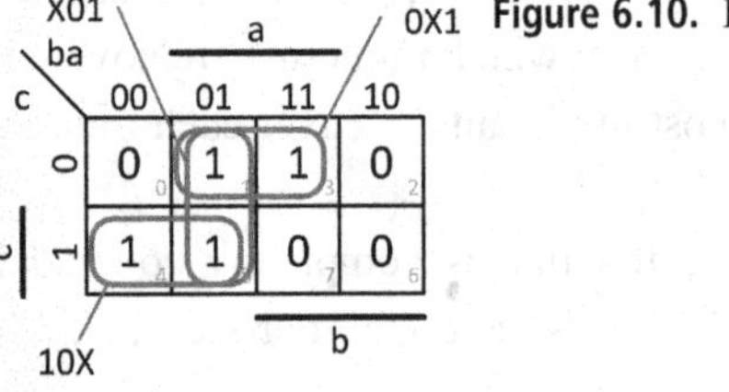

Figure 6.10. K-map used to find a cover for Example 6.4.

6.7 FROM A COVER TO GATES

Once we have a minimum-cost cover of a logic function, the cover can be directly converted to gates by instantiating an AND gate for each implicant in the cover and using a single OR

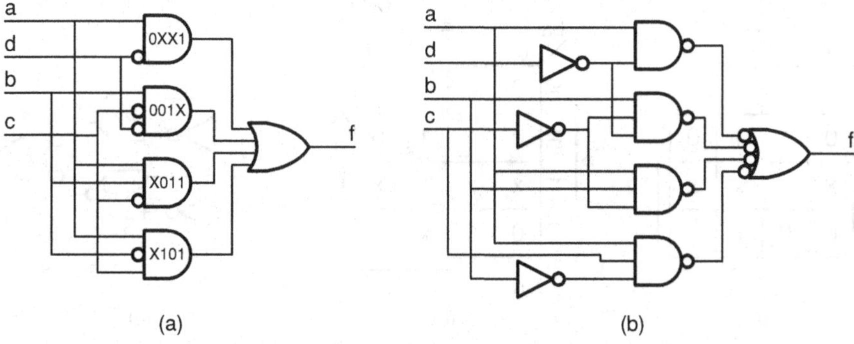

Figure 6.11. Logic circuit for the four-bit prime number function. (a) Logic circuit using AND and OR gates with arbitrary inversion bubbles on inputs. Each AND gate corresponds to a prime implicant in the cover of the function. (b) Logic circuit using CMOS NAND gates and inverters. NAND gates are used for both the AND and OR functions. Inverters complement inputs as required.

gate to sum the outputs of the AND gates. Such an AND-OR realization of the four-bit prime number function is shown in Figure 6.11(a).

With CMOS logic we are restricted to inverting gates, so we use NAND gates for both the AND and the OR functions as shown in Figure 6.11(b). Because CMOS gates have all inputs of the same polarity (all bubbles or no bubbles), we add inverters as needed to invert inputs. We could just as easily have designed the function using all NOR gates; NANDs are preferred, however, because they have lower logical effort for the same fan-in (see Section 5.3).

CMOS gates are also restricted in their fan-in (see Section 5.3). In typical cell libraries, the maximum fan-in of a NAND or NOR gate is 4. If a larger fan-in is needed, a tree of gates (e.g., two NANDs into a NOR) is used to build a large AND or OR, adding inverters as needed to correct the polarity.

6.8 INCOMPLETELY SPECIFIED FUNCTIONS

Often our specification guarantees that a certain set of input states (or minterms) will never be used. Suppose, for example, we have been asked to design a one-digit decimal prime number detecting circuit that need only accept inputs in the range from 0 to 9. That is, for an input between 0 and 9 our circuit must output 1 if the number is a prime and 0 otherwise. However, for inputs between 10 and 15 our circuit can output either 0 or 1 – the output is unspecified.

We can simplify our logic by taking advantage of these don't care input states as shown in Figure 6.12. Figure 6.12(a) shows a K-map for the decimal prime number function. We place an X in each square of the K-map that corresponds to a don't care input state. In effect we are dividing the input states into three sets: f_1 – those input combinations for which the output must be 1; f_0 – those input combinations for which the output must be 0; and f_X – those input combinations where the output is not specified and may be either 0 or 1. In this case, f_1 is the set of five minterms labeled with 1 (1, 2, 3, 5, and 7), f_0 contains the five minterms labeled 0 (0, 4, 6, 8, and 9), and f_X contains the remaining minterms (10–15).

An implicant of an incompletely specified function is any product term that includes at least one minterm from f_1 and does not include any minterms in f_0. Thus we can expand our

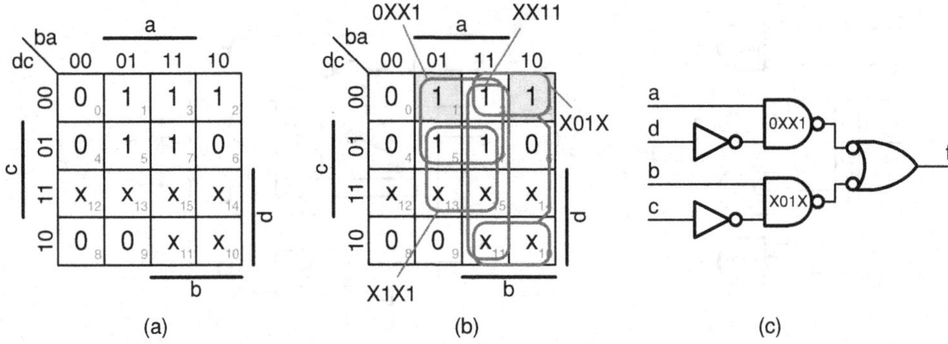

Figure 6.12. Design of a decimal prime number circuit illustrating the use of don't cares in a K-map. (a) The K-map for the decimal prime number circuit. Input states 10–15 labeled with X are don't care states. (b) The K-map with prime implicants shown. The circuit has four prime implicants: 0XX1, X01X, XX11, and X1X1. The first two are essential since they are the only implicants that cover 0001 and 0010, respectively; the last two (XX11 and X1X1) are not essential and in fact are not needed. (c) A CMOS logic circuit derived from the K-map. The two NAND gates correspond to the two essential prime implicants.

implicants by including minterms in f_X. Figure 6.12(b) shows the four prime implicants of the decimal prime number function. Note that implicant 001X of the original prime number function has been expanded to X01X to include two minterms from f_X. Two new prime implicants, X1X1 and XX11, have been added, each by combining two minterms from f_1 with two minterms from f_X. Note that products 11XX and 1X1X, which are entirely in f_X, are not implicants even though they contain no minterms of f_0. To be an implicant, a product must contain at least one minterm from f_1.

Using the notation of Equation (6.2) we can write a function with don't cares as follows:

$$f = \sum_{\text{in}} m(1, 2, 3, 5, 7) + D(10, 11, 12, 13, 14, 15). \tag{6.4}$$

That is, the function is the sum of five minterms plus six don't care terms.

We form a cover of a function with don't cares using the same procedure described in Section 6.6. In the example of Figure 6.12 there are two essential prime implicants: 0XX1 is the only prime implicant that includes 0001, and X01X is the only prime implicant that includes 0010. These two essential prime implicants cover all five of the minterms in f_1, so they form a cover of the function. The resulting CMOS gate circuit is shown in Figure 6.12(c).

EXAMPLE 6.5 Incompletely specified function

Design a circuit that detects when its four-bit input equals 7. Take advantage of the fact that you know the input is a prime number.

We fill in the Karnaugh map as shown in Figure 6.13 to divide our input space into f_1, f_X, and f_0. Input combination 7 is labeled 1, all non-prime input combinations are labeled X – because we know these combinations won't occur on our inputs, and the remaining input combinations are labeled 0. The resulting circuit can be implemented with a single two-input AND gate.

$$f(c, b, a) = b \wedge c.$$

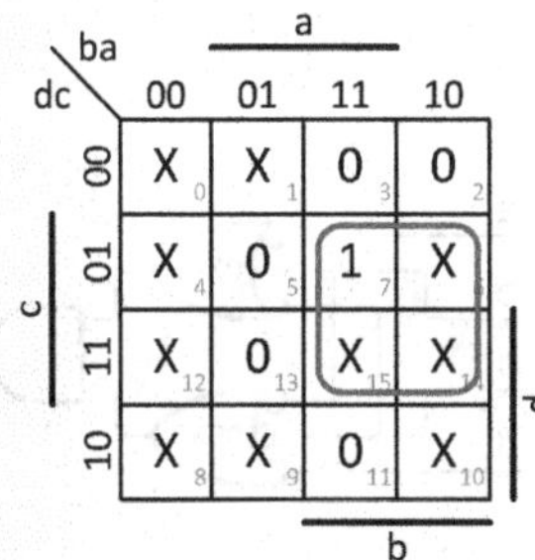

Figure 6.13. Karnaugh map and cover for Example 6.5, an incompletely specified function.

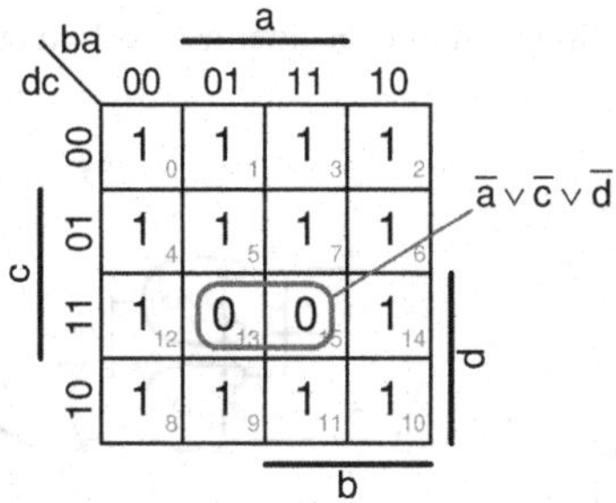

Figure 6.14. K-map for a function with two maxterms OR(0000) and OR(0010) that can be combined into a single sum, OR(00X0).

6.9 PRODUCT-OF-SUMS IMPLEMENTATION

So far we have focused on the input states where the truth table is a 1 and have generated sum-of-products logic circuits. By duality we can also realize product-of-sums logic circuits by focusing on the input states where the truth table is 0. With CMOS implementations we generally prefer the sum-of-products implementations because NAND gates have a lower logical effort than NOR gates with the same fan-in. However, there are some functions where the product-of-sums implementation is less expensive than the sum-of-products. Often both are generated and the better circuit selected.

A *maxterm* is a sum (OR) that includes every variable or its complement. Each 0 in a truth table or K-map corresponds to a maxterm. For example, the logic function shown in the K-map of Figure 6.14 has two maxterms: $\bar{a} \vee \bar{b} \vee \bar{c} \vee \bar{d}$ and $\bar{a} \vee b \vee \bar{c} \vee \bar{d}$. For simplicity, we refer to these as OR(0000) and OR(0010). Note that a maxterm corresponds to the complement of the input state in the K-map, so maxterm 0, OR(0000), corresponds to a 0 in square 15 of the K-map. We can combine adjacent 0s in the same way we combined adjacent 1s, so OR(0000) and OR(0010) can be combined into sum OR(00X0) $= \bar{a} \vee \bar{c} \vee \bar{d}$.

The design process for a product-of-sums circuit is identical to a sum-of-products design except that 0s in the K-maps are grouped instead of 1s. Figure 6.15 illustrates the process for a function with three maxterms. Figure 6.15(a) shows the K-map for the function. Two prime sums (OR terms that cannot be made any larger without including 1s) are identified in Figure 6.15(b): OR(00X0) and OR(0X10). Both of these sums are needed to cover all 0s in the K-map. Finally, Figure 6.15(c) shows the product-of-sums logic circuit that computes this function. The circuit consists of two OR gates, one for each of the prime sums, and an AND gate that combines the outputs of the OR gates so that the output of the function is 0 when the output of either OR gate is 0.

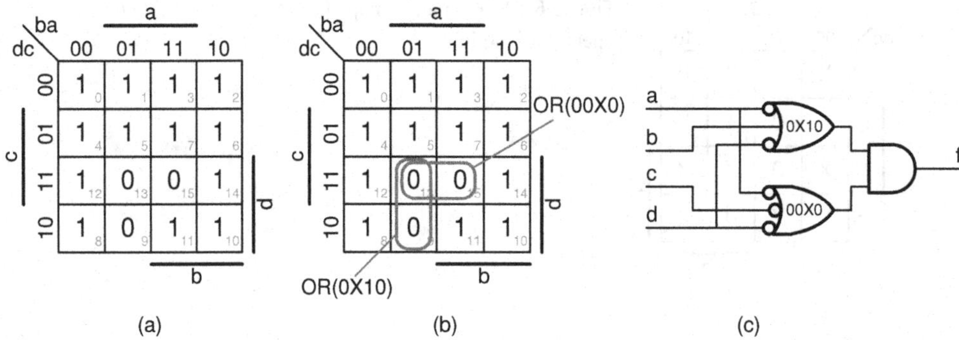

Figure 6.15. Product-of-sums synthesis. (a) K-map of a function with three maxterms. (b) Two prime sums. (c) Product of sums logic circuit.

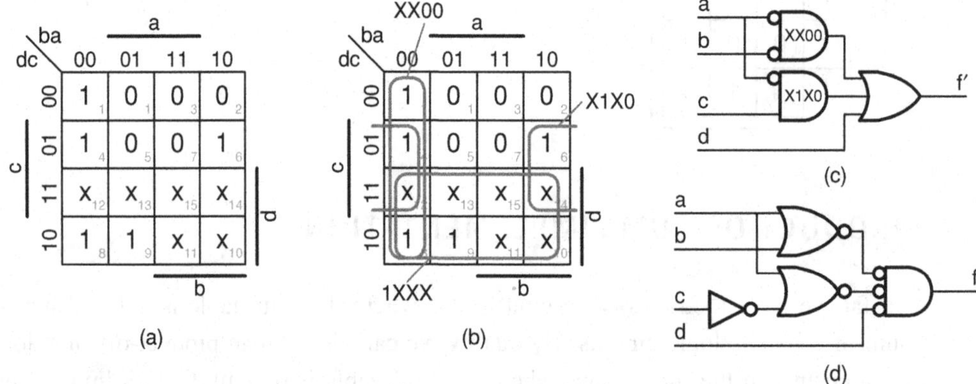

Figure 6.16. Implementation of the decimal prime number circuit in product-of-sums form using the complement method. (a) K-map for complement of the decimal prime number function (the decimal composite number function). (b) Prime implicants of this function (XX00, X1X0, and 1XXX). (c) Sum-of-products logic circuit that computes the complement decimal prime number function. (d) Logic circuit that generates the decimal prime number function. This is derived from (c) using De Morgan's theorem.

Once you have mastered sum-of-products design, the easiest way to generate a product-of-sums logic circuit is to find the sum-of-products circuit for the complement of the logic function (the function that results from swapping f_1 and f_0 leaving f_X unchanged.) Then, to complement the output of this circuit, apply De Morgan's theorem by changing all ANDs to ORs and complementing the inputs of the circuit.

For example, consider our decimal prime number function. The K-map for the complement of this function is shown in Figure 6.16(a). We identify three prime implicants of this function in Figure 6.16(b). A sum-of-products logic circuit that realizes the complement function of this K-map is shown in Figure 6.16(c). This circuit follows directly from the three prime implicants. Figure 6.16(d) shows the product-of-sums logic circuit that computes the decimal prime number function (the complement of the K-map in (a) and (b)). We derive this logic circuit by complementing the output of the circuit of Figure 6.16(c) and applying De Morgan's theorem to convert ANDs (ORs) to ORs (ANDs).

EXAMPLE 6.6 Product of sums

Express the three-input function $f = \sum_{\text{in}} m(1,7)$ as a minimal product-of-sums expression.

Drawing the Karnaugh map (Figure 6.17) and identifying the implicants of the complement of this function $f' = \sum_{\text{in}} m(0,1,3,4,5,6)$ we can write

$$f' = \bar{a} \vee (b \wedge \bar{c}) \vee (\bar{b} \wedge c).$$

Then, applying Equation (3.9), we write

$$f = a \wedge (\bar{b} \vee c) \wedge (b \vee \bar{c}).$$

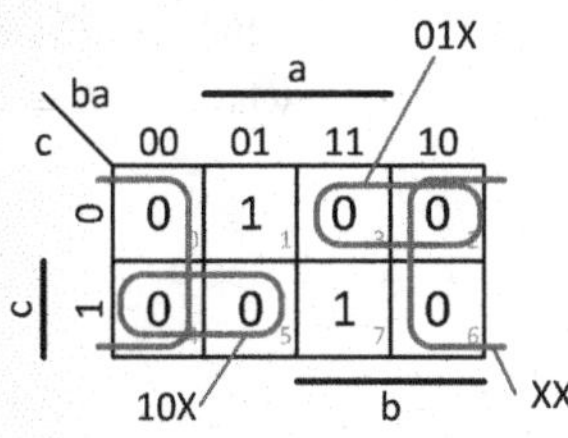

Figure 6.17. Karnaugh map and cover used to find a product-of-sums implementation of the function specified in Example 6.6.

6.10 HAZARDS

On rare occasions we are concerned with whether or not our combinational circuits generate transient outputs in response to a single transition on a single input. Most of the time this is not an issue. For almost all combinational circuits we are concerned only that the steady-state output for a given input be correct – not how the output gets to its steady state. However, in certain applications of combinational circuits, e.g., in generating clocks or feeding an asynchronous circuit, it is critical that a single input transition produces at most one output transition.

Consider, for example, the two-input multiplexer circuit shown in Figure 6.18. This circuit sets the output f equal to input a when $c = 1$ and equal to input b when $c = 0$. The K-map for this circuit is shown in Figure 6.18(a). The K-map shows two essential prime implicants, 1X1 ($a \wedge c$) and 01X ($b \wedge \bar{c}$), that together cover the function. A logic circuit that implements the function, using two AND gates for the two essential prime implicants, is shown in Figure 6.18(b). The number within each gate denotes the delay of the gate. The inverter on input c has a delay of 3, while the three other gates all have unit delay.

Figure 6.18(c) shows the transient response of this logic circuit when $a = b = 1$ and input c transitions from 1 to 0 at time 1. Three time units later, at time 4, the output of the inverter cN rises. In the meantime, the output of the upper AND gate d falls at time 2, causing output f to fall at time 3. At time 4, the rising of signal cN causes signal e to rise, which in turn causes signal f to rise at time 6. Thus, a single transition on input c causes first a falling, then a rising transition on output f.

This transient 1–0–1 on output f is called a *static-1 hazard.* The output is normally expected to be a static 1, but has a transient hazard to 0. Similarly, an output that undergoes a 0–1–0 response to a single input transition is said to have a *static-0 hazard.* More complex circuits,

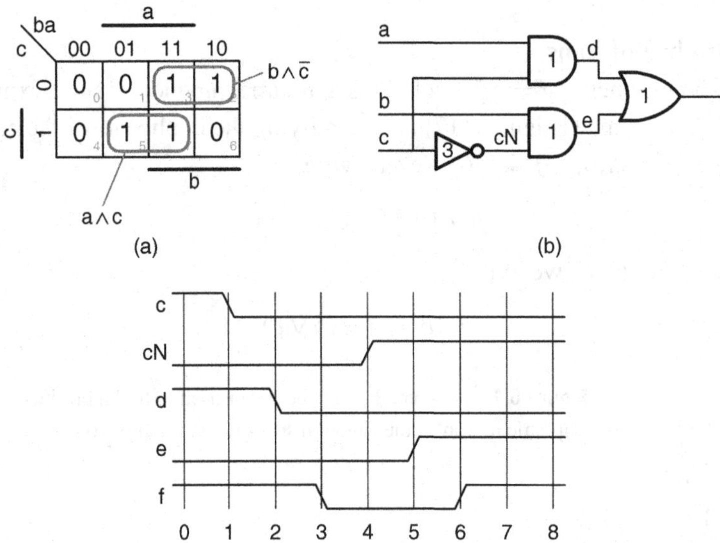

Figure 6.18. Two-input multiplexer circuit with a static-1 hazard. (a) K-map of the function showing two essential prime implicants. (b) Gate-level logic circuit for the multiplexer. The numbers denote the delay (in arbitrary units) of each gate. (c) Timing diagram showing the response of the logic circuit of (b) to a falling transition on input c when $a = b = 1$.

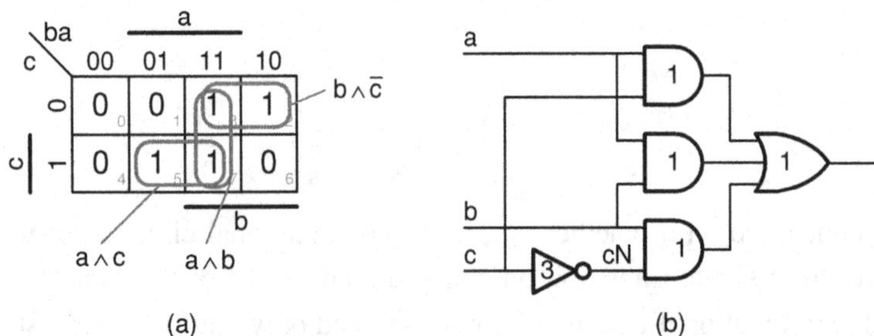

Figure 6.19. Two-input multiplexer circuit with no hazards. (a) K-map of the function showing three prime implicants. The implicant X11 is needed to cover the transition from 111 to 011, even though it is not essential. (b) Gate-level logic circuit for the hazard-free multiplexer.

with more levels of logic, may also exhibit dynamic hazards. A dynamic-1 hazard is one in which an output goes through the states 0–1–0–1: starting at 0 and ending at 1, but with three transitions instead of 1. Similarly a dynamic-0 hazard is a three-transition sequence ending in the 0 state.

Intuitively the static-1 hazard of Figure 6.18 occurs because, as the input transitions from 111 to 011, the gate associated with implicant 1X1 turns off before the gate associated with implicant 01X turns on. We can eliminate the hazard by covering the transition with an implicant of its own, X11, as shown in Figure 6.19. The third AND gate (the middle AND gate of Figure 6.19(b)), which corresponds to implicant X11, holds the output high while the other two gates switch. In general, we can make any circuit hazard-free by adding redundant implicants to cover transitions in this manner.

EXAMPLE 6.7 Hazards

Fix the hazard that occurs in the circuit shown in Figure 6.20. That is, preserve the logic function of the circuit while eliminating any hazards that occur during input transitions.

The two gates in Figure 6.20 correspond to implicants X00 and 0X1. Drawing the K-map for this circuit (in Figure 6.21), we see the need to cover the transition from 000 to 001. Otherwise,

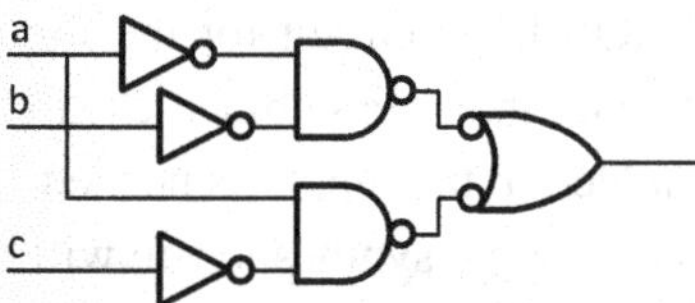

Figure 6.20. Schematic of a circuit with a hazard to be fixed in Example 6.7.

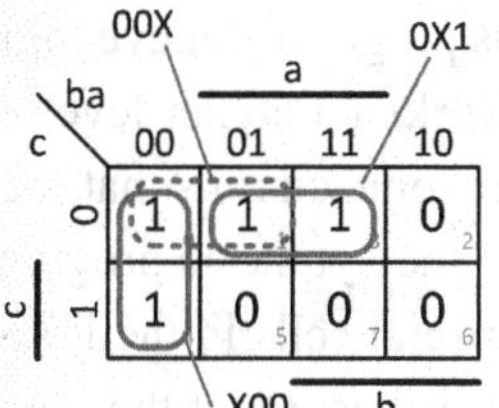

Figure 6.21. K-map showing the function implemented by Figure 6.20. While the solid implicants provide a cover, a hazard occurs when a toggles between 0 and 1 with $b = c = 0$. We fix the hazard by adding the dashed implicant.

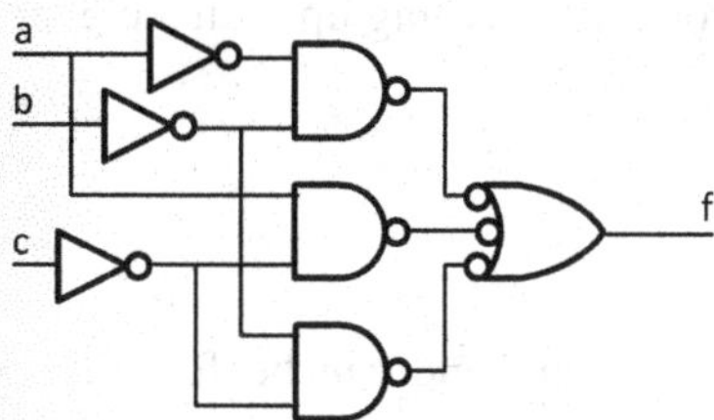

Figure 6.22. Circuit that provides the same functionality as Figure 6.20, but without any hazards.

depending on relative gate speed, the output could momentarily go to 0 during this transition when one gate turns off before the other gate turns on. Adding implicant 00X – circled with a dotted line – covers the transition. The logic circuit shown in Figure 6.22 adds a gate for implicant 00X covering the transition and eliminating the hazard.

Summary

In this chapter you have learned how to synthesize manually a combinational logic circuit. Given an English-language description of a circuit, you can generate a gate-level implementation. We start by writing a *truth table* for the circuit to precisely define the behavior of the function. Embedding the truth table in a *Karnaugh map* makes it easy to identify *implicants* of the function. Recall that *implicants* are products that include at least one minterm of f_1 and no minterms of f_0. They may or may not include minterms of f_X.

Once the implicants are identified, we generate a *cover* of the function by finding a minimal set of implicants that together contain every minterm in f_1. We start by identifying the *prime implicants*, those that are included in no larger implicant, and the *essential prime implicants*, prime implicants that cover a minterm of f_1 that is covered by no other prime implicant. We start our cover with the essential prime implicants of the function and then add prime implicants that include uncovered minterms of f_1 until all of f_1 is covered. The cover is not unique; depending on the order in which we add prime implicants to the cover, we may get a different final result.

From the cover it is straightforward to draw a CMOS logic circuit for the function. Each implicant in the cover becomes a NAND gate, their outputs are combined by a NAND gate (which performs the OR function), and inverters are added to the inputs as needed.

While it is useful to understand this process for manual logic synthesis, you will hardly ever use this procedure in practice. Modern logic design is almost always achieved using automatic logic synthesis in which a CAD program takes a high-level description of a logic function and automatically generates the logic circuit. Automatic synthesis programs relieve logic designers from the drudgery of crunching K-maps, enabling them to work at a higher level and be more productive. Also, most automatic synthesis programs produce logic circuits that are better than the ones a typical designer could easily generate manually. The synthesis program considers multi-level circuits and implementations that make use of special cells in the library, and can try thousands of combinations before picking the best one. It is best to let the CAD programs do what they are good at – finding the optimal CMOS circuit to implement a given function – and have the designer focus on what humans are good at – coming up with a clever high-level organization for the system.

BIBLIOGRAPHIC NOTES

Further details about using mapping techniques to design logic can be found in Karnaugh's original paper [63]. This paper builds on techniques proposed by Veitch in 1952 [109]. The Quine–McCluskey algorithm is used to find a minimum mapping, detailed in McCluskey's 1956 paper [74].

Exercises

6.1 *Combinational circuits.* Which of the circuits in Figure 6.23 are combinational? Each of the boxes is itself a combinational circuit.

6.2 *Fibonacci circuit.* Design a four-bit Fibonacci circuit. This circuit outputs a 1 iff its input is a Fibonacci number (i.e., 0, 1, 2, 3, 5, 8, or 13). Go through the following steps.

(a) Write a truth table for the function.

(b) Draw a Karnaugh map of the function.

(c) Identify the prime implicants of the function.

(d) Identify which of the prime implicants (if any) are essential.

(e) Find a cover of the function.

(f) Draw a CMOS gate circuit for the function.

6.3 *Minimum logic.* Draw a logic diagram that realizes a function that is true when the input (*dcba*) is 3, 4, 5, 7, 9, 13, 14, or 15. Use the minimum possible number of gate inputs. Both true and complementary inputs are available (i.e., you have both a and a').

6.4 *Decimal Fibonacci circuit.* Repeat Exercise 6.2, but for a decimal Fibonacci circuit. This circuit need only produce an output for inputs in the range 0–9. The output is a don't care for the other six input states.

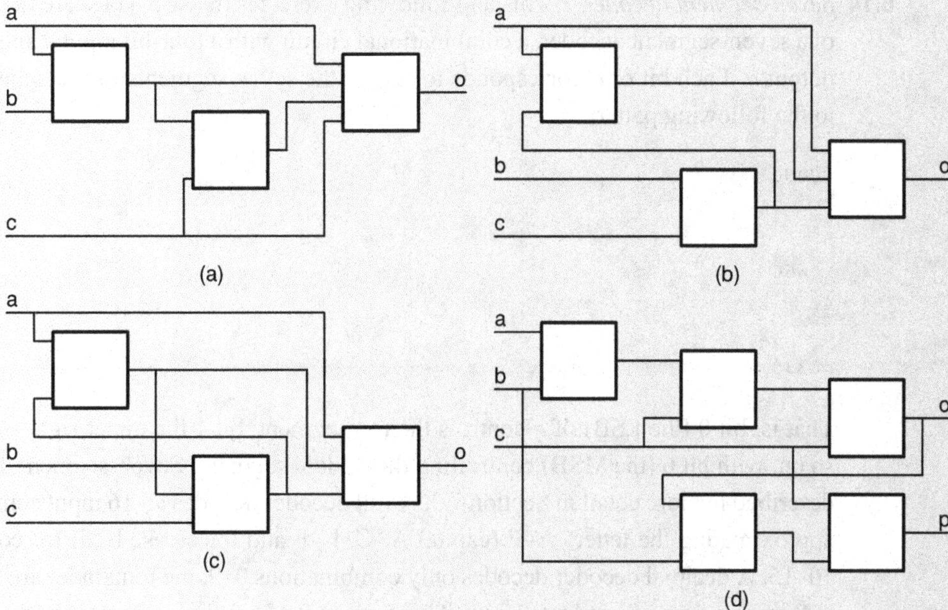

Figure 6.23. Circuits for Exercise 6.1. Each box is itself a combinational circuit.

6.5 *Prime number circuit.* Design a circuit whose output is true if its four-bit input is really a prime number – not including "1" – that is, the output is true if the input is 2, 3, 5, 7, 11, or 13. Go through the same steps as in Exercise 6.2.

6.6 *Decimal prime number circuit.* Design a circuit whose output is true if its four-bit decimal input is really a prime number – not including "1" – that is, the output is true if the input is 2, 3, 5 or 7. The output is an X (don't care) for input combinations 10–15. Go through the same steps as in Exercise 6.2.

6.7 *Multiple-of-3 circuit.* Design a four-input multiple-of-3 circuit, that is a circuit whose output is true if the input is 3, 6, 9, 12, or 15.

6.8 *Combinational design.* Design a minimal CMOS circuit that implements the function $f = \sum m(3, 4, 5, 7, 9, 13, 14, 15)$.

6.9 *Five-input prime number circuit.* Design a five-input prime number circuit. The output is true if the input is a prime number (not including "1") between 0 and 31.

6.10 *Six-input prime number circuit.* Design a six-input prime number circuit. This circuit must also recognize the primes between 32 and 63 (neither of which is prime).

6.11 *Non-unique cover.* Design a four-input circuit that implements the function $f = \sum m$ $(0, 1, 2, 9, 10, 11)$.

6.12 *Product of sums, I.* Design the decimal Fibonnaci circuit of Exercise 6.4 in product-of-sums form.

6.13 *Product of sums, II.* Design the decimal prime number circuit of Exercise 6.6 in product-of-sums form.

6.14 *Seven-segment decoder, I.* This and following exercises (6.14–6.41) share the description of a seven-segment decoder, a combinational circuit with a four-bit input *a* and a seven-bit output *q*. Each bit of *q* corresponds to one of the seven segments of a display according to the following pattern:

```
 0000
5    1
5    1
 6666
4    2
4    2
 3333
```

That is, bit 0 (the LSB) of *q* controls the top segment, bit 1 the upper right segment, and so on, with bit 6 (the MSB) controlling the middle segment. Seven-segment decoders are described in more detail in Section 7.3. A full decoder decodes all 16 input combinations – approximating the letters A–F (capital A, C, E, F and lowercase b, d) for combinations 10–15. A decimal decoder decodes only combinations 0–9, the remainder are don't cares.

Design a sum-of-products circuit for segment 0 of a full seven-segment decoder.

6.15 *Seven-segment decoder, II.* Design a sum-of-products circuit for segment 1 of the full seven-segment decoder described in Exercise 6.14.

6.16 *Seven-segment decoder, III.* Design a sum-of-products circuit for segment 2 of the full seven-segment decoder described in Exercise 6.14.

6.17 *Seven-segment decoder, IV.* Design a sum-of-products circuit for segment 3 of the full seven-segment decoder described in Exercise 6.14.

6.18 *Seven-segment decoder, V.* Design a sum-of-products circuit for segment 4 of the full seven-segment decoder described in Exercise 6.14.

6.19 *Seven-segment decoder, VI.* Design a sum-of-products circuit for segment 5 of the full seven-segment decoder described in Exercise 6.14.

6.20 *Seven-segment decoder, VII.* Design a sum-of-products circuit for segment 6 of the full seven-segment decoder described in Exercise 6.14.

6.21 *Decimal seven-segment decoder, I.* Design a sum-of-products circuit for segment 0 of the decimal seven-segment decoder described in Exercise 6.14.

6.22 *Decimal seven-segment decoder, II.* Design a sum-of-products circuit for segment 1 of the decimal seven-segment decoder described in Exercise 6.14.

6.23 *Decimal seven-segment decoder, III.* Design a sum-of-products circuit for segment 2 of the decimal seven-segment decoder described in Exercise 6.14.

6.24 *Decimal seven-segment decoder, IV.* Design a sum-of-products circuit for segment 3 of the decimal seven-segment decoder described in Exercise 6.14.

6.25 *Decimal seven-segment decoder, V.* Design a sum-of-products circuit for segment 4 of the decimal seven-segment decoder described in Exercise 6.14.

6.26 *Decimal seven-segment decoder, VI.* Design a sum-of-products circuit for segment 5 of the decimal seven-segment decoder described in Exercise 6.14.

6.27 *Decimal seven-segment decoder, VII.* Design a sum-of-products circuit for segment 6 of the decimal seven-segment decoder described in Exercise 6.14.

6.28 *Product-of-sums seven-segment decoder, I.* Design a product-of-sums circuit for segment 0 of the full seven-segment decoder described in Exercise 6.14.

6.29 *Product-of-sums seven-segment decoder, II.* Design a product-of-sums circuit for segment 1 of the full seven-segment decoder described in Exercise 6.14.

6.30 *Product-of-sums seven-segment decoder, III.* Design a product-of-sums circuit for segment 2 of the full seven-segment decoder described in Exercise 6.14.

6.31 *Product-of-sums seven-segment decoder, IV.* Design a product-of-sums circuit for segment 3 of the full seven-segment decoder described in Exercise 6.14.

6.32 *Product-of-sums seven-segment decoder, V.* Design a product-of-sums circuit for segment 4 of the full seven-segment decoder described in Exercise 6.14.

6.33 *Product-of-sums seven-segment decoder, VI.* Design a product-of-sums circuit for segment 5 of the full seven-segment decoder described in Exercise 6.14.

6.34 *Product-of-sums seven-segment decoder, VII.* Design a product-of-sums circuit for segment 6 of the full seven-segment decoder described in Exercise 6.14.

6.35 *Product-of-sums decimal seven-segment decoder, I.* Design a product-of-sums circuit for segment 0 of the decimal seven-segment decoder described in Exercise 6.14.

6.36 *Product-of-sums decimal seven-segment decoder, II.* Design a product-of-sums circuit for segment 1 of the decimal seven-segment decoder described in Exercise 6.14.

6.37 *Product-of-sums decimal seven-segment decoder, III.* Design a product-of-sums circuit for segment 2 of the decimal seven-segment decoder described in Exercise 6.14.

6.38 *Product-of-sums decimal seven-segment decoder, IV.* Design a product-of-sums circuit for segment 3 of the decimal seven-segment decoder described in Exercise 6.14.

6.39 *Product-of-sums decimal seven-segment decoder, V.* Design a product-of-sums circuit for segment 4 of the decimal seven-segment decoder described in Exercise 6.14.

6.40 *Product-of-sums decimal seven-segment decoder, VI.* Design a product-of-sums circuit for segment 5 of the decimal seven-segment decoder described in Exercise 6.14.

6.41 *Product-of-sums decimal seven-segment decoder, VII.* Design a product-of-sums circuit for segment 6 of the decimal seven-segment decoder described in Exercise 6.14.

6.42 *Multiple outputs, I.* Design a sum-of-products circuit that generates outputs for segments 0, 1, and 2 of the decimal seven-segment decoder described in Exercise 6.14. Share logic between the outputs where possible.

6.43 *Multiple outputs, II.* Design a sum-of-products circuit that generates outputs for segments 3, 4, 5, and 6 of the decimal seven-segment decoder described in Exercise 6.14. Share logic between the outputs where possible.

6.44 *Hazards, I.* Fix the hazard that may occur in Figure 6.24(a).

6.45 *Hazards, II.* Fix the hazard that may occur in Figure 6.24(b).

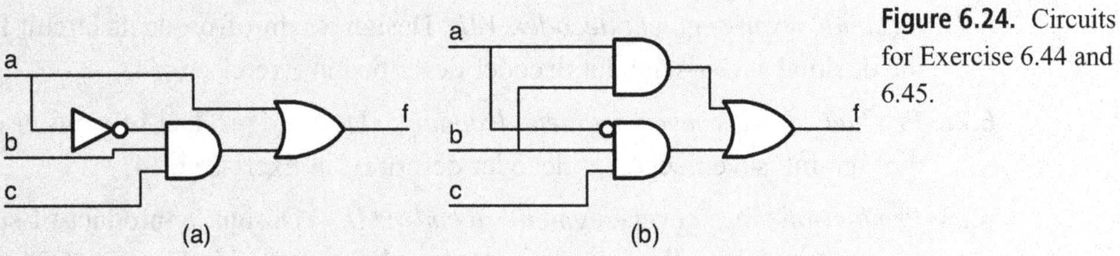

Figure 6.24. Circuits for Exercise 6.44 and 6.45.

6.46 *Adder Karnaugh maps, I.* A *half adder* is a circuit which takes in one-bit binary numbers *a* and *b* and outputs a sum *s* and a carry out *co*. The concatenation of *co* and *s* `co,s` is the two-bit value that results from adding *a* and *b* (e.g., if $a = 1$ and $b = 1$, $s = 0$ and $co = 1$). Half adders are described in more detail in Chapter 10.

(a) Write out truth tables for the *s* and *co* outputs of a half adder.

(b) Draw Karnaugh maps for the *s* and *co* outputs of the half adder.

(c) Circle the prime implicants and write out the logic equations for the *s* and *co* outputs of the half adder.

6.47 *Adder Karnaugh maps, II.* A *full adder* is a circuit which takes in one-bit binary numbers *a*, *b*, and *ci* (carry in), and outputs *s* and *co*. The concatenation of *co* and *s* `{co,s}` is the two-bit value that results from adding *a*, *b*, and *ci* (e.g., if $a = 1$, $b = 0$, and $ci = 1$ then $s = 0$ and $co = 1$). Full adders are described in more detail in Chapter 10.

(a) Write out the truth tables for the *s* and *co* outputs for the full adder.

(b) Draw Karnaugh maps for the *s* and *co* outputs of the full adder.

(c) Circle the prime implicants and write out the logic equations for the *s* and *co* outputs of the full adder.

(d) How would the use of an XOR gate help in the half adder? How would it help in the full adder?

7 VHDL descriptions of combinational logic

In Chapter 6 we saw how to synthesize combinational logic circuits manually from a specification. In this chapter we show how to describe combinational circuits in the VHDL hardware description language, building on our discussion of Boolean expressions in VHDL (Section 3.6) and the initial discussion of VHDL (Section 1.5). Once the function has been described in VHDL, it can be automatically synthesized, eliminating the need for manual synthesis.

Because all optimization is done by the synthesizer, the main goal in writing synthesizable VHDL is to make it easily readable and maintainable. For this reason, descriptions that are close to the function of a design (e.g., a truth table specified with a `case` statement) are preferable to those that are close to the implementation (e.g., equations using a concurrent assignment statement, or a structural description using gates). Descriptions that specify just the function tend to be easier to read and maintain than those that reflect a manual implementation of the function.

To verify that a VHDL design entity is correct, we write a *testbench*. A testbench is a piece of VHDL code that is used during simulation to instantiate the design entity to be tested, generate input stimulus, and check the design entity's outputs. While design entities must be coded in a strict synthesizable subset of VHDL, testbenches, which are not synthesized, can use the full VHDL language, including looping constructs. In a typical modern digital design project, at least as much effort goes into *design verification* (writing testbenches) as goes into doing the design itself.

7.1 THE PRIME NUMBER CIRCUIT IN VHDL

In describing combinational logic using VHDL we restrict our use of the language to constructs that can easily be synthesized into logic circuits.

Specifically, we restrict combinational circuits to be described using only concurrent signal assignment statements, **`case`** statements, **`if`** statements, or by the structural composition of other combinational design entities.

In this section we look at four ways of implementing the prime number (plus 1) circuit we introduced in Chapter 6 as combinational VHDL.

7.1.1 A VHDL design entity

Before diving into our four implementations of the prime number circuit, we quickly review the structure of a VHDL design entity. A design entity is a block of logic with specified input

```
entity <entity_name> is
  port( <port_declarations> );
end <entity_name>;

architecture <implementation_name> of <entity_name> is
  <component_declarations>
  <internal_signal_declarations>
begin
  <concurrent_statements>
end <implementation_name>;
```

Figure 7.1. A VHDL design entity consists of an entity declaration containing a list of input and output signal port declarations followed by an architecture body containing internal signal declarations followed by a set of concurrent statements. The logic of the design entity is implemented by the concurrent statements.

```
---------------------------------------------------------------------------
-- prime
--    input   - 4 bit binary number
--    isprime - true if "input" is a prime number 1, 2, 3, 5, 7, 11, or 13
---------------------------------------------------------------------------

library ieee;
use ieee.std_logic_1164.all;

entity prime is
  port( input : in std_logic_vector(3 downto 0);
        isprime : out std_logic );
end prime;

architecture case_impl of prime is
begin
  process(input) begin
    case input is
      when x"1" | x"2" | x"3" | x"5" | x"7" | x"b" | x"d" => isprime <= '1';
      when others => isprime <= '0';
    end case;
  end process;
end case_impl;
```

Figure 7.2. VHDL description of the four-bit prime-or-one number function using a `case` statement to encode the truth table directly.

and output ports. Logic within the design entity computes the outputs based on the inputs – on just the current state of the inputs for a design entity implementing combinational logic. After declaring a design entity, we can instantiate one or more copies, or *instances*, of the design entity within a higher-level design entity.

The basic form of a VHDL design entity is shown in Figure 7.1, and a design entity that implements the four-bit prime number function using a VHDL **`process`** statement is shown in Figure 7.2. All design entities are composed of two parts – an entity declaration and an architecture body. The entity declaration starts with the keyword **`entity`**. The design entity name is given between the keyword **`entity`** and the keyword **`is`** (e.g., `prime` in Figure 7.2). Following the keyword **`port`** is a list of input and output signal declarations in parentheses.

Each signal includes the name of the signal, followed by a colon (`:`) followed by the signal direction specified using **`in`** for inputs and **`out`** for outputs. VHDL also defines signal directions **`buffer`** and **`inout`**. Use **`buffer`** for outputs that are read inside a design entity.[1] Signal direction **`inout`** should be used only for interface signals that are bidirectional and is rarely used in practice. The direction is followed by the type of the signal. For example, in Figure 7.2 the input signal `input` has type `std_logic_vector` and the output signal `isprime` has type `std_logic` (these types were introduced in Section 1.5).

The architecture body statements perform the logic that computes the design entity outputs. The architecture body starts with the keyword **`architecture`** followed by an identifier, then the keyword **`of`** (e.g., `case_impl` in Figure 7.2), then the name of the entity declaration, then the keyword **`is`**. Next optional lists of components and internal signals are given before the keyword **`begin`** (none in the example in Figure 7.2). Next is a list of concurrent statements. Unlike statements in a software programming language such as C or Java, in VHDL each concurrent statement operates in parallel. In the subset of VHDL we will use in this book, the concurrent statements consist of one or more process statements, concurrent signal assignments, component instantiations, or generate statements. Examples of the first three of these are given in the four implementations of the prime number circuit in Sections 7.1.2 to 7.1.8. We defer discussion of generate statements to Section 9.1.

7.1.2 The `case` statement

As shown in Figure 7.2, a VHDL **`case`** statement allows us to specify directly the truth table of a logic function. The **`case`** statement must appear inside a process and it allows us to specify the output value of a logic function for each input combination. The signal after the **`case`** keyword and before the keyword **`is`** indicates the input of the lookup table. Inside a case statement the keyword **`when`** is used to specify the rows of the lookup table. In this example, to save space we specify all the input states where the output, `isprime`, is 1 using a list of different input states separated by a pipe "`|`" symbol. To specify a default rule that applies for inputs that do not match we use the keyword **`others`**. In the default case we assign the output, `isprime`, to 0. The arrow compound delimiter, `=>` , separates the list of input states from a statement used to assign the output of the truth table.

The general form of an `std_logic_vector` value in VHDL-2008 is `<size><base>"<value>"` (also known as a *bit string literal*). Here `<size>` is a decimal number that describes the width of the number in bits. In each constant definition, the size of the number is 7, specifying that each constant is seven bits wide. Note that `3b"0"` and `7b"0"` are different numbers; both have the value 0, but the first is three bits wide while the second is seven bits wide. The `<base>` portion of a number is `b` for binary, `d` for decimal, `o` for octal (base 8), or `x` for hexadecimal (base 16). Finally, the `"<value>"` portion of the number is the value in the specified base. Versions of VHDL prior to VHDL-2008 do not support `<size>`, or `"d"` for `<base>`. If `<size>` is omitted, the width is inferred from the number of digits in `<value>`.

[1] Using the direction **`buffer`** used to be discouraged in favor of declaring a separate internal signal that is copied to the output. With recent CAD tools and especially when using VHDL-2008, declaring an extra internal signal is no longer required.

For example, x"00" is eight bits wide since a single hexadecimal digit is four bits. If <size> and <base> are both omitted, then in addition the base is assumed to be binary. For example, "0101" is a four-bit binary number having a value of five.

Note that case statements must be contained within a VHDL process statement. This syntax in Figure 7.2 specifies that the case statement will be evaluated each time the signals listed in the parenthesis after the **process** keyword change state. In this example, the case statement is evaluated each time the four-bit input signal input changes state.

Whenever a process statement is used to describe a *combinational* circuit, it is critical that all inputs be included in the *sensitivity list*. The sensitivity list is a comma separated list of signals in parentheses after the **process** keyword. If an input is omitted from the sensitivity list, the body of the process statement will not be evaluated when that input changes. The result will be sequential, not combinational, logic. Another way of saying this is that omitting signals from the sensitivity list will often result in odd behavior that can be difficult to debug. In the latest version of VHDL, VHDL-2008, this can be avoided by using the keyword **all** for the sensitivity list, i.e., **process**(**all**). This causes the process to evaluate any time any signal read inside the process changes state. We use this syntax throughout the book.

The result of synthesizing the VHDL description of Figure 7.2 with the Synopsys Design Compiler® using a typical CMOS standard cell library is shown in Figure 7.3. The synthesizer has converted the *behavioral* VHDL description of Figure 7.2, which specifies what is to be done (i.e., a truth table), to a *structural* VHDL description, which specifies how to do it (i.e., five gates and the connections between them). The structural VHDL instantiates five gates: two OR-AND-invert gates (OAI), two inverters (INV), and one exclusive-OR gate (XOR). The four wires connecting the gates are declared as n1 through n4. For each gate, the Design Compiler® output instantiates the gate by using a component instantiation statement by providing a label for the component followed by a colon (:), followed by the type of the gate (e.g., OAI13), and then specifying which signal is connected to each gate input and output using a **port map** aspect with *named association* using the arrow compound delimiter. For example, "A1 => n2" means that signal n2 is connected to gate input A1).

Note that a design entity can be instantiated either with named association, which explicitly connects input and output signals, or with a simplified notation. For example, if the ports were declared in the order shown, we could instantiate the XOR gate with the following syntax, known as *positional association*:

```
U4: XOR2 port map (input(2), input(1), n4);
```

Using positional association, the above statement implicitly connects input(2) with input A because input(2) is listed first in the association list and input A is declared first in the component declaration. Named and positional association are equivalent. For complex components, the named association syntax avoids getting the order wrong (and prevents errors when the underlying order changes). For simple components, the positional association syntax is more compact and easier to read. Also, the notation input(2) indicates that we want to access only the bit with index "2" from input.

Figure 7.4 shows a schematic of the synthesized circuit to show how the synthesizer has optimized the logic. Unlike the two-level synthesis method we employed in Chapter 6, the synthesizer has used four levels of logic (not counting inverters), including an XOR gate in

```
library IEEE;
use IEEE.std_logic_1164.all;

entity prime is
  port( input : in std_logic_vector (3 downto 0);
        isprime : out std_logic );
end prime;

architecture SYN_case_impl of prime is
  component OAI13 is
    port( A1, B1, B2, B3: in std_logic; Y: out std_logic );
  end component;
  component OAI12 is
    port( A1, B1, B2: in std_logic; Y: out std_logic );
  end component;
  component INV is
    port( A: in std_logic; Y: out std_logic );
  end component;
  component XOR2 is
    port( A, B: in std_logic; Y: out std_logic );
  end component;
  signal n1, n2, n3, n4 : std_logic;
begin
  U1: OAI13 port map( A1=>n2, B1=>n1, B2=>input(2), B3=>input(3), Y=>isprime);
  U2: INV port map( A=>input(1), Y=>n1);
  U3: INV port map( A=>input(3), Y=>n3);
  U4: XOR2 port map( A=>input(2), B=>input(1), Y=>n4);
  U5: OAI12 port map( A1=>input(0), B1=>n3, B2=>n4, Y=>n2);
end SYN_case_impl;
```

Figure 7.3. Result of synthesizing the VHDL description of Figure 7.2 with the Synopsys Design Compiler® using a typical standard cell library. A schematic of this synthesized circuit is shown in Figure 7.4.

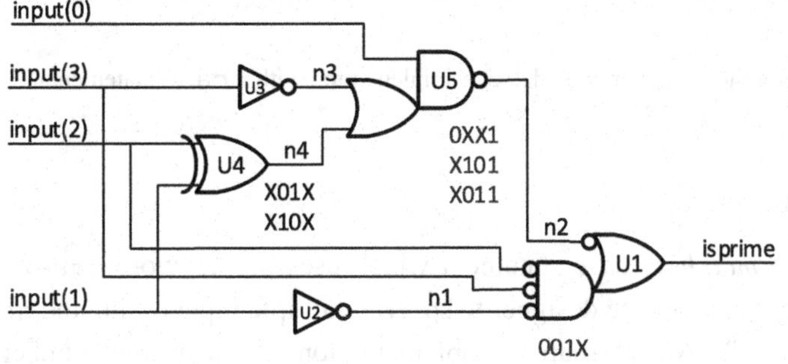

Figure 7.4. Schematic showing the circuit of Figure 7.3.

addition to the ANDs and ORs. However, this circuit still implements the same four prime implicants (0XX1, 001X, X101, and X011). As shown in the figure, the bottom part of gate U1 directly implements implicant 001X. Gate U5 implements the other three implicants – factoring `input(0)` out of the implicants so that this AND can be shared across all three. The top input to the OR of U5 (`n3`) ANDed with `input(0)` gives 0XX1. The output of the XOR gate gives products X01X and X10X, which, when ANDed with `input(0)` in U5, give the remaining two implicants X101 and X011.

This synthesis example illustrates the power of modern computer-aided design tools. A skilled designer would have to spend considerable effort to generate a circuit as compact as this one. Moreover, the synthesis tool can (via a constraint file) be asked to reoptimize this circuit for speed rather than area with minimum effort. With modern synthesis tools, the primary role of the logic designer has changed from one of optimization to one of specification. This specification task, however, continues to increase in complexity as systems become larger.

EXAMPLE 7.1 Thermometer code detector with case

Write a design entity that detects whether a four-bit input is a legal thermometer encoded signal (0000, 0001, 0011, 0111, 1111). Implement the logic using a `case` statement.

The design entity, shown in Figure 7.5, is implemented by first declaring the input (`input`) and output (`output`). The `case` statement is written in two lines: a list of minterms that output 1 and a default statement to output 0.

```
library ieee;
use ieee.std_logic_1164.all;

entity therm is
  port( input: in std_logic_vector(3 downto 0);
        output: out std_logic );
end therm;

architecture case_impl of therm is
begin
  process(all) begin
    case input is
      when "0000" | "0001" | "0011" | "0111" | "1111" => output <= '1';
      when others => output <= '0';
    end case;
  end process;
end case_impl;
```

Figure 7.5. The VHDL for a thermometer code detector implemented with a `case` statement.

7.1.3 The case? statement

VHDL-2008 includes a *matching case* statement which uses the keyword **`case?`** instead of **`case`**. The matching case enables the designer to specify multiple inputs using the `std_logic` "don't care" symbol ("`-`"). An alternative implementation of the prime number function using `case?` to specify implicants that cover the function is shown in Figure 7.6. Since we have already provided an entity declaration for `prime` in Figure 7.2 we need only provide the architecture body, though we must give the architecture body a different name (`match_case_impl`). This implementation is similar to the one in Figure 7.2 except that we use the `case?` statement in place of the `case` statement. The `case?` statement allows don't cares (-s) in the cases. This allows us to put implicants, rather than just minterms, on the left side of each case. For example, the first case `"0-1"` corresponds to implicant 0XX1 and covers minterms 1, 3, 5, and 7. A restriction of the matching `case?` statement is that each `when`

```
architecture mcase_impl of prime is
begin
  process(all) begin
    case? input is
      when "0--1" => isprime <= '1';
      when "0010" => isprime <= '1';
      when "1011" => isprime <= '1';
      when "1101" => isprime <= '1';
      when others => isprime <= '0';
    end case?;
  end process;
end mcase_impl;
```

Figure 7.6. VHDL architecture body of the four-bit prime number function using a matching case statement to describe the implicants in a cover. Note entity declaration is in Figure 7.2.

clause must not overlap. For example `"0-1"` overlaps with `"001-"` because both contain the minterm 3.

The matching case statement is useful in describing combinational design entities where one input overrides the others; for example, when a disable input causes all outputs to go low regardless of the other inputs, or for a priority encoder (see Section 8.5). For the prime number function, however, the implementation in Figure 7.2 is preferred because it more clearly describes the function being implemented and is easier to maintain. There is no need to reduce the function to implicants manually, the synthesis tools do this.

EXAMPLE 7.2 Thermometer code detector with case?

Write a design entity that detects whether a four-bit input is a legal thermometer encoded signal using a matching case statement. Assume the entity declaration from Figure 7.5 has already been given so that you need only provide a new architecture body definition.

This is a scenario where the synthesizer is better suited to do the logic reduction. Nevertheless, the thermometer code function can be written as follows:

$$f(a_3, a_2, a_1, a_0) = (\overline{a_3} \wedge \overline{a_2} \wedge \overline{a_1}) \vee (\overline{a_3} \wedge \overline{a_2} \wedge a_1 \wedge a_0) \vee (a_2 \wedge a_1 \wedge a_0).$$

Implementing this function in a matching case statement yields Figure 7.7.

```
architecture mcase_impl of therm is
begin
  process(all) begin
    case? input is
      when "000-" => output <= '1';
      when "0011" => output <= '1';
      when "-111" => output <= '1';
      when others => output <= '0';
    end case?;
  end process;
end mcase_impl;
```

Figure 7.7. The VHDL architecture body for a thermometer code detector with a `case?` statement. Note entity declaration is in Figure 7.5.

```
architecture if_impl of prime is
begin
  process(all) begin
    if input = 4d"1" then isprime <= '1';
    elsif input = 4d"2" then isprime <= '1';
    elsif input = 4d"3" then isprime <= '1';
    elsif input = 4d"5" then isprime <= '1';
    elsif input = 4d"7" then isprime <= '1';
    elsif input = 4d"11" then isprime <= '1';
    elsif input = 4d"13" then isprime <= '1';
    else isprime <= '0';
    end if;
  end process;
end if_impl;
```

Figure 7.8. VHDL description of the four-bit prime-or-one number function using an `if` statement to encode the truth table directly. Note entity declaration is in Figure 7.2.

7.1.4 The if statement

It is possible to describe combinational logic using *if* statements as illustrated in Figure 7.8 for the prime-or-one circuit. An *if* statement must appear inside a process and begins with the VHDL keyword **`if`** followed by an expression that returns a value of type boolean followed by the keyword **`then`**. After the keyword **`then`** one provides a list of sequential statements. In this example we include a single signal assignment statement, "`isprime <= '1';`". Multiple conditions can be considered in a single if statement by using the VHDL keyword **`elsif`**, which stands for "else if". In this example we include an **`elsif`** test for each prime number between 2 and 13. Finally, one can include a default action in case none of the prior conditions evaluated to true by using the keyword **`else`**. In this example we assign `isprime <= '0'` when `input` is not a prime number by using **`else`**.

While if statements are quite intuitive to anyone who has spent time writing software, we suggest avoiding using them as much as possible. The reason is that it is too easy to accidentally generate a sequential circuit by excluding an `else` clause, or by forgetting to assign to *every* output variable in *every* branch of the `if` statement (see Rule C2 in Section B.10.1). In this textbook, we use `if` statements only when specifying the next state function of state machines (Chapter 14).

7.1.5 Concurrent signal assignment statements

VHDL statements appearing inside of a process are known as sequential statements. Statements not in a process are known as *concurrent* statements. Figure 7.9 shows a fourth VHDL description of the prime number circuit. This version uses a concurrent signal assignment statement statement to describe the logic function using an equation. There is little advantage to describing the prime number circuit with an equation. The truth table description is easier to write, easier to read, and easier to maintain. The synthesizer reduces the truth table to an equation and optimized set of gates.

```
architecture logic_impl of prime is
begin
  isprime <= (input(0) AND (NOT input(3))) OR
             (input(1) AND (NOT input(2)) AND (NOT input(3))) OR
             (input(0) AND (NOT input(1)) AND input(2)) OR
             (input(0) AND input(1) AND NOT input(2)) ;
end logic_impl;
```

Figure 7.9. VHDL architecture body description of the four-bit prime number function using a concurrent assignment statement. Note entity declaration is in Figure 7.2.

EXAMPLE 7.3 Thermometer code detector with concurrent signal assignment statement

Write a design entity that detects whether a four-bit input is a legal thermometer encoded signal using a concurrent signal assignment statement. Assume the entity declaration from Figure 7.5 has already been given so that you need only provide a new architecture body definition.

This is a scenario where the synthesizer is better suited to do the logic reduction. Nevertheless, the thermometer code function can be written as follows:

$$f(a_3, a_2, a_1, a_0) = (\overline{a_3} \wedge \overline{a_2} \wedge \overline{a_1}) \vee (\overline{a_3} \wedge \overline{a_2} \wedge a_1 \wedge a_0) \vee (a_2 \wedge a_1 \wedge a_0).$$

Writing this equation as a concurrent signal assignment statement yields Figure 7.10.

```
architecture assign_impl of therm is
begin
  output <= ((NOT input(3)) AND (NOT input(2)) AND (NOT input(1))) OR
            ((NOT input(3)) AND (NOT input(2)) AND input(1) AND input(0)) OR
            (input(2) AND input(1) AND input(0));
end assign_impl;
```

Figure 7.10. The VHDL architecture body for a thermometer code detector implemented with a concurrent signal assignment statement.

7.1.6 Selected signal assignment statements

Figure 7.11 illustrates the prime function using a *selected signal assignment* statement. A selected signal assignment statement begins with the keyword **with**, followed by the name of a signal that is read, followed by the keyword **select**, followed by the name of the signal that will be assigned to. The value assigned to, `isprime` in this example, is determined by matching the value of the expression between the keywords **with** and **select**, `input` in this example, against the choices listed *after* the keyword **when**. Choices are separated by a pipe delimiter (|). The choice on the right side of **when** must be distinct constants. The keyword **others** can be used as a shorthand for values not explicitly listed. Multiple possible assignment values, indicated by multiple **when** clauses, are separated by commas. A selected signal assignment can be used to implement truth tables more compactly than a case statement, but cannot appear inside a process.[2]

[2] VHDL-2008 actually removes this confusing restriction, but as of early 2015 few CAD tools have implemented the change.

```
architecture select_impl of prime is
begin
  with input select
    isprime <= '1' when 4d"1" | 4d"2" | 4d"3" | 4d"5" | 4d"7" | 4d"11" | 4d"13",
               '0' when others;
end select_impl;
```

Figure 7.11. VHDL architecture body description of the four-bit prime number function using a selected signal assignment statement. Note entity declaration is in Figure 7.2.

```
architecture cond_impl of prime is
begin
  isprime <= '1' when input = 4d"1" else
             '1' when input = 4d"2" else
             '1' when input = 4d"3" else
             '1' when input = 4d"5" else
             '1' when input = 4d"7" else
             '1' when input = 4d"11" else
             '1' when input = 4d"13" else
             '0';
end cond_impl;
```

Figure 7.12. VHDL description of the four-bit prime number function using a conditional signal assignment statement.

7.1.7 Conditional signal assignment statements

Figure 7.12 illustrates the prime function using a *conditional signal assignment* statement. Here the value assigned to `isprime` is determined by evaluating the expression after each `when` statement. Conditional signal assignment statements are useful for expressing simple comparison logic. They are also useful when one condition takes precedence over another. If multiple `when` conditions evaluate to true the first one, in the order listed in the statement, takes priority. Similarly to the selected signal assignment statement, the conditional signal assignment statement cannot be used inside a process.

7.1.8 Structural description

Our final description of the prime number function, shown in Figure 7.13, is a structural description that, much like the output of the synthesizer in Figure 7.3, describes the function by instantiating components and describing the connections between them. Unlike the synthesizer output in Figure 7.3, this description does not instantiate components like `OAI13`. Instead we have constructed the circuit using implicants found from the K-map in Figure 6.6 in Section 6.5. In addition Figure 7.13 includes a specification of the design entities for the `and_gate` components that are instantiated and an example of a configuration statement used to select which architecture body to bind to each component instantiation. Instead of instantiating gates for the or gate and inverters we use four concurrent assignment statements, e.g., "`n1 <= not input(1);`". We do this to emphasize that concurrent assignment statements and component instantiations can both be employed together. Also important to note is that,

```
architecture struct_impl of prime is
  component and_gate is
    port( a, b, c  : in std_logic := '1'; y : out std_logic );
  end component;
  signal a1, a2, a3, a4, n1, n2, n3: std_logic;
begin
  -- Note that the order in which component instantiations and
  -- concurrent assignment statements appear has no effect.
  AND1: and_gate port map( input(1), n2, n3, a1 ); -- positional association
  AND2: and_gate port map( y=>a2, a=>input(0), b=>n3 ); -- named association
  AND3: and_gate port map( y=>a3, a=>input(0), b=>n1, c=>input(2) );
  AND4: and_gate port map( y=>a4, a=>input(0), b=>input(1), c=>n2 );
  isprime <= a1 or a2 or a3 or a4;
  n1 <= not input(1);
  n2 <= not input(2);
  n3 <= not input(3);
end struct_impl;

-- Each entity declaration must include packages used in its architecture bodies.
library ieee;
use ieee.std_logic_1164.all;

entity and_gate is
  port( a, b, c : in std_logic := '1'; y : out std_logic );
end and_gate;

architecture logic_impl of and_gate is
begin
  y <= a and b and c;
end logic_impl;

architecture alt_impl of and_gate is
begin
  y <= not (not a or not b or not c);
end alt_impl;

-- Without the optional configuration declaration below all and_gate component
-- instantiations in work.prime(struct_impl) will use work.and_gate(alt_impl).
configuration my_config of prime is
  for struct_impl
    for AND1, AND3 : and_gate
      use entity work.and_gate(alt_impl);
    end for;
    for others : and_gate
      use entity work.and_gate(logic_impl);
    end for;
  end for;
end configuration my_config;
```

Figure 7.13. VHDL description of the four-bit prime number function using structural description including an (optional) configuration declaration.

since both are concurrent statements, the order of component instantiation and concurrent assignment statement lines does not matter. Unlike in a software language, the multiple component instantiations and concurrent assignment statements operate in parallel. We illustrate both positional (AND1) and named association (AND2, AND3, AND4) for component instantiation. The component and entity declarations for the `and_gate` include a default input assignment for unused inputs with the syntax `:= '1'`. We do this to enable us to specify both three-input (AND1, AND3, AND4) and gates and, using positional association, a two-input (AND2) and gate with a single entity declaration.

We include two architecture bodies for `and_gate` with different implementations identified by `logic_impl` and `alt_impl`. We can optionally specify which implementation to use for each component instantiation using a *configuration declaration*. Configuration declarations are usually omitted in simple designs where there is a single architecture body for each entity declaration.

The optional configuration declaration starts with the keyword **`configuration`** followed by a configuration name (e.g., `my_config` in Figure 7.13), the keyword **`of`**, the name of the entity being configured (e.g., `prime`) and the keyword **`is`**. Next, since there may be multiple architecture bodies, we specify the specific one being configured (e.g., `struct_impl`) after the first **`for`**. Then, for each component instantiation, we specify a *binding* to a particular entity and architecture body combination. This binding step is performed by first specifying the instances to bind using **`for`** statements nested inside the first **`for`** (e.g., "**`for`** `AND1, AND3 : and_gate`"), then the design entity they are bound to (e.g., **`use entity`** `work.and_gate(alt_impl)`").

We show this structural description of the prime number circuit to illustrate the range of the VHDL language. For a design as simple as the prime number function the designer should use one of the earlier representations and let the synthesizer do the synthesis and optimization. For a simple circuit like the prime number example, a synthesis program like Synopsys Design Compiler® will generate the same circuit (Figure 7.3) for all of the specifications in Figures 7.2, 7.6, 7.8, 7.9, 7.11, 7.12, and 7.13. More typically, a structural description would be used only when instantiating much larger components than the `and_gate` component used here.

EXAMPLE 7.4 Structural thermometer code detector

Write a structural VHDL design entity which detects whether a four-bit input is a legal thermometer encoded signal. Assume the entity declaration from Figure 7.5 has already been given so that you need only provide a new architecture body definition.

Rewritten from Example 7.3, the reduced logic function is given by

$$f(a_3, a_2, a_1, a_0) = (\overline{a_3} \wedge \overline{a_2} \wedge \overline{a_1}) \vee (\overline{a_3} \wedge \overline{a_2} \wedge a_1 \wedge a_0) \vee (a_2 \wedge a_1 \wedge a_0).$$

We directly instantiate AND gates in the VHDL, assigning the AND gates' outputs to intermediate signals (`t2, t1, t0`). The configuration statement uses the keyword **`all`** to indicate that all component instantiations of `and_gate` are bound to `work.and_gate(impl)`. The resulting VHDL is shown in Figure 7.14.

```
architecture struct_impl of therm is
  component and_gate is
    port( a, b, c, d  : in std_logic :='1'; Y : out std_logic );
  end component;
  signal i1, i2, i3, t2, t1, t0 : std_logic;
begin
  i1 <= not input(1);
  i2 <= not input(2);
  i3 <= not input(3);
  AND1: and_gate port map(y=>t0, a=>i3, b=>i2, c=>i1 );
  AND2: and_gate port map(y=>t1, a=>i3, b=>i2, c=>input(1), d=>input(0) );
  AND3: and_gate port map(y=>t2, a=>input(2), b=>input(1), c=>input(0) );
  output <= t2 or t1 or t0;
end struct_impl;

library ieee;
use ieee.std_logic_1164.all;

entity and_gate is
  port( a, b, c, d : in std_logic := '1'; y : out std_logic );
end and_gate;

architecture impl of and_gate is
begin
  y <= a and b and c and d;
end impl;

configuration my_config of therm is
  for struct_impl
    for all: and_gate
      use entity work.and_gate(impl);
    end for;
  end for;
end;
```

Figure 7.14. Structural VHDL for a thermometer code detector.

7.1.9 The decimal prime number function

Figure 7.15 illustrates how incompletely specified functions (see Section 6.8) can be described in VHDL. While a matching case (Section 7.1.3) allows simplifying a specification by indicating the output for an implicant larger than a minterm, here we instead wish to enable the synthesis tool to further optimize the circuit. We again use the VHDL `case` statement to specify a truth table. In this example we specify that input states 10 to 15 have a don't care output by using the default (i.e., **`others =>`**) case alternative to assign the `std_logic` value `'-'` (i.e., don't care) to `isprime`. Because we can have only a single default statement, and here we choose to use it to specify don't cares, we must explicitly include the five input states for which the output is 0.

The result of synthesizing the VHDL description of Figure 7.15 using the Synopsys Design Compiler® is shown in Figure 7.16, and a schematic diagram of the synthesized circuit is

```
library ieee;
use ieee.std_logic_1164.all;

entity prime_dec is
  port( input  : in std_logic_vector(3 downto 0);
        isprime : out std_logic );
end prime_dec;

architecture impl of prime_dec is
begin
  process(input) begin
    case input is
      when x"0" | x"4" | x"6" | x"8" | x"9" => isprime <= '0';
      when x"1" | x"2" | x"3" | x"5" | x"7" => isprime <= '1';
      when others => isprime <= '-';
    end case;
  end process;
end impl;
```

Figure 7.15. VHDL description of the four-bit decimal prime number function using a `case` statement with don't care on the default output.

```
library ieee;
use ieee.std_logic_1164.all;

entity prime_dec is
   port( input  : in std_logic_vector (3 downto 0);  isprime : out std_logic);
end prime_dec;

architecture SYN_impl of prime_dec is

   component OAI21X1
      port( A, B, C : in std_logic;  Y : out std_logic);
   end component;
   component INVX1
      port( A : in std_logic;  Y : out std_logic);
   end component;
   component AND2X1
      port( A, B : in std_logic;  Y : out std_logic);
   end component;

   signal n4, n5, n6, n7 : std_logic;
begin

   U7 : AND2X1 port map( A => input(0), B => n5, Y => n6);
   U8 : INVX1 port map( A => n6, Y => n4);
   U9 : INVX1 port map( A => input(1), Y => n7);
   U10 : INVX1 port map( A => input(3), Y => n5);
   U11 : OAI21X1 port map( A => n7, B => input(2), C => n4, Y => isprime);

end SYN_impl;
```

Figure 7.16. Results of synthesizing the VHDL description of Figure 7.15 using the Synopsys Design Compiler®. A schematic of this synthesized circuit is shown in Figure 7.17. The resulting circuit is considerably simpler than the fully specified circuit of Figure 7.4.

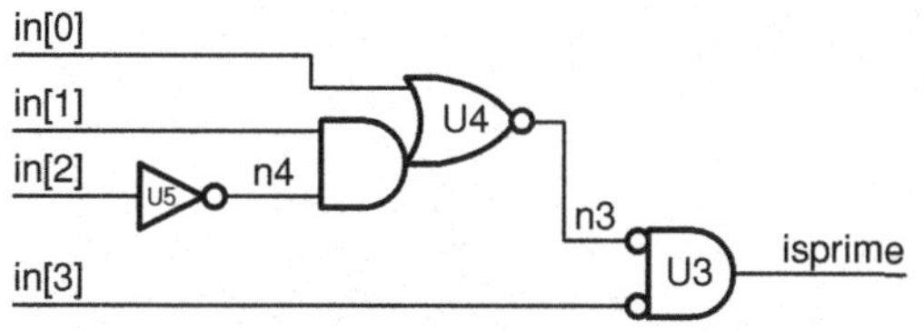

Figure 7.17. Schematic showing the circuit of Figure 7.16.

shown in Figure 7.17. With the don't cares specified, the logic is reduced to one two-input gate, one three-input gate, and one inverter, as compared with one four-input gate, one three-input gate, an XOR, and two inverters for the fully specified circuit.

7.2 A TESTBENCH FOR THE PRIME NUMBER CIRCUIT

To test via simulation that the VHDL description of a design entity is correct, we write a VHDL *testbench* that exercises the design entity. A testbench is itself a VHDL design entity. It is a design entity that is never synthesized to hardware and is used only to facilitate testing of the design entity under test. The testbench design entity instantiates the design entity under test, generates the input signals to exercise the design entity, and checks the output signals of the design entity for correctness. The testbench is analogous to the instrumentation you would use on a lab bench to generate input signals and observe output signals from a circuit.

Figure 7.18 shows a simple testbench for the prime number circuit. As in the first testbench we saw in Figure 3.8, we surround the testbench VHDL with a `--pragma translate_off` and `--pragma translate_on` pair to prevent synthesis while allowing the testbench in the same file as the synthesizable design entity being tested. As an alternative we could put the testbench in a separate file, which is used for simulation but not for synthesis. The testbench is itself a VHDL design entity, but it has no inputs or outputs. Internal signals are used as the inputs and outputs of the design entity under test, in this case `prime`. The testbench declares the input of the prime design entity, `count`, as a `signal` of type `std_logic_vector`. When we know we want to instantiate a specific architecture we can use *direct instantiation* rather than component and configuration declarations. Hence, the testbench instantiates an instance of design entity `prime` with architecture body `case_impl` using

```
DUT: entity work.prime(case_impl) port map(input, isprime);
```

The syntax to directly instantiate a design entity is given by

```
<inst_label>: entity <entity_name>(<arch_identifier>)  port map (<assoc_list>);
```

where `<inst_label>` is a label for the instance, `<entity_name>` is the name of the design entity from the entity declaration, `<arch_identifier>` is the name of the architecture body, and `<assoc_list>` is an association list using either position or named association format. In our example `<entity_name>` is `work.prime`. We preface `prime` with "`work.`" because as it analyzes each entity declaration and architecture body a VHDL compiler records the resulting design unit in the working library, which by convention is called "`work`".

The actual test code for the testbench is contained within a **`process`** statement. This process statement is like the ones we saw earlier (e.g., Figure 7.2 and Figure 7.15) except that, instead of being executed every time an input signal changes, it is executed repeatedly, starting at the beginning of the simulation. This is because it lacks a *sensitivity list* containing signals

```
-- pragma translate_off
library ieee;
use ieee.std_logic_1164.all;
use ieee.std_logic_unsigned.all;
use ieee.numeric_std.all;

entity test_prime is
end test_prime;

architecture test of test_prime is
  signal input: std_logic_vector(3 downto 0);
  signal isprime: std_logic;
begin
  -- instantiate module to test
  DUT: entity work.prime(case_impl) port map(input, isprime);

  process begin
    for i in 0 to 15 loop
      input <= std_logic_vector(to_unsigned(i,4));
      wait for 10 ns;
      report "input = " & to_string(to_integer(unsigned(input))) &
             " isprime = " & to_string(isprime);
    end loop;
    std.env.stop(0);
  end process;
end test;
-- pragma translate_on
```

Figure 7.18. VHDL testbench for prime number design entity.findexto_integer

(e.g., `input` in Figure 7.2) to trigger the statements inside the **`process`** to be evaluated. That is, in Figure 7.18 there are no signals listed in parentheses after the **`process`** keyword. During each iteration of the for loop, the simulator sets `input` to the iteration count. To do this, we convert `i` to `unsigned`, which is like `std_logic_vector` except that the bits are interpreted as an unsigned number (i.e., zero or positive). We perform the conversion using the function `to_unsigned` declared in package `ieee.numeric_std`. The first argument to the function `to_unsigned` is an integer. The second argument specifies the bit width of the result. We then convert the unsigned result to `std_logic_vector` using a *type conversion*, which is similar to a type cast in a programming language like C or C++. Next the line "`wait for 10 ns;`" specifies that during simulation the process should wait for 10 nanoseconds of simulated time before executing the next statement. This allows time for the output of the design under test to settle. The next line displays the input and output using **`report`**. To print out `input`, which is of type `std_logic_vector` in decimal, we first apply a type conversion to `unsigned`, then call the function `to_integer` defined in `ieee.numeric_std`, then call the function `to_string`. After 16 iterations, the loop completes. The call to the VHDL-2008 function `std.env.stop()` after the loop stops simulation. Without it the process would immediately restart after the for loop exits. An alternative is to use an unconditional wait statement, i.e., "**`wait;`**".

In Figure 7.18 the `DUT` component operates in parallel with the `process` that generates inputs and prints the outputs. The inputs to `DUT` are transfered from the `process` via the signal

```
input = 0 isprime = 0
input = 1 isprime = 1
input = 2 isprime = 1
input = 3 isprime = 1
input = 4 isprime = 0
input = 5 isprime = 1
input = 6 isprime = 0
input = 7 isprime = 1
input = 8 isprime = 0
input = 9 isprime = 0
input = 10 isprime = 0
input = 11 isprime = 1
input = 12 isprime = 0
input = 13 isprime = 1
input = 14 isprime = 0
input = 15 isprime = 0
```

Figure 7.19. Output from testbench of Figure 7.18 on the design entity described in Figure 7.2.

`input`. Similarly, the output of `DUT` is transfered to the `process` via the signal `isprime`. Note that `isprime` can be read inside the process even though the process does not have a sensitivity list. The sensitivity list of a process is *not* the same as the arguments of a function in a software language.

The testbench does not describe a piece of our design, but rather is just a source of input stimuli, or *test patterns*, and a monitor of output results. Because the testbench design entity does not have to be synthesized, it can use VHDL constructs that are not permitted in synthesizable designs. For example, the `process` without a sensitivity list, `for-loop`, and `wait for 10 ns` statements in Figure 7.18 are not allowed in synthesizable VHDL design entities, but are quite useful in testbenches.[3] When writing VHDL, it is important to keep in mind whether one is writing synthesizable code or a testbench. Very different styles are used for each.

The output of a VHDL simulation of the testbench of Figure 7.18 and the prime number design entity of Figure 7.2 is shown in Figure 7.19. Each iteration of the for loop, the **`report`** statement in the testbench, prints one line of output. The argument to the **`report`** command must be a string. The string we print is formed by joining several strings using the concatenation operator '`&`'. VHDL-2008 includes the `to_string` function in the `ieee.std_logic_1164` package.[4] As its name suggests, `to_string` converts a value to a string. The output is decimal format string for integer data types and binary for `std_logic_vector`. There are also functions for printing in hexadecimal (`to_hstring`) and octal (`to_ostring`). To print `input` in decimal, we first convert it to type `unsigned`, which is essentially a `std_logic_vector` that is interpreted as 2's complement unsigned number. We then use the function `to_integer`, defined in `ieee.numeric_std`, to convert from unsigned to an integer. In contrast to `std_logic_vector` and `unsigned`, an `integer` is at least 32 bits in VHDL.

[3] Technically, the **`for-loop`** statement is synthesizable, but its use is discouraged. While some forms of wait statement are synthesizable, a wait with a "timeout clause," such as the one in the above test script, is ignored during synthesis.

[4] Prior to VHDL-2008 converting `std_logic_vector` values to strings was more complicated.

```
-- pragma translate_off
library ieee;
use ieee.std_logic_1164.all;
use ieee.std_logic_unsigned.all;
use ieee.numeric_std.all;

entity test_prime_mcase is
end test_prime_mcase;

architecture test of test_prime_mcase is
  signal input: std_logic_vector(3 downto 0);
  signal check: std_logic; -- set to 1 on mismatch
  signal isprime0, isprime1: std_logic ;
begin
  -- instantiate both implementations
  p0: entity work.prime(case_impl) port map(input, isprime0) ;
  p1: entity work.prime(mcase_impl) port map(input, isprime1) ;

  process begin
    check <= '0';
    for i in 0 to 15 loop
      input <= std_logic_vector(to_unsigned(i,4));
      wait for 10 ns;
      if isprime0 /= isprime1 then
        check <= '1';
      end if;
    end loop;
    wait for 10 ns;
    if check /= '1' then report "PASS"; else report "FAIL"; end if;
    std.env.stop(0);
  end process;
end test;
-- pragma translate_on
```

Figure 7.20. Go/no-go testbench that checks results using a second implementation of the prime number design entity.

By examining the output, we can see that the prime number design entity is operating correctly.

Checking a VHDL design entity by manually examining its output works fine for small design entities that need to be checked just once. However, for larger design entities, or repeated testing,[5] manual checking is tedious and error-prone. In such cases, the testbench must check results in addition to generating inputs.

One approach to a self-checking testbench is to instantiate two separate implementations of the design entity and compare their outputs, as shown in Figure 7.20. (Another approach is to use an inverse function as shown in Section 7.3.) In Figure 7.20, the testbench creates one instance of design entity `prime(case_impl)` (Figure 7.2) and one instance of design entity

[5] It is common practice to rerun a large test suite on an entire design on a periodic basis (e.g., every night). This *regression* testing catches many errors that result from the unintended consequences of making a change to one part of the design on a different, and often distant, part.

prime(mcase_impl) (Figure 7.9).[6] All 16 input patterns are then applied to both design entities. We compare the outputs using the inequality operator /= which evaluates to true if isprime0 does not equal isprime1. The comparison is done with respect to the underlying std_logic type. This type includes special values that a VHDL simulator can use to indicate conditions such as disconnected wires (e.g., 'U' for uninitialized). Thus, using std_logic and std_logic_vector can catch errors that might otherwise go undetected if values are constrained to being either logic value one or zero.[7] If the outputs of the design entities don't match for any pattern, the variable check is set equal to 1. After all cases have been tried, a PASS or FAIL is indicated, depending on the value of check.

EXAMPLE 7.5 Thermometer code detector testbench

Write a VHDL testbench that simultaneously checks the design entities written for Examples 7.1–7.4.

Our testbench, shown in Figure 7.21, requires the user to check manually that the output of therm(assign_impl) is correct. We use pass/fail testing to verify the other three entity instantiations, comparing their outputs with the output of therm(assign_impl). The output with our test (Figure 7.22) confirms that our VHDL is correct.

```
-- pragma translate_off
library ieee;
use ieee.std_logic_1164.all;
use ieee.std_logic_unsigned.all;
use ieee.numeric_std.all;

entity therm_test is
end therm_test;

architecture test of therm_test is
  signal count: std_logic_vector(3 downto 0);
  signal t0, t1, t2, t3, check: std_logic;
begin
  M0: entity work.therm(assign_impl) port map( count, t0 );
  M1: entity work.therm(case_impl)   port map( count, t1 );
  M2: entity work.therm(mcase_impl)  port map( count, t2 );
  M3: configuration work.my_config   port map( count, t3 );

  process begin
    count <= "0000";
    check <= '0';
    for i in 0 to 15 loop
      wait for 10 ns;
```

Figure 7.21. Testbench to verify the four thermometer code detectors from Examples 7.1–7.4.

[6] In this example, there is little advantage to comparing these two implementations, since they are of roughly the same complexity. In other situations, however, there is often a very simple non-synthesizable description that can be used for comparison.

[7] For example, using the VHDL types bit and bit_vector.

```
      report "input = " & to_string(count) & " therm = " & to_string(t0);
      if t0 /= t1 then check <= '1'; end if;
      if t0 /= t2 then check <= '1'; end if;
      if t0 /= t3 then check <= '1'; end if;
      count <= count + 1;
    end loop;
    if check = '0' then report "PASS";
    else report "FAIL"; end if;
    std.env.stop(0);
  end process;
end test;
-- pragma translate_on
```

Figure 7.21. *(cont.)*

```
input = 0000 therm = 1
input = 0001 therm = 1
input = 0010 therm = 0
input = 0011 therm = 1
input = 0100 therm = 0
input = 0101 therm = 0
input = 0110 therm = 0
input = 0111 therm = 1
input = 1000 therm = 0
input = 1001 therm = 0
input = 1010 therm = 0
input = 1011 therm = 0
input = 1100 therm = 0
input = 1101 therm = 0
input = 1110 therm = 0
input = 1111 therm = 1
PASS
```

Figure 7.22. The output from the testbench of Figure 7.21. We must check manually whether each output is correct.

7.3 EXAMPLE: A SEVEN-SEGMENT DECODER

In this section we examine the design of a seven-segment decoder to introduce the concepts of constant definitions, signal concatenation, and checking with inverse functions.

A seven-segment display depicts a single decimal digit by illuminating a subset of seven light-emitting segments. The segments are arranged in the form of the numeral 8, as shown in the top part of Figure 7.23, numbered from 0 to 6. A seven-segment decoder is a hardware block that accepts a four-bit binary-coded input signal, `bin(3 downto 0)`, and generates a seven-bit output signal, `segs(6 downto 0)`, that indicates which *segments* of a seven-segment display should be illuminated to display the number encoded by `bin`. For example, if the binary code for "4," 0100, is input to a seven-segment decoder, the output is 0110011, which indicates that segments 0, 1, 4, and 5 are illuminated to display a 4.

The first order of business in describing our seven-segment decoder is to define ten constants that each describe which segments are illuminated to display a particular numeral. Figure 7.23

```
------------------------------------------------------------------------
-- define segment codes
-- seven bit code - one bit per segment, segment is illuminated when
-- bit is low.  Bits 6543210 correspond to:
--
--          0000
--         5    1
--         5    1
--          6666
--         4    2
--         4    2
--          3333
--
------------------------------------------------------------------------
library ieee;

package sseg_constants is
  use ieee.std_logic_1164.all;
  subtype sseg_type is std_logic_vector(6 downto 0);

  constant SS_0 : sseg_type := 7b"1000000";
  constant SS_1 : sseg_type := 7b"1111001";
  constant SS_2 : sseg_type := 7b"0100100";
  constant SS_3 : sseg_type := 7b"0110000";
  constant SS_4 : sseg_type := 7b"0011001";
  constant SS_5 : sseg_type := 7b"0010010";
  constant SS_6 : sseg_type := 7b"0000010";
  constant SS_7 : sseg_type := 7b"1111000";
  constant SS_8 : sseg_type := 7b"0000000";
  constant SS_9 : sseg_type := 7b"0010000";
  constant SOFF : sseg_type := 7b"1111111";

  component sseg is
    port( bin : in std_logic_vector(3 downto 0);
          segs : out sseg_type );
  end component;
end package;
```

Figure 7.23. Defining the constants for the seven-segment decoder.

illustrates how we can declare our own package that serves this purpose. Specifically, it shows a *package declaration* beginning with the VHDL keyword **`package`**, followed by an identifier used to refer to the package and then the keyword **`is`** followed by definitions for ten constants `SS_0` through `SS_9` and finally a component declaration. We also define a constant `SOFF` which has a value of all zeros – a blank display – in addition to a component declaration for our seven-segment decoder. A package declaration is identified by the **`package`** keyword followed by the name of the package (`sseg_constants`) followed by the keyword `is`. We can use the constants and component declaration from our `sseg_constants` package by writing "**`use`** `work.sseg_constants.`**`all`**`;`" before the entity declaration of the design entity where we wish to use them.

The constants are declared using the VHDL **`constant`** keyword. Each `constant` declaration maps a constant name to a constant value and has a type. For example, the constant named `SS_4` is defined to have the seven-bit string 0011001 as its value and type `sseg_type`. We define the type `sseg_type` using the keyword **subtype** where we set it equal to `std_logic_vector(6 downto 0)`. Constants can also be declared within an architecture body as follows:

```
architecture <architecture_identifier> of <entity_name> is
  <constant_declarations>
  <signal_declarations>
begin
  ...
```

We choose to define our constants inside a package since we want to use them in multiple design entities.

We define constants for two reasons. First, using constant names rather than values makes our code more readable and easier to maintain. Second, defining a constant allows us to change all uses of the constant by changing a single value. For example, suppose we decide to drop the bottom segment on the "9." To do this, we would simply change the definition of `SS_9` to be

```
------------------------------------------------------------------------
-- sseg - converts a 4-bit binary number to seven segment code
--
-- bin  - 4-bit binary input
-- segs - 7-bit output, defined above
------------------------------------------------------------------------

library ieee;
use ieee.std_logic_1164.all;
use work.sseg_constants.all;

entity sseg is
  port( bin : in std_logic_vector(3 downto 0); segs : out sseg_type );
end sseg;
architecture impl of sseg is
begin
  process(all) begin
    case bin is
      when x"0" => segs <= SS_0;
      when x"1" => segs <= SS_1;
      when x"2" => segs <= SS_2;
      when x"3" => segs <= SS_3;
      when x"4" => segs <= SS_4;
      when x"5" => segs <= SS_5;
      when x"6" => segs <= SS_6;
      when x"7" => segs <= SS_7;
      when x"8" => segs <= SS_8;
      when x"9" => segs <= SS_9;
      when others => segs <= SOFF;
    end case;
  end process;
end impl;
```

Figure 7.24. Seven-segment decoder implemented with a `case` statement.

```
library ieee;
use ieee.std_logic_1164.all;
use ieee.numeric_std.all;
use work.sseg_constants.all;

entity invsseg is
  port( segs : in sseg_type;
        bin  : out std_logic_vector(3 downto 0);
        valid : out std_logic );
end invsseg;

architecture impl of invsseg is
begin
  process(all) begin
    case segs is
      when SS_0 =>  valid <= '1'; bin <= x"0";
      when SS_1 =>  valid <= '1'; bin <= x"1";
      when SS_2 =>  valid <= '1'; bin <= x"2";
      when SS_3 =>  valid <= '1'; bin <= x"3";
      when SS_4 =>  valid <= '1'; bin <= x"4";
      when SS_5 =>  valid <= '1'; bin <= x"5";
      when SS_6 =>  valid <= '1'; bin <= x"6";
      when SS_7 =>  valid <= '1'; bin <= x"7";
      when SS_8 =>  valid <= '1'; bin <= x"8";
      when SS_9 =>  valid <= '1'; bin <= x"9";
      when SOFF =>  valid <= '0'; bin <= x"0";
      when others =>valid <= '0'; bin <= x"1";
    end case;
  end process;
end impl;
```

Figure 7.25. VHDL description of an *inverse* seven-segment decoder, used to check the output of the seven-segment decoder.

0011000 rather than 0010000, and this change would propagate automatically to every use of `SS_9`. Without the definition, we would have to edit manually every use of the constant – and we would be likely to miss at least one.

The constant definitions give another example of the syntax used to describe numbers in VHDL (the syntax for numbers was described in Section 7.1.2). In the constant definitions of Figure 7.23 all numbers are in binary. The inverse seven-segment design entity in Figure 7.25 uses hexadecimal numbers.

Now that we have defined the constants, writing the VHDL code for the seven-segment decoder design entity `sseg` is straightforward. Along with the standard packages we include our constant definitions with the line "`use work.sseg_constants.all;`". As shown in Figure 7.24, we use a `case` statement to describe the truth table of the design entity, just as we did for the prime number function in Section 7.1.2. The output values are defined using our defined constants. It is much easier to read this code with the mnemonic constant names than if the right side of the case statement were all bit strings of 1s and 0s. When an input value is not in the range 0–9, the `sseg` design entity outputs `SOFF`.

To aid in testing our seven-segment decoder, we will also define an inverse seven-segment decoder design entity as shown in Figure 7.25. Design entity `invsseg` accepts a seven-bit input string `segs`. If the input is one of the ten codes defined in Figure 7.23, the circuit outputs the corresponding binary code on output `bin` and a "1" on output `valid`. If the input is `SOFF`, which is declared as all zeros (corresponding to the output of the decoder when the input is

```
-- pragma translate_off
library ieee;
use ieee.std_logic_1164.all;
use ieee.std_logic_unsigned.all;
use ieee.std_logic_arith.all;
use work.sseg_constants.all;

entity test_sseg is
end test_sseg;

architecture test of test_sseg is
  signal bin_in: std_logic_vector(3 downto 0);  -- binary code in
  signal segs: sseg_type;                        -- segment code
  signal bin_out: std_logic_vector(3 downto 0); -- binary code out of inverse coder
  signal valid: std_logic;                       -- valid out of inverse coder
  signal err: std_logic;
begin
  -- instantiate decoder and checker
  SS:  sseg port map(bin_in,segs);
  ISS: entity work.invsseg(impl) port map(segs,bin_out,valid);

  process begin
    err <= '0';
    for i in 0 to 15 loop
      bin_in <= conv_std_logic_vector(i,4);
      wait for 10 ns;
      report to_hstring(bin_in) & " " & to_string(segs) & " " &
             to_hstring(bin_out) & " " & to_string(valid);
      if bin_in < 4d"10" then
        if (bin_in /= bin_out) or (valid /= '1') then
          report "ERROR"; err <= '1';
        end if;
      else
        if (bin_out /= "0000") or (valid /= '0') then
          report "ERROR"; err <= '1';
        end if;
      end if;
    end loop;
    if err = '0' then report "TEST PASSED";
    else report "TEST FAILED"; end if;
    std.env.stop(0);
  end process;
end test;
-- pragma translate_on
```

Figure 7.26. Testbench for the seven-segment decoder using the inverse function to test the output.

out of range), the output is `valid = 0, bin = 0`. If the input is any other code, the output is `valid = 0, bin = 1`. Again, our inverse seven-segment decoder uses a case statement to describe a truth table. In this case there are two outputs of the truth table corresponding to `valid` and `bin`.

Now that we have defined the seven-segment decoder design entity `sseg` and its inverse design entity `invsseg`, we can write a testbench that uses the inverse design entity to check the functionality of the decoder itself. Figure 7.26 shows the testbench. The testbench instantiates the decoder and its inverse. The decoder accepts input `bin_in` and generates output `segs`. The inverse circuit accepts `segs` and generates outputs `valid` and `bin_out`.

After instantiating and connecting the design entities, the testbench contains a `for loop` that loops through the 16 possible inputs. For inputs that are in range (between 0 and 9), it checks that `bin_in = bin_out` and that `valid` is 1. If these two conditions don't hold, an error is flagged. Similarly, for inputs that are out of range it checks that `bin_out` and `valid` are both zero.

Using an inverse design entity to check the functionality of a combinational design entity is a common technique in writing testbenches. It is particularly useful in checking arithmetic circuits (see Chapter 10). For example, in writing a testbench for a square-root unit, we can square the result (a much simpler operation) and check that we get the original value.

The use of an inverse design entity in a testbench is also an example of the more general technique of using *checking modules*. Checking modules in testbenches are like *assertions* in software. They are redundant logic that is inserted to check *invariants*, conditions that we know should always be true (e.g., two modules should not drive the bus at the same time). Because the checking modules are in the testbench, they cost us nothing. They are not included in the synthesized logic and consume zero chip area. However, they are invaluable in detecting bugs during simulation.

Summary

In this chapter you have learned how to write VHDL descriptions of combinational logic functions. A truth table can be directly converted to a VHDL `case` statement:

```
process(input)
begin
  case input is
    when x"1" | x"2" | x"3" | x"5" | x"7" | x"b" | x"d" => isprime <= '1';
    when others => isprime <= '0';
  end case;
end process;
```

Here, we use one case to handle all input combinations where the output is 1, and the default case makes the function 0 elsewhere. The VHDL `case?` statement is like the `case` statement, but it allows our cases to include -s (don't cares) in some of the bits. It is most useful for specifying functions where a particular pattern on some input bits makes the remaining input bits irrelevant.

The VHDL concurrent assignment statement allows us to write a logic equation directly as in

```
majority <= (a AND b) OR (a AND c) or (b AND c) ;
```

A VHDL logic function can be encapsulated in a design entity. A design entity is composed of an `entity` declaration followed by an `architecture` body declaration. The entity declaration includes a list of input and output signals in a `port` declaration. Larger functions can then *instantiate* design entities and connect the output of one design entity to the input of another by connecting them to the same signal.

We verify a VHDL design entity by writing a *testbench* for that design entity. Unlike a design entity, which is ultimately synthesized into hardware, a testbench is not synthesized and is written in a different style of VHDL. A testbench instantiates the design entity being tested, declares the input and output signals of the design entity, and then specifies a series of *test patterns* within a `process` statement. Ideally, a testbench verifies proper behavior and gives a *pass/fail* indication. We will discuss verification in more detail in Chapter 20.

BIBLIOGRAPHIC NOTES

More information about VHDL can be found in Ashenden *The Designer's Guide to VHDL* [3] or one of many other textbooks on the language. Several good VHDL references are available online.

Exercises

7.1 *Fibonacci:* `case`. Write a VHDL description for a circuit that accepts a four-bit input and outputs true if the input is a Fibonacci number (0, 1, 2, 3, 5, 8, or 13). Your implementation must be done via a `case` statement.

7.2 *Fibonacci: circuit: concurrent signal assignment.* Write a VHDL description for a circuit that accepts a four-bit input and outputs true if the input is a Fibonacci number (0, 1, 2, 3, 5, 8, or 13). Your implementation must be done via a concurrent assignment statement using a minimized logic function.

7.3 *Fibonacci: circuit: structural.* Write a VHDL description for a circuit that accepts a four-bit input and outputs true if the input is a Fibonacci number (0, 1, 2, 3, 5, 8, or 13). Your implementation must be done via structural VHDL, directly instantiating AND and OR gates. You will need to write separate AND and OR design entities.

7.4 *Fibonacci: testbench.* Write a testbench and verify that your design entities from Exercises 7.1–7.3 all work correctly. Which of the four design entities did you find easiest to write and maintain?

7.5 *Fibonacci: logic synthesis.* Use a synthesis tool to synthesize the Fibonacci circuits you wrote in Exercises 7.1–7.3. Draw the resulting logic diagram for each synthesis outputs circuit. Compare and contrast the output circuits from each design entity.

7.6 *Five-bit prime number circuit.* Write a VHDL description for a circuit that accepts a five-bit input and outputs true if the input is a prime number (2, 3, 5, 7, 11, 13, 17, 19, 23, 29, or 31). Describe why the approach you chose (`case`, concurrent assignment, structural) is the right approach.

7.7 *Multiple-of-3 circuit.* Write a VHDL description for a circuit that accepts a four-bit input and outputs true if the input is a multiple of 3 (3, 6, 9, 12, or 15). Describe why the approach you chose (`case`, concurrent assignment, structural) is the right approach.

7.8 *Testbench.* Write a VHDL testbench for the multiple-of-3 circuit of Exercise 7.7.

7.9 *Decimal Fibonacci circuit.* Write a VHDL description for a circuit that accepts a four-bit input that is guaranteed to be in the range 0–9 and outputs true if the input is a Fibonacci number (0, 1, 2, 3, 5, or 8). The output is a don't care for input states 10–15. Describe why the approach you chose (`case`, concurrent assignment, structural) is the right approach.

7.10 *Multiple-of-5 circuit.* Write a VHDL description for a circuit that outputs true if its five-bit input is a multiple of 5.

7.11 *Square circuit.* Write a VHDL description for a circuit that outputs true if its eight-bit input is a square number, i.e., 1, 4, 9, ...

7.12 *Cube circuit.* Write a VHDL description for a circuit that outputs true if its eight-bit input is a cube, i.e., 1, 8, 27, 64, ...

7.13 *Bit reversal:* `case`*.* Write a VHDL design entity that takes a five-bit input, `input`, and outputs a five-bit value that is equal to the input with its bits reversed. For example, input 01100 gives output 00110, and input 11110 gives output 01111. You must implement your design entity using a `case` statement.

7.14 *Bit reversal:* `assign`*.* Write a VHDL design entity that takes a five-bit input, `input`, and outputs a five-bit value that is equal to the input with its bits reversed. For example, input 01100, gives output 00110 and input 11110 gives output 01111. You must implement your design entity using a single concurrent assignment statement. You will need to use the concatenation operator.

7.15 *Next Fibonacci number, I.* Write a VHDL design entity that takes in a four-bit input and outputs a five-bit number representing the next Fibonacci number. The mappings of inputs to outputs are as follows:

$$f(0001) = 00010,$$
$$f(0010) = 00011,$$
$$f(0011) = 00101,$$
$$f(0101) = 01000,$$
$$f(1000) = 01101,$$
$$f(1101) = 10101.$$

You may assume that the input is a legal Fibonacci number and output don't cares (`'-'`) if the input is not valid. Remember that if you do not test your circuit, it does not work. For the sake of this problem (and Exercises 7.16 and 7.17) we are ignoring the leading 0, 1 of the Fibonacci sequence and using only 1, 2, 3, 5, ...

7.16 *Next Fibonacci number, II.* Modify your Fibonacci circuit of Exercise 7.15 to include an output signal `valid` that is 1 when the input is a valid Fibonacci number and 0 when it is not. Your design entity should still also output the next five-bit number.

7.17 *Next Fibonacci number, III.* Modify your Fibonacci circuit of Exercise 7.16 to include two new inputs: `rst` and `ivalid`. If `rst` is equal to 1, then your design entity outputs (both `valid` and the next number) must be 0. If `ivalid` is 0, then the output `valid` signal must be 0, regardless of the input number. If `rst` is 0 and `ivalid` is 1, then the circuit works as described in Exercise 7.16. Your logic should be implemented in a `case?` statement with no more than eight statements and a default.

7.18 *FPGA implementation.* Use an FPGA mapping tool (such as Altera Quartus or Xilinx Foundation) to map the seven-segment decoder of Figure 7.24 to an FPGA. Use the floorplanning tools to view the layout of the FPGA. How many CLBs did the synthesis use?

7.19 *Seven-segment decoder.* Modify the seven-segment decoder to output the characters "A" through "F" (lowercase "b" and "d") for input states 10 to 15, respectively.

7.20 *Inverse seven-segment decoder.* Modify the inverse seven-segment decoder to accept two possible codes for "9," either `"1111011"` or `"1110011"`– i.e., with the bottom segment (segment 3) on or off.

7.21 *Testbench.* Modify the testbench of Figure 7.18 to check the output and indicate only pass or fail for the test.

7.22 *Case multiplication.* Write a VHDL design entity using a `case` statement that takes two two-bit numbers and outputs a four-bit number that is the product of the inputs (e.g., $10_2 \times 11_2 = 0110_2$). You will learn how to design a real multiplier in Chapters 10–12.

7.23 *Find-first-one.* A find-first-one unit will detect and output the bit position of the most significant 1 in an input. Write the VHDL, using a `case?` statement, to implement a find-first-one that takes a 16-bit input and outputs a four-bit signal that notes the position of the first 1. Also include an output that is 1 if there are no 1s in the input. Test your circuit. We discuss a different implementation of this circuit in Chapter 8.

7.24 *Factorization circuit.* Write a VHDL description for a circuit that takes a four-bit input `input` and produces outputs `two, three, five, seven, eleven,` and `thirteen.` A given output is high if the number corresponding to the output evenly divides the input. For example, when the input is 6, outputs `two` and `three` are high and the other outputs are low.

8 Combinational building blocks

A relatively small number of modules, decoders, multiplexers, encoders, etc., are used repeatedly in digital designs. These building blocks are the idioms of modern digital design. Often, we design a module by composing a number of these building blocks to realize the desired function, rather than writing its truth table and directly synthesizing a logical implementation.

In the 1970s and 1980s most digital systems were built from small integrated circuits that each contained one of these building block functions. The popular 7400 series of TTL logic [106], for example, contained many multiplexers and decoders. During that period, the art of digital design largely consisted of selecting the right building blocks from the TTL databook and assembling them into modules. Today, with most logic implemented as ASICs or FPGAs, we are not constrained by what building blocks are available in the TTL databook. However, the basic building blocks are still quite useful elements from which to build a system.

8.1 MULTI-BIT NOTATION

Throughout this book we use *bus notation* in figures to denote multi-bit signals with a single line. For example, in Figure 8.1, we represent the eight-bit signal $b_{7:0}$ with a single line. The diagonal slash across the signal indicates that this line represents a multi-bit signal. The number "8" below the slash indicates that the width of the bus is eight bits.

Single bits and subfields are selected from a multi-bit signal using diagonal connectors, as shown for bits b_7 and b_5, and the three-bit subfield $b_{5:3}$. Each diagonal connector is labeled with the bits being selected. The bit selections may overlap – as is the case with b_5 and $b_{5:3}$. The subfield $b_{5:3}$ is itself a multi-bit signal and is labeled accordingly – with a slash and the number "3."

8.2 DECODERS

In general, a *decoder* converts symbols from one code to another. We have already seen an example of a binary to seven-segment decoder in Section 7.3. When used by itself, however, the term decoder means a binary to *one-hot* decoder. That converts a symbol from a binary code (each bit pattern represents a symbol) to a one-hot code (at most one bit can be high at a time and each bit represents a symbol). In Section 8.4 we will discuss *encoders* that reverse this process. That is, they are one-hot to binary decoders.

The schematic symbol for an $n \rightarrow m$ decoder is shown in Figure 8.2. Input signal a is an n-bit binary signal and output signal b is an m-bit ($m \leq 2^n$) one-hot signal. A truth table for

Table 8.1. **Truth table for a 3 → 8 decoder: the decoder converts a three-bit binary input, *bin*, to an eight-bit one-hot output, *ohout***

bin	*ohout*
000	00000001
001	00000010
010	00000100
011	00001000
100	00010000
101	00100000
110	01000000
111	10000000

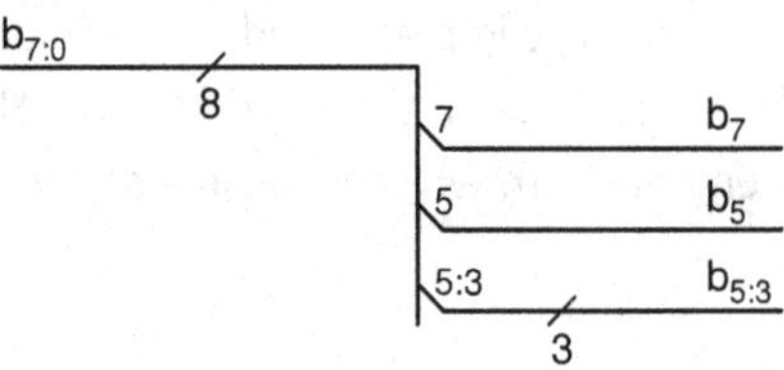

Figure 8.1. We denote a multi-bit signal or a bus with a diagonal slash across the signal labeled with the width of the bus. We select (possibly overlapping) individual signals and subfields from a multi-bit signal using diagonal connectors labeled with the bit or subfield being extracted.

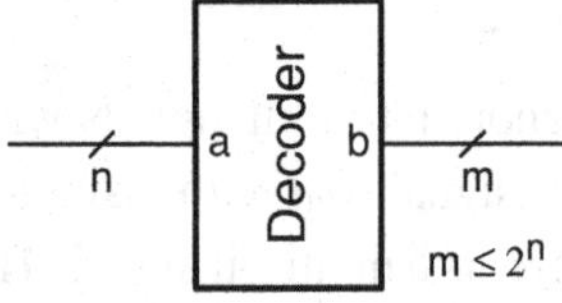

Figure 8.2. Schematic symbol for an $n \to m$ decoder.

a 3 → 8 decoder is shown in Table 8.1. If we think of both the input and the output as binary numbers, then, if the input has value i, the output has value 2^i.

A VHDL description of an $n \to m$ decoder is shown in Figure 8.3. This design entity introduces the use of VHDL *generic constants*. The design entity uses generic constants n and m to allow this single design entity to be used to instantiate component decoders of arbitrary input and output width. In the entity declaration description, the statement `n : integer := 2;` after the keyword **`generic`** declares that `n` (the input signal width) is an integer constant with a default value of 2. Similarly, `m` (the output signal width) is an integer constant with a default value of 4.

If we instantiate the component as usual, it will be created with the default values for all generic constants. For example, the following code creates a 2 → 4 decoder since the default values are `n=2` and `m=4`:

```
Dec24: Dec port map(x,y);
```

Here `x` and `y` are signals (declaration not shown) that connect to input `a` and output `b` via positional association.

```
----------------------------------------------------------------------
-- n -> m  Decoder
-- a - binary input   (n bits wide)
-- b - one-hot output (m bits wide)
----------------------------------------------------------------------
library ieee;
use ieee.std_logic_1164.all;
use ieee.numeric_std.all;

entity Dec is
  generic( n : integer := 2; m : integer := 4 );
  port( a : in std_logic_vector(n-1 downto 0);
        b : out std_logic_vector(m-1 downto 0) );
end Dec;

architecture impl of Dec is
  signal one: unsigned(m-1 downto 0);
  signal shift: integer;
begin
  one   <= to_unsigned(1,m);
  shift <= to_integer(unsigned(a));
  b     <= std_logic_vector(one sll shift);
end impl;
```

Figure 8.3. VHDL description of an *n* to *m* decoder.

We can override the default generic constant values when we instantiate a component. The general form for such a parameterized component instantiation is given by

```
<instance_label>: <component_identifier> generic map(<assoc_list>)
                                         port map(<assoc_list>);
```

For example, to instantiate a $3 \rightarrow 8$ decoder, the appropriate VHDL code is given by

```
Dec38: Dec generic map(n=>3, m=>8) port map(a=>x,b=>y);
```

Here `generic map(n=>3, m=>8)` sets `n=3` and `m=8` (using named association) for this instance of the `Dec` component with instance label `Dec38`. We can also use positional association:

```
Dec38b: Dec generic map(3,8) port map(x,y);
```

where `generic map(3,8)` sets `n=3` and `m=8` (using positional association) for another instance of the `Dec` component with instance label `Dec38b`. Similarly, a $4 \rightarrow 10$ decoder is created with

```
Dec410: Dec generic map(4,10) port map(x,y);
```

Note that the output width m need not be equal to 2^n for input width n. In many cases (where not all input states occur) it is useful to instantiate decoders that have less than full-width outputs. The design entity of Figure 8.3 uses the left shift operator[1] "`sll`" to shift a 1 over to the position specified by binary input `a` to create one-hot output `b`. An alternative to writing "`one sll shift`" is to write "`shift_left(one,shift)`".

[1] As you would expect, VHDL also has a right shift operator, `srl`.

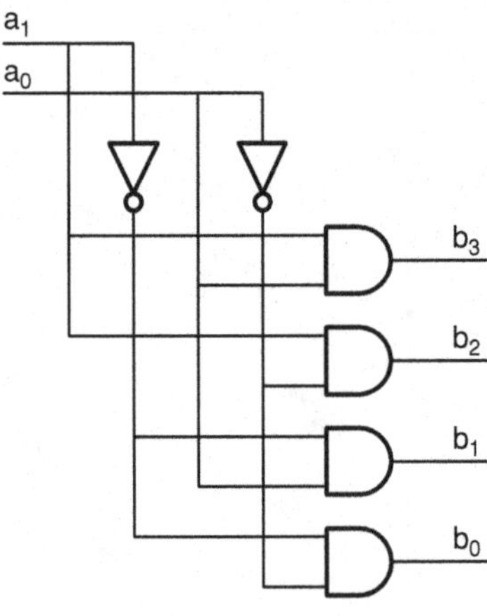

Figure 8.4. Schematic of a 2 → 4 decoder. An array of inverters creates the complements of the inputs and an array of AND gates generate the outputs.

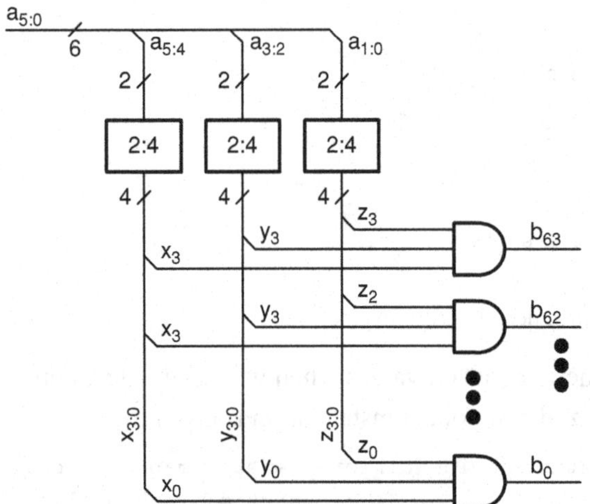

Figure 8.5. Schematic of a 6 → 64 decoder. Three 2 → 4 decoders predecode each pair of input bits into three four-bit one-hot signals, *x*, *y*, and *z*. An array of 64 three-input AND gates generates the outputs using these predecoded signals.

A small decoder is constructed using an AND gate to generate each output, as shown for a 2 → 4 decoder in Figure 8.4. Each input is complemented by an inverter. An AND gate for each output then selects the true or complement for each input to form the product corresponding to that output. Output b_1, for example, is generated by an AND gate with inputs a_0 and $\overline{a_1}$, so $b_1 = a_0 \wedge \overline{a_1}$.

Large decoders can be constructed from small decoders, as shown for a 6 → 64 decoder in Figure 8.5. The six-bit input $a_{5:0}$ is divided into three two-bit fields, and each is decoded by a 2:4 decoder generating three four-bit signals *x*, *y*, and *z*. In effect, this *predecoding* stage converts the six-digit binary input into a three-digit quaternary (base 4) input. Each of the four-bit signals *x*, *y*, and *z* represents one quaternary number. Each bit of the one-hot representation of the quaternary number corresponds to a particular value – 0, 1, 2, or 3. The 64 outputs are generated by 64 three-input AND gates that combine one bit from each of the quaternary digits. The AND gate for output b_i selects the bits that correspond to the quaternary representation of the output number, *i*. For example, output b_{27} (not shown) combines x_1, y_2, and z_3 because $27_{10} = 123_4$.

Building a large decoder using predecoders as shown in Figure 8.5 reduces logical effort (see Section 5.3) by factoring a large AND gate into two stages of smaller AND gates. For the 6:64 decoder in Figure 8.5, a six-input AND gate would be required to realize the decoder in a single stage. The implementation with 2:4 predecoders replaces the six-input AND gate with three two-input AND gates (one in each predecoder) followed by a three-input AND gate.

Efficiency is also gained by sharing the two-input AND gates across several outputs. In the figure, each two-input AND gate (each predecoder output) is shared across 16 outputs (one for each combination of the other two quaternary digits).

The design of a large decoder is a compromise between wiring density and logical efficiency. A single-level $n \rightarrow 2^n$ decoder requires $2n$ wiring tracks to run the inputs and their complements to all 2^n AND gates. Using 2:4 predecoders requires exactly the same number of wires, since the four wires required to carry two binary bits and their complements are replaced by a four-wire one-hot quaternary digit. This is clearly a win because the fan-in of the output gates is halved without any increase in wire tracks. Moreover, power is reduced since at most two of the bits of the quaternary signal change state (one up, one down) each time the input is changed, while all four wires of the true/complement binary signal may change state.

Going from 2:4 to 3:8 predecoders involves more of a tradeoff. The number of wiring tracks increases by 33% (from $2n$ to $8n/3$) in exchange for reducing the fan-in of the output gates by 33% (from $n/2$ to $n/3$). This is still usually a good trade. However, larger predecoders (e.g., 4:16) are rarely used because of the excessive number of wiring tracks required. An i-input predecoder requires $2^i n/i$ wiring tracks and (n/i)-input AND gates.

For very large decoders, the upper digits of the predecoder are often distributed to eliminate the need to run all output wires across the entire AND gate array. In Figure 8.5 (which is not really a *very* large decoder), for example, we could distribute the four AND gates of the predecoder that generate $x_{3:0}$ so that the AND gate that generates x_0 is next to the output AND gates that generate $b_{15:0}$, and the AND gate that generates x_1 is next to the AND gates that generate $b_{31:16}$, and so on. For wide input decoders, distributing decoders in this manner reduces wiring tracks. When the second most significant decoder is distributed, it is typically repeated for each output of the most significant decoder – negating some of the gate-sharing advantage of predecoding to reduce wiring complexity.

An $n \rightarrow 2^n$ decoder can be used to build an arbitrary n-input logic function. The decoder generates all 2^n minterms of n inputs. An OR gate can be used to combine the minterms that are implicants of the function to be implemented. For example, Figure 8.6 shows how a three-bit prime number function can be realized with a 3:8 decoder. The decoder generates all eight minterms $b_{7:0}$. An OR gate combines the minterms b_1, b_2, b_3, b_5, and b_7 that are implicants of the function.

A VHDL design entity that describes a three-bit prime number function using a decoder in this manner is shown in Figure 8.7. While it would be inefficient to implement the prime number function in this manner, it is a very compact and readable way to describe the function – it is very close to the notation $f = \sum m(1, 2, 3, 5, 7)$, and a good synthesizer will reduce this description to efficient logic.

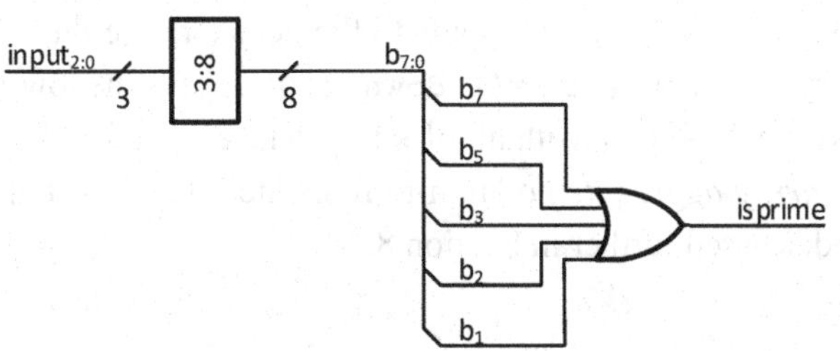

Figure 8.6. Three-bit prime-or-one number function implemented with a 3:8 decoder.

```
library ieee;
use ieee.std_logic_1164.all;
use work.ch8.all; -- ch8 package (not shown) includes component declaration for Dec

entity Primed is
  port( input: in std_logic_vector(2 downto 0);
        output: out std_logic );
end Primed;

architecture impl of Primed is
  signal b: std_logic_vector( 7 downto 0 );
begin
  -- instantiate a 3->8 decoder
  d: Dec generic map(n=>3,m=>8) port map(a=>input,b=>b);

  -- compute the output as the OR of the required minterms
  output <= b(1) or b(2) or b(3) or b(5) or b(7);
end impl;
```

Figure 8.7. VHDL design entity that implements the three-bit prime number function using a 3:8 decoder.

Note that to reuse the decoder VHDL from Figure 8.3 in our prime number circuit in Figure 8.7 we must either use a component declaration or use the direct instantiation syntax introduced in Section 7.2. It is generally preferrable to use a component declaration for synthesizable code because it allows us to change the binding later using a configuration declaration like that illustrated in Section 7.1.8. We could put the component declaration in the architecture body for `Primed`, as we have seen earlier (e.g., Figure 7.13). However, for frequently used components such as the combinational building blocks introduced in this chapter, it saves time and reduces VHDL code size to instead group the component declarations into a package. Hence, in Figure 8.7 and future examples we reference a package called "`ch8`" that includes a component declaration for each entity in this chapter. Such a package can be declared using the syntax introduced in Section 7.3 and illustrated in Figure 7.23. The VHDL code for the `ch8` package and similar packages used in subsequent chapters is included alongside the VHDL examples available on the website for this book.

EXAMPLE 8.1 Large decoder

Write a VHDL design entity that implements a 4 → 16 decoder using the 2 → 4 decoder as a building block.

The VHDL is shown in Figure 8.8. We use the hierarchical scheme shown in Figure 8.5. We first instantiate two 2 → 4 decoders, `d0` and `d1`, that decode the high and low halves of `a` to produce four-bit one-hot signals `x` and `y`. The inputs to the decoders use the VHDL *slice* notation to select a subset of bits from a bus (e.g., `a(1 downto 0)` selects the low-order two-bit fields from `a`). We then AND each bit of `y` with all of `x` to produce each four-bit field of `b`. Our design entity uses VHDL *array aggregate* notation to replicate bits of `y`. Both the array aggregate and slice syntax are discussed further in Section 8.3.

```
library ieee;
use ieee.std_logic_1164.all;
use work.ch8.all;

entity Dec4to16 is
  port( a: in std_logic_vector(3 downto 0);
        b: out std_logic_vector(15 downto 0) );
end Dec4to16;

architecture impl of Dec4to16 is
  signal x, y : std_logic_vector(3 downto 0); -- output of predecoders
begin
  -- instantiate predecoders
  d0: Dec port map(a(1 downto 0),x);
  d1: Dec port map(a(3 downto 2),y);

  -- combine predecoder outputs with AND gates
  b(3 downto 0)   <= x and (3 downto 0 => y(0));
  b(7 downto 4)   <= x and (3 downto 0 => y(1));
  b(11 downto 8)  <= x and (3 downto 0 => y(2));
  b(15 downto 12) <= x and (3 downto 0 => y(3));
end impl;
```

Figure 8.8. Implementation of a 4 → 16 decoder using two 2 → 4 decoders and 16 AND gates.

8.3 MULTIPLEXERS

Figure 8.9 shows the schematic symbol for a k-bit $n \rightarrow 1$ multiplexer. This circuit accepts n distinct k-bit wide input data signals $a_0, \ldots, a_{n-1}$ and an n-bit one-hot select signal s. The circuit selects the input signal a_i that corresponds to the high bit of s and outputs this value of a_i on the single k-bit wide output signal, b. In effect, the multiplexer acts as a k-pole n-throw switch to select one of the n input data signals under control of the select signal.

Multiplexers are commonly used in digital systems as *data selectors*. For example, a multiplexer on the input of an ALU (see Section 18.6) selects the source of data to feed the ALU, another selects an ALU's output, and a multiplexer on the address lines of a RAM selects the data source to provide a memory address each cycle.

Figure 8.10 shows two implementations of a one-bit 4:1 multiplexer. The implementation shown in Figure 8.10(a) uses AND and OR gates. Each data input a_i is ANDed with its corresponding select bit s_i, and the outputs of the ANDs are ORed together. Because the select signal is one-hot, only the select bit s_i corresponding to the selected input is true; the output of this AND gate will be a_i, and the output of all other AND gates will be zero. Thus the output of the OR will be the selected input, a_i. An alternative design using tri-state buffers is shown in Figure 8.10(b). A tri-state buffer (see Section 4.3.4) is a logic circuit with output equal to its data input (left input) if its control input (bottom input) is high, and disconnected (open circuit) if the control input is low. The high bit of the select input s_i enables one of the tri-state buffers to transmit a_i to the output; all other tri-state buffers are disabled – effectively disconnected from the output. The advantage of the tri-state implementation is that it can be distributed, with each

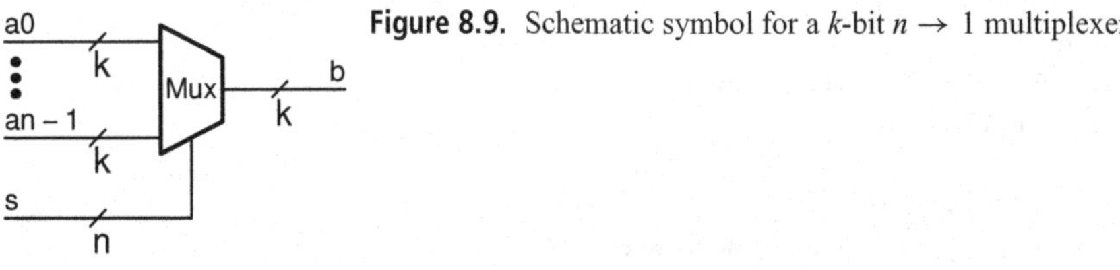

Figure 8.9. Schematic symbol for a k-bit $n \rightarrow 1$ multiplexer.

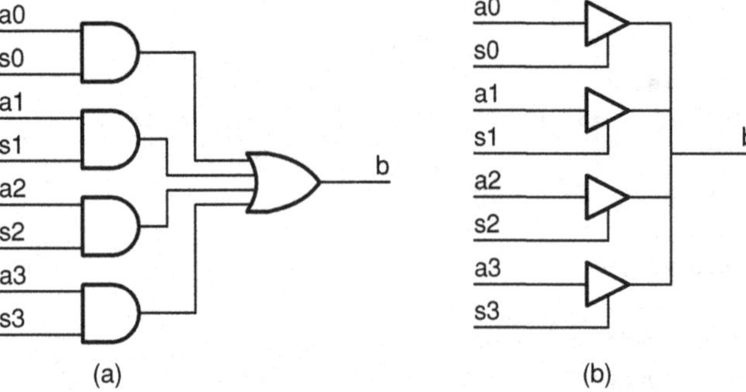

Figure 8.10. Schematic of a $4 \rightarrow 1$ multiplexer. (a) Using AND and OR gates. (b) Using tri-state buffers.

```
-- three-input mux with one-hot select (arbitrary width)
library ieee;
use ieee.std_logic_1164.all;

entity Mux3 is
  generic( k : integer := 1 );
  port( a2, a1, a0 : in std_logic_vector( k-1 downto 0 ); -- inputs
           s : in std_logic_vector( 2 downto 0 ); -- one-hot select
           b : out std_logic_vector( k-1 downto 0 ) );
end Mux3;

architecture logic_impl of Mux3 is
begin
  b <= ((k-1 downto 0 => s(2)) and a2) or
         ((k-1 downto 0 => s(1)) and a1) or
         ((k-1 downto 0 => s(0)) and a0);
end logic_impl;
```

Figure 8.11. VHDL description of an arbitrary-width $3 \rightarrow 1$ multiplexer.

buffer placed near its corresponding data source, and with only a single output line connecting the tri-state buffers (Exercise 8.7 asks you to build this structure out of CMOS logic gates). The AND/OR implementation, on the other hand, is more difficult to distribute because of the wiring required to connect to the final OR.

A VHDL description of an arbitrary-width 3:1 multiplexer is shown in Figure 8.11. This design entity takes three k-bit data inputs, `a0`, `a1`, and `a2`, and a three-bit one-hot select input `s`, and generates a k-bit output `b`. The implementation uses a concurrent signal assignment statement that matches the gate implementation of Figure 8.10(a), with two differences. First, since this is a three-input multiplexer, there are three ANDs rather than four. Second, and more importantly, since this multiplexer is k-bits wide, each AND is k-bits wide – that is, k copies of a two-input AND.

To feed each bit of the select signal, e.g., `s(0)`, into a k-bit-wide AND gate, it must first be replicated to make a k-bit-wide signal, each bit of which is `s(0)`. This is accomplished using VHDL *array aggregate* notation to perform *signal replication*. Writing `(k-1 downto 0 => x)` makes `k` copies of signal `x` concatenated end-to-end. Thus, in this design entity we make k copies of select bit 0, `s(0)`, by writing `(k-1 downto 0 => s(0))`.

The `std_logic_vector` type is an example of a VHDL array type. Specifically, `std_logic_vector` is an array of elements of type `std_logic`. As shown above, one way to specify an array value is to use array aggregate notation. The syntax we employed above for forming an array aggregate was

```
(<range_specification> => <expression>)
```

Here `<range_specification>` can be a *discrete range* such as `k-1 downto 0` or the keyword **`others`**, and `<expression>` can be either a signal or a constant.

An alternative VHDL description of the 3:1 multiplexer using a `case` statement is given in Figure 8.12. This description is equivalent to that of Figure 8.11 for the cases where `s` is a one-hot signal, but is easier to read and understand. For the default case alternative, "`when others =>`", we use an array aggregate, "`(others => '-')`", to specify a k-bit-wide `std_logic_vector` value. The use of **`others`** for the range specification in the array aggregate indicates that all elements of the array take on the don't care value (`'-'`). The width of `(others => '-')` will be inferred to match the width signal `b` which is assigned to. We use this form rather than a bit string literal as the width of signal `b` can be changed by assigning a different value for generic parameter `k` which can be modified using a **`generic map`** clause during component instantiation.

```
-- three-input mux with one-hot select (arbitrary width)
library ieee;
use ieee.std_logic_1164.all;

entity Mux3a is
  generic( k : integer := 1 );
  port( a2, a1, a0 : in std_logic_vector( k-1 downto 0 ); -- inputs
        s : in std_logic_vector( 2 downto 0 ); -- one-hot select
        b : out std_logic_vector( k-1 downto 0 ) );
end Mux3a;

architecture case_impl of Mux3a is
begin
  process(all) begin
    case s is
      when "001" => b <= a0;
      when "010" => b <= a1;
      when "100" => b <= a2;
      when others => b <= (others => '-');
    end case;
  end process;
end case_impl;
```

Figure 8.12. VHDL description of an arbitrary-width 3 → 1 multiplexer using a `case` statement.

```
architecture select_impl of Mux3a is
begin
  with s select
    b <= a0 when "001",
         a1 when "010",
         a2 when "100",
         (others => '-') when others;
end select_impl;
```

Figure 8.13. VHDL description of an arbitrary-width $3 \rightarrow 1$ multiplexer using a selected signal assignment statement. Note entity declaration is in Figure 8.12.

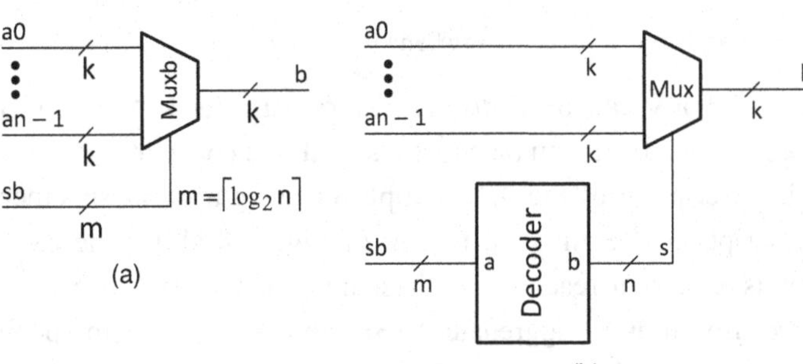

Figure 8.14. Binary-select multiplexer. (a) A k-bit-wide $n \rightarrow 1$ binary-select multiplexer selects input a_i, where i is the value of m-bit-wide binary select signal *sb*. (b) We can implement the binary-select multiplexer using a decoder and a normal (one-hot select) multiplexer.

A third alternative using a selected signal assignment is shown in Figure 8.13.[2]

Most standard-cell libraries provide multiplexers with one-hot select signals, and in most cases this is what we want – because our select signal is already in one-hot form. However, in some cases, it is desirable to have a multiplexer with a binary select signal. This may be because our select signal is in binary rather than one-hot form, or because we have to transmit our select signal over a long distance (or through a narrow pin interface) and want to economize on wiring.

Figure 8.14(a) shows the symbol for a binary-select multiplexer. This circuit takes an $m = \lceil \log_2 n \rceil$ bit binary select signal, *sb*, and selects one of the n input signals according to the binary value of *sb* – i.e., if $sb = i$, then a_i is selected.

We can implement a binary-select multiplexer using two blocks that we have already designed, as shown in Figure 8.14(b). We use an $m \rightarrow n$ decoder to decode binary select signal *sb* into a one-hot select signal *s* and then use a normal multiplexer (with one-hot select) to select the desired input.

The VHDL description of a k-bit wide 3:1 binary-select multiplexer is shown in Figure 8.15. The description exactly matches that shown in Figure 8.14(b). A 2:3 decoder is instantiated to convert a two-bit binary select *sb* to a three-bit one-hot select *s*. The one-hot select is then used in a normal (one-hot select) 3:1 multiplexer to select the desired input.

A schematic for a 4:1 binary-select multiplexer is shown in Figure 8.16. Each two-input AND gate at the decoder output has been combined with the two-input AND gate at the multiplexer input to form a single three-input AND gate. This fused implementation is more efficient than

[2] Depending on how the don't cares in Figure 8.12 and Figure 8.13 are interpreted during synthesis, these two descriptions can have different behavior than that given by Figure 8.11 in the case where `s` is not one-hot – i.e., when `s` is all zeros or has more than one bit set.

```
-- 3:1 multiplexer with binary select (arbitrary width)
library ieee;
use ieee.std_logic_1164.all;
use work.ch8.all;

entity Muxb3 is
  generic( k : integer := 1 );
  port( a2, a1, a0 : in std_logic_vector( k-1 downto 0 ); -- inputs
        sb : in std_logic_vector( 1 downto 0 ); -- binary select
        b : out std_logic_vector( k-1 downto 0 ) );
end Muxb3;

architecture struct_impl of Muxb3 is
  signal s: std_logic_vector(2 downto 0);
begin
  -- decoder converts binary to one-hot
  d: Dec generic map(2,3) port map(sb,s);
  -- multiplexer selects input
  mx: Mux3 generic map(k) port map(a2,a1,a0,s,b);
end struct_impl;
```

Figure 8.15. VHDL description of a binary-select 3:1 multiplexer using a decoder and a normal multiplexer.

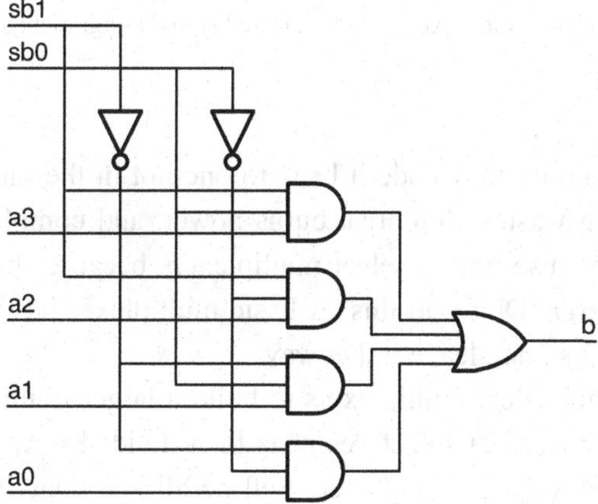

Figure 8.16. Schematic of a 4 → 1 binary-select multiplexer.

literally combining the two components. However, a VHDL description that combines two components, as in Figure 8.15, does not necessarily result in an inefficient implementation. A good synthesis program will generate a very efficient gate-level implementation from such a description. Again, the goal of the VHDL description is to be readable, maintainable, and synthesizable – optimization should be left to the synthesis tools.

VHDL descriptions of a binary-select multiplexer using `case` and selected signal assignment statements are shown in Figures 8.17 and 8.18, respectively. These are *behavioral* descriptions of the multiplexer, as opposed to the *structural* description in Figure 8.15, and are easier to read.

A common design error is to use a binary-select multiplexer in a situation where the select signal is originally in one-hot form (e.g., the output of an arbiter that determines which bidder gets access to a shared resource). All too often a designer takes such a one-hot select signal and

```
architecture case_impl of Muxb3 is
begin
  process(all) begin
    case sb is
      when "00" => b <= a0;
      when "01" => b <= a1;
      when "10" => b <= a2;
      when others => b <= (others => '-');
    end case;
  end process;
end case_impl;
```

Figure 8.17. VHDL description of a binary-select multiplexer using a `case` statement. Note entity declaration is in Figure 8.15.

```
architecture select_impl of Muxb3 is
begin
  with sb select
    b <= a0 when "00",
         a1 when "01",
         a2 when "10",
        (others => '-') when others;
end select_impl;
```

Figure 8.18. VHDL description of a binary-select multiplexer using a selected signal assignment statement. Note entity declaration is in Figure 8.15.

encodes it into binary (see Section 8.4) only to decode it back to one-hot in the multiplexer. Such gratuitous encoding and decoding wastes chip area, burns power, and complicates the VHDL description. Many designers overuse binary-select multiplexers because they equate *multiplexer* with *binary-select multiplexer*. Don't do this. A basic multiplexer has a one-hot select input. If you have a one-hot select signal, leave it that way.

We can combine several small one-hot select multiplexers to build a larger multiplexer by ORing their outputs together. The large one-hot select vector is then divided over the small multiplexers. The select signal for most of the multiplexers will be all zeros, giving a zero output. Only the small multiplexer with the selected input gets a one-hot select signal, which enables the selected input to propagate all the way to the output. For example, Figure 8.19 shows how a 6:1 multiplexer can be constructed from two 3:1 multiplexers. Note that for this ORing of small multiplexers to work, each small multiplexer must output a zero when its select input comprises all zeros. This may not be the case for tri-state multiplexers (Figure 8.10(b)) or with the don't care conditions in Figures 8.12 and 8.13.

In Figure 8.19 the select inputs to the 3:1 multiplexers use the VHDL *slice* notation to select a subset of bits from a bus (e.g., `s(2 downto 0)` selects the low-order three-bit fields from `s`). The syntax we used is

```
<array_signal_name>( <discrete_range> )
```

where `<array_signal_name>` is the name of a signal with array type (e.g., `std_logic_vector`) and `<discrete_range>` is an expression giving a range of index values such as

```
library ieee;
use ieee.std_logic_1164.all;
use work.ch8.all;

entity Mux6a is
  generic( k : integer := 1 );
  port( a5, a4, a3, a2, a1, a0 : in std_logic_vector(k-1 downto 0);
        s: in std_logic_vector(5 downto 0); -- one-hot select
        b: out std_logic_vector(k-1 downto 0) );
end Mux6a;

architecture impl of Mux6a is
  signal bb, ba : std_logic_vector(k-1 downto 0);
begin
  b <= ba or bb;
  MA: Mux3 generic map(k) port map(a2,a1,a0, s(2 downto 0), ba);
  MB: Mux3 generic map(k) port map(a5,a4,a3, s(5 downto 3), bb);
end impl;
```

Figure 8.19. Six-input multiplexer created by ORing the output of two three-input multiplexers.

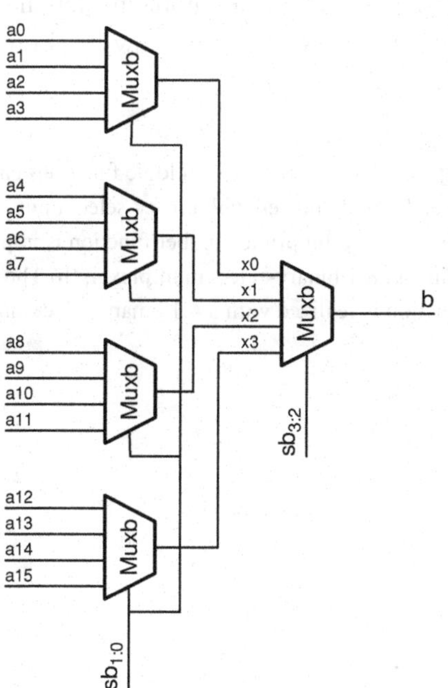

Figure 8.20. A large binary-select multiplexer is constructed from a tree of smaller binary-select multiplexers.

"`2 downto 0`" or "`8 to 15`". The *direction* (i.e., `downto` or `to`) of the slice must match the direction in the signal's declaration.

A large binary multiplexer is constructed as a tree of multiplexers as shown in Figure 8.20. The figure shows a 16:1 multiplexer constructed from five 4:1 multiplexers. A single 4:1 multiplexer at the output uses the most significant bits (MSBs) of the select signal $s_{3:2}$ to select between the four four-bit input groups. Within each input group, the least significant bits (LSBs) of the select signal $s_{1:0}$ select one of the four inputs in that group. For example, to select input $a11$, the select input is set to 1011, the binary code for 11. The MSBs are 10, which selects the $x2$ input on the output multiplexer and hence the signal group from $a8$ to $a11$. The

LSBs are 11, which selects $a11$ from within this group. One downside of this tree multiplexer structure, as compared with the one-hot approach of Figure 8.19, is that if the LSBs of the select signal switch, all four of the $x_{3:0}$ signals switch, and so dissipate energy, even though only one is needed.

An arbitrary n-input combinational logic function can be implemented using a binary-select $2^n \rightarrow 1$ multiplexer by placing the truth table of the function on the inputs of the multiplexer. The binary select input of the multiplexer acts as the input to the logic function and selects the appropriate entry of the truth table. This is shown in Figure 8.21(a) for the three-bit prime number function. We can actually realize any n-input logic function using only a 2^{n-1}-input multiplexer by factoring the function about the last input. This is shown for the three-bit prime number function in Figure 8.21(b). In effect, we split the truth table into two pieces – for $sb_2 = 0$ and $sb_2 = 1$. For each combination of the remaining inputs $sb_{1:0}$ we compare the two truth table halves. If the truth table is 0 (1) in both halves, we put a 0 (1) on the corresponding multiplexer input. However, if the function is 0 (1) when $sb_2 = 0$ and 1 (0) when $sb_2 = 1$, we put sb_2 ($\overline{sb_2}$) on the corresponding multiplexer input.

A VHDL design entity that implements the three-bit prime number function using a multiplexer is shown in Figure 8.22. The 2^n-input multiplexer implementation is chosen because it is easier to write, read, and maintain. Again, we leave the low-level optimization to the synthesis tools.

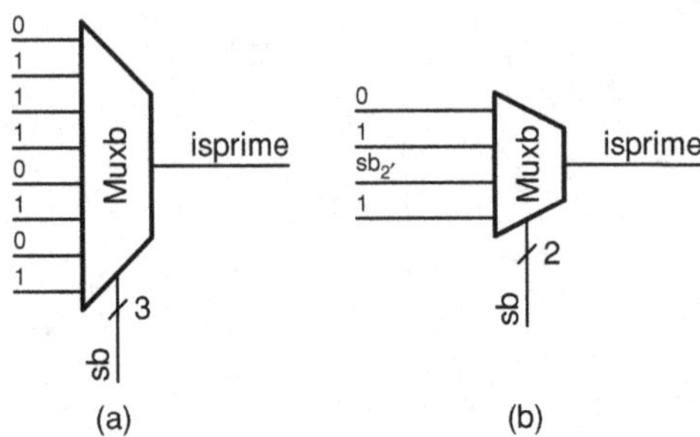

Figure 8.21. Combinational logic functions can be directly implemented with binary-select multiplexers. (a) The three-bit prime number function is implemented using an 8:1 binary-select multiplexer. (b) The same function is realized with a 4:1 binary-select multiplexer.

```
library ieee;
use ieee.std_logic_1164.all;
use work.ch8.all;

entity Primem is
  port( input : in std_logic_vector(2 downto 0);
        isprime : out std_logic );
end Primem;

architecture impl of Primem is
  signal b : std_logic_vector(0 downto 0);
begin
  M: Muxb8 generic map(1) port map("1","0","1","0","1","1","1","0",input,b);
  isprime <= b(0);
end impl;
```

Figure 8.22. VHDL description of the three-bit prime number function implemented using an 8:1 binary-select multiplexer with its data inputs set to the prime number truth table.

8.4 ENCODERS

The schematic symbol for an encoder, a logic module that converts a one-hot input signal into a binary-encoded output signal, is shown in Figure 8.23. An encoder is the inverse function of a decoder. It accepts an n-bit one-hot signal and generates an $m = \lceil \log_2 n \rceil$ bit binary output signal. The encoder will work properly only if its input signal is one-hot.

An encoder is implemented with an OR gate for each output as shown in Figure 8.24 for a 4:2 encoder. Each OR gate has as inputs all of those one-hot input bits that correspond to binary numbers with that output set. In Figure 8.24, for example, output $b1$ is true for binary codes 2 and 3. Hence it combines one-hot input bits $a2$ and $a3$.

A VHDL design entity for a 4:2 encoder described using a logic equation is shown in Figure 8.25. This description follows the schematic shown in Figure 8.24. The two-bit output signal `b` is assigned in a single statement using the concatenation operator '`&`' to combine two separate logic equations, one for each bit of `b`. We have seen the concatenation operator applied to strings in our testbench scripts (e.g., Figures 3.8, 7.18, 7.21, and 7.26). Here we apply the concatenation operator to two expressions, `(a(3) or a(2))` and `(a(3) or a(1))`, each of which produces a value of type `std_logic`. The result of this concatenation operation is a `std_logic_vector` of length two that can be assigned to the output `b`. The line

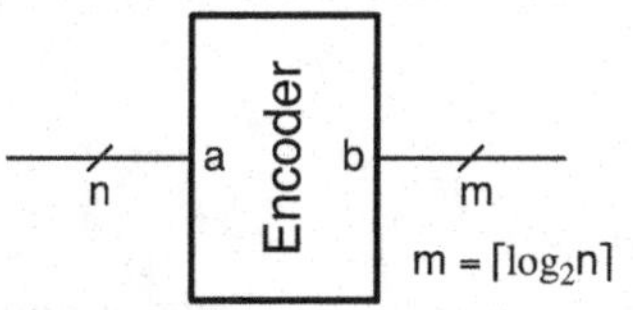

Figure 8.23. Schematic symbol for an $n \rightarrow m$ encoder that converts an n-bit one-hot signal to an $m = \lceil \log_2 n \rceil$ bit binary signal.

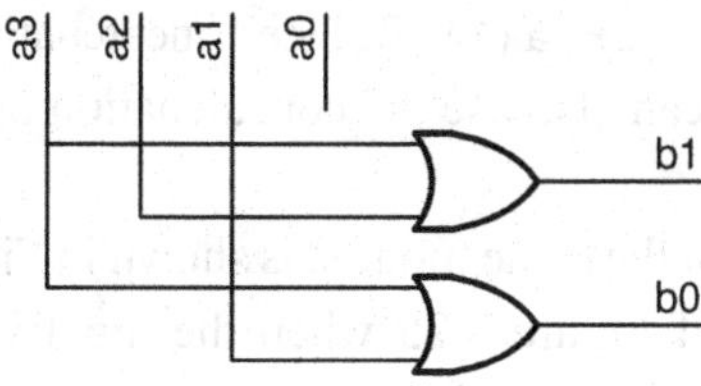

Figure 8.24. Schematic of a $4 \rightarrow 2$ encoder.

```
-- 4:2 encoder
library ieee;
use ieee.std_logic_1164.all;
entity Enc42 is
  port( a : in std_logic_vector(3 downto 0);
        b : out std_logic_vector(1 downto 0) );
end Enc42;
architecture impl of Enc42 is
begin
  b <= (a(3) or a(2)) & (a(3) or a(1));
end impl;
```

Figure 8.25. VHDL description of a 4:2 encoder.

```
-- 4:2 encoder
library ieee;
use ieee.std_logic_1164.all;

entity Enc42b is
  port( a : in std_logic_vector( 3 downto 0 );
        b : out std_logic_vector( 1 downto 0 ) );
end Enc42b;

architecture impl of Enc42b is
begin
  process(all) begin
    case a is
      when "0001" => b <= "00";
      when "0010" => b <= "01";
      when "0100" => b <= "10";
      when "1000" => b <= "11";
      when "0000" => b <= "00"; -- to facilitate large encoders
      when others => b <= "--";
    end case;
  end process;
end impl;
```

Figure 8.26. Behavioral VHDL description of a 4:2 encoder.

"`b <= (a(3) or a(2)) & (a(3) or a(1));`" is equivalent to, but more compact than, writing

```
b(1) <= a(3) or a(2);
b(0) <= a(3) or a(1);
```

The leftmost bit in the concatenation expression, "`(a(3) or a(2))`", is assigned to the leftmost bit of the signal `b` irrespective of the index value for the leftmost bit. We put parentheses around "`(a(3) or a(2))`" and "`(a(3) or a(1))`" since concatenation (`&`) has higher precedence than the logical operator **`or`**. We can also use the concatenation operator on `std_logic_vector` values.

For comparison, a behavioral VHDL description of the same module is shown in Figure 8.26. This design entity has the same behavior as that of Figure 8.25 when the input is one-hot. For other non-zero inputs, the behavioral design entity sets the output to a don't care state (`'-'`) while the logical design entity generates an output by ORing together the input bits. The behavioral description is preferred – both to detect the illegal input state during simulation and to give the synthesizer the freedom to exploit the don't cares to minimize the resulting logic.

We have coded this encoder to output 0 when the input comprises all zeros. This facilitates construction of large encoders from multiple small encoders as described below. If this feature is not needed, the zero line can be deleted from the case statement.

A large encoder can be constructed from a tree of smaller encoders, as shown in Figure 8.27. The figure shows a 16:4 encoder constructed from 4:2 encoders. To allow this composition, a summary output must be added to each 4:2 encoder. This output is true if any input of the encoder is true. The LSBs of the output are generated by ORing together[3] the outputs of the

[3] This is why we assign the output `"00"` instead of `"-"` to the input `"0000"` in Figure 8.26.

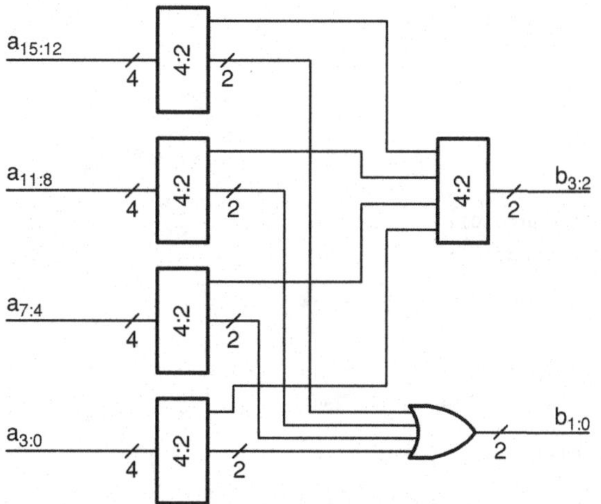

Figure 8.27. Large encoders can be constructed from a tree of smaller encoders. Each small encoder requires an additional output that indicates whether any of its inputs are true.

4:2 encoders connected directly to the inputs. The MSBs of the outputs are generated by an additional 4:2 encoder that encodes the summary outputs of the input encoders. The summary output of this final encoder (not shown) could be used as a summary output of the entire 16:4 encoder. Still larger encoders can be constructed by building additional levels on the tree.

To understand how the tree encoder works, consider the case where input a_9 is true and all other inputs are false. The third (from the bottom) input encoder has an input of 0010, and hence outputs 01, and sets its summary output true. All other input encoders have zero inputs and hence generate 00 outputs. Since all other outputs are 00, the output of the third encoder passes through the two OR gates to generate directly the two LSBs of the output, $b_{1:0} = 01$. The MSB encoder has an input of 0100, since only the third summary output is true. Hence it generates an output of $b_{3:2} = 10$, giving an overall output of $b_{3:0} = 1001$, which is 9 in decimal.

VHDL code for the tree encoder is shown in Figure 8.28. In the code for the 4:2 encoder with summary output `Enc42a`, the statement that generates summary output `c` uses the function `or_reduce(x)` defined in `std_logic_misc`. This function returns the result of ORing all of the bits of signal `x` together.

EXAMPLE 8.2 Thermometer encoder

Write a VHDL design entity that encodes a four-bit thermometer code to a binary number reflecting the number of 1s in the code. The output should be specified as don't care if the input is not a thermometer code. See Figure 8.29.

8.5 ARBITERS AND PRIORITY ENCODERS

Figure 8.30 shows the schematic symbol for an *arbiter*, which is sometimes called a *find-first-one* (FF1) unit. This circuit accepts an arbitrary input signal and outputs a one-hot signal that has its sole 1 in the position of the least significant 1 in the input signal. For example, if the

```
-- 4 to 2 encoder - with summary output
library ieee;
use ieee.std_logic_1164.all;
use ieee.std_logic_misc.all;

entity Enc42a is
  port( a : in std_logic_vector(3 downto 0);
        b : out std_logic_vector(1 downto 0);
        c : out std_logic );
end Enc42a;

architecture impl of Enc42a is
begin
  b <= (a(3) or a(2)) & (a(3) or a(1));
  c <= or_reduce(a);
end impl;

-- factored encoder
library ieee;
use ieee.std_logic_1164.all;
use work.ch8.all; -- for Enc42a component declaration

entity Enc164 is
  port( a : in std_logic_vector( 15 downto 0 );
        b : out std_logic_vector( 3 downto 0 ) );
end Enc164;

architecture impl of Enc164 is
  signal c: std_logic_vector(7 downto 0); -- intermediate result of first stage
  signal d: std_logic_vector(3 downto 0); -- if any set in group of four
begin
  --  four LSB encoders each include 4-bits of the input
  E0: Enc42a port map( a(3 downto 0), c(1 downto 0), d(0) );
  E1: Enc42a port map( a(7 downto 4), c(3 downto 2), d(1) );
  E2: Enc42a port map( a(11 downto 8), c(5 downto 4), d(2) );
  E3: Enc42a port map( a(15 downto 12), c(7 downto 6), d(3) );

  --  MSB encoder takes summaries and gives msb of output
  E4: Enc42 port map( d(3 downto 0), b(3 downto 2) );

  --  two OR gates combine output of LSB encoders
  b(1) <= c(1) or c(3) or c(5) or c(7);
  b(0) <= c(0) or c(2) or c(4) or c(6);
end impl;
```

Figure 8.28. VHDL code for a 16:4 encoder built as a tree of 4:2 encoders with a summary output (component Enc42a). The summary output is true if any input of the design entity is true.

input to an eight-bit arbiter were 01011100, the output would be 00000100, since the least significant 1 in the input is in bit 2. In some applications we reverse the arbiter and look for the most significant 1. For the remainder of this section, however, we focus on arbiters that look for the least significant 1 in the input signal.

```
library ieee;
use ieee.std_logic_1164.all;

entity ThermometerEncoder is
  port( a: in std_logic_vector(3 downto 0); -- thermometer coded input
        b: out std_logic_vector(2 downto 0)); -- # of 1s in input (if legal)
end ThermometerEncoder;

architecture impl of ThermometerEncoder is
begin
  process(all) begin
    case a is
      when "0000" => b <= 3d"0";
      when "0001" => b <= 3d"1";
      when "0011" => b <= 3d"2";
      when "0111" => b <= 3d"3";
      when "1111" => b <= 3d"4";
      when others => b <= "---";
    end case;
  end process;
end impl;
```

Figure 8.29. VHDL for a thermometer to binary encoder.

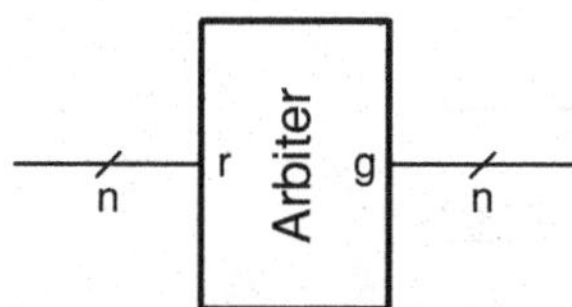

Figure 8.30. Schematic symbol for an arbiter.

Arbiters are used in digital systems to arbitrate requests for shared resources. For example, if n units share a bus that only one can use at a time, an n-input arbiter is used to determine which unit gets access to the bus during a given cycle (see Section 24.2).[4] Another use of an arbiter is in arithmetic circuits where to *normalize* numbers we need to find the position of the most significant 1 (see Section 11.3.3). In this application, they are called find-first-one units, because there is no arbitration going on and they are reversed (compared with what we discuss here) to find the most significant 1.

An arbiter can be constructed as an *iterative circuit*. That is, we can design the logic for one bit of the arbiter and repeat (or iterate) it. Figure 8.31 shows the logic for one bit (bit i) of the arbiter. One AND gate generates the grant output for this bit, g_i. The grant is set high if the request, r_i, is high and no 1 has been found so far, as signaled by a 1 on the top input. The second AND gate signals downstream bits if no 1 has been found by this stage or any previous stage. We will see many examples of iterative circuits in our study of digital design. They are widely used in arithmetic circuits (see Chapter 10).

To build a four-bit arbiter, for example, we connect four copies of the bit cell of Figure 8.31 and connect the top input of the first cell to "1." The resulting circuit is shown

[4] In this application we would typically use an arbiter with rotating priority so that the resource is shared fairly. With a fixed-priority arbiter, the unit connected to the least significant input gets an unfair advantage.

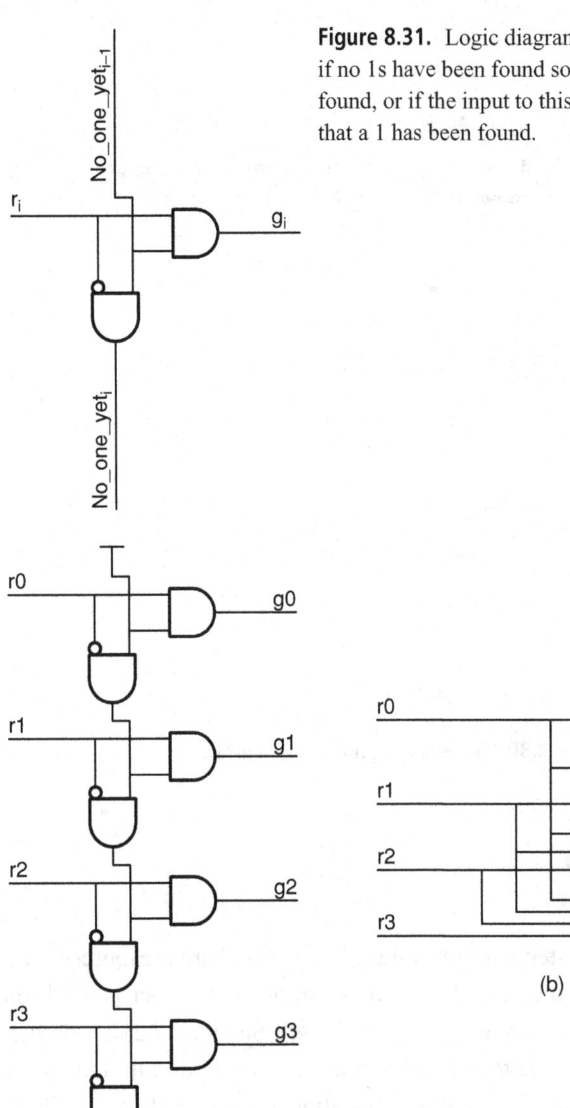

Figure 8.31. Logic diagram for one bit of an arbiter. The output is true only if no 1s have been found so far, and the input is 1. If a 1 has previously been found, or if the input to this stage is 1, an output signal informs other stages that a 1 has been found.

Figure 8.32. Two implementations of a four-bit arbiter: (a) using the bit-cell of Figure 8.31; (b) using *look-ahead*.

in Figure 8.32(a). The vertical chain of AND gates scans over the request inputs until the first 1 is found and disables all outputs below this input. The output AND gates pass the first input 1 to the output and force zeros on all outputs below this first 1.

The linear chain of AND gates in this circuit gives a delay that increases linearly with the number of inputs. In some applications this delay can be prohibitive. We can shorten the delay somewhat by flattening the logic, as shown in Figure 8.32(b). This technique is often called *look-ahead* since the gate that generates output `g3` is looking ahead at inputs `r0`, `r1`, and `r2` rather than waiting for their effects to propagate through a chain of gates. We will look at more scalable approaches to look-ahead for iterative circuits in Section 12.1.

```
-- arbiter (arbitrary width) - LSB is highest priority
library ieee;
use ieee.std_logic_1164.all;
entity Arb is
  generic( n: integer := 8 );
  port( r: in std_logic_vector(n-1 downto 0); g: out std_logic_vector(n-1 downto 0) );
end Arb;
architecture impl of Arb is
  signal c : std_logic_vector(n-1 downto 0);
begin
  c <= (not r(n-2 downto 0) and c(n-2 downto 0)) & '1';
  g <= r and c;
end impl;
---------------------------------------------------------------------------------
-- arbiter (arbitrary width) - MSB is highest priority
library ieee;
use ieee.std_logic_1164.all;
entity RArb is
  generic( n: integer := 8 );
  port( r: in std_logic_vector(n-1 downto 0); g: out std_logic_vector(n-1 downto 0) );
end RArb;
architecture impl of RArb is
  signal c : std_logic_vector(n-1 downto 0);
begin
  c <= '1' & (not r(n-1 downto 1) and c(n-1 downto 1));
  g <= r and c;
end impl;
```

Figure 8.33. VHDL description of two arbitrary-width fixed-priority arbiters. `Arb` finds the least significant 1 and `RArb` finds the most significant 1.

VHDL descriptions of two arbiters are shown in Figure 8.33. One arbiter, `Arb`, finds the least significant 1 and the other, `RArb`, finds the most significant 1. The implementation of `Arb` directly follows Figure 8.32(a). Signal `g` is generated by ANDing the 'no 1s so far' signal `c` with the request input `r`. The 'no 1s so far' signal, `c`, is generated by using a concatenation to set its LSB to 1 (always enabling output `g0`) and to set the remaining bits to `c(i) = not r(i-1) and c(i-1)`. At first this definition looks circular because it appears that we are defining `c` in terms of itself. On closer examination we realize that each bit of `c` depends only on less significant bits of `c`. The definition is not circular.

For comparison, Figure 8.34 shows a behavioral description of a four-bit arbiter that finds the least significant 1 in input `a`. This description uses a **`case?`** statement to compress the 16-line truth table to five lines. The default case ("**`when others`** `=>`") should never occur. However, we must include a default case because the `std_logic` type includes non-synthesizable values (e.g., `'U'` for "uninitialized" and `'X'` for "unknown value") and VHDL requires that all possible input combinations (including those containing non-synthesizable values) are covered in a **`case`** or **`case?`** statement.

One use of an arbiter is to build a *priority encoder*, as shown in Figure 8.35. A priority encoder takes an n-bit input signal `a` and outputs an $m = \lceil \log_2 n \rceil$ bit binary signal `b` that

```
-- arbiter four bits wide - LSB is highest priority
library ieee;
use ieee.std_logic_1164.all;

entity Arb_4b is
  port( r : in std_logic_vector( 3 downto 0 );
        g : out std_logic_vector( 3 downto 0 ) );
end Arb_4b;

architecture impl of Arb_4b is
begin
  process(r) begin
    case? r is
      when "0000" => g <= "0000";
      when "---1" => g <= "0001";
      when "--10" => g <= "0010";
      when "-100" => g <= "0100";
      when "1000" => g <= "1000";
      when others => g <= "----";
    end case?;
  end process;
end impl;
```

Figure 8.34. Behavioral VHDL description of a four-bit arbiter that finds the least significant 1.

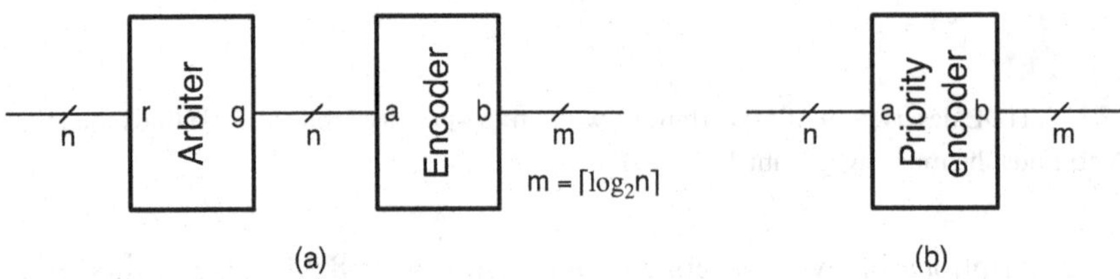

Figure 8.35. (a) A priority encoder realized by connecting an arbiter to an encoder. (b) Schematic symbol for a priority encoder.

indicates the position of the first one bit in a. The priority encoder operates in two steps, as shown in Figure 8.35(a). First, an arbiter finds the first one bit in input a and outputs a one-hot signal, g, with only this single bit set. Then, one-hot signal g is converted into a binary signal b by an encoder. A VHDL description of the priority encoder is shown in Figure 8.36.

When input a is 0, the arbiter will output g = 0, which is not a one-hot signal. In this case, the encoder, if it is constructed from OR gates as described in Section 8.4, will also output b = 0, which is often acceptable. In some applications, however, this all-zero state must be detected and a special code output to distinguish the case when the input is 0 from the case where the first bit set is bit zero. This is easily accomplished by making the arbiter one bit wider than otherwise needed and applying a constant 1 signal to the last input bit.

A behavioral VHDL description of a priority encoder is shown in Figure 8.37. This circuit outputs don't cares "---", not 0, in the case where the input is r = 8b"0". It can easily be modified by adding one line to the **case?** statement to assign any value desired to b (or to an auxiliary output) for the case where r = 8b"0".

```
-- 8:3 priority encoder
library ieee;
use ieee.std_logic_1164.all;
use work.ch8.all;

entity PriorityEncoder83 is
  port( r: in std_logic_vector(7 downto 0);
        b: out std_logic_vector(2 downto 0) );
end PriorityEncoder83;

architecture impl of PriorityEncoder83 is
  signal g: std_logic_vector(7 downto 0);
begin
  A: Arb port map(r,g);
  E: Enc83 port map(g,b);
end impl;
```

Figure 8.36. VHDL description of a priority encoder.

```
-- 8:3 priority encoder
library ieee;
use ieee.std_logic_1164.all;

entity PriorityEncoder83b is
  port( r: in std_logic_vector( 7 downto 0 );
        b: out std_logic_vector( 2 downto 0 ) );
end PriorityEncoder83b;

architecture impl of PriorityEncoder83b is
begin
  process(all) begin
    case? r is
      when "-------1" => b <= 3d"0";
      when "------10" => b <= 3d"1";
      when "-----100" => b <= 3d"2";
      when "----1000" => b <= 3d"3";
      when "---10000" => b <= 3d"4";
      when "--100000" => b <= 3d"5";
      when "-1000000" => b <= 3d"6";
      when "10000000" => b <= 3d"7";
      when others => b <= (others => '-');
    end case?;
  end process;
end impl;
```

Figure 8.37. Behavioral VHDL description of a priority encoder.

EXAMPLE 8.3 Programmable-priority arbiter

Using slice notation, write a VHDL expression for a programmable-priority arbiter. Your arbiter should accept a one-hot priority signal `p` that determines the bit of `r` having the highest priority.

Priority should decrease leftward from this bit position, wrapping around from bit $n-1$ to bit 0. For example, for an eight-bit arbiter, if bit 6 of p is set, r(6) has the highest priority, r(7) has the second highest priority, r(0) has the third highest priority, and so on. To facilitate timing verification, your design should not include cyclic logic.

We can achieve the desired function with cyclic logic by writing the following two statements:

```
c <= p or (not (r(n-2 downto 0) & r(n-1)) and c(n-2 downto 0) & c(n-1));
g <= r and c;
```

Note that in VHDL **not** has higher precedence than concatenation (&), which in turn has higher precedence than **and**. The first statement computes a per-bit carry signal c by starting with a "1" bit in the position specified by p and then propagating this carry to the left, cyclically, as long as the corresponding bit of r is low. This finds the highest priority request. Unfortunately this logic is cyclic and hence not compatible with modern timing verification tools.

To make the logic acyclic, we duplicate the carry chain, making it $2n$ long, so we can propagate a carry out of bit (n-1) without creating a cycle.

```
c <= ((n-1 downto 0 => '0') & p) or
     (not (r(n-2 downto 0) & r & r(n-1)) and c(2*n-2 downto 0) & '0');
g <= r and (c(2*n-1 downto n) or c(n-1 downto 0));
```

8.6 COMPARATORS

Figure 8.38 shows the schematic symbol for an *equality comparator*. This component accepts two n-bit binary inputs a and b and outputs a one-bit signal that indicates whether a = b; i.e., whether each bit of a is equal to the corresponding bit of b.

Figure 8.39 shows a logic diagram for a four-bit equality comparator. An array of exclusive-NOR (XNOR) gates compares individual bits of the input signals. The output of each XNOR

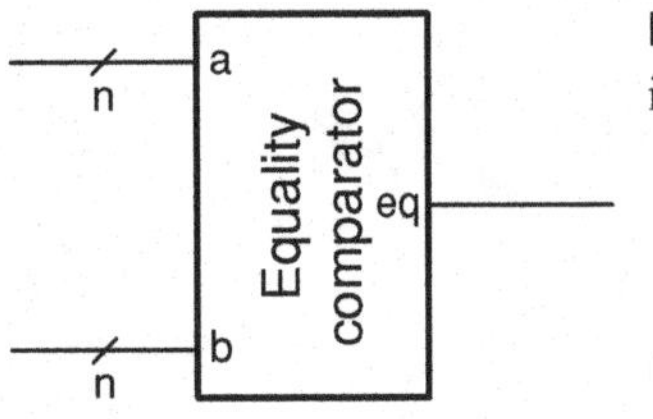

Figure 8.38. Schematic symbol for an equality comparator. The eq output is true if a = b.

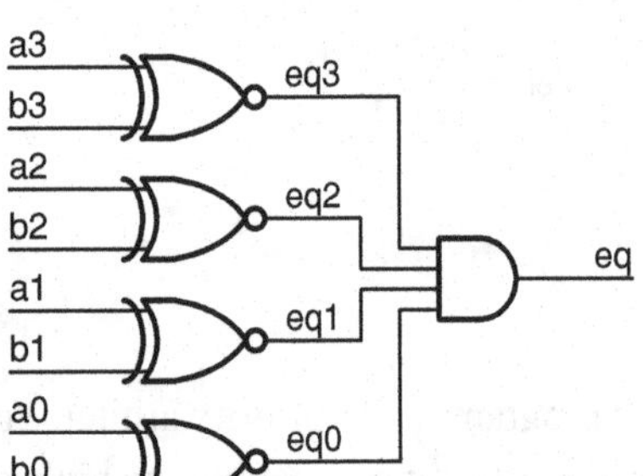

Figure 8.39. Logic diagram of a four-bit equality comparator. Exclusive-NOR (XNOR) gates are used to compare individual bits of inputs a and b. An AND gate combines the bit-by-bit comparisons and signals true if all bits are equal.

```
-- Equality comparator
library ieee;
use ieee.std_logic_1164.all;

entity EqComp is
  generic( k: integer := 8 );
  port( a, b: in std_logic_vector(k-1 downto 0);
        eq: out std_logic );
end EqComp;

architecture impl of EqComp is
begin
  eq <= '1' when a = b else '0';
end impl;
```

Figure 8.40. VHDL description of an equality comparator using a conditional signal assignment statement.

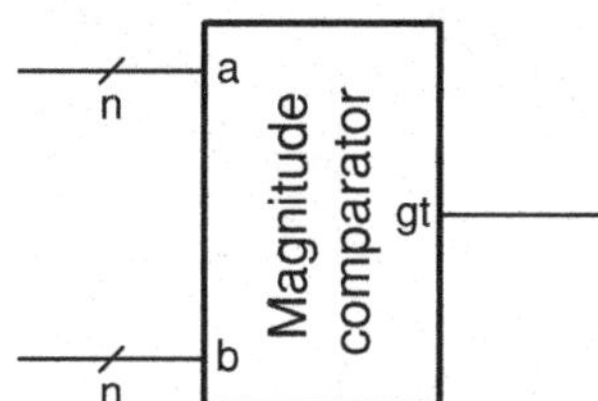

Figure 8.41. Schematic symbol for an magnitude comparator. The `gt` output is true if `a` > `b`.

gate is high if its two inputs are equal, so signal eq_i is true if $a_i = b_i$. An AND gate combines the eq_i signals and outputs true only if all bits are equal. Alternatively, we can design the equality comparator as an iterative circuit by linearly scanning the bits to determine that all are equal.

A VHDL description of an equality comparator is shown in Figure 8.40. Here we use a VHDL conditional signal assignment statement combined with the equality operator defined for `std_logic_vector`. An alternative implementation of equality would use the XNOR operator to perform a bitwise comparison as follows:

```
eq <= and_reduce(a xnor b);
```

Here we use the `and_reduce(x)` function defined in `ieee.std_logic_misc` to combine the bits of the XNOR without the need to declare an intermediate signal.

A *magnitude comparator* is a component that compares the relative magnitude of two binary numbers. Strictly speaking, this is an arithmetic circuit, since it treats its inputs as numbers, and thus we should defer its treatment until Chapter 10. We present it here because it is an excellent example of an iterative circuit.

Figure 8.41 shows a schematic symbol for a magnitude comparator. The single bit output `gt` is true if n-bit input `a` is greater than n-bit input `b`. One binary signal is greater than another if it has a 1 in the most significant position in which the two numbers are not equal.

We can structure two different iterative circuits for the magnitude comparator, as shown in Figure 8.42. In Figure 8.42(a) we scan from LSB to MSB to find the most significant bit in which the two numbers disagree. In this design we propagate a signal *gtb* (greater than below). Signal gtb_i, if set, indicates that from the LSB through bit $i-1$ signal `a` is greater than b, i.e., gtb_i implies $a_{i-1:0} > b_{i-1:0}$. Bit gtb_{i+1} is set if $a_i > b_i$ or if $a_i = b_i$ and $a_{i-1:0} > b_{i-1:0}$.

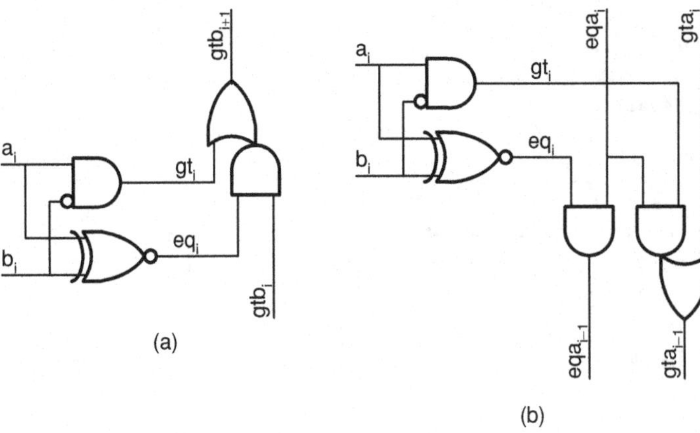

Figure 8.42. Two iterative implementations of the magnitude comparator. (a) LSB first; a greater than below, *gtb*, signal is propagated upward. (b) MSB first, two signals: greater than above, *gta*, and equal above, *eqa*, are propagated downward.

```
library ieee;
use ieee.std_logic_1164.all;

entity MagComp is
  generic( k: integer := 8 );
  port( a, b: in std_logic_vector(k-1 downto 0);
        gt: out std_logic );
end MagComp;

architecture impl of MagComp is
  signal eqi, gti : std_logic_vector(k-1 downto 0);
  signal gtb: std_logic_vector(k downto 0);
begin
  eqi <= a xnor b;
  gti <= a and not b;
  gtb <= (gti or (eqi and gtb(k-1 downto 0))) & '0';
  gt <= gtb(k);
end impl;
```

Figure 8.43. VHDL description of an LSB-first magnitude comparator.

The gtb_n signal out of the most significant bit gives the required answer since it indicates that $a_{n-1:0} > b_{n-1:0}$. A VHDL description of this LSB-first magnitude comparator is shown in Figure 8.43.

An alternative iterative implementation of a magnitude comparator that operates MSB first is shown in Figure 8.42(b). Here we have to propagate two signals between each bit position. Signal gta_i (greater than above) indicates that $a > b$ for bits more significant than the current bit, i.e., $a_{n-1:i+1} > b_{n-1:i+1}$. Similarly, eqa_i (equal above) indicates that $a_{n-1:i+1} = b_{n-1:i+1}$. These two signals scan the bits from MSB to LSB. As soon as a difference is found, we know the answer. If the first difference is a bit where $a > b$, we set gta_{i-1} and clear eqa_{i-1} out of this bit position, and these values propagate all the way to the output. On the other hand, if $b > a$ in the first bit that differs, we clear eqa_{i-1} but leave gta_{i-1} low. These signals also propagate all the way to the output. The output is the signal gta_{-1}.

A behavioral VHDL description of a magnitude comparator is given in Figure 8.44. This design entity is much simpler to understand than that of Figure 8.43. Most synthesis

```
-- Behavioral Magnitude comparator
library ieee;
use ieee.std_logic_1164.all;

entity MagComp_b is
  generic( k: integer := 8 );
  port( a, b: in std_logic_vector(k-1 downto 0);
        gt: out std_logic );
end MagComp_b;

architecture impl of MagComp_b is
begin
  gt <= '1' when a > b else '0';
end impl;
```

Figure 8.44. Behavioral VHDL description of a magnitude comparator.

tools will generate very good logic from this code, using look-ahead techniques when needed to meet the specified timing constraints. The equality comparator of Figure 8.40 is already in behavioral form.

EXAMPLE 8.4 Three-way equality comparator

Write a VHDL expression for a three-way equality comparator that outputs true if three inputs, a, b, and c, are equal to one another.

The VHDL code is shown below. It suffices to compare two pairs of inputs. Note that equality (=) has higher operator precedence than **and**. The VHDL code is

```
eq <= '1' when a = b and a = c else '0';
```

8.7 SHIFTERS

A shifter is a combinational block that takes a bit field a and a shift count n and outputs a shifted by n positions. Left-shifters shift left, right-shifters shift right, *barrel-shifters* rotate input *a* (wrapping the bits that fall off the left (or right) so they appear entering on the right (or left)) and *funnel-shifters* select a small bit field from a specified position of a larger bit field.

Figure 8.45 shows the VHDL code for a left-shifter. With the default parameters, the design entity accepts an eight-bit input a and a three-bit shift count n and produces a 15-bit output b equal to a shifted left by n bit positions. With the maximum shift count n = 7, the input bit field is left-aligned in output b. The bit field is right-aligned in b with the minimum shift count.

Figure 8.46 shows the VHDL code for a barrel-shifter. This design entity performs a left shift of a on signal x and then ORs the high bits of x with the low bits of x to generate the wrapped output b.

We leave the description of a funnel-shifter to Exercise 8.17.

```
library ieee;
use ieee.std_logic_1164.all;
use ieee.numeric_std.all;

entity ShiftLeft is
  generic( k: integer := 8; lk : integer := 3 );
  port( n: in std_logic_vector(lk-1 downto 0); -- how much to shift
        a: in std_logic_vector(k-1 downto 0); -- number to shift
        b: out std_logic_vector(2*k-2 downto 0) ); -- the output
end ShiftLeft;

architecture impl of ShiftLeft is
  signal input: unsigned(2*k-2 downto 0);
  signal shift: integer;
begin
  input <= unsigned((2*k-2 downto k => '0') & a);
  shift <= to_integer(unsigned(n));
  b <= std_logic_vector( input sll shift );
end impl;
```

Figure 8.45. VHDL description of a left-shifter.

```
library ieee;
use ieee.std_logic_1164.all;
use ieee.numeric_std.all;

entity BarrelShift is
  generic( k: integer := 8; lk: integer := 3 );
  port( n: in std_logic_vector(lk-1 downto 0); -- how much to shift
        a: in std_logic_vector(k-1 downto 0); -- number to shift
        b: out std_logic_vector(k-1 downto 0) ); -- the output
end BarrelShift;

architecture impl of BarrelShift is
  signal shift_amt: integer; -- amount to shift
  signal input: unsigned(2*k-2 downto 0); -- zero padded input
  signal shift_out: std_logic_vector(2*k-2 downto 0); -- output before wrapping
begin
  input <= unsigned( (2*k-2 downto k => '0') & a );
  shift_amt <= to_integer(unsigned(n));
  shift_out <= std_logic_vector( input sll shift_amt );
  b <= shift_out(k-1 downto 0) or ('0' & shift_out(2*k-2 downto k));
end impl;
```

Figure 8.46. VHDL description of a barrel-shifter.

8.8 READ-ONLY MEMORIES

A *read-only memory*, or ROM, is a component that implements a look-up table. It accepts an address as input, and outputs the value stored in the table at that address. The ROM is *read-only* because the values stored in the table are predetermined – hard-wired at the time the ROM is manufactured – and cannot be changed. In Section 8.9 we will examine *read–write memories*, where the table entries can be changed. A discussion of memories from a systems perspective is given in Chapter 25.

The schematic symbol for a ROM is shown in Figure 8.47. For an N-word × b-bit ROM, an $n = \lceil \log_2 N \rceil$ bit address signal a selects a word of the table. The b-bit value stored in that word is output on data output d.

A ROM can implement an arbitrary logic function by storing the truth table of that function in the ROM. For example, we can implement a seven-segment decoder with a ten-word × seven-bit ROM. The value 1111110, the segment pattern for 0, is placed in the first location (location 0), the value 0110000, the segment pattern for 1, is placed in the second location (location 1), and so on.

A simple implementation of a ROM using a decoder and tri-state buffers is shown in Figure 8.48. An $n \rightarrow N$ decoder decodes the n-bit binary address a into an N-bit one-hot word select signal, w. Each bit of this word select signal is connected to a tri-state gate. When an address $a = i$ is applied to the ROM, word select signal w_i goes high and enables the corresponding tri-state buffer to drive table entry d_i onto the output.

For large ROMs, the one-dimensional ROM structure of Figure 8.48 becomes unwieldy and inefficient. The decoder becomes very large – requiring N AND gates. Above a certain size, it is more efficient to construct a ROM as a two-dimensional array of cells, as shown in

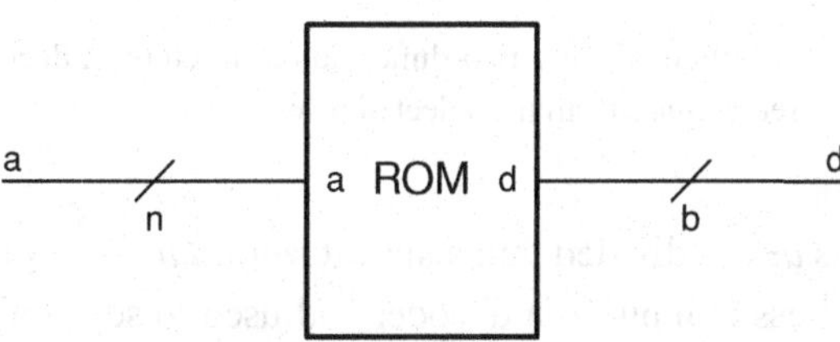

Figure 8.47. Schematic symbol for a ROM. The n-bit address a selects a location in a table. The value stored in that location is output on the b-bit data output d.

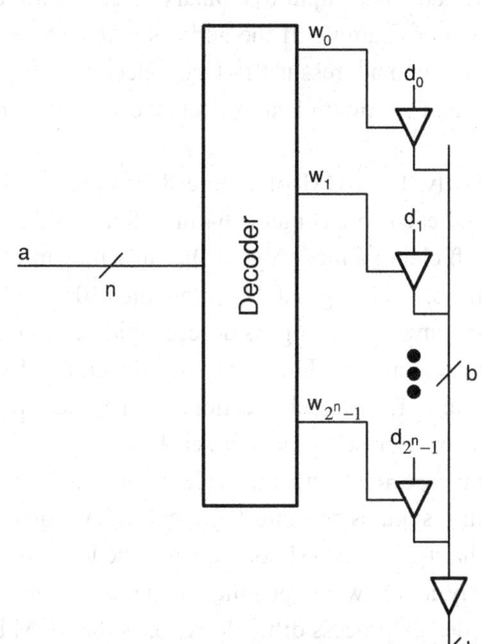

Figure 8.48. A ROM can be implemented with a decoder and a set of tri-state gates with constants connected to their inputs. The address is decoded to select one of the tri-state gates. That gate drives its corresponding value onto the output.

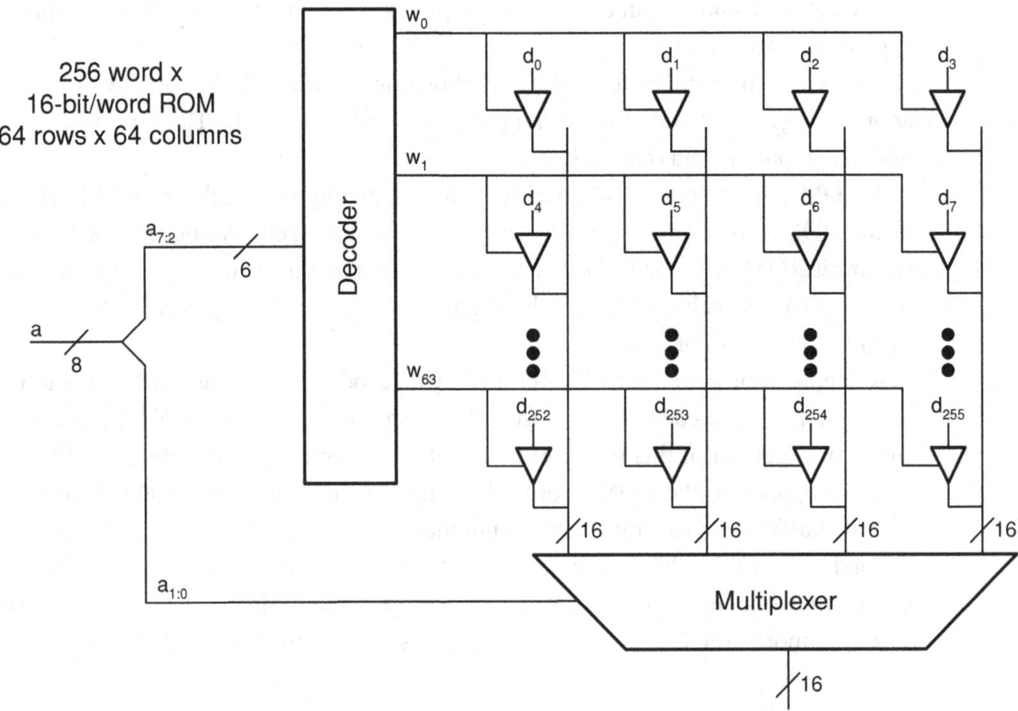

Figure 8.49. A ROM can be implemented more efficiently as a two-dimensional structure. A decoder selects a *row* of words. A multiplexer selects the desired *column* from the selected row.

Figure 8.49. Here the eight-bit address $a_{7:0}$ is divided into a six-bit *row address* $a_{7:2}$ and a two-bit *column address* $a_{1:0}$. The row address is input to a decoder and used to select a row via a 64-bit one-hot select signal w. The column address is input to a binary-select multiplexer that selects the appropriate word from the row. For example, if the address is $a = 5 = 000101$, then the row address is $0001 = 1$, and the column address is 01. Thus select line w_1 goes high to select the row containing d_4 through d_7, and the multiplexer selects the second word of this line, d_5.

While the address bits are not split evenly, the ROM of Figure 8.49 is realized using a *square* array of bits. There are 64 rows, and each row contains 64 bits; four words of 16 bits each. Square arrays tend to give the most efficient memory layouts because they minimize the peripheral (decoder and multiplexer) overhead. Very large ROMs (more than $10^5 - 10^6$ bits) are built from multiple arrays using either bit-slicing or banking, as described in Section 25.2.

In practice, ROMs employ highly optimized circuits. For clarity, we illustrate a ROM here with conventional logic symbols – a tri-state buffer at each location. In fact, most ROMs use circuits that require only a single transistor (or its absence) to store each bit.

A ROM can be modeled in VHDL using a `case` statement, as shown in Figure 8.50. Such a block can either be implemented by being synthesized into logic gates that implement the ROM functionality, or be replaced with a hard-wired ROM component. The logic synthesized for this block will have a different area, delay, and power depending on the ROM contents. This can make modifying ROM data late in the design process difficult. Also, as the ROM becomes larger, the size of the `case` statement and the potential for designer errors increases.

```
-- rom(fixed width) built with a case statement
library ieee;
use ieee.std_logic_1164.all;

entity rom_case is
  port( a : in std_logic_vector(3 downto 0);
        d : out std_logic_vector(7 downto 0) );
end rom_case;

architecture impl of rom_case is
begin
  process(all) begin
    case a is
      when x"0" => d <= x"00";
      when x"1" => d <= x"11";
      when x"2" => d <= x"22";
      when x"3" => d <= x"33";
      when x"4" => d <= x"44";
      when x"5" => d <= x"12";
      when x"6" => d <= x"34";
      when x"7" => d <= x"56";
      when x"8" => d <= x"78";
      when x"9" => d <= x"9a";
      when x"a" => d <= x"bc";
      when x"b" => d <= x"de";
      when x"c" => d <= x"f0";
      when x"d" => d <= x"12";
      when x"e" => d <= x"34";
      when x"f" => d <= x"56";
      when others => d <=x"00";
    end case;
  end process;
end impl;
```

Figure 8.50. VHDL description of a ROM built using a `case` statement. Because it is not parameterized or easily modified, we recommend using the ROM of Figure 8.51.

Alternatively, a ROM can be modeled using an array of `std_logic_vector` values that are initialized from a file, as shown in Figure 8.51. Unfortunately, VHDL as defined in IEEE Standard 1076 does not provide a standardized approach for initializing a ROM from a file. As a consequence, different tool vendors have implemented support for this important feature using different approaches. Figure 8.51 includes support for two common approaches, which are described below. Before discussing how the ROM is initialized, we first discuss the syntax used to declare the ROM.

The ROM itself is declared as a VHDL array using the following three lines:

```
subtype word_t is std_logic_vector(data_width-1 downto 0);
type mem_t is array(0 to (2**addr_width-1)) of word_t;
...
signal rom_data: mem_t ...
```

```
-- rom (arbitrary width, size)
library ieee;
use ieee.std_logic_1164.all;
use ieee.numeric_std.all;
use ieee.std_logic_textio.all;
use std.textio.all;

entity ROM is
  generic( data_width: integer := 32;
           addr_width: integer := 4;
           filename: string := "dataFile" );
  port( addr: in std_logic_vector(addr_width-1 downto 0);
        data: out std_logic_vector(data_width-1 downto 0) );
end ROM;

architecture impl of ROM is
  subtype word_t is std_logic_vector(data_width-1 downto 0);
  type mem_t is array(0 to (2**addr_width-1)) of word_t;

  -- ModelSim and Vivado will initialize RAM/ROMs using the following function
  impure function init_rom (filename: in string) return mem_t is
    file init_file: text open read_mode is filename;
    variable init_line: line;
    variable result_mem: mem_t;
  begin
    for i in result_mem'range loop
      readline(init_file,init_line);
      ieee.std_logic_textio.read(init_line, result_mem(i));
    end loop;
    return result_mem;
  end init_rom;

  signal rom_data: mem_t := init_rom(filename);

  -- Quartus initializes RAM/ROMs via ram_init_file synthesis attribute
  -- filename must be in MIF format (different format than used by init_rom)
  attribute ram_init_file : string;
  attribute ram_init_file of rom_data : signal is filename;
begin
  data <= rom_data(to_integer(unsigned(addr)));
end impl;
```

Figure 8.51. VHDL for an arbitrarily sized ROM. At the beginning of simulation or synthesis, the ROM is initialized with the data loaded from the file with parameterized name `filename`.

The first line declares `word_t` to be a *subtype* of `std_logic_vector` with a prespecified width of `data_width` bits. The next line, beginning with the keyword **`type`**, declares a new type `mem_t` as an *array* containing 2^{addr_width} elements, each with the subtype `word_t`. The last line above declares `rom_data` as a signal with type `mem_t`.

To access a value in the ROM at address `addr` we use the following line:

```
data <= rom_data(to_integer(unsigned(addr)));
```

As a consequence of its strong typing, VHDL requires that we explicitly convert `addr` to a value of integer type before using it to index an array.

One approach to initialize the ROM, which is supported by, e.g., Mentor Graphics ModelSim® and Xilinx Vivado®, uses a VHDL `function`. In Figure 8.51 this function is declared starting at the line:

```
impure function init_rom (filename: in string) return mem_t ...
```

and it is invoked at the line:

```
signal rom_data: mem_t := init_rom(filename);
```

This invocation of the function is performed once to provide the contents of memory either at the start of simulation or during synthesis.

The second approach to initializing the ROM, which works with, e.g., Altera Quartus II®, employs a VHDL attribute declaration and specification to enable the designer to indicate the name of a file containing the ROM contents (in `.mif` format):

```
attribute ram_init_file : string;
attribute ram_init_file of rom_data : signal is filename;
```

During synthesis Quartus II® recognizes the `ram_init_file` attribute on `rom_data` as a special (Altera specific) *synthesis attribute*.

However the ROM is modeled, with a `case` or an array, ROMs above a critical size (a few thousand bytes) should generally be implemented as a custom ROM component rather than as a synthesized logic block. First, with optimized circuit design and layout, a ROM is typically much smaller, and often much faster, than an equivalent logic circuit. Second, the regular structure allows us to change the contents of the ROM without changing its overall layout – as long as its size doesn't change. Only small changes to the internals of the ROM are required. Some ROMs are programmable by changing only a single metal layer, making changing ROM contents relatively inexpensive.

The contents of most ROMs are determined at the time the ROM is manufactured – by the presence or absence of a transistor. Programmable ROMs, or PROMs, are manufactured without a pattern and are programmed electrically later – by blowing a fuse or placing charge on a floating gate. Using PROMs makes low-volume applications more economical by removing the tooling costs otherwise required to configure a ROM. Some PROMs are one-time programmable – once programmed, they cannot be changed. Erasable programmable ROMs, or EPROMs, can be erased and reprogrammed multiple times. Some EPROMs are erased by exposure to UV light (UV-EPROMs), while others are electrically erasable (EEPROMs).

8.9 READ–WRITE MEMORIES

A *read–write memory*, or RWM, is like a ROM, but also allows the contents of the table to be changed or *written*. For historical reasons, read–write memories are commonly called

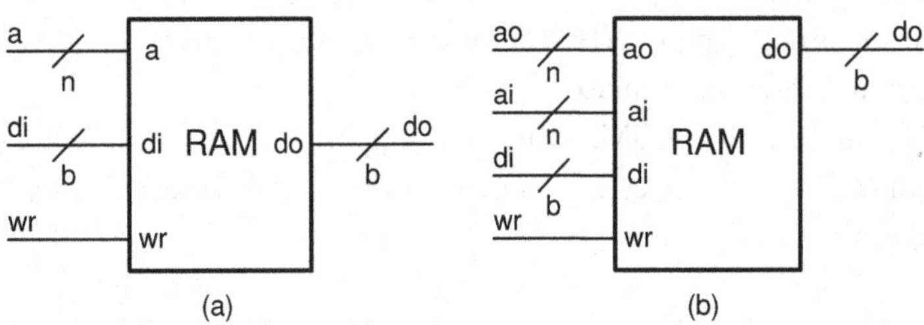

Figure 8.52. Schematic symbols for RAMs. The data input *di* is written to the selected location if the write line *wr* is true. (a) A single-port RAM shares address lines for both read and write. (b) A dual-port RAM has separate address lines: *ao* for read and *ai* for write.

RAMs.[5] The use of the term RAM to refer to a RWM is almost universal, so we will adopt it as well. Strictly speaking, a RAM is a sequential logic device – it has state, and hence its outputs depend on its input history – and so we should defer discussing RAMs until Chapter 14. However, because they are common building blocks, we will go ahead and discuss the basics of RAMs here.

A schematic symbol for a single-port RAM is shown in Figure 8.52(a). If the write signal *wr* is low, the RAM performs just like the ROM. Applying an address to input *a* results in the contents of the corresponding location being output on data output *do*. When signal *wr* goes high, a write takes place. The value on data input *di* is written to the location specified by *a*. We can use this RAM to store a value in a location – by addressing the location and setting *wr* high. At a later time we can read out the stored value – by addressing the location again with *wr* low.

In the single-port RAM of Figure 8.52(a), only one location, specified by the single address *a*, can be addressed at a time. If we are writing at the location specified by *a* we cannot read a different location at the same time. A dual-port RAM overcomes this limitation. The schematic symbol for a dual-port RAM is shown in Figure 8.52(b). With the dual-port RAM, the read port (signals *ao* and *do*) is independent from the write port (signals *ai*, *di*, and *wr*). A location specified by *ao* can be read onto data output *do* at the same time as data input *di* is being written to a different location specified by *ai*. Dual-port RAMs are often used to interface two subsystems, with one subsystem writing the RAM and the other subsystem reading it. RAMs can be constructed with an arbitrary number of read and write ports, but the cost of the RAM increases quadratically with the number of ports.

A simple implementation of a dual-port RAM using two decoders, latches, and tri-state buffers is shown in Figure 8.53. The read decoder and tri-state buffers form a structure identical to the ROM of Figure 8.48. The read address *ao* is decoded to N read word select lines $wo_0, \ldots, wo_{N-1}$. Each read word select line enables the corresponding location onto the read output *do*.

The difference between the read port and the ROM is that the data stored at each location are obtained from a latch in the RAM (rather than being a constant in the ROM). A latch is a simple storage element that copies its input D to its output Q when its enable G input is high.

[5] The acronym RAM stands for *random-access memory*. Both ROMs and RAMs as we have defined them allow random access – i.e., any location can be accessed in any order. In contrast, a tape is a sequential access memory – words stored on the tape must be accessed in sequential order.

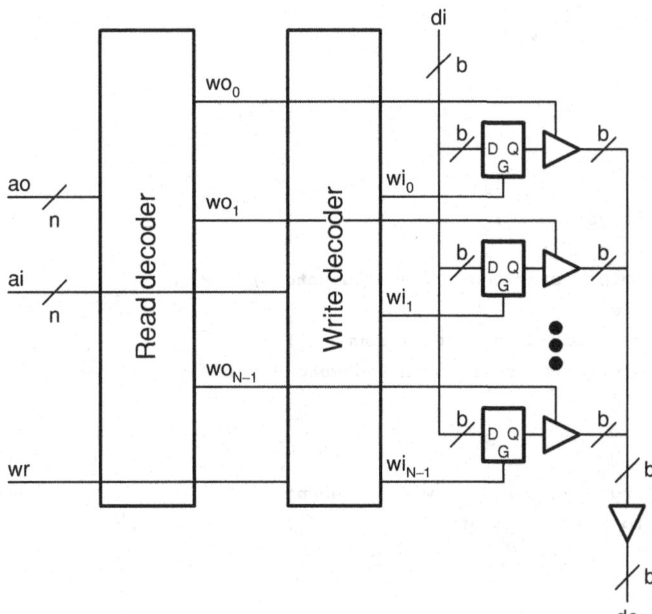

Figure 8.53. A dual-port RAM can be implemented with two decoders, latches to store the data, and an array of tri-state buffers for reading. Actual RAMs use a two-dimensional structure and more efficient circuit design for the storage element.

When G is low, the output Q holds its previous value – giving us a simple one-bit memory. Latches are described in more detail in Section 27.1.

The write port of the RAM uses the write decoder to decode the write address *ai* to write word select lines $wi_0, \ldots, wi_{N-1}$ when *wr* is asserted. When *wr* is not asserted, all write word select lines remain low. When location i is written, write word select line wi_i goes high, storing the input data on *di* into the ith latch. When the address changes or *wr* goes low, wi_i goes low and the latch retains the stored data.

As with the ROM, actual RAM implementations are much more efficient than the simple one we illustrate here. Most RAMs employ a two-dimensional structure, like the one shown for a ROM in Figure 8.49. The column "multiplexing" for writes is a bit more involved and will not be discussed further here. Most practical RAMs also use a much more efficient bit cell than the latch plus tri-state buffer shown here. Most static RAMs (SRAMs) use a six-transistor storage cell, and modern dynamic RAMs (DRAMSs) use a cell composed of a single transistor and a storage capacitor. The circuit details of these RAM cells are beyond the scope of this book.

We show the VHDL code to model a dual-port RAM in Figure 8.54. The read function is identical to that of the ROM of Figure 8.51. The write function is implemented within the `process` and updates the selected location any time signal `write` is high. When `write` is low, the last value written to an address is saved. As mentioned above, because it saves state, a RAM is not a combinational circuit, but rather a sequential circuit (see Chapter 14). Unlike our ROM in Figure 8.51, the RAM is not initialized to any values. Each address must be written before it is read to avoid getting undefined data values.[6] Many RAM components have a synchronous interface where a clock edge is required to perform a read or write operation. We discuss RAM blocks further in Chapter 25.

[6] Some SRAMs provide either a reset signal or logic to write 0 into every register location.

```
-- RAM of parameterized size and width
library ieee;
use ieee.std_logic_1164.all;
use ieee.numeric_std.all;

entity RAM is
  generic( data_width: integer := 32;
           addr_width: integer := 4 );
  port( ra, wa: in std_logic_vector(addr_width-1 downto 0);
        write: in std_logic;
        din: in std_logic_vector(data_width-1 downto 0);
        dout: out std_logic_vector(data_width-1 downto 0) );
end RAM;

architecture impl of RAM is
  subtype word_t is std_logic_vector(data_width-1 downto 0);
  type mem_t is array(0 to (2**addr_width-1)) of word_t;
  signal data: mem_t;
begin
  dout <= data(to_integer(unsigned(ra)));

  process(all) begin
    if write = '1' then
      data(to_integer(unsigned(wa))) <= din;
    end if;
  end process;
end impl;
```

Figure 8.54. The VHDL for a RAM block. Whenever the write signal is high, *din* is written into location *wa*. We continuously assign the output to be the value stored at location *ra*.

8.10 PROGRAMMABLE LOGIC ARRAYS

A programmable logic array, or PLA, is a regular structure that can be configured to realize an arbitrary set of sum-of-products logic functions. As shown in Figure 8.55, a PLA consists of an AND plane and an OR plane. The AND plane is a two-dimensional structure with literals (inputs and their complements) running vertically and product terms running horizontally. In each row, an arbitrary set of literals is selected as inputs to an AND gate to realize an arbitrary product term. The literals connected to each AND gate are denoted by squares in the figure. For example, the top AND gate takes as input three literals, a_0, $\overline{a_1}$, and $\overline{a_2}$, and generates their product $a_0 \wedge \overline{a_1} \wedge \overline{a_2}$.

The AND plane is much like the decoder in a ROM, except that the product term for each row is arbitrary, whereas the product term for each row of a decoder is the minterm corresponding to the address of that row. Several rows of a PLA may go high at the same time, while only one row of a ROM's decoder will be activated at a time.

The OR plane is another two-dimensional structure: the product terms run horizontally and the outputs (sums) run vertically. In each column, arbitrary product terms are ORed together to form an output. In Figure 8.55, the products combined by each OR are denoted by squares. For example, the rightmost column ORs together the bottom three products.

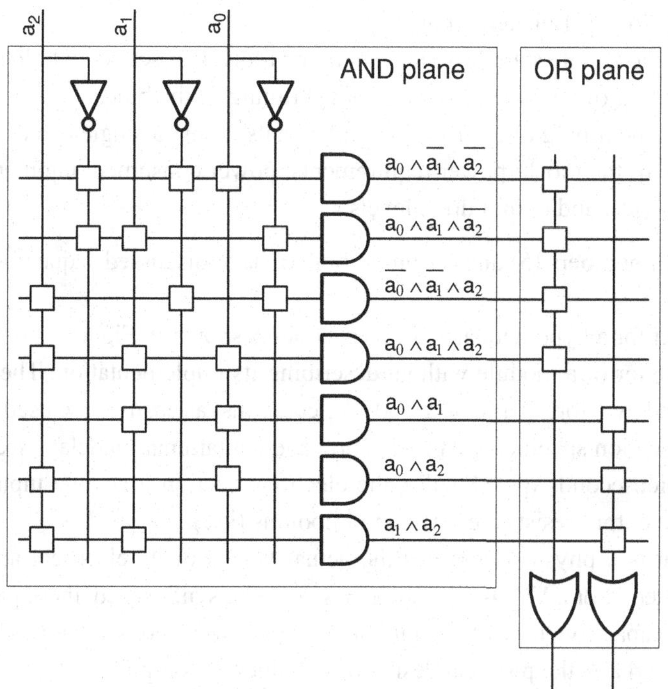

Figure 8.55. A programmable logic array (PLA) consists of an AND plane and an OR plane. By programming the connections, an arbitrary set of product terms can be realized by the AND plane. These product terms are then combined into sum-of-products realizations of logic functions by the OR plane. This figure shows a full adder implemented as a PLA with seven product terms in the AND plane and two sums (for sum and carry) in the OR plane.

In practice, PLAs usually use the same structure both for the AND plane and for the OR plane – either a NAND or a NOR gate. By De Morgan's law, A NAND–NAND PLA is equivalent to the AND–OR PLA shown in Figure 8.55. A NOR–NOR PLA realizes the complement of a function and can be followed by an inverter to realize a function directly. A highly optimized circuit structure in which each crosspoint in a plane requires only a single transistor (or its absence) is often used.

Most PLAs are hard-wired at manufacture. The literals in each product and the products in each sum are selected by the presence or absence of a transistor. Some PLAs are configurable with a bit of storage controlling whether a literal is included in a product or a product in a sum. Such configurable PLAs are many times larger than an equivalent hard-wired PLA.

8.11 DATA SHEETS

We will often use a building block, or an entire subsystem, in a larger design without understanding its implementation. When we use a building block in this manner, we rely on the specification of the block. This specification, often called a *data sheet*, gives enough information for us to use the block, but omits internal details of how the block is constructed. A data sheet typically contains the following.

(1) A functional description of the block – what the block does. This should be in sufficient detail as to specify completely the block's behavior. For a combinational block, a truth table or equation is often used to specify the block's function.

(2) A detailed description of the inputs and outputs of the block: a signal-by-signal description that gives the signal name, width, direction, and a brief description.

(3) A description of all block parameters, if any.
(4) A description of all the visible state and registers in the block (for sequential blocks).
(5) The synchronous timing of the block: the cycle-level timing of the block.
(6) The detailed timing: the timing of input and output signals within a single cycle.
(7) Electrical properties of the block: power requirements, power consumed, input and output signal levels, input loads, and output drive levels.

We defer discussion of numbers (5) and (6) until after we have discussed sequential circuits and timing.

An example data sheet for a hypothetical 4 → 16 decoder is shown in Figure 8.56. This data sheet describes the behavior of a module without describing its implementation. The function of the module is specified by a formula (`b = 1<<a`). We could just as easily have used a 16-row truth table. The timing section specifies the propagation and contamination delay (see Chapter 15) of the module in picoseconds (ps). Finally, the electrical section gives the input load in femtofarads (fF) and the output resistance (drive) in kiloohms (kΩ).

A building block that is a physical chip has the actual values of its electrical and timing parameters in the data sheet. For a VHDL block that has yet to be synthesized, these parameters are not yet known. The capacitive load of each input, for example, is not known until after the block has been synthesized and the physical design of the block is complete.

```
Name: decode_4_16

Description: 4 to 16 decoder

Inputs:
  Name  Width  Direction  Description
  a     4      in         binary input
  b     16     out        one-hot output

Function:
  b = 1<<a

Timing:
  Parameter Min    Max    Units Description
  t_dab            300    ps    Delay from a to b - no load on b
  t_cab     100           ps    Contamination delay from a to b - no load on b

Electrical
  Parameter Min    Max    Units Description
  c_a              20     fF    Capacitance of each bit of a
  r_b        5            kOhms Effective output resistance of each bit of b
```

Figure 8.56. Example data sheet for a 4 → 16 decoder.

```
set_max_delay 0.2 -from {a} -to {b}
set_driving_cell -lib_cell INV4 {a}
set_load -pin_load 5 {b}
```

Figure 8.57. Example constraint file for 4 → 16 decoder.

A *constraint file* is used to specify targets for timing and electrical parameters. These targets (or constraints) are then used to direct the synthesis and physical design tools. A very simple constraint file for our 4 $\rightarrow$ 16 decoder is shown in Figure 8.57. This file is in a form suitable for use by the Synopsys Design Compiler®. The file specifies that the delay of the decoder from a to b must not exceed 0.2 ns. Rather than specify the input load, the file specifies that input a is driven by a cell equivalent in drive to a `INV4` cell. If the synthesizer makes the input capacitance too large, the delay of this cell driving a, which is included in the total delay, will make it hard to meet the timing constraint. Finally, the file specifies that the output load, on each bit of b, is 5 (capacitive units). The synthesizer must size the output driver for the decoder large enough to drive this load without excessive delay.

8.12 INTELLECTUAL PROPERTY

A design team builds a chip by combining design entities they design themselves with design entities that they obtain from other sources. The design entities that are obtained elsewhere are often called IP, for *intellectual property*.[7]

IP blocks are available from vendors and as *open-source* gateware. Some vendors specialize in particular types of IP. For example, ARM and MIPS specialize in selling microprocessors as IP. The microprocessors in most cell phones are licensed as IP from ARM.

The open-source movement that has revolutionized software has its parallel in the hardware world. Many useful pieces of VHDL IP are available for free under an open-source license at `http://www.opencores.org`. Modules available include processors, interfaces (e.g., Ethernet, PCI, USB, etc.), encryption/decryption blocks, compression/decompression blocks, and others. While these building blocks are more complex than the ones described in this chapter, the concepts involved in using them are the same. The design team builds a system by combining a number of blocks. The blocks themselves are described by data sheets (and constraint files) that specify their function, interfaces, and parameters.

As with all things, caveat emptor (buyer beware) applies to IP. Purchased IP does not always meet its specification. The cautious designer thoroughly tests every piece of acquired IP.

Summary

In this chapter you have learned about some common design patterns, or *idioms*, of digital design – common circuits that we use repeatedly in our designs.

Decoders convert a binary encoding of a signal to a *one-hot* representation of the same signal. We use them, for example, to select one of many rows in a memory.

Encoders do the reverse, converting a *one-hot* signal to a binary representation.

[7] The term *intellectual property (IP)* is much broader than its use here. IP encompasses anything of value which is independent of a physical object. That is, the value has been created by intellectual effort, rather than by manufacturing. For example, all software, books, movies, music, designs – including VHDL designs – are IP.

Arbiters find the first high bit (searching from right or left) in an input word. They are used, for example, to control access to shared resources and to *normalize* floating-point numbers. Combining an *arbiter* with an *encoder* gives a *priority encoder*. The arbiter finds the first one bit in the input and the encoder converts this one-hot signal to binary – outputting the binary encoding of the position of the first 1.

Multiplexers select one of many inputs to be driven onto their output under control of a *one-hot* select signal. They are widely used for data steering in all types of datapaths. Combining a multiplexer with a decoder gives a *binary-select* multiplexer. The decoder converts a binary select signal to one-hot, and the multiplexer uses the resulting signal to select an input.

Comparators compare two binary numbers and indicate whether they are equal or whether one is greater than the other. Comparators are often implemented as iterative circuits.

Shifters shift or rotate an input signal to produce an output signal. They are used, for example, to align floating-point numbers for addition.

Memories are tables that may be read-only (ROMs) or read-write (RAMs). Given an address, the memory returns the value stored at that address. For a RAM, the value at that address can also be written. Memories are discussed in detail in Chapter 25.

Large versions of building block circuits such as decoders, encoders, and multiplexers can be constructed hierarchically, as trees of smaller versions of these circuits. These hierarchical circuits have a smaller gate count, are faster, and consume less energy than a *flat* implementation when the size of the block is large.

Building blocks (and other design entities) are often used in a design without understanding their internal implementation. The external characteristics of such *IP* blocks are specified by a *data sheet*.

BIBLIOGRAPHIC NOTES

The *TTL Data Book* [106], first published in the 1970s, describes the building block functions available as separate chips in the classic 7400-series TTL logic family. The chips include simple gates, multiplexers, decoders, seven-segment decoders, arithmetic functions, registers, counters (see Chapter 16), and many others. The *TTL Data Book* also gives many good examples of data sheets. The function, interfaces, electrical parameters, and timing parameters are listed for each part.

FPGA vendors often provide a library of building block design entities to designers. Altera's library of parameterized components [1] is one such example. It includes decoders, multiplexers, and many of the arithmetic circuits discussed in Chapter 10.

The design of RAMs, ROMs, and PLAs is explored in digital integrated circuits textbooks, such as refs. [94] and [112].

Exercises

8.1 *Decoder.* Write a structural VHDL description of a 3 → 8 decoder.

8.2 *Decoder logic.* Implement a seven-segment decoder using a 4 → 16 decoder and OR gates.

8.3 *Two-hot decoder.* Consider the alphabet of two-hot signals, that is binary signals with only two bits equal to 1. There are $(n(n-1))/2$ n-bit two-hot symbols. Assume these symbols are ordered by their binary value; i.e., for $n = 5$ the order is 00011, 00101, 00110, ..., 11000. Design a $4 \rightarrow 5$ binary to two-hot decoder.

8.4 *Large decoder, I.* Write a VHDL design entity that implements a $5 \rightarrow 32$ decoder using a $2 \rightarrow 4$ and a $3 \rightarrow 8$ decoder as building blocks.

8.5 *Large decoder, II.* Write a VHDL design entity that implements a $6 \rightarrow 64$ decoder using $3 \rightarrow 8$ decoders as building blocks.

8.6 *Large decoder, III.* Write a VHDL design entity that implements a $6 \rightarrow 64$ decoder using $2 \rightarrow 4$ decoders as building blocks.

8.7 *Distributed multiplexer.* Implement a large (32-input) multiplexer in which each multiplexer input and its associated select signal is in a different part of a large chip. The 32 inputs and selects are located along a 0.4 mm long line. Show how this can be implemented using static CMOS gates (e.g., NANDs, NORs, and inverters – no tri-states) with only a single wire running along the line between adjacent input locations.

8.8 *Multiplexer logic.* Implement a four-bit Fibonacci circuit (output true if the input is a Fibonacci number) using an $8 \rightarrow 1$ binary-select multiplexer.

8.9 *Decoder testbench.* Write a testbench for a $4 \rightarrow 16$ decoder using an encoder as a checker.

8.10 *Two-hot encoder.* Design a $5 \rightarrow 4$ two-hot encoder using the conventions of Exercise 8.3.

8.11 *Programmable priority encoder.* Write a VHDL design entity for a priority encoder with programmable priority – an input (one-hot) selects which bit is highest priority. The priority rotates rightward from that bit position.

8.12 *Binary-priority arbiter.* Write a VHDL design entity for an arbiter with programmable priority – a binary input selects which bit is highest priority. The priority rotates rightward from that bit position.

8.13 *Round-robin arbiter.* Design an arbiter when the highest-priority input in each cycle is one input to the right (cyclically) of the input that last won an arbitration. Assume that the previous winner is an input to your module.

8.14 *Comparator.* Write a VHDL design entity for an arbitrary-width magnitude comparator that propagates information downward from MSB to LSB, as shown in Figure 8.42.

8.15 *Three-way magnitude comparator, I.* Write a VHDL design entity for a three-way magnitude comparator that outputs true if its three inputs are in strict order: $a > b > c$.

8.16 *Three-way magnitude comparator, II.* Write a VHDL design entity for a three-way magnitude comparator that outputs true if its three inputs are not out of order: $a \geq b \geq c$.

8.17 *Funnel-shifter.* Write a VHDL design entity for an i to j funnel-shifter, a block that accepts an i-bit input `a` and an $l = \log_2(i - j)$ bit shift count `n` and generates a $j < i$ bit output `b = a(n+j-1 downto n)`.

8.18 *Using building blocks, I.* Using building blocks such as binary adders, comparators, multiplexers, decoders, encoders, and arbiters, as well as logic gates, design an 8×2 *popularity* circuit – a circuit that accepts eight two-bit binary numbers and outputs the number of times each of the four two-bit numbers appears on the input. Modify your circuit to output also the two-bit number that appears the most often (high number wins ties).

8.19 *Using building blocks, II.* Design a combinational circuit with three eight-bit inputs that outputs the minimum value of the three.

8.20 *Using building blocks, III.* Design a combinational circuit with three eight-bit inputs that outputs the median value of the three (the one that is neither the maximum nor the minimum).

8.21 *ROM logic – prime number function.* Implement the four-bit prime number function using a ROM. How large a ROM is needed (what are N and b)? What is stored in each location?

8.22 *ROM logic – seven-segment decoder.* Implement a seven-segment decoder with a ROM. How large a ROM is needed (what are N and b)? What is stored in each location?

8.23 *PLA – prime number function.* Implement the four-bit prime number function using a PLA. How many product terms and sum terms are needed? What are the connections for each term?

8.24 *PLA – seven-segment decoder.* Implement a seven-segment decoder with a PLA. How many product terms and sum terms are needed? What are the connections for each term?

9 Combinational examples

In this chapter we work through several examples of combinational circuits to reinforce the concepts in the preceding chapters. A *multiple-of-3* circuit is another example of an iterative circuit. The *tomorrow* circuit from Section 1.4 is an example of a counter circuit with subcircuits for modularity. A priority arbiter is an example of a building-block circuit – built using design entities described in preceding chapters. Finally, a circuit designed to play *tic-tac-toe* gives a complex example combining many concepts.

9.1 MULTIPLE-OF-3 CIRCUIT

In this section we develop a circuit that determines whether an input number is a multiple of 3. We implement this function using an iterative circuit (like the magnitude comparator of Section 8.6). A block diagram of an iterative multiple-of-3 circuit is shown in Figure 9.1. Each stage performs, in binary, a step of long division of the input number by 3, passing along the remainder but discarding the quotient. The circuit checks the input number one bit at a time starting at the MSB. At each bit, we compute the remainder so far (0, 1, or 2). At the LSB we check whether the overall remainder is 0. Each bit cell takes the remainder so far to its left, and one bit of the input, and computes the remainder so far to its right.

The VHDL design entity for the bit cell of our iterative multiple-of-3 circuit is shown in Figure 9.2. The remainder in `remin` represents the remainder from the neighboring bit to the left, and hence has a weight of 2 relative to the current bit position. In our neighboring bit this signal represented a remainder of 0, 1, or 2. However, in the present bit, this value is shifted to the left by one bit and it represents a value of 0, 2, or 4. Hence we can concatenate `remin` with the current bit of the input, `input`, to form a three-bit binary number and then take the remainder (mod 3) of this number. A case statement is used to compute the new remainder.

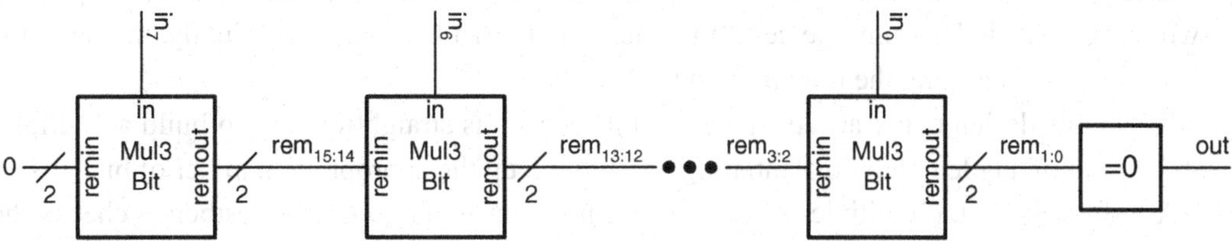

Figure 9.1. Block diagram of a multiple-of-3 circuit. The circuit computes the remainder, mod 3, of the input one bit at a time working from the left (MSB) to the right (LSB). Each bit cell computes the remainder of a three-bit number formed by concatenating the two-bit remainder in (`remin`) with the current bit of the input (`in`). If the remainder out of the low bit is zero, the number is a multiple of 3.

```
--------------------------------------------------------------------------------
-- Multiple_of_3_bit
-- Cell for iterative multiple-of-3 circuit.
-- Determines the remainder (mod 3) of the number from this bit to the MSB.
-- Input:
--    input - the current bit of the number being checked
--    remin - the remainder after the last bit checked (2 bits)
-- Output:
--    remout - the remainder after checking this bit (2 bits).
--
-- remin has weight 2 since it is from the bit to the left, thus remin & input
-- forms a 3-bit number.  We divide this number by 3 and produce the remainder
-- on remout.
--------------------------------------------------------------------------------

library ieee;
use ieee.std_logic_1164.all;

entity Multiple_of_3_bit is
  port( input: in std_logic;
        remin: in std_logic_vector(1 downto 0);
        remout: out std_logic_vector(1 downto 0) );
end Multiple_of_3_bit;

architecture impl of Multiple_of_3_bit is
begin
  process(all) begin
    case remin & input is
      when "000" => remout <= 2d"0";
      when "001" => remout <= 2d"1";
      when "010" => remout <= 2d"2";
      when "011" => remout <= 2d"0";
      when "100" => remout <= 2d"1";
      when "101" => remout <= 2d"2";
      when others => remout <= "--";
    end case;
  end process;
end impl;
```

Figure 9.2. VHDL description of bit cell for multiple-of-3 circuit.

The top-level multiple-of-3 design entity is shown in Figure 9.3. This design entity instantiates eight copies of the bit cell of Figure 9.2. The cells are connected together by passing two bits of remainder from one cell to the next via the 16-bit signal `re`. Finally, the output is generated by comparing the remainder out to zero.

While this design entity accepts an eight-bit input, it is straightforward to build a multiple-of-3 circuit of any length by instantiating and linking up the appropriate number of bit cells.

A testbench for the multiple-of-3 circuit is shown in Figure 9.4. The testbench checks the result of the circuit under test by checking the remainder (mod 3) using the VHDL modulo operator **mod** and comparing it with zero. The result of this comparison has the VHDL type `boolean`. We convert `output` from `std_logic` to `boolean` using the VHDL-2008 *condition operator* `??` since both operands of the inequality operator must have the same type. Note that

```
-----------------------------------------------------
-- Multiple_of_3
-- Determines whether input is a multiple of 3
-- Input:
--     input - an 8-bit binary number
-- Output:
--     output - true if in is a multiple of 3
-----------------------------------------------------

library ieee;
use ieee.std_logic_1164.all;
use work.ch9.all;

entity Multiple_of_3 is
  port( input: in std_logic_vector( 7 downto 0 );
        output: out std_logic );
end Multiple_of_3;

architecture impl of Multiple_of_3 is
  signal re: std_logic_vector(17 downto 0); -- two bits of remainder per cell
begin
  -- instantiate 8 copies of the bit cell
  b7: Multiple_of_3_bit port map(input(7),"00",re(15 downto 14));
  b6: Multiple_of_3_bit port map(input(6),re(15 downto 14),re(13 downto 12));
  b5: Multiple_of_3_bit port map(input(5),re(13 downto 12),re(11 downto 10));
  b4: Multiple_of_3_bit port map(input(4),re(11 downto 10),re(9 downto 8));
  b3: Multiple_of_3_bit port map(input(3),re(9 downto 8),re(7 downto 6));
  b2: Multiple_of_3_bit port map(input(2),re(7 downto 6),re(5 downto 4));
  b1: Multiple_of_3_bit port map(input(1),re(5 downto 4),re(3 downto 2));
  b0: Multiple_of_3_bit port map(input(0),re(3 downto 2),re(1 downto 0));

  -- output is true if remainder out is zero
  output <= '1' when re(1 downto 0) = "00" else
            '0';
end impl;
```

Figure 9.3. VHDL code for eight-bit multiple-of-3 circuit. This design entity instantiates eight copies of the design entity from Figure 9.2 and checks the final output for zero.

we do not want to use the **mod** operator in our circuit itself because use of this operator will cause the synthesis program to instantiate a prohibitively expensive divider. However, use of the **mod** operator in a testbench, which is not synthesized, causes no problems.

The testbench declares the device under test's input, and output signals, instantiates the multiple-of-3 design entity, and then walks through all possible input states. In each input state the output of the device under test is compared with an output computed using the **mod** operator. If there is a mismatch, an error is flagged. If all states are tested with no mismatch, the test is passed.

9.2 TOMORROW CIRCUIT

In Section 1.4 we introduced a calendar circuit. The key design entity of this circuit was a *tomorrow* circuit that, given today's date in month, day-of-month, day-of-week format,

```
-- pragma translate_off
library ieee;
use ieee.std_logic_1164.all;
use ieee.numeric_std.all;

entity testMul3 is
end testMul3;

architecture test of testMul3 is
  signal input: std_logic_vector( 7 downto 0 );
  signal output, err: std_logic;
begin
  DUT: entity work.Multiple_of_3(impl) port map(input,output);

  process begin
    err <= '0';
    for i in 0 to 255 loop
      input <= std_logic_vector(to_unsigned(i,8));
      wait for 10 ns;
      if (?? output) /= ((i mod 3) = 0) then
          err <= '1';
      end if;
    end loop;
    if not err then report "PASS"; end if;
    std.env.stop(0);
  end process;
end test;
-- pragma translate_on
```

Figure 9.4. VHDL description of the testbench for the multiple-of-3 circuit.

computes tomorrow's date in the same format. In this section we present the VHDL implementation of this tomorrow circuit.

A key step in designing a digital circuit is dividing a large problem into simpler subproblems. We can then design simple design entities to solve these subproblems and compose these design entities to solve our larger problem. For the tomorrow circuit we can define two subproblems:

(1) increment the day of the week (this is completely independent of the month or day of the month);
(2) determine the number of days in the current month.

Figure 9.5 shows a VHDL design entity that increments the day of the week. If the current day is `SATURDAY` (which is defined to be 7 in a package called `calendar` that we would define, which is not shown), this design entity sets tomorrow's day to be `SUNDAY` (which is defined to be 1). If the current day is other than `SATURDAY`, the design entity just increments today to get tomorrow.

In Section 16.1 we will see that the `NextDayOfWeek` design entity is the combinational part of a *counter*, a circuit that increments its state. The variation here is that when the counter reaches `SATURDAY` it resets back to `SUNDAY`.

```
library ieee;
use ieee.std_logic_1164.all;
use ieee.std_logic_unsigned.all;
use work.calendar.all;  -- our definition of constants SUNDAY..SATURDAY (not shown)

entity NextDayOfWeek is
  port( today: in std_logic_vector( 2 downto  0 );
        tomorrow: out std_logic_vector( 2 downto 0 ) );
end NextDayOfWeek;

architecture behav of NextDayOfWeek is
begin
  tomorrow <= SUNDAY when today = SATURDAY else today + 1;
end behav;
```

Figure 9.5. VHDL description of the `NextDayOfWeek` design entity which increments the day of the week.

```
library ieee;
use ieee.std_logic_1164.all;

entity DaysInMonth is
  port( month : in std_logic_vector(3 downto 0); -- month of the year 1=Jan, 12=Dec
        days : out std_logic_vector(4 downto 0) ); -- number of days in month
end DaysInMonth;

architecture impl of DaysInMonth is
begin
  process(all) begin
    case month is
      -- thirty days have September...
      -- all the rest have 31
      -- except for February which has 28
      when 4d"4" | 4d"6" | 4d"9" | 4d"11" => days <= 5d"30";
      when 4d"2" => days <= 5d"28";
      when others => days <= 5d"31";
    end case;
  end process;
end impl;
```

Figure 9.6. VHDL description of a design entity that computes the number of days in a month when it is not a leap year.

This design entity is coded using definitions for `SATURDAY` and `SUNDAY`. However, it will work only if the days are represented by consecutive three-bit integers starting with `SUNDAY` and ending with `SATURDAY`. In Exercise 9.9 we explore writing a more general version of this design entity that will work with arbitrary representations for the days of the week.

Figure 9.6 shows a VHDL design entity that computes the number of days in a given month. This is just a simple case statement that uses a default case (**`when others`** `=>`) to handle the common case of a month with 31 days. This design entity would be easier to read if we defined constants for the month names. However, we use numbers for months often enough in our daily lives that little is lost by using the numbers here.

```
library ieee;
use ieee.std_logic_1164.all;
use ieee.std_logic_unsigned.all;
use work.calendar.all;

entity Tomorrow is
  port( todayMonth: in std_logic_vector(3 downto 0);
        todayDoM: in std_logic_vector(4 downto 0);
        todayDoW: in std_logic_vector(2 downto 0);
        tomorrowMonth: out std_logic_vector(3 downto 0);
        tomorrowDoM: out std_logic_vector(4 downto 0);
        tomorrowDoW: out std_logic_vector(2 downto 0) );
end Tomorrow;

architecture impl of tomorrow is
  signal daysinmonth : std_logic_vector(4 downto 0);
  signal lastday, lastmonth : std_logic;
begin
  -- compute next day of week
  ndow: entity work.nextdayofweek port map(todaydow,tomorrowdow);

  -- compute month and day of month
  dim: entity work.daysinmonth port map(todaymonth,daysinmonth);

  -- compute month and day of month
  lastday <= '1' when todaydom = daysinmonth else '0';
  lastmonth <= '1' when todaymonth = december else '0';
  tomorrowmonth <= january when lastday and lastmonth else
                   todaymonth+1 when lastday else
                   todaymonth;
  tomorrowdom <= 5d"1" when lastday else todaydom+1;
end impl;
```

Figure 9.7. VHDL description of a tomorrow circuit. The circuit accepts today's date in month, day-of-month, day-of-week format and outputs tomorrow's date in the same format.

The astute reader will have observed that this `DaysInMonth` design entity is not quite right. We have not considered leap years – when February has 29 days. We leave it as an exercise for the reader (Exercise 9.10) to remedy this situation.

With our two components defined, we can now develop the full `Tomorrow` design entity. Figure 9.7 shows the code for our full tomorrow design entity. After the design entity, input/output, and signal declarations, it starts by instantiating our two components. The `NextDayOfWeek` design entity directly generates the `tomorrowDoW` output. This is also the only code to use the `todayDoW` input. The day-of-week function is completely independent of the month and day-of-month functions.

Next, the circuit instantiates the `DaysInMonth` component. This component generates an internal signal `daysInMonth` that encodes the last day of the current month. The tomorrow component then generates two other internal signals: `lastDay` is true if today is the last day of the month, and `lastMonth` is true if the current month is December. Using these two internal signals, the component then computes `tomorrowMonth` and `tomorrowDoM` using conditional assignment statements.

Verifying the tomorrow circuit efficiently is a bit of a challenge. A brute force enumeration of all states requires simulating the circuit for seven years of input dates – 2555 inputs. We can reduce this to 365 inputs by observing that the day-of-week function is completely independent and can be verified independently. We can further collapse the set of tests by simulating only the beginning and end of each month. We also must unit test the `DaysInMonth` design entity to ensure that it outputs the correct amount of days for each month.

9.3 PRIORITY ARBITER

Our next example is a four-input *priority arbiter*, a circuit which accepts four inputs and outputs the index of the input with the highest value. In the event of a tie, the circuit outputs the lowest index that has the highest value. As an example of the arbiter's function, suppose the four inputs are 28, 32, 47, and 19. The arbiter will output 2 because input 2 has the highest value, 47. If the four inputs are 17, 23, 19, 23, the arbiter will output 1 because, of the two inputs (1 and 3) with the high value of 23, input 1 has the lowest index.

This circuit is used, for example, in networking equipment where the next packet to send is selected according to a quality-of-service (QoS) policy that gives each packet a score. The packet with the highest score is sent first. In this application, the score for each packet is an input to the priority arbiter, which selects the packet to transmit.

Our implementation of the priority arbiter is shown in Figure 9.8 and a VHDL implementation is given in Figure 9.9. The implementation performs a tournament to select the winning input. In the first round, inputs 0 and 1 and inputs 2 and 3 are compared. The second round compares the winners of the first round.

Each match in the tournament is performed using a magnitude comparator (see Section 8.6). To break ties in favor of the lower number input, the magnitude comparator computes a signal

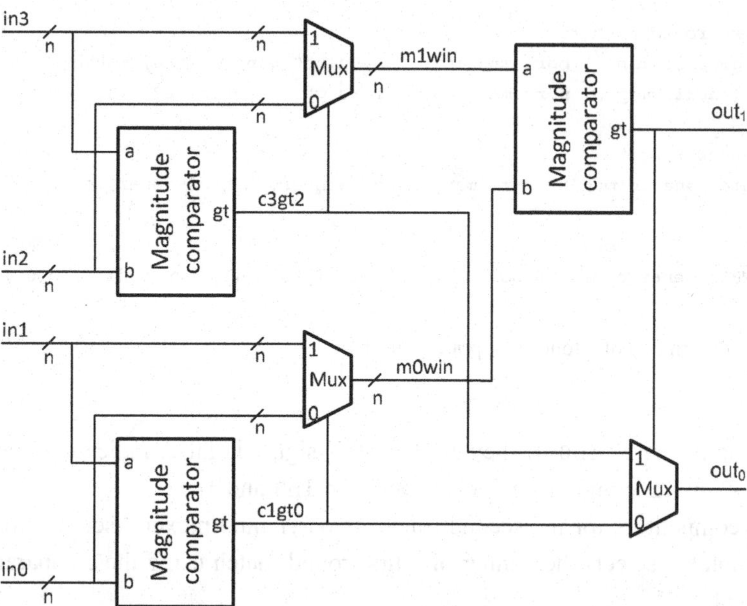

Figure 9.8. A four-input priority arbiter. This circuit accepts four inputs and outputs the index of the input with the largest value. It operates by performing a *tournament* on the inputs to find the highest value and then selecting match results to compute the index.

```
----------------------------------------------------------------------
-- 4-input Priority Arbiter
-- Outputs the index of the input with the highest value
-- Inputs:
--   in0, in1, in2, in3 - n-bit binary input values
-- Out:
--   o - 2-bit index of the input with the highest value
--
-- We pick the "winning" output via a tournament.
-- In the first round we compare in0 against in1 and in2 against in3
-- The second round compares the winners of the first round.
-- The MSB comes from the final round, the LSB from the selected first round.
--
-- Ties are given to the lower numbered input.
----------------------------------------------------------------------

library ieee;
use ieee.std_logic_1164.all;
use work.ch8.all;

entity PriorityArbiter is
  generic( n : integer := 8 );
  port( in3, in2, in1, in0: in std_logic_vector(n-1 downto 0);
        o: buffer std_logic_vector( 1 downto 0 ) );
end PriorityArbiter;

architecture impl of PriorityArbiter is
  signal match0winner, match1winner: std_logic_vector(n-1 downto 0);
  signal c1gt0, c3gt2: std_logic_vector(0 downto 0);
begin
  -- first round of tournament
  round0match0: MagComp generic map(n) port map(in1,in0,c1gt0(0)); -- compare in0 and in1
  round0match1: MagComp generic map(n) port map(in3,in2,c3gt2(0)); -- compare in2 and in3

  -- select first round winners
  match0: Mux2 generic map(n) port map(in1, in0, c1gt0 & not c1gt0, match0winner);
  match1: Mux2 generic map(n) port map(in3, in2, c3gt2 & not c3gt2, match1winner);

  -- compare round0 winners
  round1: MagComp generic map(n) port map( match1winner, match0winner, o(1));

  -- select winning LSB index
  winningLSB: Mux2 generic map(1) port map(c3gt2, c1gt0, o(1) & not o(1), o(0 downto 0) );
end impl;
```

Figure 9.9. VHDL description of a four-input priority arbiter.

`c1gt0` that is true if `in1` > `in0`. If they are tied, this signal is false, indicating that `in0` has won the match. A similar comparison is made between `in3` and `in2`.

To select the competitors for the second round, two 2:1 multiplexers (see Section 8.3) are used. Each multiplexer selects the winner of a first-round match using the comparator output as the select signal.

A third magnitude comparator performs the second-round match – comparing the two winning outputs from the first-stage multiplexers. The output of this second-round comparator is the MSB of the priority arbiter. If this signal is true, the winner is `in2` or `in3`; if it is false, the winner is `in0` or `in1`.

To get the LSB of the priority arbiter output, we select the output of the winning first-round comparator. This is accomplished with a single-bit wide 2:1 multiplexer controlled by the output of the final comparator.

9.4 TIC-TAC-TOE

In this section we develop a combinational circuit that plays the game of tic-tac-toe. Given a starting board position, it selects the square on which to play its next move. Being a combinational circuit, it can play only one move. However, it can easily be transformed into a sequential circuit (see Chapter 14) that plays an entire game. We implement this sequential version in Section 19.3.

Our first task is to decide how to represent the playing board. We represent the input board position as two nine-bit vectors: one `xin` encodes the position of the Xs and the other `oin` encodes the position of the Os. We map each nine-bit vector to the board as shown in Figure 9.10(a). The upper left corner is the LSB and the bottom right corner is the MSB. For example, the board shown in Figure 9.10(b) is represented by `xin` = 100000001 and `oin` = 000011000. For a legal board position, `xin` and `oin` must be orthogonal, that is `xin` ∧ `oin` = 0.

Strictly speaking, playing as X, a legal board should also have $N_X + 1 \geq N_O \geq N_X$, where N_O is the number of bits set in `oin` and N_X is the number of bits set in `xin`. If X goes first, the input should always have equal numbers of bits set in the two inputs. If O goes first, the input will always have one more bit set in `oin` than in `xin`.

Our output will also be a nine-bit, one-hot vector `xout` that indicates which position our circuit will be playing. A legal move must be orthogonal to both input vectors. On the next turn, `xin` will be replaced by the OR of the old `xin` and `xout`, and the opponent will have added a bit to `oin`.

Now that we have represented the board, our next step is to structure our circuit. A useful structure is as a set of ordered strategy components that each apply a strategy to generate the next move, a *divide-and-conquer* approach. The highest-priority that is able to generate a move is selected. For example, a good set of strategy components is as follows.

(1) *Win:* if a move will complete three-in-a-row, do it.
(2) *Don't lose:* if a move will block an opponent with two in a row, do it.
(3) *Pick first open square:* traversing the board in a particular order, pick the first square that is open.

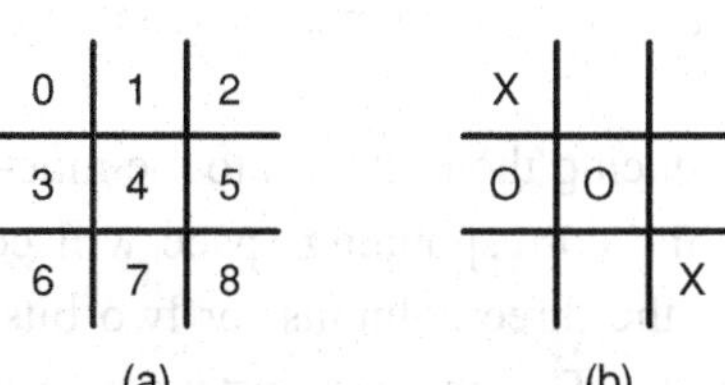

Figure 9.10. Representation of tic-tac-toe board. (a) Mapping a bit vector to the board. (b) Board represented by `xin` = 100000001, `oin` = 000011000.

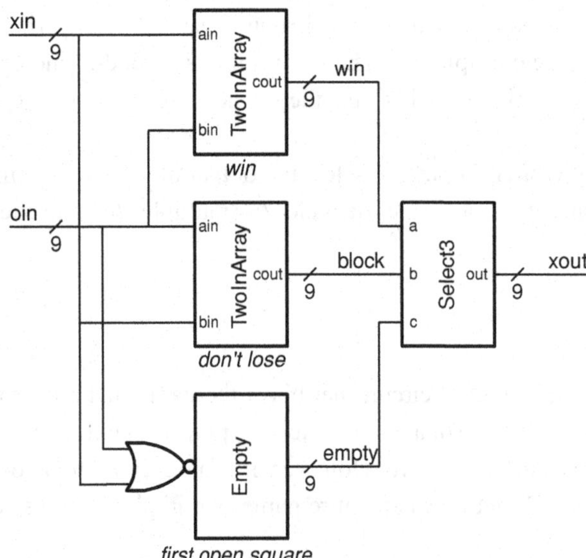

Figure 9.11. High-level design of the tic-tac-toe design entity. Three strategy components accept the inputs, `xin` and `oin`, and compute possible moves to win, not lose, or pick an empty square. The `Select3` component then picks the one with the highest priority of these possible moves to be the next move.

A selection circuit combines the inputs from our components and selects the highest-priority component with an output. With this modular design, we can easily add more strategy components later to refine the ability of our circuit.

The top-level design entity for our tic-tac-toe move generator is shown in Figure 9.11, and the VHDL for this design entity is given in Figure 9.12. It instantiates four components: two instances of `TwoInArray` and one instance each of `Empty` and `Select3`. The first `TwoInArray` component finds spaces (if any) where a play would cause us to win – that is, spaces where there is a row, column, or diagonal with two Xs and no Os. The second `TwoInArray` component finds spaces (if any) where, if we didn't play, the opponent could win on their next play – spaces where a row, column, or diagonal has two Os and no Xs. We use the same component both for the win and for the block strategies because they require the same function – just with the Xs and Os reversed. The next component, `Empty`, finds the first empty space according to a particular ordering of the spaces. The ordering picks empty spaces in order of their strategic value. Finally, the component `Select3` takes the three outputs of the previous components and selects the highest-priority move.

Most of the work in our tic-tac-toe implementation is done by the `TwoInArray` design entity shown in Figure 9.13. This design entity creates eight instances of the `TwoInRow` design entity (Figure 9.14). Each `TwoInRow` design entity checks one line (row, column, or diagonal). If the line being checked has two bits of `a` true and no bits of `b` true, a 1 is generated in the position of the open space. The design entity consists of three four-input AND gates, one for each of the three positions being checked. Note that we check only one bit of `b` in each AND gate since we are assuming that the inputs are legal, so that if a bit of `a` is true, the corresponding bit of `b` is false.

Three instances of `TwoInRow` check the rows producing their result into the nine-bit vector `rows`. If a bit of `rows` is true, playing an `a` into the corresponding space will complete a row. Similarly, three instances of `TwoInRow` check the three columns for two bits of `a` and no bits of `b`, producing results into nine-bit vector `cols`. The final two instances of `TwoInRow`

```
--------------------------------------------------------------------------------
-- TicTacToe
-- Generates a move for X in the game of tic-tac-toe
-- Inputs:
--    xin, oin - (9-bit) current positions of X and O.
-- Out:
--    xout - (9-bit) one-hot position of next X.
--
-- Inputs and outputs use a board mapping of:
--
--    0 | 1 | 2
--   ---+---+---
--    3 | 4 | 5
--   ---+---+---
--    6 | 7 | 8
--
-- The top-level circuit instantiates strategy components that each generate
-- a move according to their strategy and a selector component that selects
-- the highest-priority strategy component with a move.
--
-- The win strategy component picks a space that will win the game if any exists.
--
-- The block strategy component picks a space that will block the opponent
-- from winning.
--
-- The empty strategy component picks the first open space - using a particular
-- ordering of the board.
--------------------------------------------------------------------------------

library ieee;
use ieee.std_logic_1164.all;
use work.tictactoe_declarations.all;

entity TicTacToe is
  port( xin, oin: in std_logic_vector( 8 downto 0 );
        xout: out std_logic_vector( 8 downto 0 ) );
end TicTacToe;

architecture impl of TicTacToe is
  signal win, blk, emp : std_logic_vector( 8 downto 0 );
begin
  WINX: TwoInArray port map(xin, oin, win);
  BLOCKX: TwoInArray port map(oin, xin, blk);
  EMPTYX: Empty port map( not (oin or xin), emp);
  COMB: Select3 port map(win,blk,emp,xout);
end impl;
```

Figure 9.12. Top-level VHDL description for our tic-tac-toe move generator.

check the two diagonals producing results into three-bit vectors `ddiag` and `udiag` for the downward-sloping and upward-sloping diagonals.

After checking the rows, columns, and diagonals, the final assign statement combines the results into a single nine-bit vector by ORing together the individual components. The `rows`

```
-----------------------------------------------------------------------
-- TwoInArray
-- Indicates whether any row or column or diagonal in the array has two pieces of
-- type a and no pieces of type b. (a and b can be x and o or o and x)
-- Inputs:
--   ain, bin - (9 bits) array of types a and b
-- Output:
--   cout - (9 bits) location of space to play in to complete row, column
--          or diagonal of a.
-- If more than one space meets the criteria the output may have more than
-- one bit set.
-- If no spaces meet the criteria, the output will be all zeros.
-----------------------------------------------------------------------

library ieee;
use ieee.std_logic_1164.all;
use work.tictactoe_declarations.all;

entity TwoInArray is
  port( ain, bin: in std_logic_vector( 8 downto 0 );
        cout: out std_logic_vector( 8 downto 0 ) );
end TwoInArray;

architecture impl of TwoInArray is
  signal rows, cols, cc: std_logic_vector( 8 downto 0 );
  signal ddiag, udiag : std_logic_vector( 2 downto 0 );
begin
  -- check each row
  TOPR: TwoInRow port map( ain(2 downto 0), bin(2 downto 0), rows(2 downto 0) );
  MIDR: TwoInRow port map( ain(5 downto 3), bin(5 downto 3), rows(5 downto 3) );
  BOTR: TwoInRow port map( ain(8 downto 6), bin(8 downto 6), rows(8 downto 6) );

  -- check each column
  LEFTC:  TwoInRow port map( ain(6) & ain(3) & ain(0),
                             bin(6) & bin(3) & bin(0),
                             cc(8 downto 6) );
  MIDC:   TwoInRow port map( ain(7) & ain(4) & ain(1),
                             bin(7) & bin(4) & bin(1),
                             cc(5 downto 3) );
  RIGHTC: TwoInRow port map( ain(8) & ain(5) & ain(2),
                             bin(8) & bin(5) & bin(2),
                             cc(2 downto 0) );
  (cols(6),cols(3),cols(0),cols(7),cols(4),cols(1),cols(8),cols(5),cols(2)) <= cc;

  -- check both diagonals
  DNDIAGX: TwoInRow port map( ain(8)&ain(4)&ain(0), bin(8)&bin(4)&bin(0), ddiag );
  UPDIAGX: TwoInRow port map( ain(6)&ain(4)&ain(2), bin(6)&bin(4)&bin(2), udiag );

  cout <= rows or cols or (ddiag(2) & "000" & ddiag(1) & "000" & ddiag(0)) or
          ("00" & udiag(2) & "0" & udiag(1) & "0" & udiag(0) & "00");
end impl;
```

Figure 9.13. VHDL description of the TwoInArray design entity.

```
-----------------------------------------------------------------------------
-- TwoInRow
-- Indicates whether a row (or column, or diagonal) has two pieces of type a
-- and no pieces of type b. (a and b can be x and o or o and x)
-- Inputs:
--   ain, bin - (3 bits) row of types a and b.
-- Outputs:
--   cout - (3 bits) location of empty square if other two are type a.
-----------------------------------------------------------------------------
library ieee;
use ieee.std_logic_1164.all;

entity TwoInRow is
  port( ain, bin : in std_logic_vector( 2 downto 0 );
        cout : out std_logic_vector( 2 downto 0 ) );
end TwoInRow;

architecture impl of TwoInRow is
begin
  cout(0) <= not bin(0) and not ain(0) and ain(1) and ain(2);
  cout(1) <= not bin(1) and ain(0) and not ain(1) and ain(2);
  cout(2) <= not bin(2) and ain(0) and ain(1) and not ain(2);
end impl;
```

Figure 9.14. VHDL description of the `TwoInRow` design entity. This design entity outputs a 1 in the empty position of a row that contains two bits of `a` and no bits of `b`.

```
-----------------------------------------------------------------------------
-- Empty
-- Pick first space not in input.  Permute vector so middle comes first,
-- then corners, then edges.
-- Inputs:
--   i - (9 bits) occupied spaces
-- Outputs:
--   o - (9 bits) first empty space
-----------------------------------------------------------------------------
library ieee;
use ieee.std_logic_1164.all;
use work.ch8.all; -- for RArb

entity Empty is
  port( i: in std_logic_vector(8 downto 0);
        o: out std_logic_vector(8 downto 0) );
end Empty;

architecture impl of Empty is
  signal op: std_logic_vector(8 downto 0);
begin
  RA: RArb generic map(9) port map( i(4)&i(0)&i(2)&i(6)&i(8)&i(1)&i(3)&i(5)&i(7),op );
  (o(4),o(0),o(2),o(6),o(8),o(1),o(3),o(5),o(7)) <= op;
end impl;
```

Figure 9.15. VHDL description of an `Empty` design entity. This design entity uses an arbiter to find the first empty space by searching first the middle space, then the four corners, then the four edges.

```
--------------------------------------------------------------------------------
-- Select3
-- Picks the highest-priority bit from 3 9-bit vectors
-- Inputs:
--    a, b, c - (9 bits) Input vectors
-- Outputs:
--    output - (9 bits) One-hot output has a bit set (if any) in the highest
--             position of the highest-priority input.
--------------------------------------------------------------------------------

library ieee;
use ieee.std_logic_1164.all;
use work.ch8.all;

entity Select3 is
  port( a, b, c: in std_logic_vector( 8 downto 0 );
        output: out std_logic_vector( 8 downto 0 ) );
end Select3;

architecture impl of Select3 is
  signal x: std_logic_vector(26 downto 0);
begin
  RA: RArb generic map(27) port map(a & b & c, x);
  output <= x(26 downto 18) or x(17 downto 9) or x(8 downto 0);
end impl;
```

Figure 9.16. VHDL description of the `Select3` design entity. A 27-input arbiter is used to find the first set bit of the highest-priority strategy component. A three-way OR combines the arbiter outputs.

and `cols` vectors are combined directly. The three-bit diagonal vectors are first expanded to nine bits to place their active bits in the appropriate positions.

The `Empty` design entity, shown in Figure 9.15, uses an arbiter (see Section 8.5) to find the first non-zero bit in its input vector. Note that the top-level has ORed the two input vectors together and taken the complement so each one bit in the input to this design entity corresponds to an empty space. The input vector is permuted, using a concatenation statement, to give the priority order we want (middle first, then corners, then edges). The output is permuted in the same order to maintain correspondence.

The `Select3` design entity, shown in Figure 9.16, is also just an arbiter. In this case, a 27-bit arbiter scans all three inputs to find the first bit. This both selects the highest-priority non-zero input and also selects the first set bit of this input. The 27-bit output of the arbiter is reduced to nine bits by ORing the bits corresponding to each input together.

It is worth pointing out that the entire tic-tac-toe design entity is, at the bottom level, built entirely from just two component types: `TwoInRow` and `RArb`. This demonstrates the utility of combinational building blocks.

A simple testbench for the tic-tac-toe design entity is shown in Figure 9.17. The testbench instantiates two copies of the `TicTacToe` design entity. One plays X and the other plays O. The testbench starts by checking the copy that plays X, called `dut` in the testbench, with some directed testing. The five vectors check empty, win, and block strategies and check row, column, and diagonal patterns.

```
-- pragma translate_off
library ieee;
use ieee.std_logic_1164.all;

entity TestTic is
end TestTic;

architecture test of TestTic is
  signal x, o, xo, oo: std_logic_vector( 8 downto 0 );
begin
  DUT: entity work.TicTacToe(impl) port map(x,o,xo);
  OPPONENT: entity work.TicTacToe(impl) port map(o,x,oo);

  process begin
    -- all zeros, should pick middle
    x <= "000000000"; o <= "000000000";
    wait for 10 ns; report to_string(x) & " " & to_string(o) & " -> " & to_string(xo);
    -- can win across the top
    x <= "000000101"; o <= "000000000";
    wait for 10 ns; report to_string(x) & " " & to_string(o) & " -> " & to_string(xo);
    -- near-win: can't win across the top due to block
    x <= "000000101"; o <= "000000010";
    wait for 10 ns; report to_string(x) & " " & to_string(o) & " -> " & to_string(xo);
    -- block in the first column
    x <= "000000000"; o <= "000100100";
    wait for 10 ns; report to_string(x) & " " & to_string(o) & " -> " & to_string(xo);
    -- block along a diagonal
    x <= "000000000"; o <= "000010100";
    wait for 10 ns; report to_string(x) & " " & to_string(o) & " -> " & to_string(xo);
    -- start a game - x goes first
    x <= "000000000"; o <= "000000000";
    for i in 0 to 6 loop
      wait for 10 ns;
      report to_hstring(x(0)&o(0))&" "&to_hstring(x(1)&o(1))&" "&to_hstring(x(2)&o(2));
      report to_hstring(x(3)&o(3))&" "&to_hstring(x(4)&o(4))&" "&to_hstring(x(5)&o(5));
      report to_hstring(x(6)&o(6))&" "&to_hstring(x(7)&o(7))&" "&to_hstring(x(8)&o(8));
      report "";
      x <= x or xo;
      wait for 10 ns;
      report to_hstring(x(0)&o(0))&" "&to_hstring(x(1)&o(1))&" "&to_hstring(x(2)&o(2));
      report to_hstring(x(3)&o(3))&" "&to_hstring(x(4)&o(4))&" "&to_hstring(x(5)&o(5));
      report to_hstring(x(6)&o(6))&" "&to_hstring(x(7)&o(7))&" "&to_hstring(x(8)&o(8));
      report "--------";
      o <= o or oo;
    end loop;
    std.env.stop(0);
  end process;
end test;
-- pragma translate_on
```

Figure 9.17. VHDL testbench for the tic-tac-toe design performs directed testing and then plays one instance against another.

```
. . .     . . .     O . .     O . X     O . X
. . .     . X .     . X .     . X .     . X .
. . .     . . .     . . .     . . .     O . .

O . X     O . X     O . X     O O X     O O X
X X .     X X O     X X O     X X O     X X O
O . .     O . .     O . X     O . X     O X X
```

Figure 9.18. Results of playing one `TicTacToe` instance against another.

After the five directed patterns, the testbench plays a game of tic-tac-toe by ORing the outputs of each instance into its input to compute the input for the next round. The results of the game (obtained by writing a script to massage the output of the `report` statements) are shown in Figure 9.18.

The game starts with an empty board. The *empty* rule applies and X plays to the center – our highest-priority empty space. The *empty* rule applies for the next two turns as well, and O and X take the top two corners. At this point X has two in a row, so the *block* rule applies and O plays to the bottom left corner (position 6), completing the first row of the figure.

The second row of the figure starts with the *block* rule, causing X to play on the left edge (position 3); O then blocks X in the middle row. At this point *empty* causes X to take the remaining corner. In the last two moves, *empty* causes O and X to fill the two remaining open spaces. The game ends in a draw.

The verification performed by this testbench is by no means adequate to verify proper operation. Many combinations of inputs have not been tried. To verify the design thoroughly a checker is required. This would typically be implemented in a high-level programming language (like "C") and interfaced to the simulator. Proper operation would then be verified by comparing the simulation results with the high-level language model. One hopes that the same mistake would not be made in both models.

Once a checker is in place, we still need to pick the test vectors. After more directed testing (e.g., win, block, near-win, and near-block on all eight lines), we could take two approaches. We could exhaustively test the design (there are 2^{18} input cases). Depending on how fast our simulator runs, we may have time to try them all. Alternatively, if we don't have time for exhaustive testing, we could apply random testing, randomly generating input patterns and checking the resulting outputs.

Summary

In this chapter you have seen four extended examples that bring together much of what we have learned up to this point in the book. The *multiple-of-3* circuit is an example of an iterative circuit. It is built from eight `Multiple_of_3_bit` components, each of which is an example of a combinational design entity specified with a `case` statement. The top-level design entity is a good example of structural VHDL. The testbench for this circuit performs exhaustive testing and is self-checking.

Our *tomorrow* circuit implements a function specified in Section 1.4 that computes the day and date of tomorrow given the day and date of today as input. The components of this circuit give examples of defining combinational circuits using both `case` and concurrent assignment statements. The top-level design entity mixes structural VHDL – instantiating and connecting the components – with the use of concurrent assignment statements at the same level.

The *priority arbiter* gives an example of realizing a function by composing combinational building blocks – in this case comparators and multiplexers. The circuit performs a *tournament* among the inputs using a comparator to select a winner of each match and using a multiplexer to route the winner to the next level of the tournament.

Finally, our *tic-tac-toe* circuit illustrates how complex functionality can be built from simple circuits and how the *divide-and-conquer* technique can be used to break a complex task down into simple parts. The top-level circuit instantiates a *strategy* component and then uses an arbiter to select the highest-priority result. Each strategy component, in turn, is implemented from simple logic components. At the bottom level, the entire circuit is implemented by *arbiters* and `TwoInRow` components.

Exercises

9.1 *Voting circuit.* Using combinational building blocks such as adders, comparators, multiplexers, decoders, encoders, and arbiters, as well as logic gates, design a circuit that accepts five three-bit one-hot numbers and outputs the three-bit one-hot number that occurred most often on the inputs. Ties can be broken in any manner. For example, if the inputs are 100, 100, 100, 010, and 001, the output will be 100.

9.2 *Middle circuit.* Using building blocks such as binary adders, comparators, multiplexers, decoders, encoders, and arbiters, as well as logic gates, design a circuit that accepts three one-hot eight-bit numbers, $a2_{7:0}$, $a1_{7:0}$, and $a0_{7:0}$, and outputs the input with the middle of the three values. For example, if the inputs are $a2 = 10000000$, $a1 = 00010000$, and $a0 = 00000001$, the output should be 00010000, the middle of the three one-hot values.

9.3 *Multiple-of-5 circuit, design.* Using an approach similar to the multiple-of-3 circuit of Section 9.1, design a multiple-of-5 circuit that outputs true iff its eight-bit input is a multiple of 5.

9.4 *Multiple-of-5 circuit, implementation.* Code your design from Exercise 9.3 in VHDL and exhaustively verify it with a testbench.

9.5 *Multiple-of-10 circuit, design.* Design a circuit that outputs true if its eight-bit input is a multiple of 10. (Hint: think about how few bits of remainder you need to implement this function.)

9.6 *Multiple-of-10 circuit, implementation.* Code your design from Exercise 9.5 in VHDL and exhaustively verify it with a testbench.

9.7 *Modulo-3 circuit, design.* Modify the multiple-of-3 circuit of Section 9.1 to output `in %3`, that is the input modulo 3.

9.8 *Modulo-3 circuit, implementation.* Code your design from Exercise 9.7 in VHDL and exhaustively verify it with a testbench.

9.9 *Modifying the calendar circuit, I.* Recode the `NextDayOfWeek` design entity so it will work with arbitrary definitions of the constants `'SUNDAY, 'MONDAY, ..., 'SATURDAY`.

9.10 *Modifying the calendar circuit, II.* Modify the calendar circuit to work correctly in leap years. Assume your input includes the year – in 12-bit binary format.

9.11 *Calendar representations.* Design a combinational logic circuit that takes a date as the number of days since January 1, 0000, and returns the date in month, day-of-month format. (Optional: also generate day-of-week format.)

9.12 *Ties in the priority arbiter, design.* The priority arbiter of Section 9.3 currently breaks ties in favor of the lower-numbered input. Modify the circuit so that it breaks ties in favor of the higher-numbered input.

9.13 *Ties in the priority arbiter, implementation.* Code your design from Exercise 9.12 in VHDL and verify it with selected test cases.

9.14 *Five-input priority arbiter.* Modify the priority arbiter of Section 9.3 to take five inputs.

9.15 *Eight-input priority arbiter.* Modify the priority arbiter of Section 9.3 to take eight inputs.

9.16 *Inverted priority.* Modify the priority arbiter of Section 9.3 to pick the input with the lowest value.

9.17 *Winning value.* Modify the priority arbiter of Section 9.3 to output the value of the winning input.

9.18 *Tic-tac-toe strategy, I.* Extend the tic-tac-toe design entity Section 9.4 by adding a strategy design entity to play to a space that creates two in a row. Create a design entity called `OneInARow` that finds rows, columns, and diagonals with one X and no Os. Use this component to build a design entity `OneInArray` that implements this strategy.

9.19 *Tic-tac-toe strategy, II.* Extend the tic-tac-toe design entity of Section 9.4 by adding a strategy component that, on an empty board, plays to space 0 (upper left corner).

9.20 *Tic-tac-toe strategy, III.* Extend the tic-tac-toe design entity of Section 9.4 by adding a strategy component that, on a board that is empty except for an opponent O in two opposite corners and your X in the middle, plays to an adjacent edge space. (Play to the space marked H in the diagram below.[1])

```
O . .
H X .
. . O
```

9.21 *Tic-tac-toe, input check.* Add a component to the tic-tac-toe design entity that checks whether the inputs are legal.

[1] If you play to a corner in this situation, your opponent O will win in two moves.

9.22 *Tic-tac-toe, game over.* Add a component to the tic-tac-toe design entity that outputs a signal when the game is over and indicates the outcome. The signal should encode the options: playing, win, lose, draw.

9.23 *Verification.* Build a checker for the tic-tac-toe design entity and write a testbench that performs random testing on the design.

Part III

Arithmetic circuits

10 Arithmetic circuits

Many digital systems operate on numbers, performing arithmetic operations such as addition and multiplication. For example, a digital audio system represents a waveform as a sequence of numbers and performs arithmetic to filter and scale the waveform.

Digital systems internally represent numbers in binary form. Arithmetic functions, including addition and multiplication, are performed as combinational logic functions on these binary numbers. In this chapter, we introduce binary representations for positive and negative integers and develop the logic for simple addition, subtraction, multiplication, and division operations. In Chapter 11, we expand on these basics by looking at floating-point number representations that approximate real numbers. In Chapter 12, we look at methods for accelerating arithmetic operations. Finally, Chapter 13 presents several examples of designs that use these arithmetic operations.

10.1 BINARY NUMBERS

As human beings we are used to representing numbers in *decimal*, or base-10, notation. That is, we use a positional notation in which each digit is weighted by ten times the weight of the digit to its right. For example, the number 1234_{10} (the subscript implies base 10) represents $1 \times 1000 + 2 \times 100 + 3 \times 10 + 4$. It is likely that we use the decimal system because we have ten fingers on which to count.

With digital electronics, we do not have ten fingers; instead, we have two states, 1 and 0, with which to represent values. Thus, while computers can (and sometimes do) represent numbers in base 10, it is more natural to represent numbers in base 2, or *binary*, notation. With binary notation, each digit is weighted by two times the weight of the digit to its right. For example, the number 1011_2 (the subscript implies base 2) represents $1 \times 8 + 0 \times 4 + 1 \times 2 + 1 = 11_{10}$.

More formally, a number $a_{n-1}, a_{n-2}, \ldots, a_1, a_0$ in base b represents the value

$$v = \sum_{i=0}^{n-1} a_i b^i. \tag{10.1}$$

For a binary number, $b = 2$, and we have

$$v = \sum_{i=0}^{n-1} a_i 2^i. \tag{10.2}$$

Bit a_{n-1} is the leftmost or *most significant bit* (MSB) of the binary representation, while bit a_0 is the rightmost or *least significant bit* (LSB).

We can convert from one base to another by evaluating either Equation (10.1) or Equation (10.2) in the target base, as we did above to convert 1011_2 to 11_{10}. Converting from decimal to binary can be done by applying this technique. For example, $1234_{10} = 1 \times 1111101000_2 + 10_2 \times 1100100_2 + 11_2 \times 1010_2 + 100_2 \times 1 = 1111101000_2 + 11001000_2 + 11110_2 + 100_2 = 10011010010_2$. However, this procedure is a bit tedious, requiring lots of binary calculations.

It is usually more convenient to subtract repeatedly the highest power of 2 less than the number, and then add these powers of 2 to form the representation in base 2. For example, in Equation (10.3) we convert the number 1234_{10} to binary. We start with 1234_{10} in the left column and repeatedly subtract the largest power of 2 that is smaller than the remaining number. Each time we subtract a value from the left column, we add the same number – but with binary representation – to the right column. At the bottom, the left column is 0; we have subtracted away the entire value of 1234_{10}, and the right column is 10011010010_2, the binary representation of 1234_{10}. We have added the entire value of 1234_{10} in this column, one bit at a time:

$$
\begin{array}{rr}
1234_{10} & 0_2 \\
\underline{-1024_{10}} & \underline{+10,000,000,000_2} \\
210_{10} & 10,000,000,000_2 \\
\underline{-128_{10}} & \underline{+10,000,000_2} \\
82_{10} & 10,010,000,000_2 \\
\underline{-64_{10}} & \underline{+1,000,000_2} \\
18_{10} & 10,011,000,000_2 \\
\underline{-16_{10}} & \underline{+10,000_2} \\
2_{10} & 10,011,010,000_2 \\
\underline{-2_{10}} & \underline{+10_2} \\
0_{10} & 10,011,010,010_2
\end{array}
\tag{10.3}
$$

Because binary numbers can be quite long – representing a four-digit decimal number takes 11 digits – we sometimes display them using *hexadecimal*, or base-16, notation. Because $16 = 2^4$, it is easy to convert between binary and hexadecimal. We simply break a binary number into four-bit chunks and convert each chunk to base 16. For example 1234_{10} is 10011010010_2 and $4D2_{16}$. As shown below, we simply take each four-bit group of 10011010010_2, starting from the right, and convert the group to hexadecimal. We use the characters A–F to represent single digits with values 10–15, respectively. So $10_{10} = A_{16}$, $11_{10} = B_{16}$, $12_{10} = C_{16}$, $13_{10} = D_{16}$, $14_{10} = E_{16}$, and $15_{10} = F_{16}$. So, the character D in $4D2_{16}$ implies that the second digit (with weight 16) has a value of 13:

$$
\begin{array}{cccc}
0100 & 1101 & 0010 & {}_2 \\
4 & D & 2 & {}_{16}
\end{array}
\tag{10.4}
$$

Some digital systems use a *binary-coded decimal*, or BCD, representation to encode decimal numbers. This is a representation where each decimal digit is represented by a four-bit binary number. In BCD, the value is given by

$$v = \sum_{i=0}^{n-1} d_i 10^i \tag{10.5}$$

$$= \sum_{i=0}^{n-1} \left(10^i \times \sum_{j=0}^{3} a_{ij} 2^j \right). \tag{10.6}$$

That is, each decimal group d_i of four bits is weighted by a power of 10, and each binary digit b_{ij} within decimal group d_i is additionally weighted by a power of 2. For example, the number $1234_{10} = 0001001000110100_{\mathrm{BCD}}$:

$$\begin{array}{ccccc} 1 & 2 & 3 & 4 & {}_{10} \\ 0001 & 0010 & 0011 & 0100 & {}_{\mathrm{BCD}} \end{array} \tag{10.7}$$

The reason why we use binary notation to represent numbers in digital systems is that it makes common operations (addition, subtraction, multiplication, etc.) easy to perform. As always, we pick a representation suitable for the task at hand. If we had a different set of operations to perform, we might pick a different representation.

EXAMPLE 10.1 Binary conversion

Convert the number 5961 to the following notations: binary, hexadecimal, and BCD.

We can follow the same process as in Equation (10.3) to convert to binary. There will be a total of 13 bits in the answer:

$$\begin{array}{rr}
5961_{10} & 0_2 \\
\underline{-4096_{10}} & \underline{+1,000,000,000,000_2} \\
1865_{10} & 1,000,000,000,000_2 \\
\underline{-1024_{10}} & \underline{+10,000,000,000_2} \\
841_{10} & 1,010,000,000,000_2 \\
\underline{-512_{10}} & \underline{+1,000,000,000_2} \\
329_{10} & 1,011,000,000,000_2 \\
\underline{-256_{10}} & \underline{+100,000,000_2} \\
73_{10} & 1,011,100,000,000_2 \\
\underline{-64_{10}} & \underline{+1,000,000_2} \\
9_{10} & 1,011,101,000,000_2 \\
\underline{-8_{10}} & \underline{+1,000_2} \\
1_{10} & 1,011,101,001,000_2 \\
\underline{-1_{10}} & \underline{+1_2} \\
0_{10} & 1,011,101,001,001_2
\end{array}$$

The number 5961_{10} is 1011101001001_2 in binary. To write the value in hexadecimal, we group the binary digits into sets of four:

$$\begin{array}{ccccc} 0001 & 0111 & 0100 & 1001 & {}_{2} \\ 1 & 7 & 4 & 9 & {}_{16} \end{array}$$

Finally, in BCD, the number becomes 0101 1001 0110 0001.

Table 10.1. **Truth table for a half adder**

a	b	r	c	s
0	0	0	0	0
0	1	1	0	1
1	0	1	0	1
1	1	2	1	0

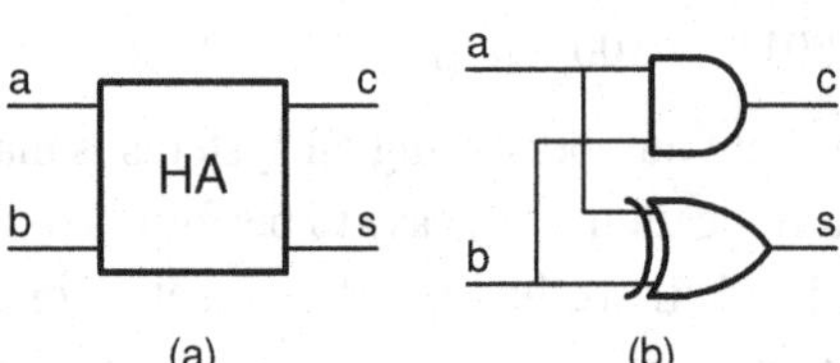

Figure 10.1. Half adder: (a) symbol, (b) logic circuit.

10.2 BINARY ADDITION

The first operation we will consider is addition. We add binary numbers the same way we add decimal numbers – one digit at a time, starting from the right. The only difference is that the digits are binary, not decimal. This actually simplifies addition considerably since we only have to remember four possible combinations of digits (rather than 100).

To add two bits, a and b, together, there are only four possibilities for the result, r, as shown in Table 10.1. In the first row, we add $0 + 0$ to get $r = 0$. In the second and third rows we add $0 + 1$ (or equivalently $1 + 0$) to get $r = 1$. Finally, in the last row, if both a and b are 1, we get $r = 1 + 1 = 2$.

To represent results r ranging from 0 to 2 requires two bits, s and c, as shown in Table 10.1. The LSB, s, we refer to as the *sum*, and the MSB, c, we refer to as the *carry*. (The reason for these names will become clear shortly when we discuss multi-bit addition.)

A circuit that adds two bits together to produce a sum and carry is called a *half adder*. The reader will have noticed that the truth table for the sum is the same as the truth table for an XOR gate and the truth table for the carry is just the truth table for an AND gate. Thus we can realize a half adder with just these two gates, as shown in Figure 10.1.

To handle a carry input, we require a circuit that accepts three input bits: a, b, and *cin* (for carry in), and generates a result r that is the sum of these bits. Now r can range from 0 to 3, but it can still be represented by two bits, s and *cout* (for carry out). A circuit that adds three equally weighted bits together to generate a sum and a carry is called a *full adder*, and a truth table for this circuit is shown in Table 10.2.

A full adder circuit is shown in Figure 10.2. From Table 10.2, we observe that the sum output has the truth table of a three-input exclusive-or (i.e., the output is true when an odd number of the inputs are true). The carry output is true whenever a majority of the inputs are

Table 10.2. **Truth table for a full adder**

a	*b*	*cin*	*r*	*cout*	*s*
0	0	0	0	0	0
0	0	1	1	0	1
0	1	0	1	0	1
0	1	1	2	1	0
1	0	0	1	0	1
1	0	1	2	1	0
1	1	0	2	1	0
1	1	1	3	1	1

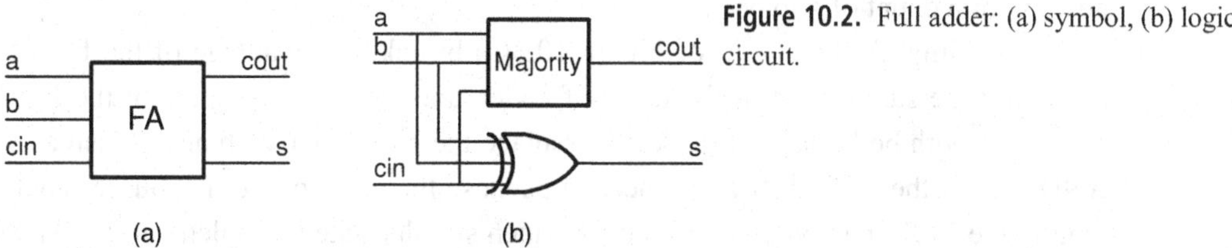

Figure 10.2. Full adder: (a) symbol, (b) logic circuit.

true (two or three inputs true out of three), and thus can be implemented using a majority circuit (Equation (3.6)).

The astute reader will have observed by now that adder circuits are really counters. A half adder, or a full adder, just counts the number of 1s on its inputs (all inputs are equivalent) and reports the count in binary form on its output. For a half adder the count is in the range of 0 to 2, and for a full adder the count ranges from 0 to 3.

We can use this counting property to construct a full adder from half adders as shown in Figure 10.3(a). In the figure the numbers in parentheses indicate the weight of a signal. The inputs are all weighted (1). The result output is a binary number with the sum weighted (1) and the carry out weighted (2). Because an adder counts the 1s on its inputs, they should all be of equal weight – otherwise one input would count more than another. We use one half adder to count two of the original inputs, producing a sum which we call p (for propagate) and a carry which we call g (for generate). If the propagate signal is true, a one bit on carry in (*cin*) will cause carry out to go high. That is, the carry in *propagates* to the carry out. If the generate signal is true, the carry out will be true regardless of the carry in. We say that bits a and b *generate* the carry out. We will see in Section 12.1 how the generate and propagate signals are used to build very fast adders. For now, however, we will continue with our simple adder.

A second half adder combines p (of weight (1)) with the carry input (also of weight (1)) to produce the sum output s (of weight (1)) and a carry output, which we call cp (for propagated carry) (of weight (2)). At this point we have a single weight (1) signal, s, and two weight (2) signals, cp and g. We use a third half adder to combine the two weight (2) signals. The sum output of this third half adder is the carry out (of weight (2)). The carry output (which would

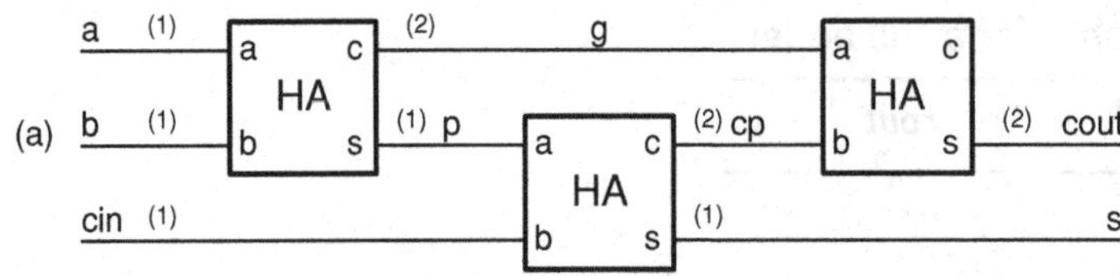

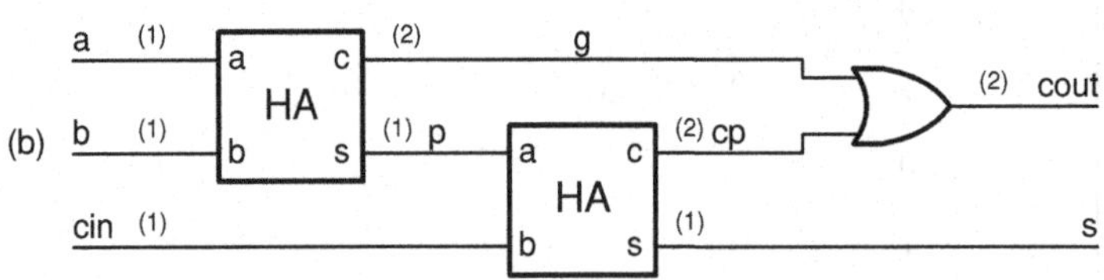

Figure 10.3. A full adder can be constructed using (a) three half adders or (b) two half adders and an OR gate. The numbers in parentheses show the weight of each signal.

be of weight (4)) of this half adder is unused. This is because, with only three inputs, there is no way for a count of 4 to occur.

We can simplify the circuit of Figure 10.3(a) by taking advantage of the facts that (a) we need only the sum output of the last half adder, and (b) the two inputs of the last half adder will never both be high. Fact (a) lets us replace the half adder with an exclusive-or gate. The AND gate in the half adder is not needed because the carry output is unused. Fact (b) lets us replace the XOR gate with an OR gate (a much simpler gate to implement in CMOS) because their truth tables are identical, except for the state where both inputs are high. The result is the circuit of Figure 10.3(b). The VHDL description for this circuit is shown in Figure 10.4.

Figure 10.5 shows an optimized CMOS logic circuit for a full adder. The circuit consists of an inverter and five CMOS gate circuits, Q1–Q5, including two two-input NAND gates, Q1 and Q3, and three three-input OR-AND-invert (OAI) gates, Q2, Q4, and Q5. Gates Q1 and Q2 form the first half adder – providing complemented outputs p and g. Gates Q3 and Q4 form an exclusive-NOR (XNOR) gate which acts as the XOR of the second half adder – producing output s. The input OR (low-true AND) part of gate Q5 performs the AND part of the second half adder. Signal *cp* is not produced, however. It remains internal to gate Q5. The output AND (low-true OR) part of gate Q5 performs the OR to combine g with *cp* and generates the carry out.

It is illustrative to see how these arithmetic circuits are realized in CMOS gates. Fortunately, with modern logic synthesis, you will very rarely have to work with arithmetic circuits at the gate level.

Now that we have circuits that can add single bits, we can move on to multi-bit addition. To add multi-bit numbers, we simply apply this single-bit binary addition one bit at a time from right to left. For example, suppose we are working with four-bit binary numbers. To add 3_{10} 0011 to 6_{10} 0110 we compute as shown below:

```
  1 1 0
 0 1 1 0
+0 0 1 1
--------
 1 0 0 1
```

```
------------------------------------------------------------------------
-- half adder
library ieee;
use ieee.std_logic_1164.all;
entity HalfAdder is
  port( a, b: in std_logic;
        c, s: out std_logic ); -- carry and sum
end HalfAdder;
architecture impl of HalfAdder is
begin
  s <= a xor b;
  c <= a and b;
end impl;
------------------------------------------------------------------------
-- full adder - from half adders
library ieee;
use ieee.std_logic_1164.all;
use work.ch10.all;
entity FullAdder is
  port( a, b, cin: in std_logic;
        cout, s: out std_logic ); -- carry and sum
end FullAdder;
architecture impl of FullAdder is
  signal g, p: std_logic; -- generate and propagate
  signal cp: std_logic;
begin
  HA1: HalfAdder port map(a,b,g,p);
  HA2: HalfAdder port map(cin,p,cp,s);
  cout <= g or cp;
end impl;
```

Figure 10.4. VHDL description of a full adder constructed from half adders.

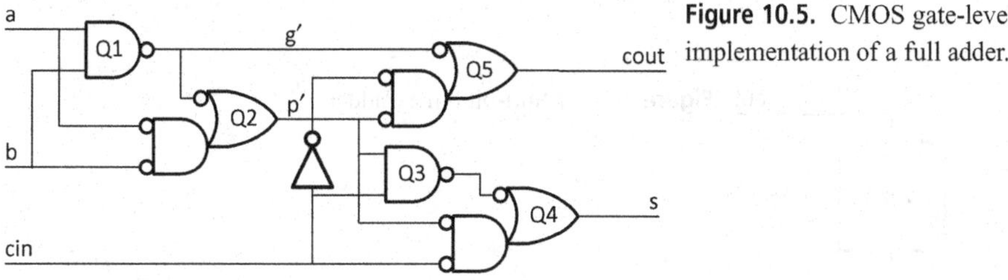

Figure 10.5. CMOS gate-level implementation of a full adder.

We start in the rightmost column by adding the two LSBs $0 + 1$ to give 1 as the LSB of the result. The result can be represented in a single bit (i.e., it is less than 2), so the carry into the next column (denoted by the small gray number on the top row) is 0. With the carry, we have three bits to sum for the second column, 0, 1, and 1, and the result is 2 – so the second bit of the sum is 0 and we carry a 1 to the top of the third column. In the third column the bits are 1, 1, and 0 – again two 1s, so the sum and carry are again 0 and 1, respectively. In the fourth and final column, only the carry is a 1, so the sum is 1 and the carry (not shown) is 0. The result is $0110 + 0011 = 1001$ or $6_{10} + 3_{10} = 9_{10}$.

EXAMPLE 10.2 Binary addition

Perform, in binary, the following addition: $71_{10} + 51_{10}$.

First, we convert the numbers to binary: $71_{10} = 1000111_2$ and $51_{10} = 110011$. Next, we add (the carry outputs on the top row, sum outputs on the bottom):

$$\begin{array}{rr} c & 0001110 \\ & 1000111 \\ + & \underline{0110011} \\ s & 1111010 \end{array}$$

When we convert the answer back to decimal, we can check that $1111010_2 = 122_{10}$.

We can build a multi-bit adder circuit from full adders by operating in the same manner – starting from the LSB and working toward the MSB. Such a circuit is shown in Figure 10.6. The bottom full adder, FA0, sums a carry in, *cin*, with the LSBs of the two inputs a_0 and b_0 generating the LSB of the sum s_0 and a carry into bit 1, c_1. We could have used a half adder for this bit; however, we chose to use a full adder to allow us to accept a carry in. Each subsequent full adder bit, FA*i*, sums the carry into that bit, c_i, with that bit's inputs a_i and b_i to generate that bit of the sum, s_i, and the carry into the next bit, c_{i+1}.

This circuit is often referred to as a *ripple-carry adder* because, if the inputs are set correctly (exactly one of a_i or b_i true for all *i*), the carry will *ripple* from *cin* to *cout*, propagating through all *n* full adders. For large *n* (more than 8) this can be quite slow. We will see in Section 12.1 how to build adders with delay proportional to $\log(n)$ rather than *n*.

For most applications, the appropriate way to describe an adder in VHDL is behaviorally, as shown in Figure 10.7. After declaring the inputs and outputs, the actual description here is a single line that uses the "+" operator to add single-bit `cin` to *n*-bit `a` and `b`. The concatenation of `cout` and `s` accepts the output.

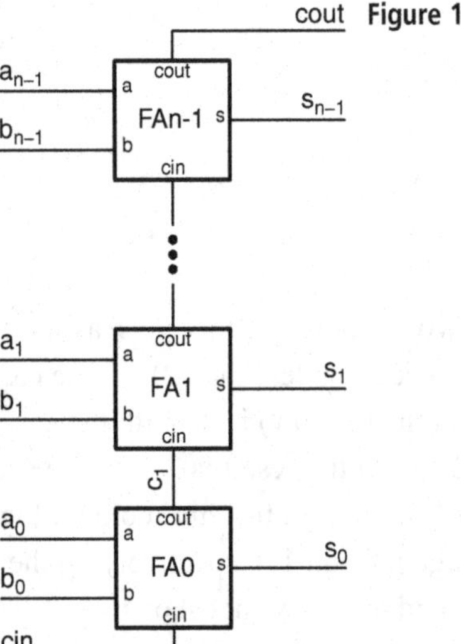

Figure 10.6. Multi-bit binary adder.

```
-- multi-bit adder - behavioral
library ieee;
use ieee.std_logic_1164.all;
use ieee.std_logic_unsigned.all;

entity Adder is
  generic( n: integer := 8 );
  port( a, b : in std_logic_vector(n-1 downto 0);
        cin: in std_logic;
        cout: out std_logic;
        s: out std_logic_vector(n-1 downto 0));
end Adder;

architecture impl of Adder is
  signal sum: std_logic_vector(n downto 0);
begin
  sum <= ('0' & a) + ('0' & b) + cin;
  cout <= sum(n);
  s <= sum(n-1 downto 0);
end impl;
```

Figure 10.7. Behavioral VHDL description of a multi-bit adder. This description uses the definition of "+" from `std_logic_unsigned` to describe addition.

```
-- multi-bit adder - bit-by-bit logical
architecture ripple_carry_impl of Adder is
  signal p, g: std_logic_vector(n-1 downto 0);
  signal c: std_logic_vector(n downto 0);
begin
  p <= a xor b; -- propagate
  g <= a and b; -- generate
  c <= (g or (p and c(n-1 downto 0))) & cin; -- carry = g or (p and c)
  s <= p xor c(n-1 downto 0); -- sum
  cout <= c(n);
end ripple_carry_impl;
```

Figure 10.8. Bit-wise logical VHDL description of a ripple-carry adder.

Modern synthesis tools are quite good at taking a behavioral description, like the one shown here, and generating a very efficient logic netlist. In fact, many tools come with libraries of optimized versions of various arithmetic units. There is rarely any reason to describe an adder in more detail.

For illustrative purposes, an alternative VHDL description of an adder is shown in Figure 10.8. This design entity describes the bit-by-bit logic of a ripple-carry adder in terms of AND, OR, and XOR operations. The description defines n-bit propagate and generate variables and then uses them to compute the carry. The definition of the carry uses a concatenation and subfield specification to make bit i of the carry a function of bit $i - 1$. This is not a circular definition.

While this description is useful for showing the logical definition of an adder, it may generate a logic netlist that is inferior to the behavioral description of Figure 10.7. This is because the synthesis tool may not recognize it as an adder – and hence not perform its special adder synthesis. In contrast, when you use the "+" operator, there is no doubt that the circuit being described is an adder. The logical description is also harder to read and maintain. Without the design entity name, signal names, and comments, you would have to study this design entity for a while to discern its function. In contrast, using "+," it is immediately obvious to a reader (as well as a synthesis tool) what you mean.

Our n-bit adder (of Figures 10.6–10.8) accepts two n-bit inputs and produces an $(n + 1)$-bit output. This ensures that we have enough bits to represent the largest possible sum. For example with a three-bit adder, adding binary 111 to 111 gives a four-bit result, 1110. In many applications, however, we need an n-bit output. For example, we may want to use the output as a later input. In these cases we need to discard the carry out and retain just the n-bit sum. Restricting ourselves to an n-bit output raises the specter of *overflow* – a condition that occurs when we compute an output that is too large to be represented as n bits.

Overflow is usually an error condition. It is easily detected; any time carry out is 1, an overflow has occurred. Most adders perform modulo arithmetic on an overflow condition – they compute $a + b \pmod{2^n}$. For example, with a three-bit adder, adding $111 + 010$ gives 001 ($7_{10} + 2_{10} = 1_{10} \pmod{8_{10}}$). In Exercise 10.20 we will look at a saturating adder which takes a different approach to producing an output during an overflow condition.

10.3 NEGATIVE NUMBERS AND SUBTRACTION

With n-bit binary numbers, using Equation (10.2) we can represent only non-negative integers up to a maximum value of $2^n - 1$. We often refer to binary numbers that represent only positive integers as *unsigned* numbers (because they don't have a + or − sign). In this section we will see how to use binary numbers to represent both positive and negative integers, often referred to as *signed* numbers. To represent signed numbers, we have three main choices: 2's complement, 1's complement, and sign-magnitude.

The simplest system conceptually is sign-magnitude. Here we simply add a sign bit, s, to the number, with the convention that if $s = 0$ the number is positive and if $s = 1$ the number is negative. By convention we place the sign bit in the leftmost (MSB) position. Consider the numbers $+23_{10}$ and -23_{10}. In sign-magnitude representation, $+23_{10} = 010111_2$ and $-23_{10} = 110111_{2SM}$. All that changes between the two numbers is the sign bit. Our value function becomes

$$v = -1^s \times \sum_{i=0}^{n-1} a_i 2^i. \tag{10.8}$$

To negate a 1's complement number, we complement all of the bits of the number. So, to negate our example number, $+23_{10} = 010111_2$, we get $-23_{10} = 101000_{2OC}$. The value function becomes

$$v = -a_{n-1}(2^{n-1} - 1) + \sum_{i=0}^{n-2} a_i 2^i. \tag{10.9}$$

Here the sign bit, a_{n-1}, is weighted, but by $-(2^{n-1}-1)$ (a number that is all 1s in binary representation – hence the name 1's complement).

Finally, to negate a 2's complement number, we complement all of the bits of the number and then add 1. For our example number, $+23_{10} = 010111_2$ and $-23_{10} = 101001_2$. The value function becomes

$$v = -a_{n-1}2^{n-1} + \sum_{i=0}^{n-2} a_i 2^i. \tag{10.10}$$

Compared with 1's complement, the weight on the sign bit has been decreased by 1 to -2^{n-1}.

So which of these three formats should we use in a given system? The answer depends on the system. However, the vast majority of digital systems use a 2's complement number system because it simplifies addition and subtraction. We can add positive or negative 2's complement numbers directly, using binary addition, and get the correct answer. The same is not true of sign-magnitude or 1's complement numbers.

Consider, for example, adding +4 and −3 to get a result of +1 represented as four-bit signed binary numbers. The inputs and outputs of this computation are shown for the three number systems below. For 2's complement, adding +4 (0100) to −3 (1101) gives +1 (10001) – the correct answer if we ignore the carry (there is more on carry and overflow below). In contrast, just adding the 1's complement numbers gives 10000,[1] and adding the sign-magnitude numbers gives 1111.[2] Here are the inputs and outputs:

	2's comp	1's comp	sign-mag
+4	0100	0100	0100
−3	1101	1100	1011
+1	0001	0001	0001

To see why 2's complement numbers make adding negative numbers easy, it is instructive to review how we generate a 2's complement integer. We complement the bits and add 1. Complementing the bits of a number x gives $2^n - 1 - x$ ($15 - x$ for four-bit integers); for example, $15 - 3 = 12$, which is 1100 in binary (the 1's complement of 3). The 2's complement of x is one more than this, or $2^n - x$ ($16 - x$ for four-bit integers); for example, $16 - 3 = 13$, which is 1101 in binary (the 2's complement of 3). Because all addition is performed mod 2^n, the 2's complement of a number, $2^n - x$, is the same as $-x$. Hence we get the correct result. Returning to our example, we have

$$\begin{aligned} 4 - 3 &= 4 + (16 - 3) \pmod{16} \\ &= 17 \pmod{16} \\ &= 1. \end{aligned} \tag{10.11}$$

It is often helpful when thinking about 1's complement or 2's complement arithmetic to visualize the numbers on a wheel, as shown in Figure 10.9. Here we show the four-bit numbers from 0000 (at the 12 o'clock position) to 1111 incrementing in a clockwise direction around a circle. In Figures 10.9(a), (b), and (c) we show the values assigned to these bit patterns by the 2's complement, 1's complement, and sign-magnitude number systems, respectively.

[1] In Exercise 10.33 we see how a 1's complement adder can be built by using an end-around carry.

[2] Sign-magnitude addition of negative numbers is performed by first converting to 1's complement or 2's complement

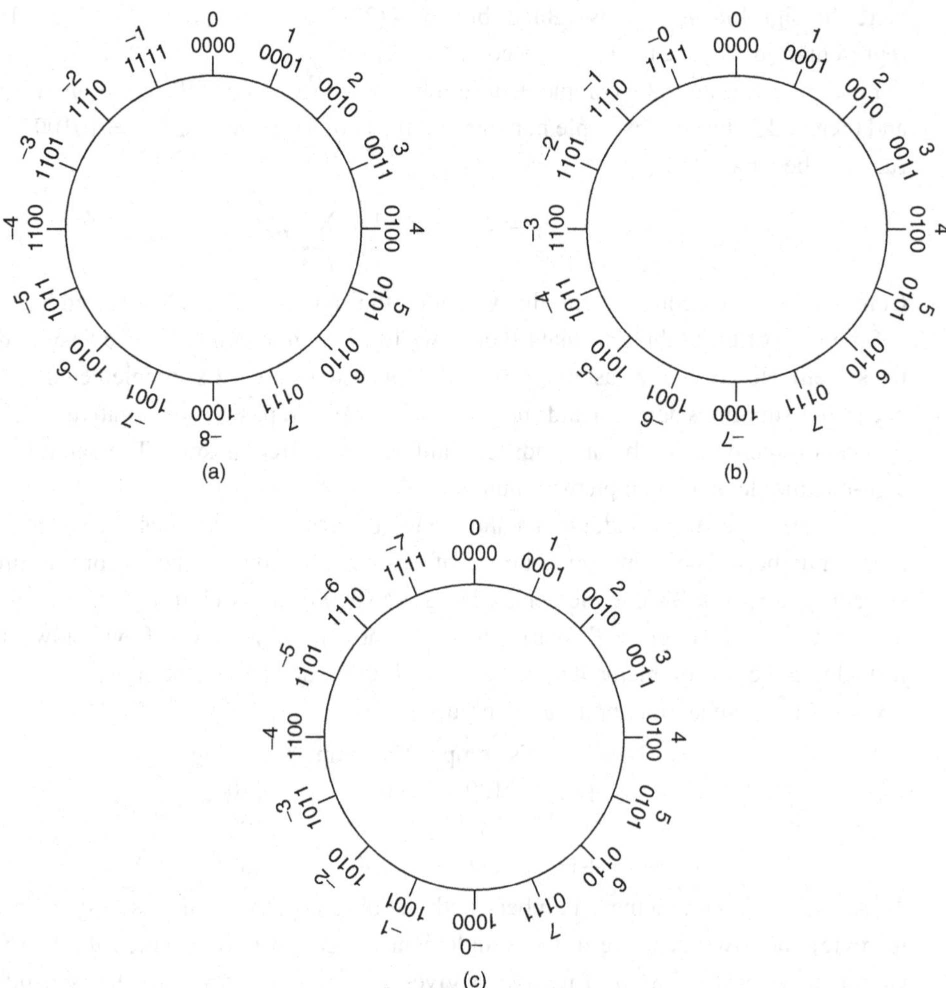

Figure 10.9. Number wheels showing three encodings of negative numbers: (a) 2's complement, (b) 1's complement, and (c) sign-magnitude.

One thing that becomes immediately apparent from the figure is that 1's complement and sign-magnitude do not have a unique representation for 0. In 1's complement, for example, both 0000 and 1111 represent the value 0. This makes comparison difficult. An equality comparator (see Section 8.6) cannot by itself determine whether two 1's complement or sign-magnitude numbers are equal because one may be $+0$ and the other -0, which are equivalent.

More importantly, the circle lets us see the effect of modular arithmetic. Adding $-x$ to a number has the effect of moving $16 - x$ steps clockwise around the circle, which is exactly the same as moving x steps counterclockwise around the circle. For example, -3 is equivalent to moving 13 steps clockwise or three steps counterclockwise, so adding -3 to any value between -5 and $+7$ gives the correct result. (Adding -3 to a value between -6 and -8 results in an overflow because we cannot represent results less than -8.)

How do we detect overflow when adding 2's complement numbers? We saw above that we may generate a carry as a result of modular arithmetic and get the correct answer. However, we can still generate results that are out of range. For example, what happens when we add -3 to

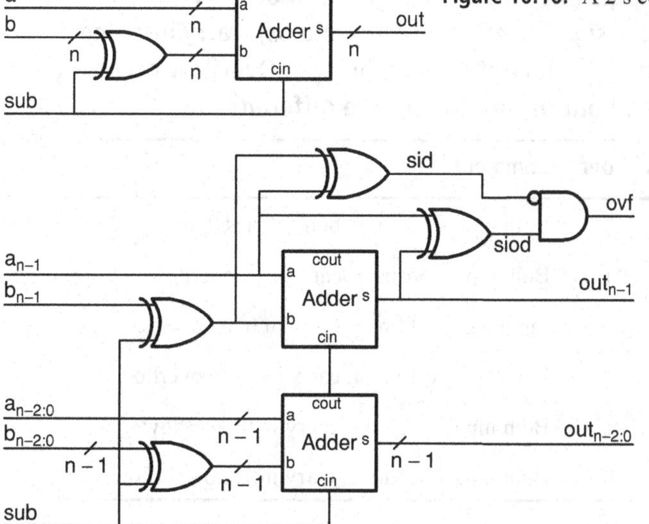

Figure 10.10. A 2's complement add/subtract unit.

Figure 10.11. A 2's complement add/subtract unit with overflow detection based on comparison of sign bits.

-6 or $+4$ to $+4$? We get $+7$ and -8, respectively – both incorrect. How do we detect this to signal an overflow?

The key thing to observe here is that the signs changed. We can always add a positive number to a negative number (or vice versa) and get a result that is in range. An overflow will occur only if we add two numbers of the same sign and get a result of the opposite sign. Thus we can detect overflows by comparing the signs of the inputs and outputs.[3]

Now that we can add negative numbers, we can build a circuit to subtract. A subtractor accepts two 2's complement numbers, a and b, as input and outputs $q = a - b$. A circuit to both add and subtract is shown in Figure 10.10. In add mode, the *sub* input is low, so the XORs pass the b input unchanged and the adder generates $a + b$. When the *sub* input is high, the XORs complement the b input and the carry into the adder is high, so the adder generates $a + \bar{b} + 1 = a - b$.

Figure 10.11 shows how we can augment our add/subtract circuit to detect overflow with three gates. The first XOR gate detects whether the two input signs are different (*sid*); the second XOR determines whether an input sign is different than the output sign (*siod*). The AND gate checks whether the two input signs are the same (*sid* = 0) and different than the output sign (*siod* = 1). If so, then overflow has occurred.

We can simplify the overflow detection to a single XOR gate as shown in Figure 10.12. This simplification is based on an observation about the carries into and out of the sign bit. Table 10.3 enumerates the six cases – inputs positive, different, or negative, and carry in 0 or 1. When the input signs are different ($p = 1, g = 0$), the carry into the sign bit will propagate. Thus the carry in and out are the same in this case. When the inputs are both positive ($p = 0, g = 0$) a carry into the sign bit indicates an overflow and will not propagate. Finally, if the inputs are both negative ($g = 1$), an overflow will occur unless there is a carry into the sign bit. Thus, we see that overflow occurs iff the carry into the sign bit (*cis*) and the carry out of the sign bit (*cos*) are different.

[3] We shall see below that we can accomplish the same function by comparing the carry into the last bit with the carry out of the last bit.

Table 10.3. **Cases for inputs and carry into sign bit of adder to detect overflow. Columns show sign bit of *a* and *b* (*as* and *bs*), carry into and out of sign bit (*cis* and *cos*), and output sign bit (*qs*). Overflow occurs only if the carry into and out of the sign bit are different.**

as	*bs*	*cis*	*qs*	*cos*	*ovf*	Comment
0	0	0	0	0	0	Both inputs positive, both carries 0, no overflow
0	0	1	1	0	1	Both inputs positive, carry in 1, overflow
0	1	0	1	0	0	Input signs different, carry in 0, no overflow
0	1	1	0	1	0	Input signs different, carry in 1, no overflow
1	1	0	0	1	1	Both inputs negative, carry in 0, overflow
1	1	1	1	1	0	Both inputs negative, carry in 1, no overflow

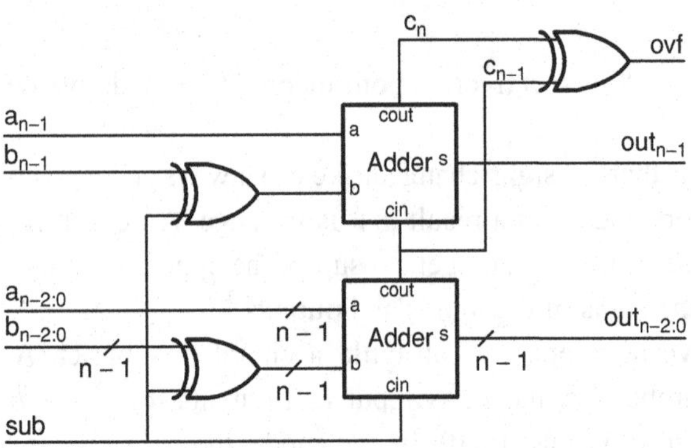

Figure 10.12. A 2's complement add/subtract unit with overflow detection based on carry in and out of the last bit.

VHDL code for the add/subtract unit is shown in Figure 10.13. This code instantiates a one-bit adder to add the sign bits and an $n - 1$ bit adder to add the remaining bits. The XOR of the `b` input with the `sub` input is performed in the argument list for each adder.

An alternative VHDL implementation is shown in Figure 10.14. This code uses assign statements with the "+" operator in place of instantiating predefined adders. It still performs an explicit XOR on the `b` input before the add.

One might be tempted to avoid the explicit XOR and instead code an add/subtract unit (ignoring carry out and overflow) using a statement like

```
process(a,b,sub) begin
  if sub then
     s <= a - b;
  else
     s <= a + b;
  end if;
end process;
```

Don't do this! Almost all synthesis systems will generate two separate adders for this code: one to do the "+" and a second to do the "−." While this code is quite clear and easy to read, it does not synthesize well – generating twice the logic of the alternative version.

```
-- add a+b or subtract a-b, check for overflow
library ieee;
use ieee.std_logic_1164.all;
use ieee.std_logic_unsigned.all;
use work.ch10.all;

entity AddSub is
  generic( n: integer := 8 );
  port( a, b: in std_logic_vector(n-1 downto 0);
        sub: in std_logic; -- subtract if sub=1, otherwise add
        s: out std_logic_vector(n-1 downto 0);
        ovf: out std_logic ); -- 1 if overflow
end AddSub;

architecture structural_impl of AddSub is
  signal c1, c2: std_logic; -- carry out of last two bits
begin
  ovf <= c1 xor c2; -- overflow if signs don't match

  -- add non sign bits
  Ai: Adder generic map(n-1)
        port map(a(n-2 downto 0),b(n-2 downto 0) xor (n-2 downto 0 => sub),sub,c1,
             s(n-2 downto 0));
  -- add sign bits
  As: Adder generic map(1)
        port map(a(n-1 downto n-1), b(n-1 downto n-1) xor (0 downto 0 => sub),c1,c2,
             s(n-1 downto n-1));
end structural_impl;
```

Figure 10.13. Structural VHDL code for add/subtract unit with overflow detection. This implementation instantiates the adder component.

```
-- add a+b or subtract a-b, check for overflow
architecture behavioral_impl of AddSub is
  signal c1, c2: std_logic; -- carry out of last two bits
  signal c1n: std_logic_vector(n-1 downto 0);
  signal c2s: std_logic_vector(1 downto 0);
begin
  -- overflow if signs don't match
  ovf <= c1 xor c2;
  -- add non-sign bits
  c1n <= ('0'&a(n-2 downto 0)) + ('0'&(b(n-2 downto 0)xor(n-2 downto 0=>sub))) + sub;
  s(n-2 downto 0) <= c1n(n-2 downto 0);
  c1 <= c1n(n-1);
  -- add sign bits
  c2s <= ('0'&a(n-1)) + ('0'&(b(n-1) xor sub)) + c1;
  s(n-1) <= c2s(0);
  c2 <= c2s(1);
end behavioral_impl;
```

Figure 10.14. Behavioral VHDL description of add/subtract unit with overflow detection. This implementation uses the "+" operator to perform the 2's complement addition. Entity declaration is shown in Figure 10.13.

Once we have a subtractor, then, with the addition of a zero-checker on the output, we also have a comparator. If we subtract, computing $s = a - b$, then if $s = 0$ then $a = b$, and if the sign bit of s is true, then $(a - b) < 0$, so $a < b$.

When adding 2's complement signed numbers of different lengths, one must first *sign extend* the shorter number. It is required that the sign bits – which have negative weight – be in the same position. If the numbers are added without sign extension, the negative-weight sign bit of the shorter number will be incorrectly added to a positive-weight bit of the longer number. For example, if we add 1010, a four-bit representation of -6_{10}, to 001000, a six-bit representation of $+8_{10}$, we get $010010 = 18_{10}$. This is because 1010 is misinterpreted as 10_{10}.

A 2's complement number can be sign extended by just copying the sign bit into the new positions to the left. For example, sign extending 1010 to six bits gives 111010. Our addition now becomes $111010 + 001000 = 000010 = 2_{10}$, which is the correct result.

In hardware, sign extension is accomplished with no additional gates, just wiring to repeat the sign bit. In VHDL it is easily expressed using the concatenate operator. For example, if `a` is n bits long and `b` is $m < n$ bits long, we sign extend `b` to n bits by writing

```
...
generic( n: integer := 6;   m: integer := 4 );

signal a: std_logic_vector(n-1 downto 0);
signal b: std_logic_vector(m-1 downto 0);

... (n-m downto 0 => b(m-1)) & b(m-2 downto 0)  ...  -- sign extend b to n bits
```

When shifting a 2's complement signed number it is important that the sign bit be duplicated when right-shifting and that overflow be checked when left-shifting. Overflow on a left-shift occurs when the sign of the result differs from the sign of the input. To right-shift an `n`-bit 2's complement number `b` correctly by three places we can write

```
use ieee.numeric_std.all;
...
signal b: signed(n-1 downto 0);
...
shift_right(b,3)
```

To shift `b` by a variable amount `s`, $0 \leq$ `s` $\leq$ `m`, to the right we can write

```
use ieee.numeric_std.all;
...
signal b: signed(n-1 downto 0);
...
shift_right(b,to_integer(unsigned(s)));
```

EXAMPLE 10.3 Negative numbers

Convert the number -82 to sign-magnitude, 1's complement, and 2's complement eight-bit binary numbers.

We note that 82 in binary is 1010010_2. To convert to sign-magnitude format, we simply append the sign (1) to the beginning of the number: 11010010. The 1's complement form is found by inverting all the bits of the positive number (01010010): 10101101. Finally, we find the 2's complement number by adding 1 to the 1's complement: 10101110.

EXAMPLE 10.4 Subtraction

Subtract, in 2's complement, 82_{10} from 72_{10}.

We must first convert both 72 and −82 to binary: 01001000 and 10101110 (see Example 10.3), respectively. Next, we perform the addition:

$$\begin{array}{rr} c & 00010000 \\ & 01001000 \\ + & \underline{10101110} \\ s & 11110110 \end{array}$$

Finally, we verify that 11110110_2 does indeed equal -10_{10}.

10.4 MULTIPLICATION

We multiply binary numbers the same way we multiply decimal numbers: by multiplying digits, shifting and adding. Multiplying one bit by another is simpler than multiplying decimal digits. There are only four cases to consider: $0 \times 0 = 0$, $1 \times 0 = 0$, $0 \times 1 = 0$, $1 \times 1 = 1$. Hence, multiplying two one-bit numbers can be performed by an AND gate. Multiplying multi-bit binary number n by a single bit b results in two cases: $n \times b = n$ if $b = 1$ and $n \times b = 0$ if $b = 0$. To multiply two multi-bit binary numbers we employ shifting. Shifting a binary number left by one position is the same as multiplying it by 2. For example, the number $101_2 = 5_{10}$; if we shift it left by one position, we get $1010_2 = 10_{10}$; another left-shift gives $10100_2 = 20_{10}$, and so on.

To multiply two unsigned binary numbers $a_{n-1}, \ldots, a_0$ and $b_{n-1}, \ldots, b_0$, we add a copy of a shifted to the appropriate position for each position in which b is 1. That is, we compute $b_0 a + b_1(a << 1) + \cdots + b_{n-1}(a << (n-1))$.

For example, consider multiplying $a = 101_2$ by $b = 110_2$ ($5_{10} \times 6_{10}$). In long notation we write

$$\begin{array}{r} 1\,0\,1 \\ \times\,1\,1\,0 \\ \hline 0\,0\,0 \\ 1\,0\,1 \\ 1\,0\,1 \\ \hline 1\,1\,1\,1\,0 \end{array}$$

Here $b_0 = 0$, hence the row of 0s in the unshifted position. We add in 101 shifted by 1 since $b_1 = 1$ and 101 shifted by 2 since $b_2 = 2$. Summing these three *partial products* gives $11110 = 30_{10}$.

A circuit to perform multiplication on two four-bit unsigned binary numbers is shown in Figure 10.15. An array of 16 AND gates forms four four-bit partial products. The first row of four AND gates forms b_0a. The second row forms b_1a shifted one position left, and so on. An array of 12 full adders then sums the partial products by columns to produce the eight-bit product $p_7, \ldots, p_0$. Partial product bit pp_{00}, formed by $b_0 \wedge a_0$, is the only partial product of weight (1), so it directly becomes p_0. Partial product bits pp_{01} and pp_{10} are both of weight (2) and are summed by a full adder to give p_1. Product bit p_2 is computed by summing pp_{02}, pp_{11}, and pp_{20} along with the carry out of the weight (1) adder. The remaining bits are computed in a similar manner – by summing the partial products in their column along with carries from the previous column.

Note that all partial products in a column have indices that sum to the weight of that column: e.g., 02, 11, and 20 all sum to (2). This is because the weight of a partial product is equal to the sum of the indices of the input bits from which it's derived. To see this, consider that multiplication can be expressed as

$$p = \sum_{i=0}^{n-1} \sum_{j=0}^{n-1} (a_i \wedge b_j) \times 2^{i+j}. \tag{10.12}$$

VHDL code for the four-bit multiplier is shown in Figure 10.16. Four assignments form the partial products `pp0` to `pp3`, each a four-bit vector. Three four-bit adders are then instantiated to add up the partial products. The second input to each of these adders is a concatenation of the high three bits out of the previous adder and either 0 (for the first adder) or the carry out of the previous adder.

The multiplier of Figures 10.15 and 10.16 multiplies unsigned numbers. It will not produce the correct result with a 2's complement signed number on the a input because the partial products are not sign extended to full width before being added. Also, it will not produce correct results with a 2's complement negative number on the b input. This is because the multiplication counts on b_3 are being weighted by 8 rather than -8. We leave the modification of the multiplier to handle a 2's complement number as an exercise (Exercise 10.50). Also, the use of Booth recoding (see Section 12.2) results in a multiplier that naturally handles signed numbers.

EXAMPLE 10.5 Binary multiplication

Multiply the unsigned hexadecimal number E_{16} by the number D_{16}.

The final result of the multiplication is $B6_{16}$. We show our work below:

```
          1 1 1 0
×         1 1 0 1
-----------------
          1 1 1 0
        0 0 0 0
      1 1 1 0
    1 1 1 0
-----------------
  1 0 1 1 0 1 1 0
```

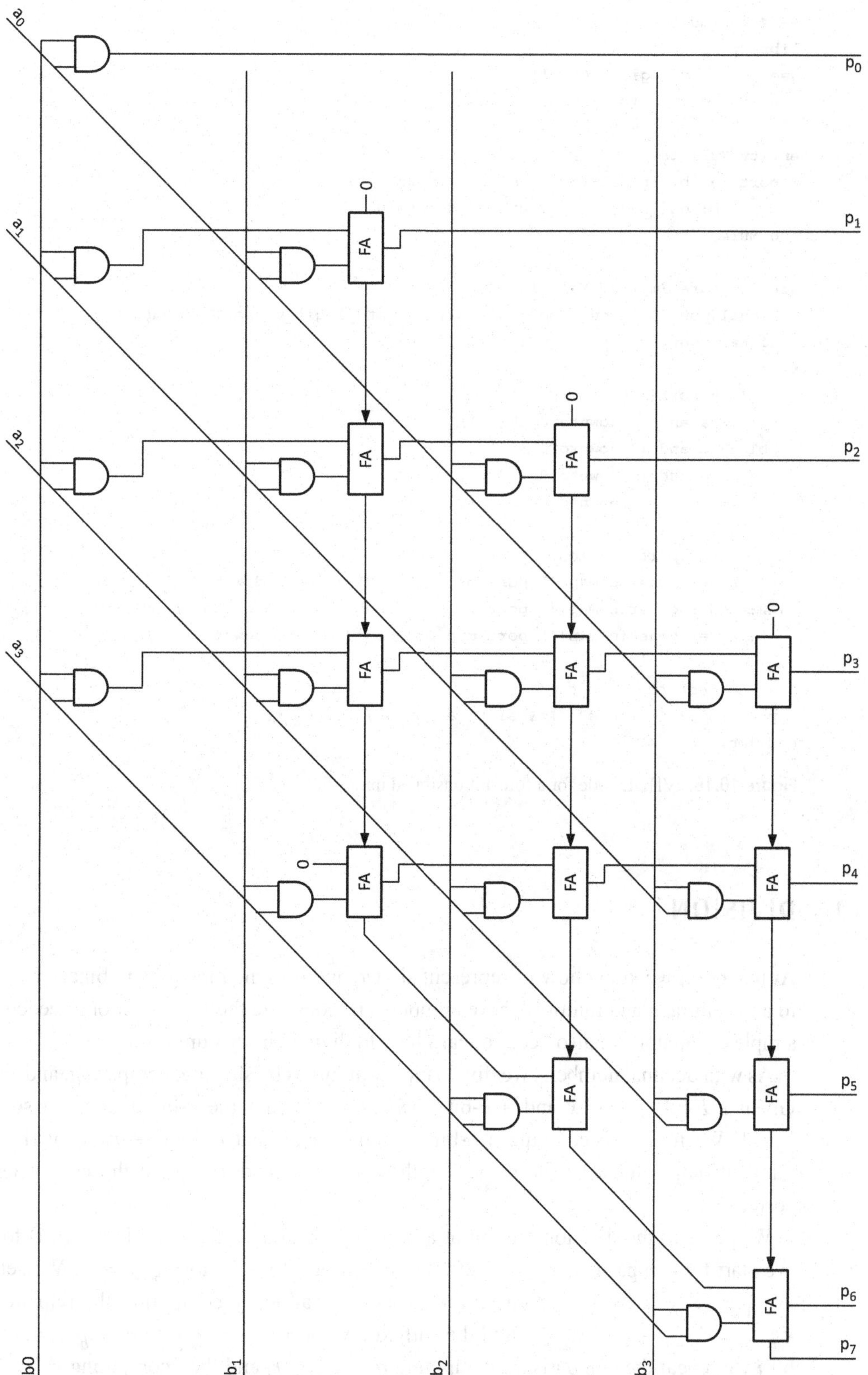

Figure 10.15. Four-bit unsigned binary multiplier.

```
-- 4-bit multiplier
library ieee;
use ieee.std_logic_1164.all;
use work.ch10.all;

entity Mul4 is
  port( a, b: in std_logic_vector(3 downto 0);
        p: out std_logic_vector(7 downto 0) );
end Mul4;

architecture impl of Mul4 is
  signal pp0, pp1, pp2, pp3, s1, s2, s3: std_logic_vector(3 downto 0);
  signal cout1, cout2, cout3: std_logic;
begin
  -- form partial products
  pp0 <= a and (3 downto 0=>b(0));
  pp1 <= a and (3 downto 0=>b(1));
  pp2 <= a and (3 downto 0=>b(2));
  pp3 <= a and (3 downto 0=>b(3));

  -- sum up partial products
  A1: Adder generic map(4) port map(pp1, '0' & pp0(3 downto 1),'0',cout1,s1);
  A2: Adder generic map(4) port map(pp2,cout1 & s1(3 downto 1),'0',cout2,s2);
  A3: Adder generic map(4) port map(pp3,cout2 & s2(3 downto 1),'0',cout3,s3);

  -- collect the result
  p <= cout3 & s3 & s2(0) & s1(0) & pp0(0);
end impl;
```

Figure 10.16. VHDL code for a four-bit unsigned multiplier.

10.5 DIVISION

At this point we know how to represent signed and unsigned integers in binary form and how to add, subtract, and multiply these numbers. To complete the four functions needed to build a simple calculator, we also need to learn how to divide binary numbers.

As with decimal numbers, we divide binary numbers by shifting, comparing, and subtracting. Given a b-bit *divisor* x and a c-bit *dividend* y we find the c-bit quotient q so that $q = \lfloor y/x \rfloor$. We may also compute a b-bit remainder r so that $r = y - qx = y \pmod{x}$. The quotient output must be the same length as the dividend to handle the case where $x = 1$ correctly.

We perform the division one bit at a time – generating q from the left (MSB) to the right. We start by comparing $x'_{2b-1} = 2^{2b-1}x = x << (2b - 1)$ to $r'_{2b-1} = y$. We set $q_{2b-1} = x'_{2b-1} \leq r'_{2b-1}$. We then prepare for the next iteration by computing the remainder so far, $r'_{2b-2} = r'_{2b-1} - q_{2b-1}x'_{2b-1}$, and the shifted divisor, $x'_{2b-2} = 2^{2b-2}x = x'_{2b-1} >> 1$. At each bit i we repeat the comparison, computing $q_i = x'_i \leq r'_i$ and then computing $r'_{i-1} = r'_i - q_i x'_i$ and $x'_{i-1} = x'_i >> 1$.

For example, consider dividing $132_{10} = 10000100_2$ by $11_{10} = 1011_2$. The process is shown below:

```
         1100
1011|10000100  y
    -1011000   x'3
      101100   r'2
     -101100   x'2
           0
```

For the first four iterations, $i = 7, \ldots, 4$ (not shown), $x'_i > y$, and no subtractions are performed. Finally, on the fifth iteration, $i = 3$, we have $x'_3 = 1011000 < r'_3 = y$, so we set bit $q_3 = 1$ and subtract to compute $r'_2 = y - x'_3 = 101100$. We shift x'_3 right to get $x'_2 = 101100$. These two values are equal, so bit $q_2 = 1$. Subtracting gives $r'_1 = 0$, and all subsequent bits of q are 0.

A six-bit by three-bit divider is shown in Figure 10.17. The circuit consists of six nearly identical stages. Stage i generates bit q_i of the quotient by comparing an appropriately shifted version of the input x'_i with the remainder from the previous stage r'_i. The remainder from stage i, r'_{i-1}, is generated by a subtractor and a multiplexer. The subtractor subtracts the shifted input from the previous remainder. If q_i is true, the multiplexer selects the result of the subtraction, $r'_i - x'_i$, to be the new remainder. If q_i is false, the previous remainder is passed unchanged. Stage 0 generates the LSB of the quotient q_0 and the final remainder, r.

A divider is quite costly in terms of gate count, but not quite as costly as Figure 10.17 indicates. For clarity, the figure shows six subtractors and six comparators. In practice, the subtractor can be used as the comparator. The carry out of the subtractor can be used as the quotient bit. If the carry out of the subtractor is 1, then $a >= b$. With this optimization, the circuit can be realized with six subtractors and six multiplexers.

A further optimization can be made by reducing the width of the subtractors. The first subtractor can be made a single bit wide. To see this, note that $x << 5$ has zeros in its low five bits. Thus, the low five bits of the result will be equal to the low five bits of y, $y_{4:0}$, and we don't need to subtract these bits. Also note that if x has any non-zeros anywhere other than its LSB, then $x << 5$ is guaranteed to be larger than y, and the result of the subtraction is not needed. Thus, we need not feed the upper bits of x into the subtractor. We do, however, need to check whether the upper bits of x are non-zero as part of computing the MSB of the quotient q_5. If an upper bit of x is non-zero, then $x << 5 > y$ and $q_5 = 0$. With these observations we see that the first subtractor needs only to subtract the LSB of x from the MSB of y. Hence a one-bit subtractor suffices. By a similar set of arguments, we can use a two-bit subtractor for the stage that computes q_4, a three-bit subtractor for the q_3 stage, and a four-bit subtractor for the q_2 stage. The q_1 and q_0 stages can also use a four-bit subtractor. The remainder into the q_1 stage is guaranteed to be no more than five bits in length, and the remainder into the q_0 stage is at most four bits long. Hence the upper bits of these subtractors can be omitted.

VHDL code for a six-bit by three-bit divider is shown in Figure 10.18. This code uses the subtractors to perform the comparison and optimizes the width of the subtractors for each stage. Each subtractor is realized using an `Adder` component (Figure 10.7) with the second input complemented and the carry in set to `1'b1`. Each multiplexer is implemented with an assign statement using the `? :` operation. For the first stage, the upper two bits of input `x` are checked

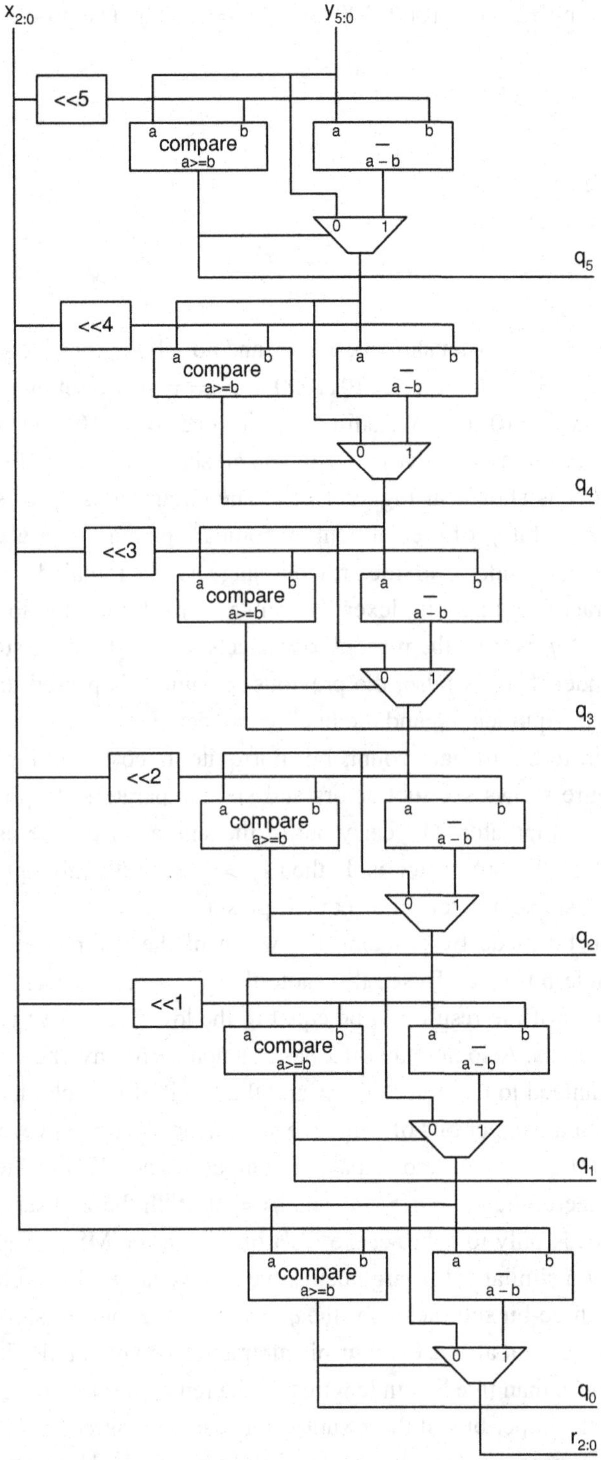

Figure 10.17. Binary divider.

```
-- Six-bit by three-bit divider
--    At each stage we use an adder to both subtract and compare.
--    The adders start 1 bit wide and grow to 4 bits wide.
--    We check the bits of x to the left of the adder as part of
--    the comparison.
--    Starting with the fourth iteration (that computes q[2]) we
--    drop a bit of the remainder each iteration.  It is guaranteed
--    to be zero.
library ieee;
use ieee.std_logic_1164.all;
use work.ch10.all;

entity Divide is
  port( y: in std_logic_vector( 5 downto 0 ); -- dividend
        x: in std_logic_vector( 2 downto 0 ); -- divisor
        q: buffer std_logic_vector( 5 downto 0 ); -- quotient
        r: out std_logic_vector( 2 downto 0 ) ); -- remainder
end Divide;
architecture impl of Divide is
  signal co5, co4, co3, co2, co1, co0: std_logic;  -- carry out of adders
  signal sum5: std_logic_vector(0 downto 0); -- sum out of adder - stage 1
  signal sum4: std_logic_vector(1 downto 0); -- sum out of adder - stage 2
  signal sum3: std_logic_vector(2 downto 0); -- sum out of adder - stage 3
  signal sum2, sum1, sum0: std_logic_vector(3 downto 0); -- sum out of adder -
         stage 4, 5, 6
  signal r4, r3, r2: std_logic_vector(5 downto 0);
  signal r1: std_logic_vector(4 downto 0);
  signal r0: std_logic_vector(3 downto 0);
begin
  SUB5: Adder generic map(1) port map(y(5 downto 5),not x(0 downto 0),'1',co5,sum5);
  q(5) <= co5 and not (x(2) or x(1)); -- if x<<5 bigger than y, q(5) is 0
  r4 <= (sum5 & y(4 downto 0)) when q(5) else y;

  SUB4: Adder generic map(2) port map(r4(5 downto 4),not x(1 downto 0),'1',co4,sum4);
  q(4) <= co4 and not x(2); -- compare
  r3 <= (sum4 & r4(3 downto 0)) when q(4) else r4;

  SUB3: Adder generic map(3) port map(r3(5 downto 3),not x(2 downto 0),'1',co3,sum3);
  q(3) <= co3; -- compare
  r2 <= (sum3 & r3(2 downto 0)) when q(3) else r3;

  SUB2: Adder generic map(4) port map(r2(5 downto 2),'1' & not x,'1',co2,sum2);
  q(2) <= co2; -- compare
  r1 <= (sum2(2 downto 0) & r2(1 downto 0)) when q(2) else r2(4 downto 0); -- msb is zero,
         drop it

  SUB1: Adder generic map(4) port map(r1(4 downto 1),'1' & not x,'1',co1,sum1);
  q(1) <= co1; -- compare
  r0 <= (sum1(2 downto 0) & r1(0)) when q(1) else r1(3 downto 0); -- msb is zero, drop it

  SUB0: Adder generic map(4) port map(r0(3 downto 0),'1' & not x,'1',co0,sum0);
  q(0) <= co0; -- compare
  r <= sum0(2 downto 0) when q(0) else r0(2 downto 0); -- msb is zero, drop it
end impl;
```

Figure 10.18. VHDL code for a six-bit by three-bit divider.

as part of the comparison, and in the second stage the MSB of x is checked. The remainder of the VHDL code follows directly from the schematic.

In addition to consuming a great deal of space, dividers are also very slow. This is because the subtract in one stage must be completed to determine the multiplexer command – and hence the intermediate remainder before the subtract in the next stage can be started. Thus, the delay through a c-bit by b-bit divider using ripple-carry adders is proportional to $c \times b$, since c subtracts, of $b + 1$ bits in length (after the first b stages), must be performed. This is in contrast to a multiplier where the partial products can be summed in parallel.

EXAMPLE 10.6 Division

Divide EA_{16} by 12_{16}.

Our work in arriving at the final answer, D_{16}, is shown below:

```
          00001101
         ---------
    10010|11101010
         -10010000
         ---------
           1011010
          -1001000
         ---------
             10010
            -10010
         ---------
                 0
```

Summary

In this chapter you have learned how to perform arithmetic using combinational logic circuits. We have seen how to represent positive and negative numbers and how to perform the four functions found on a typical calculator: $+$, $-$, $\times$, and $\div$ on integers.

We represent numbers with multi-bit digital signals using a *weighted* representation, where the rightmost bit (the least significant bit, or LSB) is weighted 1 and every other bit has a weight that is $2\times$ that of its right neighbor. In other words, bit i, if it is 1, represents a value of 2^i.

We represent negative numbers with *2's complement* notation in which the weight of the leftmost bit (the most significant bit or MSB) is negated. In a signed number, we refer to this bit as the *sign bit*. A number with a sign bit of 0 is positive. A negative number has a sign bit of 1. We negate a number by complementing each bit (the *1's complement*) and adding 1. When we operate on signed numbers, we must often *sign extend* them, copying the sign bit multiple times, so that all numbers being operated on have their sign bit in the same position.

With this weighted binary notation, we can perform addition of multi-bit signals one bit at a time – performing a carry operation when the sum of two bits generates an output that carries into the next bit position. By writing out the truth table for summing three bits of equal weight into a two-bit binary output, we derive the *full adder*. Multi-bit addition is performed by an iterative circuit of full adders. To perform subtraction we take the 2's complement of one input (negating that input) and then add.

Overflow occurs when the result of an arithmetic operation is too large to be represented in the width of the output signal. For an unsigned addition, overflow is detected when there is a carry out of the last bit. For signed addition, an overflow occurs when the two inputs have one sign and the output has the opposite sign – or equivalently when the carry into the sign bit is different than the carry out of the sign bit. Some systems use *saturating arithmetic* to limit the error caused by an overflow.

Multiplication and *division* are performed exactly as with decimal numbers. With multiplication, every bit of one input is multiplied (ANDed) by every bit of the other input, forming a matrix of *partial products*. Each partial product has a weight given by the product of the weights of the inputs from which it was derived. That is, $pp_{ij} = a_i \wedge b_j$ has weight 2^{i+j}. An array of adders is used to sum the partial products to give the final answer. We will see in Chapter 12 how to do this faster.

A divider performs long division in binary by subtracting shifted versions of the *divisor* from the current remainder – the *dividend* for the first iteration.

Exercises

10.1 *Decimal to binary conversion, I.* Convert 817 from decimal to binary notation. Use the minimum number of bits possible. Express the result in hexadecimal.

10.2 *Decimal to binary conversion, II.* Convert 1492 from decimal to binary notation. Use the minimum number of bits possible. Express the result in hexadecimal.

10.3 *Decimal to binary conversion, III.* Convert 1963 from decimal to binary notation. Use the minimum number of bits possible. Also express the result in hexadecimal.

10.4 *Decimal to binary conversion, IV.* Convert 2012 from decimal to binary notation. Use the minimum number of bits possible. Also express the result in hexadecimal.

10.5 *Binary to decimal conversion, I.* Convert 0011 0011 0001 from binary to decimal notation.

10.6 *Binary to decimal conversion, II.* Convert 0111 1111 from binary to decimal notation.

10.7 *Binary to decimal conversion, III.* Convert 0100 1100 1011 0010 1111 from binary to decimal notation.

10.8 *Binary to decimal conversion, IV.* Convert 0001 0110 1101 from binary to decimal notation.

10.9 *Hexadecimal to decimal conversion, I.* Convert 2C from hexadecimal to decimal notation. Also express the number in binary-coded decimal (BCD) notation.

10.10 *Hexadecimal to decimal conversion, II.* Convert BEEF number from hexadecimal to decimal notation. Also express the number in binary-coded decimal (BCD) notation.

10.11 *Hexadecimal to decimal conversion, III.* Convert 2015 from unsigned hexadecimal to decimal notation. Also express the number in binary-coded decimal (BCD) notation.

10.12 *Hexadecimal to decimal conversion, IV.* Convert F00D from unsigned hexadecimal to decimal notation. Also express the number in binary-coded decimal (BCD) notation.

10.13 *Hexadecimal to decimal conversion, V.* Convert DEED from unsigned hexadecimal to decimal notation. Also express the number in binary-coded decimal (BCD) notation.

10.14 *Binary addition, I.* Add the following pair of binary numbers:

```
 1010
+0111
```

10.15 *Binary addition, II.* Add the following pair of binary numbers:

```
 011 1010
+110 1011
```

10.16 *Binary addition, III.* Add the following pair of hexadecimal numbers:

```
 2A
+3C
```

10.17 *Binary addition, IV.* Add the following pair of hexadecimal numbers:

```
 BC
+AD
```

10.18 *Bit counting circuit, design.* Using full adders, design a circuit that accepts a seven-bit input and outputs the number of inputs that are 1 as a three-bit binary number.

10.19 *Bit counting circuit, implementation.* Write VHDL code to represent your circuit from Exercise 10.18 and demonstrate its correct operation via simulation.

10.20 *Saturating adder, design.* In some applications, particularly signal processing, it is desirable to have an adder *saturate*, producing a result of $2^n - 1$ on an overflow condition rather than producing a modular result. Design a saturating adder. You may use n-bit adders and n-bit multiplexers as basic components.

10.21 *Saturating adder, implementation.* Write VHDL code for your saturating adder of Exercise 10.20 and demonstrate its correct operation by simulating it on representative test cases. Your code should take the width of the adder as a parameter.

10.22 *Vector adder.* Design and write VHDL for a circuit that can function as a 32-bit adder, two 16-bit adders, or four eight-bit adders. You may have no more than 32 full adders in your circuit. The inputs to your design entity are `a(31 downto 0)`, `b(31 downto 0)`, `add2x16`, **and** `add4x8`, and the output is `s(31 downto 0)`. Refer to Table 10.4 for operation; you do not need to handle subtraction or overflow.

Table 10.4. Output specification for the vector adder of Exercise 10.22

`add2x16`	`add4x8`	Result
0	0	s(31*downto*0) = a(31*downto*0) + b(31*downto*0)
1	0	s(31*downto*16) = a(31*downto*16) + b(31*downto*16)
		s(15*downto*0) = a(15*downto*0) + b(15*downto*0)
0	1	s(31*downto*24) = a(31*downto*24) + b(31*downto*24)
		...
		s(7*downto*0) = a(7*downto*0) + b(7*downto*0)
1	1	s(31*downto*0) = 32sb"-"

You may not simply implement Table 10.4 as a `case` or `case?` statement.

10.23 *BCD addition, design.* Design a circuit that accepts two three-digit (12-bit) BCD numbers (Equation (10.6)) and outputs their sum as a BCD number.

10.24 *BCD addition, implementation.* Code your BCD adder from Exercise 10.23 in VHDL and verify its operation with a testbench.

10.25 *Negative numbers, I.* Express +17 as an eight-bit sign-magnitude, 1's complement, and 2's complement binary number.

10.26 *Negative numbers, II.* Express −17 as an eight-bit sign-magnitude, 1's complement, and 2's complement binary number.

10.27 *Negative numbers, III.* Express −31 as an eight-bit sign-magnitude, 1's complement, and 2's complement binary number.

10.28 *Negative numbers, IV.* Express −32 as an eight-bit sign-magnitude, 1's complement, and 2's complement binary number.

10.29 *Subtraction, I.* Subtract the following pairs of 2's complement binary numbers:

$$\begin{array}{r} 0101 \\ \underline{-0110} \end{array}$$

10.30 *Subtraction, II.* Subtract the following pairs of 2's complement binary numbers:

$$\begin{array}{r} 0101 \\ \underline{-1110} \end{array}$$

10.31 *Subtraction, III.* Subtract the following pairs of 2's complement binary numbers:

$$\begin{array}{r} 1010 \\ \underline{-0010} \end{array}$$

10.32 *Subtraction, IV.* Subtract the following pairs of 2's complement binary numbers:

$$\begin{array}{r} 0101 \\ \underline{-0111} \end{array}$$

10.33 *1's complement adder, design.* Design an adder for 1's complement numbers. (Hint: first add the two numbers normally. If there is a carry out of this first add, you need to increment the result to give the correct answer. While the straightforward solution requires an adder and an incrementer, it can be done with a single adder.)

10.34 *1's complement adder, implementation.* Write VHDL code for your 1's complement adder (Exercise 10.33) and demonstrate its correct operation by simulating it on representative test cases.

10.35 *Saturating 2's complement adder, design.* In Exercise 10.20 we saw how to build a saturating adder for positive numbers. In this exercise you are to extend this design so that it handles negative numbers – saturating both in the positive direction and in the negative direction. On overflows in the positive direction your adder is to generate $2^{n-2} - 1$, and on overflows in the negative direction it should generate -2^{n-2}.

10.36 *Saturating 2's complement adder, implementation.* Write VHDL code for your saturating 2's complement adder (Exercise 10.35) and demonstrate its correct operation by simulating it on representative test cases.

10.37 *Sign-magnitude adder, design.* Design a circuit that accepts two sign-magnitude binary numbers and outputs their sum, also in sign-magnitude form.

10.38 *Sign-magnitude adder, implementation.* Write VHDL code for your sign-magnitude adder of Exercise 10.37 and demonstrate its correct operation by simulating it on representative test cases.

10.39 *Non-standard signed representation.* Consider a four-bit representation that can represent numbers from −4 to 11. Each negative number is represented by its normal 2's complement representation, e.g., −4 = 1100. Similarly, each positive number is represented by its normal binary representation, e.g., 11 = 1011.

(a) Draw a number wheel (see Figure 10.9) for this representation.

(b) Draw an adder that operates on these numbers and detects overflow.

(c) Explain how to negate one of these numbers.

10.40 *Multiplication, I.* Multiply the following pair of unsigned binary numbers:

$$\begin{array}{r} 0101 \\ \underline{\times\ 0101} \end{array}$$

10.41 *Multiplication, II.* Multiply the following pair of unsigned binary numbers:

$$\begin{array}{r} 0110 \\ \underline{\times\ 0011} \end{array}$$

10.42 *Multiplication, III.* Multiply the following pair of unsigned binary numbers:

$$\begin{array}{r} 1001 \\ \underline{\times\ 1001} \end{array}$$

10.43 *Multiplication, IV.* Multiply the following pair of unsigned hexadecimal numbers:

$$\begin{array}{r} A \\ \underline{\times\ C} \end{array}$$

10.44 *Five-times circuit.* Using adders, combinational building blocks, and gates, design a circuit that accepts a four-bit 2's complement binary input `a(3 downto 0)` and outputs a seven-bit 2's complement output `b(6 downto 0)` that is five times the value input. You cannot use a multiplier building block. Use the minimum number of adder bits possible.

10.45 *Fifteen-times circuit.* Using adders, combinational building blocks, and gates, design a circuit that accepts a four-bit 2's complement binary input `a(3 downto 0)` and outputs an eight-bit 2's complement output `b(7 downto 0)` that is 15 times the value input. You cannot use a multiplier building block. Use the minimum number of adder bits possible.

10.46 *Sixteen-times circuit.* Design a circuit that accepts an eight-bit 2's complement binary input `a(7 downto 0)` and outputs a 12-bit 2's complement output `b(11 downto 0)` that is 16 times the value input. Use as little logic as possible.

10.47 *BCD multiplication, design.* Design a circuit that accepts two three-digit (12-bit) BCD numbers (see Equation (10.6)) and outputs their product as a BCD number.

10.48 *BCD multiplication, implementation.* Code your design from Exercise 10.47 in VHDL and verify its operation with a testbench.

10.49 *Circuit design.* Using adders, combinational building blocks, and gates, design a circuit that accepts the four inputs a, b, c, and d and outputs the expression $a-b+(c\times d)$. Input a

is a four-bit 2's complement number. Input b is a four-bit 1's complement number. Input c is a two-bit unsigned number. Input d is a four-bit unsigned number.

10.50 *2's complement multiplier, design.* Design a multiplier for 2's complement binary numbers. Consider two approaches:

(a) Sign extend the partial products to deal with input a being negative and add a "complementer" to negate the last set of partial products if b is negative.

(b) Convert the two inputs to sign-magnitude notation, multiply unsigned numbers, and convert the result back to 2's complement.

Compare the cost and performance (delay) of the two approaches. Select the approach that gives the lowest cost and show its design in terms of basic components (gates, adders, etc.)

10.51 *2's complement multiplier, implementation.* Write VHDL code for your 2's complement multiplier from Exercise 10.50 and demonstrate its correct operation by simulating it on representative test cases.

10.52 *Binary division, I.* Divide the following pair of unsigned binary numbers (show each step of the process): $101110_2 \div 101_2$.

10.53 *Binary division, II.* Divide the following pair of unsigned binary numbers (show each step of the process): $101110_2 \div 011_2$.

10.54 *Binary division, III.* Divide the following pair of unsigned hexadecimal numbers (show each step of the process): $AE_{16} \div E_{16}$.

10.55 *Binary division, IV.* Divide the following pair of unsigned hexadecimal numbers (show each step of the process): $F7_{16} \div 6_{16}$.

10.56 *Subtractor widths for dividers.* For each of the following pairs of argument widths, determine the width of subtractor needed in each stage of a divider:

(a) dividend four bits, divisor four bits;

(b) dividend six bits, divisor four bits;

(c) dividend four bits, divisor three bits.

11 Fixed- and floating-point numbers

In Chapter 10 we introduced the basics of computer arithmetic: adding, subtracting, multiplying, and dividing binary integers. In this chapter we continue our exploration of computer arithmetic by looking at number representation in more detail. Often integers do not suffice for our needs. For example, suppose we wish to represent a pressure that varies between 0 (vacuum) and 0.9 atmospheres with an error of at most 0.001 atmospheres. Integers don't help us much when we need to distinguish 0.899 from 0.9. For this task we will introduce the notion of a *binary point* (similar to a decimal point) and use *fixed-point* binary numbers.

In some cases, we need to represent data with a very large dynamic range. For example, suppose we need to represent time intervals ranging from 1 ps (10^{-12} s) to one century (about 3×10^9 s) with an accuracy of 1%. To span this range with a fixed-point number would require 72 bits. However, if we use a *floating-point* number – in which we allow the position of the binary point to vary – we can get by with 13 bits: six bits to represent the number and seven bits to encode the position of the binary point.

11.1 REPRESENTATION ERROR: ACCURACY, PRECISION, AND RESOLUTION

With digital electronics, we represent a number, x, as a string of bits, b. Many different *number systems* are used in digital systems. A number system can be thought of as two functions R and V. The representation function R maps a number x from some set of numbers (e.g., real numbers, integers, etc.) into a bit string b: $b = R(x)$. The value function V returns the number (from the same set) represented by a particular bit string: $y = V(b)$.

Consider mapping to and from the set of real numbers in some range. Because there are more possible real numbers than there are bit strings of a given length, many real numbers necessarily map to the same bit string. Thus, if we map a real number to a bit string with R and then back with V we will almost always get a slightly different real number than we started with. That is, if we compute $y = V(R(x))$ then y and x will differ. The difference is the error of the representation. We can express error either in an absolute sense (e.g., the representation has an error of 2 mm), or relative to the magnitude of the number (e.g., the representation has an error of 3%). We express the *absolute error* of a representation at a point x as

$$e_a = |V(R(x)) - x| , \tag{11.1}$$

and the *relative error* as

$$e_{\mathrm{r}} = \left| \frac{V(R(x)) - x}{x} \right|. \tag{11.2}$$

The quality of a number representation is given by its *accuracy* or *precision*,[1] the maximum error over its input range X. The absolute accuracy is given by

$$a_{\mathrm{a}} = \max_{x \in X} |V(R(x)) - x|, \tag{11.3}$$

and the relative accuracy is

$$a_{\mathrm{r}} = \max_{x \in X} \left| \frac{V(R(x)) - x}{x} \right|. \tag{11.4}$$

Naturally, the relative accuracy is not defined near $x = 0$. When we want to represent numbers economically with a given relative accuracy, floating-point numbers are often used. When we want to represent numbers economically with a given absolute accuracy, fixed-point numbers are more efficient. We describe these two representations in the following sections.

Sometimes people refer to the number of bits used in a number system, its *length* (the term *precision* is often misused for length; e.g., saying a system has 32-bit precision). At other times, people refer to the smallest difference that can be distinguished by the number system, the *resolution* of the system. When determining the quality of the representation, neither length nor resolution is useful. What matters is accuracy.

For example, suppose we represent real numbers over the range $X = [0, 1000]$ as ten-bit binary integers by representing each real number with the nearest integer. Picking the nearest integer to a real number is often referred to as *rounding* the real number to an integer. We would then represent 512.742 as 513 or 1000000001_2, and the error of representing this number would be $e_{\mathrm{a}}(512.742) = |512.742 - 513| = 0.258$. The error over the entire range is $a_{\mathrm{a}}(X) = 0.5$ since a value halfway between two integers, e.g., 512.500, has this much error whether it is rounded up or down. Note that the error here depends on the representation function R. If we choose R so that each number x is represented by the nearest integer *less than* x, then we get $e_{\mathrm{a}}(512.742) = 0.742$ and $a_{\mathrm{a}}(X) = 1$. For positive real numbers, applying this second representation function is often referred to as *truncating* the real number to an integer.

Again, one should not confuse *accuracy* with *resolution*. The resolution of both the rounding and truncating representations discussed above is 1.0 – integers are spaced one unit apart. However, the accuracy of rounding is 0.5, and the accuracy of truncation is 1.0.

EXAMPLE 11.1 Calculating accuracy

In Section 1.3 we represented temperatures between 68 and 82 as a three-bit number, where

$$T = 68 + 2\sum_{i=0}^{2} 2^i \mathrm{TempA}_i.$$

For this representation, find the absolute accuracy, relative accuracy, and resolution of this scheme. Compute these values only for numbers between 68 and 82 and assume rounding.

[1] We use the terms accuracy and precision interchangeably in this book.

The resolution, or weighted value of the LSB, in this scheme is 2. The absolute accuracy a_{a} is 1. For example, the value 79 is rounded up to 80 and $|79 - 80| = 1$. Numbers between 68 and 70, specifically 69, have the lowest relative accuracy:

$$a_{\mathrm{r}} = \left| \frac{V(R(69)) - 69}{69} \right|;$$

$$a_{\mathrm{r}} = \left| \frac{70 - 69}{69} \right|;$$

$$a_{\mathrm{r}} = 1.4\%.$$

This representation has 1.4% accuracy for numbers between 68 and 82.

EXAMPLE 11.2 Representation design

What is the resolution needed to represent numbers from 54 500 000 km to 4 500 000 000 km with 3% accuracy? This is the range of numbers representing the distance between the Sun and Mercury, and the Sun and Neptune, respectively. Assume a rounding scheme.

The maximum error will be halfway between 54 500 000 and 54 500 000 + r, where r represents our resolution. So

$$a_{\mathrm{r}} = \left| \frac{V(R(x)) - x}{x} \right|;$$

$$3\% = \left| \frac{54\,500\,000 - 54\,500\,000 - 0.5r}{54\,500\,000 + 0.5r} \right|;$$

$$r = 3\,370\,000.$$

To achieve our goal, the LSB in our representation needs to represent 3 370 000 km or less. The value function for calculating the represented distance D from an 11-bit representation with an LSB weight of 3 000 000 km is

$$D = 54.5 \times 10^6 + 3 \times 10^6 \sum_{i=0}^{10} 2^i.$$

11.2 FIXED-POINT NUMBERS

11.2.1 Representation

A b-bit binary fixed-point number is a representation where the value of the number $a_{n-1}, a_{n-2}, \ldots, a_1, a_0$ is given by

$$v = 2^p \sum_{i=0}^{n-1} a_i 2^{i-n}, \tag{11.5}$$

where p is a constant giving the position of the binary point – in bits from the left end of the number.

Consider, for example, a fixed-point number system with $n = 4$ bit numbers, with the binary point to the right of the most significant bit at $p = 1$. That is, there are three bits to the right of the binary point – the fractional part of the number – and one bit to the left of the binary point – the integral part of the number. We often use the shorthand $p.f$ to refer to the integral and fractional bits of a number. Using this shorthand, the system with $n = 4$ and $p = 1$ is a 1.3 fixed-point system. If we add an additional sign bit to the left of the integral bits, we will refer to the resulting $(p + f + 1)$-bit system as an $sp.f$ system. We use 2's complement numbers in $sp.f$ systems.

The number of fractional bits, $f = n - p$, determines the resolution of our number system. The resolution, or the smallest interval we can distinguish, is $r = 2^{-f}$. With $f = 3$, for example, our 1.3 fixed-point system has a resolution of 1/8, or 0.125. Each increment of the binary number changes the value represented by 1/8. The number of integral bits, p, determines the range of our number system. The largest number we can represent with our system is $2^p - r$. For a signed number system, the lowest (most negative, not closest to zero) number we can represent is -2^p. The range and precision are sometimes easier to see if we rewrite Equation (11.5) as follows:

$$v = r \sum_{i=0}^{n-1} a_i 2^i. \tag{11.6}$$

To convert a binary fixed-point number to decimal, we just convert it into an integer and multiply by r. Table 11.1 shows some example fixed-point numbers and their conversion to decimal and fractional representations.

To convert a decimal number to a fixed-point binary number, the easiest approach is to (a) multiply the decimal number by 2^f, (b) round the resulting product to the nearest whole integer, and (c) convert the resulting decimal integer to a binary integer. For example, suppose we want to convert 1.389 to our 1.3 fixed-point format. We first multiply by 8, giving 11.112. Then we round to 11, and convert to binary, giving 1.011, which represents 1.375. Hence our *error* in this representation (the difference between the represented value and the actual value) is $1.389 - 1.375 = 0.014$, or just over 1% of the actual value. If we always round to the nearest value, the largest error over all values in the range (the accuracy of the representation) should be $r/2$ – in this case, 0.0625. As we get closer to zero, this error as a percentage of the value being represented grows. For numbers close to zero, the error is 100%.

While a decimal integer can be converted into a binary integer with zero error, a decimal fraction cannot, in general, be converted into a finite-length binary fraction without error. Because 5 does not evenly divide any power of 2, 0.1_{10} cannot be exactly represented as a

Table 11.1. **Example fixed-point numbers**

Format	Number	r	Integer	Value
1.3	1.011	0.125	11	1.375 (11/8)
$s1.3$	01.011	0.125	11	1.375 (11/8)
$s1.3$	11.011	0.125	−5	−0.625 (−5/8)
2.4	10.0111	0.0625	39	2.4375 (39/16)

finite-length binary fraction. Decimal fractions that represent powers of 2, like 0.25, 0.125, etc., can be exactly represented as binary fractions. However, other decimal fractions, such as 0.1 or 0.389, cannot be exactly represented. As you add more bits, the error becomes smaller, but it is never zero. If zero error is required, the number can be scaled (e.g., by 1000), or a BCD representation can be used.

Fixed-point binary numbers are often used in signal processing applications – for example, to process audio and video streams. In these applications, the range and precision are well known, and the binary point can be placed so that the full range of the number system is used while eliminating (or minimizing) the possibility of an overflow. Typically the values being represented are scaled so they fall between -1 and 1 so they can be represented in an $s0.f$ format. For most signal processing, 16 bits suffice and an $s0.15$ format is used.

Consider our example of representing a voltage between 0 and 10 V with 10 mV precision. Suppose we would like this representation to use the fewest bits possible. It is clear that we will need four bits to the left of the binary point to represent 10. To achieve 10 mV precision, we will need 20 mV resolution. Hence we need six bits to the right of the binary point – giving us a resolution of $2^{-6} = 0.015625$ and a precision of $2^{-7} = 0.0078125$. Thus, a 4.6 fixed-point format can directly represent this range of voltages to the specified precision using ten bits.

An alternative representation would be to use a *scaled* number. If we use a nine-bit binary number, we can represent values from 0 to 511. If we then scale this number by 20 mV – i.e., a count of 1 corresponds to 20 mV – we can then represent our 10 V range with 10 mV precision with just nine bits.

EXAMPLE 11.3 Conversion to binary fixed-point

Convert the number 4.23 into each of the following fixed-point schemes, and then convert back to decimal. Assume full rounding.

(1) $s4.2$;
(2) $s4.5$;
(3) 12-bit, where we represent the number $\times 100$.

The conversion of the integer portion of the number to decimal is trivial: $4_{10} = 100_2$. We find the fractional portion of the number using a methodology similar to how we convert integer numbers:

$$\begin{array}{rr}
0.23_{10} & 0.000_2 \\
\underline{-0.125_{10}} & \underline{+0.001_2} \\
0.105_{10} & 0.0010_2 \\
\underline{-0.0625_{10}} & \underline{+0.0001_2} \\
0.0425_{10} & 0.00110_2 \\
\\
\underline{-0.03125_{10}} & \underline{+0.00001_2} \\
0.01125_{10} & 0.0011100_2 \\
\underline{-0.0078125_{10}} & \underline{+0.0000001_2} \\
0.0034375_{10} & 0.0011101_2
\end{array}$$

$$\ldots$$

Note that we cannot precisely represent the fractional 0.23 in a finite amount of binary fraction bits. To represent this value as an $s4.2$ number, we round the fractional portion of the number to 0.01_2, giving an answer of 00100.01. Converting this number back to decimal gives 4.25_{10}. The $s4.5$ version of the number is 00100.00111_2 or 4.21875_{10}. Finally, scaling $4.23 \times 100 = 423_{10} = 00110100111_2$.

EXAMPLE 11.4 Designing a fixed-point system

Describe the number of bits needed to represent the distances from 0 to 31 AU with a precision of 0.05 AU.

Since all the numbers in this representation are positive, a sign bit is not necessary. We need only four bits to represent the integral numbers 0–31. With a rounding scheme, we need a resolution of 0.1 AU or lower. Since $2^{-4} < 0.1$, we use four bits for the fractional portion of our number. Thus, our final format is eight bits: 4.4.

11.2.2 Operations

We can perform the four basic operations on fixed-point binary numbers just as if they were integers. The same arithmetic circuits as those described in Chapter 10 can be used. However, we need to be careful to consider that the range and precision of the results of arithmetic operations may be different than the range and precision of the inputs.

Adding two $p.f$ fixed-point numbers gives a result that is a $(p+1).f$ fixed-point number. If we wish to force the result to be a $p.f$ fixed-point number, we may encounter an overflow condition in which the result is outside of our representable range. For example, consider our 4.6 fixed-point representation for representing voltages. If we add two voltages together, we will get a result between 0 and 20 V. A 5.6 fixed-point representation is required in order to represent this full range.

When adding sequences of fixed-point numbers, the numbers are often added using a greater range and then scaled and rounded to fit into the desired range and precision for the result. For example, suppose we have 16 values we wish to sum up, each in $s4.6$ fixed-point format and each representing a voltage between -10 V and 10 V. However, we know the sum will result in a number between -10 and 10. We perform the summation using $s8.6$ format to avoid any overflows on intermediate results and then convert back to $s4.6$ format at the end. In some cases, this final conversion is performed using *saturation*, where the value is clamped to the maximum representable value if it is out of range. (See Exercise 10.20.)

To add two fixed-point numbers with different representations, it is necessary first to align the binary points of the two numbers. This is most often achieved by converting both numbers to a fixed-point representation that has both p and f large enough to overlap both representations. For example, consider adding the 2.3 format number 01.101 to the 3.2 format number 101.01 We first convert both numbers to 3.3 format and then add 001.101 + 101.010, giving 110.111.

When we multiply two fixed-point numbers, the result has twice as many bits on both sides of the binary point as the inputs. For example, if we multiply two 4.6 fixed-point numbers, the result will be an 8.12 fixed-point number. For example, suppose we multiply a voltage signal with a range of 10 V and a precision of 10 mV by a current signal with a range of 10 A and a precision of 10 mA – both of these signals are in 4.6 format. The result is a power signal with a range of 100 W and a precision of 100 μW – in 8.12 format.

Many signal processors scale numbers to a 0.16 format (or $s0.15$ for signed numbers). Multiplying two 0.16 numbers gives a 0.32 number. A common operation is to take a dot product of two vectors in 0.16 format. To allow this operation to take place with no loss of precision, many popular signal processors have 40-bit accumulators. They accumulate up to 256 0.32 multiplication results, giving a sum in 8.32 format. (For signed numbers, the result is in $s8.30$ format.) This sum is then usually scaled and rounded to get a final result back in 0.16 format.

In most cases, a result calculated to a high precision eventually must be *rounded* to the original precision. Rounding is the process of reducing the precision of a number by discarding some of the rightmost bits of the number. When rounding decimal numbers to the nearest integer, we know that we should always round up if the next digit is a 5 or more and down if it is a 4 or less. Binary rounding works the same way. We round up if the most significant bit discarded is a 1 and down if it is a 0. For example, the number .10001000 in 0.8 format is rounded to .1001 in 0.4 format, while .10000111 in 0.8 format is rounded to .1000 in 0.4 format. Rounding requires an add (or at least an increment) to increment the result when rounding up – hence its not a free operation. The rounding can potentially change all of the remaining bits. For example, rounding .01111000 in 0.8 to 0.4 yields .1000.

EXAMPLE 11.5 Fixed-point operations

Add, subtract, and multiply the $s2.3$ number 010.001 and $s0.4$ number 0.1011, leaving the result in full precision.

First, we add after aligning the binary point:

$$\begin{array}{r} 010.0010 \\ +\underline{000.1011} \\ 010.1101 \end{array}$$

To subtract, we add the 2's complement of 0.1011: 1.0101. We must sign extend this number when aligning the values:

$$\begin{array}{r} 010.0010 \\ +\underline{111.0101} \\ 001.0111 \end{array}$$

Finally, we compute the multiplication:

$$\begin{array}{r} 00010.001 \\ \times\underline{0000.1011} \\ 000010001 \end{array}$$

$$\begin{array}{r} 000100010 \\ +\underline{010001000} \\ 010111011 \end{array}$$

The final format of the product is $s2.7$: 001.0111011_2.

11.3 FLOATING-POINT NUMBERS

11.3.1 Representation

High-dynamic-range numbers are often represented in *floating-point format.* In particular, a floating-point format is efficient for representing a number when we need a fixed proportional (not absolute) precision.

A floating-point number has two components; the exponent e and the mantissa m. The value represented by a floating point number is given by

$$v = m \times 2^{e-x}, \tag{11.7}$$

where m is a binary fraction, e is a binary integer, and x is a *bias* on the exponent that is used to center the dynamic range. The *mantissa, m,* is a fraction – meaning that the binary point is left of the MSB of m. The *exponent, e,* is an integer – its binary point is to the right of its LSB. If the bits of m are $m_{n-1}, \ldots, m_0$ and the bits of e are $e_{k-1}, \ldots, e_0$, the value is given by

$$v = \sum_{i=0}^{n-1} m_i 2^{i-n} \times 2^{\left(\sum_{i=0}^{k-1} e_i 2^k - x\right)}. \tag{11.8}$$

We refer to a floating-point number system with an a-bit mantissa and a b-bit exponent as an aEb format. For example, a system with a five-bit mantissa and a three-bit exponent is a 5E3 system. We will also use the "E" notation to write numbers. For example, the 5E3 number with a mantissa of 10010 and an exponent of 011 is 10010E011. Assuming zero bias, this number has a value of $v = 18/32 \times 8 = 4.5$.

We could also represent 4.5 as 01001E100 ($9/32 \times 16$). Most floating-point number systems disallow this second representation of 4.5 by insisting that all floating-point numbers be *normalized* by shifting the mantissa left (and decrementing the exponent) until either there is a 1 in the MSB of the mantissa or the exponent is 0. With normalized numbers we can quickly check for equality by simply comparing two numbers bit-by-bit. If numbers are unnormalized they must first be normalized (or at least aligned) before they can be compared. Some number systems take advantage of normalization by omitting the MSB of the mantissa, since it is almost always 1 (see Exercise 11.23).

Typically, when a floating-point number is stored, the exponent is stored to the left of the mantissa. For example, 11001E011 would be stored in eight bits as 01111001. Storing the exponent to the left allows integer comparison to work on floating-point numbers as long as numbers are normalized. That is, for two floating-point numbers a and b, if $a > b$ then $i_a > i_b$, where i_a and i_b are the integer interpretations of the bits of a and b, respectively.

If we want to represent signed values, we typically add a sign bit to the left of the exponent. For example, in eight bits we can represent an S4E3 number which, from left to right, would contain a sign bit, a three-bit exponent, and then a four-bit mantissa (SEEEMMMM). In this representation, the bit string 11001001 represents −9E4 or (with zero bias) $-9/16 \times 2^4 = -9$.

Floating-point numbers are just *scientific notation* applied to binary numbers. Like scientific notation, floating-point numbers have an error that is proportional to the magnitude of the number. For this reason, floating-point numbers are an efficient way to represent values with a specified proportional accuracy – particularly when the values in question have a high dynamic range.

For example, suppose we need to represent times from 1 ns to 1000 s with an accuracy of 1%. At the low end of this range we need an accuracy of 10 ps and at the high end of the range we need to represent 1000 s, i.e., 10^{14} times the required accuracy at the low end. A fixed-point representation would require 46 bits (10.36) to represent 1000 s with 10 ps accuracy (20 ps resolution). With a floating-point number, we can take advantage of the fact that we need only an accuracy of 10 s (a resolution of 20 s) at the high end of the range. Hence our mantissa needs only six bits. We can cover the large dynamic range (of $10^{12} < 2^{40}$) by using a six-bit exponent that can represent a range of 2^{64}. We set our exponent offset (x in Equation (11.7)) to 54 so we can represent numbers up to 2^{10}. Hence we can achieve the same relative accuracy with a 12-bit 6E6 floating-point number as we can with a 46-bit 10.36 fixed-point number.

Just like unscaled fixed-point numbers, binary floating-point numbers cannot *exactly* represent arbitrary decimal numbers – because 1/10 cannot be exactly represented as a finite-length binary fraction. Thus, the value 0.3, for example, can only be approximated in binary. This approximation can be made arbitrarily precise by adding mantissa bits, but an error will always remain. In applications where the error must be zero, BCD or a *scaled* representation must be used.

EXAMPLE 11.6 Floating-point design

Design a floating-point scheme to represent a measurement of 1×10^{-6} to 1×10^{7} with 5% error. Represent the value 4.5 in this format.

We need a total of five mantissa bits in our scheme to represent the most significant 1 and the rest of the mantissa to the needed precision. The smallest number to represent is 1×2^{-20}, while the largest is 1×2^{24}. We need a range in our exponent of at least 44, requiring six bits. The exponent bias is 20.

Note that 4.5 is equal to 0.10010×2^3. In our format, this is represented as 10010E010111.

11.3.2 Denormalized numbers and gradual underflow

If we disallow all denormalized numbers, we have a large gap in our representation function in that the closest number to zero we can represent (in 4E3, for example) is 1000E000, which with no bias represents 0.5. This gives a large relative error for numbers smaller than 0.5. We can reduce this relative error by allowing denormalized numbers only with an exponent of 0. We can then represent 1/4 as 0100E000, 1/8 as 0010E000, and 1/16 as 0001E000. In this case, the magnitude of the error for small numbers is reduced by a factor of 8. In general, the error for small numbers is reduced by a factor of 2^{n-1} for n-bit mantissas.

This representation is often referred to as *gradual underflow* because it reduces the error due to *underflow* – when an arithmetic operation gives a result closer to zero than the smallest number that can be represented. This solves the problem of having multiple representations for the same number. Because these denormalized numbers are restricted to have an exponent of 0, there is only one representation for each value.

To simplify the presentation, the arithmetic units we describe here do not support gradual underflow. We leave their extension to support this representation as Exercises 11.29 and 11.30.

11.3.3 Floating-point multiplication

Multiplying floating-point numbers is simple: we just multiply the mantissas and add the exponents. This doubles the number of mantissa bits and increments the number of exponent bits. We typically produce a result in the same format as the inputs by rounding the mantissa (as described in Section 11.2.2) and discarding the extra bit generated by the exponent add. When the mantissa is rounded, the exponent must be adjusted to account for the bits being dropped. If there is an exponent bias, the exponent must also be adjusted to compensate for the effect of applying the bias twice. If the number cannot be represented without an extra exponent bit, an overflow is signaled. The increment required to round the mantissa may itself result in a carry into the next mantissa bit position. If this occurs, the mantissa is shifted right again, and the exponent incremented accordingly.

For example, consider multiplying 101E011 (5) by 101E100 (10), both in 3E3 format with no bias. Our goal is to produce a normalized result in the same format. Our inputs are 101E011 and 101E100. Multiplying the mantissas 101 and 101 gives 011001 (25/64), and adding the exponents 011 and 100 gives 111 (7_{10}). This is, in fact, the correct answer since $25/64 \times 2^7 = 50$. We now need to convert this result back to 3E3 format.

To give a three-bit normalized mantissa, we shift the mantissa left once and discard the lowest two bits. The exponent is adjusted by decrementing it once to 110. Because the most significant bit discarded is 0, no increment is required for the rounding. Thus, our result in the original format is 110E110 ($6/8 \times 2^6 = 48$). The error of 2 here was caused by dropping the LSB of the mantissa during rounding.

EXAMPLE 11.7 Floating-point multiplication

Multiply the following two numbers written in 4E3 with an exponent bias of 4:

$$1100\text{E}010 \times 1100\text{E}110.$$

We first find the initial output exponent of 100_2 by summing the two input exponents (1000_2) and subtracting off the bias (100_2). Next, we multiply the two mantissas, giving a product of 0.10010000_2 or 0.1001_2. No normalization is needed because our final answer is 1001E100.

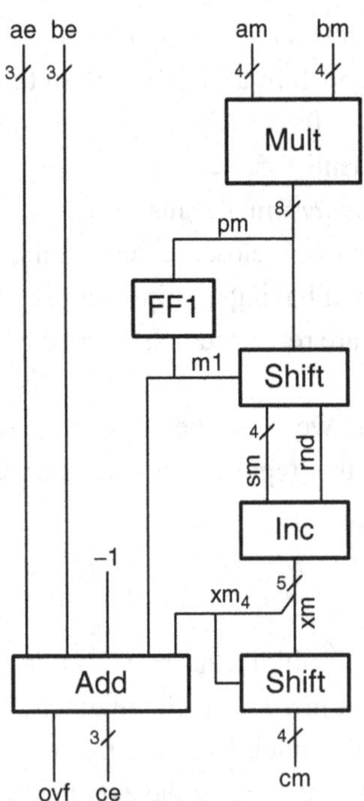

Figure 11.1. Floating-point multiplier with 3E4 inputs and output.

Figure 11.1 shows a block diagram of a floating-point multiplier, and the VHDL description of this multiplier is given in Figure 11.2. The FF1 block in the figure finds the leftmost one bit in `pm`, the product out of the multiplier. Because both inputs are normalized, this bit is guaranteed to be one of the left two bits of the product. Hence, we can use `pm[7]` directly to select which group of four bits from `pm` to use as `sm`, the shifted product.[2] Signal `rnd`, the first discarded bit of `pm`, is used to determine whether a rounding increment is needed. Signal `xm`, the rounded version of `sm`, is a five-bit signal. Like `pm`, it is guaranteed to have a 1 in one of its most significant two bits. Thus, we use its MSB `xm(4)` to select which group of four bits to output as the normalized mantissa. Note that we are guaranteed not to need another round after this final shift, since, if `xm(4)` is a 1, `xm(0)` is guaranteed to be a 0 (see Exercise 11.26).

11.3.4 Floating-point addition/subtraction

Floating-point addition is slightly more involved than multiplication because of the need to align the inputs and normalize the output. The process has three steps: alignment, addition, and normalization. During the alignment step the mantissa of the number with the smaller exponent is shifted right to align it with the mantissa of the number with the larger exponent – so bits of the same weight are aligned with one another. Once the two mantissas are aligned, they can be added or subtracted as if they were integers. This addition may produce a result that is

[2] The FF1 block in this case simply selects one bit out of pm as shown in the VHDL. With denormalized numbers, however, a full priority encoder function is required.

```
library ieee;
use ieee.std_logic_1164.all;
use ieee.std_logic_unsigned.all;
use work.ch10.all;

entity FP_Mul is
  generic( e: integer := 3 );
  port( ae, be: in std_logic_vector(e-1 downto 0); -- input exponents
        am, bm: in std_logic_vector(3 downto 0); -- input mantissas
        ce: out std_logic_vector(e-1 downto 0); -- result exponent
        cm: out std_logic_vector(3 downto 0); -- result mantissa
        ovf: out std_logic ); -- overflow indicator
end FP_Mul;

architecture impl of FP_Mul is
  signal pm: std_logic_vector(7 downto 0); -- result of initial multiply
  signal sm: std_logic_vector(3 downto 0); -- after shift
  signal xm: std_logic_vector(4 downto 0); -- after inc
  signal rnd: std_logic; -- true if MSB shifted off was one
  signal oece: std_logic_vector(e+1 downto 0); -- to detect exponent ovf
begin
  -- multiply am and bm
  MULT: Mul4 port map(am,bm,pm);

  -- Shift/Round: if MSB is 1 select bits 7:4 otherwise 6:3
  sm <= pm(7 downto 4) when pm(7) else pm(6 downto 3);
  rnd <= pm(3) when pm(7) else pm(2);

  -- Increment
  xm <= ('0' & sm) + ("000" & rnd);

  -- Final shift/round
  cm <= xm(4 downto 1) when xm(4) else xm(3 downto 0);

  -- Exponent add
  oece <= ("00" & ae) + be + (pm(7) or xm(4)) - 1;

  ce <= oece(e-1 downto 0);
  ovf <= oece(e-1) or oece(e-2);
end impl;
```

Figure 11.2. VHDL description of a floating-point multiplier.

unnormalized. A carry out of the add may result in a mantissa that must be shifted right by one bit for the most significant 1 to be placed into the MSB of the result. Alternatively, a subtraction may leave many of the MSBs of the result 0, requiring a left shift of an arbitrary number of bits to place the most significant 1 into the MSB of the result. The normalization step finds the most significant 1 of the result, shifts the result to place this 1 into the MSB of the mantissa, and adjusts the exponent accordingly. If the normalization is a right shift, it may discard one LSB of the result. A rounding increment is needed to round, rather than truncate, when discarding this bit.

As an example of floating-point addition, suppose we wish to add the numbers 5 and 11, both represented in 5E3 notation. In this notation, the number 5 is 10100E011 and the number 11 is 10110E100. During the alignment step, we shift the mantissa of 5 to the right by one bit to align it with the mantissa of 11. In effect, we are rewriting 5 as 01010E100 – denormalizing the mantissa to make the exponent agree with the other argument. With the two arguments aligned, we can now add the mantissas. There is a carry out of the mantissa add, giving a result of 100000E100 in 6E3 format. To normalize this result so it fits into 5E3 format, we shift the mantissa to the right by one position and increment the exponent, giving the final result of 10000E101, or 16.

As a second example, consider subtracting 9 from 10, both represented in *s*5E3 format. Here, 9 is +10010E100 and 10 is +10100E100. The two numbers have the same exponent, so they are already aligned. No shifting is required before the subtraction. Subtracting the two numbers gives 00010E100, which is unnormalized. To normalize this number, we shift the mantissa three places to the left and decrement the exponent by 3, giving 10000E001, or 1.

Aligning floating-point numbers before addition and subtraction causes bits beyond the LSB of the larger number to be dropped. Because of this, floating-point operations are not associative. For example, consider adding a small value b to a large value A. If b is small enough, $b + A = A$. If we then subtract A from this sum, we find $(b + A) - A = 0$. Subtracting first gives a different answer: $b + (A - A) = b$.

EXAMPLE 11.8 Floating-point addition and subtraction

Add and subtract the following floating-point numbers written in 4E3 format with a bias of 4:

$$1100\text{E}100 + 1110\text{E}011; \qquad 1100\text{E}100 - 1110\text{E}011.$$

To perform the addition, we must first shift the smaller number to align the binary points:

$$\begin{aligned} &0.1100\text{E}100 \\ &+\underline{0.0111\text{E}100} \\ &= 1.0011\text{E}100 \\ &\approx 1010\text{E}101 \end{aligned}$$

After doing the addition, we had to normalize the sum, rounding up after shifting out a 1.

The subtraction follows a similar methodology, though we add the 2's complement of the shifted value being subtracted:

$$\begin{aligned} &0.1100\text{E}100 \\ &+\underline{1.1001\text{E}100} \\ &0.0101\text{E}100 \\ &= 1010\text{E}011 \end{aligned}$$

This time, the difference must be left-shifted to obtain the correct answer.

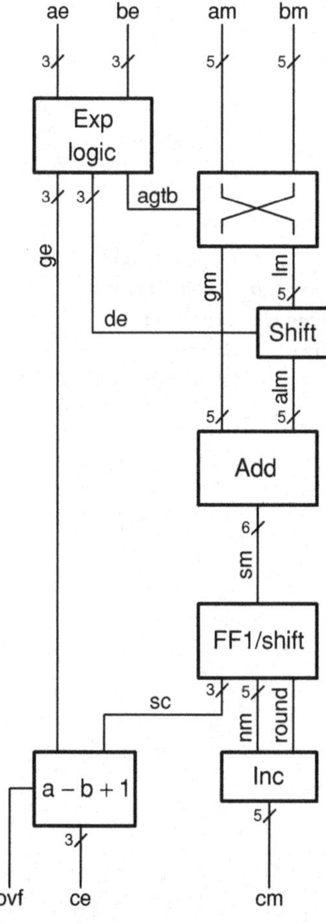

Figure 11.3. Floating-point adder.

A block diagram of a floating-point adder is shown in Figure 11.3, and a VHDL description of this adder is shown in Figure 11.4. The input exponent logic compares the two exponents, generating signal `agtb`, to determine which mantissa needs to be shifted, and signal `de`, which gives the number of bits to shift. The input switch uses `agtb` to switch the two mantissas so the mantissa with the greater exponent is on signal `gm` and the mantissa with the smaller exponent is on signal `lm`. Mantissa `lm` is then shifted by `de` to align the mantissas.[3] The aligned mantissas are added, producing signal `sm`, which is one bit wider than the mantissas. A reverse-priority encoder is then used to find the most significant 1 in `sm`. A shift is then performed to move this bit to the most significant position of the result, giving signal `nm`. This shift ranges from a one-bit right-shift to a full-width left-shift. Signal `rnd` captures the bit discarded on a right-shift by 1. The exponent is adjusted to reflect the shift amount. If the exponent cannot be represented in the given number of bits, an overflow occurs.

[3] The astute reader will notice that we have potentially shifted off and lost a round bit by making `alm` 5 bits instead of 6. Exercise 11.24 will ask you to fix that.

```
library ieee;
use ieee.std_logic_1164.all;
use ieee.std_logic_unsigned.all;
use ieee.numeric_std.all;
use work.ch8.all;

entity FP_Add is
  generic( e: integer := 3; m: integer := 5 );
  port( ae, be: in std_logic_vector(e-1 downto 0); -- input exponents
        am, bm: in std_logic_vector(m-1 downto 0); -- input mantissas
        ce: out std_logic_vector(e-1 downto 0); -- result exponent
        cm: out std_logic_vector(m-1 downto 0); -- result mantissa
        ovf: out std_logic );
end FP_Add;

architecture impl of FP_Add is
  signal ge, le, de, sc: std_logic_vector(e-1 downto 0);
  signal gm, lm, alm: std_logic_vector(m-1 downto 0);
  signal sm, nmrnd: std_logic_vector(m downto 0);
  signal ovfce: std_logic_vector(e downto 0);
  signal agtb: std_logic;
begin
  -- input exponent logic
  agtb <= '1' when ae >= be else '0';
  ge <= ae when agtb else be;
  le <= be when agtb else ae;
  de <= ge - le;

  -- select input mantissa
  gm <= am when agtb else bm;
  lm <= bm when agtb else am;

  -- shift mantissa to align
  alm <= std_logic_vector(shift_right(unsigned(lm),to_integer(signed(de))));

  -- add
  sm <= ('0' & gm) + ('0' & alm);

  -- find first one
  FF1: RevPriorityEncoder generic map(6,3) port map(sm, sc);

  -- shift first 1 to MSB
  nmrnd <= std_logic_vector(shift_left(unsigned(sm),to_integer(signed(sc))));

  -- adjust exponent
  ovfce <= ('0' & ge) - ('0' & sc) + 1;
  ovf <= ovfce(e);

  -- round result
  cm <= nmrnd(m downto 1) + ((m-1 downto 1 => '0') & nmrnd(0));
end impl;
```

Figure 11.4. VHDL description of a floating-point adder.

Summary

In this chapter you have learned how to select a number system that has adequate range and accuracy for a particular application. We can express the error of a representation in either *absolute* or *relative* terms. For a given representation function $R(x)$ we can compute the absolute and relative errors of the representation at any point x. The *accuracy* of a representation is the maximum error over the required range.

Fixed-point numbers have a fixed *binary point* with fractional bits to the right of the binary point and integral bits to the left. For example, an $s1.14$ number is a 16-bit number with a sign bit, one integral bit to the left of the binary point, and 14 fractional bits to the right of the binary point. Bits to the right of the binary point have fractional value. In the $s1.14$ representation, the LSB is weighted 2^{-14}. Operations on fixed-point numbers are simpler than operations on floating-point numbers and, if properly scaled, they have good absolute accuracy.

The value represented by a *floating-point* number is $v = m \times 2^{e-x}$, where m is the *mantissa*, e is the *exponent*, and x is the *bias*. Multiplying m by 2^{e-x} has the effect of allowing the binary point in m to *float*, hence the name. Operations on floating-point numbers are more complex than on integer numbers because of the need to align mantissas – so that similar bits have similar weights – before adding. The advantage of floating-point numbers is that they have greater relative accuracy than fixed-point numbers for a given number of bits and required range. Floating-point numbers also allow designers to be lazy and not *scale* their variables. The large range of the floating-point number allows designers to defer the scaling to the hardware.

To have a single representation for each value, we generally insist that floating-point numbers be *normalized* by shifting the mantissa left (and decrementing the exponent accordingly) so that the MSB of the mantissa is 1. Since the MSB of the mantissa is always 1, we often omit it – we call this an *implied 1*. To allow small numbers to be represented accurately, we allow denormalized numbers when the exponent is 0 – this is called *gradual underflow*.

Floating-point multiplication is performed by adding the exponents and multiplying the mantissas. The result of the multiply must be rounded – to fit into the required number of output bits – and normalized – so that the most significant bit of the output is 1.

A floating-point addition operation requires that the mantissas first be *aligned* – by shifting the mantissa with the smaller exponent right – so that similar bits have similar weights. After the addition, the result must be *normalized* – the result shifted so that it has a 1 in its MSB and the exponent adjusted accordingly.

BIBLIOGRAPHIC NOTE

For more information about floating-point formats, consult the IEEE floating-point standard [58] and [54].

Exercises

11.1 *Fixed-point representation, I.* Convert the following fixed-point number to decimal: 1.0101 in 1.4.

11.2 *Fixed-point representation, II.* Convert the following fixed-point number to decimal: 11.0101 in *s*1.4.

11.3 *Fixed-point representation, III.* Convert the following fixed-point number to decimal: 101.011 in 3.3.

11.4 *Fixed-point representation, IV.* Convert the following fixed-point number to decimal: 101.011 in *s*2.3.

11.5 *Fixed-point representation, I.* Convert 1.5999 to the nearest fixed-point *s*1.5 representation. Give the absolute and relative errors.

11.6 *Fixed-point representation, II.* Convert 0.3775 to the nearest fixed-point *s*1.5 representation. Give the absolute and relative errors.

11.7 *Fixed-point representation, III.* Convert 1.109375 to the nearest fixed-point *s*1.5 representation. Give the absolute and relative errors.

11.8 *Fixed-point representation, IV.* Convert −1.171875 to the nearest fixed-point *s*1.5 representation. Give the absolute and relative errors.

11.9 *Fixed-point representation, absolute error.* Find a decimal value between −1 and 1 for which the absolute value of the error of representation as an *s*1.5 fixed-point number is maximum.

11.10 *Fixed-point representation, relative error.* Find a decimal value between 0.1 and 1 for which the percentage of the error of representation as an *s*1.5 fixed-point number is maximum.

11.11 *Selecting a fixed-point representation, I.* You need to represent a relative pressure signal with a range from −10 PSI to 10 PSI with an accuracy of 0.1 PSI. Select a fixed-point representation that covers this range with the specified accuracy with a minimum number of bits.

11.12 *Selecting a fixed-point representation, II.* Select a fixed-point representation that covers a range from 0.001 to 1 with an accuracy of 1% across the range and uses a minimum number of bits.

11.13 *Floating-point representation, I.* Convert the following floating-point number with a bias of 3 to decimal: 1111E111 in 4E3.

11.14 *Floating-point representation, II.* Convert the following floating-point number with a bias of 3 to decimal: 1010E100 in 4E3.

11.15 *Floating-point representation, III.* Convert the following floating-point number with a bias of 3 to decimal: 1100E001 in *s*3E3.

11.16 *Floating-point representation, IV.* Convert the following floating-point number with a bias of 3 to decimal: 0101E101 in *s*3E3.

11.17 *Floating-point representation, I.* Convert −23 to *s*3E5 floating-point format with a bias of 8. Give the relative and absolute errors.

11.18 *Floating-point representation, II.* Convert 100 000 to *s*3E5 floating-point format with a bias of 8. Give the relative and absolute errors.

11.19 *Floating-point representation, III.* Convert 999 to *s*3E5 floating-point format with a bias of 16. Give the relative and absolute errors.

11.20 *Floating-point representation, IV.* Convert 64 to *s*3E5 floating-point format with a bias of 16. Give the relative and absolute errors.

11.21 *Selecting a floating-point representation, I.* Select a floating-point representation that covers a range from -10 to 10 with an accuracy of 0.1 for numbers with magnitudes $> 1/32$ using a minimum number of bits.

11.22 *Selecting a floating-point representation, II.* Select a floating-point representation that covers a range from 0.001 to 100 000 000 with an accuracy of 1% across the range and uses a minimum number of bits.

11.23 *Implied 1.* Many floating-point formats omit the MSB of the mantissa. That is, they don't bother to store it. The IEEE single-precision floating-point standard, for example, stores a 24-bit mantissa in 23 bits by omitting the MSB of the mantissa. This is referred to as an *implied 1*. Some formats insist that this missing MSB is always 1. However, this leads to interesting error behavior near zero. Better error characteristics can be achieved (at some cost in complexity) by implying that the MSB of the mantissa is 0, and that the exponent is $1 - x$ when e is 0 (the same exponent as when e is 1). A number system having this feature is said to provide *gradual underflow*.

(a) Suppose you have a 5E3 floating-point number system with a bias x of 0 and an implied 1. (The mantissa consists of an implied 1 followed by four bits.) Plot the error curve over the interval $[-2, 2]$ for a system without gradual underflow.

(b) On the same axes as (a), plot the error curve for the same number system but with gradual underflow.

(c) At what value is the percentage error of the system without gradual underflow largest?

(d) Is there a value range in which the gradual underflow system has a larger error than the system without gradual underflow? If so, what is this range?

11.24 *Floating-point add with round bit.* Fix the block diagram of Figure 11.3 and VHDL of Figure 11.4 to account for the potentially lost round bit of *lm*. For example, adding $1.0000 \times 2^0 + 1.1111 \times 2^{-1}$ should output 1.0000×2^1 (rounded from 1.11111×2^0) instead of 1.1111×2^0. Create a new wire, *guard*, that represents the bit to the right of *alm*'s LSB and add it into the sum *sm* if set.

11.25 *Floating-point subtract.* Extend the floating-point adder shown in Figures 11.3 and 11.4 to handle signed floating-point numbers and to perform floating-point subtractions. Assume that the sign bit for each input operand is provided on separate lines `as` and `bs`, and that the sign of the result is to be output on line `cs`.

11.26 *Floating-point multiply.* In Section 11.3.3 we stated that if the MSB of the rounded product `xm` was a 1, then its LSB must be a 0. However, we gave no justification for this claim. Prove that this is true.

11.27 *Floating-point multiply with denormalized numbers.* Modify the design of the floating-point multiplier of Section 11.3.3 to work with denormalized inputs.

11.28 *Floating-point addition with underflow.* A floating-point add of two normalized numbers may result in a number that cannot be represented with a 1 in the MSB of the mantissa, but yet is not 0. This situation is an *underflow*. Modify the adder of Section 11.3.4 to detect and signal an underflow condition.

11.29 *Add with gradual underflow.* Extend the adder design of Section 11.3.4 to handle gradual underflow; that is, to handle denormalized inputs when the input exponent is 0.

11.30 *Multiply with gradual underflow.* Extend the multiplier design of Section 11.3.3 to handle gradual underflow.

11.31 *Gradual underflow and implied 1.* Consider a system that uses a representation with an implied 1 in the MSB of the mantissa, as described in Exercise 11.23. Extend this system to allow numbers with gradual underflow. Make sure that you do not create any gaps or redundant representations in your system. (Hint: let exponents of 0 and 1 represent the same value, but one with an implied 1, and one without.) Convert the following numbers to a 4E3 version of your representation:

(a) 1/8,

(b) 4,

(c) 1/16,

(d) 32.

11.32 *Add with gradual underflow and implied 1.* Extend the adder design of Section 11.3.4 to handle gradual underflow with an implied 1 (Exercise 11.31).

11.33 *Multiply with gradual underflow and implied 1.* Extend the multiplier design of Section 11.3.3 to handle gradual underflow with an implied 1 (Exercise 11.31).

11.34 *Logarithmic representation.* Consider a number system where $v = b^{e-x}$ for a fixed *base* b. This system is like a floating-point representation where the mantissa is always 1 and hence is omitted. Consider the specific case where $b = 2^{1/8}$. Suppose you have to represent voltages from 1 μV to 1 MV with a relative accuracy of 5%. How does this representation compare with fixed- and floating-point representations in terms of the number of bits required?

12 Fast arithmetic circuits

In this chapter, we look at three methods for improving the speed of arithmetic circuits, and in particular multipliers. We start in Section 12.1 by revisiting binary adders and see how to reduce their delay from $O(n)$ to $O(\log(n))$ by using hierarchical carry-look-ahead circuits. This technique can be applied directly to build fast adders and is also used to accelerate the summation of partial products in multipliers. In Section 12.2 we see how the number of partial products that need to be summed in a multiplier can be greatly reduced by *recoding* one of the inputs as a sequence of higher-radix, signed digits. Finally, in Section 12.3 we see how the partial products can be accumulated with $O(\log(n))$ delay by using a tree of full adders. The combination of these three techniques into a fast multiplier is left as Exercises 12.17 to 12.20.

12.1 CARRY LOOK-AHEAD

Recall that the adder developed in Section 10.2 is called a *ripple-carry adder* because a transition on the carry signal must *ripple* from bit to bit to affect the final value of the MSB of the sum. This ripple-carry results in an adder delay that increases linearly with the number of bits in the adder. For large adders, this linear delay becomes prohibitive.

We can build an adder with a delay that increases logarithmically, rather than linearly, with the width of the adder by using a dual-tree structure as shown in Figure 12.1. This circuit works by computing carry *propagate* and carry *generate* across groups of bits in the upper tree and then using these signals to generate the carry signal into each bit in the lower tree. The propagate signal p_{ij} is true if a carry into bit i will propagate from bit i to bit j and generate a carry out of bit j. The generate signal g_{ij} is true if a carry will be generated out of bit j regardless of the carry into bit i. We can define p and g recursively as follows:

$$p_{ij} = p_{ik} \wedge p_{(k+1)j} \ (\forall k : i <= k < j), \tag{12.1}$$

$$p_{ii} = p_i = a_i \oplus b_i, \tag{12.2}$$

$$g_{ij} = (g_{ik} \wedge p_{(k+1)j}) \vee g_{(k+1)j} \ (\forall k : i <= k < j), \tag{12.3}$$

$$g_{ii} = g_i = a_i \wedge b_i. \tag{12.4}$$

The first two equations define the propagate signals. A carry signal will propagate across a range of bits from i to j if it propagates from i to k (p_{ik}) and then from $k+1$ to j ($p_{(k+1)j}$). This works for any choice of k from i to $j-1$. Of course, we usually split our intervals evenly,

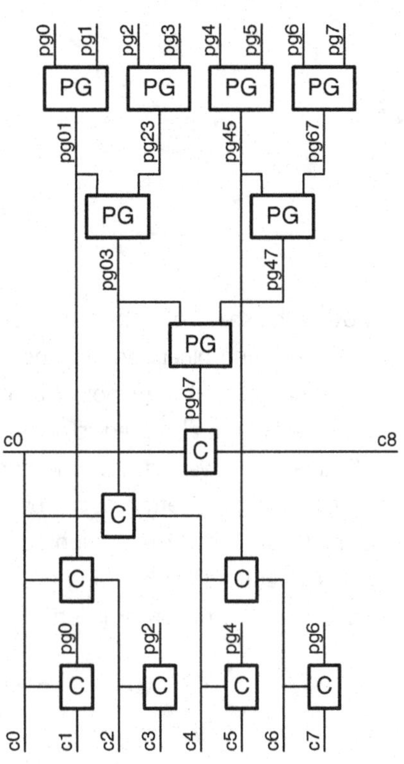

Figure 12.1. Block diagram of a carry-look-ahead circuit. Each *pg* signal represents two bits, a *p* bit, which specifies that the carry propagates across the specified bit range, and a *g* bit, which specifies that the carry is generated out of the specified bit range. The *pg* signals are combined in a tree to span increasing ranges of bits. The *pg* signals are then used in a second tree to generate the carry, *c*, signals for each bit.

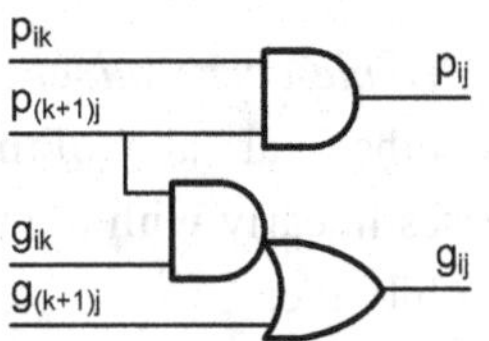

Figure 12.2. Logic to recursively generate propagate, p_{ij}, and generate, g_{ij}, signals across a group of bits from i to j as a function of the p and g signals across adjacent subranges ik and $(k+1)j$.

choosing $k = \lfloor (i+j)/2 \rfloor$. When our interval is down to a single bit, we compute p_{ii}, or just p_i, as discussed in conjunction with Figure 10.3. The carry propagates across a single bit i when exactly one input to that bit is true (i.e., when $a_i \oplus b_i$).

The first generate equation states that a carry signal will be generated out of bit j regardless of the carry into bit i if either (1) it is generated out of k regardless of the carry into i and then propagated across bits $k+1$ to j, or (2) it is generated out of bit j regardless of the carry into bit $k+1$. The base case for generate is as discussed in Figure 10.3. The carry is generated out of a single bit only when both inputs to that bit are high.

It is easy to construct the top part of the carry circuit of Figure 12.1 by using Equations (12.1)–(12.4). For the eight-bit carry circuit in the figure, we would like to know what the carry out of bit 7, c_8, is. Hence we would like to compute p_{07} and g_{07}. To simplify the drawing, we refer to these two signals collectively as pg_{07}. We choose $k = 3$ and compute pg_{07} from pg_{03} and pg_{47}. The logic of the block labeled PG is that of Equations (12.1) and (12.3), and is shown in Figure 12.2. We then recursively subdivide each of these intervals until we bottom out at the single-bit p and g terms.

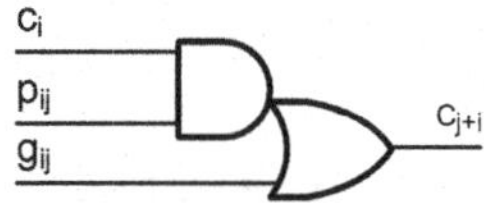

Figure 12.3. Logic to generate carry signals from upper-level carry signals and *pg* signals.

Once we have generated the *pg* signals recursively, we use these signals to generate the carries. We proceed to build a tree, starting with the carry across all eight bits, then groups of four, two, and one. At each level, we compute the carry from carries at the previous levels and the *pg* signals as follows:

$$c_{j+1} = g_{ij} \vee (c_i \wedge p_{ij}). \tag{12.5}$$

The logic in each of the C blocks of Figure 12.1 is that of Equation (12.5), and is shown in Figure 12.3.

VHDL code for the eight-bit carry-look-ahead circuit of Figure 12.1, along with logic that generates p and g from adder inputs a and b, is shown in Figure 12.4. The code includes the equations for the input, PG, and C blocks directly, rather than defining modules for each block type and instantiating each of them separately. In this case, coding with equations makes the code easier both to write and to understand.

The VHDL code in Figure 12.4 generates many unused signals. For example, pairwise propagate signals $p_{i(i+1)}$, named `p2(i)` in the code, are generated for all eight bits even though only the even bits are needed. Similarly, four-bit wide propagate and generate signals `p4` and `g4` are generated for all eight bits, even though only bits 0 and 4 are used, and eight-bit wide propagate and generate signals `p8` and `g8` are generated, and only bit 0 is used. This coding style makes it easier to write the equations, and it is not wasteful because the synthesizer optimizes away the unused signals and any logic needed only to generate them.

The circuit of Figures 12.1 and 12.4 uses a fan-in of 2. That is, p and g signals at each level are combined pairwise to form the p and g signals at the next level. Depending on the optimal fan-in and fan-out of one stage of logic in the technology at hand (see Sections 5.2 and 5.3), it may be faster to build a carry-look-ahead circuit with a larger fan-in. For example, Figure 12.5 shows a 16-bit carry-look-ahead circuit with a fan-in of 4.

VHDL for the circuit of Figure 12.5 is shown in Figure 12.6. In contrast to Figure 12.4, which implements the look-ahead equations directly, this module is coded using submodules (shown in Figure 12.7) to implement the four-bit wide PG and carry functions. For the radix-4 case, this makes the code more readable.

Of course, the fan-in, or *radix*, of a carry-look-ahead circuit need not be a power of 2. One can build carry-look-ahead circuits with fan-ins of 3, 5, 6, or any other value. In some cases, better performance is achieved with a mixed-radix design, in which a larger fan-in is used in the earlier stages (where wires are short) and a smaller fan-in is used in the later stages (where wires are longer) – to balance the larger electrical effort associated with the long wires with a smaller logical effort due to the fan-in (see Section 5.3).

While we have illustrated carry look-ahead in the context of an adder, the technique can be applied to any one-dimensional iterative function. Arbiters (see Section 8.5) and comparators (see Section 8.6), for example, can be realized with delay that increases as the log of the number of inputs using a look-ahead tree. All that is required is to write the logic of the iterative circuit

```
-- 8-bit carry-look-ahead
-- takes 8-bit inputs a and b, ci and produces co ;
-- this module generates many unused signals which the synthesizer optimizes away
library ieee;
use ieee.std_logic_1164.all;

entity Cla8 is
  port( a, b: in std_logic_vector(7 downto 0);
        ci: in std_logic;
        co: out std_logic_vector(8 downto 0) );
end Cla8;

architecture impl of Cla8 is
  signal p, g, p2, g2, p4, g4, p8, g8: std_logic_vector(7 downto 0);
  signal coi: std_logic_vector(8 downto 0);
begin
  -- input stage of PG cells
  p <= a xor b;
  g <= a and b;

  -- p and g across multiple bits
  -- px(i)/gx(i) is propagate/generate across x bits starting at bit i
  p2 <= p and ('0' & p(7 downto 1)); -- across pairs - only 0,2,4,6 used
  g2 <= ('0' & g(7 downto 1)) or (g and ('0' & p(7 downto 1)));
  p4 <= p2 and ("00" & p2(7 downto 2)); -- across nybbles - only 0,4 used
  g4 <= ("00" & g2(7 downto 2)) or (g2 and ("00" & p2(7 downto 2)));
  p8 <= p4 and ("0000" & p4(7 downto 4)); -- across byte - only 0 used
  g8 <= ("0000" & g4(7 downto 4)) or (g4 and ("0000" & p4(7 downto 4)));

  -- first level of output, derived from ci
  coi(0) <= ci;
  coi(8) <= g8(0) or (ci and p8(0));
  coi(4) <= g4(0) or (ci and p4(0));
  coi(2) <= g2(0) or (ci and p2(0));
  coi(1) <= g(0) or (ci and p(0));

  -- second level of output, derived from first level
  coi(6) <= g2(4) or (coi(4) and p2(4));
  coi(5) <= g(4) or (coi(4) and p(4));
  coi(3) <= g(2) or (coi(2) and p(2));

  -- final level of output derived from second level
  coi(7) <= g(6) or (coi(6) and p(6));
  co <= coi;
end impl;
```

Figure 12.4. VHDL description of an eight-bit carry-look-ahead circuit. The code generates many unused intermediate signals, which are optimized out by the synthesizer.

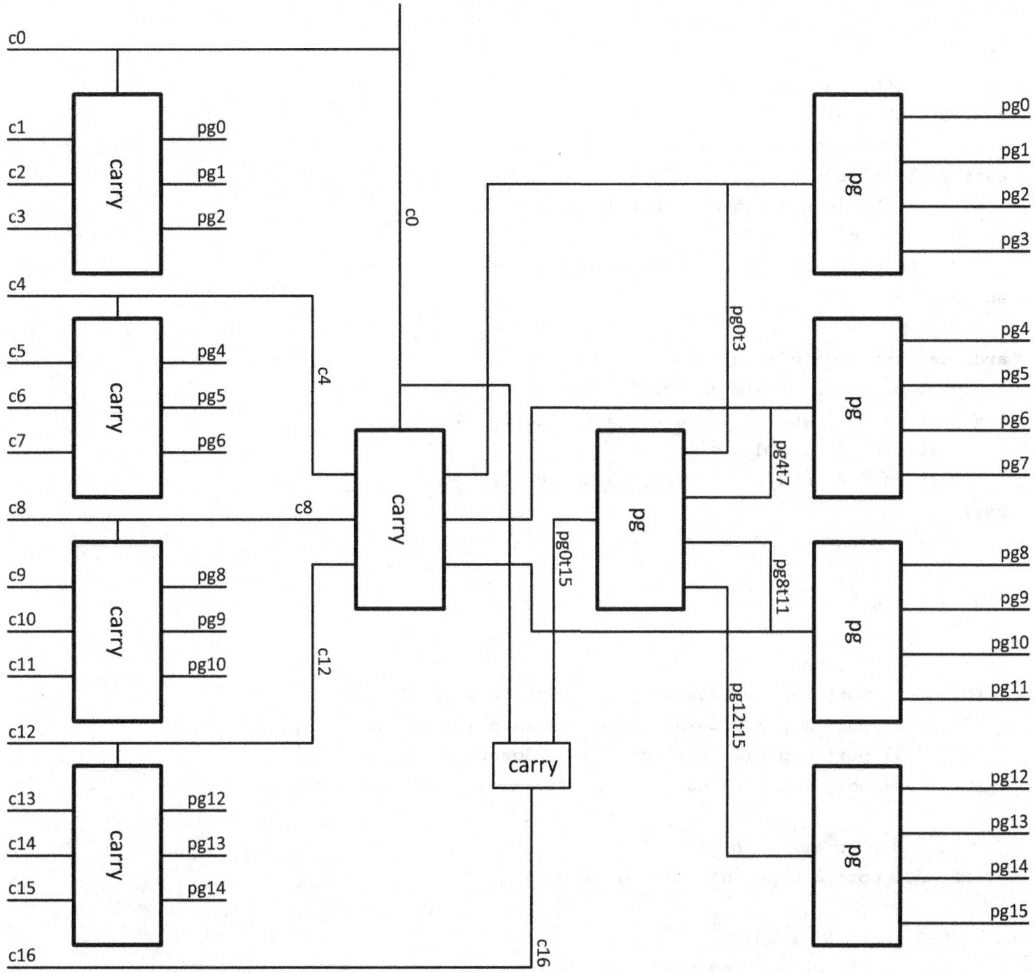

Figure 12.5. Sixteen-bit carry-look-ahead circuit with a fan-in, or *radix*, of 4.

in propagate–generate form:

$$p_i = f_p(a_i, b_i, \ldots), \tag{12.6}$$

$$g_i = f_g(a_i, b_i, \ldots), \tag{12.7}$$

$$c_{i+1} = g_i \vee (p_i \wedge c_i), \tag{12.8}$$

$$o_i = f_o(c_i, a_i, b_i, \ldots). \tag{12.9}$$

For example, a magnitude comparator with carry propagating from LSB to MSB in propagate–generate can be put in propagate–generate form:

$$p_i = \overline{(a_i \oplus b_i)}, \tag{12.10}$$

$$g_i = a_i \wedge \overline{b_i}, \tag{12.11}$$

$$o = c_N. \tag{12.12}$$

```
-- 16-bit radix-4 carry-look-ahead
library ieee;
use ieee.std_logic_1164.all;
use work.ch12.all;

entity Cla16 is
  port( a, b: in std_logic_vector(15 downto 0);
        ci: in std_logic;
        co: out std_logic_vector(16 downto 0) );
end Cla16;

architecture impl of Cla16 is
  signal p, g: std_logic_vector(15 downto 0);
  signal p4, g4: std_logic_vector(3 downto 0);
  signal p16, g16: std_logic;
  signal co1284: std_logic_vector(2 downto 0);
begin
  -- input stage of PG cells
  p <= a xor b;
  g <= a and b;

  --  input PG stage
  PG10: PG4 port map(p(3 downto 0),g(3 downto 0),p4(0),g4(0));
  PG11: PG4 port map(p(7 downto 4),g(7 downto 4),p4(1),g4(1));
  PG12: PG4 port map(p(11 downto 8),g(11 downto 8),p4(2),g4(2));
  PG13: PG4 port map(p(15 downto 12),g(15 downto 12),p4(3),g4(3));

  -- p and g across 16 bits
  PG2: PG4 port map(p4, g4, p16, g16);

  -- MSB and LSB of carry
  co(16) <= g16 or (ci and p16);
  co(0) <= ci;

  -- first level of carry
  C10: Carry4 port map(ci,p4(2 downto 0), g4(2 downto 0),co1284);
  co(12) <= co1284(2);
  co(8) <= co1284(1);
  co(4) <= co1284(0);

  -- second level of carry
  C20: Carry4 port map(ci,p(2 downto 0),g(2 downto 0),co(3 downto 1));
  C21: Carry4 port map(co1284(0),p(6 downto 4),g(6 downto 4),co(7 downto 5));
  C22: Carry4 port map(co1284(1),p(10 downto 8),g(10 downto 8),co(11 downto 9));
  C23: Carry4 port map(co1284(2),p(14 downto 12),g(14 downto 12),co(15 downto 13));
end impl;
```

Figure 12.6. VHDL description of the 16-bit, radix-4 carry-look-ahead unit of Figure 12.5.

```
-- four-bit PG module
library ieee;
use ieee.std_logic_1164.all;
use ieee.std_logic_misc.all;
entity PG4 is
  port( pi, gi: in std_logic_vector(3 downto 0);
        po, go: out std_logic );
end PG4;
architecture impl of PG4 is
begin
  po <= and_reduce(pi);
  go <= gi(3) or (gi(2) and pi(3)) or (gi(1) and pi(3) and pi(2)) or
        (gi(0) and pi(3) and pi(2) and pi(1));
end impl;
------------------------------------------------------------------------
-- four-bit carry module
library ieee;
use ieee.std_logic_1164.all;
entity Carry4 is
  port( ci: in std_logic; p, g: in std_logic_vector(2 downto 0);
        co: out std_logic_vector(2 downto 0) );
end Carry4;
architecture impl of Carry4 is
  signal gg: std_logic_vector(3 downto 0);
begin
  gg <= g & ci;
  co <= gg(3 downto 1) or (gg(2 downto 0) and p) or
        ((gg(1 downto 0) & '0') and p and (p(1 downto 0) & '0')) or
        ((gg(0) & "00") and p and (p(1 downto 0)&'0') and (p(0) & "00"));
end impl;
```

Figure 12.7. Four-bit `PG` and `Carry` modules for the carry-look-ahead module of Figure 12.6.

Using this formulation, we can build a 16-bit magnitude comparator with only four levels of logic using the circuit of Figure 12.5. Because the magnitude comparator needs only the MSB of the carry for its output, the `Carry` blocks in Figure 12.5 can be omitted.

When writing VHDL, it is more important to describe other functions using a carry-look-ahead formulation than it is to describe an adder with this formulation. Modern synthesis tools are very good at taking the VHDL code

```
s <= a + b;
```

and expanding this into a highly optimized adder, including carry look-ahead if appropriate for the constraints. Synthesis tools will not perform this optimization for other functions such as priority encoders and comparators.

We look at some other example applications of carry-look-ahead circuits in Exercises 12.2 to 12.4.

12.2 BOOTH RECODING

The unsigned binary multiplier described in Section 10.4 generates m n-bit partial products and requires $m \times (n-1)$ full adder cells to sum these partial products into a final result. We can reduce the number of partial products by a factor of 2 (or more) using Booth recoding. As an added bonus, the recoding easily handles 2's complement signed inputs.

Radix-2^i recoding works by reinterpreting an n-bit multiplier as an (n/i)-digit radix-2^i number. For example, we can reinterpret the six-bit binary number $b = 011011_2$ as the four-digit quaternary (radix-4) number 123_4. To multiply this number by another binary number, $a = 010011_2$, we need to sum only three partial products, as shown in Figure 12.8.

This simple recoding requires that the partial products be selected from shifted multiples of a up to $2^i - 1$. For example, multiples of a, $2a$, and $3a$ are required for the radix-4 multiplication discussed above. The value of $2a$ is available with a simple shift, but $3a$ requires a shift and an addition. Even with this addition to pre-compute $3a$, the total number of full adders needed to compute the product is reduced by almost a factor of 2.

Booth recoding eliminates the need for the pre-addition (for radix-4) and naturally handles 2's complement signed numbers. It does this by using overlapping digits that are interpreted as signed numbers. Each digit is an $(i+1)$-bit field and overlaps adjacent digits by one bit. Consider the bits of a digit as a bit vector $b_{i-1}, b_{i-2}, \ldots, b_0, b_{-1}$. The MSB of each digit b_{i-1} is weighted $-(2^{i-1})$, the middle bits b_j of each digit are each weighted 2^j, and the LSB of each digit b_{-1} is weighted 1. With the overlap, this weighting maintains the correct total weighting of each bit in the recoded number.

Consider radix-4 Booth recoding an eight-bit 2's complement number $b_7, \ldots, b_0$. We interpret this as a four-digit radix-4 number $d_3, \ldots, d_0$, where digit i consists of the three bits b_{2i+1}, b_{2i}, and b_{2i-1} weighted by -2, 1, and 1, respectively (and the whole digit is weighted by 4^i). Summing the weights for the overlapping bits gives the correct weight for a 2's complement number. For example, bit b_1 is the MSB of digit d_0, where it has weight -2, and the LSB of digit d_1, where it has weight 4; summing these values gives a total weight of 2, the correct weight for b_1. Table 12.1 illustrates how this overlap also works for bits b_3 and b_5. The sum of the weights of each bit below the line equals the weight above the line. To make the three-bit digit d_0 work out evenly, we assume a b_{-1} which is always zero.

As illustrated in Table 12.2, each radix-4 digit d_i in the Booth recoded multiplier can take on one of five values: $-2, -1, 0, 1$, or 2. Thus, we can build a multiplier that generates half

```
     010011
   x    123
   ----
    0111001
  0100110
0010011
------
01000000001
```

Figure 12.8. An example of multiplying a binary number by a quaternary number. Only three binary partial products must be summed, compared with six for binary × binary multiplication.

Table 12.1. **Weights for radix-4 Booth recoding of an eight-bit binary number**

Bit	b_7	b_6	b_5	b_4	b_3	b_2	b_1	b_0	b_{-1}
weight	−128	64	32	16	8	4	2	1	n/a
d_3	−128	64	64						
d_2			−32	16	16				
d_1					−8	4	4		
d_0							−2	1	1

Table 12.2. **Possible values for a radix-4 digit d_i**

b_{2i+1}	b_{2i}	b_{2i-1}	d_i
0	0	0	0
0	0	1	1
0	1	0	1
0	1	1	2
1	0	0	−2
1	0	1	−1
1	1	0	−1
1	1	1	0

the number of partial products by using each Booth recoded digit to select one of these five multiples of the multiplicand as the partial product for that digit. All of these multiples of the multiplicand can be generated with simple shifts (to multiply by 2) and logical inversions (to perform a 2's complement negation).

Figure 12.9 shows a six-bit × four-bit 2's complement multiplier using radix-4 Booth recoding. A VHDL implementation of this design is shown in Figures 12.10 and 12.11.

At the left side of Figure 12.9, three recoder (R) blocks are used to recode three-bit overlapping fields of input $b_{5:0}$ into three signed quaternary digits $d_{0:2}$. Note that the least significant digit d_0 is computed using b_1, b_0, and b_{-1} (which is always 0). The details of the recoding are shown in Figure 12.11. Each recoded digit is represented as a three-bit field. The MSB encodes whether the digit is negative, and the lowest two bits encode whether the digit is 2 or 1. If the digit is 0, all three bits are low.

The output of each R block drives a row of select (S) blocks. The S blocks, driven by d_i, perform a signed multiply of $a_{3:0}$ by d_i by selectively shifting (if d_i is ± 2) and inverting (if d_i is -1 or -2). When $d_i = \pm 1$, the a input must be sign extended to five bits to give a five-bit

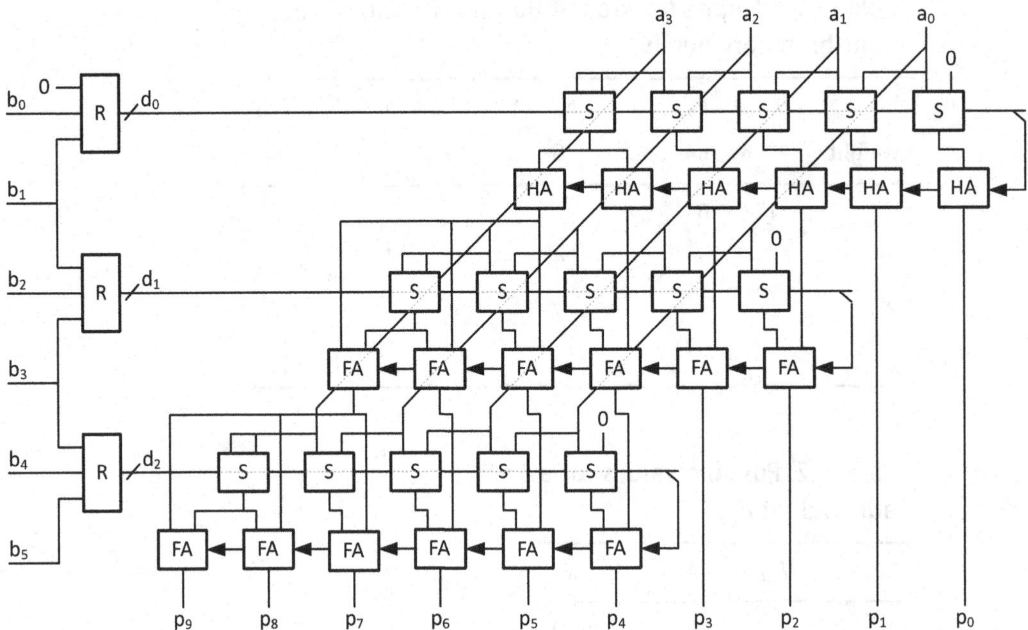

Figure 12.9. A six-bit × four-bit 2's complement multiplier using radix-4 Booth recoding.

S block output with the correct sign. The negation bit, $d_{i,2}$, must be added to the output of each row of S blocks to complete the multiply – since 2's complement negation requires inverting a number and adding 1.

Even though operand a is only four bits wide, and the output of the S blocks is only five bits wide, the addition used to complete the negation must be six bits wide to handle all cases. Consider, for example, when $a = -8_{10} = 1000_2$ and $d_i = -2$. The resulting partial product $pp_i = +16_{10} = 010000$ requires six bits. Both inputs to the six-bit add must be sign extended to get the correct result.

For each row of adders, the low two bits are used directly as outputs and the upper four bits are sign extended to six bits and added with the next row's partial product, which is also sign extended to six bits. The final output is a ten-bit 2's complement number.

While our example performs radix-4 recoding, it is possible to recode with a radix that is any power of 2. A radix-8 recoder uses overlapping four-bit fields with bits weighted $-4, 2, 1, 1$ to generate octal digits with values ranging from -4 to 4. A pre-adder is needed to generate $3a$ to be one of the inputs to the S block.

A radix-16 recoder uses overlapping five-bit fields with weights $-8, 4, 2, 1, 1$ to generate hexadecimal digits with values ranging from -8 to 8. Pre-adders are needed to generate values of $3a$, $5a$, and $7a$. The $6a$ is generated by the S blocks by shifting $3a$. Radix-32, and even radix-64, recoders are also possible and may be of interest for very large multipliers.

12.3 WALLACE TREES

The simple, unsigned multiplier of Figure 10.15 and the recoded multiplier of Figure 12.9 both propagate the carry linearly through a series of adders proportional to the size of the input. For

```
-- 6-bit x 4-bit radix-4 Booth multiplier
library ieee;
use ieee.std_logic_1164.all;
use ieee.std_logic_signed.all;
use work.ch12.all;

entity R4Mult64 is
  port( a: in std_logic_vector(3 downto 0);
        b: in std_logic_vector(5 downto 0);
        s: out std_logic_vector(9 downto 0) );
end R4Mult64;

architecture impl of R4Mult64 is
  -- recoded digits - {negate, select2, select1}
  signal d2, d1, d0: std_logic_vector(2 downto 0);
  signal pp0, pp1, pp2: std_logic_vector(4 downto 0);
  signal ps0, ps1, ps2: std_logic_vector(5 downto 0);
  signal bi: std_logic_vector(6 downto 0);
begin
  bi <= b & '0';
  -- Recoders
  R0: Recode4 port map(bi(2 downto 0), d0);
  R1: Recode4 port map(bi(4 downto 2), d1);
  R2: Recode4 port map(bi(6 downto 4), d2);

  -- Selectors - in equation form - sign extend on select 1 (d(0))
  pp0 <= (4 downto 0 => d0(2)) xor (((4 downto 0 => d0(1)) and (a & '0'))
           or ((4 downto 0 => d0(0)) and (a(3) & a)));
  pp1 <= (4 downto 0 => d1(2)) xor (((4 downto 0 => d1(1)) and (a & '0'))
           or ((4 downto 0 => d1(0)) and (a(3) & a)));
  pp2 <= (4 downto 0 => d2(2)) xor (((4 downto 0 => d2(1)) and (a & '0'))
           or ((4 downto 0 => d2(0)) and (a(3) & a)));

  -- Adders - behavioral - sign extend partial sums
  ps0 <= (pp0(4) & pp0) + ("0000" & d0(2));
  ps1 <= (pp1(4) & pp1) + ((2 downto 0 => ps0(5)) & ps0(4 downto 2))
           + ("0000" & d1(2)); -- second row of adders
  ps2 <= (pp2(4) & pp2) + ((2 downto 0 => ps1(5)) & ps1(4 downto 2))
           + ("0000" & d2(2)); -- third row of adders

  -- Output
  s <= ps2 & ps1(1 downto 0) & ps0(1 downto 0);
end impl;
```

Figure 12.10. VHDL code for the radix-4 Booth recoded multiplier of Figure 12.9.

a simple n-bit $\times$ m-bit multiplier, the length of the carry chain is $n + m - 2$. For an n-bit $\times$ m-bit radix-4 Booth multiplier, the length of the carry chain is $n + m/2 + 1$. We can reduce this $O(n)$ delay in accumulating partial products to an $O(\log(n))$ delay by organizing the adders in a tree, rather than linear arrays, to reduce the partial products to at most two of each weight. A carry-look-ahead adder (see Section 12.1) can then be used for the final summation.

```
-- Radix-4 recode block
-- Output is invert, select 2, select 1.
library ieee;
use ieee.std_logic_1164.all;

entity Recode4  is
  port( b: in std_logic_vector(2 downto 0);
        d: out std_logic_vector(2 downto 0) );
end Recode4;

architecture impl of Recode4 is
begin
  process(all) begin
    case b is
      when "000" | "111" => d <= "000"; -- no select, no invert
      when "001" | "010" => d <= "001"; -- select 1
      when "011"         => d <= "010"; -- select 2
      when "100"         => d <= "110"; -- select 2, invert
      when "101" | "110" => d <= "101"; -- select 1, invert
      when others        => d <= "000"; -- should never be selected
    end case;
  end process;
end impl;
```

Figure 12.11. Recode block for the Booth recoded multiplier of Figure 12.10.

Figure 12.12 shows a circuit that accumulates the partial products (labeled pp_{ij}) for the four-bit by four-bit multiplier of Figure 10.15. The partial products at the left of the figure are the outputs of the AND gates in Figure 10.15, signal $pp_{ij} = a_i \wedge b_j$. The Wallace tree, consisting of five full adders (FA) and a half adder, reduces the partial products until there are at most two of each weight. A carry-look-ahead adder then forms the final product with delay logarithmic in the word length.

In the figure, partial products are grouped vertically by weight. Dashed lines separate the groups. Groups 0 and 1, weights of 1 and 2, have two or fewer partial products, hence no reduction is required. These partial products can be directly input to the carry-look-ahead adder.

Group 2, with weight 4, has three partial product bits (pp_{02}, pp_{11}, and pp_{20}) which are input to a full adder to produce one signal of weight 4, $w1a2$, which is input to the carry-look-ahead adder, and one of weight 8, $w1a3$ which is passed to the next group.

The intermediate signal names specify the stage of the signal and the log of the signal weight, and identify the particular signal of this stage and weight. For example, $w1a3$ is in stage 1, has weight 8, 2^3, and is the first signal with these two properties (hence the label a).

Group 3, with weight 8, has four partial products. Three of these are input to one full adder, and the fourth is passed to a second full adder in stage 2. The second full adder combines two group 3, stage 1 signals, $w1a3$ and $w1b3$, and the remaining group 3 input, pp_{30}, to generate one group 3, stage 2 signal, $w2a3$, and one group 4, stage 2 signal, $w2a4$, both of which are inputs to the carry-look-ahead adder.

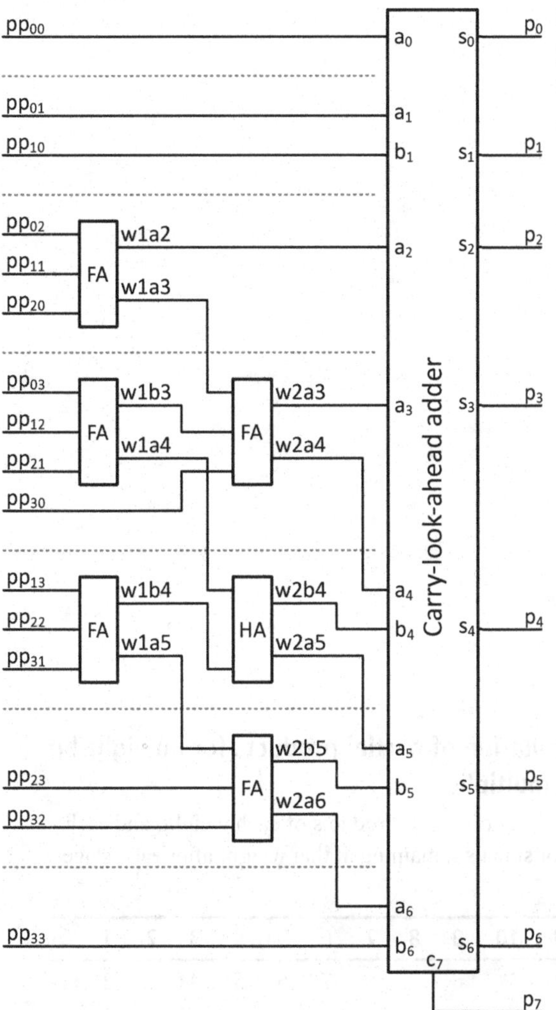

Figure 12.12. A Wallace tree to accumulate the partial products for the four-bit by four-bit simple, unsigned multiplier of Figure 10.15.

The process continues with the remaining groups. Partial products and intermediate signals in each group are reduced by full adders and one half adder, until the number of signals in each group is two or fewer, at which point they are input to the final carry-look-ahead adder.

Using full adders and half adders, each stage of our adder reduces partial products, as shown in Table 12.3. The table shows how many outputs of weights i and $2i$ are produced by one stage of full and half adders for each number of inputs of weight i. For example, with six inputs of weight i, we use two full adders to generate two outputs, each of weights i and $2i$. In some cases we have a choice as to whether to reduce or not. With two partial products of a given weight, for example, we can just pass these inputs directly through – perhaps to be input to the carry-look-ahead adder as in group 1 of Figure 12.12 – or we can input them to a half adder to generate one output each of weights i and $2i$. Where it makes sense, this alternative reduction is shown in the last two columns of the table.

Using the reduction numbers from Table 12.3, Table 12.4 shows how the partial products for an eight-bit by eight-bit simple, unsigned multiplier can be reduced in four stages of full adders and half adders. The first row shows the group number, $\log_2$ of the weight. The second

Table 12.3. **Reduction of number of partial products in one stage of a Wallace tree**

	Out		Alternative	
In	*i*	2*i*	*i*	2*i*
1	1	0		
2	1	1	2	0
3	1	1		
4	2	1		
5	2	2		
6	2	2		
7	3	2		
8	3	3	4	2

Table 12.4. **Wallace-tree accumulation of partial products for an eight-bit by eight-bit simple, unsigned multiplier**

The row labeled "pps" shows the number of partial products of each weight, and each subsequent row shows the number of signals remaining at that weight after each stage of full adders.

$\log_2$ weight	14	13	12	11	10	9	8	7	6	5	4	3	2	1	0
pps	1	2	3	4	5	6	7	8	7	6	5	4	3	2	1
Stage 1	1	3	2	4	4	4	6	5	5	4	3	3	1		
Stage 2	2	1	3	3	3	4	4	4	3	3	2	1			
Stage 3	2	2	2	2	2	3	3	3	2	1					
Stage 4				2	2	2	2	1							

row shows the number of partial products (pps) in each group: the number of bits of a given weight that need to be summed. The remaining rows show the number of signals of each weight remaining after each stage of full adders. The entries are calculated using the reduction rules of Table 12.3. Each number depends only on the number directly above it and the number above and one column to the right. For example, group 9 has four signals after stage 1. This comprises two signals reduced from the six partial products in group 9 and two additional signals that carry out of the reduction of seven partial products in group 8.

At stage 2 in group 12 we have a choice as to whether to pass the two stage 1 signals through to stage 3, or to use a half adder to pass one signal through and carry a signal to group 13. By passing the signals through we are able to use a full adder in stage 3 to reduce three signals to

Table 12.5. **Wallace-tree accumulation of partial products for an eight-bit by eight-bit signed radix-4 Booth multiplier**

Bits 10–15 each have four inputs because all partial products are sign extended.

$\log_2$ weight	15	14	13	12	11	10	9	8	7	6	5	4	3	2	1	0
pps	4	4	4	4	4	4	4	4	4	5	3	4	2	3	1	2
Stage 1	3	3	3	3	3	3	3	3	4	3	2	3	2	1		
Stage 2	2	2	2	2	2	2	2	2	3	2	2	1				
Stage 3	2	2	2	2	2	2	2	2	1							

two – one each in groups 12 and 13. This strategy of avoiding the use of half adders allows us to accumulate enough signals to use a full adder, which reduces the total number of signals in the next stage.

Table 12.5 shows how Booth recoding reduces the number of partial products and hence the number of stages needed to reduce them. Since we are using Booth recoding to do a *signed* multiplication, we must sign extend all partial products to full bit width. We must also account for the carry-in bits of each input (bits 0, 2, 4, and 6) in our tree. With recoding, a 16-bit multiplier can be realized with just three stages of adders before the carry-look-ahead adder, compared with four stages without recoding.

In addition to summing the partial products in multipliers, a tree of full-adder cells can also be applied to summing N b-bit numbers in $O(\log(N)\log(b))$ time. Figure 12.13, for example, shows how a three-level tree of full adders can be used to sum six numbers. The VHDL code for this circuit is shown in Figure 12.14.

Because each stage of adders reduces the number of signals of a given weight by 2/3, the number of stages S required to sum N numbers down to two adder inputs is bounded by

$$S \geq \lceil \log_{3/2}(N/2) \rceil . \tag{12.13}$$

Unfortunately this bound is not quite tight. It suggests, for example, that ten inputs can be summed in four stages, while five are required. A tight bound can be arrived at via the recurrence

$$S(N) = 0 \qquad \text{if} \quad 0 \leq N \leq 2, \tag{12.14}$$

$$S(N) = 1 + S(N - \lfloor N/3 \rfloor) \quad \text{if} \quad N \geq 3. \tag{12.15}$$

Each stage with N input signals of equal weight uses $a = \lfloor N/3 \rfloor$ full adders to reduce the input to $N - a$ signals until the output is just two signals. We enumerate the first 16 values of $S(N)$ in Table 12.6.

In this section we have looked at compressing bits of equal weight using full adders which compress three bits of equal weight to a two-bit binary number that reflects the number of 1s in the three-bit input. Hence, when used in this context a full adder is sometimes referred to as a *3–2 compressor* or a *3–2 counter*. It is possible to build reduction trees with larger radix compressors. For example, we can design a *7–3 compressor* that accepts seven bits of equal

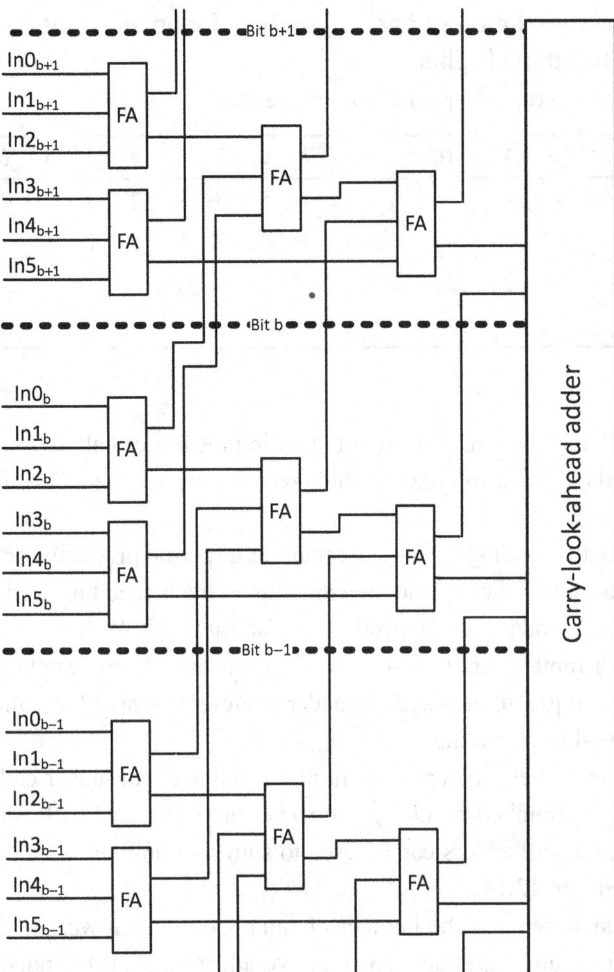

Figure 12.13. Multiple-input adder where every bit is compressed via a series of full adders from the number of input bits to two. The final two numbers are then summed via a conventional adder.

weight and outputs a three-bit binary number, which is a count of the 1s in the input. Exercise 12.12 explores building a Wallace tree with this block. We can also combine two full adders in a way that (modulo a carry in and carry out) gives us a *4–2 compressor*. This concept is explored in Exercise 12.13.

12.4 SYNTHESIS NOTES

While there are times when one needs to code multipliers and adders explicitly to achieve particular goals, in most cases modern synthesis tools do a sufficiently good job and one can simply generate very efficient signed and unsigned multipliers and adders by just using the VHDL `std_logic_signed` operators * and +, respectively. The synthesis tools have a large library of adder and multiplier designs and will choose the cheapest design that meets the specified timing constraints.

```
library ieee;
use ieee.std_logic_1164.all;
use ieee.std_logic_signed.all;
use work.ch10.all;

entity MultiAdder_FA_Tree is
  port( in0, in1, in2, in3, in4, in5: in std_logic_vector(3 downto 0);
        output: out std_logic_vector(6 downto 0) );
end MultiAdder_FA_Tree;

architecture impl of MultiAdder_FA_Tree is
  signal se_in0, se_in1, se_in2, se_in3, se_in4, se_in5: std_logic_vector(6 downto 0);
  signal s00, s01, s10, s20: std_logic_vector(6 downto 0); -- s(level)(unit)
  signal c00, c01, c10, c20: std_logic_vector(6 downto 0); -- c(level)(unit)
  signal toss: std_logic_vector(3 downto 0); -- outputs thrown away as bit 7 not needed
begin
  -- sign extend the inputs
  se_in0 <= in0(3) & in0(3) & in0(3) & in0;
  se_in1 <= in1(3) & in1(3) & in1(3) & in1;
  se_in2 <= in2(3) & in2(3) & in2(3) & in2;
  se_in3 <= in3(3) & in3(3) & in3(3) & in3;
  se_in4 <= in4(3) & in4(3) & in4(3) & in4;
  se_in5 <= in5(3) & in5(3) & in5(3) & in5;

  --  Set lower bit carry ins to 0
  c00(0) <= '0'; c01(0) <= '0';
  c10(0) <= '0'; c20(0) <= '0';

  FA01_0: FullAdder port map(se_in0(0),se_in1(0),se_in2(0),c00(1),s00(0));
  FA02_0: FullAdder port map(se_in3(0),se_in4(0),se_in5(0),c01(1),s01(0));
  FA10_0: FullAdder port map(s00(0),c00(0),c01(0),c10(1),s10(0));
  FA20_0: FullAdder port map(s01(0),s10(0),c10(0),c20(1),s20(0));

  -- Array adders for bits 1, 2, 3, 4, 5 to reduce code length
  FAA: for i in 1 to 5 generate
    FA00i: FullAdder port map(se_in0(i),se_in1(i),se_in2(i),c00(i+1),s00(i));
    FA01i: FullAdder port map(se_in3(i),se_in4(i),se_in5(i),c01(i+1),s01(i));
    FA10i: FullAdder port map(s00(i),c00(i),c01(i),c10(i+1),s10(i));
    FA20i: FullAdder port map(s01(i),s10(i),c10(i),c20(i+1),s20(i));
  end generate;

  FA01_6: FullAdder port map(se_in0(6),se_in1(6),se_in2(6),toss(0),s00(6));
  FA02_6: FullAdder port map(se_in3(6),se_in4(6),se_in5(6),toss(1),s01(6));
  FA10_6: FullAdder port map(s00(6),c00(6),c01(6),toss(2),s10(6));
  FA20_6: FullAdder port map(s01(6),s10(6),c10(6),toss(3),s20(6));

  output <= s20 + c20;
end impl;
```

Figure 12.14. The VHDL design entity for a six-input signed adder, made up of single-bit full-adder blocks (`fa`).

Table 12.6. **Number of stages required to reduce N inputs to two outputs using a Wallace tree of full adders**

N	*S*
1	0
2	0
3	1
4	2
5	3
6	3
7	4
8	4
9	4
10	5
11	5
12	5
13	5
14	6
15	6
16	6

Summary

In this chapter you have learned how to design fast arithmetic circuits. We have covered three techniques used in virtually all high-performance arithmetic circuits today: carry look-ahead, Booth recoding, and Wallace trees.

By computing the carry out of a group of bits with a *tree* of logic modules, we can perform an n-bit add in time proportional to $\log_2(n)$ rather than n. To build a *carry-look-ahead* tree, we formulate the add (or other iterative function) by specifying functions p and g that *propagate* the carry across a bit and *generate* the carry out of a bit – regardless of what the carry was coming in. We compute the p and g functions for a group of bits recursively from the p and g functions for smaller groups – and ultimately individual bits. Once we have computed the carry out of a group of bits, we use an inverse tree to compute the carry out of each bit in the group.

We can reduce the number of partial products that need to be summed in a multiplier by a factor of 2 (or more) by *Booth recoding* one of the inputs to the multiplier. With radix-4 Booth recoding, for example, we recode an n-bit input into an $(n/2)$-digit radix-4 number where each digit takes on integer values between -2 and 2. The AND used to compute each partial product is extended with an optional shift (for 2 and -2) and negation (for -2 and -1). The resulting, much smaller, set of partial product bits are then summed normally. As an added benefit, Booth recoding handles 2's complement numbers with little additional effort.

We can use a tree of adders to sum the partial products of a multiplier (whether it uses recoding or not). Such *Wallace trees* reduce the time required to sum up the partial products to be proportional to $\log_{3/2}(N/2)$ for N partial products.

While it is important to understand these techniques for building fast arithmetic circuits, modern synthesis tools draw on a library of optimized arithmetic unit designs that include these techniques – and others. Thus, it is not necessary to instantiate these methods manually when implementing standard arithmetic functions. The tools handle these cases well. It is necessary to implement these methods manually when implementing other functions, for example, other iterative circuits that can benefit from carry look-ahead.

BIBLIOGRAPHIC NOTES

The classic 1961 paper by MacSorley [75] includes all of the techniques described in this chapter, as well as a few more. It shows how mature the field of computer arithmetic was by the early 1960s. Carry look-ahead was first described in 1956 by Weinberger and Smith [111]. For more information about fast adders, refer to Harris's taxonomy of parallel prefix adders [46] or Ling's paper on fast addition [70].

Recoding derives from Booth's algorithm [15], which was originally developed to perform multiplication sequentially in software. The parallel form we describe here is similar to that described by MacSorley [75]. The Wallace tree was first described in ref. [110].

The past three chapters have barely scratched the surface of the interesting topic of computer arithmetic. The interested reader is referred to one of the many excellent textbooks and monographs on the subject, including refs. [25], [42], [43], and [53].

Exercises

12.1 *Mixed-radix look-ahead.* Write the VHDL code for a 32-bit magnitude comparator using carry look-ahead using PG blocks with a fan-in of 5 or 6.

12.2 *Reverse carry propagation.* Write the VHDL code for a 32-bit magnitude comparator formulated by propagating carry from MSB to LSB and using carry look-ahead with a fan-in of 4. (Hint: look at Figure 8.42(b).)

12.3 *Look-ahead arbitration.* Write the VHDL code for a 32-bit arbiter using carry look-ahead with a fan-in of 4.

12.4 *Look-ahead priority encoder.* Combine carry look-ahead with the technique shown in Figure 8.24 to build a 32-bit priority encoder. Use a fan-in of 4.

12.5 *Unsigned recoded multiplier.* Redesign the radix-4 Booth recoded multiplier of Figures 12.9 and 12.10 to handle unsigned numbers.

12.6 *Radix-8 Booth recoder.* Design a radix-8 Booth recorder. Start by writing out a table like those shown in Tables 12.1 and 12.2. Write the VHDL for doing a six-bit by six-bit multiply, only generating two partial products.

12.7 *Radix-16 Booth recoder.* Repeat Exercise 12.6, but instead use a radix-16 Booth encoding and an eight-bit by eight-bit multiply.

12.8 *Optimal Booth recoding.* For a 64-bit signed multiply determine the recoding radix that requires the fewest full adders. Include both the adders used to pre-compute multiples of a and the adders used to sum partial products.

12.9 *Optimal Booth recoding.* Repeat Exercise 12.8 for a 128-bit signed multiply.

12.10 *Are all the adders needed?* The leftmost two adders in each row of Figure 12.9 have identical input signals. Hence, one might think that the leftmost adder can be eliminated and its output connected to the carry out of the penultimate adder in the same row. Is this true? If so, explain why; if not, explain why not, and suggest an alternative way to simplify this logic. (Hint: make sure you consider the case where $a = -8$ and $b = 2, 8,$ or 12.)

12.11 *Wallace tree for* 16×16*-bit radix-4 Booth multiplier.* Draw a table like that of Tables 12.4 and 12.5 for a Wallace tree to sum the partial products of a 16-bit $\times$16-bit radix-4 Booth recoded multiplier. Remember to sign extend your partial products.

12.12 *Wallace trees with seven-input counter cells.* Suppose that, in addition to a full adder, which takes three inputs of weight i and produces one output with weight i and one with weight $2i$, we also have a seven-input "double adder" cell which takes seven inputs of weight i and generates one output each with weights i, $2i$, and $4i$. Draw a table like that of Tables 12.4 and 12.5 for a 16-bit $\times$16-bit radix-4 Booth recoded multiplier using these seven-input cells (as well as full and half adders).

12.13 *4–2 compressors.* Figure 12.15 shows a 4–2 compressor that takes four bits of input and produces two bits of output by using two adders.

(a) Draw a table like that of Tables 12.4 and 12.5 for a 16-bit $\times$16-bit radix-8 Booth recoded multiplier using these "four"-input cells. You may assume that pre-added partial products are available.

(b) What is the bound on the delay of using 4–2 compressors to reduce n partial products to two? Write your answer in terms of the delay of one full adder. Is this faster or slower than using a basic 3–2 Wallace tree?

(c) What must the delay of one 4–2 compressor be in order for a 4–2 Wallace tree to have a delay equal to a 3–2 Wallace tree, for the same number of inputs?

12.14 *Wallace tree choices.* Redo Table 12.5 but using a half adder in stage 1 of group 12 rather than passing both group 12 partial products through to stage 2.

12.15 *Wallace tree with two bits per partial product.* Draw a table like that of Tables 12.4 and 12.5 for a Wallace tree to sum the partial products of a 15-bit radix-8 Booth recoded multiplier. Assume that each partial product consists of two bits to allow the $3a$ case to be handled without doing a pre-add.

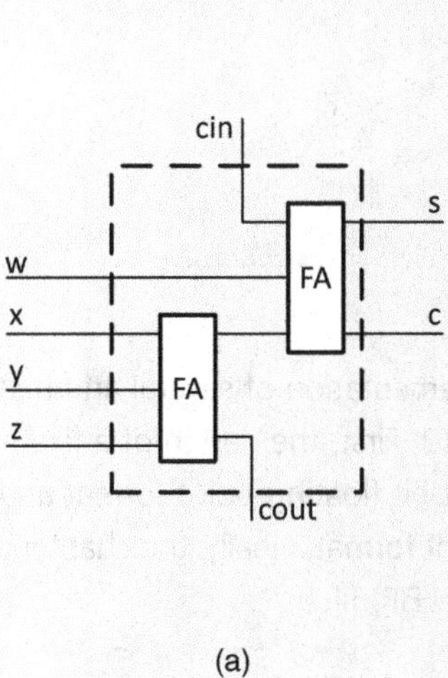

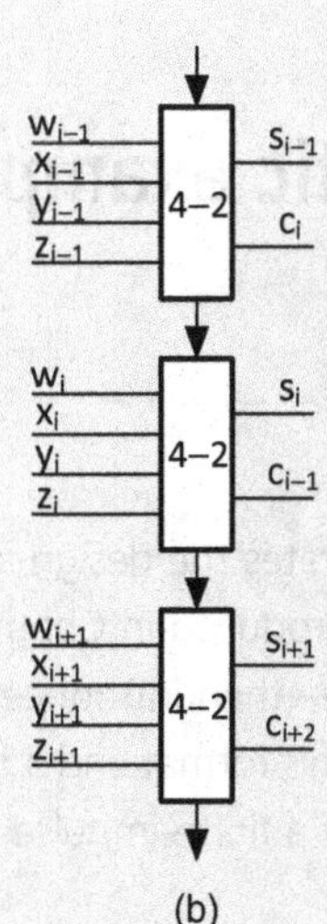

Figure 12.15. A 4–2 compressor composed of two full adders (a) takes four bits and produces a carry and sum. As shown in (b), the intermediate carries are passed from one bit position to the next. We use this structure in Exercise 12.13.

12.16 *Simplified Wallace tree.* Two of the full adders in Figure 12.12 can be replaced by half adders. Redraw the figure with this change, showing how the now extra input for each of these adders is handled. Is this design, with three half adders and three full adders, better than the original design? Explain why it is, or why not.

12.17 *Fast multiplication, design.* Design a 32-bit ×32-bit 2's complement multiplier with minimum delay. Your design should include Booth recoding (with optimal radix) of partial products, a Wallace tree to reduce the partial products, and a carry-look-ahead adder (with optimal radix) to perform the final summation. Be careful to handle sign extension properly through your summation tree. Show an analysis of the delay through your design.

12.18 *Fast multiplication, implementation.* Write the VHDL code for the multiplier of Exercise 12.17 and validate it with a testbench.

12.19 *Fast unsigned multiplication, design.* Design a 32-bit ×32-bit unsigned multiplier with minimum delay. Your design should include Booth recoding (with optimal radix) of partial products, a Wallace tree to reduce the partial products, and a carry-look-ahead adder (with optimal radix) to perform the final summation. Show an analysis of the delay through your design.

12.20 *Fast unsigned multiplication, implementation.* Write the VHDL code for the multiplier of Exercise 12.19 and validate it with a testbench.

13 Arithmetic examples

This chapter demonstrates the design and implementation of several arithmetic circuits that use the techniques introduced in Chapters 10–12. First, the design of a fixed-point complex multiplier is shown. We then introduce an eight-bit floating-point format and develop units that switch between this format and a fixed-point format. Finally, the chapter concludes with the implementation of a finite-impulse-response (FIR) filter.

13.1 COMPLEX MULTIPLICATION

Multiplying two complex numbers $a + ib$ and $c + id$ gives

$$(a + ib) \times (c + id) = (ac - bd) + i(bc + da). \tag{13.1}$$

To produce the real and imaginary parts of the product requires four multipliers and two adders.[1] Since $i^2 = -1$, the product of the two imaginary input components must be subtracted from the product of the real input components to give the real part of the output.

Our complex multiplier uses an $s1.14$ fixed-point format (see Section 11.2), as is common in signal processing. To avoid triggering an overflow error or losing precision until after the final summation, we keep intermediate values at full bit width. Multiplying two $s1.14$ numbers gives an $s2.28$ product, and adding two of these products gives an $s3.28$ result. The final stage of our module checks for overflow and rounds the $s3.28$ result back to an $s1.14$ number. An overflow has occurred when the three most significant bits of the $s3.28$ result are not the same value. A leading 1 indicates a negative overflow, while a leading 0 indicates positive overflow. Our complex multiplier uses *saturating arithmetic*, clamping the result on an overflow to minimize the error (see Exercise 10.20).

The block diagram and VHDL for this implementation are shown in Figures 13.1 and 13.2, respectively.

The design of Figure 13.1 is simple, since neither the multiplier nor the adders are specialized units, but it is not the fastest. Each of the multipliers has its own partial product generation and reduction tree, producing one result via a 31-bit adder (see Section 12.3). To remove the delay of this adder, we can modify the multipliers to omit this final add stage and instead output reduced partial products. Two more full-adder blocks can be used to reduce these four values to the two inputs to the final addition stage. The real portion of the output requires a subtraction operation, which is performed by inverting `a_img` before inputting it to the modified multiplier.

[1] It is also possible to build this circuit with three multipliers (ad, bc, $(a - b)(c + d)$), but we focus on the implementation with four products.

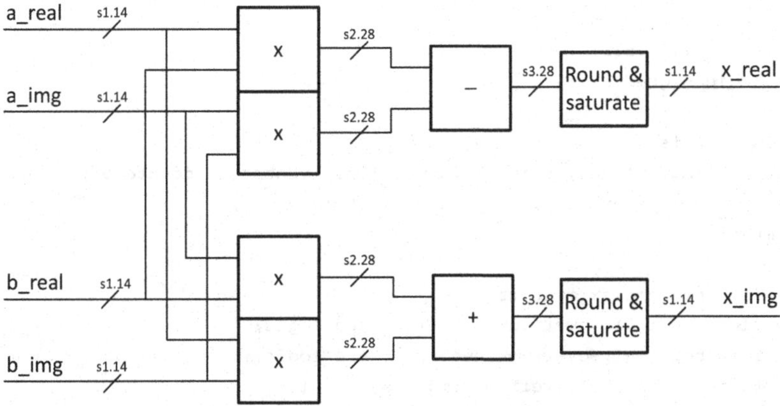

Figure 13.1. An $s1.14$ fixed-point complex number multiplier. Composed of two 16-bit multipliers and two 31-bit adders, this circuit takes two complex inputs a and b and produces complex output x.

We have supplied the block diagram in Figure 13.3, but leave sizing the intermediate wires and finalizing the design to the reader in Exercise 13.1.

13.2 CONVERTING BETWEEN FIXED- AND FLOATING-POINT FORMATS

In this section we introduce an eight-bit floating-point format and explore how to convert between a 12-bit signed number and this new format. The format that we will use is custom, but resembles that used in A-law and μ-law audio compression. Like these formats, it compresses a 12-bit signed integer to an eight-bit representation. It does not represent fractional magnitudes.

13.2.1 Floating-point format

Our eight-bit format, shown in Figure 13.4, represents values ranging from −2047 to 2047 with five bits of precision. The format contains a single sign bit, three bits of exponent, and four bits of mantissa. We augment the four bits of mantissa with an implied MSB of "1" (see Exercise 11.23) to represent the five most significant bits of each fixed-point number. In order to represent numbers less than 16 accurately, we include gradual underflow (see Section 11.3.2).

Zero is simply represented as 00_{16}, and the value 80_{16} (negative 0) represents an arithmetic error code.[2] When an overflow occurs when doing arithmetic, we will set the result to 80_{16} to indicate an error has occurred. When the error code is the input to any floating-point function, such as addition, the output is automatically set to that code.

The value function of our floating-point format is given by

$$v = \begin{cases} -1^s m & \exp = 0, \\ -1^s 2^{e-1}(10000_2 + m) & \exp \neq 0. \end{cases} \tag{13.2}$$

To provide gradual underflow, exponents of 0 and 1 both encode a binary point just to the right of the LSB of the mantissa. The difference is that an exponent > 0 implies a 1 left of the MSB

[2] Similar to NaN in the IEEE floating-point standard.

```
library ieee;
use ieee.std_logic_1164.all;
use ieee.std_logic_signed.all;

entity complex_mult is
  port( a_real, a_img, b_real, b_img : in std_logic_vector(15 downto 0);
        x_real, x_img : out std_logic_vector(15 downto 0) );
end complex_mult;

architecture impl of complex_mult is
  signal overflow_pos_real, overflow_pos_img : std_logic;
  signal overflow_neg_real, overflow_neg_img : std_logic;
  signal no_overflow_real, no_overflow_img : std_logic;
  signal p_ar_br, p_ai_bi, p_ai_br, p_ar_bi : std_logic_vector(31 downto 0);
  signal s_real, s_img : std_logic_vector(31 downto 0);
  signal s_real_rnd, s_img_rnd: std_logic_vector(17 downto 0);
begin
  -- s2.28
  p_ar_br <= a_real * b_real;
  p_ai_bi <= a_img * b_img;
  p_ai_br <= a_img * b_real;
  p_ar_bi <= a_real * b_img;

  -- s3.28
  s_real <= p_ar_br - p_ai_bi;
  s_img <= p_ar_bi + p_ai_br;

  -- Round up on half, s3.14
  s_real_rnd <= s_real(31 downto 14) + s_real(13);
  s_img_rnd <= s_img(31 downto 14) + s_img(13);

  -- check for overflow & clamp  (bits 17, 16, 15 not equal)
  overflow_pos_real <= (not s_real_rnd(17)) and (s_real_rnd(16) or s_real_rnd(15));
  overflow_neg_real <= (s_real_rnd(17)) and not(s_real_rnd(16) and s_real_rnd(15));
  no_overflow_real <= not (overflow_pos_real or overflow_neg_real);

  overflow_pos_img <= (not s_img_rnd(17)) and (s_img_rnd(16) or s_img_rnd(15));
  overflow_neg_img <= (s_img_rnd(17)) and not(s_img_rnd(16) and s_img_rnd(15));
  no_overflow_img <= not (overflow_pos_img or overflow_neg_img);

  x_real <= ((15 downto 0 => overflow_pos_real) and x"7fff") or
            ((15 downto 0 => overflow_neg_real) and x"8000") or
            ((15 downto 0 => no_overflow_real) and s_real_rnd(15 downto 0));

  x_img <= ((15 downto 0 => overflow_pos_img) and x"7fff") or
           ((15 downto 0 => overflow_neg_img) and x"8000") or
           ((15 downto 0 => no_overflow_img) and s_img_rnd(15 downto 0));
end impl;
```

Figure 13.2. VHDL for the complex multiplier.

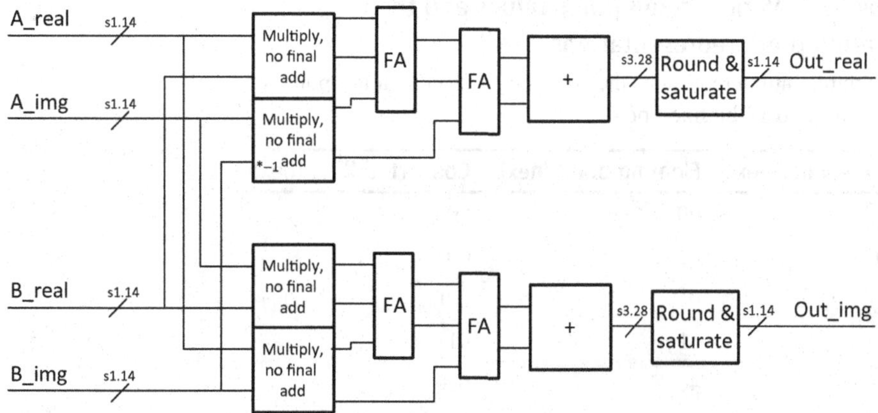

Figure 13.3. Faster $s1.14$ fixed-point complex number multiplier. This implementation removes the final adder in each multiplier and instead reduces the four output values to two via two full adders.

Sign	Exponent	Mantissa
7	6:4	3:0

Figure 13.4. The floating-point format used throughout this chapter.

of the mantissa while an exponent of 0 does not. Exponents of 1 to 7 have a bias of 1 – i.e. the mantissa is shifted left $e - 1$ positions to give the value.

The equations to compute the floating-point representation, omitting the potential mantissa round, are as follows:

$$s = v < 0; \tag{13.3}$$

$$e = \begin{cases} 0 & \log_2(|x|) < 4, \\ \lfloor \log_2(|x|) \rfloor - 3 & \log_2(|x|) \geq 4; \end{cases} \tag{13.4}$$

$$m = |x| 2^{-\min(e-1,0)}. \tag{13.5}$$

To encode values in floating-point, the mantissa is implicitly the most significant five bits of the input. The exponent encodes the bit position of the most significant bit.

The conversion to floating-point also includes a step to round the new mantissa. We use a simple rounding scheme that rounds up on one-half. Figures 13.5 and 13.6 show the represented values and relative error, respectively. The relative error of this representation is never over 3.1%, and the first value to have any error is 33. Error-free representation occurs only when the magnitude bits shifted out are all 0.

Table 13.1 shows several example numbers and their floating-point format.

13.2.2 Fixed- to floating-point conversion

Figure 13.7 shows a block diagram of a 12-bit fixed-point to an eight-bit floating-point converter. The input is first converted from 2's complement to sign magnitude using a negation block and a multiplexer. Next, a find-first-one unit finds the most significant 1 (see Section 8.5).

This one-hot signal is immediately input into a priority encoder to encode the three-bit exponent and the mantissa shift amount. The logic shifts the least significant ten bits of the

Table 13.1. **Various fixed-point values and their floating-point representations**

The third column represents the conversion of the floating-point value back to 12-bit fixed-point.

Fixed-point (hex)	Floating-point (hex)	Converted float (hex)
000	00	000
003	03	003
00f	0f	00f
011	11	011
0de	4c	0e0
59d	76	580
7ff	7f	7C0
fff	81	fff

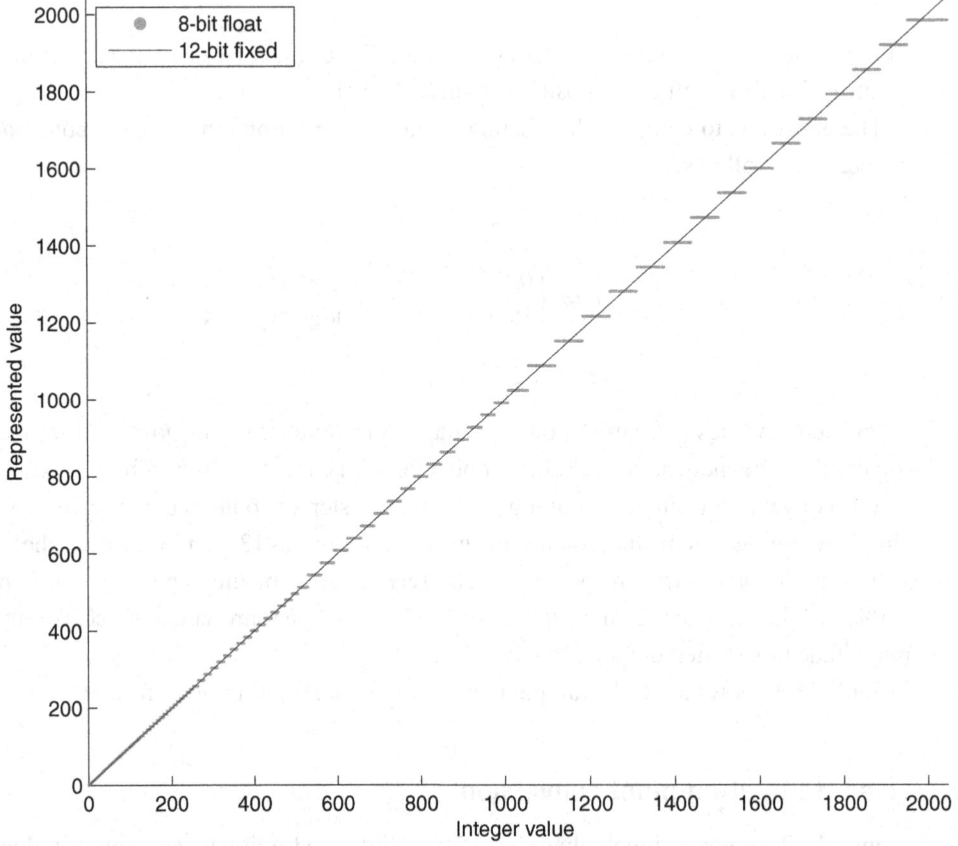

Figure 13.5. Fixed-point numbers and their floating-point representation. A truncation scheme causes the floating-point number to be less than or equal to the initial value.

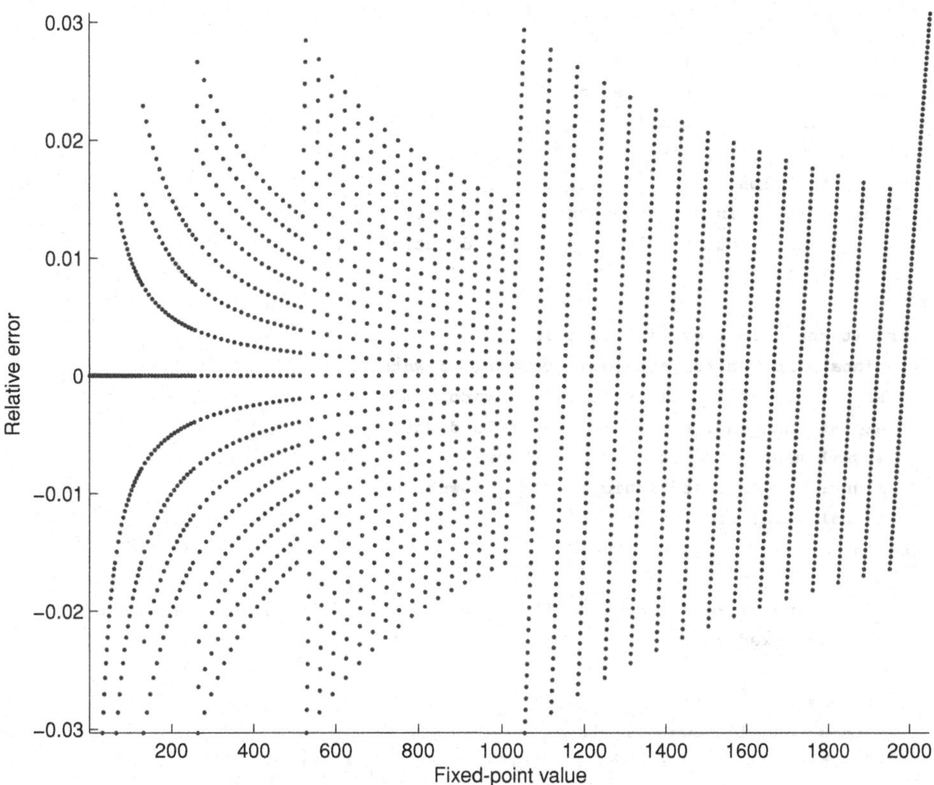

Figure 13.6. The representation error of our eight-bit floating-point scheme.

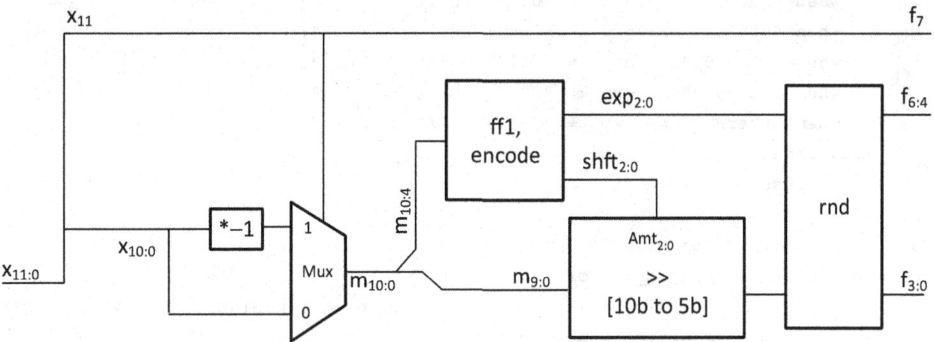

Figure 13.7. Block diagram for converting fixed-point to floating-point numbers.

input into the mantissa. Rounding a number can result in needing to increment the exponent (and potentially causing overflow).

The VHDL code for the fixed- to floating-point converter can be found in Figure 13.8.

When shifting, the design entity saves a round bit (`mant_lng(0)`) and adds it back to the mantissa. If the mantissa had been `4b"1111"` before the round operation, it will overflow and we add 1 to the exponent. If the exponent overflows, we saturate the output value. In the design entity provided, the input of 800_{16} (-2048) does not convert to ff_{16}, but rather to our error code of 0×80. Exercise 13.4 will ask you to fix this flaw.

```
library ieee;
use ieee.std_logic_1164.all;
use ieee.std_logic_unsigned.all;
use ieee.numeric_std.all;

entity fix2float is
  port( fixed: in std_logic_vector(11 downto 0);
        float: out std_logic_vector(7 downto 0) );
end fix2float;

architecture impl of fix2float is
  signal exp, shift: std_logic_vector(2 downto 0);
  signal mag: std_logic_vector(10 downto 0);
  signal mant_lng: std_logic_vector(10 downto 0);
  signal mant: std_logic_vector(4 downto 0);
  signal new_exp: std_logic_vector(3 downto 0);
  signal sign: std_logic;
begin

  mag <= (not fixed(10 downto 0))+1 when fixed(11)='1' else fixed(10 downto 0);
  sign <= fixed(11);

  process(all) begin
    case? mag(10 downto 4) is
      when "1------" => exp <= "111"; shift <= "110";
      when "01-----" => exp <= "110"; shift <= "101";
      when "001----" => exp <= "101"; shift <= "100";
      when "0001---" => exp <= "100"; shift <= "011";
      when "00001--" => exp <= "011"; shift <= "010";
      when "000001-" => exp <= "010"; shift <= "001";
      when "0000001" => exp <= "001"; shift <= "000";
      when "0000000" => exp <= "000"; shift <= "000";
      when others    => exp <= "---"; shift <= "---";
    end case?;
  end process;

  -- Shift the mantissa and round
  mant_lng <= std_logic_vector(shift_right(unsigned(mag(9 downto 0) & '0'),
                                          to_integer(unsigned(shift))));
  mant <= ('0' & mant_lng(4 downto 1)) + mant_lng(0);

  -- Check for round overflow
  new_exp <= ('0' & exp) + mant(4);

  -- If the exponent overflowed, saturate
  float <= (sign & "1111111") when new_exp(3) = '1' else
           (sign & new_exp(2 downto 0) & mant(3 downto 0));
  -- Using mant(3 downto 0) is correct even with the round overflow,
  -- since in that case mant(4 downto 1)=mant(3 downto 0)="0000"
end impl;
```

Figure 13.8. The top-level VHDL design entity for fixed- to floating-point conversion.

13.2.3 Floating- to fixed-point conversion

A floating- to fixed-point converter block diagram and VHDL are shown in Figures 13.9 and 13.10. The first step in the conversion process is to determine whether the implicit 1 is present by checking whether the exponent is equal to 0. If the 1 is present, we subtract 1 from the

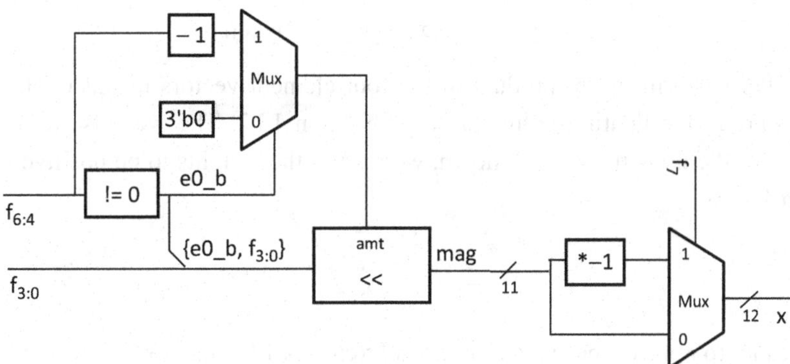

Figure 13.9. Block diagram of a floating- to fixed-point converter.

```
library ieee;
use ieee.std_logic_1164.all;
use ieee.std_logic_unsigned.all;
use ieee.numeric_std.all;

entity float2fix is
  port( float: in std_logic_vector(7 downto 0);
        fixed: out std_logic_vector(11 downto 0) );
end float2fix;

architecture impl of float2fix is
  signal sign, implied_one: std_logic;
  signal exponent, shift: std_logic_vector(2 downto 0);
  signal mant: std_logic_vector(3 downto 0);
  signal mag: std_logic_vector(11 downto 0);
begin
  sign <= float(7);
  exponent <= float(6 downto 4);
  mant <= float(3 downto 0);

  shift <= "000" when exponent = "000" else exponent-1;
  implied_one <= '1' when not (exponent = "000") else '0';

  mag <= std_logic_vector( shift_left( unsigned("0000000" & implied_one & mant),
                                        to_integer(unsigned(shift)) ) );
  fixed <= (not mag)+1 when sign='1' else mag;
end impl;
```

Figure 13.10. VHDL design entity for floating-point to fixed-point conversion.

exponent and then shift the mantissa and implicit bit to the left by the prescribed amount. Finally, the negative value of the magnitude is taken, if necessary.

13.3 FIR FILTER

This section details the design of a four-tap finite-impulse-response (FIR) filter. Given a set of four input values and weights, the output of the module is given by

$$y = w_0x_0 + w_1x_1 + w_2x_2 + w_3x_3. \tag{13.6}$$

That is, the filter performs a dot product of the four-element vectors w and x. The input and output values are in the floating-point format of Section 13.2. The weights are in 1.4 fixed-point format. For the implementation shown, we restrict the weights to be positive and sum to no more than 1:

$$0 \leq w_i \leq 1, \qquad \sum_{i=0}^{3} w_i \leq 1. \tag{13.7}$$

We ask the reader to remove these restrictions in Exercises 13.7 and 13.8.

Since our floating-point scheme represents only a 12-bit dynamic range ($s11.0$), we opt to do all arithmetic in fixed-point. These fixed-point units will be smaller and faster than the floating-point counterparts. The weight values are input as unsigned 1.4 values (16ths). To avoid losing intermediate precision, the output of the $s11.0 \times 1.4$ multiplier is kept in $s11.4$ fixed-point. Because of the restriction that the sum of the weights is less than 1, the adder outputs an $s11.4$ value. The last stage of the FIR filter rounds the remaining fractional portion and converts back to eight-bit floating-point format. The block diagram and numerical representations are shown in Figure 13.11.

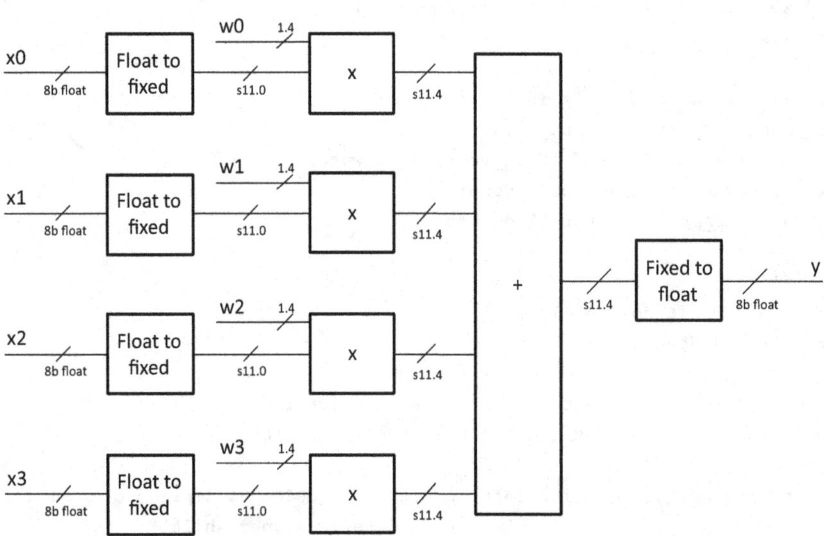

Figure 13.11. FIR filter block diagram. All weights must be positive and sum to no more than 1. The filter converts the eight-bit floating-point inputs into a 12-bit fixed-point format. All intermediate values are all fixed-point and sized so as to not lose precision.

```
-- A four-input floating-point FIR filter
library ieee;
use ieee.std_logic_1164.all;
use ieee.std_logic_signed.all;
use work.ch13.all;

entity fir is
  -- A four-input floating-point FIR filter
  port( x0, x1, x2, x3: in std_logic_vector(7 downto 0);
        -- The output will only be an error if any of the 4 inputs is an error code
        -- As the weights are restricted to be no greater than one
        w0, w1, w2, w3: in std_logic_vector(4 downto 0);
        -- In 1.4 format, max value = 16/16
        output: out std_logic_vector(7 downto 0) );
end fir;

architecture impl of fir is
  signal fix0, fix1, fix2, fix3, fixed_out: std_logic_vector(11 downto 0);
  -- The weighted floating point numbers, s11.4
  signal weighted0, weighted1, weighted2, weighted3, shift1: std_logic_vector(16 downto 0);
  signal w_sum: std_logic_vector(16 downto 0);
  signal float_out: std_logic_vector(7 downto 0);
  signal err: std_logic;
begin
  CONV0: float2fix port map(x0, fix0);
  CONV1: float2fix port map(x1, fix1);
  CONV2: float2fix port map(x2, fix2);
  CONV3: float2fix port map(x3, fix3);

  weighted0 <= fix0 * w0;
  weighted1 <= fix1 * w1;
  weighted2 <= fix2 * w2;
  weighted3 <= fix3 * w3;

  w_sum <= weighted0 + weighted1 + weighted2 + weighted3;
  fixed_out <= w_sum(15 downto 4)+w_sum(3);

  CONVOUT: fix2float port map(fixed_out,float_out);

  err <= '1' when (x0 = x"80") or (x1 = x"80") or (x2 = x"80") or (x3 = x"80") else '0';
  output <= x"80" when err = '1' else float_out;
end impl;
```

Figure 13.12. VHDL design entity for the floating-point FIR filter.

The FIR VHDL is shown in Figure 13.12. Each floating-point input is first converted to fixed-point using the conversion block of Figure 13.10. Next, we sign extend the fixed-point values (`fixi`) and multiply by the weights (giving `weightedi`). The sum of the weighted numbers is rounded and converted back to floating-point. The last step of the process is to output an error code if an error code was input.

Summary

In this chapter you have seen three extended examples that bring together much of what you have learned from Chapters 10–12. Our complex multiplier illustrates the issues of overflow and precision that come when working with fixed-point numbers and how rounding and saturation are used to deal with these issues.

Our discussion of fixed- to floating-point conversion, which is based loosely on popular audio compression standards, shows the advantage of floating-point representation for relative precision and illustrates many details of floating-point representation. The representation includes an *implied 1* and *gradual underflow*. Careful rounding and normalization are required during the conversion.

Our final example is a finite-impulse-response filter. This module illustrates the use of multiple number systems. It has floating-point inputs and outputs, fixed-point weights, and performs its internal calculations in fixed-point. (It uses our fixed- to floating-point conversion design entities to convert back and forth.) Careful attention is given to the range and precision of internal representations to avoid overflow or loss of precision.

BIBLIOGRAPHIC NOTE

Our version of μ-law comes from the G.711 standard [59].

Exercises

13.1 *Faster complex multiplication, design.* Give a detailed design (including wire widths and Wallace tree) of the complex multiplier found in Figure 13.3. You do not need to write the VHDL, but should rather describe the module in such a way that doing so would be simple.

13.2 *Faster complex multiplication, implementation.* Write and verify the VHDL to implement your design from Exercise 13.1.

13.3 *More complex multiplication.* Design, write, and verify the VHDL for a module that multiplies eight different $s1.14$ complex numbers by each other. You should not lose precision before the final output step.

13.4 *Fixed- to floating-point conversion.* The fixed-point conversion of Figure 13.8 does not work properly given an input value of 800_{16}; fix this.

13.5 *Fixed- to floating-point conversion, truncation.* Modify the fixed-point to floating-point converter of Figure 13.8 to use a truncation scheme (drop the bits that get shifted out), instead of rounding. What is the maximum representation error in this format? Plot the error over all input numbers.

13.6 *3E5 floating-point.* Write both a fixed- to floating-point and a floating- to fixed-point converter that can convert between a 32-bit signed integer and a 3E5 floating-point value. This format will have a sign bit, five bits of exponent (maximum value 29), an implicit 1, and two explicit mantissa bits. Use an implicit 1, gradual underflow, and truncation. This representation allows for a 4× data compression by representing only the three MSBs of the input value. What is the maximum representation error?

13.7 *Extending the FIR filter, I.* Modify the FIR filter of Figure 13.12 to accept weights in *s*2.5 format, subject only to the following constraint:

$$-1 \leq \sum_{i=0}^{3} w_i \leq 1.$$

13.8 *Extending the FIR filter, II.* Modify the FIR filter of Exercise 13.7 and remove the constraint on the sum of the weights. Be sure to check for overflow.

13.9 *Extending the FIR filter, III.* Using, without modification, the FIR filter block of Figure 13.12, design a 16-tap FIR function that finds the average of all 16 inputs. Take care to minimize precision loss. What parts of the filter need to be modified to eliminate precision loss completely?

13.10 *Putting it all together.* Draw the block diagram for a complex four-tap FIR filter. The value inputs and final outputs will be 3E5 complex numbers. The weights will be *s*1.4 fixed-point complex numbers. Be sure to denote the number formatting and bit widths of all intermediate wires. No precision loss should occur until the final output stage.

13.11 *Cross products.* The cross product of a vector (a_x, a_y, a_z) with another vector (b_x, b_y, b_z) results in a third vector (c_x, c_y, c_z), where

$$c_x = a_y b_z - a_z b_y,$$

$$c_y = a_z b_x - a_x b_z,$$

$$c_z = a_x b_y - a_y b_x.$$

Design a module with inputs of two 3-vectors of *s*3.14 numbers that outputs a 3-vector. Your output vector will not be composed of *s*3.14 numbers, but rather the minimum bit-width number with no precision loss.

13.12 *Approximate square root.* An approximation[3] for finding the square root of numbers between 0.5 and 2 is found on computing the value

$$\sqrt{x} \approx 1 + \frac{x-1}{2} - \frac{(x-1)^2}{8} + \frac{(x-1)^3}{16}.$$

Design and write a VHDL design entity that computes the approximate square root of a 1.8 number using the above formula. You may assume that the input is between 0.5 and 2. You should output a 1.8 number, but not suffer any intermediate precision loss. What is the worst-case error for all numbers between 0.5 and 2?

13.13 *Approximate division.* An approximation[4] for finding the reciprocal of a number between 0.5 and 1 is given by

$$\frac{1}{x} \approx 1 + (1-x) + (1-x)^2 + (1-x)^3 + (1-x)^4.$$

[3] Found by computing the Taylor series of $\sqrt{x}$ about $x = 1$.

[4] Found by computing the Taylor series of $1/x$ about $x = 1$.

(a) Plot the error of this approximation from $x = 0.5$ to $x = 1$.

(b) Design a floating-point divider module (23E8 format) that multiplies the dividend by the reciprocal of the divisor. You should provide a block diagram at least as detailed as that of Figure 11.1.[5]

13.14 *BCD to binary.* Design and code a VHDL design entity to convert a four-digit unsigned BCD whole number into a 14-bit binary number.

13.15 *Binary to BCD.* Design and code a VHDL design entity to convert a 14-bit unsigned binary number into a four-digit BCD value.

[5] Beware that few systems can tolerate having any error on a division operation.

Part IV

Synchronous sequential logic

14 Sequential logic

The output of sequential logic depends not only on its input, but also on its *state*, which may reflect the history of the input. We form a sequential logic circuit via feedback – feeding state variables computed by a block of combinational logic back to its input. General sequential logic, with asynchronous feedback, can become complex to design and analyze due to multiple state bits changing at different times. We simplify our design and analysis tasks in this chapter by restricting ourselves to *synchronous* sequential logic, in which the state variables are held in a register and updated on each rising edge of a clock signal (*clk*).[1]

The behavior of a synchronous sequential logic circuit, or *finite-state machine* (FSM), is completely described by two logic functions: one that computes its next state as a function of its input and present state, and one that computes its output – also as a function of its input and present state. We describe these two functions by means of a *state table*, or graphically with a *state diagram*. If states are specified symbolically, a *state assignment* maps the symbolic states onto a set of bit vectors – both binary and one-hot state assignments are commonly used.

Given a state table (or state diagram) and a state assignment, the task of implementing a finite-state machine is a simple one of synthesizing the next-state and output logic functions. For a one-hot state encoding, the synthesis is particularly simple because each state maps to a separate flip-flop and all edges in the state diagram leading to a state map into a logic function on the input of that flip-flop. For binary encodings, Karnaugh maps for each bit of the state vector are written and reduced to logic equations.

Finite-state machines can be implemented in VHDL by creating a state register to hold the current state, and describing the next-state and output functions with combinational logic descriptions, such as `case` statements as described in Chapter 7. State assignments should be specified using `constants` to allow them to be changed without altering the machine description itself. Special attention should be given to resetting the FSM to a known state at startup.

14.1 SEQUENTIAL CIRCUITS

Recall that a combinational circuit produces an output that depends only on the current state of its input. Recall also that combinational circuits must be acyclic. If we add feedback to a combinational circuit, creating a cycle as shown in Figure 14.1, the circuit becomes sequential. The output of a sequential circuit depends not only on its current input, but also on the history

[1] We revisit asynchronous sequential circuits in Chapter 26.

Table 14.1. **State table for RS flip-flop**

r	s	q_{old}	q_{new}
0	0	0	0
0	0	1	1
0	1	X	1
1	X	X	0

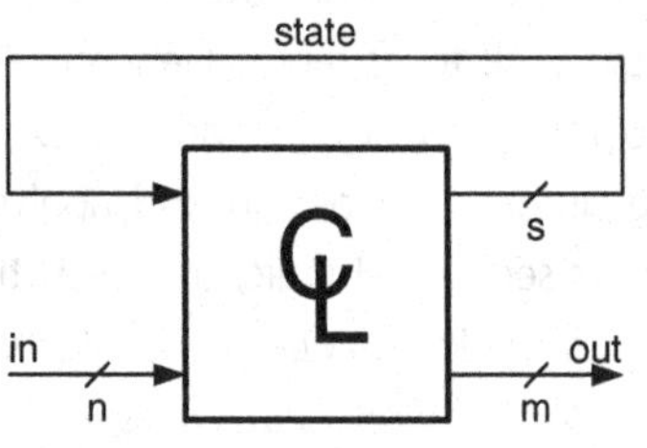

Figure 14.1. Sequential circuits are formed when feedback paths carrying state information are added to combinational circuits. The output of a sequential circuit depends both on the current input and on the state, which is a function of previous inputs.

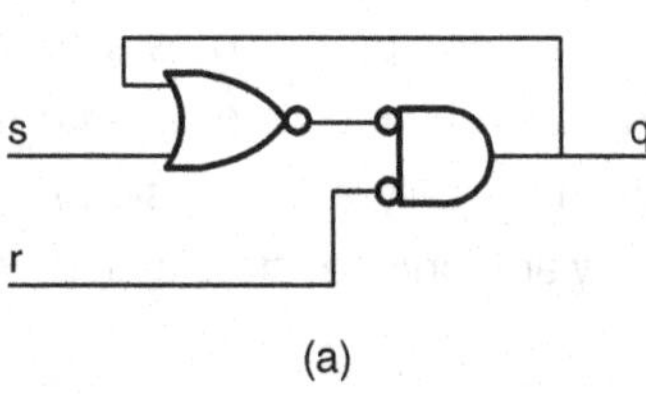

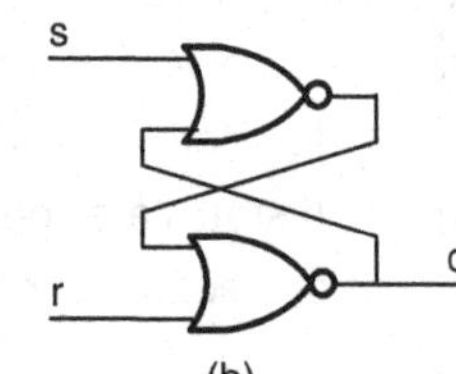

Figure 14.2. An RS flip-flop is a simple example of a sequential circuit: (a) schematic; (b) alternative schematic that does not obey the *bubble rule.*

of previous inputs. The cycle created by the feedback allows the circuit to store information about its previous input. We refer to the information stored on the feedback signals as the *state* of the circuit.

A sequential circuit generates an output that is a function both of its input and of its present state. It also generates its next state, also as a function of its input and its present state.

Figure 14.2 shows a reset–set (RS) flip-flop, a very simple sequential logic circuit that is composed of two NOR gates.[2] The output q is fed back to the input as a state variable. The circuit's behavior is described by the equation $q = \bar{r} \wedge (s \vee q)$. The state variable q appears on both sides of the equation. To make the dynamics clearer, we rewrite this as $q_{\text{new}} = \bar{r} \wedge (s \vee q_{\text{old}})$. That is, the equation tells us how to derive the *new* state of q as a function of the inputs and the *old* state of q.

From the equation (and the schematic), it is easy to see that if $r = 1$, $q = 0$ and the flip-flop is reset; if $s = 1$ and $r = 0$, $q = 1$ and the flip-flop is set; and if $s = 0$ and $r = 0$, the output q stays in whatever state it was in. The output q reflects the last input to be high. If r was high last, $q = 0$. If s was high last, $q = 1$. We summarize this behavior in the state table of Table 14.1.

Because the function of sequential circuits depends on the evolution of signals over time, we often describe their behavior using a *timing diagram*. A timing diagram illustrating operation of the RS flip-flop is shown in Figure 14.3. The figure shows the waveforms, signal levels as a

[2] We draw the circuit as shown in Figure 14.2(a). Many people (who do not obey the bubble rule) draw it as shown in Figure 14.2(b).

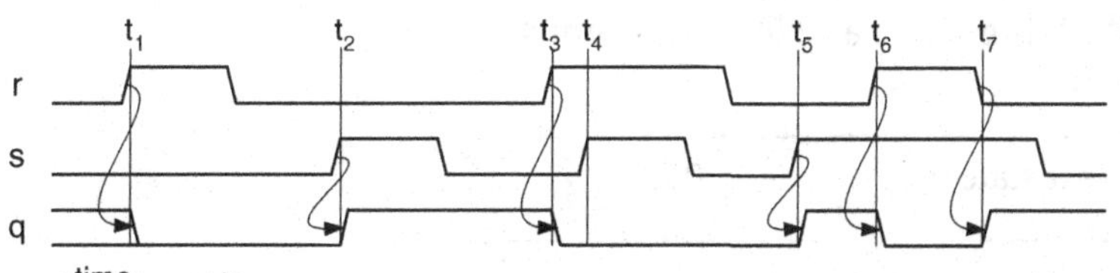

Figure 14.3. Timing diagram showing operation of the RS flip-flop. The value of signals is shown as time advances from left to right.

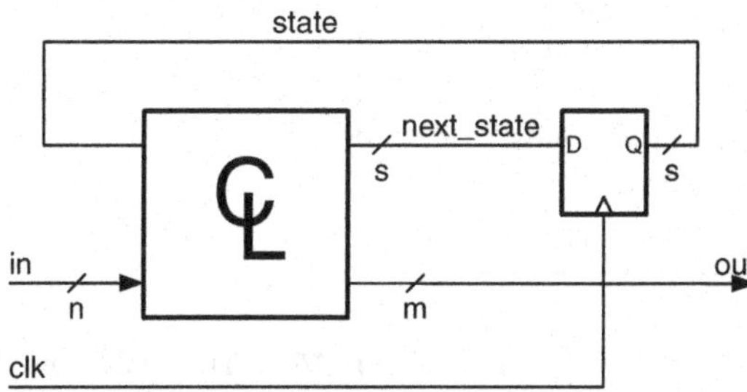

Figure 14.4. A *synchronous* sequential circuit breaks the state feedback loop with a clocked storage element (in this case a D-type flip-flop). The flip-flop ensures that all state variables change value at the same time – when the clock signal rises.

function of time, for the signals r, s, and q. Time advances from left to right. Arrows from one signal to another show cause and effect.

Initially, q is in an unknown state; it could be either high or low, denoted by both high and low lines. At time t_1, r goes high, causing q to fall – resetting the flip-flop. Signal s goes high at t_2 causing q to rise – setting the flip-flop. The flip-flop is reset again at t_3. Signal s goes high at t_4, but this has no effect on the output, since r is also high. Signal s going high with r low at t_5 does set the flip-flop. It is reset again at t_6 when r goes high – even though s is still high. The flip-flop is set for a final time at t_7 when r goes low.

14.2 SYNCHRONOUS SEQUENTIAL CIRCUITS

While the RS flip-flop of Figure 14.2 is simple enough to understand, arbitrary sequential circuits, with many bits of state feedback, can give complex behavior. Part of the complexity is due to the fact that the different bits of the next-state signal may change at different times. Such *races* can lead to a next-state output that depends on circuit delay. We defer discussion of general *asynchronous* sequential circuits until Chapter 26. Until then, we restrict our attention to *synchronous* sequential circuits in which clocked storage elements are used to ensure that all state variables change state at the same time, synchronized to a clock signal. Synchronous sequential circuits are sometimes called *finite-state machines*, or FSMs.

A block diagram of a synchronous sequential logic circuit is shown in Figure 14.4. The circuit is synchronous because the state feedback loop is broken by an s-bit wide D flip-flop (where s is the number of state bits). This flip-flop circuit, described in Section 15.2, updates its output with the value of its input on the rising edge of the clock signal. At all other times, the output remains stable. Inserting the D flip-flop into the feedback loop constrains all of the state bits to change at the same time – eliminating the possibility of races.

Operation of a synchronous sequential circuit is illustrated in the timing diagram of Figure 14.5. During each clock cycle, the time from one rising edge of the clock to the next, a block of combinational logic computes the next state and output (not shown) as a combinational

Table 14.2. **State table for example synchronous logic circuit**

State	Next state in = 0	Next state in = 1	Out in = 0	Out in = 1
00	00	01	0	0
01	00	11	0	0
11	01	10	0	0
10	11	00	0	1

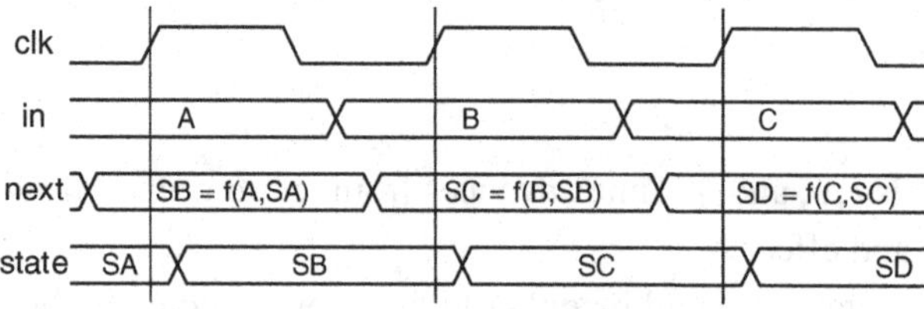

Figure 14.5. Timing diagram showing operation of a synchronous sequential circuit. The state advances on each rising edge of the clock signal, *clk*.

function of the input and current state. At each rising edge of the clock, the current state (*state*) is updated with the next state computed during the previous clock cycle.

During the first clock cycle of Figure 14.5, for example, the current state is *SB*, and the input changes to *B* before the end of the cycle. The combinational logic then computes the next state $SC = f(B, SB)$. At the end of the cycle the clock rises again, updating the current state to be *SC*. The state will be *SC* until the clock rises again.

We can analyze synchronous sequential circuits on a clock-by-clock basis. The next state and output during a given clock cycle depend only on the current state and input during that clock cycle. At each clock, the current state advances to the next state.

For example, suppose our next state and output logic are as given by Table 14.2. If our circuit starts in state 00 and has an input sequence of 011011011 for the first nine cycles, what will its state and output be each cycle?

Operation of our example circuit is shown in Table 14.3. In cycle 0 we start in state 00. The input and output are both 0 this cycle. With an input of 0 in state 00, the next state is also 00, so in cycle 1 we remain in state 00, but now with an input of 1. A state of 00 and an input of 1 gives us a next state of 01 for cycle 2. With the input high in state 01, we get a next state of 11 for cycle 3. The input goes low in state 11 in cycle 3, taking us back to state 01 for cycle 4. The input is high for the next two cycles, taking us to 11 and 10 in cycles 5 and 6, respectively. A low input in cycle 6 takes us back to 11 for cycle 7. High inputs for cycles 7 and 8 take us to states 10 and 00 for cycles 8 and 9. The output goes high in cycle 8 since the state is 10 and the input is 1.

One representation of a finite-state machine is a *state table*, such as Table 14.2, that gives the next-state and output functions in tabular form. An equivalent graphical representation is a state diagram, as shown in Figure 14.6.

Table 14.3. **State sequence for the sequential logic circuit described by Table 14.2 on the input sequence 011011011**

Cycle	State	In	Out
0	00	0	0
1	00	1	0
2	01	1	0
3	11	0	0
4	01	1	0
5	11	1	0
6	10	0	0
7	11	1	0
8	10	1	1
9	00		0

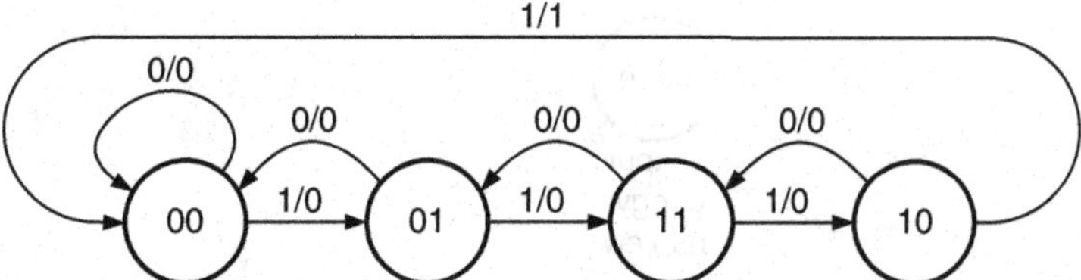

Figure 14.6. State diagram for finite-state machine of Table 14.2. The circles represent the four states. Each arrow represents a *state transition* from a current state to a next state and is labeled *input*/*output* – the input that causes the transition, and the output in the current state for that input.

Each circle in Figure 14.6 represents one state. It is labeled with the name of the state. Here we use the value of the state variables as the state name. Later we will introduce symbolic state names, independent of the state coding. The next-state function is shown by the arrows. Each arrow represents a *state transition* and is labeled with the input and output values during that transition. For example, the arrow from state 00 to state 01, labeled 1/0, implies that in state 00 when the input is 1 the next state is 01 and the output is 0. Note that an arrow may go from a state to itself, as in the case of the input 0 transition from state 00 to state 00. Also, a transition may go quite a distance in the diagram, as with the input 1 transition from state 10 to state 00.

14.3 TRAFFIC-LIGHT CONTROLLER

As a second example of an FSM, consider the problem of controlling the traffic lights at an intersection of a north–south road with an east–west road, as shown in Figure 14.7. There are

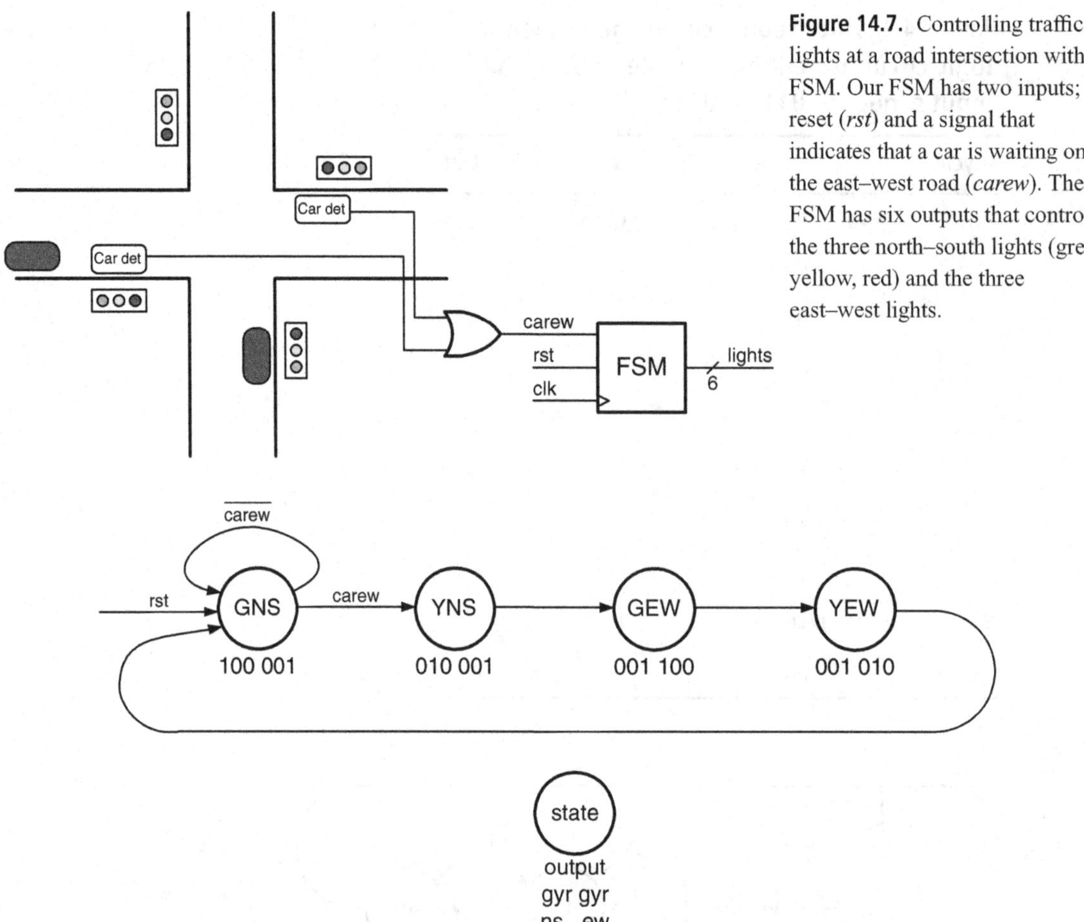

Figure 14.7. Controlling traffic lights at a road intersection with an FSM. Our FSM has two inputs; a reset (*rst*) and a signal that indicates that a car is waiting on the east–west road (*carew*). The FSM has six outputs that control the three north–south lights (green, yellow, red) and the three east–west lights.

Figure 14.8. State diagram for a traffic-light controller FSM. The states are labeled with symbolic names. The outputs are given under each state (green–yellow–red (three-bits) for north–south followed by green–yellow–red for east–west). The reset arrows are omitted. The FSM resets to state GNS.

six lights to control: green, yellow, and red both for the north–south road and for the east–west road. Our FSM will take as input signal *carew* that indicates that a car is waiting on the east–west road. A second input, *rst*, resets the FSM to a known state.

We start with an English-language description of our FSM.

(1) Reset to a state where the light is green in the north–south direction and red in the east–west direction.
(2) When a car is detected in the east–west direction (*carew* = 1), go through a sequence that makes the light go green in the east–west direction and then return to green in the north–south direction.
(3) A direction with a green light must first transition to a state where the light is yellow before going to a state where the light goes red.
(4) A direction can have a green light only if the light in the other direction is red.

A state diagram for an FSM that meets our specification is shown in Figure 14.8. Compared with the state diagram of Figure 14.6, there are two major differences. First, the states are

Table 14.4. **State table for the traffic-light controller FSM; the FSM resets to state GNS**

	Next state		
State	*carew* = 0	*carew* = 1	Out
GNS	GNS	YNS	100 001
YNS	GEW	GEW	010 001
GEW	YEW	YEW	001 100
YEW	GNS	GNS	001 010

labeled with symbolic names. Second, the output values are placed under the states rather than on the transitions. This is because the output is a function only of the state, not of the input.[3]

The FSM resets to state GNS (for green–north–south). In this state the output is 100 001. The first 100 represents the north–south lights (green–yellow–red). The 001 represents the east–west lights (also green–yellow–red). Hence the light is green in the north–south direction and red in the east–west direction. Resetting to this state satisfies specification number 1. The arrow labeled *carew* keeps us in state GNS until a car is detected in the east–west direction.

When a car is detected in the east–west direction, the signal *carew* becomes true, and the next rising edge of the clock causes the machine to enter state YNS (yellow–north–south). In this state the output is 010 001. The 010 implies yellow in the north–south direction, and 001 implies red in the east–west direction. Transitioning to this state before making east–west green satisfies specification 3 on the transition from GNS to GEW. The arrow out of state YNS has no label. This implies that this state transition always occurs (unless the FSM is reset).

State YNS is always followed by state GEW (green, east–west). In this state the output is 001 100 – red in the north–south direction and green in the east–west direction. This state, and the sequence it is part of, satisfies specification 2. State GEW is always followed by state YEW. State YEW (yellow, east–west) has output 001 010 – red in the north–south direction and yellow in the east–west direction. This state satisfies specification 3 on the transition between GEW and GNS.

A state table for the traffic-light controller is shown in Table 14.4. Reset is not shown. We explore some variations of this basic traffic-light controller in Exercises 14.3 to 14.9.

EXAMPLE 14.1 State machine

Draw a state diagram for a state machine that fills missing pulses in a pulse train. The input *a* is a signal that normally goes high for one cycle every five cycles. When input *a* goes high in the expected cycle, or one cycle early, the output *q* should go high for the following cycle. If the input *a* is one cycle early or late, timing is reset so *a* is expected to go high again five cycles

[3] An FSM where the output depends only on the current state and not on the input is sometimes called a *Moore machine*. An FSM where the output depends on both the current state and the input is sometimes called a *Mealy machine*.

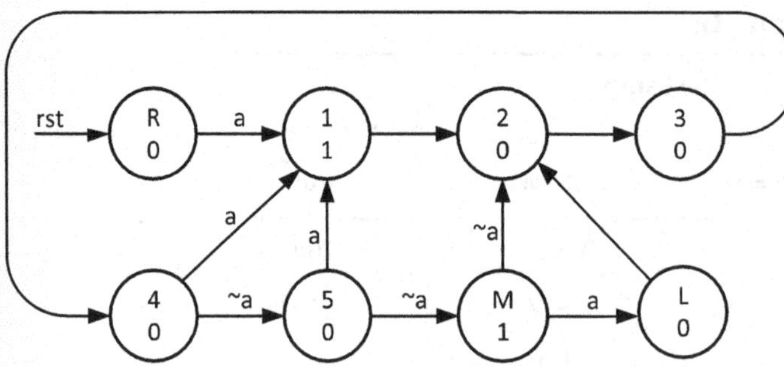

Figure 14.9. State diagram for the pulse-filler FSM, as described in Example 14.1.

later. If a goes high two cycles early or late, it is ignored. If a fails to appear in the expected cycles, the output q goes high anyway in the cycle after a was expected to appear.

The state diagram, shown in Figure 14.9, has eight states. The output for each state is shown under the state name. On reset, the machine starts in state R, awaiting the first pulse on a. When this pulse arrives, the machine moves to state 1 and outputs a 1. The machine advances to states 2, 3, and 4, ignoring any inputs on a that arrive in states 1, 2, or 3. If a is high in state 4 (one cycle early), the machine goes to state 1 – resetting the timing. A pulse arriving at the expected time, in state 5, causes a transition to state 1 – restarting the sequence. If a is zero in state 5, control transfers to state M, where the output goes high to handle a missing or late pulse. If a goes high in state M, control transfers to state L to reset timing for the late pulse. Otherwise, control transfers to state 2 to handle a missing pulse without resetting timing.

14.4 STATE ASSIGNMENT

When an FSM is specified with symbolic state names as in Figure 14.8 or Table 14.4, we need to assign actual binary values to the states before we can implement the FSM. This process of assigning values to states is called *state assignment.*

With a synchronous machine, we can assign the states to any set of values, as long as the value for each state is unique.[4] It takes at least $s_{\min} = \log_2(N)$ bits to represent N states; however, the best state assignment is not always the one with the fewest bits. We refer to each bit of a state vector as a *state variable.*

A *one-hot* state assignment uses N bits to represent N states. Each state gets its own bit. When the machine is in the ith state, the corresponding bit b_i of the state variable is 1. In all other states, b_i is 0. A one-hot state assignment for the traffic-light controller FSM is shown in Table 14.5. It takes four bits to represent the four states. In any state, only one bit is set. A one-hot assignment makes the logic design of a finite-state machine particularly simple, as we shall see below.

A binary state assignment uses the minimum number of bits, $s_{\min} = \log_2(N)$, to represent N states. Of the $N!$ possible binary state assignments (24 for four states), it does not really matter which one you choose. While many academic papers have been written about choosing state assignments to minimize implementation logic, in practice it does not really matter. Except in

[4] This is not true of an asynchronous machine, where careful state assignment is required in order to avoid races.

Table 14.5. **One-hot state assignment for the traffic-light controller FSM**

State	Encoding
GNS	0001
YNS	0010
GEW	0100
YEW	1000

Table 14.6. **Binary state assignment for the traffic-light controller FSM**

State	Encoding
GNS	00
YNS	01
GEW	11
YEW	10

very rare cases, the number of gates saved by optimizing the state assignment is unimportant. Don't waste much time on state assignment. Design time is more important than a few gates.

One possible binary state assignment for the traffic-light controller FSM is shown in Table 14.6. This particular state assignment uses a Gray code so that only one bit of state changes on each state transition. This sometimes reduces power and minimizes logic. We could just as easily have chosen a straight binary count (GEW = 10, YEW = 11), and it wouldn't make much difference.

14.5 IMPLEMENTATION OF FINITE-STATE MACHINES

Given a state table (or state diagram) and a state assignment, the implementation of an FSM is reduced to the problem of designing two combinational logic circuits, one for the next state and one for the output. These combinational logic circuits are combined with an s-bit wide D flip-flop to update the current state from the next state on each rising edge of the clock. A multi-bit D flip-flop like this is often called a *register*, and when it is used to hold the state of an FSM it is called the *state register*.

With a one-hot state assignment, the implementation of the next-state logic is a direct translation of the state diagram, as shown in Figure 14.10 for our traffic-light controller FSM. Four flip-flops correspond to the four states: GNS, YNS, GEW, and YEW. When the first flip-flop is set, the FSM is in state GNS. The logic feeding the D input of each flip-flop is an OR of the transition arrows feeding the corresponding state in the state diagram. For states GEW and YEW this is just a wire. These states always follow the preceding state. For state YNS, the input logic is an AND gate that ANDs the previous state (GNS) with the condition for a

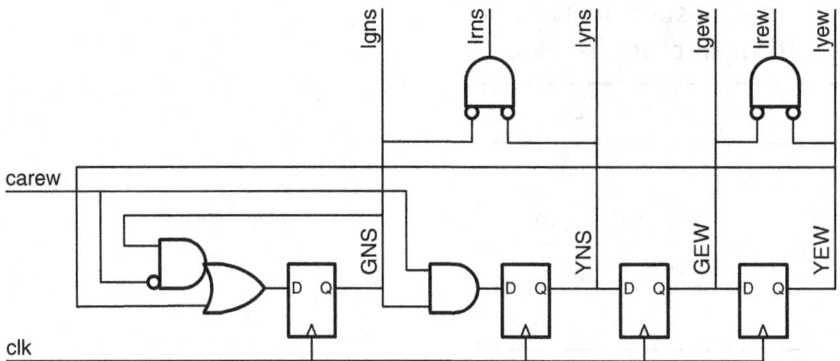

Figure 14.10. Implementation of the traffic-light controller FSM using a one-hot state encoding. Four flip-flops are used, one corresponding to each state. The state transition arrows into a state directly translate into the logic preceding the corresponding flip-flop. The outputs, shown at the top, represent the state of each individual light. For example, *lrns* is the red light in the north–south direction.

transition from GNS to YNS (*carew*). State GNS is the target of two destination arrows and hence requires an OR gate to combine them. Its input logic is the OR of a wire from state YEW, and an AND gate combining state GNS and condition $\overline{carew}$ – this corresponds to the back edge from state GNS to itself.

It is always possible to implement directly a one-hot FSM in this manner. This made design and maintenance of FSMs very easy in the era before logic synthesis. One would simply instantiate a flip-flop for each state, and appropriate input gates for each transition arrow. The function is immediately apparent from the logic. Adding, deleting, or changing the condition on a transition arrow was straightforward, and affected only the part of the logic associated with the transition arrow. With modern logic synthesis, the advantage of this approach is greatly diminished.

The output logic for the circuit of Figure 14.10 consists of two NOR gates. States GNS, YNS, GEW, and YEW drive the green and yellow light outputs directly. The red light outputs are generated by observing that for each direction, if the yellow and green lights are off, the red light should be on: $r = \overline{y} \wedge \overline{g}$.

To implement our FSM with a binary state encoding, we proceed with logic synthesis of each of the state variables. First, we convert our state table into a truth table – showing each next-state variable as a function of the current state variable and all inputs. For example, a truth table for the traffic-light controller FSM with the state encoding of Table 14.6 is shown in Table 14.7.

From this state table we draw the two Karnaugh maps shown in Figure 14.11. The left K-map shows the truth table for the MSB of the next state (ns_1), and the right K-map shows the truth table for the LSB of the next stage (ns_0). The next-state logic here is very simple. The ns_1 function has a single prime implicant, and the ns_0 function has two. All three are essential. The logic here is very simple:

$$ns_1 = s_0, \tag{14.1}$$

$$ns_0 = (s_0 \vee carew) \wedge \overline{s_1}, \tag{14.2}$$

where ns_1, ns_0 are the next-state variables and s_1, s_0 are the current-state variables.

Now that we have the next-state function, what remains is to derive the output function. To do this we write down the truth table for the outputs – a function of only the current state. This

Table 14.7. **Truth table for the next-state function of the traffic-light controller FSM with the state assignment of Table 14.6**

State	*carew*	Next state (ns_1, ns_0)	Comment
00	0	00	green north/south *carew* = 0
00	1	01	green north/south *carew* = 1
01	0	11	yellow north/south *carew* = 0
01	1	11	yellow north/south *carew* = 1
11	0	10	green east/west *carew* = 0
11	1	10	green east/west *carew* = 1
10	0	00	yellow east/west *carew* = 0
10	1	00	yellow east/west *carew* = 1

Table 14.8. **Truth table for the output function of the traffic-light controller FSM with the state assignment of Table 14.6**

State	Output
00	100 001
01	010 001
11	001 100
10	001 010

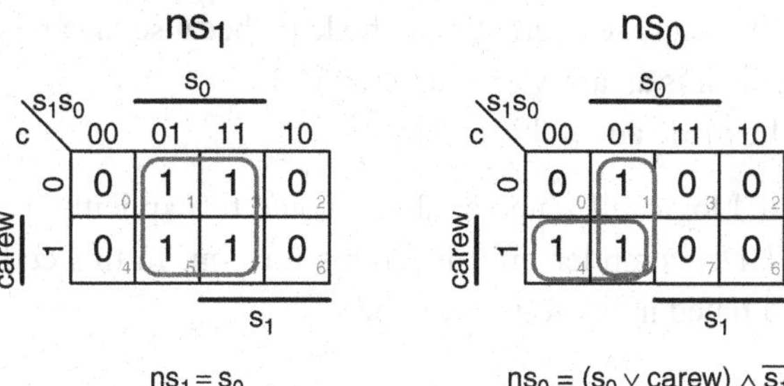

Figure 14.11. Karnaugh maps for computing the next state (ns_1 ns_0) based on the current state and input *carew*.

is shown in Table 14.8. The logic functions for the output variables can be derived directly from this table as follows:

$$g_{\mathrm{ns}} = \overline{s_1} \wedge \overline{s_0}, \tag{14.3}$$

$$y_{\mathrm{ns}} = \overline{s_1} \wedge s_0, \tag{14.4}$$

$$r_{\mathrm{ns}} = s_1, \tag{14.5}$$

$$g_{\mathrm{ew}} = s_1 \wedge s_0, \tag{14.6}$$

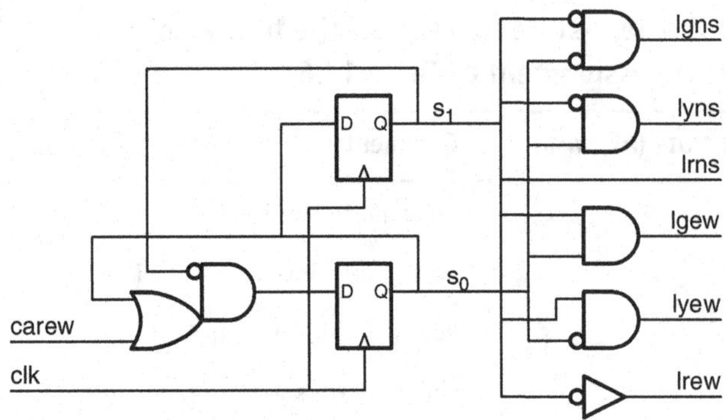

Figure 14.12. Logic diagram for traffic-light controller FSM implemented with the state assignment of Table 14.6.

$$y_{\mathrm{ew}} = s_1 \wedge \overline{s_0}, \tag{14.7}$$

$$r_{\mathrm{ew}} = \overline{s_1}. \tag{14.8}$$

By combining the next-state and logic equations, we obtain the logic diagram of Figure 14.12.

14.6 VHDL IMPLEMENTATION OF FINITE-STATE MACHINES

With VHDL, designing an FSM is a simple matter of specifying the next-state and output functions and selecting a state assignment. Logic synthesis does all the work of generating the next-state and output logic. A VHDL description of the traffic-light controller FSM is shown in Figure 14.13. The bulk of the logic is in a single case statement that defines both the next-state and output functions. The syntax `std_logic_vector'(X)` used inside the case statement is a VHDL qualified expression. A qualified expression has the syntax `<type_mark>'(<expression>)` and is used when the type of `<expression>` is ambiguous. In this case we need to make it clear that the array aggregate has type `std_logic_vector` for the assignment to `next1` in the last case alternative to be legal because the array aggregate could also be of type `string` (which is an array of characters).

There are three key points to be made about this code.

(1) In implementing all sequential logic, all state variables should be explicitly declared as D flip-flops. Do not let the VHDL compiler infer flip-flops for you. In this code, the state flip-flops are explicitly instantiated in the following code:

```
-- instantiate state register
STATE_REG: vDFF generic map(SWIDTH) port map(clk,next_state,current_state);
```

This code instantiates an `SWIDTH`-wide D flip-flop that is clocked by `clk`, has input `next_state`, and outputs `current_state`. The component `vDFF` is defined in the package `ff` in Figure 14.16 which will be described below.

(2) In designing finite-state machines, use VHDL **constant** declarations to define all constants. Do not hard-code any constants in the code. Constants that should be declared in this way include the width of the state vector `SWIDTH`, the state encodings (e.g., `GNS`), and input and output encodings (e.g., `GNSL`). In particular, defining symbolic names for the

```
----------------------------------------------------------
-- Traffic_Light
-- Inputs:
--    clk - system clock
--    rst - reset - high true
--    carew - car east/west - true when car is waiting in east-west direction
-- Outputs:
--    lights - (6 bits) {gns, yns, rns, gew, yew, rew}
-- Waits in state GNS until carew is true, then sequences YNS, GEW, YEW
-- and back to GNS.
----------------------------------------------------------
library ieee;
use ieee.std_logic_1164.all;
use work.Traffic_Light_Codes.all;
use work.ff.all;

entity Traffic_Light is
  port( clk, rst, carew: in std_logic;
        lights: out lights_type );
end Traffic_Light;

architecture impl of Traffic_Light is
  signal current_state, next_state, next1: state_type;
begin
  -- instantiate state register
  STATE_REG: vDFF generic map(SWIDTH) port map(clk,next_state,current_state);

  -- next state and output equations - this is combinational logic
  process(all) begin
    case current_state is
      when GNS =>
        if carew then next1 <= YNS;
        else next1 <= GNS; end if;
        lights <= GNSL;
      when YNS => next1 <= GEW; lights <= YNSL;
      when GEW => next1 <= YEW; lights <= GEWL;
      when YEW => next1 <= GNS; lights <= YEWL;
      when others =>
        next1 <= std_logic_vector'(SWIDTH-1 downto 0 => '-');
        lights <= "------";
    end case;
  end process;

  -- add reset
  next_state <= GNS when rst else next1;
end impl;
```

Figure 14.13. VHDL description of traffic-light controller FSM.

state encodings enables you to change a state assignment by changing the definitions. We will see an example of this in the following.

(3) Make sure you reset your FSM. Here we declare two next-state vectors, `next1` and **`next`**. The **`case`** statement computes `next1` as the next state, ignoring reset `rst`. A final assign statement overrides this next state with the reset state `GNS` if `rst` is asserted:

```
-- add reset
next_state <= GNS when rst else next1;
```

Factoring the reset out of the next state function in this manner greatly improves the readability of the code. If we did not do this, we'd have to repeat the reset logic in every state, rather than doing it just once.

The VHDL constant definitions for the traffic-light controller FSM are shown in Figure 14.14. Using the keyword **constant** in this manner enables us to introduce symbolic names in our code, improving readability, and also makes it easy to change encodings. For example, substituting the one-hot state encodings of Figure 14.15 changes our FSM from a binary to a one-hot state assignment without changing any other lines of code.

The VHDL code for the `vDFF` component is shown in Figure 14.16 along with a declaration for the package `ff`. The edge-sensitive behavior of the flip-flop is described by the **process** statement:

```
process(clk) begin
  if rising_edge(clk) then
    Q <= D;
  end if;
end process;
```

```
library ieee;
use ieee.std_logic_1164.all;

package Traffic_Light_Codes is
  ----------------------------------------------
  -- define state assignment - binary
  ----------------------------------------------
  constant SWIDTH: integer := 2;
  subtype state_type is std_logic_vector(SWIDTH-1 downto 0);
  constant GNS: state_type := "00";
  constant YNS: state_type := "01";
  constant GEW: state_type := "10";
  constant YEW: state_type := "11";
  ----------------------------------------------
  -- define output codes
  ----------------------------------------------
  subtype lights_type is std_logic_vector(5 downto 0);
  constant GNSL: lights_type := "100001";
  constant YNSL: lights_type := "010001";
  constant GEWL: lights_type := "001100";
  constant YEWL: lights_type := "001010";
end package;
```

Figure 14.14. VHDL definitions for traffic-light controller state variables and output encodings.

```
constant GNS: state_type := "1000";
constant YNS: state_type := "0100";
constant GEW: state_type := "0010";
constant YEW: state_type := "0001";
```

Figure 14.15. VHDL definitions for a one-hot state assignment for the traffic-light controller FSM.

```
library ieee;
package ff is
  use ieee.std_logic_1164.all;
  component vDFF is -- multi-bit D flip-flop
    generic( n: integer := 1 ); -- width
    port( clk: in std_logic;
          D: in std_logic_vector( n-1 downto 0 );
          Q: out std_logic_vector( n-1 downto 0 ) );
  end component;
  component sDFF is -- single-bit D flip-flop
    port( clk, D: in std_logic; Q: out std_logic );
  end component;
end package;

library ieee;
use ieee.std_logic_1164.all;

entity vDFF is
    generic( n: integer := 1 );
    port( clk: in std_logic;
          D: in std_logic_vector( n-1 downto 0 );
          Q: out std_logic_vector( n-1 downto 0 ) );
end vDFF;

architecture impl of vDFF is
begin
  process(clk) begin
    if rising_edge(clk) then
      Q <= D;
    end if;
  end process;
end impl;

library ieee;
use ieee.std_logic_1164.all;

entity sDFF is
  port( clk, D: in std_logic;
        Q: out std_logic );
end sDFF;

architecture impl of sDFF is
begin
  process(clk) begin
    if rising_edge(clk) then
      Q <= D;
    end if;
  end process;
end impl;
```

Figure 14.16. VHDL description of a D flip-flop.

This statement performs the update of the output `Q <= D` on every rising edge of `clk`. This is indicated by the condition "`rising_edge(X)`" which is true whenever signal `X` has a rising edge. Most synthesis tools allow more complex expressions than "`Q <= D`" inside the "**`if`** `rising\_edge(clk)` **`then`**" statement to be synthesized into gates. However, such code can be more difficult to understand.

Our testbench for the traffic-light controller is shown in Figure 14.17. To test an FSM thoroughly, we would like to visit every state *and* traverse every edge of the state diagram. Achieving this coverage is not particularly difficult for our traffic-light controller FSM.

The testbench has three parts. First, it instantiates a `Traffic_Light` design entity – the unit being tested. The second part is a **`process`** that performs clock generation and output. The code displays some variables and generates a clock with a 10 nanosecond period. The output is done in the middle of the clock cycle – just after `clk` goes low.

```
-- pragma translate_off
library ieee;
use ieee.std_logic_1164.all;
use work.Traffic_Light_Codes.all;

entity Test_Fsm1 is
end Test_Fsm1;

architecture test of Test_Fsm1 is
  signal clk, rst, carew: std_logic;
  signal lights: std_logic_vector(5 downto 0);
begin
  DUT: entity work.Traffic_Light(impl) port map(clk,rst,carew,lights);

  -- clock with period of 10 ns
  process begin
    clk <= '1'; wait for 5 ns;
    clk <= '0'; wait for 5 ns;
    report to_string(rst) & " " & to_string(carew) & " " &
           to_string( <<signal DUT.current_state: state_type>> ) & " " &
           to_string(lights);
  end process;

  -- input stimuli
  process begin
    rst <= '0'; carew <= '0';        -- start w/o reset to show x state
    wait for 15 ns; rst <= '1';      -- reset
    wait for 10 ns; rst <= '0';      -- remove reset
    wait for 20 ns; carew <= '1';  -- wait 2 cycles, then car arrives
    wait for 30 ns; carew <= '0';  -- car leaves after 3 cycles (green)
    wait for 20 ns; carew <= '1';  -- wait 2 cycles then car comes and stays
    wait for 60 ns;
    std.env.stop(0);
  end process;
end test;
-- pragma translate_on
```

Figure 14.17. VHDL testbench for the traffic-light controller FSM.

Notice that to print out the value of `current_state` inside of `Traffic_Light` the second part of the testbench used the syntax "`<<`**`signal`**` DUT.current_state: state_type>>`". This is an example of a VHDL 2008 *external name*. It enables us to read the value of a signal deep inside our design under test without having to add signals to the top level interface. An external name begins with the compound delimiter "`<<`" followed by the type of object accessed, which can be **`signal`**, **`variable`**, or **`constant`**, followed by an external path name, followed by a colon (`:`), followed by the subtype of the object, followed by the compound delimiter "`>>`". In this case `DUT.current_state` is the external path name for the signal we would like to print out. The *external path name* is composed of the instance labels of the instantiated component, in this case, `DUT`, and the internal signal name, each separated by a period symbol (`.`).

The final part is process containing a test script that generates the inputs for the module being tested. We use the VHDL-2008 `std.env.stop()` function defined in the package `std.env`. The function `stop()` causes simulation to stop. If we did not call this function our simulation would not terminate since the first process, generating the clock, never reaches an unconditional wait statement (i.e., "**`wait`**`;`").

The inputs and the response of the device under test (the component labeled "DUT") are shown in textual form in Figure 14.18, which shows the signals on each falling edge of the clock, and as waveforms in Figure 14.19. Initially, `state` and `next` are both in an unknown state (x in the text output, and a line midway between a 1 and 0 in the waveform). Signal `rst` is asserted in the second clock cycle to reset to a known state. Next-state signal `next` responds immediately, and `state` follows on the rising edge of the clock. The FSM stays in state 00 until `carew` rises on clock 5, causing the FSM to go to state 01 on clock 6. This starts a sequence through states 01, 11, 10, and back to 00 on clock 9. It stays in state 00 for two cycles this time, until `carew` going high in clock 10 causes the FSM to start the sequence again in clock 11. This time `carew` stays high, and the sequence repeats until the simulation ends.

```
0 0 UU ---
1 0 - ---
0 0 00 100001
0 0 00 100001
0 1 00 100001
0 1 01 010001
0 1 11 001100
0 0 10 001010
0 0 00 100001
0 1 00 100001
0 1 01 010001
0 1 11 001100
0 1 10 001010
0 1 00 100001
0 1 01 010001
```

Figure 14.18. Results of simulating the traffic-light controller FSM of Figure 14.13 using the testbench of Figure 14.17. Each line shows the values of `rst`, `carew`, `state`, and `lights` on a falling edge of the clock.

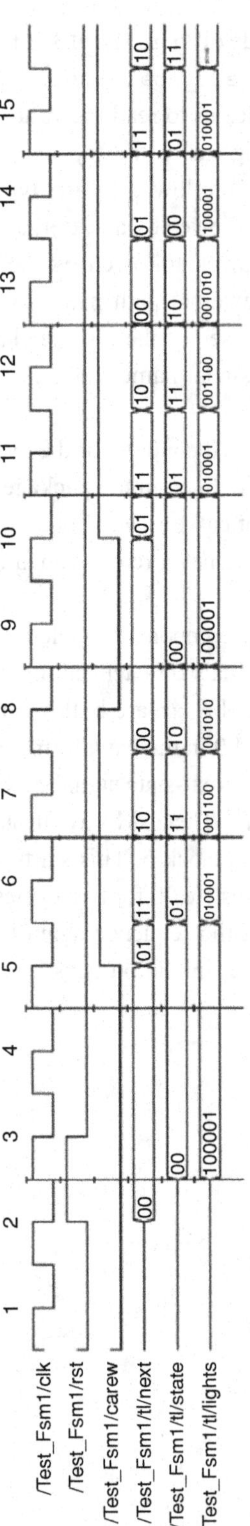

Figure 14.19. Waveforms from simulation of the traffic-light controller of Figure 14.13 using the testbench of Figure 14.17.

EXAMPLE 14.2 VHDL FSM

Write a VHDL design entity for the pulse-filler FSM of Example 14.1.

```
library ieee;
use ieee.std_logic_1164.all;
use work.ff.all;

entity PulseFiller is
  port( clk, rst, a: in std_logic;
        q: out std_logic );
end PulseFiller;

architecture impl of PulseFiller is
  constant SWIDTH: integer := 3;
  constant SR: std_logic_vector(SWIDTH-1 downto 0) := "000";
  constant S1: std_logic_vector(SWIDTH-1 downto 0) := "001";
  constant S2: std_logic_vector(SWIDTH-1 downto 0) := "011";
  constant S3: std_logic_vector(SWIDTH-1 downto 0) := "010";
  constant S4: std_logic_vector(SWIDTH-1 downto 0) := "110";
  constant S5: std_logic_vector(SWIDTH-1 downto 0) := "111";
  constant SM: std_logic_vector(SWIDTH-1 downto 0) := "101";
  constant SL: std_logic_vector(SWIDTH-1 downto 0) := "100";
  constant SX: std_logic_vector(SWIDTH-1 downto 0) := "---";

  signal state, nxt: std_logic_vector(SWIDTH-1 downto 0);
begin
  sreg: vDFF generic map(SWIDTH) port map(clk, nxt, state);

  process(all) begin
    case? rst & a & state is
      when '1' & '-' & SX => q <= '0'; nxt <= SR;
      when '0' & '0' & SR => q <= '0'; nxt <= SR;
      when '0' & '1' & SR => q <= '0'; nxt <= S1;
      when '0' & '-' & S1 => q <= '1'; nxt <= S2;
      when '0' & '-' & S2 => q <= '0'; nxt <= S3;
      when '0' & '-' & S3 => q <= '0'; nxt <= S4;
      when '0' & '0' & S4 => q <= '0'; nxt <= S5;
      when '0' & '1' & S4 => q <= '0'; nxt <= S1;
      when '0' & '0' & S5 => q <= '0'; nxt <= SM;
      when '0' & '1' & S5 => q <= '0'; nxt <= S1;
      when '0' & '0' & SM => q <= '1'; nxt <= S2;
      when '0' & '1' & SM => q <= '1'; nxt <= SL;
      when '0' & '-' & SL => q <= '0'; nxt <= S2;
      when others => q <= '-'; nxt <= SX;
    end case?;
  end process;
end impl;
```

Figure 14.20. VHDL implementing the pulse-filler FSM of Example 14.2.

The code is shown in Figure 14.20. In this example we declare constants for the state machine inside the architecture body instead of placing them in a separate package. We declare two internal signals, one for the current state, called `state`, and one for the next state, called `nxt`. We instantiate a register to hold the state. A process and a matching `case` statement are used to encode the state table as combinational logic.

Summary

In this chapter you have added the dimension of time to your study of digital systems. By adding feedback to combinational logic circuits, we create *sequential* logic circuits: circuits whose outputs are a function not only of their inputs, but also of their present *state* and thus indirectly of their history. While combinational logic circuits are static, always producing the same output given the same input, the behavior of sequential logic circuits unfolds over time.

To tame potential races between state variables, *synchronous* sequential logic circuits include a clocked *flip-flop* or *register* in all feedback paths. This causes all state variables to be updated simultaneously on the rising edge of a *clock* signal. The state of a synchronous sequential logic circuit evolves in discrete steps. Each clock cycle, the state is updated to a value computed from the previous state and the inputs on the rising clock edge. This step-by-step behavior of synchronous sequential logic circuits makes them easy to analyze and design.

We design a synchronous sequential circuit, or *finite-state machine*, by starting with a *state diagram* or *state table* that describes the function of the machine. A *state assignment* assigns a unique bit pattern to each state of our machine. From the assignment and the diagram or table we can write down the combinational logic functions for the next state and output using the methods described earlier in this book.

We can implement finite-state machines in VHDL by *explicitly* instantiating a *state register* and then designing combinational logic using `case` or concurrent signal assignment statements to compute the next state and output. It is important to make sure the state register is initialized to a known state at reset.

BIBLIOGRAPHIC NOTES

One of the first descriptions of FSMs and sequential logic can be found in Huffman's "The synthesis of sequential switching circuits" [52]. Moore [83] and Mealy's [79] papers on sequential FSMs also provide background on the subject. For a recent exploration of FSM theory, refer to Kohavi's text in ref. [68]. Many textbooks, such as Brown's *Fundamentals of Digital Logic* [20], explore the hand synthesis of state machine logic.

Readers interested in the tangential topic of traffic lights can read two papers from 1927 [71, 76] discussing the use of this now ubiquitous invention.

Exercises

14.1 *Homing sequences, I.* The finite-state machine described by Table 14.2 does not have a reset input. Explain how you can get the machine in a known state regardless of its initial starting state by providing a fixed input sequence. An input sequence that always takes an FSM to the same state is called a *homing sequence*.

14.2 *Homing sequences, II.* Suppose the traffic-light controller FSM of Table 14.4 did not reset to state GNS. Find a homing sequence for the machine that will get it to state GNS.

14.3 *Modified traffic-light controller, I-I.* Modify the traffic-light controller FSM of Table 14.4 so that it makes lights go red in both directions for one cycle before turning a light green. Show a state table and state diagram for your new FSM.

14.4 *Modified traffic-light controller, I-II.* Choose a state assignment for the modified traffic-light controller of Exercise 14.3 and derive the logic to compute the next-state and output values. Show Karnaugh maps for the next-state variables and output variables and a gate-level schematic for the FSM.

14.5 *Modified traffic-light controller, I-III.* Write and verify the VHDL that implements your state machine from Exercise 14.3.

14.6 *Modified traffic-light controller, II-I.* Modify the traffic-light controller FSM of Table 14.4 so that it takes an additional input, *carns*, that indicates when there is a car waiting in the north–south direction. Change the logic so that, once the light has changed to east–west, it stays with east–west green until a car waiting in the north–south direction is detected. Show a state table and state diagram for your new FSM.

14.7 *Modified traffic-light controller, II-II.* Choose a state assignment for the modified traffic-light controller of Exercise 14.6 and derive the logic to compute the next-state and output values. Show Karnaugh maps for the next-state variables and output variables and a gate-level schematic for the FSM.

14.8 *Modified traffic-light controller, II-III.* Write and verify the VHDL that implements your state machine from Exercise 14.6.

14.9 *Modified traffic-light controller, III-I.* Modify the traffic-light controller FSM of Table 14.4 so that the FSM stays in state GEW as long as *carew* is true. Show a state table and state diagram for your new FSM.

14.10 *Modified traffic-light controller, III-II.* Choose a state assignment for the modified traffic-light controller of Exercise 14.9 and derive the logic to compute the next-state and output values. Show Karnaugh maps for the next-state variables and output variables and a gate-level schematic for the FSM.

14.11 *Modified traffic-light controller, III-III.* Write and verify the VHDL that implements your state machine from Exercise 14.9.

14.12 *Modified pulse filler I.* Modify the FSM of Example 14.1 so that it expects pulses on input *a* every six cycles, rather than five. Draw the state diagram for your modified FSM.

14.13 *Pulse-filler state table.* Write a state table for the FSM of Example 14.1.

14.14 *Pulse-filler state assignment.* Design a state assignment for the pulse-filler FSM of Example 14.1 using three state variables. The *R* state should have the encoding 000, and state 1 should have the encoding 001. Assign the remaining states so that as few state bits as possible change on each transition.

14.15 *Pulse-filler implementation.* Write equations for the next-state logic for the pulse-filler FSM of Example 14.1. Use the following state assignment: $R = 000$, $1 = 001$, $2 = 010$, $3 = 011$, $4 = 100$, $5 = 101$, $M = 110$, $L = 111$.

14.16 *Pulse-filler one-hot implementation.* Draw a schematic diagram for an implementation of the pulse-filler FSM of Example 14.1 using a one-hot state assignment.

14.17 *FSM implementation.* Implement the traffic-light controller FSM with a state encoding where GNS = 00, YNS = 01, GEW = 10, and YEW = 11. Show Karnaugh maps for the next-state variables and output variables and a gate-level schematic for the FSM.

14.18 *Digital lock, I.* Draw a state diagram and a state table for a digital lock. The lock has two inputs, *a* and *b*, and one output, *unlock*. The output is asserted only if the sequence *a, b, a, a* is observed. Each element of the sequence must last for one or more cycles, and there must be one or more cycles of both inputs low between the sequence elements. After unlocking, either input going high causes unlock to go low.

14.19 *Digital lock, II.* Implement, in VHDL, your digital lock state machine from Exercise 14.18.

14.20 *Basic vending machine, I.* Exercises 14.20–14.22 will focus on designing an FSM for a simple vending machine. This machine vends a single \$0.40 item and accepts only nickels and dimes. The input signals are *nickel* and *dime*, and the outputs are *vend* and *change*. The two input signals are pulsed high when a nickel (\$0.05) or dime (\$0.10) is inserted into the machine (only one will be high at a time). When enough money has been added, the *vend* signal goes high for a single cycle. If \$0.45 had been inserted, *change* also pulses high for one cycle. After vending an item, the state returns to the initial state where no money has been inserted. We will explore a more flexible vending machine in Section 16.3.1. Draw a state diagram and state table for this machine.

14.21 *Basic vending machine, II.* Using the state diagram from Exercise 14.20 and a binary state assignment, derive the output signals and next-state logic.

14.22 *Basic vending machine, III.* Implement the vending machine of Exercise 14.20 in VHDL. Show a waveform of the outputs and state when the user inserts consecutively two nickels, six dimes, and three nickels.

14.23 *Vending machine with quarters, I.* Modify your state table and diagram from Exercise 14.20 to include a *quarter* (\$0.25) input. Assume the vending machine will output one nickel for every cycle that *change* remains high.

14.24 *Vending machine with quarters, II.* Write the VHDL that implements the state machine of Exercise 14.23.

14.25 *Airplane indicator lights, I.* Draw the state diagram and table for an FSM that controls the seat-belt and no-electronics signs for a commercial airliner. The state machine has three inputs: *alt10k*, *alt25k*, and *smooth*. Whenever the airplane passes 10 000 (25 000) feet moving in either direction, *alt10k* (*alt25k*) will pulse high for one cycle. If the plane is not climbing, descending, or experiencing turbulence, the *smooth* signal will be set to high. The state machine should set the *noelectronics* signal to high when the plane is below 10 000 feet, and low otherwise. The *seatbelt* signal should be low only when the plane is above 25 000 feet and *smooth* has been asserted for at least five cycles. Assume that the plane is initially on the ground.

14.26 *Airplane indicator lights, II.* Using a one-hot state encoding, derive the logic needed to compute the next-state and output logic of your state diagram from Exercise 14.25.

14.27 *Airplane indicator lights, III.* Write the VHDL to implement your state machine from Exercise 14.25.

14.28 *Anti-lock brakes, I.* An FSM for an anti-lock brake system accepts two inputs (*wheel* and *time*), and generates a single output (*unlock*). The *wheel* input pulses high for one clock cycle each time the wheel rotates a small amount. The *time* input pulses high for one clock cycle every 10 ms. If the machine detects two *time* pulses since the last *wheel* pulse, it concludes that the wheel is locked, and *unlock* is asserted for one clock cycle to "pump" the brakes. After *unlock* goes high, the machine waits for two *time* pulses before resuming normal operation. Thus, there is a minimum of four *time* pulses between *unlock* pulses. Draw a state diagram (bubble diagram) for this state machine.

14.29 *Anti-lock brakes, II.* Write and verify the VHDL to implement the anti-lock brake state machine of Exercise 14.28.

14.30 *Direction sensor, I.* A direction sensor is used to detect the direction of a rotating gear. An input pulses each time a gear tooth is over the left side or right side of the sensor. This machine has two inputs, *il* and *ir*, and two outputs, *ol* and *or*. The FSM should output a one-cycle pulse on *ol* any time a high level on *il* for one or more cycles is followed, after zero or more cycles, by a high level on *ir* for zero or more cycles. Similarly, a one-cycle pulse on *or* is output if a high level on *ir* is followed by a high level on *il*. We have provided an example waveform in Figure 14.21. Draw the state diagram and table.

14.31 *Direction sensor, II.* Write and verify the VHDL to implement the direction sensor of Exercise 14.30.

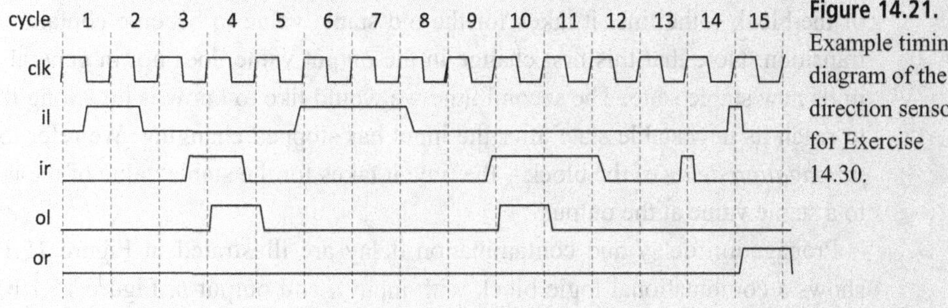

Figure 14.21. Example timing diagram of the direction sensor for Exercise 14.30.

15 Timing constraints

How fast will an FSM run? Could making our logic too fast cause our FSM to fail? In this chapter, we will see how to answer these questions by analyzing the timing of our finite-state machines and the flip-flops used to build them.

Finite-state machines are governed by two timing constraints – a maximum delay constraint and a minimum delay constraint. The maximum speed at which we can operate an FSM depends on two flip-flop parameters (the setup time and propagation delay) along with the maximum propagation delay of the next-state logic. On the other hand, the minimum delay constraint depends on the other two flip-flop parameters (hold time and contamination delay) and the minimum contamination delay of the next-state logic. We will see that if the minimum delay constraint is not met, our FSM may fail to operate at any clock speed due to hold-time violations. Clock skew, the delay between the clocks arriving at different flip-flops, affects both maximum and minimum delay constraints.

15.1 PROPAGATION AND CONTAMINATION DELAY

In a synchronous system, logic signals advance from the stable state at the end of one clock cycle to a new stable state at the end of the next clock cycle. Between these two stable states, they may go through an arbitrary number of transitions.

In analyzing timing of a logic block we are concerned with two times. First, we would like to know for how long the output retains its initial stable value (from the last clock cycle) after an input first changes (in the new clock cycle). We refer to this time as the *contamination delay* of the block – the time it takes for the old stable value to become contaminated by an input transition. Note that this first change in the output value does not in general leave the output in its new stable state. The second time we would like to know is how long it takes the output to reach its new stable state after the input has stopped changing. We refer to this time as the *propagation delay* of the block – the time it takes for the stable value of the input to propagate to a stable value at the output.

Propagation delay and contamination delay are illustrated in Figure 15.1. Figure 15.1(a) shows a combinational logic block with input a and output b. Figure 15.1(b) shows how the output b responds when input a changes state. Up to time t_1, both input a and output b are in their stable state from the last clock cycle. At time t_1 input a first changes. If a is a multi-bit signal, this is the time when the first bit of a to change state toggles – other bits may change at later times. Whether single-bit or multi-bit, t_1 is the time of the first transition on a. A given bit of a may toggle more than once before reaching its new stable state. At time t_2, a contamination delay of t_{cab} after t_1, this first change on a may affect output b, and b may change state. Up

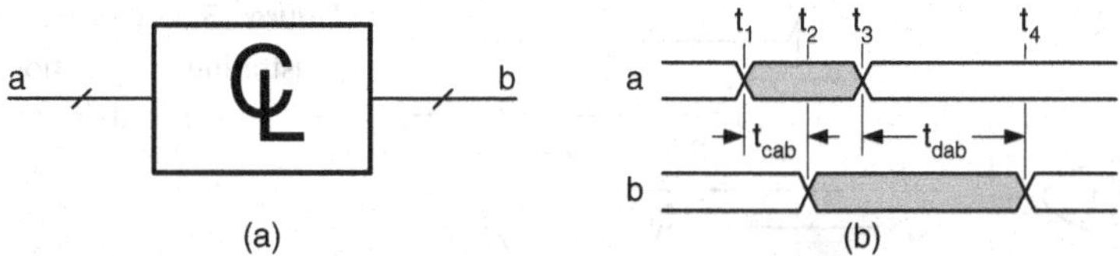

Figure 15.1. Propagation delay t_{dab} and contamination delay t_{cab}. The *contamination delay* of a logic block is the time from when the *first* input signal *first* changes to when the *first* output signal *first* changes. The *propagation delay* of a logic block is the time from when the *last* input signal *last* changes to when the *last* output signal *last* changes.

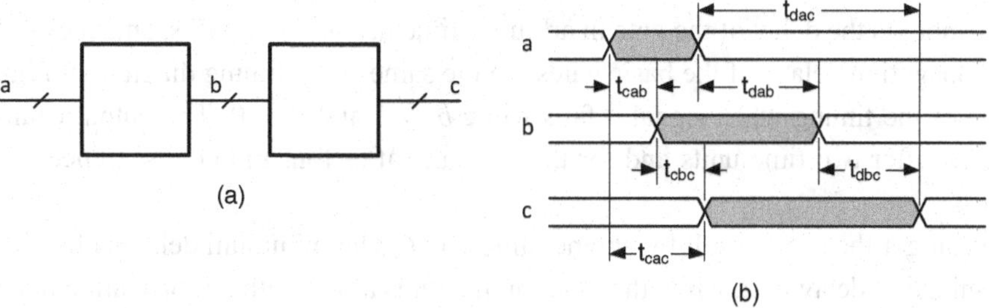

Figure 15.2. Propagation and contamination delay sum over a linear path. (a) Two modules in series with input a, intermediate signal b, and output c. (b) Timing diagram showing that $t_{cac} = t_{cab} + t_{cbc}$ and similarly for propagation delay, $t_{dac} = t_{dab} + t_{dbc}$.

until t_2, output b was guaranteed to have the steady-state value from the previous clock cycle. The first bit of b to change toggles for the first time at time t_2 as with a at t_1; this bit of b may toggle again before reaching a steady state, and other bits of b may change state later.

At time t_3, input a stops changing state. From t_3 until at least the end of the current clock cycle, signal a is guaranteed to be in its stable state. Time t_3 represents the time at which the last bit of a to toggle toggles for the last time. At time t_4, a propagation delay t_{dab} after t_3, the last change of input a has its final effect on output b. From this point to at least the end of the clock cycle, output b is guaranteed to be in its stable state for this clock cycle.

We denote a propagation (contamination) delay from a signal a to a signal b as t_{dab} (t_{cab}). The "d" or "c" in the subscript denotes propagation or contamination. The rest of the subscript gives the source signal and destination signal of the delay. That is, t_{dxy} is the delay starting with a transition on signal x to a transition on signal y.

As described in Section 5.1, we measure delay from the point where an input signal crosses 50% of its signal swing to the point where an output signal crosses 50% of its signal swing. Measuring delay in this manner lets us sum propagation and contamination delays over linear paths, as shown in Figure 15.2. The timing diagram in the figure shows that when two modules are composed in series, their delays sum:

$$t_{cac} = t_{cab} + t_{cac}, \tag{15.1}$$

$$t_{dac} = t_{dab} + t_{dac}. \tag{15.2}$$

To handle circuits with parallel paths, we simply enumerate all possible single-bit paths. The overall contamination delay is the *minimum* contamination delay over all paths, and the overall propagation delay is the *maximum* propagation delay over all paths.

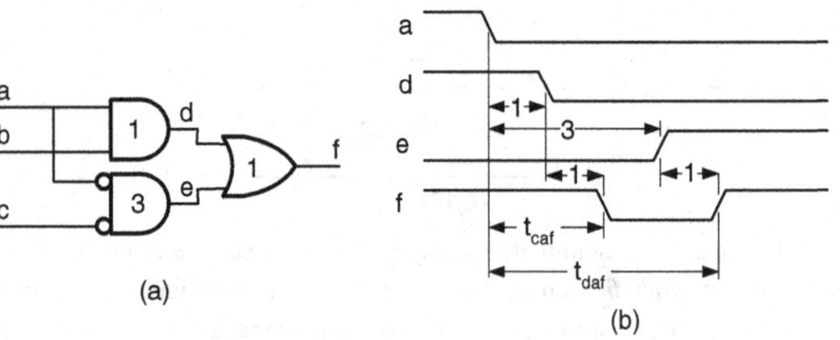

Figure 15.3. Circuit with a hazard illustrating propagation and contamination delay.

Figure 15.3(a) shows a circuit with a static-1 hazard (recall Section 6.10). The value in each gate symbol is the delay of the gate in arbitrary time units. (Here we assume the contamination and propagation delays of the basic gates are the same.) The timing diagram in Figure 15.3(b) illustrates the timing when signal a falls while $b = 1$ and $c = 0$. The output changes for the first time after two time units and for the last time after four time units. Hence, $t_{caf} = 2$ and $t_{daf} = 4$.

We can get the same result by enumerating paths. The minimum delay path is *a–d–f* with a contamination delay of 2, while the maximum path is *a–e–f* with a propagation delay of 4.

Contamination and propagation delay are independent of input state. The contamination delay of the circuit in Figure 15.3(a) from a to f is 2, regardless of the state of signals b and c. This delay represents the possibility that the output *may* change two time units after a transition on a, not a guarantee that it will change.

Many people confuse contamination delay with minimum propagation delay. They are not the same thing. The minimum propagation delay is the minimum value (over some range of parameters: voltage, temperature, process variation, input combinations) of the time for the correct steady-state value to appear on the output of a circuit after a transition on the input. In contrast, the contamination delay is the time until the output first changes from its old steady-state value after a transition on its input. These are not the same thing. The transition that sets the contamination delay is not, in general, a change of the output to its steady-state value, but rather a change to some intermediate value – for example, the transition of a to 0 in the hazard of Figure 15.3.

EXAMPLE 15.1 Propagation and contamination delay

Compute the propagation and contamination delays from input a to output q in the circuit shown in Figure 15.4. The numbers above each gate give its delay in picoseconds.

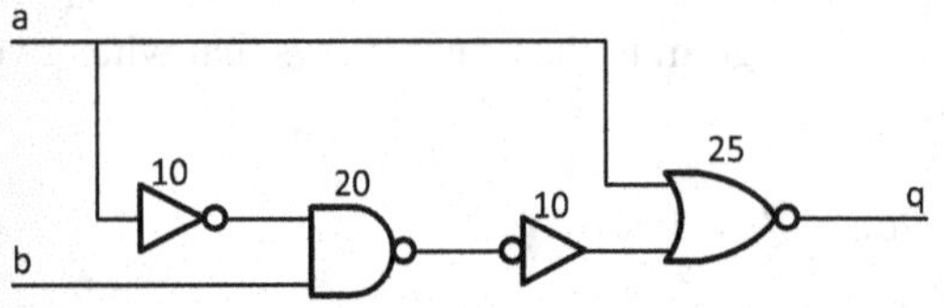

Figure 15.4. Circuit for the delay calculation in Example 15.1.

The minimum delay is the path directly from a to the NOR gate to q. Thus, $t_{caq} = 25$ ps. The maximum delay includes the two inverters and the NAND gate, giving $t_{daq} = 65$ ps.

15.2 THE D FLIP-FLOP

The timing constraints that determine whether an FSM will operate or not, and at what speed it will operate, are governed by the clocked storage elements used to construct the FSM – in our case, the D flip-flop. A schematic symbol for a D flip-flop is shown in Figure 15.5. A multi-bit D flip-flop is sometimes called a *register*. Here we consider the D flip-flop to be a *black box* – that is, we look at its external behavior without looking inside to see how this behavior is achieved. We defer exploring inside the D flip-flop until Chapter 27.

A D flip-flop samples its input on the rising edge of the clock signal and updates its output with the value sampled. This sampling and update is illustrated in the timing diagram of Figure 15.5(b). For the sampling to take place correctly, the input data (shown in the top waveform of the timing diagram) must be stable for a period before and after the rising edge of the clock. Specifically, the data must have reached its correct value (labeled x in the figure) at least a *setup time* t_s before the clock reaches its 50% point, and the data must be held stable at this value until a *hold time* t_h after the clock has reached its 50% point.[1] During the gray areas in the data waveform, D can take on any value. However, it must remain stable with a value of x during the setup- and hold-intervals for the flip-flop to sample the value x correctly.

If the input meets its setup- and hold-time constraints, the flip-flop will update the output with the sampled value x, as shown on the bottom waveform of Figure 15.5(b). The old value (which was sampled on the previous rising edge of the clock) will remain stable on the output until a *contamination delay* t_{cCQ} after the rising edge of the clock. The circuit may continue to rely on the old value being stable up until this point in time. After the contamination delay the output of the flip-flop may change, but not necessarily to the correct value. This period where the output value is not guaranteed is shaded gray in the figure. At a *propagation delay* t_{dCQ} after the rising edge of the clock, the output is guaranteed to have the value x sampled from the input. It will then hold this value stable until t_{cCQ} after the next rising edge of the clock.

15.3 SETUP- AND HOLD-TIME CONSTRAINTS

Now that we have introduced the nomenclature, the timing constraints on a finite-state machine are quite simple. To ensure that the clock cycle t_{cy} is long enough for the longest path to satisfy

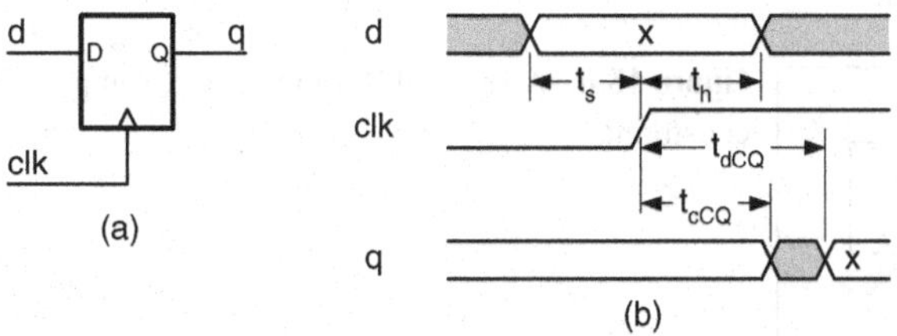

Figure 15.5. D flip-flop: (a) schematic symbol; (b) timing diagram. The D flip-flop samples its input on the rising edge of the clock and updates the output with the value sampled. For correct sampling the input must be *stable* from t_s before the clock rises to t_h after the clock rises. The output may change as soon as t_{cCQ} after the clock. The output takes on the correct value no later than t_{dCQ} after the clock.

[1] Note that t_s or t_h may be negative, but $t_s + t_h$ will always be positive.

the setup time of the D flip-flop, we must satisfy the following:

$$t_{cy} \geq t_{dCQ} + t_{dMax} + t_s, \tag{15.3}$$

where t_{dMax} is the maximum propagation delay from the output of a D flip-flop to the input of a D flip-flop.

We also must ensure that no signal is contaminated so quickly as to violate the hold-time constraint on the input of a D flip-flop by satisfying

$$t_h \leq t_{cCQ} + t_{cMin}, \tag{15.4}$$

where t_{cMin} is the minimum contamination delay from the output of a D flip-flop to the input of a D flip-flop.

The two constraints (15.3) and (15.4) govern system timing. The setup-time constraint (15.3) determines performance by giving the minimum cycle time t_{cy} at which the circuit will operate. The hold-time constraint, on the other hand, is a correctness constraint. If Equation (15.4) is violated, the circuit may not meet its hold time constraint – and hence may malfunction – regardless of the cycle time.

Figure 15.6 shows a simple finite-state machine that we shall use to illustrate setup- and hold-time constraints. The FSM consists of two flip-flops. The upper flip-flop generates state bit *a* that propagates through a maximum-length (maximum propagation delay) logic path (Max) to generate signal *b*. Signal *b* is in turn sampled by the lower flip-flop. The lower flip-flop generates signal *c*, which propagates through a minimum-length (minimum contamination delay) block to generate signal *d*, which is sampled by the upper flip-flop. Note that the destination flip-flop of a minimum-delay path is not necessarily the source flip-flop of a maximum-delay path (and vice versa). In general, we need to test all possible paths from all flip-flops to all flip-flops to find the minimum and maximum paths.

The maximum-delay path from the upper flip-flop to the lower flip-flop of Figure 15.6 stresses the setup time of the lower flip-flop. If this path is too slow, the next clock edge may arrive at the lower flip-flop before its input signal *b* has settled at its final value for the cycle. Figure 15.7 repeats Figure 15.6 with this path highlighted. A timing diagram corresponding to this path is shown in Figure 15.8. Suppose the rising edge of the clock samples value *x* on signal *d*, then, after a flip-flop propagation delay t_{dCQ}, flip-flop output *a* will take on value *x* and hold this value through the remainder of the clock cycle. Signal *a* is input to combinational block

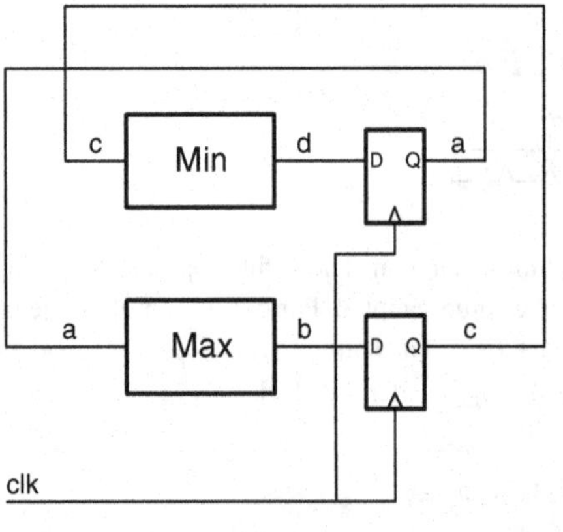

Figure 15.6. A simple FSM to illustrate setup and hold constraints.

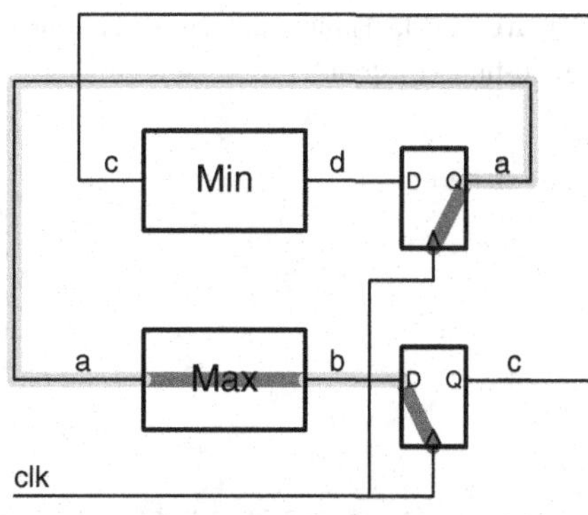

Figure 15.7. Setup-time constraint. The maximum path from the clock on a source flip-flop to the clock on a destination flip-flop is shaded. From the rising edge of the clock, the signal must propagate to the Q output of the flip-flop (t_{dCQ}) and propagate through the maximum-delay logic path (t_{dab}) at least a setup time (t_s) before the next clock edge.

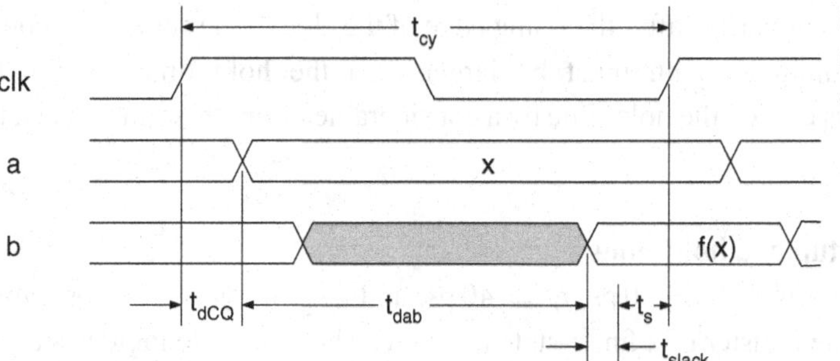

Figure 15.8. Timing diagram illustrating the setup-time constraint.

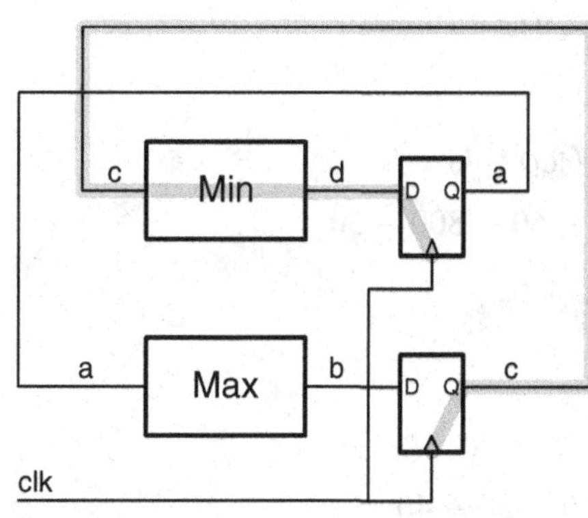

Figure 15.9. Hold-time constraint. The minimum contamination delay path from the clock on a source flip-flop to the clock on a destination flip-flop is shaded. From the rising edge of the clock, the contamination delay must be long enough for signal d to remain stable until a hold time t_h after this clock edge.

Max, which generates signal b. After an additional propagation delay from a to b, t_{dab} (which corresponds to t_{dMax} in constraint (15.3)), signal b takes on its final value for the clock cycle $f(x)$. Signal b must settle at this final value at least t_s before the rising edge of the next clock for constraint (15.3) to be satisfied. The sum of the propagation delays along the maximum path and the setup time must be less than the cycle time. In the timing diagram, signal b settles slightly early, leaving a timing margin, or *slack* time, of t_{slack}. The clock cycle t_{cy} could be reduced by t_{slack} and the setup constraint would still be met.

The minimum-delay path from the lower flip-flop to the upper flip-flop of Figure 15.6 stresses the hold time of the upper flip-flop. If this path is too fast, signal d might change before a hold time after the rising edge of the clock. Figure 15.9 shows this timing path highlighted. A timing diagram illustrating the signals along this path is shown in Figure 15.10. A flip-flop contamination delay t_{cCQ} after the rising edge of the clock, signal c may first change. A contamination delay of the logic block t_{ccd} (which corresponds to t_{cMin} in constraint (15.4))

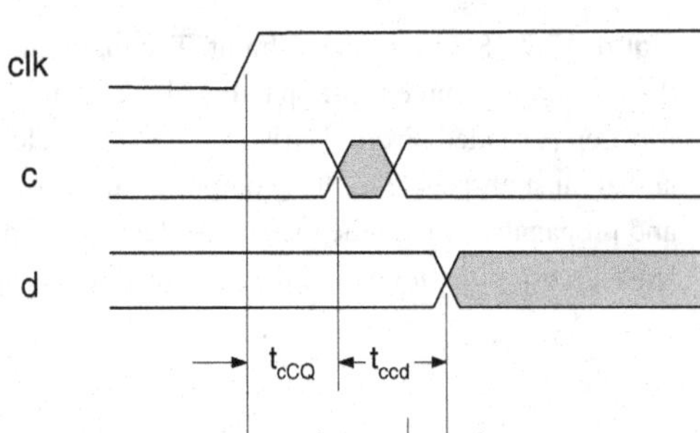

Figure 15.10. Timing diagram illustrating the hold-time constraint.

later, signal d may change. For the hold-time constraint to be satisfied, this first change on signal d is not allowed to occur until t_{h} after the rising edge of the clock. The sum of the contamination delays along the minimum path must be larger than the hold time. In the figure, the contamination delays exceed the hold time by a considerable timing margin or slack time, t_{slack}.

EXAMPLE 15.2 Setup and hold time

Consider a D flip-flop with $t_{\mathrm{s}} = 50$ ps, $t_{\mathrm{h}} = 40$ ps, and $t_{\mathrm{c}CQ} = t_{\mathrm{d}CQ} = 60$ ps, which is used to implement the state register of a finite-state machine. The next-state logic has a propagation delay of $t_{\mathrm{d}} = 800$ ps and a contamination delay of $t_{\mathrm{c}} = 50$ ps. Compute the setup- and hold-time *slack* for this arrangement operating at $f_{cy} = 1$ GHz.

For the setup constraint, we write

$$\begin{aligned} t_{\mathrm{sslack}} &= t_{\mathrm{cy}} - t_{\mathrm{d}CQ} - t_{\mathrm{d}} - t_{\mathrm{s}} \\ &= 1000 - 60 - 800 - 50 \\ &= 90 \text{ ps.} \end{aligned}$$

The hold-time slack is given by

$$\begin{aligned} t_{\mathrm{hslack}} &= t_{\mathrm{c}CQ} + t_{\mathrm{c}} - t_{\mathrm{h}} \\ &= 60 + 50 - 40 \\ &= 70 \text{ ps.} \end{aligned}$$

15.4 THE EFFECT OF CLOCK SKEW

On an ideal chip, the clock signal would change at the input of all flip-flops at the same time. In practice, device variations and wire delays in the clock distribution network cause the timing of the clock signal to vary slightly from flip-flop to flip-flop. We refer to this spatial variation in clock timing as *clock skew*. Clock skew adversely affects both the setup- and hold-time constraints. With a skew of t_{k}, these two constraints become

$$t_{\mathrm{cy}} \geq t_{\mathrm{d}CQ} + t_{\mathrm{dMax}} + t_{\mathrm{s}} + t_{\mathrm{k}} \quad (15.5)$$

and

$$t_{\mathrm{h}} \leq t_{\mathrm{c}CQ} + t_{\mathrm{cMin}} - t_{\mathrm{k}}. \quad (15.6)$$

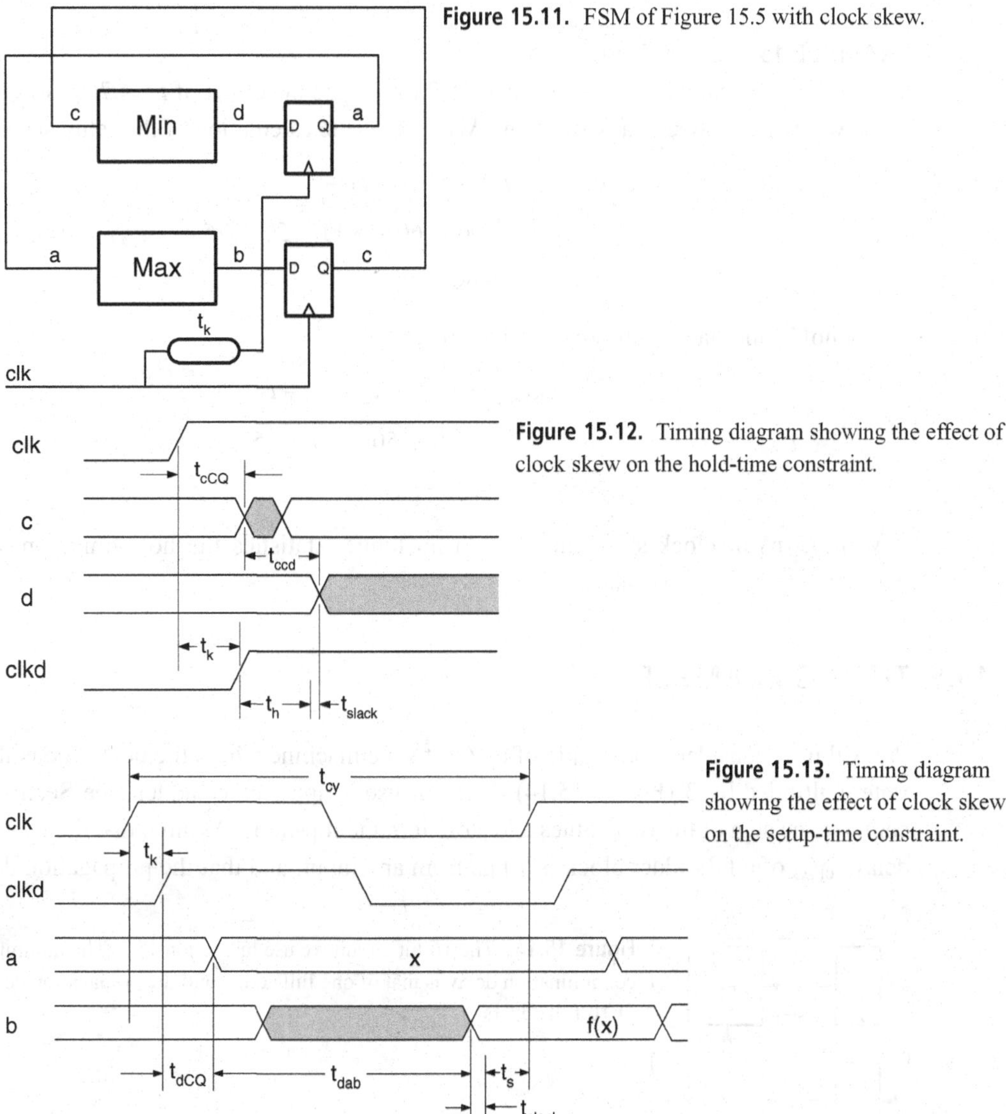

Figure 15.11. FSM of Figure 15.5 with clock skew.

Figure 15.12. Timing diagram showing the effect of clock skew on the hold-time constraint.

Figure 15.13. Timing diagram showing the effect of clock skew on the setup-time constraint.

Figure 15.11 shows the FSM of Figure 15.5 with clock skew added. A delay line (the oval-shaped block) with a delay of t_k (the magnitude of the skew) is connected between the clock input and the clock to the upper flip-flop. Hence each edge of the clock arrives at the upper flip-flop a time t_k later than it arrives at the lower flip-flop. Delaying the clock to the source of the maximum-length path causes this path effectively to get longer. In a similar manner, delaying the clock to the destination of the minimum-length path effectively makes this path shorter.

The effect of skew on the minimum-length path, and hence on the hold-time constraint, is shown in Figure 15.12. The contamination delays from the clock to c to d add as above. However, now signal d must stay stable until t_h after the delayed clock *clkd* or $t_h + t_k$ after the original clock *clk*. The effect is the same as increasing the hold time by t_k.

The timing diagram of Figure 15.13 illustrates the effect of clock skew on the setup-time constraint. Delaying the clock to the upper flip-flop delays the transition of a by t_k, effectively adding t_k to the maximum path.

EXAMPLE 15.3 Clock skew

Repeat the slack calculation of Example 15.2 to include the effect of $t_k = 75$ ps of clock skew.

Skew always reduces slack (margin). With skew, the calculation of the setup slack becomes

$$\begin{aligned} t_{\text{sslack}} &= t_{\text{cy}} - t_{\text{d}CQ} - t_{\text{d}} - t_{\text{s}} - t_{\text{k}} \\ &= 1000 - 60 - 800 - 50 - 75 \\ &= 15\,\text{ps}. \end{aligned}$$

The hold-time slack with skew is given by

$$\begin{aligned} t_{\text{hslack}} &= t_{\text{c}CQ} + t_{\text{c}} - t_{\text{h}} - t_{\text{k}} \\ &= 60 + 50 - 40 - 75 \\ &= -5\,\text{ps}. \end{aligned}$$

With 75 ps of clock skew, the system no longer statisfies the hold-time constraint – as indicated by the negative slack.

15.5 TIMING EXAMPLES

We will now consider an example of a 16-bit state machine where the next state is the current state multiplied by 3 (Figure 15.14). We will use a ripple-carry adder (see Section 10.2) to compute the sum of the two values and store it in the flip-flops. Assume that the contamination delay, t_{cFA}, of a full-adder block is 10 ps from any input, and that the propagation delay, t_{dFA},

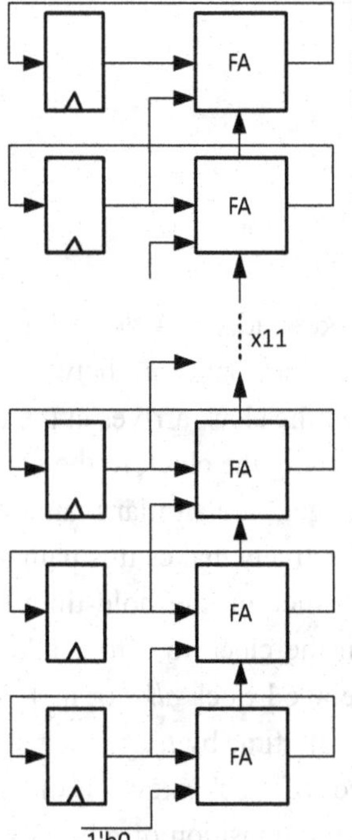

Figure 15.14. The 16-bit circuit we use in Section 15.5. The minimum logic contamination delay is that of one full adder, and the propagation delay is that of 16 full adders.

is 30 ps. The minimum contamination delay between flip-flops is 10 ps, and the maximum propagation delay is $16t_{\mathrm{dFA}} = 480\,\mathrm{ps}$.

Assume that the flip-flops have a $t_{\mathrm{c}CQ}$ and $t_{\mathrm{d}CQ}$ of 10 ps and 20 ps, respectively. The flip-flops also have a 20 ps setup time and 10 ps hold time. First, we will check that our circuit meets hold-time constraints:

$$t_{\mathrm{h}} \le t_{\mathrm{c}CQ} + t_{\mathrm{cFA}},$$

$$10\,\mathrm{ps} \le 10\,\mathrm{ps} + 10\,\mathrm{ps}.$$

Next, we compute the maximum cycle time:

$$t_{\mathrm{cy}} \ge t_{\mathrm{d}CQ} + t_{\mathrm{d}} + t_{\mathrm{s}},$$

$$t_{\mathrm{cy}} = 20 + 480 + 20 = 520\,\mathrm{ps}.$$

These equations are still valid even if any of the flip-flop parameters are negative. The steps to perform skew-free timing analysis remain the same in any situation: find the minimum and maximum logic delay, and check Equations (15.3) and (15.4).

To build a more robust circuit, we wish to check that we have no timing violations, even in the presence of clock skew of up to 20 ps. This clock skew can be in either direction between any two flip-flops. Again, we begin with the hold time:

$$t_{\mathrm{h}} \le t_{\mathrm{c}CQ} + t_{\mathrm{cFA}} - t_{\mathrm{k}},$$

$$10 \le 10 + 10 - 20.$$

We have an error: hold time has been violated. Unlike setup violations, we cannot simply increase our cycle time to fix the problem. We must redesign the flip-flop, modify the clock distribution to have less skew, or add extra logic at the input of the flip-flop. The easiest solution, and the one we choose, is to add extra logic. To do so, we must insert logic with a *contamination* delay of 10 ps immediately before the input of each flip-flop (or immediately after the output).

We must include this new delay in our cycle time computation (we assume the propagation and contamination delay of this logic to be equal):

$$t_{\mathrm{cy}} \ge t_{\mathrm{d}CQ} + t_{\mathrm{d}} + t_{\mathrm{extra}} + t_{\mathrm{s}} + t_{\mathrm{k}},$$

$$t_{\mathrm{cy}} = 20 + 480 + 10 + 20 + 20 = 550\,\mathrm{ps}.$$

Our example circuit is able to run at a frequency of 1.8 GHz.

15.6 TIMING AND LOGIC SYNTHESIS

For every design, the timing analysis we performed in Section 15.6 must be repeated over all logic paths between all possible combinations of flip-flops at every operating condition.[2] With a small number of gates, timing analysis performed by hand is time-consuming and error-prone. Timing checks become near impossible to do by hand with large amounts of logic. Thankfully, synthesis and timing tools do this analysis for us.

2 Hold-time violations are more frequent with best-case logic delay: high voltage and cold temperature. Setup-time violations are most common in worst-case conditions with a low operation voltage and a hot chip.

```
set top pc_28bit_top
set src_files [list\
 ./rtl/pc_28bit_top.vhd\
 ./rtl/pc_28bit.vhd ]
read_vhdl -vhdl2008 ${src_files}
current_design ${top}
# Clocks
set clk_name   CLK
set clk_period 1
create_clock -name ${clk_name} -period ${clk_period} \
 [get_ports ${clk_name}]

set_input_delay .2 -clock CLK $all_inputs
set_output_delay .5 -clock CLK $all_outputs
```

Figure 15.15. A section of a Tcl script that sets up a clock with a 1 ns period. We also set the input and output delay of the system to be 200 ps and 500 ps, respectively.

Logic synthesis tools, which transform VHDL into logic gates, have timing models for each gate. Given an operating condition, logic, and constraints, this tool will determine whether timing has been violated on any path. If the tool finds a timing violation, it will try to replace the combinational logic with an implementation that meets the timing constraints. Instead of always starting with the fastest implementation of a design, tools use an iterative approach. The tool first generates minimum area (or energy) logic. The tool then generates faster implementations only for the paths that violate setup time. For example, a VHDL addition may first be implemented as a ripple-carry adder. If the ripple-carry adder is too slow, it will be replaced by a faster, larger carry-look-ahead adder.

The standard-cell library is characterized to provide timing models, including flip-flop constraints, for each cell at each of several *process corners*.[3] The designer is responsible for specifying the name of the clock signal, desired cycle time, and input/output delays in a *constraint file*. An example script, written in the Tcl language, specifying these constraints and invoking a synthesis tool, is shown in Figure 15.15. The script imports the RTL files – a state machine for controlling a program counter – sets the clock name and period (ns), creates the clock, and specifies the input and output delays. The output of the synthesis tool includes a listing of all logic paths that violate timing. For example, Figure 15.16 shows a failing path from our example. This path runs from bit 11 to bit 26 of the PC and does not arrive within the setup window of the flip-flop.

In addition to the timing analysis performed by the synthesis tool, a separate static timing analysis (STA) tool is typically used to verify that the final design – including interconnect parasitics – meets all timing constraints.

We can perform VHDL simulations with timing models. However, such simulations can find timing errors only if the input vectors (the cases simulated) activate the critical path. Because it is very difficult to *prove* that all paths have been tested, timing simulation is not sufficient to

[3] These models and other characterization data are provided by the supplier of the standard-cell library. Characterizing a standard-cell library often requires more effort than designing the library.

```
Des/Clust/Port      Wire Load Model        Library
  ------------------------
  pc_28bit_top        area_1Kto2K            CORE

  Point                                    Incr       Path
  ------------------------------
  clock CLK (rise edge)                    0.00       0.00
  clock network delay (ideal)              0.00       0.00
  I_pc_28bit/PC_reg[11]/CP (DFPQX9)        0.00       0.00 r
  I_pc_28bit/PC_reg[11]/Q (DFPQX9)         0.20       0.20 f
  U382/Z (BFX53)                           0.08       0.28 f
  U197/Z (NAND2X7)                         0.06       0.34 r
  U458/Z (OAI12X18)                        0.05       0.39 f
  U267/Z (AOI21X12)                        0.03       0.42 r
  U265/Z (OAI21X12)                        0.03       0.45 f
  U257/Z (IVX18)                           0.06       0.51 r
  U256/Z (AND2X35)                         0.10       0.61 r
  U358/Z (NAND2X7)                         0.05       0.66 f
  U628/Z (XNOR2X18)                        0.10       0.76 f
  U430/Z (NAND2X14)                        0.05       0.81 r
  U533/Z (NAND3X13)                        0.11       0.92 f
  I_pc_28bit/PC_reg[26]/D (DFPQX9)         0.00       0.92 f
  data arrival time                                    0.92

  clock CLK (rise edge)                    1.00       1.00
  clock network delay (ideal)              0.00       1.00
  I_pc_28bit/PC_reg[26]/CP (DFPQX9)        0.00       1.00 r
  library setup time                      -0.12       0.88
  data required time                                  0.88
  ----------------------------
  data required time                                  0.88
  data arrival time                                  -0.92
  ----------------------------
  slack (VIOLATED)                                   -0.04
```

Figure 15.16. An example path that violates setup constraints after synthesis. This particular path, one of many failed paths, runs from the 11th bit of the program counter to the 26th bit. Often, the names of the logic signals are ambiguous when converting from behavioral VHDL to gates.

verify that a chip meets all timing constraints. Static timing analysis, which finds all paths that violate timing constraints, is required in order to prove that a chip meets timing.

Summary

In this chapter you have learned how to analyze the timing of synchronous logic circuits. The timing of a combinational logic module is described by two numbers. The *contamination delay*, t_{cab}, of a circuit is the time from any change on input a until the previous steady-state value is changed or contaminated on output b. Contamination delay is important for analyzing hold-time violations. The *propagation delay*, t_{dab}, of a circuit is the time from the last change on input a to the output b settling at its steady-state value for the remainder of the clock cycle. Propagation delay is used to analyze setup-time violations.

The timing of a D flip-flop or register is governed by four numbers. The input to the register must be stable a *setup time* t_s before the rising edge of the clock and must remain stable until a *hold time* t_h after the rising edge of the clock. If the setup- and hold-time constraints of a register are met, the output of the register will be contaminated (the old value is no longer stable) a contamination delay t_{ccq} after the rising edge of the clock, and will be stable with the new value a propagation delay t_{dcq} after the rising edge of the clock. For most flip-flops, $t_{ccq} = t_{dcq}$, and the flip-flop changes to its final state in one step.

From these timing properties of combinational logic and registers, we can derive that the hold-time constraint of a register will be met if

$$t_h \leq t_{cCQ} + t_{cMin} - t_k,$$

and that the setup-time constraint of a register will be met if

$$t_{cy} \geq t_{dCQ} + t_{dMax} + t_s + t_k,$$

where t_k is the clock skew, t_{cMin} is the minimum contamination delay over all combinational paths, and t_{dMax} is the maximum propagation delay over all combinational paths.

In practice, we use *static timing analysis* tools to verify that our designs meet setup and hold constraints. These tools compute the delays of all possible combinational paths in a design and check that the setup- and hold-time constraints are met for every flip-flop.

BIBLIOGRAPHIC NOTES

Dally and Poulton [33] and Weste and Harris [112] detail timing analysis with both our simple flip-flop-based clocking scheme and more complex latch-based schemes. A description of the algorithms used in synthesis and optimization tools is presented in ref. [80]. A recent book from Brunvand [21] details the design process, including using commercial timing analysis tools.

Exercises

15.1 *Propagation and contamination delays, I.* Calculate the propagation and contamination delays of each input to the output in Figure 15.17. Assume that each gate has a 10 ps delay.

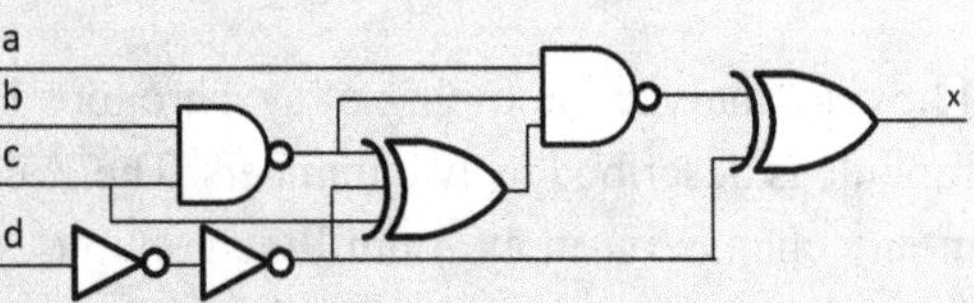

Figure 15.17. The simple combinational circuit used in Exercise 15.1.

15.2 *Propagation and contamination delays, II.* Compute the contamination and propagation delays of the circuit in Figure 15.18 from flip-flop A to the output. Assume that the delay through each gate is 10 ps.

15.3 *Propagation and contamination delays, III.* Compute the contamination and propagation delays of the circuit in Figure 15.18 from flip-flop B to the output. Assume that the delay through each gate is 10 ps.

15.4 *Propagation and contamination delays, IV.* Compute the contamination and propagation delays of the circuit in Figure 15.18 from flip-flop C to the output. Assume that the delay through each gate is 10 ps.

15.5 *Propagation and contamination delays, V.* Compute the contamination and propagation delays of the circuit in Figure 15.18 from flip-flop D to the output. Assume that the delay through each gate is 10 ps.

15.6 *Propagation and contamination delays, VI.* What are the overall contamination and propagation delays of the circuit in Figure 15.18? Assume that the delay through each gate is 10 ps.

15.7 *Setup and hold, I.* For flip-flop A in Table 15.1, draw a waveform showing the flip-flop's input going high just before t_s and low just after t_h. Include *clk*, *D*, and *Q* (previously 0). Label all constraints on your diagram.

15.8 *Setup and hold, II.* For flip-flop B in Table 15.1, draw a waveform showing the flip-flop's input going high just before t_s and low just after t_h. Include *clk*, *D*, and *Q* (previously 0). Label all constraints on your diagram.

Table 15.1. Three different flip-flop specifications that we use throughout the exercises

Parameter (ps)	FF A	FF B	FF C
t_s	20	100	−30
t_h	10	−20	80
t_{cCQ}	10	2	40
t_{dCQ}	20	30	50

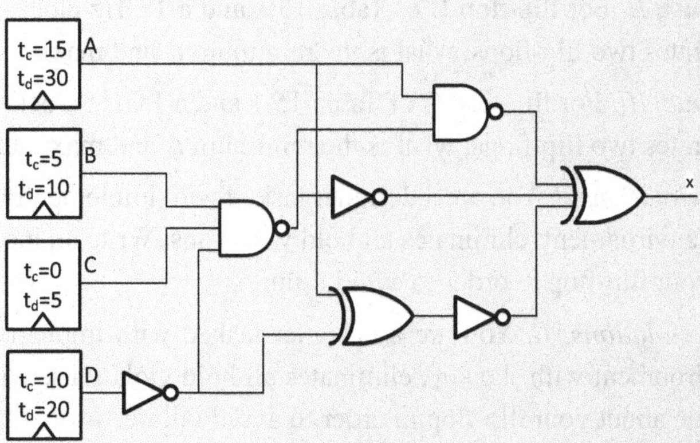

Figure 15.18. The simple circuit used in Exercises 15.2–15.6. Labels for each flip-flop are in picoseconds, and each gate has a 10 ps delay.

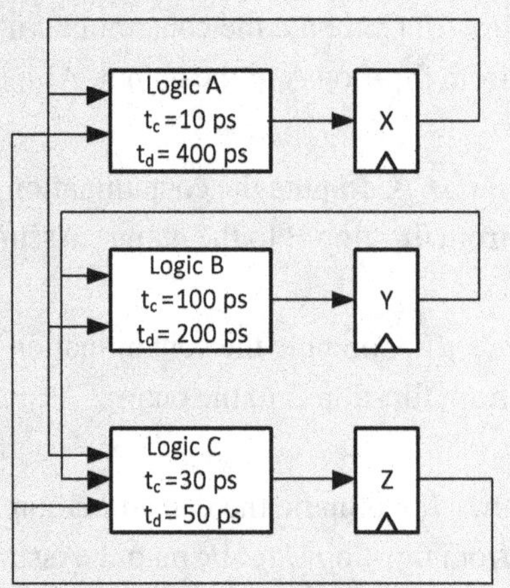

Figure 15.19. The simple logic circuit used in Exercises 15.10–15.12 and 15.22–15.24.

15.9 *Setup and hold, III.* For flip-flop C in Table 15.1, draw a waveform showing the flip-flop's input going high just before t_s and low just after t_h. Include *clk*, *D*, and *Q* (previously 0). Label all constraints on your diagram.

15.10 *Setup and hold violations, I.* For flip-flop A of Table 15.1 and a 2 GHz clock, check Figure 15.19 for timing violations. If there is a hold violation, indicate where delay must be added, specify the necessary delay, and recheck for setup violations. If there is a setup violation, calculate the maximum error-free frequency.

15.11 *Setup and hold violations, II.* For flip-flop B of Table 15.1 and a 2 GHz clock, check Figure 15.19 for timing violations. If there is a hold violation, indicate where delay must be added, specify the necessary delay, and recheck for setup violations. If there is a setup violation, calculate the maximum error-free frequency.

15.12 *Setup and hold violations, III.* For flip-flop C of Table 15.1 and a 2 GHz clock, check Figure 15.19 for timing violations. If there is a hold violation, indicate where delay must be added, the necessary delay, specify and recheck for setup violations. If there is a setup violation, calculate the maximum error-free frequency.

15.13 *Logic constraints, I.* For flip-flop A of Table 15.1 and a 1 GHz clock, for any block of logic that separates two flip-flops, what is the minimum t_c and maximum t_d?

15.14 *Logic constraints, II.* For flip-flop B of Table 15.1 and a 1 GHz clock, for any block of logic that separates two flip-flops, what is the minimum t_c and maximum t_d?

15.15 *Logic constraints, III.* For flip-flop C of Table 15.1 and a 1 GHz clock, for any block of logic that separates two flip-flops, what is the minimum t_c and maximum t_d?

15.16 *Avoiding hold violations, I.* You are a designer tasked with implementing a flip-flop that, in a skew-free environment, eliminates all hold violations. Write an inequality that must be true about your flip-flop in order to avoid failure.

15.17 *Avoiding hold violations, II.* You are a designer tasked with implementing a flip-flop that, in an environment with skew t_k, eliminates all hold violations. Write an inequality that must be true about your flip-flop in order to avoid failure.

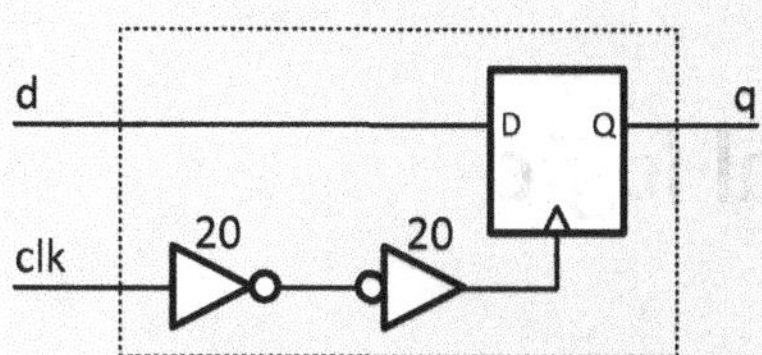

Figure 15.20. An *outer* flip-flop is made by inserting two inverters with a combined delay of 40 ps between the clock input and an *inner* flip-flop. This figure is used by Exercise 15.19.

15.18 *Post-fabrication violations.* After taping out a chip, it comes back from the fabrication plant and does not work. How can you test whether the failure is either a setup violation or a hold violation? Describe the tests to run.

15.19 *Clock delay.* Suppose the *inner* flip-flop in Figure 15.20 has $t_s = t_h = 50$ ps and $t_{dCQ} = t_{cCQ} = 80$ ps. What are t_s, t_h, t_{dCQ}, and t_{cCQ} for the *outer* flip-flop.

15.20 *Data delay.* Repeat Exercise 15.19, but instead of having two inverters on the *clk* input, place them on the *d* input.

15.21 *Output delay.* Repeat Exercise 15.19, but instead of having two inverters on the *clk* input, place them on the *q* output.

15.22 *Clock skew, I.* Calculate the maximum allowable clock skew in Figure 15.19. Indicate between which pair of flip-flops the skew occurs and whether it will trigger a setup or hold violation. Use flip-flop specification A and a 2 ns clock. You may want to enumerate all possible paths and write the setup and hold equations.

15.23 *Clock skew, II.* Repeat Exercise 15.22, instead using flip-flop specification B and a 2 ns clock.

15.24 *Clock skew, III.* Repeat Exercise 15.22, instead using flip-flop specification C and a 4 ns clock.

15.25 *Timing analysis.* Multiplying two 64-bit numbers gives a 128-bit result.

(a) Assuming the inputs and outputs are terminated by flip-flops, how many combinations of start and end flip-flops exist in the system?

(b) It takes 30 s to check for a violation on a register–register path by hand. If you wanted to check for setup and hold violations on each of these paths in nine different processes, estimate the time it would take to perform a complete timing analysis on the multiplier.

16 Datapath sequential logic

In Chapter 14 we saw how a finite-state machine can be synthesized from a state diagram by writing down a table for the next-state function and synthesizing the logic that realizes this table. For many sequential functions, however, the next-state function can be more simply described by an expression rather than by a table. Such functions are more efficiently described and realized as *datapaths*, where the next state is computed as a logical function, often involving arithmetic circuits, multiplexers, and other building block circuits.

16.1 COUNTERS

16.1.1 A simpler counter

Suppose you want to build a finite-state machine with the state diagram shown in Figure 16.1. This circuit is forced to state 0 whenever input r is true. Whenever input r is false, the machine counts through the states from 0 to 31 and then cycles back to 0. Because of this counting behavior, we refer to this finite-state machine as a *counter*.

We could design the counter employing the methodology developed in Chapter 14. A VHDL description taking this approach for a three-bit counter (eight states) is shown in Figure 16.2.[1] A three-bit wide bank of flip-flops holds the current state, `count`, and updates it from the next state, `nxt`, on each rising edge of the clock. The matching `case` statement captures the state table, specifying the next state for each input and current state combination.

While this method for generating counters works, it is verbose and inefficient. The lines of the state table are repetitive. The behavior of the machine can be captured entirely by the single line

```
nxt <= (others => '0') when rst else count+1;
```

We use array aggregate notation, "`(others => '0')`", to specify a `std_logic_vector` value with all elements equal to `'0'`. This is a *datapath* description of the finite-state machine in which we specify the next state as a function of the current state and inputs.

A VHDL description of an n-bit counter design entity using such a datapath description is shown in Figure 16.3. An n-bit wide bank of flip-flops holds the current state, `count`, and updates it from the next state, `nxt`, on each rising edge of the clock. A single concurrent signal assignment statement describes the next-state function. The next state is 0 if `rst` is high, or `count + 1` otherwise.

[1] A five-bit counter (32 states) using this approach would require 32 lines in the state table.

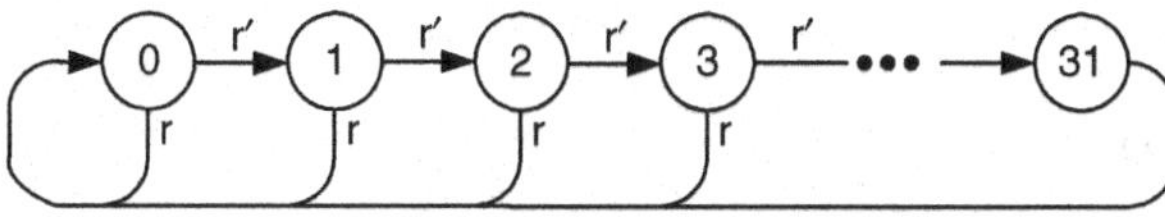

Figure 16.1. State diagram for a five-bit counter.

```
library ieee;
use ieee.std_logic_1164.all;
use work.ff.all;

entity Counter1 is
  port( clk, rst: in std_logic;
        count: buffer std_logic_vector(2 downto 0) );
end Counter1;

architecture impl of Counter1 is
  signal nxt: std_logic_vector( 2 downto 0 );
begin
  COUNTER: vDFF generic map(3) port map(clk,nxt,count);

  process(all) begin
    case? rst & count is
      when "1---" => nxt <= 3d"0";
      when  4d"0" => nxt <= 3d"1";
      when  4d"1" => nxt <= 3d"2";
      when  4d"2" => nxt <= 3d"3";
      when  4d"3" => nxt <= 3d"4";
      when  4d"4" => nxt <= 3d"5";
      when  4d"5" => nxt <= 3d"6";
      when  4d"6" => nxt <= 3d"7";
      when others => nxt <= "000";
    end case?;
  end process;
end impl;
```

Figure 16.2. Three-bit counter FSM specified as a state table.

A block diagram of this counter is shown in Figure 16.4. The figure illustrates the datapath nature of this implementation. The next state is computed by data flowing through paths involving combinational building blocks – including arithmetic blocks and multiplexers. In this case, the two building blocks are a multiplexer, which selects between reset (0) and increment (*count* +1), and an incrementer, which increments *count* to give *count* +1.

In general, a sequential datapath circuit takes the form shown in Figure 16.5. As with any synchronous sequential logic circuit, the state is held in a state register. An output logic circuit computes the output signals as a function of inputs and current state. The next state is computed by a next-state circuit as a function of inputs and current state. What distinguishes a datapath is that the next-state logic and output logic are described by functions rather than tables. For our simple counter, the output logic is just the current state, and the next-state logic is a selection between incrementing or setting to zero. We will see more complex examples of sequential datapath circuits in the following sections. However, they all retain this functional description of the next-state and output functions.

```
library ieee;
use ieee.std_logic_1164.all;
use ieee.std_logic_unsigned.all;
use work.ff.all;

entity Counter is
  generic( n: integer := 5 );
  port( clk, rst: in std_logic;
        count: buffer std_logic_vector(n-1 downto 0) );
end Counter;

architecture impl of Counter is
  signal nxt: std_logic_vector(n-1 downto 0);
begin
  COUNTER: vDFF generic map(n) port map(clk,nxt,count);

  nxt <= (others => '0') when rst else count+1;
end impl;
```

Figure 16.3. An *n*-bit counter FSM specified with a single concurrent signal assignment statement.

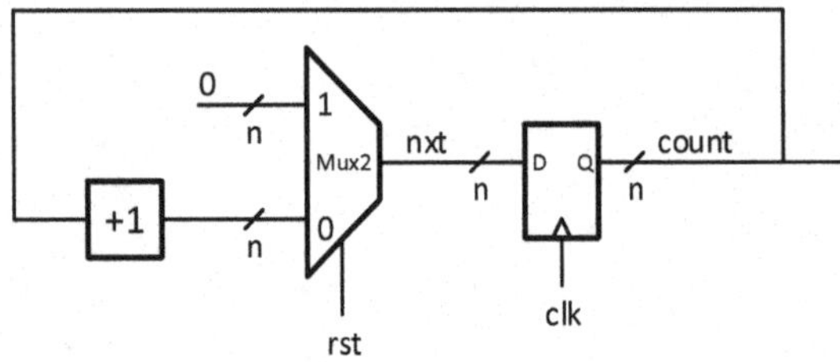

Figure 16.4. Block diagram of a simple counter. The counter state is held in a state register (flip-flops). The next state is selected by a multiplexer to be zero if *rst* is asserted, or the output of an incrementer (*count* +1) otherwise.

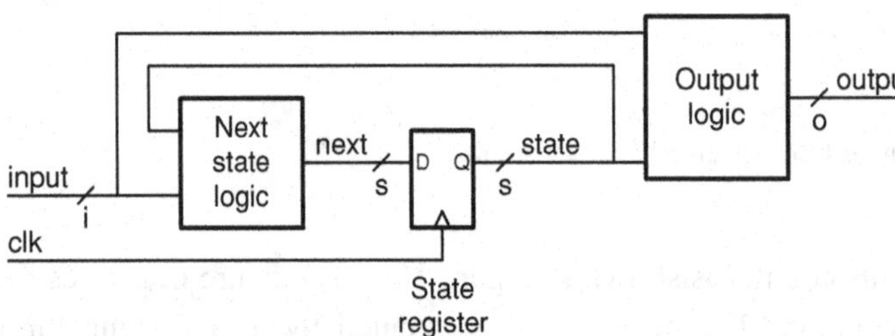

Figure 16.5. In general, a sequential datapath circuit consists of a state register, next-state logic, and output logic. The next-state and output logic are described by functions rather than tables.

16.1.2 Up/down/load counter

Our simple counter had only two choices for next state: reset or increment. Often we require a counter with more options for the next state. A counter may need to count down (decrement) as well as count up; there may be cases where it needs to hold its value; and, on occasion, we may need to load an arbitrary value into the counter.

Figure 16.6 shows the VHDL description of such an up/down/load (UDL) counter. The counter has four control inputs (`rst`, `up`, `down`, and `load`), and one *n*-bit wide data input (`input`). The next-state function is described by a matching case statement. Some CAD tools may have difficulty recognizing that the result of concatenating the four `std_logic` signals

```
library ieee;
use ieee.std_logic_1164.all;
use ieee.std_logic_unsigned.all;
use work.ff.all;

entity UDL_Count1 is
  generic( n: integer := 4 );
  port( clk, rst, up, down, load: in std_logic;
        input: in std_logic_vector(n-1 downto 0);
        output: buffer std_logic_vector(n-1 downto 0) );
end UDL_Count1;

architecture impl of UDL_Count1 is
  signal nxt: std_logic_vector(n-1 downto 0);
begin
  COUNT: vDFF generic map(n) port map(clk,nxt,output);

  process(all) begin
    case? std_logic_vector'(rst & up & down & load) is
      when "1---" => nxt <= (others => '0');
      when "01--" => nxt <= output + 1;
      when "001-" => nxt <= output - 1;
      when "0001" => nxt <= input;
      when others => nxt <= output;
    end case?;
  end process;
end impl;
```

Figure 16.6. VHDL description of an up/down/load (UDL) counter.

rst, up, down, and output has type std_logic_vector. Hence, we use the *qualified expression* std_logic_vector'(rst & up & down & load). If reset (rst) is asserted, the next state is 0. Otherwise, if up is asserted, the next state is out + 1. If down is asserted (and not up or rst), the counter decrements by setting the next state to output - 1. The counter is loaded when load is asserted by setting the next state to input. Finally, if none of the control inputs are asserted, the counter holds its present value by setting next to output.

This description accurately captures the function of the UDL counter and is adequate for most purposes. However, it is somewhat inefficient in that it will result in the instantiation of both an incrementer (to compute output + 1) and a decrementer (to compute output - 1). From our brief study of computer arithmetic (see Chapter 10), we know that these two circuits could be combined.

If we are in an operating mode where saving a few gates matters (which is unlikely), we can describe a more economical counter circuit as shown in Figure 16.7. This circuit factors the increment and decrement operations out of the case statement. Instead, it realizes them in signal outpm1 (out plus or minus 1). This signal is generated by a concurrent signal assignment statement that adds 1 to output if down is false, and adds −1 to output if down is true. The code is otherwise identical to that of Figure 16.6.

A block diagram of the UDL counter is shown in Figure 16.8. Like the simple counter of Figure 16.4, and like most datapath circuits, the UDL counter uses a multiplexer to

```
library ieee;
use ieee.std_logic_1164.all;
use ieee.std_logic_unsigned.all;
use work.ff.all;

entity UDL_Count2 is
  generic( n: integer := 4 );
  port( clk, rst, up, down, load: in std_logic;
        input: in std_logic_vector(n-1 downto 0);
        output: buffer std_logic_vector(n-1 downto 0) );
end UDL_Count2;

architecture impl of UDL_Count2 is
  signal outpm1, nxt: std_logic_vector(n-1 downto 0);
begin
  COUNT: vDFF generic map(n) port map(clk,nxt,output);

  outpm1 <= output + ((n-2 downto 0 => down) & '1');

  process(all) begin
    case? std_logic_vector'(rst & up & down & load) is
      when "1---" => nxt <= (others => '0');
      when "01--" => nxt <= outpm1;
      when "001-" => nxt <= outpm1;
      when "0001" => nxt <= input;
      when others => nxt <= output;
    end case?;
  end process;
end impl;
```

Figure 16.7. VHDL description of an up/down/load counter with a shared incrementer/decrementer.

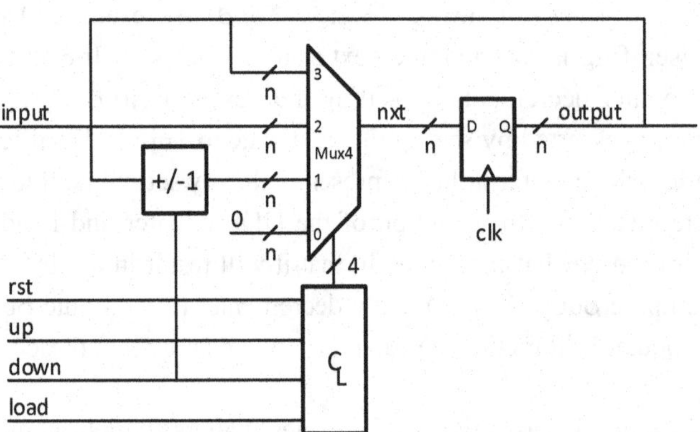

Figure 16.8. Block diagram of an up/down/load counter.

select different options for the next state. Some of the options are created by function units. Here the multiplexer has four inputs – to select the input (*load*), the output of the incrementer/decrementer (*up* or *down*), 0 (*reset*), and count (*hold*). The single function unit is an incrementer/decrementer that can either add or subtract 1 from the current count. The *down* line controls whether to increment or decrement. A block of combinational logic generates the

```
library ieee;
use ieee.std_logic_1164.all;
use ieee.std_logic_unsigned.all;
use work.ff.all;
use work.ch8.all; -- for Mux4

entity UDL_Count3 is
  generic( n: integer := 4 );
  port( clk, rst, up, down, load: in std_logic;
        input: in std_logic_vector(n-1 downto 0);
        output: buffer std_logic_vector(n-1 downto 0) );
end UDL_Count3;

architecture impl of UDL_Count3 is
  signal outpm1, nxt: std_logic_vector(n-1 downto 0);
  signal sel: std_logic_vector(3 downto 0);
begin
  REG: vDFF generic map(n) port map(clk,nxt,output);

  outpm1 <= output + ((n-2 downto 0 => (not up)) & '1');

  MUX: Mux4 generic map(n) port map(output, input, outpm1, (n-1 downto 0 => '0'),
        ((not rst) and (not up) and (not down) and (not load)) &
        ((not rst) and load) &
        ((not rst) and (up or down)) &
        rst,
        nxt);
end impl;
```

Figure 16.9. VHDL description of an up/down/load counter using a shared incrementer/decrementer and an explicit multiplexer.

select signals for the multiplexer by decoding the control inputs. VHDL code corresponding to this block diagram is shown in Figure 16.9.

16.1.3 A timer

In many applications, for example a more involved version of our traffic-light controller, we would like a timer that can be set with an initial time *t*, and that, after *t* cycles have passed, signals us that the time is complete. This is analogous to a kitchen timer that you set with an interval, and it signals you audibly when the interval is complete.

A block diagram of an FSM timer is shown in Figure 16.10. It follows our familiar theme of using a multiplexer to select the next state from among constants, inputs, and the outputs of function units (in this case a decrementer). What is different about this block diagram is that it includes an output function unit. A zero checker asserts signal *done* when the count has reached zero.

To operate the timer, the time interval is applied to *input* and control signal *load* is asserted to load the interval. Each cycle after this load, the internal state *count* counts down. When *count* reaches zero, output *done* is asserted and counting stops. The reset input *rst* is used only to initialize the timer on power-up.

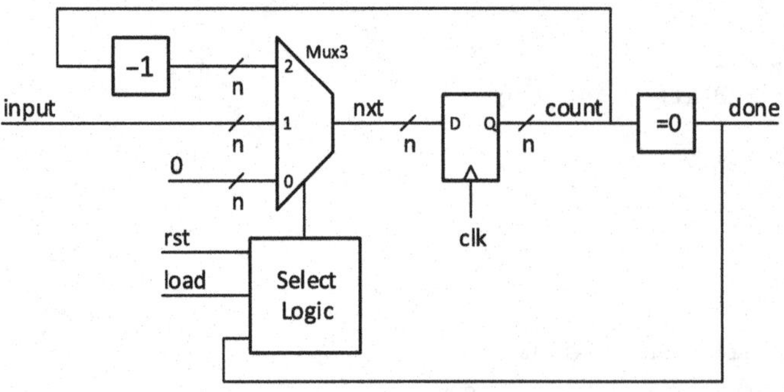

Figure 16.10. Block diagram of a timer FSM.

```
--------------------------------------------------------------------------
-- Timer design entity
-- rst sets count to zero
-- load sets count to input
-- Otherwise count decrements and saturates at zero (doesn't wrap)
-- Done is asserted when count is zero
--------------------------------------------------------------------------

library ieee;
use ieee.std_logic_1164.all;
use ieee.std_logic_unsigned.all;
use work.ch8.all;
use work.ff.all;

entity Timer is
  generic( n: integer := 4 );
  port( clk, rst, load: in std_logic;
        input: in std_logic_vector(n-1 downto 0);
        done: buffer std_logic );
end Timer;

architecture impl of Timer is
  signal count, next_count, cntm1, zero: std_logic_vector(n-1 downto 0);
  signal sel: std_logic_vector(2 downto 0);
begin
  CNT: vDFF generic map(n) port map(clk,next_count,count);
  MUX: Mux3 generic map(n) port map(count - 1,input, (n-1 downto 0 => '0'),
         ((not rst) and (not load) and (not done)) &
         (load and (not rst)) &
         (rst or (done and (not load))),
         next_count);
  done <= '1' when count = 0 else '0';
end impl;
```

Figure 16.11. VHDL description of a timer.

A VHDL description of the timer is shown in Figure 16.11. The style is similar to our simple counter and structural UDL counter with a multiplexer and a state register. The decrementer is implemented in the argument list of the multiplexer. To keep the timer from continuing to decrement, we select the zero input of the multiplexer when `done` is asserted and `load` is not. The conditional signal assignment to `done` implements the zero checker.

EXAMPLE 16.1 Increment/decrement-by-3 counter

Write a VHDL design entity for an arbitrary-width counter that, after being reset, increments each clock by 3 if an `inc` input is asserted, decrements by 3 if a `dec` input is asserted, and holds its value otherwise.

The VHDL code is shown in Figure 16.12. We define signal `pm3` that is `-3` if `dec` is true and `3` otherwise. We use the function `conv_std_logic_vector` defined in the standard package `ieee.std_logic_arith`. The function takes two integer arguments. The first is a value to be converted to a `std_logic_vector` and the second is the width of the resulting `std_logic_vector`. The signal `pm3` is input to a single adder to achieve both increment and decrement. A matching `case` statement is used to implement the multiplexer that selects the appropriate value for `nxt`. Note that if `inc` and `dec` are asserted simultaneously `nxt` is set to an unspecified (don't care) value, "`(others => '-')`".

```
library ieee;
use ieee.std_logic_1164.all;
use ieee.std_logic_unsigned.all;
use ieee.std_logic_arith.all;
use work.ff.all;

entity IncDecBy3 is
  generic( n: integer := 8 );
  port( clk, rst, inc, dec: in std_logic;
        output: buffer std_logic_vector(n-1 downto 0) );
end IncDecBy3;

architecture impl of IncDecBy3 is
  signal outpm3, pm3, nxt: std_logic_vector(n-1 downto 0);
begin
  SR: vDFF generic map(n) port map(clk,nxt,output);

  pm3 <= conv_std_logic_vector(-3,n) when dec else conv_std_logic_vector(3,n);
  outpm3 <= output + pm3;

  process(all) begin
    case? rst & inc & dec & outpm3 is
      when "1--" => nxt <= (others => '0'); -- reset
      when "010" => nxt <= outpm3;          -- increment
      when "001" => nxt <= outpm3;          -- decrement
      when "000" => nxt <= output;          -- hold
      when others=> nxt <= (others => '-');
    end case?;
  end process;
end impl;
```

Figure 16.12. VHDL implementation of an increment/decrement-by-3 counter.

16.2 SHIFT REGISTERS

In addition to incrementing, decrementing, and comparing, another popular datapath function is shifting. Shifters are particularly useful in serializers and deserializers that convert data from parallel to serial form and back again.

16.2.1 A simple shift register

A block diagram of a simple shift register is shown in Figure 16.13 and a VHDL description of this circuit is given in Figure 16.14. Unless the shift register is reset, the next state is the current state shifted one bit to the left, with the `sin` input providing the rightmost bit (LSB). In the VHDL implementation, the 2:1 multiplexer is performed using a conditional signal assignment statement, and the left shift is performed by the expression

```
output(n-2 downto 0) & sin
```

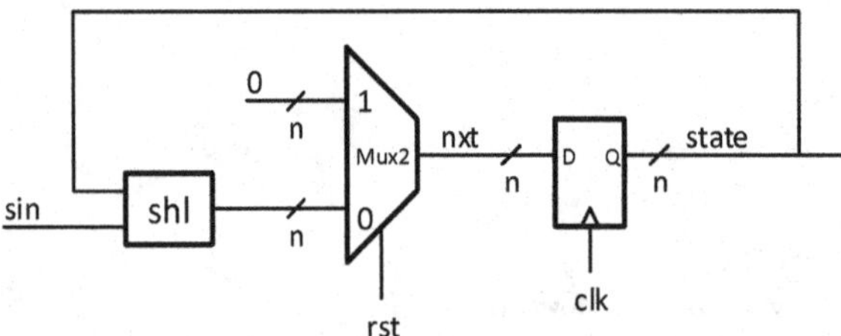

Figure 16.13. Block diagram of a simple shift register.

```
----------------------------------------------------------------------
-- Basic shift register
-- rst - sets out to zero, otherwise out shifts left - sin becomes lsb
----------------------------------------------------------------------

library ieee;
use ieee.std_logic_1164.all;
use work.ff.all;

entity Shift_Register1 is
  generic( n: integer := 4 );
  port( clk, rst, sin: in std_logic;
        output: buffer std_logic_vector(n-1 downto 0) );
end Shift_Register1;

architecture impl of Shift_Register1 is
  signal nxt: std_logic_vector(n-1 downto 0);
begin
  nxt <= (others => '0') when rst else output(n-2 downto 0) & sin;
  CNT: vDFF generic map(n) port map(clk, nxt, output);
end impl;
```

Figure 16.14. VHDL description of simple shift register. If `rst` is asserted, the register is set to all 0s; otherwise it shifts left, with input `sin` filling the LSB.

Here the concatenate operation is used to concatenate the rightmost $n-1$ bits of `output` with `sin`, effectively shifting the bits of `output` one position to the left and inserting `sin` into the LSB.

This simple shift register could be used, for example, as a deserializer. A serial input is received on input `sin` and, every n clocks, the parallel output is read from `output`. A framing protocol of some type is necessary to determine where each parallel symbol starts and stops; i.e., during which clock the parallel output should be read.

16.2.2 Left/right/load (LRL) shift register

Analogously to our complex counter, we can make a complex shift register that can be loaded and shifts in either direction. The block diagram of this left/right/load (LRL) shift register is shown in Figure 16.15 and the VHDL description is shown in Figure 16.16. The design entity uses a matching case statement to select between zero, a left shift (done with concatenation), a right shift (also done with concatenation), the input, and the output. Note that the shift expressions

```
output(n-2 downto 0) & sin
sin & output(n-1 downto 1)
```

do not actually generate any logic. This shift operation is just wiring – from `sin` and the selected bits of `output` to the appropriate input bits of the multiplexer.

16.2.3 Universal shifter/counter

If we made the LRL shift register design entity also increment, decrement, and check for zero, it could serve in place of any of the design entities we have discussed so far in this section. This idea is not as far-fetched as it may seem. At first you may think that using a full-featured design entity when all we need is a simple counter is wasteful. However, most synthesis systems given constant inputs will eliminate logic that is never used (i.e., if up and down are always zero the incrementer/decrementer will be eliminated). Thus, the unused logic should not cost

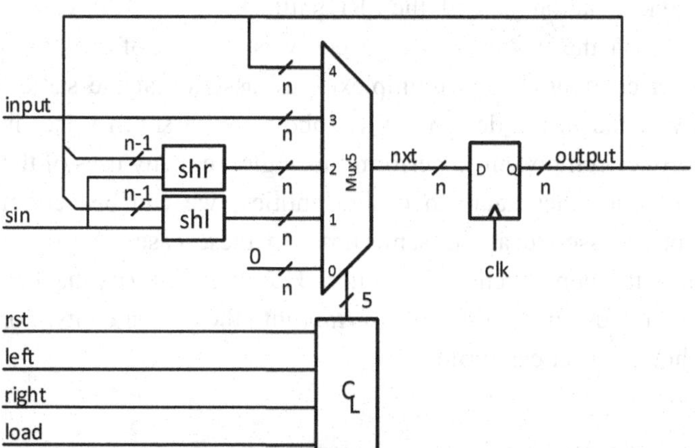

Figure 16.15. Block diagram of a left/right/load shift register.

```
--------------------------------------------------------------------------
-- Left/Right SR with Load
--------------------------------------------------------------------------

library ieee;
use ieee.std_logic_1164.all;
use work.ff.all;

entity LRL_Shift_Register is
  generic( n: integer := 4 );
  port( clk, rst, left, right, load, sin: in std_logic;
        input: in std_logic_vector(n-1 downto 0);
        output: buffer std_logic_vector(n-1 downto 0) );
end LRL_Shift_Register;

architecture impl of LRL_Shift_Register is
  signal nxt: std_logic_vector(n-1 downto 0);
begin
  CNT: vDFF generic map(n) port map(clk,nxt,output);
  process(all) begin
    case? std_logic_vector'(rst & left & right & load) is
      when "1---" => nxt <= (others => '0');              -- reset
      when "01--" => nxt <= output(n-2 downto 0) & sin; -- left
      when "001-" => nxt <= sin & output(n-1 downto 1); -- right
      when "0001" => nxt <= input;                        -- load
      when others => nxt <= output;                       -- hold
    end case?;
  end process;
end impl;
```

Figure 16.16. VHDL description of a left/right/load shift register.

us anything in practice.[2] The advantage of the universal shifter/counter is that it comprises a single design entity to remember and maintain.

VHDL for a universal shifter/counter design entity is shown in Figure 16.17. This code largely combines the code of the individual design entities above. It uses a seven-input multiplexer to select between input, current state, increment, decrement, left shift, right shift, and zero. The increment/decrement is done with an assignment (as with the UDL counter), and the shifts are done with concatenation (as with the LRL shifter).

One thing that is new with the universal design entity is the use of an arbiter `RArb` to make sure that no two select inputs to the multiplexer are asserted at the same time. This arbitration was explicitly coded in the design entities above. With seven select inputs, it is easier to instantiate the arbiter component to perform this logic. In many uses of the universal shifter/counter (or any of our other datapath design entities) we can be certain that two command inputs will not be asserted at the same time. In these cases, we don't need the arbitration (whether done with a component or explicitly). However, to verify that the command inputs are in fact one-hot, it is useful to code an *assertion* into the design entity. For example, to assert a signal is one-hot, the user can write

[2] Check that this is true with your synthesis system before using this approach.

```
----------------------------------------------------------------------
-- Universal Shifter/Counter
-- inputs take priority in order listed
-- rst - resets state to zero
-- left - shifts state to the left, sin fills LSB
-- right - shifts state to the right, sin fills MSB
-- up - increments state
-- down - decrements state - will not decrement through zero.
-- load - load from in
--
-- Output done indicates when state is all zeros.
----------------------------------------------------------------------
library ieee;
use ieee.std_logic_1164.all;
use ieee.std_logic_unsigned.all;
use work.ff.all;
use work.ch8.all;

entity UnivShCnt is
  generic( n: integer := 4 );
  port( clk, rst, left, right, up, down, load, sin: in std_logic;
        input: in std_logic_vector(n-1 downto 0);
        output: buffer std_logic_vector(n-1 downto 0);
        done: buffer std_logic );
end UnivShCnt;

architecture impl of UnivShCnt is
  signal sel: std_logic_vector(6 downto 0);
  signal nxt, outpm1: std_logic_vector(n-1 downto 0);
begin
  outpm1 <= output + ((n-1 downto 1 => down) & '1');
  CNT: vDFF generic map(n) port map(clk,nxt,output);
  ARB: RArb generic map(7)
    port map(rst & left & right & up & (down and (not done)) & load & '1', sel);
  MUX: Mux7 generic map(n)
    port map( (n-1 downto 0 => '0'), output(n-2 downto 0) & sin,
              sin & output(n-1 downto 1), outpm1, outpm1, input, output, sel, nxt);
  done <= '1' when output = 0 else '0';
end impl;
```

Figure 16.17. VHDL description of a universal shifter/counter using a shared incrementer/decrementer and an explicit multiplexer. An arbiter is used to ensure that no more than one multiplexer select line is asserted at a time.

```
count <= ("0" & input(0)) + ("0" & input(1)) + ("0" & input(2));
err <= '1' when count > 1 else '0';
```

The designer must check for this error condition on every clock edge. The assertion is a logical expression that generates no gates,[3] but flags an error during simulation if the expression is violated.

[3] Many synthesis tools have built-in commands to declare what logic should not be synthesized. Synopsys's Design Compiler®, for example, uses `- pragma translate_off` and `- pragma translate_on` to turn synthesis off and on.

EXAMPLE 16.2 Variable shift amount

Write a VHDL design entity for a shift register that shifts zero, one, two, or three bits left, as determined by a two-bit input `sh_amount`. The bits on the right are filled by a three-bit input `sin` MSB first.

The VHDL code is shown in Figure 16.18. We first concatenate `output` with `sin`, which effectively shifts `output` three bits left. We *cast* the result to type `unsigned` since the shift operation must know whether the shifted value is signed or unsigned. We then shift the concatenated value to the right by `3-sh_amount`, which puts `output` in the proper location with the new bits filled by `sin` MSB first. The right-hand operator to the **`srl`** shift operator must have type `integer`. Hence we first convert `sh_amount` to `integer` by first casting it to type `unsigned` then calling the function `to_integer` defined in `ieee.numeric_std`.

```
library ieee;
use ieee.std_logic_1164.all;
use ieee.numeric_std.all;
use work.ff.all;

entity VarShift is
  generic( n: integer := 8 );
  port( clk, rst: in std_logic;
        sh_amount: in std_logic_vector(1 downto 0);
        sin: in std_logic_vector(2 downto 0);
        output: buffer std_logic_vector(n-1 downto 0) );
end VarShift;

architecture impl of VarShift is
  signal sh_i, sh_o: unsigned(n+2 downto 0);
  signal nxt: std_logic_vector(n-1 downto 0);
begin
  SR: vDFF generic map(n) port map(clk,nxt,output);
  sh_i <= unsigned(output & sin);
  sh_o <= sh_i srl 3-to_integer(unsigned(sh_amount));
  nxt  <= (others => '0') when rst else std_logic_vector(sh_o(n-1 downto 0));
end impl;
```

Figure 16.18. VHDL for the variable shifter described in Example 16.2.

16.3 CONTROL AND DATA PARTITIONING

A common theme in digital design is the separation of a design into a control finite-state machine and a datapath, as shown in Figure 16.19. The datapath computes its next state via multiplexers and function units – like the counters and shift registers in this chapter. The control FSM, on the other hand, computes its next state via a state table. The inputs to the design entity are separated into control inputs (that affect the state of the control FSM) and data inputs (that supply values to the datapath). Design entity outputs are partitioned in a similar

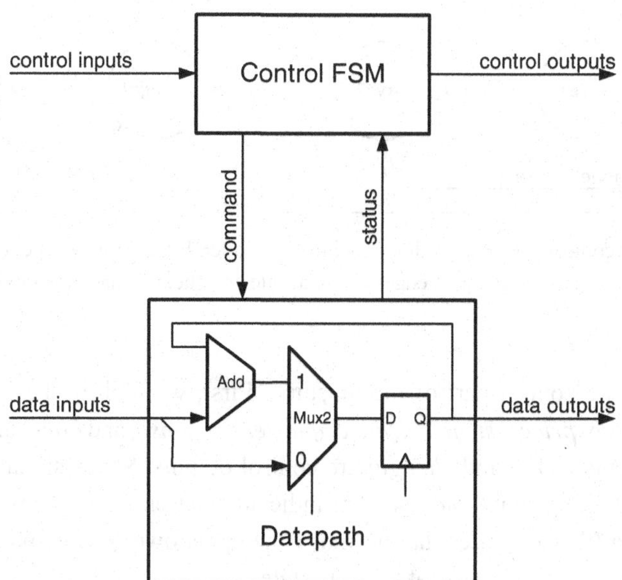

Figure 16.19. Systems are often partitioned into datapath, where the next state is determined by function units, and control, where the next state is determined by state tables.

manner. The control FSM controls the operation of the datapath via a set of command signals. The datapath communicates back to the control FSM via a set of status signals.

The counter and shifter examples we have looked at so far in this chapter are degenerate examples of this organization. They each consist of a datapath – with multiplexers and function units (shifters, incrementers, and/or adders) – and a control unit. However, their control units have been strictly combinational. The datapath commands (e.g., multiplexer selects and add/subtract controls) and control outputs have been functions solely of the current control inputs and datapath status. In this section we will examine two examples of design entities where the control section includes internal state.

16.3.1 Example: vending machine FSM

Consider the problem of designing the controller for a soft-drink vending machine. The specification is as follows. The vending machine accepts nickels, dimes, and quarters. Whenever a coin is deposited into the coin slot, a pulse appears for one clock cycle on one of three lines indicating the type of coin: *nickel*, *dime*, or *quarter*. The price of the item is set on an n-bit switch internal to the machine (in units of nickels), and is input to the controller on the n-bit signal *price*. When sufficient coins have been deposited to purchase a soft drink, the status signal *enough* is asserted. Any time *enough* is asserted and the user presses a *dispense* button, signal *serve* is asserted for exactly one cycle to serve the soft drink. After asserting *serve*, the FSM must wait until signal *done* is asserted, indicating that the mechanism has finished serving the soft drink. After serving, the machine returns change (if any) to the user. It does this one nickel at a time, asserting the signal *change*, for exactly one cycle and waiting for signal *done* to indicate that a nickel has been dispensed before dispensing the next nickel or returning to its original state. Any time the signal *done* is asserted, we must wait for *done* to go low before proceeding.

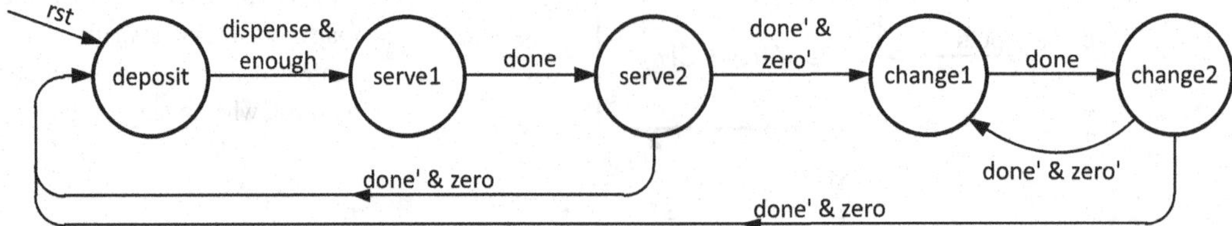

Figure 16.20. State diagram for control part of a vending-machine controller. Edges from a state to itself are omitted for clarity. If none of the logic conditions on edges out of a state are true, the machine will remain in that state.

We start by considering the control part of the machine. First, we look at the inputs and outputs. All of the inputs except *price* (*rst*, *nickel*, *dime*, *quarter*, *dispense*, and *done*) are control inputs, and both of the outputs (*serve* and *change*) are control outputs. Status signals that we need from the datapath include a signal *enough* (that indicates that enough money has been deposited) and a signal *zero* (that indicates that no more change is owed). Commands to the datapath are considered below when we look at the data state.

Now consider the states of the control portion of the machine. As illustrated in the state diagram of Figure 16.20, our vending machine operates in three main phases.[4] First, during the deposit phase (state deposit), the user deposits coins and then presses *dispense.* When signal *dispense* and *enough* are both asserted, the machine advances to the serving phase (states serve1 and serve2). In state serve1 it waits for signal *done* to indicate that the soft drink has been served, and in state serve2 it waits for *done* to be deasserted. We assert the output *serve* during the first cycle only of state serve1. If there is no change to be dispensed, *zero* is true, the FSM returns to the deposit state from serve2. However, if there is change to be dispensed, the FSM enters the change phase (states change1 and change2). The machine cycles through these states, asserting *change* and waiting for *done* in change1 and waiting for *done* to go low in change2. We assert output *change* during the first cycle of each visit to state change1. Only when all change has been dispensed do we return to the deposit state.

Now that we have the control states defined, we turn our attention to the data state. This FSM has a single piece of data state – the amount of money the machine currently owes the user – in units of nickels. We call this state variable *amount.* The different actions that affect *amount* are

reset: amount $\leftarrow$ 0;
deposit a coin: amount $\leftarrow$ amount + value, where value = 1, 2, or 5 for a nickel, dime, or quarter, respectively.
serve a drink: amount $\leftarrow$ amount – price;
return one nickel of change: amount $\leftarrow$ amount – 1.
otherwise: no change in amount.

We can now design a datapath that supports these operations. State variable *amount* can be zeroed, can be added to or subtracted from, or can hold its value. From these *register transfers* we see that we need the datapath of Figure 16.21. The next state for amount *next* is selected by

[4] In this diagram, edges from a state to itself are omitted. If the conditions on all edges leading out of the current state are not satisfied, the FSM stays in that state.

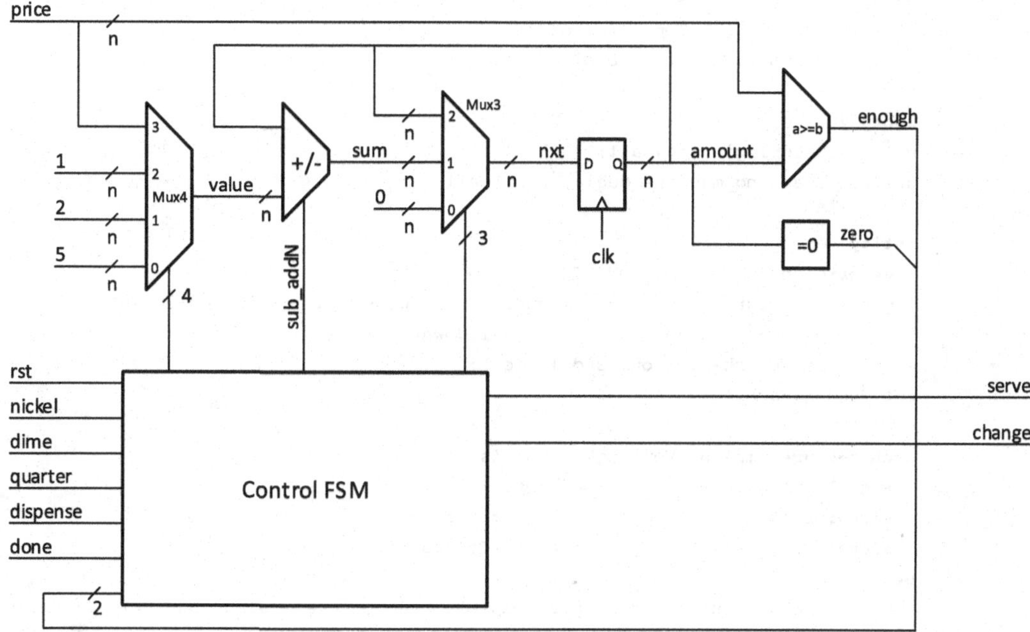

Figure 16.21. Block diagram of a vending-machine controller.

a 3:1 multiplexer that selects between 0, *amount*, or *sum*, the output of an add/subtract unit that adds or subtracts *value* from *amount*. The *value* is selected by a 4:1 multiplexer to be 1, 2, 5, or *price*. We see from the figure that the command signals needed to control the datapath are the two multiplexer select signals and the add/subtract control.

VHDL code for our vending-machine controller is shown in Figures 16.22 through 16.25. The top-level design entity is shown in Figure 16.22. This design entity just instantiates the control (`VendingMachineControl`) and data (`VendingMachineData`) components and connects them up. The command signals `sub`, `selval`, and `selnext` are declared at this level, as are the status signals `enough` and `zero`.

The VHDL for the control design entity of our vending-machine controller is split over two figures. Figure 16.23 shows the first half of the design entity, which includes the logic for generating the output and command variables. A variable `first` is defined, which is used to distinguish the first cycle of states serve1 and change1. After one cycle in either of these states, `first` goes low. This variable enables us to assert only the outputs for one cycle on each visit to these states and to decrement only `amount` once on each visit to the change1 state. Without the `first` variable, we would need to expand both serve1 and change1 into two states. The outputs `serve` and `change` are generated by ANDing `first` with signals that are true in states serve1 and change1, respectively.

The datapath control signals are determined from input and state variables. The two select signals are one-hot variables. Each bit is determined by a logic expression. Signal `selval` (which selects the value input to the add/subtract unit) selects `price` when in the deposit state (`dep` true) and `dispense` is pressed. A 1 is selected if a nickel is input in the deposit state or if we are in the change state. Inputs of 2 and 5 are selected in the deposit state if a dime or a quarter is input, respectively. Signal `selnext` selects the next state for the amount variable.

```
--------------------------------------------------------------------------
-- VendingMachine - Top level design entity
-- Just hooks together control and datapath
--------------------------------------------------------------------------
library ieee;
use ieee.std_logic_1164.all;
use work.vending_machine_declarations.all;

entity VendingMachine is
  generic( n: integer := DWIDTH );
  port( clk, rst, nickel, dime, quarter, dispense, done: in std_logic;
        price: in std_logic_vector(n-1 downto 0);
        serve, change: out std_logic );
end VendingMachine;

architecture impl of VendingMachine is
  signal enough, zero, sub: std_logic;
  signal selval: std_logic_vector(3 downto 0);
  signal selnext: std_logic_vector(2 downto 0);
begin
  VMC: VendingMachineControl port map(clk, rst, nickel, dime, quarter, dispense, done,
                              enough, zero, serve, change, selval, selnext,
                              sub);
  VMD: VendingMachineData generic map(n) port map(clk, selval, selnext, sub, price,
                              enough, zero);
end impl;
```

Figure 16.22. VHDL top-level design entity for vending-machine controller just instantiates the control and data components and connects them together.

Zero is selected if `rst` is true, and the output of the add/subtract unit is selected if variable `selv` is true. Otherwise, the current value of `amount` is selected. Variable `selv` is true in the deposit state if a coin is entered or if `dispense` and `enough` are both true, and it is true in the change state if `first` is true. Note that we can select `price` to be the value whenever `dispense` is asserted in the deposit state because we select the add/subtract output only if `enough` is also asserted. Finally, the add/subtract control signal is set to subtract on `dispense` and `change` actions. Otherwise the adder adds.

The second half of the control design entity in Figure 16.24 shows the next-state logic. This logic is implemented using a matching `case` statement, where the selection is based upon the concatenation of four input bits and the current state. Most transitions encode to a single case alternative. For example, when the machine is in the deposit state, and `dispense` and `enough` are both true, the first case alternative `"11--"& DEPOSIT` is active. It takes the next two case alternatives to encode when the machine stays in the deposit state. A separate concurrent conditional signal assignment statement is used to reset the FSM to the deposit state.

A VHDL description of the vending-machine controller datapath is shown in Figure 16.25. This code follows the datapath part of Figure 16.21 very closely. A state register holds the current amount. A three-input multiplexer feeds the state register with zero, the `sum` output of the add/subtract unit, or the current value of `amount`. An add/subtract unit adds or

```
library ieee;
use ieee.std_logic_1164.all;
use work.ff.all;
use work.vending_machine_declarations.all;

entity VendingMachineControl is
  port( clk, rst, nickel, dime, quarter, dispense, done, enough, zero: in std_logic;
        serve, change: out std_logic;
        selval: out std_logic_vector(3 downto 0);
        selnext: out std_logic_vector(2 downto 0);
        sub: out std_logic );
end VendingMachineControl;

architecture impl of VendingMachineControl is
  signal state, nxt, nxt1: std_logic_vector(SWIDTH-1 downto 0);
  signal nfirst, first, serve_1, change_1, change_int, dep, selv: std_logic;
begin
  -- outputs
  serve_1 <= '1' when state = SERVE1 else '0';
  change_1 <= '1' when state = CHANGE1 else '0';
  serve <= serve_1 and first;
  change_int <= change_1 and first;
  change <= change_int;

  -- datapath controls
  dep <= '1' when state = DEPOSIT else '0';
  selval <= (dep and dispense) &
            ((dep and nickel) or change_int) &
            (dep and dime) &
            (dep and quarter);

  -- amount, sum, 0
  selv <= (dep and (nickel or dime or quarter or (dispense and enough))) or
          (change_int and first);
  selnext <= (not (selv or rst)) & ((not rst) and selv) & rst;

  -- subtract
  sub <= (dep and dispense) or change_int;

  -- only do actions on first cycle of serve_1 or change_1
  nfirst <= not (serve_1 or change_1);
  first_reg: sDFF port map(clk, nfirst, first);
```

Figure 16.23. VHDL description of control design entity for vending-machine controller (part 1 of 2). This first half of the control design entity shows the output and command variables implemented with concurrent conditional signal assignment and concurrent simple signal assignment statements.

subtracts `value` to or from `amount`. A four-input multiplexer selects the `value` to be added or subtracted. Finally, two conditional assignment statements generate the status signals `enough` and `zero`.

A testbench for the vending-machine controller is shown in Figure 16.26 and waveforms from simulating the controller with this testbench are shown in Figure 16.27. This

```
  -- state register
  state_reg: vDFF generic map(SWIDTH) port map(clk, nxt, state);

  -- next state logic
  process(all) begin
    case? dispense & enough & done & zero & state is
      when "11--" & DEPOSIT => nxt1 <= SERVE1;   -- dispense & enough
      when "01--" & DEPOSIT => nxt1 <= DEPOSIT;
      when "-0--" & DEPOSIT => nxt1 <= DEPOSIT;
      when "--1-" & SERVE1  => nxt1 <= SERVE2;   -- done
      when "--0-" & SERVE1  => nxt1 <= SERVE1;
      when "--01" & SERVE2  => nxt1 <= DEPOSIT; -- not done and zero
      when "--00" & SERVE2  => nxt1 <= CHANGE1; -- not done and not zero
      when "--1-" & SERVE2  => nxt1 <= SERVE2;   -- done
      when "--1-" & CHANGE1 => nxt1 <= CHANGE2; -- done
      when "--0-" & CHANGE1 => nxt1 <= CHANGE1; -- done
      when "--00" & CHANGE2 => nxt1 <= CHANGE1; -- not done and not zero
      when "--01" & CHANGE2 => nxt1 <= DEPOSIT; -- not done and zero
      when "--1-" & CHANGE2 => nxt1 <= CHANGE2; -- not done and zero
      when others => nxt1 <= DEPOSIT;
    end case?;
  end process;

  nxt <= DEPOSIT when rst = '1' else nxt1;
end impl;
```

Figure 16.24. VHDL description of control design entity for vending-machine controller (part 2 of 2). This second half of the control design entity shows the next-state function implemented with a matching case statement.

testbench illustrates a new feature of VHDL introduced in VHDL-2008: external names. In the **report** statement the first call to `to_hstring` prints the value of signal `state` inside `VendingMachineControl`, which is instantiated inside `VendingMachine`, which in turn is instantiated inside the testbench. Notice that rather than having to add signals to the entity declarations for `VendingMachineControl` and `VendingMachine` we use the *external name* syntax "`<<signal DUT.VMC.STATE: std_logic_vector>>`" to access the value of `state` inside of `VendingMachineControl`. Here `DUT.VMC.STATE` is the external path name for the signal we would like to print out.

The test begins by resetting the machine. A nickel is then deposited, followed by a dime – bringing `amount` to 3 (15 cents). At this point, we go one cycle with no input to make sure `amount` stays at 3. We then assert `dispense` to make sure that trying to dispense a soft drink before enough money has been deposited doesn't work. Next we deposit two quarters in back-to-back cycles, bringing `amount` to 8 and then 13. When `amount` reaches 13, signal `enough` goes high since we have exceeded the price (which is 11). After an idle cycle, we again assert `dispense`. This time it works. The state advances to 001 (serve1) and `amount` is reduced to 2 (price of 11 deducted).

In the first cycle of the serve1 state (state = 001), `serve` is asserted. The machine remains in this state for one more cycle, waiting for `done` to go high. It spends just one cycle in state serve2 (state = 011), since `done` is already low, and continues into state change1 (state = 010).

```
library ieee;
use ieee.std_logic_1164.all;
use work.ff.all;
use work.ch8.all;
use work.ch10.all;
use work.vending_machine_declarations.all;

entity VendingMachineData is
  generic( n: integer := 6 );
  port( clk: in std_logic;
        selval: in std_logic_vector(3 downto 0); -- price, 1, 2, 5
        selnext: in std_logic_vector(2 downto 0); -- amount, sum, 0
        sub: in std_logic;
        price: in std_logic_vector(n-1 downto 0); -- price of soft drink - in nickels
        enough: out std_logic; -- amount > price
        zero: out std_logic ); -- amount = zero
end VendingMachineData;

architecture impl of VendingMachineData is
  signal sum, amount, nxt, value, z: std_logic_vector(n-1 downto 0);
  signal ovf: std_logic;
begin
  --  state register holds current amount
  AMT: vDFF generic map(n) port map(clk, nxt, amount);

  --  select next state from 0, sum, or hold
  z <= (nxt'range => '0');
  NSMUX: Mux3 generic map(n) port map(amount, sum, z, selnext, nxt);

  --  add or subtract a value from current amount
  ADD: AddSub generic map(n) port map(amount, value, sub, sum, ovf);

  --  select the value to add or subtract
  VMUX: Mux4 generic map(n) port map(price, CNICKEL, CDIME, CQUARTER, selval, value);

  -- comparators
  enough <= '1' when amount >= price else '0';
  zero <= '1' when amount = (amount'range => '0') else '0';
end impl;
```

Figure 16.25. Datapath for vending-machine controller.

On the first cycle of change1, output `change` is asserted (to return a nickel to the user), and `amount` is decremented to 1. The machine remains in the change1 state for one more cycle waiting for `done`. It then spends just one cycle in state change2 (state = 100) before returning to change1 – because `zero` is not true. Again `change` is asserted and `amount` is decremented on the first cycle of change1. This time, however, `amount` is decremented to zero and signal `zero` is asserted. After waiting a second cycle in the change1 state, the machine transitions through change2 and back to the deposit state – since `zero` is set – ready to start again.

An alternative implementation of the vending machine as a single VHDL design entity is shown in Figure 16.28. The control part of this machine is identical to that of Figures 16.23

```
-- pragma translate_off
library ieee;
use ieee.std_logic_1164.all;
use work.vending_machine_declarations.all;

entity testVend is
end testVend;

architecture test of testVend is
  signal clk, rst, n, d, q, dispense, done: std_logic;
  signal NDQd, price: std_logic_vector(3 downto 0);
  signal serve, change: std_logic;
begin
  DUT: entity work.VendingMachine(impl) generic map(4)
    port map(clk=>clk, rst=>rst, nickel=>n, dime=>d, quarter=>q, dispense=>dispense,
             done=>done, price=>price, serve=>serve, change=>change);

  process begin
    report to_string(n & d & q & dispense) & " " &
      to_hstring(<<signal DUT.VMC.STATE: std_logic_vector>>) & " " &
      to_hstring(<<signal DUT.VMD.AMOUNT: std_logic_vector>>) & " " &
      to_string(serve) & " " & to_string(change);
    wait for 5 ns; clk <= '1';
    wait for 5 ns; clk <= '0';
  end process;

  process(clk) begin
    if rising_edge(clk) then
      done <= serve or change; -- give prompt feedback
    end if;
  end process;

  process begin
    rst <= '1'; price <= CPRICE;
    (n,d,q,dispense) <= std_logic_vector'("0000");
    wait for 20 ns; rst <= '0';
    wait for 10 ns; (n,d,q,dispense) <= std_logic_vector'("1000"); -- nickel 1
    wait for 10 ns; (n,d,q,dispense) <= std_logic_vector'("0100"); -- dime 3
    wait for 10 ns; (n,d,q,dispense) <= std_logic_vector'("0000"); -- nothing
    wait for 10 ns; (n,d,q,dispense) <= std_logic_vector'("0001"); -- try dispense
    wait for 10 ns; (n,d,q,dispense) <= std_logic_vector'("0010"); -- quarter 8
    wait for 10 ns; (n,d,q,dispense) <= std_logic_vector'("0010"); -- quarter 13
    wait for 10 ns; (n,d,q,dispense) <= std_logic_vector'("0000"); -- nothing
    wait for 10 ns; (n,d,q,dispense) <= std_logic_vector'("0001"); -- dispense 2
    wait for 10 ns; (n,d,q,dispense) <= std_logic_vector'("0000");
    wait for 100 ns;
    std.env.stop(0);
  end process;
end test;
-- pragma translate_on
```

Figure 16.26. VHDL testbench for vending-machine controller.

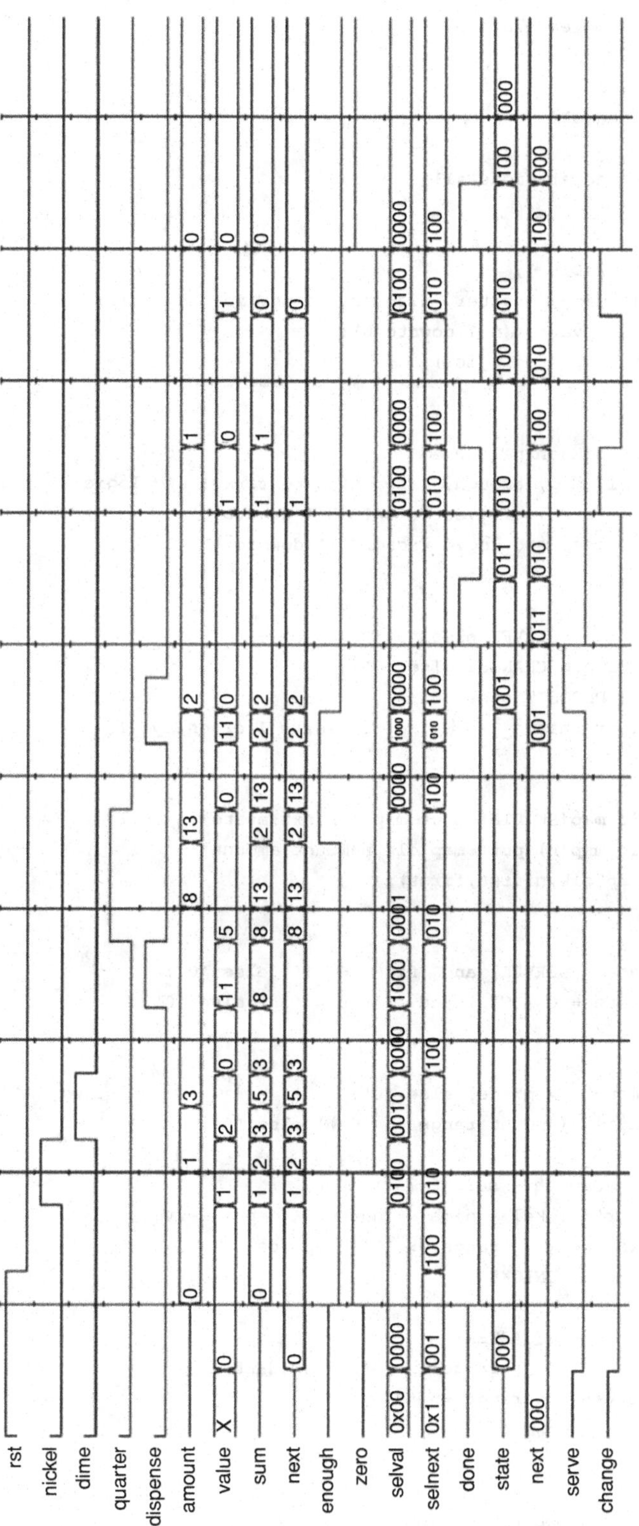

Figure 16.27. Waveforms from simulating the vending-machine controller with the testbench of Figure 16.26.

```
-- VendingMachine - Flat implementation
library ieee;
use ieee.std_logic_1164.all;
use ieee.std_logic_unsigned.all;
use work.ff.all;
use work.vending_machine_declarations.all;

entity VendingMachine1 is
  generic( n: integer := DWIDTH );
  port( clk, rst, nickel, dime, quarter, dispense, done: in std_logic;
        price: in std_logic_vector(n-1 downto 0);
        serve, change: buffer std_logic );
end VendingMachine1;

architecture impl of VendingMachine1 is
  signal serve_1, change_1, dep, enough, zero, nfirst, first: std_logic;
  signal state, nxt, nxt1: std_logic_vector(SWIDTH-1 downto 0);
  signal amount, namount, inc: std_logic_vector(n-1 downto 0);
begin
  -- decode
  serve_1 <= '1' when state = SERVE1 else '0';
  change_1 <= '1' when state = CHANGE1 else '0';
  dep <= '1' when state = DEPOSIT else '0';
  nfirst <= not (serve_1 or change_1); -- not in serve_1 or change_1

  -- state registers
  STATE_REG: vDFF generic map(SWIDTH) port map(clk,nxt,state);
  DATA_REG:  vDFF generic map(n) port map(clk,namount,amount);
  FIRST_REG: sDFF port map(clk,nfirst,first);

  -- outputs
  serve <= '1' when (state = SERVE1) and (first = '1') else '0';
  change <= '1' when (state = CHANGE1) and (first = '1') else '0';

  -- status signals
  enough <= '1' when (amount >= price) else '0';
  zero <= '1' when (amount = (amount'range => '0')) else '0';

  process(all) begin -- datapath - select increment
    case? std_logic_vector'(nickel & dime & quarter & dep & serve & change) is
      when "---010" => inc <= (inc'range => '0') - price;
      when "100100" => inc <= CNICKEL;
      when "010100" => inc <= CDIME;
      when "001100" => inc <= CQUARTER;
      when "---001" => inc <= (inc'range => '0') - CNICKEL;
      when others => inc <= (inc'range => '0');
    end case?;
  end process;

  -- datapath - select next amount
  namount <= (namount'range => '0') when rst else amount + inc;
```

Figure 16.28. An alternative version of the vending machine implemented as a single design entity. The datapath is implemented as a single process that selects an increment. Part 1 of 2; part 2 is identical to Figure 16.24.

and 16.24 except that the logic to generate signals `selval` and `selnext` is omitted. The datapath uses a single matching `case` statement to select an increment, `inc`, which is zero when `amount` stays constant.

16.3.2 Example: combination lock

As our second control and datapath example, consider an electronic combination lock that accepts input from a decimal keypad. The user must enter the code as a sequence of decimal digits and then press an *enter* key. If the user enters the correct sequence, the *unlock* output is asserted (presumably to actuate a large bolt that unlocks a door). To relock the machine, the user presses *enter* a second time. If the user enters an incorrect sequence, a *busy* output is asserted and a timer is activated to wait a predefined period before allowing the user to try again. The busy light should not come on until the user enters the entire sequence and presses *enter*. If the light were to come on at the first incorrect keypress, this would give the user information that can be used to discover the code one digit at a time.

There are three inputs to our lock system: *key*, *key_valid*, and *enter*. A four-bit code *key* indicates the current key being pressed and is accompanied by a signal *key_valid* that indicates when a key is valid. The keyboard is preprocessed so that every key press asserts key and valid for exactly one cycle. In a similar manner, the *enter* signal is preprocessed so that it is valid for exactly one cycle each time the user presses the enter key to unlock or relock the bolt. The length of the sequence is set by an internal variable *length*, and the sequence itself is stored in an internal memory.

A state diagram describing the control portion of our combination lock machine is shown in Figure 16.29. The machine resets to the enter state. In this state the machine accepts input. Each time a key is pressed, it is checked against the expected digit. If it is correct (*kmatch*) the machine stays in the enter state; if not ($valid \wedge \overline{kmatch}$), the machine transitions to the wait1 state. The machine waits in the wait1 state until *enter* is pressed, and then enters the wait2 state. The machine starts a timer in the wait2 state and remains in this state, asserting *busy* until the timer signals *done*.

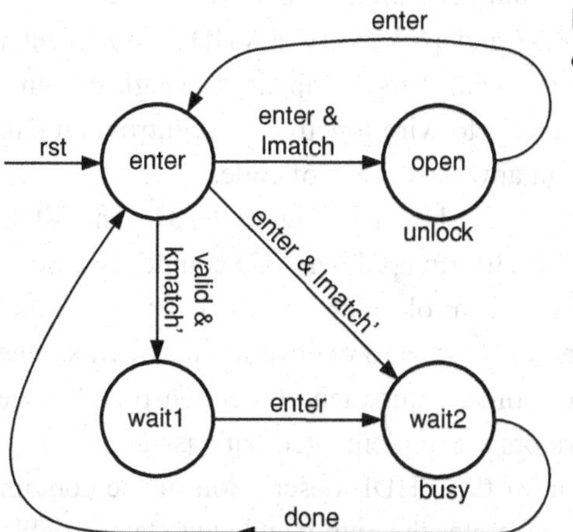

Figure 16.29. State diagram for control portion of combination lock.

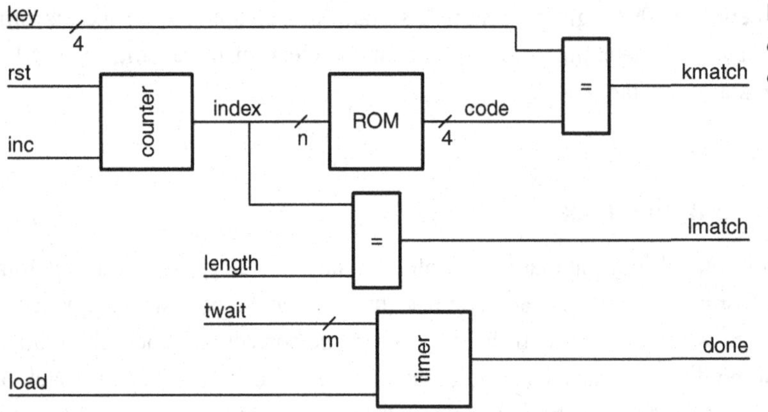

Figure 16.30. Block diagram of datapath portion of combination lock.

When the entire code has been entered correctly, the machine will still be in the enter state – an incorrect digit would have taken it to wait1 – and *lmatch* will be true; *lmatch* will be true only if the length of the code entered matches the internal *length* variable. If *enter* is pressed at this point, the machine transitions to the open state ($enter \wedge lmatch$) and the bolt is unlocked. A second *enter* in the open state returns to the reset state. If, in the enter state, the *enter* key is pressed when the length does not match (there are either too few or too many digits in the codeword) ($enter \wedge \overline{lmatch}$), the machine goes to the wait2 state.

A block diagram of the datapath portion of our combination lock is shown in Figure 16.30. The datapath has two distinct sections. The upper section compares the length and value of the code entered. The lower section times the wait period when an incorrect code is entered. This section consists of a single timer component that loads the timer interval *twait* when control signal *load* is asserted. The timer then counts down. When it reaches zero, status signal *done* is asserted.

The upper half of the datapath consists of a counter, a ROM, and two comparators. The counter keeps track of which digit of the code we are on. It is reset to zero before entering the enter state and then counts up as each key is pressed. The counter output, *index*, selects the digit to be compared next. Signal *index* is compared with length to generate status signal *lmatch*. It is also used as an address for the ROM containing the code. The current digit of the code is read from the ROM, signal *code*, and compared with the key entered by the user to generate status signal *kmatch*. Figures 16.31 and 16.32 show a VHDL implementation of our combination lock. Here we have put both the control and datapath in a single design entity. This eliminates the code that is otherwise required, to wire together the control and data sections, but for a large design entity can result in an unwieldy piece of code.

The datapath portion of the design entity is shown in Figure 16.31. The VHDL follows directly from the block diagram of Figure 16.30. An up/down/load counter (see Section 16.1.2) is used for the counter. Only the `rst` and `up` control inputs are used. The `load` and `down` are zero. The synthesizer will take advantage of these zero control inputs when synthesizing the counter and eliminate the unused logic. The timer component of Section 16.1.3 is used to count down in the wait2 state. The two comparators are implemented with assignment statements.

Figure 16.32 shows the control portion of the VHDL description of the combination lock design entity. The top portion of the code generates the output and next-state variables. Outputs

```
-----------------------------------------------------------------------
-- CombLock
-- Inputs:
--    key - (4-bit) accepts a code digit each time key_valid is true
--    key_valid - signals when a new code digit is on key
--    enter - signals when entire code has been entered
-- Outputs:
--    busy - asserted after incorrect code word entered during timeout
--    unlock - asserted after correct codeword is entered until enter
--            is pressed again.
-----------------------------------------------------------------------

library ieee;
use ieee.std_logic_1164.all;
use work.ff.all;
use work.ch16.all;
use work.comb_lock_codes.all;

entity CombLock is
  generic( n: integer := 4;    -- bits of code length
           m: integer := 4 ); -- bits of timer
  port( clk, rst, key_valid, enter: in std_logic;
        key: in std_logic_vector(3 downto 0);
        busy, unlock: buffer std_logic );
end CombLock;

architecture impl of CombLock is
  signal rstctr: std_logic; -- reset the digital counter
  signal inc: std_logic; -- increment the digit counter
  signal load: std_logic; -- load the timer
  signal done: std_logic; -- timer done
  signal kmatch, lmatch, senter, swait1: std_logic;
  signal index: std_logic_vector(n-1 downto 0);
  signal code: std_logic_vector(3 downto 0);
  signal state, nxt, nxt1: std_logic_vector(SWIDTH-1 downto 0);
begin
  ----- datapath ------------------------------------------
  CTR: UDL_Count3 generic map(n)
         port map(clk,rstctr,inc,'0','0',"0000",index);  -- counter
  MEM: ROM generic map(n,4,"comb_lock.txt") port map(index, code);
  TIM: Timer generic map(m) port map(clk,rst,load,TWAIT,done); -- wait timer
  kmatch <= '1' when code = key else '0'; -- key comparator
  lmatch <= '1' when index = LENGTH else '0'; -- length comparator
```

Figure 16.31. VHDL description of the combination lock (part 1 of 2). This section describes the datapath and closely follows Figure 16.30.

`busy` and `unlock` are generated by decoding the state variable `state` since these outputs are true whenever the machine is in the wait2 and open states, respectively. The enter and wait1 states are decoded as well for use in the command equations. The `rstctr` command signal is set to reset the digit counter on reset or in states that transition to the enter state (wait2 and open) so the counter is zeroed and ready to sequence digits in the enter state. Similarly, the

```
----- control ----------------------------------------
senter <= '1' when state = S_ENTER else '0'; -- decode state
unlock <= '1' when state = S_OPEN  else '0';
busy   <= '1' when state = S_WAIT2 else '0';
swait1 <= '1' when state = S_WAIT1 else '0';
rstctr <= rst or unlock or busy;  -- reset before returning to enter
inc <= senter and key_valid;      -- increment on each key entry
load <= senter or swait1;         -- load before entering wait2

SR: vDFF generic map(SWIDTH) port map(clk,nxt,state); -- state register

process(all) begin
  case? enter & lmatch & key_valid & kmatch & done & state is
    when "--10-" & S_ENTER => nxt1 <= S_WAIT1; -- valid and not kmatch
    when "0-11-" & S_ENTER => nxt1 <= S_ENTER; -- valid and kmatch
    when "110--" & S_ENTER => nxt1 <= S_OPEN;  -- enter and lmatch
    when "100--" & S_ENTER => nxt1 <= S_WAIT2; -- enter and not lmatch
    when "0-0--" & S_ENTER => nxt1 <= S_ENTER; -- not enter and not valid

    when "1----" & S_OPEN  => nxt1 <= S_ENTER; -- enter
    when "0----" & S_OPEN  => nxt1 <= S_OPEN;  -- not enter

    when "1----" & S_WAIT1 => nxt1 <= S_WAIT2; -- enter
    when "0----" & S_WAIT1 => nxt1 <= S_WAIT1; -- not enter

    when "----1" & S_WAIT2 => nxt1 <= S_ENTER; -- done
    when "----0" & S_WAIT2 => nxt1 <= S_WAIT2; -- not done

    when others => nxt1 <= S_ENTER;
  end case?;
end process;

nxt <= S_ENTER when rst else nxt1;
end impl;
```

Figure 16.32. VHDL description of the combination lock (part 2 of 2). This section describes the control. Concurrent assignment statements generate the command and output signals. A process describing combinational logic computes the next-state function.

`load` signal loads the timer in states that transition to the wait2 state (enter and wait1) so it can count down in wait2. The `inc` signal increments the counter in the enter state each time a key is pressed. The next-state function is implemented with a process.

Waveforms generated by simulating the combination lock design entity on a sequence of test inputs are shown in Figure 16.33. The test visits all states and traverses all edges of the state diagram of Figure 16.29. This is done with three attempts to unlock the lock; one correct attempt and two failures. After reset, the test first enters the correct sequence with a pause after the 1 and after the 7. On the cycle after entering the final 8, `enter` is pressed and the next cycle `unlock` goes high and the machine enters the open state (`state = 1`). After one cycle low, `enter` goes high again, returning the machine to the enter state.

The second attempt to unlock the lock involves entering an invalid code. As soon as the first key is entered incorrectly (7 instead of 8), the machine transits to the wait1 state (`state` = 2).

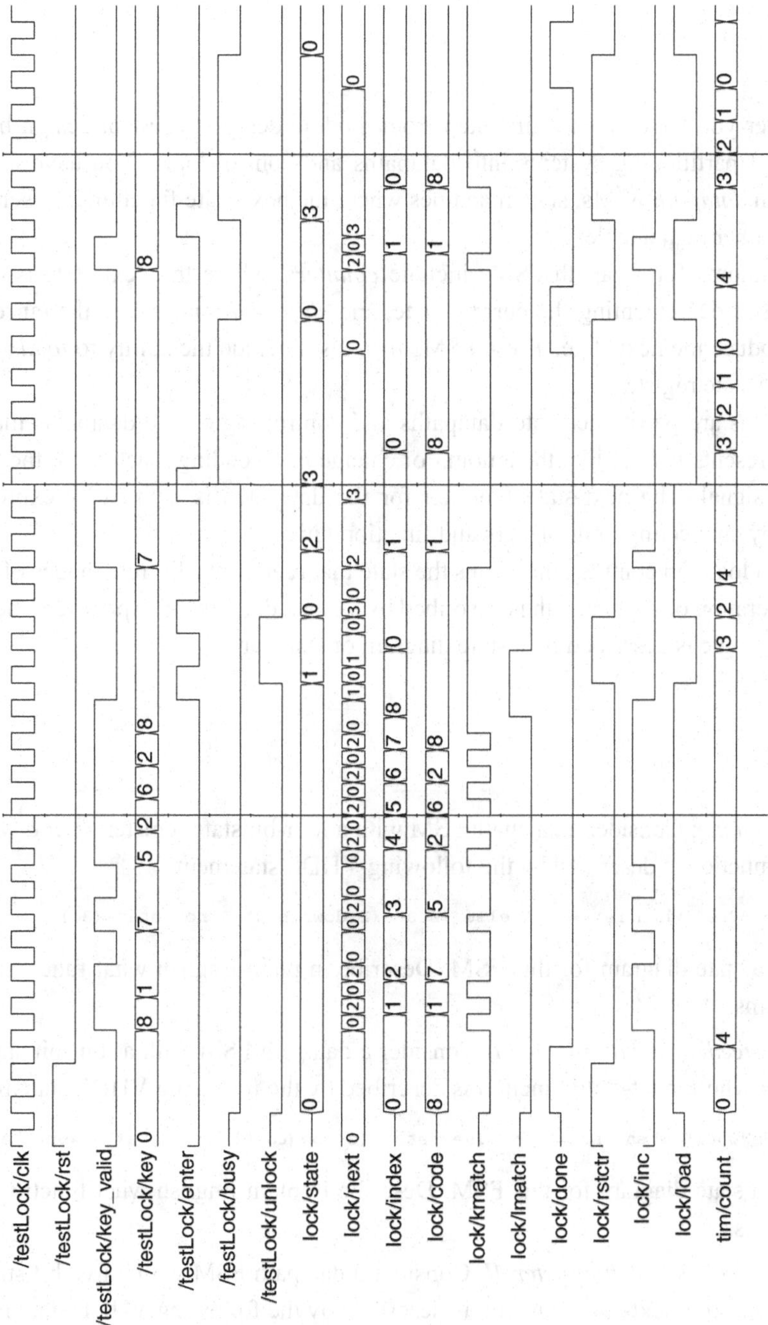

Figure 16.33. Waveforms generated by simulating the combination lock design entity of Figures 16.31 and 16.32.

It stays in this state until `enter` is pressed, at which time it goes to the wait2 state (`state` = 3). In the wait2 state the `busy` output is asserted and the timer starts counting down from 4.[5]

The third attempt to unlock the lock involves entering a code of the wrong length. After the first digit is entered correctly, `enter` is asserted. Because the correct length code has not yet been entered (`lmatch` = 0), the machine transits to the wait2 state and starts counting down the timer.

[5] In practice, we would use a much longer timeout. However, using a short timeout greatly reduces simulation time and results in an easier to read set of waveforms.

Summary

In this chapter you have taken a first step from module design to *system* design by learning the process of partitioning systems into datapaths and control logic. You have also learned how to design *datapath* FSMs, state machines where the next-state function is described by an expression, rather than a table.

Common idioms for datapath FSMs include *counters*, where the next state is derived by incrementing or decrementing the current state, and *shift registers*, where the current state is shifted to produce the next state. These FSMs may also include the ability to *load* an arbitrary value into the state register.

Most systems are partitioned into datapaths and control logic. The datapaths maintain the state that represents *values*, like the amount of change in a vending machine or the magnitude of an audio signal. The next-state function for the datapaths is defined by expressions, or structurally by connecting multiplexers and function units.

The control logic, in contrast, maintains the state that represents discrete modes of operation. While the operation of the datapath is described by a block diagram or expression, the operation of the control logic is described by a state diagram or state table.

Exercises

16.1 *Ring counter.* Consider a datapath FSM with a four-bit state register `state` whose next-state function is described by the following VHDL statement:

```
nxt <= "000" when rst = '1' else (state(2 downto 0) & not state(3));
```

Draw a state diagram for this FSM. Describe in plain English what function this FSM performs.

16.2 *Linear-feedback shift register, I.* Consider a datapath FSM with a four-bit state register `state` whose next-state function is described by the following VHDL statement:

```
nxt <= "0001" when rst = '1' else (state(2 downto 0) & (state(3) xor state(2)));
```

Draw a state diagram for this FSM. Describe in plain English what function this FSM performs.

16.3 *Linear-feedback shift register, II.* Consider a datapath FSM with a five-bit state register `state` whose next-state function is described by the following VHDL statement:

```
nxt <= "00001" when rst = '1' else (state(3 downto 0) & (state(4) xor state(2)));
```

Draw a state diagram for this FSM. Describe in plain English what function this FSM performs.

16.4 *Saturating counter, I.* Draw the block diagram for a counter that can count up, count down, and load. This counter, however, must saturate when counting up (at a programmable maximum count) and counting down (at zero).

16.5 *Saturating counter, II.* Write the VHDL for your saturating counter of Exercise 16.4.

16.6 *Multiple counters, I.* Modify the up/down/load counter of Figure 16.9 to have four separate count registers. Add a two-bit input *rd* that selects which of the four counts are to be modified on any given operation.

16.7 *Multiple counters, II.* Modify the counter of Exercise 16.6 to have both a source register input, *rs*, and a destination register input, *rd*. This allows the user to set $cnt3 = cnt0 - 1$, for example ($rd = 3, rs = 0$). Use the *rd* input as the destination of a load.

16.8 *Fibonacci numbers, I.* Draw the block diagram for a datapath circuit to compute 16-bit Fibonacci numbers. During each cycle, the circuit should output the next Fibonacci number (starting with 0 after reset). The circuit should signal when the next number is larger than 16 bits.

16.9 *Fibonacci numbers, II.* Implement your datapath FSM from Exercise 16.8 in VHDL.

16.10 *Vending machine, I.* Modify the vending machine of Section 16.3.1 in the following manner. When a coin is placed into the machine, the appropriate signal will go high for an unknown number of cycles. You should count each coin once only, and the input will go low before the next coin has been inserted.

16.11 *Vending machine, II.* Modify the vending machine of Section 16.3.1 in the following manner. If the user keeps inserting money into the current implementation, the counter could overflow. Design the vending machine to have a saturating add that returns excess coins to the user.

16.12 *Combination lock, I.* Modify the combination lock of Section 16.3.2 in the following manner. Add the capability of the lock to have multiple users (up to eight). The first digit selects the user, and the rest of the inputs are the unlock sequence. All codes must be stored in a single ROM.

16.13 *Combination lock, II.* Modify the combination lock of Section 16.3.2 in the following manner. Allow the (single) user to get one value incorrect and still unlock the lock. For example, if the code was 12345, any of the following should open the lock: 11345, 12385, or 12349. For a code of length *n*, what is the probability of guessing a correct combination, both for the original implementation and for the forgiving implementation?

16.14 *Tomato weighing, I.* A packing plant packs tomatoes into cartons so that each carton contains at least 16 oz of tomatoes. Each tomato weighs 4–6 oz, so each carton will hold three or four tomatoes. The next three questions will ask you to create a module that accepts, in sequence, the weight of each tomato and outputs the total weight for a package along with a valid signal. The inputs to your system are `rst` (clears the weight), `clk`, `weight(2 downto 0)` (the weight of the current tomato), and `valid_in` (the input weight is valid). When the stored weight (output signal `weight\_out(4 downto 0)`) is 16 oz or more raise the `valid_out` signal. After the valid output signals more than 16 oz of tomatoes, the next input weight should be added to 0, starting a new package. Draw the block diagram for this module's datapath.

16.15 *Tomato weighing, II.* Describe how to generate the control signals (using either logic equations or a block diagram) for the datapath you created in Exercise 16.14.

16.16 *Tomato weighing, III.* Write and verify the VHDL to implement the tomato weigher of Exercise 16.14.

16.17 *Calculator FSM, I.* The datapath of an octal calculator has a 24-bit input register `in_reg` and a 24-bit accumulator `acc`. The contents of both registers are displayed as eight octal (radix-8) digits. Both are cleared on reset. The calculator has buttons for C (clear), the numbers 0–7, and the functions $+$, $-$, and $\times$. Pressing C once clears `in_reg`. Pressing C a second time with no other intervening keys clears `acc`. Pressing a number shifts `in_reg` to the left three bits and puts the number pressed into the low three bits. Pressing a function performs that function on the two registers and puts the result in `acc`. Draw a block diagram for the datapath for this calculator.

16.18 *Calculator FSM, II.* Draw a block diagram for the controller component of the calculator described in Exercise 16.17.

16.19 *Calculator FSM, III.* Write the VHDL to implement the calculator of Exercise 16.17. Write the testbench and verify your VHDL code.

16.20 *Calculator FSM, IV.* Describe, in English and with a block diagram, what must change about your calculator to enable it to accept base-10 input (the numbers 0–9) instead of octal digits. Each digit must be stored and displayed separately – that is, with a module like the seven-segment display of Exercise 6.14.

16.21 *Ascending sequence detector.* Write a VHDL design entity that accepts an eight-byte sequence on eight-bit input `input`, where the first byte is signaled by a single-bit `start` signal. Your design entity should assert a single-bit `done` signal on the cycle after the last byte is input. In the same cycle, it should assert a single-bit `in_sequence` signal if the eight bytes were in ascending sequence; that is, if the $(i+1)$st byte is one more than the ith byte, $b_{i+1} = b_i + 1$, for i from 1 to 7.

16.22 *Descending sequence detector.* Write a VHDL design entity that accepts an eight-byte sequence on eight-bit input `input`, where the first byte is signaled by a single-bit `start` signal. Your design entity should assert a single-bit `done` signal on the cycle after the last byte is input. In the same cycle, it should assert a single-bit `in_sequence` signal if the eight bytes were in descending sequence; that is, if the $(i+1)$st byte is one less than the ith byte, $b_{i+1} = b_i - 1$, for i from 1 to 7.

17 Factoring finite-state machines

Factoring a state machine is the process of splitting the machine into two or more simpler machines. Factoring can greatly simplify the design of a state machine by separating orthogonal aspects of the machine into separate FSMs where they can be handled independently. The separate FSMs communicate via logic signals. One FSM provides input control signals to another FSM and senses its output status signals. Such factoring, if done properly, makes the machine simpler and also makes it easier to understand and maintain – by separating issues.

In a factored FSM, the state of each sub-machine represents one dimension of a multi-dimensional state space. Collectively the states of all of the sub-machines define the state of the overall machine – a single point in this state space. The combined machine has a number of states that is equal to the product of the number of states of the individual sub-machines – the number of points in the state space.[1] With individual sub-machines having a few tens of states, it is not unusual for the overall machine to have thousands to millions of states. It would be impractical to handle such a large number of states without factoring.

We have already seen one form of factoring in Section 16.3 where we developed a state machine with a datapath component and a control component. In effect, we factored the total state of the machine into a datapath portion and a control portion. Here we generalize this concept by showing how the control portion itself can be factored.

In this chapter, we illustrate factoring by working two examples. In the first example, we start with a flat FSM and factor it into multiple simpler FSMs. In the second example we derive a factored FSM directly from the specification, without bothering with the flat FSM. Most real FSMs are designed using the latter method. A factoring is usually a natural outgrowth of the specification of a machine. It is rarely applied to an already flat machine.

17.1 A LIGHT FLASHER

Suppose you have been asked to design a light flasher. The flasher has a single input *in* and a single output *out*. When *in* goes high (for one cycle) it initiates the flashing sequence. During this sequence output *out*, which drives a light-emitting diode (LED), *flashes* three times. For each flash, *out* goes high (LED on) for six cycles. Output *out* goes low for four cycles between flashes. After the third flash your FSM returns to the OFF state awaiting the next pulse on *in*.

[1] For most machines, not all points in the state space are reachable.

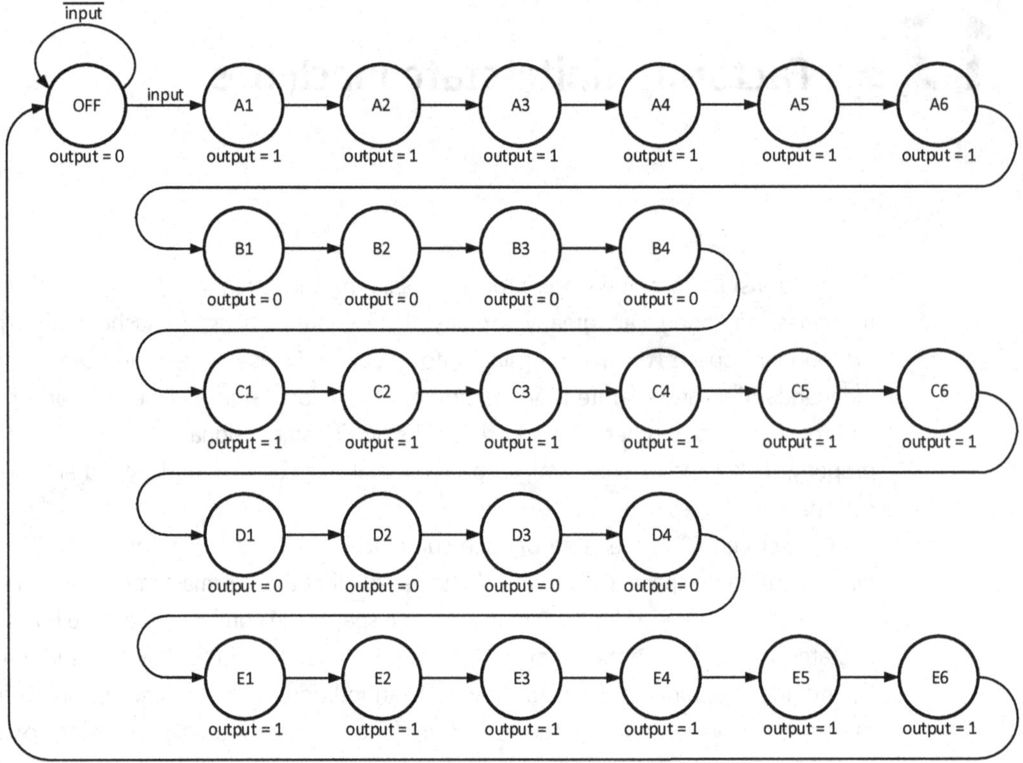

Figure 17.1. State diagram for the light flasher.

A state diagram for this light flasher is shown in Figure 17.1. The FSM contains 27 states: six for each of the three flashes, four for each of the two periods between flashes, and one OFF state. We can implement this FSM with a big `case` statement, with 27 cases. However, suppose the specifications change to require 12 cycles for each flash, four flashes, and seven cycles between flashes. Because this machine is *flat*, any one of these changes would require completely changing the `case` statement.

We will illustrate the process of factoring with our light flasher. This machine can be factored in two ways. First, we can factor out the counting of on and off intervals into a timer. This allows each of the six- or four-state sequences for a given flash or interval to be reduced to a single state. This factoring of timing both simplifies the machine and makes it considerably simpler to change the on and off intervals. Second, we can factor out the three flashes (even though the last is slightly different). This again simplifies the machine and also allows us to change the number of flashes.

Figure 17.2 shows how the flasher is factored to separate the process of counting time intervals into a separate machine. As shown in Figure 17.2(b), the factored machine consists of a *master* FSM that accepts *in* and generates *out* and a timer FSM. The timer FSM receives two control signals from the master and returns a single status signal. The *tload* control signal instructs the timer to load a value into a countdown timer (i.e., to start the timer), and *tsel* selects the value to be loaded. When *tsel* is high, the timer is set to count six cycles. The counter counts four cycles when *tsel* is low. When the counter has completed its countdown, the *done* signal

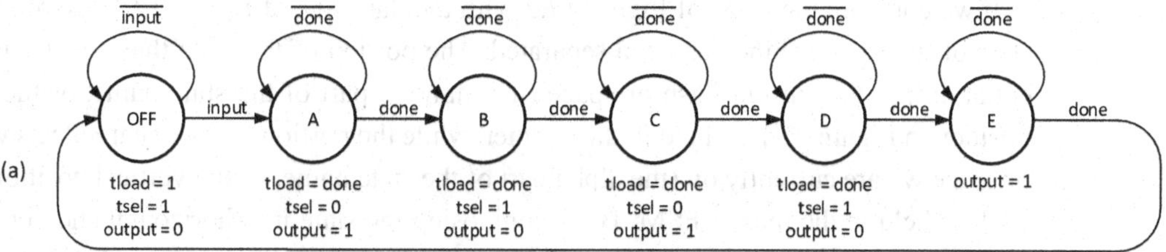

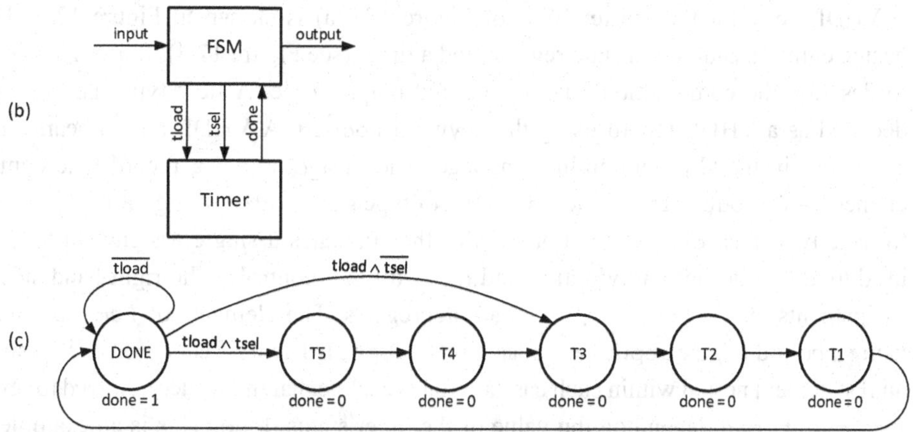

Figure 17.2. State diagram for the light flasher with on and off timing factored out into a timer. (a) State diagram of master FSM. (b) Block diagram showing connections between master FSM and timer FSM. (c) State diagram of timer FSM.

is asserted and remains asserted until the counter is reloaded. A state diagram for this timer is shown in Figure 17.2(c).[2] The timer is implemented as a datapath FSM as discussed in Section 16.1.3.

A state diagram of the master FSM is shown in Figure 17.2(a). The states correspond exactly to the states of Figure 17.1, except that each sequence of repeated states (e.g., A1 to A6) is replaced by a single state (e.g., A). The machine starts in the OFF state. This state repeatedly loads the timer with the countdown value for the *on* sequence ($tsel = 1$). When *in* goes true, the FSM enters state A, *out* goes high, and the timer begins counting down. The master FSM stays in state A, waiting for the timer to finish counting and signal *done*. It will remain in this state for six cycles. During the last cycle in state A, *done* is true, so *tload* is asserted ($tload = done$), and the counter is loaded with the countdown value for the *off* sequence ($tsel = 0$). Note that *tload* is asserted only during this final cycle of state A, when *done* is true. If it were asserted during each cycle of state A, the timer would continually reset and never reach DONE. Since *done* is true in this final cycle, the machine enters state B on the next cycle and *out* goes low. The process repeats in states B–E, with each state waiting for *done* before moving to the next state. Each of these states (except E) loads the counter for the next state during its last cycle by specifying $tload = done$.

[2] This state diagram shows only *tload* being active in state DONE. In fact, our timer can be loaded in any state. The additional edges are omitted for clarity.

If we compare the FSM of Figure 17.2 with the flat FSM of Figure 17.1, we realize that the state of the flat machine has been separated. The portion of the state that represents the cycle count during the current flash or space (the numeric part of the state name, or the horizontal position in Figure 17.1) is held in the counter, while the portion of state that reflects which flash or space we are currently on (the alpha part of the state name, or the vertical position in Figure 17.1) is held in the master FSM. By decomposing the state into horizontal and vertical in this manner, we are able to represent all 27 states of the original machine with just two six-state machines.

VHDL code for the master FSM of Figure 17.2(a) is shown in Figure 17.3. The `flash` design entity instantiates a state register and a timer (see Figure 17.4). It uses a `case` statement to describe the combinational next state and output logic. A new type, `fsm_output_t`, is declared as a VHDL record using the keyword **`record`**. A VHDL record bears similarity to a `struct` in the C programming language. The `fsm_output_t` record type contains three elements – `output`, `tload`, and `tsel` – all of type `std_logic`. A signal `fsm_out` is declared to have type `fsm_output_t`. For each of the six states a single assignment to `fsm_out` is used to assign design entity output and the two timer controls. The right-hand sides of these assignments, e.g., `('0','1','1')`, are aggregates. The elements of `fsm_out` are accessed using a period (e.g., `output <= fsm_out.output`). The next state, `nxt1`, is assigned using an if statement nested within each case alternative. Note that the values assigned to timer control `fsm_out.tload` depend on the value of the timer status `done`. This is an example of a state machine where an input `done` directly affects an output, `fsm_out.tload`, with no delay. In each state, the next state remains the same as the current state unless either `input` or `done` is asserted. A final conditional signal assignment statement resets the machine to the `OFF` state when `rst` is asserted.

For completeness, the VHDL code for the timer is shown in Figure 17.4. The approach taken is similar to that described in Section 16.1.3.

The waveform display from a simulation of the factored flasher of Figure 17.2 is shown in Figure 17.5. The output, the fourth line from the top, shows the desired three pulses with each pulse six clocks wide and spaces four clocks wide. The state of the master FSM is directly below the output, and the state of the timer is on the bottom line. The waveforms show how the master FSM stays in one state while the timer counts down – from 5 to 0 for flashes, and from 3 to 0 for spaces. The timer control lines (directly above the timer state) show how in states A–E (1–5) `tload` follows `done`.

We can factor our flasher further by recognizing that states A, C, and E are repeating the same function. The only difference is the number of remaining flashes. We can factor the number of remaining flashes out into a second counter, as shown in Figure 17.6. Here the master FSM has only three states that determine whether the machine is off, in a flash, or in a space. The state that determines the position within a flash or space is held in a timer (just as in Figure 17.2). Finally, the state that determines the number of remaining flashes is held in a counter. Collectively the three FSMs, the master, the timer, and the counter, determine the total state of the factored machine. Each of the three sub-machines determines the state along one axis of a three-dimensional state space.

The state diagram of the master FSM for the doubly factored machine is shown in Figure 17.6(b). The machine has only three states. It starts in the OFF state. In the OFF state, both the timer and the counter are loaded. The timer is loaded with the countdown value for a flash.

```
-- Flash - flashes out three times 6 cycles on, 4 cycles off
--         each time in is asserted.
library ieee;
use ieee.std_logic_1164.all;
use work.ff.all;
use work.flash_declarations.all;

entity Flash is
  port( clk, rst, input: in std_logic; -- input triggers start of flash sequence
        output: out std_logic ); -- output drives LED
end Flash;

architecture impl of Flash is
  type fsm_output_t is record output, tload, tsel : std_logic; end record;
  signal fsm_out : fsm_output_t;
  signal state: std_logic_vector(SWIDTH-1 downto 0); -- current state
  signal nxt, nxt1: std_logic_vector(SWIDTH-1 downto 0); -- next state with and w/o reset
  signal done: std_logic; -- timer output
begin
  -- instantiate state register
  STATE_REG: vDFF generic map(SWIDTH) port map(clk,nxt,state);

  -- instantiate timer
  TIMER: Timer1 port map(clk,rst,fsm_out.tload,fsm_out.tsel,done);

  process(all) begin
    case state is
      when S_OFF => fsm_out <= ('0','1','1');
        if input then nxt1 <= S_A; else nxt1 <= S_OFF; end if;
      when S_A =>   fsm_out <= ('1',done,'0');
        if done then nxt1 <= S_B; else nxt1 <= S_A; end if;
      when S_B =>   fsm_out <= ('0',done,'1');
        if done then nxt1 <= S_C; else nxt1 <= S_B; end if;
      when S_C =>   fsm_out <= ('1',done,'0');
        if done then nxt1 <= S_D; else nxt1 <= S_C; end if;
      when S_D =>   fsm_out <= ('0',done,'1');
        if done then nxt1 <= S_E; else nxt1 <= S_D; end if;
      when S_E =>   fsm_out <= ('1',done,'1');
        if done then nxt1 <= S_OFF; else nxt1 <= S_E; end if;
      when others => fsm_out <= ('1',done,'1');
        if done then nxt1 <= S_OFF; else nxt1 <= S_E; end if;
    end case;
  end process;

  nxt <= S_OFF when rst else nxt1;
  output <= fsm_out.output;
end impl;
```

Figure 17.3. VHDL description of the master FSM from Figure 17.2(a).

The counter is loaded with one fewer than the number of flashes required (i.e., the counter is loaded with a 3 for four flashes). Having input *in* go high causes a transition to the FLASH state. In the FLASH state the output *out* is true, the timer counts down, and the counter is idle. During the last cycle of the FLASH state the timer has reached its zero state and *tdone* is true.

```
-- Timer 1 - reset to done state.  Load time when tload is asserted
--    Load with T_ON if tsel, otherwise T_OFF.  If not being loaded or
--    reset, timer counts down each cycle.  Done is asserted and timing
--    stops when counter reaches 0.
library ieee;
use ieee.std_logic_1164.all;
use ieee.std_logic_unsigned.all;
use ieee.std_logic_misc.all;
use work.ff.all;
use work.flash_declarations.all;

entity Timer1 is
  generic( n: integer := T_WIDTH );
  port( clk, rst, tload, tsel: in std_logic;
        done_o: out std_logic );
end Timer1;

architecture impl of Timer1 is
  signal done: std_logic;
  signal next_count, count: std_logic_vector(n-1 downto 0);
begin
  -- state register
  STATE: vDFF generic map(n) port map(clk, next_count, count);

  -- signal done
  done <= not or_reduce(count); done_o <= done;

  -- next count logic
  process(all) begin
    case? std_logic_vector'(rst & tload & tsel & done) is
      when "1---" => next_count <= (others => '0');
      when "011-" => next_count <= T_ON;
      when "010-" => next_count <= T_OFF;
      when "00-0" => next_count <= count - '1';
      when "00-1" => next_count <= count;
      when others => next_count <= count;
    end case?;
  end process;
end impl;
```

Figure 17.4. VHDL description for the timer FSM used by the light flasher of Figure 17.2.

During this cycle the timer is reloaded with the countdown value for a space. With *tdone* true, the FSM proceeds from the FLASH state to the SPACE state if the counter is not done (*cdone* false). Otherwise, if this was the last flash (*cdone* true), the machine returns to the OFF state. In the SPACE state, the output is false, the timer counts down, and the counter is idle. In the final cycle of the SPACE state, *tdone* is true. This causes the counter to decrement, reducing the count of the number of remaining flashes, and the timer to reload with the countdown value for a flash.

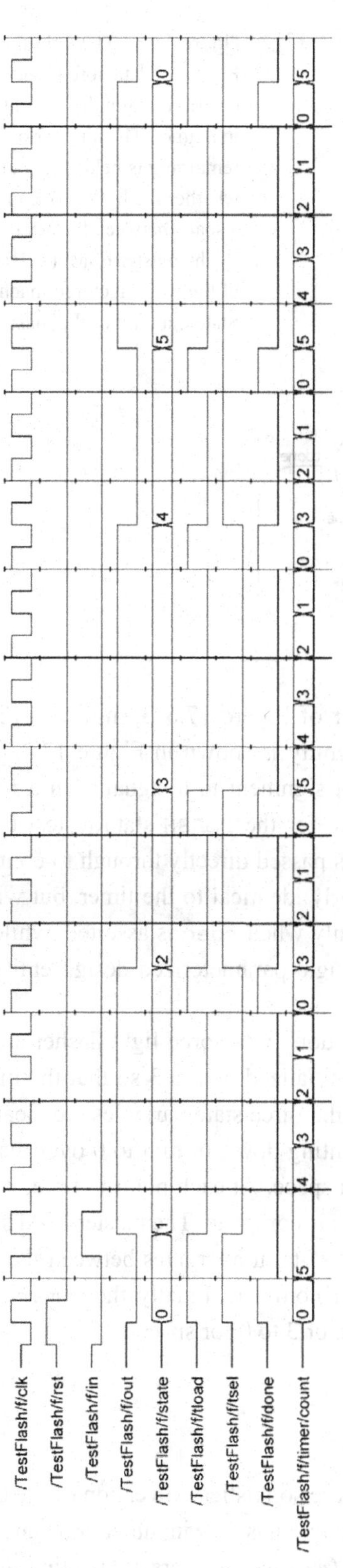

Figure 17.5. Waveform display showing a simulation of the factored light flasher of Figure 17.2.

(a)

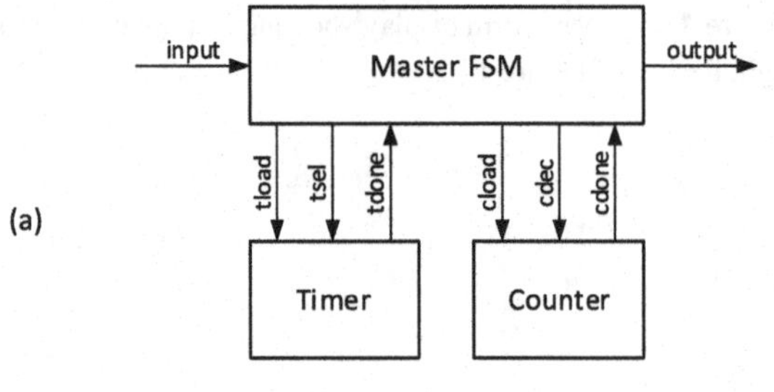

(b)

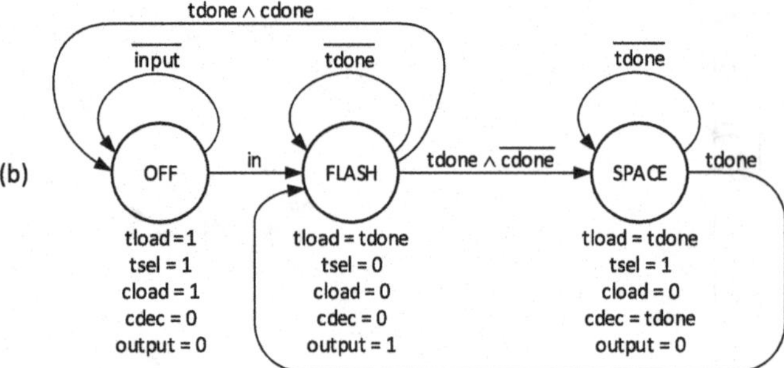

Figure 17.6. The light flasher of Figure 17.1 factored twice. The position within the current flash is held in a timer. The number of flashes remaining is held in a counter. Finally, whether the light is off, in a flash, or in a space between flashes is determined by the master FSM. (a) Block diagram of the twice-factored machine. (b) State diagram of the master FSM.

The VHDL code for the doubly factored flasher of Figure 17.6 is shown in Figure 17.7, and the VHDL description of the counter design entity is shown in Figure 17.8. The output functions for the master FSM again use a single assignment to a signal with a record type. An **if** statement is used to compute the next state for the `FLASH` state to test both `tdone` and `cdone`. In several states, status signal `tdone` is passed directly through to control signals `tload` and `cdec`. The counter design entity is nearly identical to the timer, but with slightly different control because the counter decrements only when `cdec` is asserted, while the timer always decrements. With some generalization a single parameterized design entity could be used for both functions.

The waveforms generated by a simulation of the doubly factored light flasher are shown in Figure 17.9. For this simulation, the counter was initialized with a 3 so that the master FSM produces four flashes. The different time scales of the three state variables are clearly visible. The counter (bottom line) moves most slowly, counting down from 3 to 0 over the four-flash sequence. It decrements after the last cycle of each space. At each point in time, it represents the number of flashes left to do after the current flash is complete. The master FSM state moves the next most slowly. After starting on the 0 (`OFF`) state, it alternates between the 1 (`FLASH`) state and the 2 (`SPACE`) state until the four flashes are complete. Finally, the `timer` state moves most rapidly, counting down from 5 to 0 for flashes, or 3 to 0 for spaces.

17.2 TRAFFIC-LIGHT CONTROLLER

As a second example of factoring, we consider a more sophisticated version of the traffic-light controller we introduced in Section 14.3. This machine has two inputs, *carew* and *carlt*, that indicate that cars are waiting on the east–west road (*ew*) and that cars are waiting in a left-turn lane (*lt*). The machine has nine output lines that drive three sets of three lights each, one set

```
library ieee;
use ieee.std_logic_1164.all;
use work.flash_declarations.all;
use work.ff.all;

entity Flash2 is
  port( clk, rst, input: in std_logic; -- in triggers start of flash sequence
        output: out std_logic ); -- out drives LED
end Flash2;

architecture impl of Flash2 is
  type fsm_output_t is record
    output, tload, tsel, cload, cdec : std_logic;
  end record;
  signal fsm_out : fsm_output_t;
  signal state, nxt, nxt1: std_logic_vector(XWIDTH-1 downto 0);
  signal tload, tsel, cload, cdec: std_logic;  -- timer and counter inputs
  signal tdone, cdone: std_logic;              -- timer and counter outputs
begin
  -- instantiate timer and counter
  TIMER: Timer1 port map(clk, rst, tload, tsel, tdone);
  COUNTER: Counter1 port map(clk, rst, cload, cdec, cdone);

  -- instantiate state register
  STATE_REG: vDFF generic map(XWIDTH) port map(clk, nxt, state) ;

  process(all) begin
    case state is
      when X_OFF =>  fsm_out <= ('0','1','1','1','0');
        if input then nxt1 <= X_FLASH;
        else nxt1 <= X_OFF; end if;
      when X_FLASH => fsm_out <= ('1',tdone,'0','0','0');
        if not tdone then nxt1 <= X_FLASH;
        elsif not cdone then nxt1 <= X_SPACE;
        else nxt1 <= X_OFF; end if;
      when X_SPACE => fsm_out <= ('0',tdone,'1','0',tdone);
        if not tdone then nxt1 <= X_SPACE;
        else nxt1 <= X_FLASH; end if;
      when others =>  fsm_out <= ('0',tdone,'1','0',tdone);
        if not tdone then nxt1 <= X_SPACE;
        else nxt1 <= X_FLASH; end if;
    end case;
  end process;

  nxt <= X_OFF when rst = '1' else nxt1;
end impl;
```

Figure 17.7. VHDL description of the master FSM from Figure 17.6.

each for the north–south road, the east–west road, and the left-turn lane (from the north–south road). Each set of lights consists of a red light, a yellow light, and a green light.

Normally, the light will be green for the north–south road. However, if a car is detected on the east–west road or the left-turn lane, we wish to switch the lights so that the east–west or left-turn

```
-- Counter1 - pulse counter
--    cload - loads counter with C_COUNT
--    cdec  - decrements counter by one if not already zero
--    cdone - signals when count has reached zero
library ieee;
use ieee.std_logic_1164.all;
use ieee.std_logic_misc.all;
use ieee.std_logic_unsigned.all;
use work.flash_declarations.all;
use work.ff.all;

entity Counter1 is
  generic( n: integer := C_WIDTH );
  port( clk, rst, cload, cdec: in std_logic;
        cdone: buffer std_logic );
end Counter1;

architecture impl of Counter1 is
  signal count, next_count: std_logic_vector(n-1 downto 0);
begin
  -- state register
  STATE: vDFF generic map(n) port map(clk, next_count, count);

  -- signal done
  cdone <= not or_reduce(count);

  -- next count logic
  process(all) begin
    case? std_logic_vector'(rst & cload & cdec & cdone) is
      when "1---" => next_count <= (others => '0');
      when "01--" => next_count <= C_COUNT;
      when "0010" => next_count <= count - '1';
      when "00-1" => next_count <= count;
      when others => next_count <= count;
    end case?;
  end process;
end impl;
```

Figure 17.8. VHDL description of the counter from Figure 17.6.

light is green (with priority going to the left-turn lane). Once the lights have been switched to east–west or left-turn, we leave them switched either until no more cars are detected in that direction, or until a timer expires. The lights then return to green in the north–south direction.

Each time we switch the lights, we switch the lights in the active direction from green to yellow. After a time interval, we switch them to red. Then, after a second time interval, we switch the lights in a different direction to green. The lights are not allowed to change again until they have been green for a third time interval.

Given this specification, we decide to factor the finite-state machine that implements this traffic-light controller into five component modules as shown in Figure 17.10. A master FSM accepts the inputs and decides which direction should be green, signal `dir`. To determine when it is time to force the lights back to north–south, it uses a timer, Timer1. It also receives a signal

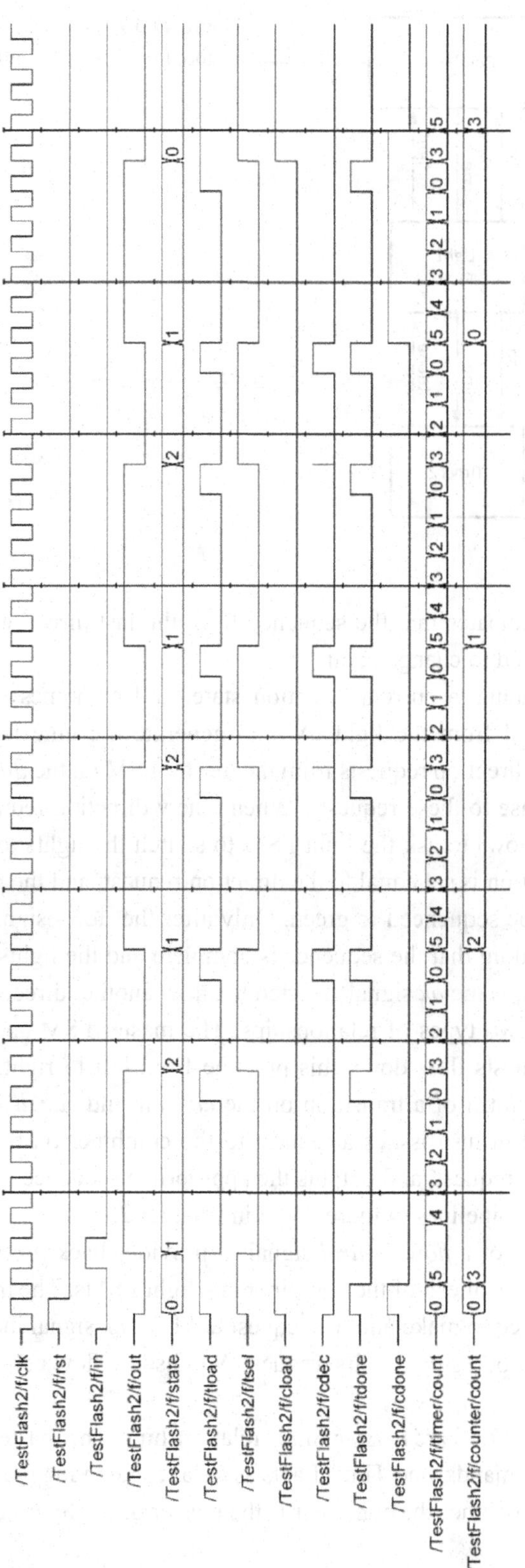

Figure 17.9. Waveform display showing a simulation of the doubly factored light flasher of Figure 17.6.

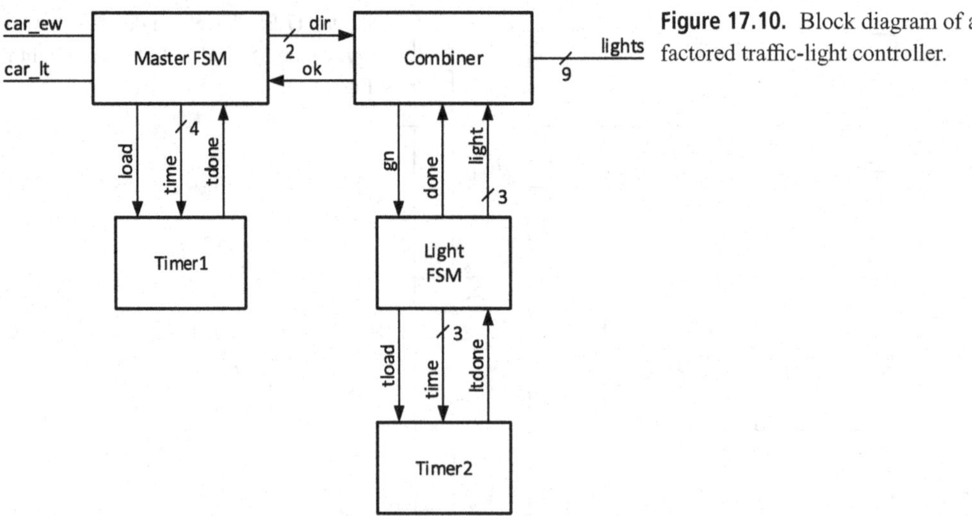

Figure 17.10. Block diagram of a factored traffic-light controller.

ok back from the combiner that indicates that the sequence from the last direction change is complete and the direction is allowed to change again.

The combiner component maintains a current direction state, and combines this current direction state with the *light* signal from the light FSM to generate the nine light outputs *lights*. The combiner also receives direction requests from the master FSM on the *dir* signal and sequences the light FSM in response to these requests. When a new direction request occurs, the combiner deasserts *gn* (sets it low) to ask the light FSM to switch the lights to red. Once the lights are red, the current direction is set equal to the direction request, and the *gn* signal is asserted to request that the lights be sequenced to green. Only after the `done` signal from the light FSM has been asserted, signaling that the sequence is complete and the lights have been green for the required time interval, is the *ok* signal asserted to allow another direction change.

The components here illustrate two types of relationships. The master FSM and combiner components form a *pipeline*. Requests flow down this pipeline from left to right. A request is input to the master FSM in the form of a transition on the *car_ew* and *car_lt* inputs. The master processes this request, and in turn issues a request to the combiner on the *dir* lines. The combiner in turn processes the request, and outputs the appropriate sequence on the *lights* output in response. We will discuss pipelines in more depth in Chapter 23.

The *ok* signal here is an example of a *flow-control* signal. It provides back-pressure on the master FSM to prevent it from getting ahead of the combiner and light FSMs. The master FSM makes a request, and it is not allowed to make another request until the *ok* signal indicates that the rest of the circuit has finished processing the first request. We discuss flow control in more detail in Section 22.1.3.

The other relationships in Figure 17.10 are *master–slave* relationships. The master FSM acts as a master to Timer1, giving it commands, and Timer1 acts as a slave, receiving the commands and carrying them out. In a similar manner the combiner is the master of the light FSM, which, in turn, is the master of Timer2.

The light FSM sequences the traffic lights from green to red and then back to green. It receives requests from the combiner on the *gn* signal and responds to these requests with the *done* signal. It also generates the three-bit *light* signal, which indicates which light in the current direction should be illuminated. When the *gn* signal is set high, it requests that the light FSM switch the *light* signal to green. When this is completed, and the minimum green time interval has elapsed, the light FSM responds by asserting *done*. When the *gn* signal is set low, it requests that the light FSM sequence the *light* signal to red – via a yellow state and observing the required time intervals. When this is completed, *done* is set low. The light FSM uses its own timer (Timer2) to count out the time intervals required for light sequencing.

For the interface between the light FSM and the combiner, the *done* signal provides flow control. The combiner can only toggle *gn* high when *done* is low, and can only toggle *gn* low when *done* is high. After toggling *gn*, it must wait for *done* to switch to the same state as *gn* before switching *gn* again.

A state diagram for the master FSM is shown in Figure 17.11. The machine starts in the `NS` state. In this state, Timer1 is loaded so it can count down in the `LT` and `EW` states, and the requested direction is `NS`. State `NS` is exited when the *ok* signal indicates that a new direction can be requested and when one of the *car* signals indicates that there is a car waiting in another direction. In the `EW` and `LT` states, the new direction is requested, and the state is exited when `ok` is true and either there is no longer a car in that direction or the timer has expired, as signaled by *tdone*.

The light FSM is a simple light sequencer similar to our light flasher of Section 17.1. A state diagram for the light FSM is shown in Figure 17.12. Like the light flasher, during the last

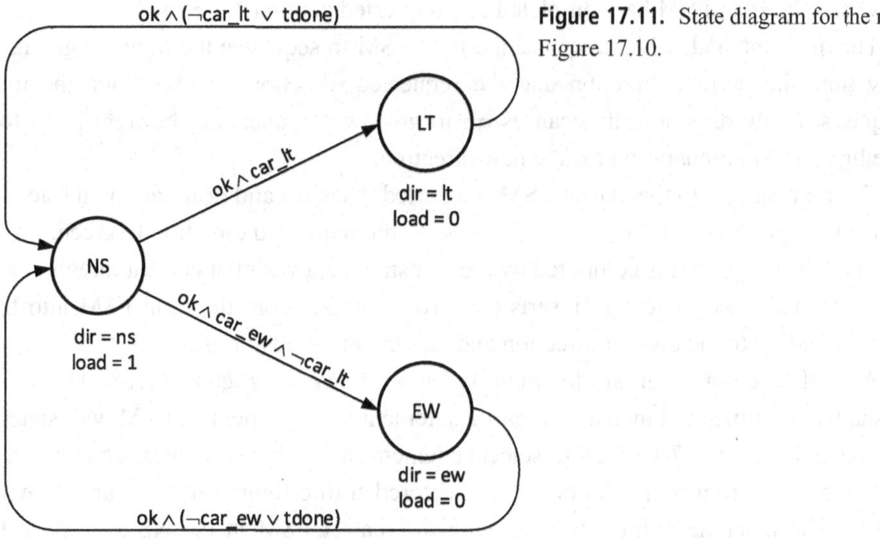

Figure 17.11. State diagram for the master FSM of Figure 17.10.

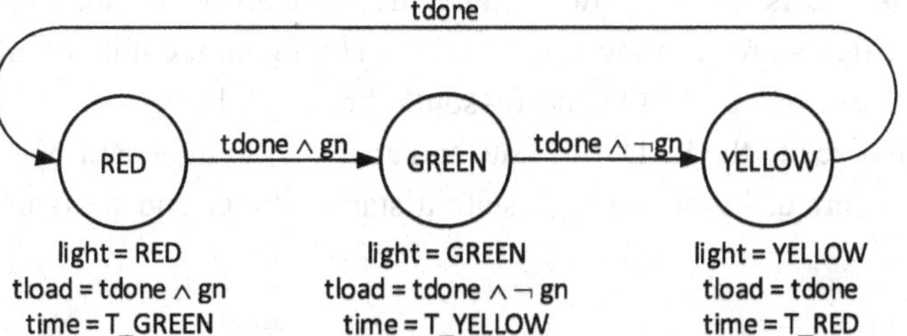

Figure 17.12. State diagram for the light FSM of Figure 17.10.

cycle in each state the timer is loaded with the timeout for the next state. The machine starts in the *RED* state. It transitions to *GREEN* when the timer is done and the `gn` signal from the combiner indicates that a green light is requested. The `done` signal is asserted after the timer in the *GREEN* state has completed its countdown. The transition to *YELLOW* is triggered by the timer being complete and the *gn* signal being low.[3] The transition from *YELLOW* to *RED* occurs on the timer alone. The *done* signal is held high during the *YELLOW* state. It is allowed to go low, signaling that the transition to *RED* is complete, only after the timer has completed its countdown in the *RED* state.

A VHDL description of the master FSM for the factored traffic-light controller is shown in Figure 17.13. This design entity instantiates a timer and then uses a `case` statement to realize a three-state FSM that exactly follows the state diagram of Figure 17.11. "**`if`**" statements are used for the next-state logic. A chain of four **`if-elsif-else`** statements is used in state `M_NS` to test in turn signals `ok`, `car_lt`, and `car_ew`.

The master FSM does not instantiate the combiner. Both the master and the combiner are instantiated as *peer* components at the top level.

A VHDL description of the combiner design entity is shown in Figure 17.14. This design entity accepts a `dir` input from the master FSM, responds to the master FSM with flow-control signal `ok`, and generates the `lights` output. The key piece of state in the combiner design entity is the current direction register `cur_dir`. This holds the direction that the light FSM is currently sequencing. It is updated with the requested direction, `dir`, when `gn` and `done` are both low. This occurs when the combiner has requested that the lights be sequenced to red (`gn` low) and the light FSM has completed this requested action (`done` low).

The `gn` command, which requests the light FSM to sequence the lights to green, is asserted any time the current direction and the requested direction match. When the master FSM requests a new direction, this causes `gn` to go low – requesting the light FSM to sequence the lights red in preparation for the new direction.

The `ok` response to the master FSM is asserted when `gn` and `done` are both true. This occurs when the light FSM has completed sequencing the requested direction to green.

The `lights` output is computed by a `case` statement with the current direction as the `case` variable. This case statement inserts the `light` output from the light FSM into the position corresponding to the current direction and sets the other positions to red.

A VHDL description of the light FSM is shown in Figure 17.15. The design entity instantiates a timer and then uses a `case` statement to implement an FSM with state transitions as shown in Figure 17.12. VHDL select statements are used for the next-state logic.

Waveforms from a simulation of the factored traffic-light controller are shown in Figure 17.16. The machine is initially reset with the master FSM in the NS state (00). Output *dir* is also NS (00). The *ok* line is initially low because the light FSM has not yet finished its sequencing to make the lights green in the NS direction. The lights are initially all red (444) but change after one cycle to be green in the north–south direction (144).

The light FSM is initialized to the RED (00) state and advances to the GREEN state because *gn* and *tdone* are both asserted. In the GREEN state it starts a timer and waits until *tdone* is

[3] The check for *tdone* should be redundant since *gn* should not go low until *done* is high.

```
------------------------------------------------------------------------
--Master FSM
--   car_ew - car waiting on east-west road
--   car_lt - car waiting in left-turn lane
--   ok     - signal that it is ok to request a new direction
--   dir    - output signaling new requested direction
------------------------------------------------------------------------
library ieee;
use ieee.std_logic_1164.all;
use work.ch17_tlc.all;
use work.ff.all;
use work.ch16.all;

entity TLC_Master is
  port( clk, rst, car_ew, car_lt, ok: in std_logic;
        dir: out std_logic_vector(1 downto 0) ); -- direction output
end TLC_Master;

architecture impl of TLC_Master is
  type fsmo_t is record dir: std_logic_vector(1 downto 0); tload: std_logic; end record;
  signal fsmout: fsmo_t;
  signal state, nxt: std_logic_vector(MWIDTH-1 downto 0); -- current and next state
  signal nxt1: std_logic_vector(MWIDTH-1 downto 0); -- next data without reset
  signal tdone: std_logic; -- timer completion
begin
  -- instantiate state register
  STATE_REG: vDFF generic map(MWIDTH) port map(clk, nxt, state);

  -- instantiate timer
  TIMERT: Timer generic map(TWIDTH) port map(clk, rst, fsmout.tload, T_EXP, tdone);

  process(all) begin
    case state is
      when M_NS => fsmout <= (M_NS,'1');
        if not ok then nxt1 <= M_NS;
        elsif car_lt then nxt1 <= M_LT;
        elsif car_ew then nxt1 <= M_EW;
        else nxt1 <= M_NS; end if;
      when M_EW => fsmout <= (M_EW,'0');
        if ok and (not car_ew or tdone) then nxt1 <= M_NS;
        else nxt1 <= M_EW; end if;
      when M_LT => fsmout <= (M_LT,'0');
        if ok and (not car_ew or tdone) then nxt1 <= M_NS;
        else nxt1 <= M_LT; end if;
      when others => fsmout <= (M_NS,'0');
        nxt1 <= M_NS;
    end case;
  end process;
  nxt <= M_NS when rst = '1' else nxt1;
end impl;
```

Figure 17.13. VHDL description of the master FSM for the traffic-light controller.

```
--------------------------------------------------------------------
-- Combiner -
--    dir - direction request from master FSM
--    ok  - acknowledge to master FSM
--    lights - 9-bits to control traffic lights {NS,EW,LT}
--------------------------------------------------------------------
library ieee;
use ieee.std_logic_1164.all;
use work.ch17_tlc.all;
use work.ff.all;

entity TLC_Combiner is
  port( clk, rst: in std_logic;
        ok: out std_logic;
        dir: in std_logic_vector( 1 downto 0 );
        lights: out std_logic_vector( 8 downto 0 ) );
end TLC_Combiner;

architecture impl of TLC_Combiner is
  signal done, gn: std_logic;
  signal light: std_logic_vector(2 downto 0);
  signal cur_dir, next_dir: std_logic_vector(1 downto 0);
begin
  -- current direction register
  DIR_REG: vDFF generic map(2) port map(clk, next_dir, cur_dir);

  -- light FSM
  LT: TLC_Light port map(clk, rst, gn, done, light);

  -- request green from light FSM until direction changes
  gn <= '1' when cur_dir = dir else '0';

  -- update direction when light FSM has made lights red
  next_dir <= "00" when rst else
              dir  when gn and done else
              cur_dir;

  -- ok to take another change when light FSM is done
  ok <= gn and done ;

  -- combine cur_dir and light to get lights
  process(all) begin
    case cur_dir is
      when M_NS => lights <= light & RED   & RED;
      when M_EW => lights <= RED   & light & RED;
      when M_LT => lights <= RED   & RED   & light;
      when others => lights <= RED & RED   & RED;
    end case;
  end process;
end impl;
```

Figure 17.14. VHDL description of the combiner for the traffic-light controller.

```
--------------------------------------------------------------------------
-- Light FSM
--------------------------------------------------------------------------
library ieee;
use ieee.std_logic_1164.all;
use work.ch17_tlc.all;
use work.ff.all;
use work.ch16.all;

entity TLC_Light is
  port( clk, rst, gn: in std_logic;
        done: out std_logic;
        light: out std_logic_vector(2 downto 0) );
end TLC_Light;

architecture impl of TLC_Light is
  type fsm_output_type is record
    tload : std_logic;
    tin   : std_logic_vector(LTWIDTH-1 downto 0);
    light : std_logic_vector(2 downto 0);
    done  : std_logic;
  end record;
  signal fsmo: fsm_output_type;
  signal state, nxt: std_logic_vector(LWIDTH-1 downto 0); -- current state, next state
  signal nxt1: std_logic_vector(LWIDTH-1 downto 0); -- next state w/o reset
  signal tdone: std_logic;
begin
  -- instantiate timer
  TIMERT: Timer port map(clk, rst, fsmo.tload, fsmo.tin, tdone);

  -- instantiate state register
  STATE_REG: vDFF generic map(LWIDTH) port map(clk, nxt, state);

  process(all) begin
    case state is
      when L_RED =>    fsmo <= ((tdone and gn), T_GREEN, RED, not tdone);
        if tdone and gn then nxt1 <= L_GREEN;
        else nxt1 <= L_RED; end if;
      when L_GREEN =>  fsmo <= ((tdone and not gn), T_YELLOW, GREEN, tdone);
        if tdone and not gn then nxt1 <= L_YELLOW;
        else nxt1 <= L_GREEN; end if;
      when L_YELLOW => fsmo <= (tdone, T_RED, YELLOW, '1');
        if tdone then nxt1 <= L_RED; else nxt1 <= L_YELLOW; end if;
      when others =>   fsmo <= (tdone, T_RED, YELLOW, '1');
        if tdone then nxt1 <= L_RED; else nxt1 <= L_YELLOW; end if;
    end case;
  end process;

  nxt <= L_RED when rst else nxt1;
  done <= fsmo.done; light <= fsmo.light;
end impl;
```

Figure 17.15. VHDL description of the light FSM for the traffic-light controller.

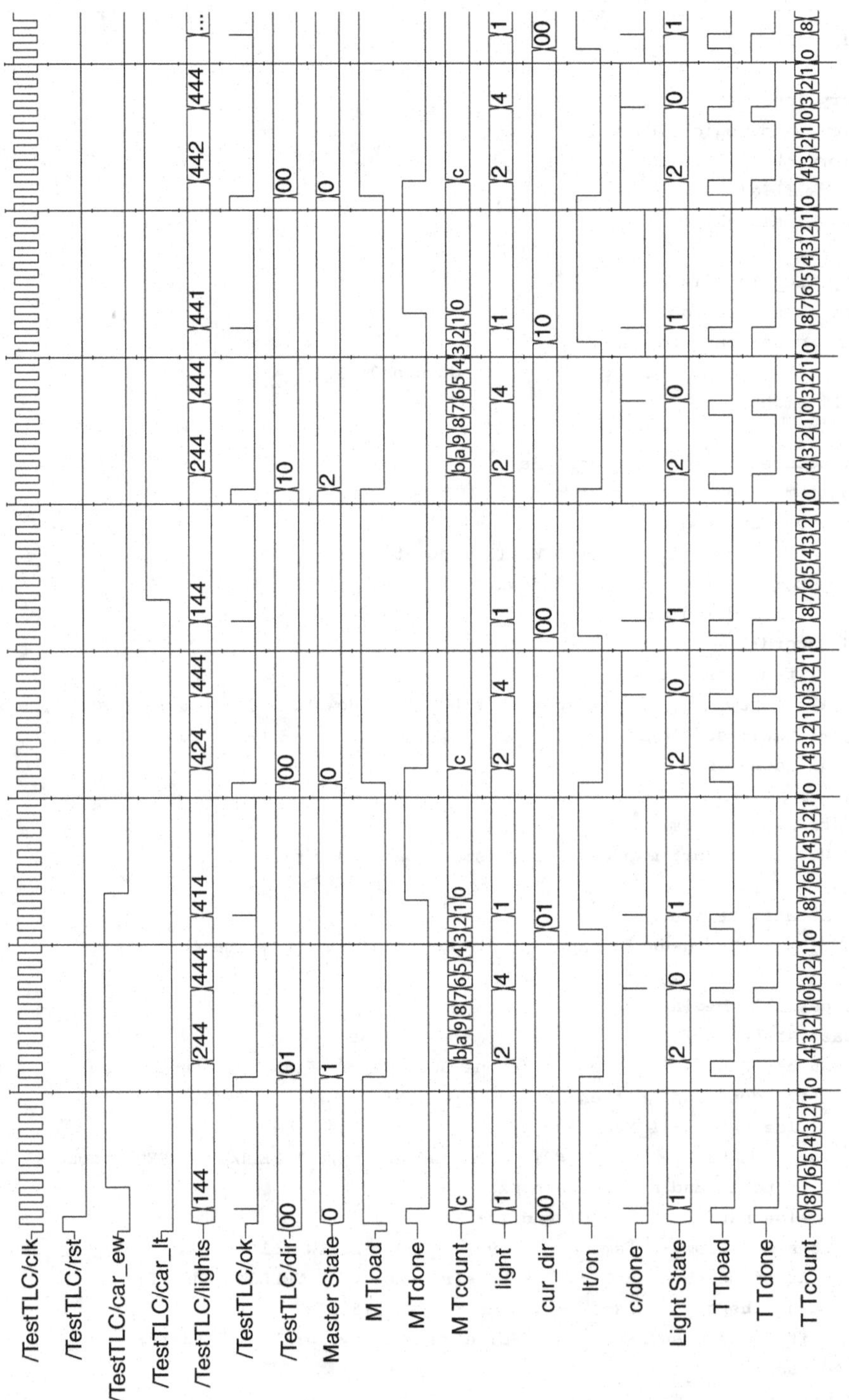

Figure 17.16. Waveform display showing a simulation of the factored traffic-light controller.

again asserted before signaling *done* to the combiner, which in turn causes the combiner to signal *ok* to the master FSM.

Because *car_ew* is asserted, once the *ok* signal goes high, the master FSM requests a change in direction to east–west by setting *dir* to 01. The combiner responds by setting *gn* low to request the light FSM to sequence the current direction's light to red. The light FSM in turn responds by transitioning to the `YELLOW` state (10), which causes *light* to go to yellow (2) and *lights* to go to 244 – yellow in the north–south direction. When the light timer completes its countdown, the light machine enters the `RED` state (00), *light* becomes 4 (RED), and *lights* becomes 444 – red in all directions. Once the light timer has counted down the minimum time in the all-red state, the light FSM sets *done* low, signaling that it has completed the transition.

With *gn* and *done* both low, the combiner updates *cur_dir* to east–west (01) and sets *gn* high to request the light FSM to sequence the lights green in the east–west direction. When the light FSM has completed this action, including timing the GREEN state, it sets *done* true. This in turn causes the combiner to set *ok* true, signaling the master FSM that it is ready to accept a new direction.

When *ok* is asserted the second time, the master FSM requests the north–south direction again ($dir = 00$). This decision is driven by the *car_ew* line being low or the master timer being done. This new direction causes *gn* to go low, sequencing the lights to the all-red state. Then, after *cur_dir* has been updated, *gn* is set high to sequence them back to the GREEN state. When this is all complete, *ok* is asserted again.

On this third assertion of *ok*, signal *car_lt* is true, so a left-turn is requested ($dir = 10$) and the lights are sequenced to red and back to green again. When the sequencing has been completed, *ok* is asserted for a fourth time. This time *car_lt* is still asserted, but the master timer is done, so a north–south direction is again requested.

Summary

In this chapter you have learned, through two examples, how to *factor* a complex finite-state machine into multiple, simpler machines. Factoring a state machine takes a large, one-dimensional state space and maps it onto a multi-dimensional state space – with one dimension for each of the machines.

Combining common state sequences is one approach to factoring. If a state diagram contains an identical (or nearly identical) sequence multiple times, this sequence can be *factored* out. One state machine implements the common sequence, and another state machine keeps track of which instance of the repeated sequence is currently taking place – so control can transfer to the proper state at the end of the sequence. Our light flasher is factored by recognizing common sequences.

Layering is a another approach to factoring, in which we build a hierarchy of machines. The top-level machine makes the top-level decisions, e.g., which direction should have the green light, and keeps the top-level state. The top-level machine calls on one or more lower-level state machines to carry out its directives, like sequencing the lights from green to red and vice versa. The lower-level machines may call on their own servant machines – timers, for example. *Flow control* may be required in order to synchronize the operations of the different layers of machines.

Exercises

17.1 *Factor a state diagram, I-I.* Consider the state diagram shown in Figure 17.17.

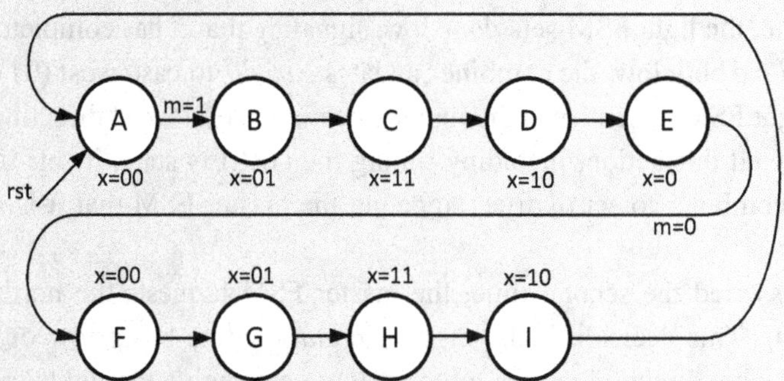

Figure 17.17. Unfactored state diagram for Exercise 17.1. It takes in a one-bit signal m and outputs a two-bit signal x. Edges not labeled with an input value transition automatically.

(a) Identify identical or nearly identical sequences of states in this FSM.

(b) Draw the state diagram for a separate FSM that implements these sequences of states – inputs should select between variations in the sequence.

(c) Draw a revised top-level state diagram that invokes your FSM from (b) to implement the repeated sequence.

17.2 *Factor a state diagram, I-II.* Implement your factored state machine from Exercise 17.1 in VHDL.

17.3 *Factor a state diagram, II-I.* Consider the state diagram shown in Figure 17.18.

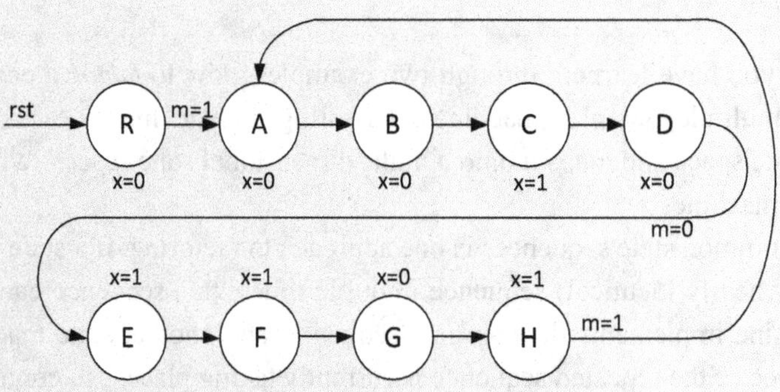

Figure 17.18. Unfactored state diagram for Exercise 17.3. It takes in a one-bit signal m and outputs a one-bit signal x. Edges not labeled with an input value transition automatically.

(a) Identify identical or nearly identical sequences of states in this FSM.

(b) Draw the state diagram for a separate FSM that implements these sequences of states – inputs should select between variations in the sequence.

(c) Draw a revised top-level state diagram that invokes your FSM from (b) to implement the repeated sequence.

17.4 *Factor a state diagram, II-II.* Implement your factored state machine from Exercise 17.3 in VHDL.

17.5 *Factor a state diagram, III-I.* Consider the state diagram shown in Figure 17.19.

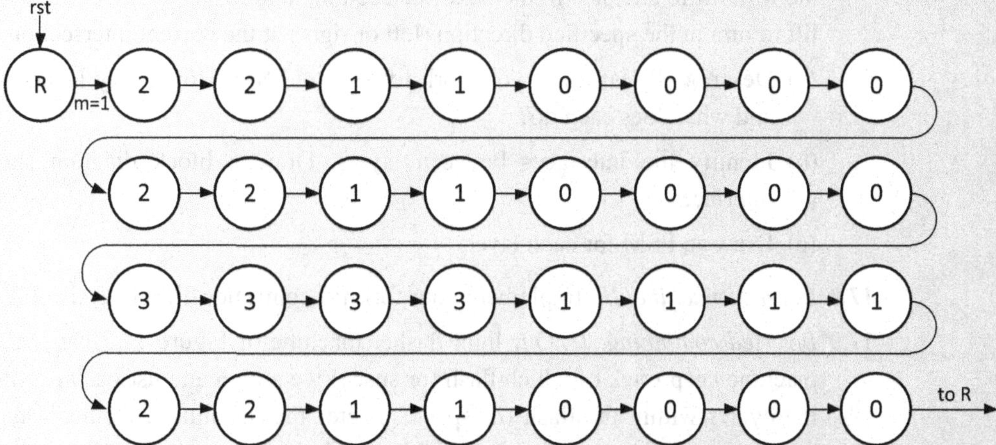

Figure 17.19. Unfactored state diagram for Exercise 17.5. It takes in a one-bit signal *m* and outputs a two-bit signal. Each state is labeled with its output value (0, 1, 2, or 3) instead of a state ID.

(a) Identify identical or nearly identical sequences of states in this FSM that can be replaced with a timer.

(b) Draw a revised top-level state diagram that invokes the timer from (a).

(c) Identify identical or nearly identical sequences of states in your new FSM.

(d) Draw the state diagram for a separate FSM that implements these sequences of states and includes your timer.

(e) Draw a revised top-level state diagram that invokes your FSMs from (e) to implement the repeated sequence. Your top-level should have, at most, five states.

17.6 *Factor a state diagram, III-II.* Implement your doubly factored state machine from Exercise 17.5 in VHDL.

17.7 *Layered machine, I.* Design an FSM to control an automated fork lift in a warehouse. The warehouse has reflective lines painted on the floor at the center of each aisle. The lines branch at each intersection. All intersections involve aisles meeting at 90 degree angles. The FSM has inputs *far_left*, *left*, *center*, *right*, and *far_right*, indicating that a reflective line is seen to the far left, just left of center, under the center, just right of center, or to the far right. It also has an input *meter* that goes high for one cycle each time the fork lift moves one meter. The FSM has outputs `go`, which causes the fork lift to move forward, and *turn_right* and *turn_left*, which cause the fork lift to change direction by 5 degrees in the specified direction. During each clock cycle of your state machine, the fork lift moves forward by about 1 cm. The sensors are spaced so that if the reflective line is near the center of the fork lift, one of the three middle sensors will be on. A 5 degree course error to the right will cause the fork lift to move from the left sensor going off to the right sensor going on in 100 clocks.

The fork lift is directed to a particular location in the warehouse by a list of instructions. There are only three types of instruction: `advance n` asks the fork lift to move `n` meters forward following the current reflective stripe; `advance next` asks the fork lift to advance to the next intersection, and `turn <direction>` asks the fork lift to turn in the specified direction (left or right) at the current intersection.

(a) Describe the layers of your fork-lift control FSM. How many layers do you have, and what does each do?

(b) Identify the interfaces between layers. Draw a block diagram showing these interfaces.

(c) Draw an FSM for each layer.

17.8 *Layered machine, II.* Implement your fork-lift controller from Exercise 17.7 in VHDL.

17.9 *Inverted combining, I.* Our light flasher machine of Figure 17.2 has the *master* state machine keep track of which flash (or space) we are on and uses a *slave* timer to count the cycles within the flash (or space). Factor the machine the other way, so that the *master* machine counts the cycles within each flash and the *slave* keeps track of which flash or space we are on.

(a) Draw a block diagram of your *inverted* light flasher.

(b) Draw state diagrams for the master and slave machines.

17.10 *Inverted combining, II.* Implement your new flasher from Exercise 17.9 in VHDL.

17.11 *SOS flasher, I.* Modify the flasher FSM of Figure 17.2 to flash an SOS sequence – three short flashes (one clock each), followed by three long flashes (four clocks each), followed by three short flashes again. Spaces within each character should be one clock long. Spaces between characters within one SOS should be three clocks long. Spaces between one SOS and the next SOS should be seven clocks long. (Hint: build a *character* FSM and an SOS FSM.)

17.12 *SOS flasher, II.* Implement your SOS flasher from Exercise 17.11 in VHDL. Be sure to implement a testbench and verify your design.

17.13 *SOS flasher, III.* Modify your SOS flasher from Exercises 17.11 and 17.12 to implement an FSM to flash the Morse code for TOSS instead of SOS. (The code for T is a single long flash.)

17.14 *Modified light flasher, I.* Modify the flasher FSM of Figure 17.6 to keep the light on for five cycles if the count is odd and 15 cycles if the count is even. You should provide a verified VHDL design entity as your solution.

17.15 *Modified light flasher, II.* Modify the flasher FSM of Figure 17.6 to keep the light on with a duration equal to the current count. You should provide a verified VHDL design entity as your solution.

17.16 *Walk, don't walk flasher.* Write an FSM in VHDL that controls two lights: a walk and a don't walk (stop) signal. Controlling the output should be a three-bit, one-hot input value *ctl*. The operation of the flasher is shown below for each possible value of *ctl*: 3'b001: walk – off, stop – on; 3'b010: walk – off, stop – flashing (10 cycles off, 15 cycles on); 3'b100: walk – on, stop – off.

17.17 *Traffic-light controller, I.* Modify the traffic-light controller of Figure 17.10 so that north–south and east–west have equal priority. Add an additional input, *car_ns* that indicates when a car is waiting in the north–south direction. In either the NS or EW states, change to the other if there is a car waiting in the other direction *and* the master timer is done.

17.18 *Traffic-light controller, II.* Modify the traffic-light controller of Figure 17.10 so that a switch from a red light to a green light is preceded by both red and yellow being on for three clocks.

17.19 *Traffic-light controller, III.* Modify the traffic-light controller of Figure 17.10 to include pedestrian signals (see Exercise 17.16) in both the north–south and east–west directions. The sequence of lights should no longer be green, yellow, red. Instead use green-walk, green-flash, yellow, and red. The walk signal should be on only in green-walk. The don't walk light should flash in green-flash, and be on in the yellow and red states.

18 Microcode

Implementing the next-state and output logic of a finite-state machine using a memory array gives a flexible way of realizing an FSM. The function of the FSM can be altered by changing the contents of the memory. We refer to the contents of the memory array as *microcode,* and a machine realized in this manner is called a *microcoded* FSM. Each word of the memory array determines the behavior of the machine for a particular state and input combination, and is referred to as a *microinstruction*.

We can reduce the required size of a microcode memory by augmenting the memory with special logic to compute the next state, and by selectively updating infrequently changing outputs. A single microinstruction can be provided for each state, rather than one for each state × input combination, by adding an instruction sequencer and a *branch* microinstruction to cause changes in control flow. Bits of the microinstruction can be shared by different functions by defining different microinstruction types for control, output, and other functions.

18.1 SIMPLE MICROCODED FSM

Figure 18.1 shows a block diagram of a simple microcoded FSM. A memory array holds the next-state and output functions. Each word of the array holds the next state and output for a particular combination of input and current state. The array is addressed by the concatenation of the current state and the inputs. A pair of registers holds the current state and current output.

In practice, the memory could be realized as a RAM or EEPROM allowing software to reprogram the microcode. Alternatively, the memory could be a ROM. With a ROM, a new mask set is required to reprogram the microcode. However, this is still advantageous because changing the program of the ROM does not otherwise alter the layout of the chip. Some ROM designs even allow the program to be changed by altering only a single metal-level mask – reducing the cost of the change. Some designs take a hybrid approach, putting most of the microcode into ROM (to reduce cost) but keeping a small portion of microcode in RAM. A method is provided to redirect an arbitrary state sequence into the RAM portion of the microcode to allow any state to be *patched* using the RAM.

A VHDL description of this FSM is shown in Figure 18.2. It follows the schematic exactly, with the addition of some logic to reset the state when `rst` is asserted. The component `ROM` is a read-only memory that takes an address "`state & input`" and returns a microinstruction `uinst`. The microinstruction is then split into its next-state and output components. The

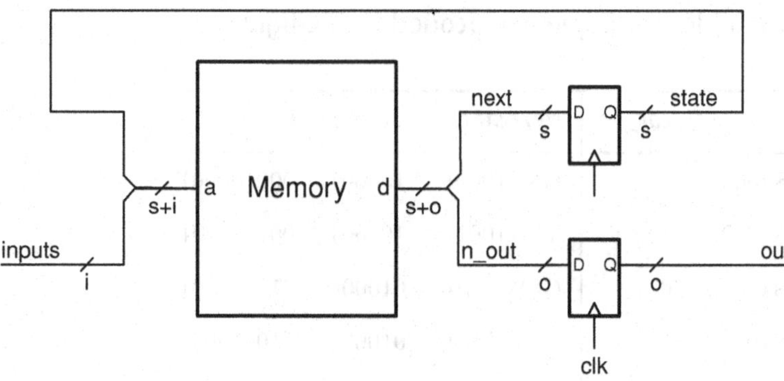

Figure 18.1. Block diagram of a simple microcoded FSM.

```
library ieee;
use ieee.std_logic_1164.all;
use work.ff.all;

entity ucode1 is
  generic( i: integer := 1; -- input width
           o: integer := 6; -- output width
           s: integer := 3; -- bits of state
           p: string := "ucode1_1.asm" );
  port( clk,rst: in std_logic;
        input: in std_logic_vector(i-1 downto 0);
        output: out std_logic_vector(o-1 downto 0) ) ;
end ucode1;

architecture impl of ucode1 is
  signal nxt, state: std_logic_vector(s-1 downto 0);
  signal uinst : std_logic_vector(s+o-1 downto 0);
begin
  STATE_REG: vDFF generic map(s) port map(clk, nxt, state);  -- state register
  OUT_REG: vDFF generic map(o) port map(clk, uinst(o-1 downto 0), output);
           -- output register
  UC: ROM generic map(s+o,s+i,p) port map(state & input, uinst); -- microcode store
  nxt <= (others => '0') when rst else uinst(s+o-1 downto o); -- reset state
end impl;
```

Figure 18.2. VHDL description of a simple microcoded FSM.

FSM of Figures 18.1 and 18.2 is very simple because it has no function until the ROM is programmed.

To see how to program the ROM to realize a finite-state machine, consider our simple traffic-light controller from Section 14.3. The state diagram of this controller is repeated in Figure 18.3. To fill our microcode ROM, we simply write down the next state and output for each current state/input combination, as shown in Table 18.1. Consider the first line of the table. Address 0000 corresponds to state GNS (green–north–south) with input $car\_ew = 0$. For this state, the output is 100001 (green–north–south, red–east–west) and the next state is GNS (000). Thus, the content of the ROM location 0000 is 000100001, the concatenation of the next state 000 with the output. For the second line of the table, address 0001 corresponds to GNS with

Table 18.1. **State table for simple microcoded traffic-light controller**

Address	State	*car_ew*	Next state	Output	Data
0000	GNS (000)	0	GNS (000)	100001	000100001
0001	GNS (000)	1	YNS (001)	100001	001100001
0010	YNS (001)	0	GEW (010)	010001	010010001
0011	YNS (001)	1	GEW (010)	010001	010010001
0100	GEW (010)	0	YEW (011)	001100	011001100
0101	GEW (010)	1	YEW (011)	001100	011001100
0110	YEW (011)	0	GNS (000)	001010	000001010
0111	YEW (011)	1	GNS (000)	001010	000001010

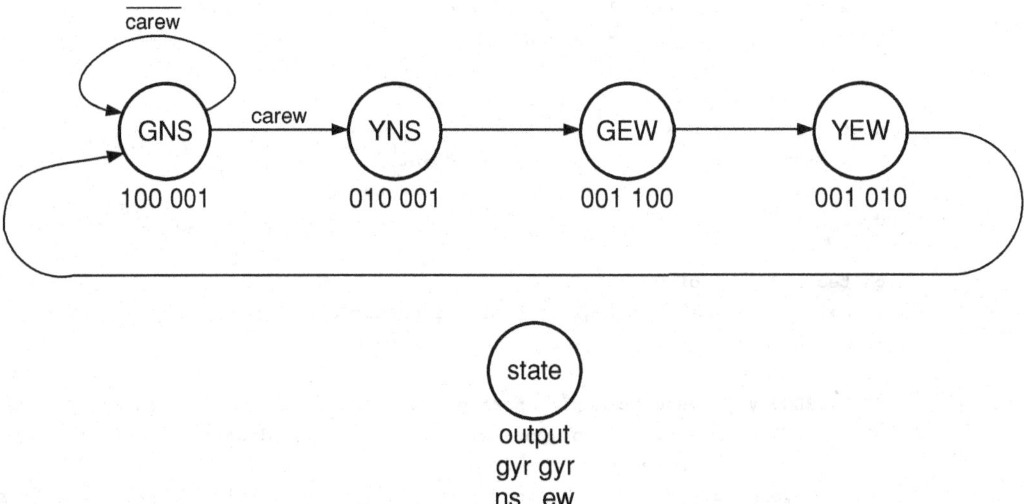

Figure 18.3. State diagram of simple traffic-light controller.

car_ew = 1. The output here is the same as for the first line, but the next state is now YNS (001), hence the content of this ROM location is 001100001. The remaining rows of the table are derived in a similar manner. The ROM itself is loaded with the contents of the column labeled "Data."

The results of simulating the microcoded FSM of Figure 18.2 using the ROM contents from Table 18.1 are shown in Figure 18.4. The output, state, and microcode ROM address, and microcode ROM data (microinstruction) (the bottom four signals) are displayed in octal (base 8). The system initializes to state 0 (GNS) with output 41 (green (4) north–south, red (1) east–west). Then the input line (*car_ew*) goes high, switching the ROM address from 00 to 01. This causes the microinstruction to switch from 041 to 141, which selects a next state of 1 (YNS) on the next clock. The machine then proceeds through states 2 (GEW) and 3 (YEW) before returning to state 0.

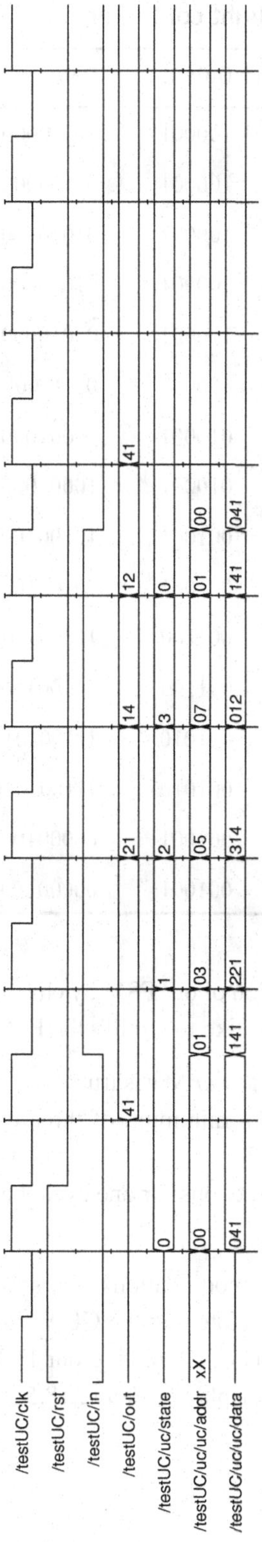

Figure 18.4. Waveform display showing a simulation of the microcoded FSM of Figure 18.2 using the microcode from Table 18.1.

Table 18.2. **State table for simple microcoded traffic-light controller**

Address	State	*car_ew*	Next state	Output	Data
0000	GNS1 (000)	0	GNS2 (001)	100001	001100001
0001	GNS1 (000)	1	GNS2 (001)	100001	001100001
0010	GNS2 (001)	0	GNS3 (010)	100001	010100001
0011	GNS2 (001)	1	GNS3 (010)	100001	010100001
0100	GNS3 (010)	0	GNS3 (010)	100001	010100001
0101	GNS3 (010)	1	YNS (011)	100001	011100001
0110	YNS (011)	0	RNS (100)	010001	100010001
0111	YNS (011)	1	RNS (100)	010001	100010001
1000	RNS (100)	0	GEW (101)	001001	101001001
1001	RNS (100)	1	GEW (101)	001001	101001001
1010	GEW (101)	0	YEW (110)	001100	110001100
1011	GEW (101)	1	GEW (101)	001100	101001100
1100	YEW (110)	0	REW (111)	001010	111001010
1101	YEW (110)	1	REW (111)	001010	111001010
1110	REW (111)	0	GNS (000)	001001	000001001
1111	REW (111)	1	GNS (000)	001001	000001001

The beauty of microcode is that we can change the function of our FSM by changing only the contents of the ROM. Suppose, for example, that we would like to modify the FSM so that

(1) the light stays green in the east–west direction as long as *car_ew* is true;
(2) the light stays green in the north–south direction for a minimum of three cycles (states GNS1, GNS2, and GNS3);
(3) after a yellow light, the lights should go red in both directions for one cycle before turning the new light green.

Table 18.2 shows the state table that implements these modifications. We split state GNS into three states and add two new states (RNS and REW). Note that the GEW state now tests *car_ew* and stays in GEW as long as it is true. The results of simulating our FSM of Figure 18.2 with this new microcode are shown in the waveform display of Figure 18.5.

18.2 INSTRUCTION SEQUENCING

A microcoded FSM can be made considerably more efficient by using a *sequencer* to generate the address of the next instruction. Performing instruction sequencing has two big advantages.

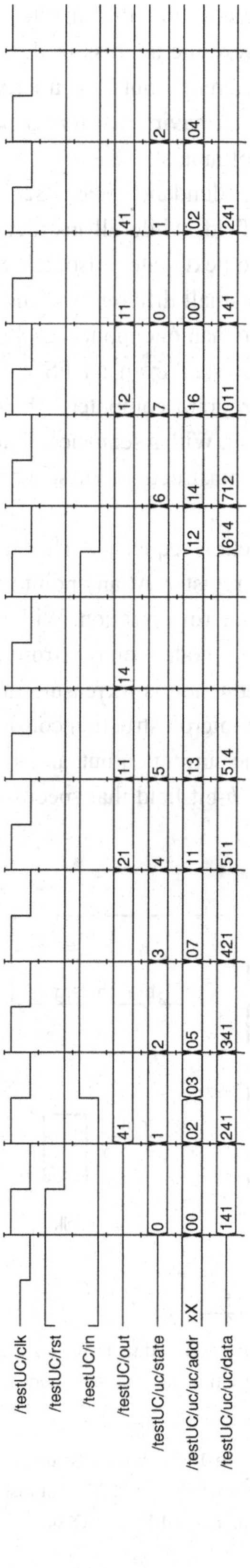

Figure 18.5. Waveform display showing a simulation of the microcoded FSM of Figure 18.2 using the microcode from Table 18.2.

First, for microinstructions that simply proceed to the next instruction, this instruction address can be generated with a counter, eliminating the need to store the address in the microcode memory. Second, by using logic to select or combine the different inputs, the microcode store can store a single microinstruction for each state, rather than having to store separate (and nearly identical) instructions for each possible combination of inputs.

A review of the microcode from Table 18.2 shows the redundancy that a sequencer can eliminate. Each instruction includes an explicit next-state field, while all instructions either select themselves or the next instruction in sequence for the next-state. Also, all instructions are duplicated for both input states, with only two having a small difference in the next-state field. This overhead would be even higher if there were more than one input signal.

Adding a sequencer to our microcoded FSM is the first step from an FSM (where the next state is determined by a logic function) to a stored-program computer, where the next instruction is determined by interpreting the current instruction. With a sequencer, like a stored-program computer, our microcoded machine *executes* microinstructions in sequence until a branch instruction redirects execution to a new address.

Figure 18.6 shows a microcoded FSM that uses an instruction sequencer. The state register here is replaced with a microprogram counter (μPC or uPC) register. At any point in time, this register represents the current state by selecting the current microinstruction. With this design we have reduced the number of microinstructions in the microcode memory from 2^{s+i} to 2^s, that is, from 2^i per state to one per state. The cost of this reduction is increasing the width of each microinstruction from $s + o$ bits to $s + o + b$ bits. Each microinstruction consists of three fields, as shown in Figure 18.7: an o-bit field that specifies the current output, an s-bit field that specifies the address to branch to (the branch target), and a b-bit field that specifies a branch instruction.

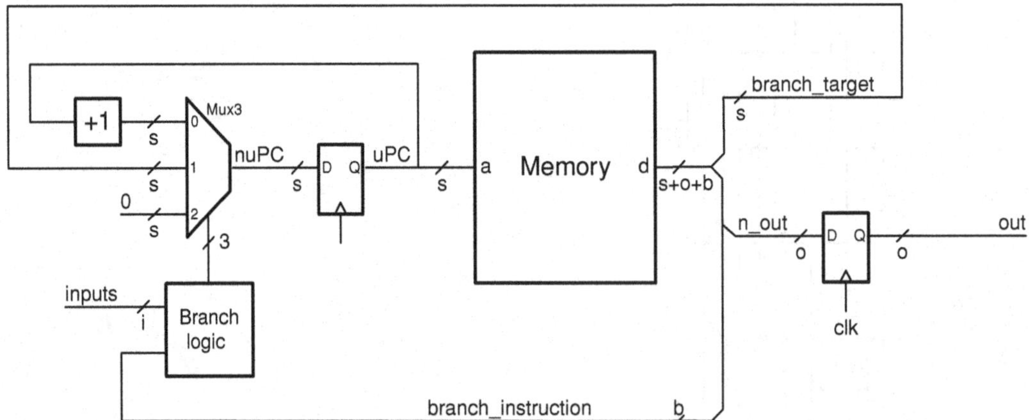

Figure 18.6. Microcoded FSM using an instruction sequencer. A multiplexer and incrementer compute the next microinstruction address (next state) on the basis of a branch instruction and the input conditions.

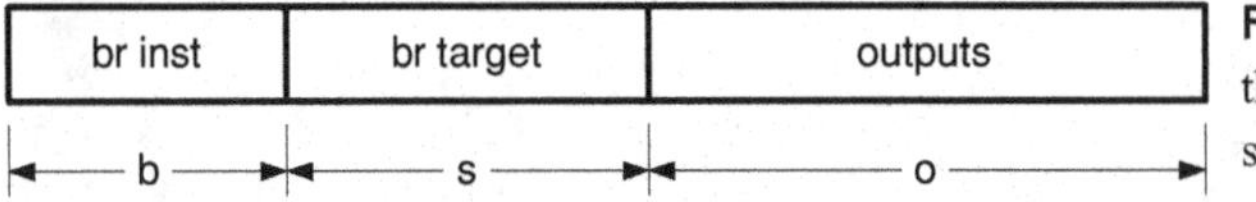

Figure 18.7. Microinstruction format for the microcoded FSM with instruction sequencer of Figure 18.6.

Table 18.3. **Branch instruction encodings**

Encoding	Opcode	Description
000	NOP	Never branch, always proceed to uPC + 1
001	B0	Branch on input 0; if `in[0]`, branch to `br_upc`, otherwise continue to uPC + 1
010	B1	Branch on input 1
011	BA	Branch any; branch if either input is true
100	BR	Always branch; select `br_upc` as the next uPC regardless of the inputs
101	BN0	Branch on not input 0; if `in[0]` is false, branch, otherwise continue to uPC + 1
110	BN1	Branch on not input 1
111	BNA	Branch only if inputs 0 and 1 are both false

With the instruction sequencer, inputs are tested by the branch logic as directed by a branch instruction from the current microinstruction. As a result of this test, the sequencer either branches (by selecting the branch target field as the next uPC) or doesn't (by selecting uPC + 1 as the next uPC).

Consider an example with a two-bit input field. We can define a three-bit branch instruction, `brinst`, as follows:

```
branch <= ((brinst(0) and input(0)) or (brinst(1) and input(1))) xor brinst(2);
```

Branch instruction bits 0 and 1 select whether we test input bits 0 or 1 (or either). Branch instruction bit 2 controls the polarity of the test. If `brinst[2]` is low, we branch if the selected bit(s) is high. Otherwise we branch if the selected bit(s) is low. Using this encoding of the branch instruction, we can perform the branchings shown in Table 18.3.

Other encodings of the branch instruction are possible. A common n-bit encoding uses $n-1$ bits to select one of 2^{n-1} inputs to test and the remaining bit to select whether to branch on the selected input high or low. One of the inputs is set always high to allow the NOP and BR instructions to be created. For this encoding the branch signal would be as follows:

```
branch = brinst(n-1) xor input( to_integer(unsigned( brinst(n-2 downto 0) )) ) ;
```

The branch instructions created by this alternative encoding (for a three-bit `brinst`) and three inputs are listed in Table 18.4. To provide the NOP and BR instructions, we use a constant 1 for the fourth input. Here each branch instruction tests exactly one input, while in the encoding of Table 18.3 the instructions may test zero, one, or two inputs. Many other possible encodings beyond the two presented here are possible.

A VHDL description of the microcoded FSM with an instruction sequencer is shown in Figure 18.8. The VHDL follows the block diagram of Figure 18.6 closely. One concurrent signal assignment statement calculates signal `branch`, which is true if the sequencer is to branch on the next cycle. A second concurrent signal assignment statement then calculates the next microprogram counter (`nupc`) on the basis of `branch` and `rst`. To make the code more readable, we use a VHDL record to hold the three fields of the microinstruction.

Table 18.4. **Alternative branch instruction encodings**

Encoding	Opcode	Description
000	B0	Branch on input 0
001	B1	Branch on input 1
010	B2	Branch on input 2
011	BR	Always branch (input 3 is the constant "1")
100	BN0	Branch on not input 0
101	BN1	Branch on not input 1
110	BN2	Branch on not input 2
111	NOP	Never branch

Consider a slightly more involved version of our traffic-light controller that includes a left-turn signal as well as north–south and east–west signals. Table 18.5 shows the microcode. Here input 0 is *car_lt* and input 1 is *car_ew* so we rename our branches BLT (branch if *car_lt*, BNEW (branch if not *car_ew*), and so on.

The microcode of Table 18.5 starts in state NS1, where the light is green in the north–south direction. In this state the left-turn sensor is checked with a BLT to LT1. If *car_lt* is true, control transfers to LT1. Otherwise the uPC proceeds to the next state, NS2. In NS2 the microcode branches back to NS1 if *car_ew* is false (BNEW NS1). Otherwise, control falls through to EW1, where the north–south light goes yellow. EW1 is always followed by EW2, where the east–west light is green. A BEW EW2 keeps the uPC in state EW2 as long as *car_ew* is true. When *car_ew* goes false, the uPC proceeds to state EW3, where the east–west light is yellow and a BR NS1 transfers control back to NS1. The left-turn sequence (LT1, LT2, LT3) operates in a similar manner.

Waveforms from a simulation of the microcoded sequencer of Figure 18.8 running the microcode of Table 18.5 are shown in Figure 18.9. The fifth row from the top shows the microprogram counter `upc`. The machine is reset to `upc` = 0 (NS1), and advances to 1 (NS2), then back to 0 (NS1), before branching to 5 (LT1). It proceeds from 5 (LT1) to 6 (LT2) and remains in 6 until *car_lt* goes low. At that point it advances to 7 (LT3) and back to 0 (NS1). At this point, *car_ew* = 1, and the sequence followed is 0, 1, 2, 3 (NS1, NS2, EW1, EW2). The machine stays in 3 (EW2) until *car_ew* goes low, and then proceeds to 4 (EW3) and back to 0 (NS1). The machine cycles between NS1 and NS2 a few times until both *car_ew* and *car_lt* go high at the same time. Since the machine is in NS1 when this happens, car_lt is checked first and the uPC is directed to LT1.

Because the microcoded FSM of Figure 18.6 can only branch one way in each microinstruction, it takes two states (NS1 and NS2) to perform a three-way branch among staying with north–south, going to east–west, and going to left-turn. This results in two states with the lights green in the north–south direction (NS1 and NS2) and two states with the lights yellow in the

```
library ieee;
use ieee.std_logic_1164.all;
use ieee.std_logic_unsigned.all;
use work.ff.all;

entity ucode2 is
  generic( n: integer := 2; -- input width
           m: integer := 9; -- output width
           k: integer := 4; -- bits of state
           j: integer := 3; -- bits of instruction
           p: string := "ucode2_1.asm" );
  port( clk, rst: in std_logic;
        input: in std_logic_vector(n-1 downto 0);
        output: out std_logic_vector(m-1 downto 0) );
end ucode2;

architecture impl of ucode2 is
  type inst_t is record
    brinst: std_logic_vector(j-1 downto 0);
    br_upc: std_logic_vector(k-1 downto 0);
    nout: std_logic_vector(m-1 downto 0);
  end record;
  signal nupc, upc: std_logic_vector(k-1 downto 0); -- microprogram counter
  signal ibits: std_logic_vector(j+k+m-1 downto 0); -- rom output
  signal uinst: inst_t; -- microinstruction word
  signal branch: std_logic;
begin
  -- split off fields of microinstruction
  uinst <= (ibits(j+m+k-1 downto m+k), ibits(m+k-1 downto m), ibits(m-1 downto 0));

  UPC_REG: vDFF generic map(k) port map(clk, nupc, upc);  -- microprogram counter
  OUT_REG: vDFF generic map(m) port map(clk, uinst.nout, output); -- output register
  UC: ROM generic map(m+k+j,k,p) port map(upc, ibits); -- microcode store

  -- branch instruction decode
  branch <= ((uinst.brinst(0) and input(0)) or (uinst.brinst(1) and input(1)))
            xor uinst.brinst(2);

  -- sequencer
  nupc <= (others => '0') when rst else
          uinst.br_upc when branch else
          upc + 1;
end impl;
```

Figure 18.8. VHDL description of a microcoded FSM with an instruction sequencer.

north–south direction (EW1 and LT1). The real solution to this problem is to support a multi-way branch (which we will discuss below). However, we can partially solve the problem in software by using the alternative microcode shown in Table 18.6.

In the alternative microcode of Table 18.6, the uPC stays in state NS1 as long as *car_ew* and *car_lt* are both false by using a BNA NS1 (branch on not any inputs to NS1); NS1 is

Table 18.5. **Microcode for traffic-light controller with sequencer of Figure 18.6**

Address	State	Br inst	Target	NS LT EW	Data
0000	NS1	BLT (001)	LT1 (0101)	100001001	0010101100001001
0001	NS2	BNEW (110)	NS1 (0000)	100001001	1100000100001001
0010	EW1	NOP (000)		010001001	0000000010001001
0011	EW2	BEW (010)	EW2 (0011)	001001100	0100011001001100
0100	EW3	BR (100)	NS1 (0000)	001001010	1000000001001010
0101	LT1	NOP (000)		010001001	0000000010001001
0110	LT2	BLT (001)	LT2 (0110)	001100001	0010110001100001
0111	LT3	BR (100)	NS1 (0000)	001010001	1000000001010001

now the only state with the lights green in the north–south direction. If any inputs are true, the uPC proceeds to state NS2, which is the single state with the lights yellow in the north–south direction. State NS2 tests the *car_lt* input and branches to state LT1 if true (BLT LT1). If *car_lt* is false, the uPC proceeds to EW1. The remainder of the machine is similar to that of Table 18.5, except that the EW and LT states have been renumbered.

Simulation waveforms for this alternative microcode are shown in Figure 18.10.

18.3 MULTI-WAY BRANCHES

As we saw in Section 18.2, using an instruction sequencer greatly reduces the size of our microcode memory, but at the expense of restricting each state to have at most two next states (`upc+1` and `br_upc`). This restriction can be a problem if we have an FSM that has a large number of exits from a particular state. For example, in a microcoded processor, it is typical to branch to one of tens to hundreds of next states on the basis of the *opcode* of the current instruction. Another multi-way branch is then needed on the *addressing mode* of the instruction. Implementing such a multi-way dispatch using the sequencer of Section 18.2 would result in very poor efficiency because *n* cycles would be required in order to test for *n* different opcodes.

We can overcome this limitation of two-way branches by using an instruction sequencer that supports multi-way branches, as shown in Figure 18.11. This sequencer is similar to that of Figure 18.6 except that the branch target, *br_upc*, is generated from the branch instruction, *brinst*, and inputs rather than being provided directly from the microinstruction. With this approach, we can branch to up to 2^i next states (one for each input combination) from each state.

The branch instruction encodes not just the condition to test, but also how to determine the branch target. Table 18.7 shows one possible method of encoding multi-way branch instructions. The BRx and BRNx instructions are two-way branch instructions identical to those specified in Table 18.4. The BR4 instruction is a four-way branch that selects one of four adjacent states (from *br_upc* to *br_upc+3*), depending on the input.

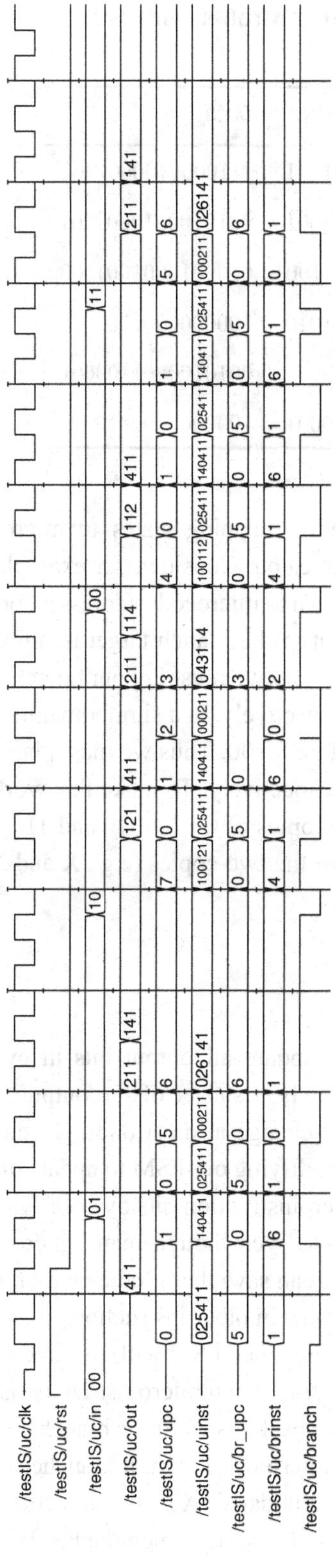

Figure 18.9. Waveform display showing a simulation of the microcoded FSM of Figure 18.8 using the microcode from Table 18.5.

Table 18.6. **Alternative microcode for traffic-light controller with sequencer of Figure 18.6**

Address	State	Br inst	Target	NS LT EW	Data
0000	NS1	BNA (111)	NS1 (0000)	100001001	1110000100001001
0001	NS2	BLT (001)	LT1 (0100)	010001001	0010100010001001
0010	EW1	BEW (010)	EW1 (0010)	001001100	0100010001001100
0011	EW2	BR (100)	NS1 (0000)	001001010	1000000001001010
0100	LT1	BLT (001)	LT1 (0100)	001100001	0010100001100001
0101	LT2	BR (100)	NS1 (0000)	001010001	1000000001010001

To use the BR4 instruction requires some care in mapping states to microinstruction addresses, and may require that some states are duplicated. Consider, for example, the state diagram of Figure 18.12. A mapping of this state diagram to microcode addresses for a machine with a four-way branch instruction that adds the input to the branch target is shown in Table 18.8. The four-way branch from X targets address 000, so we must lay out branch targets A1, B1, C1, and X in locations 000, 001, 010, and 011, respectively. In a similar manner, we locate states so that the four-way branch from C1 targets address 100. Thus we must place states C2, C3, X, and C1 at locations 100, 101, 110, and 111, respectively. To make this work, we need two copies of X, one at 011 and one at 110, and two copies of C1 (at 010 and 111). When we duplicate a state in this manner, we simply arrange for the two copies (e.g., X and X′) to have identical behavior.

18.4 MULTIPLE INSTRUCTION TYPES

So far we have considered microcoded FSMs that update all output bits in every microinstruction. In general, most FSMs need to update only a subset of the outputs in a given state. Our traffic-light controller FSMs, for example, change at most one light on each state change. We can save bits of the microinstruction by modifying our FSM to update only a single output register in any given state. We waste other microinstruction bits by specifying a branch instruction and branch target in each microinstruction, even though many microinstructions always proceed to the next state and don't branch. We can save these redundant branch bits by only branching in some instructions and updating outputs in other instructions.

Figure 18.13 shows the instruction formats for a microcoded FSM with two microinstruction types: a branch instruction and a store (output) instruction. Each microinstruction is one or the other. A branch instruction, identified by a 1 in the leftmost bit, specifies a branch condition and a branch target. When the FSM encounters a branch microinstruction, it branches (or doesn't) as specified by the condition and target. No outputs are updated. A store instruction, identified by a 0 in the leftmost bit, specifies an output register and a value. When the FSM encounters a store microinstruction, it stores the value to the specified output register and then proceeds to the next microinstruction, in sequence. No branching takes place with a store instruction.

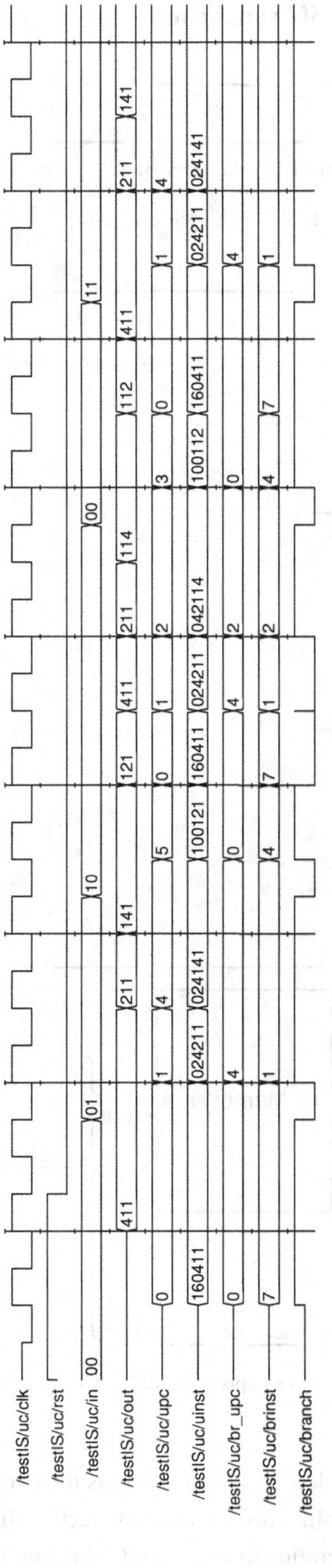

Figure 18.10. Waveform display showing a simulation of the microcoded FSM of Figure 18.8 using the microcode from Table 18.6.

Table 18.7. **Branch instructions for microcoded FSM supporting multi-way branches**

Encoding	Opcode	Description
BRx	00xx	Branch on condition x (as in Table 18.4, includes BR)
BRNx	01xx	Branch on not condition x (as in Table 18.4, includes NOP)
BR4	1000	Branch four ways; $nupc = br\_upc + in$

Table 18.8. **Mapping of state diagram of Figure 18.12 onto microcode addresses**

States X and C1 are duplicated because they each appear in two four-way branches.

Address	State	Branch instruction	Branch target
000	A1	BR	A2
001	B1	BR	C2
010	C1	BR4	C2
011	X	BR4	A1
100	C2	BR	X
101	C3	BR	X
110	*X′*	BR4	A1
111	*C1′*	BR4	C2

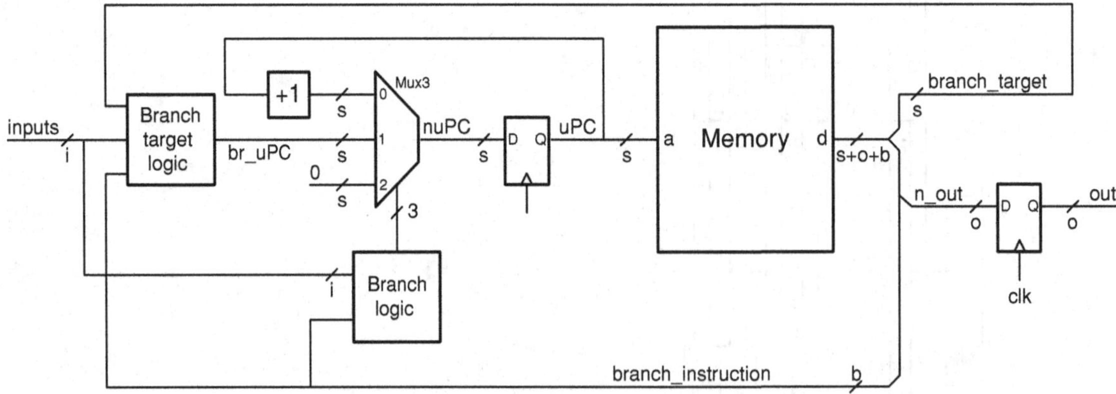

Figure 18.11. Microcoded FSM with an instruction sequencer that supports multi-way branches.

Figure 18.14 shows a block diagram of a microcoded FSM that supports the two instruction types of Figure 18.13. Each microinstruction is split into an x-bit instruction field and an s-bit value field. The instruction field holds the operation-code (opcode) bit (the leftmost bit that distinguishes between branch and store) and either the condition (for a branch) or the

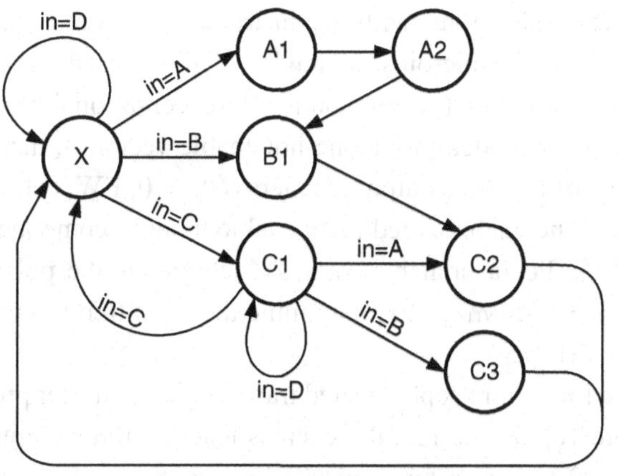

Figure 18.12. State diagram with two four-way branches.

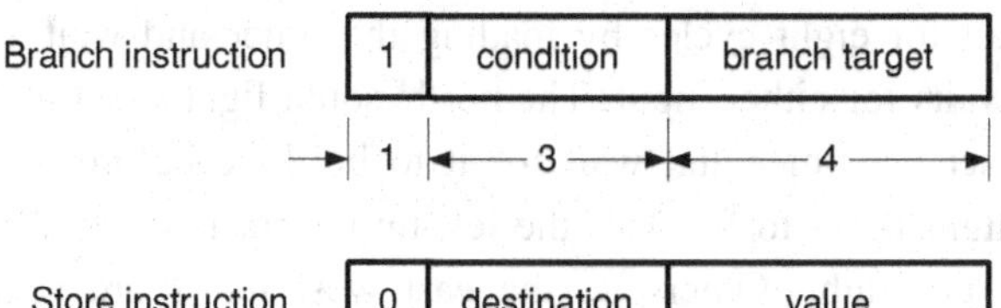

Figure 18.13. Instruction formats for a microcoded FSM with separate output and branch instructions.

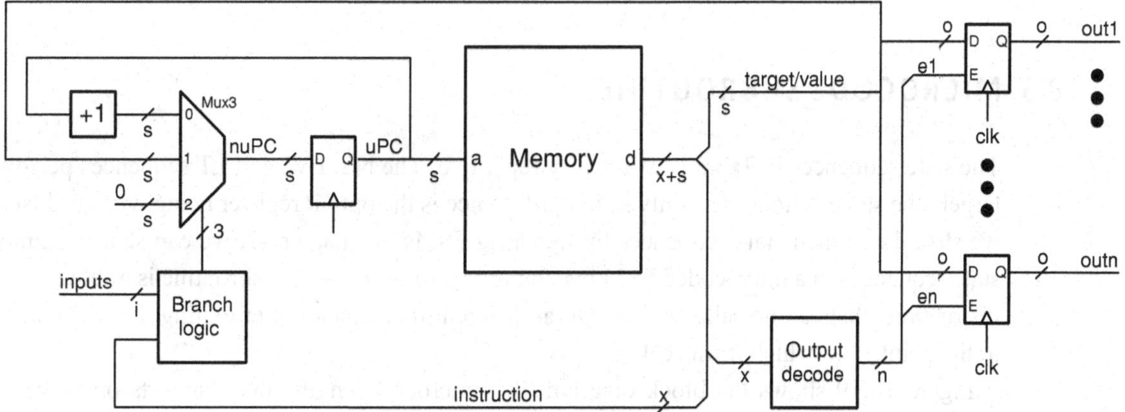

Figure 18.14. Block diagram of a microcoded FSM with output instructions. For our traffic-light controller, we replace the last output register with a timer and feed its *done* signal into the branch logic.

destination (for a store). The value field holds either the branch target (for a branch) or the new output value (for a store).

The instruction sequencer of Figure 18.14 is identical to that of Figure 18.6 except that the branch logic always selects the next microinstruction, uPC + 1, when the current microinstruction is a store instruction.[1] The major difference is with the output logic. Here a decoder enables at most one output register to receive the value field on a store instruction.

[1] One can just as easily add multiple instruction types and output registers to an FSM that supports multi-way branches (Figure 18.11).

A VHDL description for the microcode engine with two instruction types of Figure 18.14 is shown in Figures 18.15 and 18.16. Here the microinstruction is a VHDL record that contains an opcode (0 = store, 1 = branch), an instruction (destination for store, condition for branch), and a value. For a store, the destination is decoded into a one-hot enable vector, `e`, that is used to enable the value to be stored in one of the three output registers (NS = 0, EW = 1, LT = 2) or to load the timer (destination = 3). The enable is fed to an enable input to component `vDFFE`, which is described in Figure 18.17. For a branch, `inst(2)` determines the polarity of the branch, and the low two bits, `inst(1 downto 0)`, determine the condition to be tested (LT = 0, EW = 1, Lt|EW = 2, timer = 3).

Table 18.9 shows the microcode for a more sophisticated traffic-light controller programmed on the microcode engine of Figure 18.15. The first three states load the three output registers with RED in the east–west and left-turn registers and GREEN in the north–south register. Next states NS1 and NS2 wait for eight cycles by loading the timer and waiting for it to signal `done`. State NS4 then waits for either input. The north–south light is set to YELLOW in NS5; NS6 and NS7 then set the timer and wait for it to be done before advancing to NS8, where the north–south light is set to RED. If the left-turn input is true, NS9 branches to LT1 to sequence the left-turn light. Otherwise the east–west lights are sequenced in states EW1–EW9. Waveforms generated by a simulation of this microcode are shown in Figure 18.18.

18.5 MICROCODE SUBROUTINES

The state sequences in Table 18.9 are very repetitive. The NS, EW, and LT sequences perform largely the same actions. The only salient difference is the output register being written. Just as we shared common state sequences by factoring FSMs in Chapter 17, we can share common state sequences in a microcoded FSM by supporting *subroutines*. A subroutine is a sequence of instructions that can be called from several different points, and, after exiting, returns control to the point from which it was called.

Figure 18.19 shows the block diagram of a microcode engine that supports one level of subroutines. This machine is identical to that of Figure 18.14 except for two differences: (a) a return uPC register, *rupc*, and associated logic have been added to the sequencer, and (b) a select register and associated logic have been added to the output section.

The *rupc* register is used to hold the *upc* to which a subroutine should branch when it completes. When a subroutine is called, the branch target is selected as the next *upc*, and *upc+1*, the address of the next instruction in sequence, is saved in the *rupc* register. A special branch instruction CALL is used to cause the enable line to the *rupc* register, *erpc*, to be asserted. When the subroutine is complete, it returns control to the saved location using another special branch instruction RET to select the *rupc* as the source of the next *upc*.

The select register is used to allow the same state sequence to write to different output registers when called from different places. A two-bit register identifier (NS = 0, EW = 1, LT = 2) can be stored in the select register. A special store instruction SSEL can then be used to store to the register specified by the select register (rather than by the destination bits of the

```
library ieee;
use ieee.std_logic_1164.all;
use ieee.std_logic_unsigned.all;
use work.ff.all;
use work.ch16.all;
use ieee.numeric_std.all;

entity ucodeMI is
  generic( n: integer := 2; -- input width
           m: integer := 9; -- output width
           o: integer := 3; -- output sub-width
           k: integer := 5; -- bits of state
           j: integer := 4; -- bits of instruction
           p: string := "ucode.asm" );
  port( clk, rst: in std_logic;
        input: in std_logic_vector(n-1 downto 0);
        output: out std_logic_vector(m-1 downto 0) );
end ucodeMI;

architecture impl of ucodeMI is
  type inst_t is record
    opcode: std_logic; -- opcode bit
    inst: std_logic_vector(j-2 downto 0); -- condition for branch, dest for store
    value : std_logic_vector(k-1 downto 0); -- target for branch, value for store
  end record;

  signal nupc, upc: std_logic_vector(k-1 downto 0); -- microprogram counter
  signal ibits: std_logic_vector(j+k-1 downto 0); -- microinstruction raw bits
  signal ui: inst_t; -- microinstruction
  signal done: std_logic; -- timer done signal
  signal branch: std_logic;
  signal e: std_logic_vector(3 downto 0); -- enable for output registers and timer
  signal a, blt, bew, ble, btd: std_logic;
begin
  -- split off fields of microinstruction
  ui <= (opcode => ibits(j+k-1),
         inst   => ibits(j+k-2 downto k),
         value  => ibits(k-1 downto 0));

  UPC_REG: vDFF generic map(k) port map(clk, nupc, upc) ;  -- microprogram counter
  UC: ROM generic map(k+j,k,p) port map(upc, ibits) ; -- microcode store

  -- output registers and timer
  NS: vDFFE generic map(o)
         port map(clk, e(0), ui.value(o-1 downto 0), output(o-1 downto 0));
  EW: vDFFE generic map(o)
         port map(clk, e(1), ui.value(o-1 downto 0), output(2*o-1 downto o));
  LT: vDFFE generic map(o)
         port map(clk, e(2), ui.value(o-1 downto 0), output(3*o-1 downto 2*o));
  TIM: Timer generic map(k) port map(clk, rst, e(3), ui.value, done);

  e <= "0000" when ui.opcode else
      std_logic_vector( to_unsigned(1,4) sll to_integer(unsigned(ui.inst)) );
```

Figure 18.15. VHDL description of a microcoded FSM with two instruction types.

```
  -- branch instruction decode
  blt <= '1' when ui.inst(1 downto 0) = "00" else '0'; -- left turn
  bew <= '1' when ui.inst(1 downto 0) = "01" else '0'; -- east/west
  ble <= '1' when ui.inst(1 downto 0) = "10" else '0'; -- left turn or east/west
  btd <= '1' when ui.inst(1 downto 0) = "11" else '0'; -- timer done
  branch <=  (ui.inst(2) xor ((blt and input(0)) or
                              (bew and input(1)) or
                              (ble and (input(0) or input(1))) or
                              (btd and done))) when ui.opcode
                         else '0'; -- for a store opcode

  -- microprogram counter
  nupc <=  (others => '0') when rst else
           ui.value when branch else
           upc + 1;
end impl;
```

Figure 18.16. VHDL description of a microcoded FSM with two instruction types.

```
library ieee;
use ieee.std_logic_1164.all;

entity vDFFE is
  generic( n: integer := 1 ); -- width
  port( clk, en: in std_logic;
        D: in std_logic_vector( n-1 downto 0 );
        Q: buffer std_logic_vector( n-1 downto 0 ) );
end vDFFE;

architecture impl of vDFFE is
  signal Q_next: std_logic_vector(n-1 downto 0);
begin
  Q_next <= D when en else Q;

  process(clk) begin
    if rising_edge(clk) then
      Q <= Q_next;
    end if;
  end process;
end impl;
```

Figure 18.17. VHDL for flip-flop with enable.

instruction). Thus, the main program can store 0 (NS) into the select register and then call a subroutine to sequence the north–south lights on and off. The program can then store 1 (EW) into the select register and call the same subroutine to sequence the east–west lights on and off. The same subroutine can sequence different lights because it performs all of its output using the SSEL instruction.

Table 18.9. **Microcode to implement traffic-light controller on FSM with two instruction types**

Address	State	Instruction	Value	Data
00000	RST1	SLT (0010)	RED 001	001000001
00001	RST2	SEW (0001)	RED 001	000100001
00010	NS1	SNS (0000)	GREEN 100	000000100
00011	NS2	STIM (0011)	TGRN 01000	001101000
00100	NS3	BNTD (1111)	NS3 00100	111100100
00101	NS4	BNLE (1110)	NS4 00101	111000101
00110	NS5	SNS (0000)	YELLOW 010	000000010
00111	NS6	STIM (0011)	TYEL 00011	001100011
01000	NS7	BNTD (1111)	NS7 01000	111101000
01001	NS8	SNS (0000)	RED 001	000000001
01010	NS9	BLT (1000)	LT1 10100	100010100
01011	EW1	STIM (0011)	TRED 00010	001100010
01100	EW2	BNTD (1111)	EW2 01100	111101100
01101	EW3	SEW (0001)	GREEN 100	000100100
01110	EW4	STIM (0011)	TGRN 01000	001101000
01111	EW5	BNTD (1111)	EW5 01111	111101111
10000	EW6	SEW (0001)	YELLOW 010	000100010
10001	EW7	STIM (0011)	TYEL 00011	001100011
10010	EW8	BNTD (1111)	EW8 10010	111110010
10011	EW9	BTD (1011)	RST2 00001	101100001
10100	LT1	STIM (0011)	TRED 00010	001100010
10101	LT2	BNTD (1111)	LT2 10101	111110101
10110	LT3	SLT (0010)	GREEN 100	001000100
10111	LT4	STIM (0011)	TGRN 01000	001101000
11000	LT5	BNTD (1111)	LT5 11000	111111000
11001	LT6	SLT (0010)	YELLOW 010	001000010
11010	LT7	STIM (0011)	TYEL 00011	001100011
11011	LT8	BNTD (1111)	LT8 10010	111111011
11100	LT9	BTD (1011)	RST1 00000	101100000

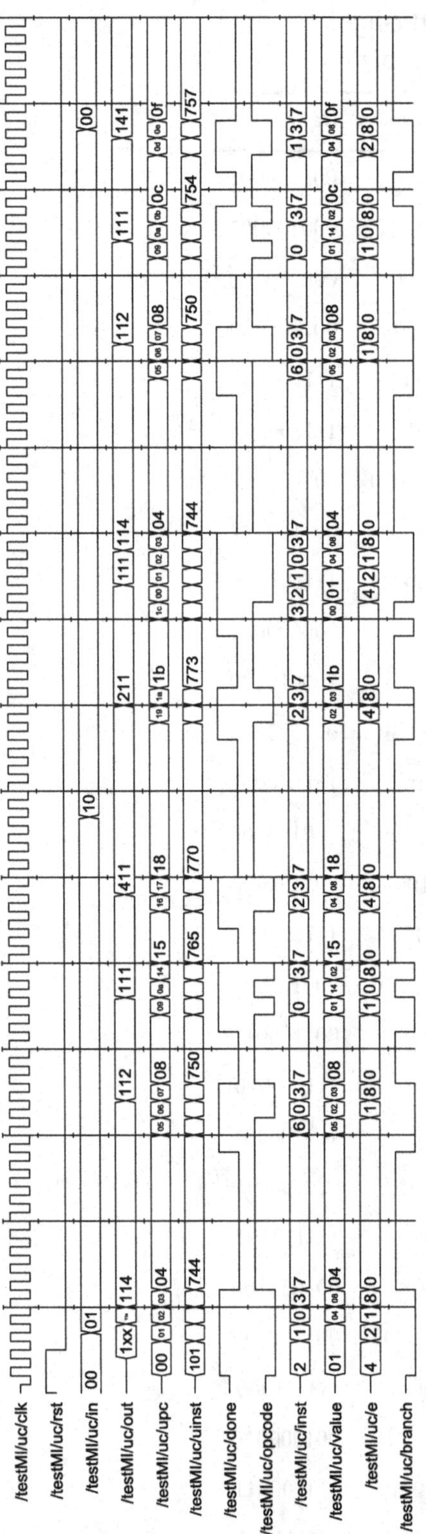

Figure 18.18. Waveform display showing a simulation of the microcoded FSM of Figures 18.15 and 18.16 using the microcode from Table 18.9.

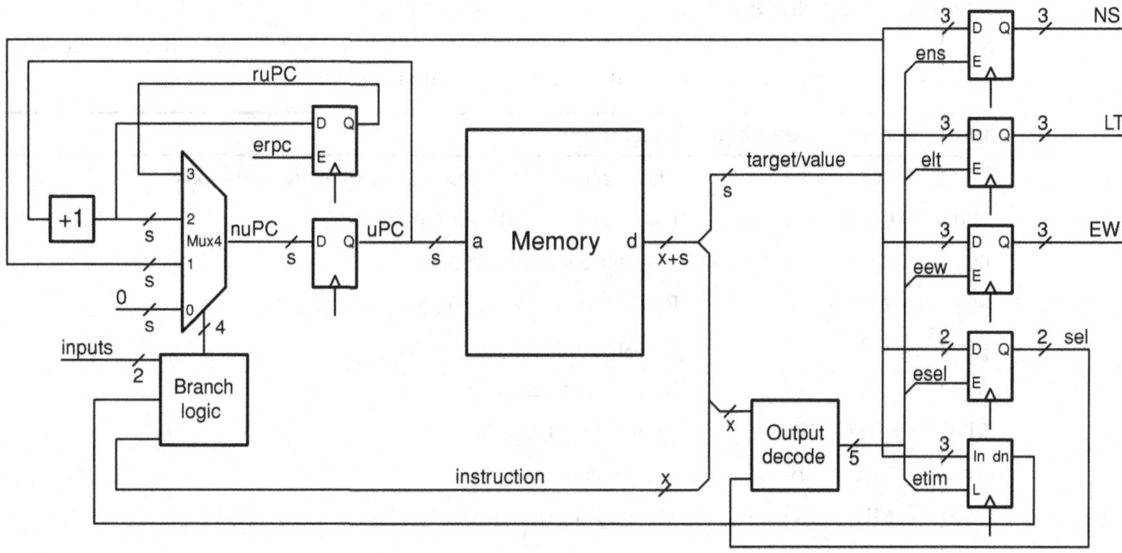

Figure 18.19. Microcoded FSM with support for one level of subroutines.

Table 18.10. **List of opcodes used in the processor**

Opcodes with a leading 1 are computed in the ALU. The lower four bits of the instruction encode RS for ALU instructions and LDA. For specific instructions, they can signify a branch condition (BR.ACC), immediate (BR.IMI, LDA.I), or destination (STA).

Opcode *i(7 downto 4)*	Instruction	Description
0000	BR	Branch to PC stored in BRD
0001	BR.S	Branch to PC stored in BRD, store PC + 1 into BRD
0010	BR.IM	Branch based on the inputs and branch instruction currently stored in IM; the destination is BRD
0011	BR.IMI	Same as BRIM, except destination is PC + *i(3 downto 0)*
0100	BR.ACC	Branch to address in BR register if ACC meets condition stored in bits *(4 downto 3)* (00: equal 0, 01: not equal 0; 10: greater than 0; 11: less than 0)
0101	LDA	ACC = RS (*i(3 downto 0)*)
0110	LDA.I	ACC = *i(3 downto 0)*
0111	STA	RD (*i(3 downto 0)*) = ACC
1000	ADD	ACC = ACC + RS(*i(3 downto 0)*)
1001	SUB	ACC = ACC − RS
1010	MUL	{ACC.H, ACC} = ACC*RS
1011	SH	{ACC.H, ACC} = {16'd0, ACC}<<RS
1100	XOR	ACC = ACC ⊕ RS
1101	AND	ACC = ACC ∧ RS
1110	OR	ACC = ACC ∨ RS
1111	NOT	ACC = $\overline{ACC}$

Table 18.11. **The state that is kept in our processor**

Not shown is the instruction ROM (accessed with the PC) and data RAM. The data RAM is accessed with address MA whenever MD is used as a source or destination operand.

ID	Register	Length (*b*)	Description
0000	ACC	16	The implicit destination of all arithmetic operations
0001	ACC.H	16	Upper 16 bits of the accumulator
0010	O0	16	Register connected to output
0011	O1	16	Register connected to output
0100	O2	16	Register connected to output
0101	O3	16	Register connected to output
0110	BRD	16	Branch target register
0111	MA	16	Memory address
1000	MD	X	Memory source register
1001	IM	10	Branch instruction register
1010	T0	16	Temporary register 0
1011	T1	16	Temporary register 1
1100	T2	16	Temporary register 2
1101	IN	8	Input values, read-only
1110	PC	16	The current program counter, read-only
1111	timer	16	Timer, write-only; loads a new start time when used as a destination

18.6 SIMPLE COMPUTER

In this chapter, we started with a simple microcoded FSM and built up to a system with branch instructions and multiple output registers. This section continues the progression by implementing a simple processor. This design is intended to demystify processors – by showing how simple they actually are. This design is not intended to be an example of what a processor should be. It has been optimized for simplicity, not performance, efficiency, or ease of programming.

The processor supports three main categories of instructions: branches, moves, and arithmetic. Regardless of type, instructions are of a fixed one-byte size. The upper four bits, *i(7 downto 4)*, indicate the *opcode* of instruction. The interpretations of the lower four bits are opcode-dependent. We summarize the 16 different instructions in Table 18.10.

Instead of storing branch targets in our instruction ROM, they are stored in a branch destination register, BRD. The next PC after a BR instruction is equal to the value of BRD. To enable more efficient calling of subroutines, the BR.S instruction stores PC + 1 into BRD. The BR.IM instruction branches according to the eight inputs and timer *done* signal, using the ten-bit branch instruction stored in register IM. The branch instructions work the same as in Table 18.3, except that there are nine input signals (eight inputs and timer) instead of two. Instruction BR.IMI also uses the IM register, but the destination is PC + *i(3 downto 0)*. Finally,

```
library ieee;
use ieee.std_logic_1164.all;
use ieee.std_logic_unsigned.all;
use ieee.numeric_std.all;
use work.alu_ch18_6.all;

entity alu is
  port( opcode: in std_logic_vector(2 downto 0);
        s0, s1: in std_logic_vector(15 downto 0);
        o_high, o_low: out std_logic_vector(15 downto 0);
        write_high: out std_logic );
end alu;

architecture impl of alu is
  type out_t is record
    high, low: std_logic_vector(15 downto 0);
    write_high: std_logic;
  end record;
  signal output: out_t;
  signal sub: std_logic;
  signal addsub_val, s1i: std_logic_vector(15 downto 0);
  signal product: std_logic_vector(31 downto 0);
  signal shft: std_logic_vector(31 downto 0);
  signal us0: unsigned(31 downto 0);
begin
  sub <= '1' when opcode = OP_SUB else '0';
  s1i <= (not s1) when sub = '1' else s1;
  addsub_val <= s0 + s1i + ((15 downto 1 => '0') & sub);
  product <= s0 * s1;

  us0 <= unsigned(std_logic_vector'(x"0000" & s0));
  shft <= std_logic_vector( us0 sll to_integer(unsigned(s1)) );

  process(all) begin
    case opcode is
      when OP_ADD => output <= (16x"0", addsub_val, '0');
      when OP_SUB => output <= (16x"0", addsub_val, '0');
      when OP_MUL => output <= (product(31 downto 16), product(15 downto 0), '1');
      when OP_SH =>  output <= (shft(31 downto 16), shft(15 downto 0), '1');
      when OP_XOR => output <= (16x"0", (s0 xor s1), '0');
      when OP_AND => output <= (16x"0", (s0 and s1), '0');
      when OP_OR =>  output <= (16x"0", (s0 or s1),  '0');
      when OP_NOT => output <= (16x"0", not s0, '0');
      when others => output <= (16x"0", 16x"0", '0');
    end case;
  end process;

  (o_high, o_low, write_high) <= output;
end impl; -- alu
```

Figure 18.20. VHDL for our simple ALU. Only shift and multiply write the upper accumulator bits.

```
library ieee;
use ieee.std_logic_1164.all;
use ieee.std_logic_unsigned.all;
use ieee.std_logic_misc.all;
use ieee.numeric_std.all;
use work.ff.all;
use work.processor_opcodes.all;
use work.alu_ch18_6.all;
use work.ch16.all;

entity processor is
  generic( programFile: string := "fib.asm" );
  port( o0, o1, o2, o3: buffer std_logic_vector(15 downto 0);
        input: in std_logic_vector(7 downto 0);
        rst, clk: in std_logic );
end processor;

architecture impl of processor is
  signal i: std_logic_vector(7 downto 0); -- the instruction
  signal pc: std_logic_vector(15 downto 0);
  signal op: std_logic_vector(3 downto 0); -- opcode
  signal alu_op: std_logic; -- alu operation?
  signal alu_opcode: std_logic_vector(2 downto 0);
  signal br_op: std_logic_vector(1 downto 0); -- branch opcode
  signal rs: std_logic_vector(3 downto 0); -- source register
  signal acc, acch, brd, ma, mout, t0, t1, t2: std_logic_vector(15 downto 0);

   --The register state
  signal tdone: std_logic;
  signal im: std_logic_vector(9 downto 0);
  signal s1: std_logic_vector(15 downto 0); -- source register
  signal imbranch, acc_eqz, accbranch, bran: std_logic;
  signal npc, npcr: std_logic_vector(15 downto 0);
  signal write_high: std_logic;
  signal o_high, o_low, acc_nxt, acch_nxt: std_logic_vector(15 downto 0);
  signal brdn, brdr, en_i, en, accr: std_logic_vector(15 downto 0);
  signal ld, lda, ldai, sta, brs, en_acc, en_acch, en_brd: std_logic;
begin
  --Instruction fetch and parse
  instStore: ROM generic map(8, 16, programFile) port map(pc, i);
  op <= i(7 downto 4);
  alu_op <= op(3);
  alu_opcode <= op(2 downto 0);
  br_op <= i(3 downto 2);
  rs <= i(3 downto 0); -- source register
```

Figure 18.21. VHDL of processor, part 1 of 3. This part of the design entity reads the instruction ROM with address `pc` and parses the instruction, `i`.

```
--Decode source register
process(all) begin
  case rs is
    when RACC => s1 <= acc;
    when RACCH => s1 <= acch;
    when RO0 => s1 <= o0;
    when RO1 => s1 <= o1;
    when RO2 => s1 <= o2;
    when RO3 => s1 <= o3;
    when RBRD => s1 <= brd;
    when RMA => s1 <= ma;
    when RMD => s1 <= mout;
    when RIM => s1 <= 6d"0" & im;
    when RT0 => s1 <= t0;
    when RT1 => s1 <= t1;
    when RT2 => s1 <= t2;
    when RIN => s1 <= 8d"0" & input;
    when RPC => s1 <= pc;
    when others => s1 <= 16d"0";
  end case;
end process;

--Compute the next PC
--im reg branch condition
imbranch <= im(9) xor or_reduce( (im(8) and tdone) & (im(7 downto 0) and input) );
--acc branch condition
acc_eqz <= '1' when acc = 16x"0" else '0';
accbranch <= '1' when (br_op = BR_EQ) and (acc_eqz = '1') else
             '1' when (br_op = BR_NEQ) and (acc_eqz = '0') else
             '1' when (br_op = BR_GZ) and (acc_eqz = '0') and (acc(15) = '0') else
             '1' when (br_op = BR_LZ) and (acc_eqz = '0') and (acc(15) = '0') else
             '0';
--Do we branch?
bran <=     '1' when (op = OP_BR) or (op = OP_BRS) or
                      (((op = OP_BRIM) or (op = OP_BRIMI)) and (imbranch = '1')) or
                      ((op = OP_BRACC) and (accbranch = '1'))
                 else '0';

--compute next PC
npc <= pc + i(3 downto 0) when bran = '1' and op = OP_BRIMI else
       brd when bran = '1' else
       pc + 1;
npcr <= 16x"0" when rst = '1' else npc;
```

Figure 18.22. VHDL of processor, part 2 of 3. The top of this code finds the correct source register using a `case` statement, and then computes the branching conditions and destination.

```
--The ALU, and next accumulator inputs
theALU: alu port map(alu_opcode, acc, s1, o_high, o_low, write_high);

lda <= '1' when op = OP_LDA else '0';
ldai <= '1' when op = OP_LDAI else '0';
acc_nxt <= (((acc_nxt'range => alu_op) and o_low) or
            ((acc_nxt'range => lda) and s1) or
            ((acc_nxt'range => ldai) and (12x"0" & rs))) and
           (acc_nxt'range => not rst);

sta <= '1' when op = OP_STA else '0';
acch_nxt <= (((acch_nxt'range => alu_op) and o_high) or
             ((acch_nxt'range => sta) and acc)) and
            (acch_nxt'range => not rst);

--The next brd register value
brdn <= pc + 1 when op = OP_BRS else acc;
brdr <= 16x"0" when rst = '1' else brdn;

--Compute the write signals for the registers
en_i <= std_logic_vector( shift_left( unsigned(std_logic_vector'(16x"1")),
                                      to_integer(unsigned(rs)) ) );
en <= (en_i and (en'range => sta)) or (en'range => rst);
ld <= lda or ldai; -- Load the acc?
en_acc <= alu_op or ld or en(to_integer(unsigned(RACC)));
en_acch <= (alu_op and write_high) or en(to_integer(unsigned(RACCH)));
brs <= '1' when op = OP_BRS else '0';
en_brd <= en(to_integer(unsigned(RBRD))) or brs;
accr <= 16x"0" when rst = '1' else acc;

ACC_REG: vDFFE generic map(16) port map(clk, en_acc, acc_nxt, acc);
ACCH_REG: vDFFE generic map(16) port map(clk, en_acch, acch_nxt, acch);
O0_REG: vDFFE generic map(16) port map(clk, en(to_integer(unsigned(RO0))), accr, o0);
O1_REG: vDFFE generic map(16) port map(clk, en(to_integer(unsigned(RO1))), accr, o1);
O2_REG: vDFFE generic map(16) port map(clk, en(to_integer(unsigned(RO2))), accr, o2);
O3_REG: vDFFE generic map(16) port map(clk, en(to_integer(unsigned(RO3))), accr, o3);
BRD_REG:vDFFE generic map(16) port map(clk, en_brd, brdr, brd);
MA_REG: vDFFE generic map(16) port map(clk, en(to_integer(unsigned(RMA))), accr, ma);
dataStore: RAM generic map(16, 16)
               port map(ma, ma, en(to_integer(unsigned(RMD))), accr, mout);
IM_REG: vDFFE generic map(10)
              port map(clk, en(to_integer(unsigned(RIM))), accr(9 downto 0), im);
TO_REG: vDFFE generic map(16) port map(clk, en(to_integer(unsigned(RT0))), accr, t0);
T1_REG: vDFFE generic map(16) port map(clk, en(to_integer(unsigned(RT1))), accr, t1);
T2_REG: vDFFE generic map(16) port map(clk, en(to_integer(unsigned(RT2))), accr, t2);
--IN, not included
PC_REG: vDFFE generic map(16) port map(clk, '1', npcr, pc);
TTIMER: Timer generic map(16)
              port map(clk, rst, en(to_integer(unsigned(RTIME))), acc, tdone);
end impl;
```

Figure 18.23. VHDL of processor, part 3 of 3. This code assigns the next values for the accumulator register and the enable signals, `en`, for our architected state, assigns the enable signals, and instantiates the registers.

```
LDAI 0111
STA BRD #Load branch target (insn 7)
LDA IN
STA 01 #01=loop count, from input
LDAI 0001
STA TO #Store 1 into TO  for dec loop count
STA T1 #Store 1 into T1 as first num
#begin loop
LDA 00 #Acc = last fib
ADD T1 #Add = 2nd to last fib
STA T2 #T2 = last fib
LDA 00
STA T1 #T1 = 2nd to last fib
LDA T2
STA 00 #00 = T2 (last fib)
LDA 01
SUB TO
STA 01 #01 = 01-1 (next loop iteration)
BRACC 0100 #Branch if no more iterations
```

Figure 18.24. Code for computing Fibonacci numbers on our simple processor.

the BR.ACC instruction branches according to the accumulator register (ACC). Bits 3 and 2 of the instruction indicate the branch condition.

Table 18.11 lists our processor's registers, many of which have unique functionality. The outputs of our system are four 16-bit registers: O0 through O3. The temporary registers (T0–T2) are used to store intermediate values. The PC holds the current program counter and is read-only for all non-branch instructions (used as RS). We also include support for a timer that is loaded when written. The eight input bits appear as a read-only register. The accumulator is broken into 16-bit high and low registers. Only the multiply and shift instructions write the high bits.

The processor includes an arithmetic-logic unit (ALU). Given an opcode and two inputs, the ALU performs the specified operation and outputs the result. Structurally, the ALU computes the eight different operations and uses an eight-input multiplexer to select the output. It resembles the universal shifter/counter of Section 16.2.3 without the internal state. The VHDL for the ALU is shown in Figure 18.20. The output function is selected by a `case` statement, and we instantiate only one adder/subtractor.

A data RAM (see Section 8.9) is accessed via the MD and MA registers. When MD is used as a source for any LD or ALU instruction, the value stored in address MA of the RAM is loaded. When MD is the destination for an STA instruction, the value in ACC is placed into memory address MA.

The VHDL for our processor design entity is shown in Figures 18.21 to 18.23. The first section of the code (Figure 18.21) loads and does some initial parsing of the current instruction from the instruction ROM. In the second and third parts (Figures 18.22 and 18.23), the code selects the correct source register from the 16 options using a `case` statement, and calculates branching conditions and the next program counter. The enable signals, `en`, are used to write the correct register. The VHDL concludes with all the state registers, including the timer and PC.

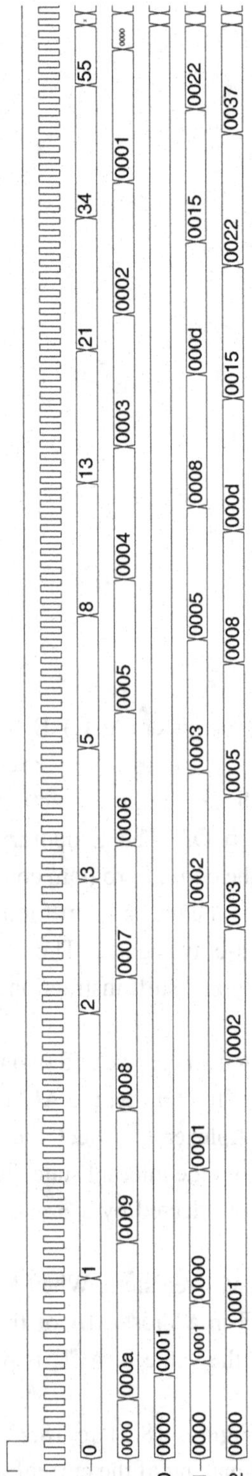

Figure 18.25. The Fibonacci code from Figure 18.24 running on our processor. Register O0 shows the current Fibonacci number and register O1 is the remaining number of iterations in the program. We also show the temporary values stored in T0, T1, and T2.

Our simple processor executes a *software* program stored in ROM. For example, a program for calculating Fibonacci numbers is shown in Figure 18.24. Waveforms resulting from executing this program are shown in Figure 18.25. The code initializes the state by loading several constants into registers as well as reading the input to determine how many numbers to compute. The loop, beginning at `PC`=9, computes the next number and transfers it into `O0`. The loop index, `O1`, is decremented, and a branch to the loop start is taken if the index does not equal 0.

Summary

In this chapter you have learned the powerful technique of *stored-program control* and how to implement finite-state machines using *microcode*.

Any finite-state machine can be implemented using a table in memory (ROM or RAM) to store the next-state and output functions. A concatenation of all input signals and the current state is used as the memory address. The memory output gives the next state and current output. This technique, while general, requires a memory with $S2^i$ words, where S is the number of states and i is the number of input signal bits.

By adding a *sequencer* to generate the next-state memory address, we can reduce the size of the required memory to S words. The sequencer implements a family of *branch* instructions that select either the next state in sequence or a *branch target*, specified by the current microcode word, depending on the value of the input bits. Multi-way branches can be performed by modifying the branch target address on the basis of the input conditions.

If only a subset of the outputs changes on each state transition, we can further reduce our memory requirements by defining *store* instructions. With this organization, the output of the microcode machine is held in a set of registers. Each store microinstruction updates the state of one output register. The other registers hold their previous states.

If our microcode has repeated sequences, we can reduce the size of our code by extending our branch instructions to include *subroutine* call and return instructions. The CALL instruction saves the address of the instruction following the call in a special `rupc` register. After the common sequence has been executed, the RET instruction jumps to the address in the `rupc`.

BIBLIOGRAPHIC NOTES

Microcode was originated by Maurice Wilkes at Cambridge University in 1951 to implement the control logic for the EDSAC computer [113]. It has been widely used since that time in many different types of digital systems. Microcode became very popular in the late 1970s for implementing processors with bipolar bit-slice chip sets [81]. Microcode is still widely used today to implement complex instruction sets such as x86 [44]. The approach taken to generate the microcode for the Motorola 680000 (the processor used in the original Apple Macintosh) is described in ref. [107].

Two of the most popular books about processor design are Patterson and Hennessy's introductory [93] and more advanced [48] texts. For another example of a relatively simple processor, O'Brien's *The Apollo Guidance Computer* [90] overviews the computer that went to the Moon.

Exercises

18.1 *Modified traffic-light controller, I.* Modify the traffic controller microcode of Table 18.1 (for the controller shown in Figure 18.1) by adding an additional input *car_ns*. Now a traffic light in any direction stays green until the input signaling a car traveling in the opposite direction goes high. This happens regardless of whether any cars remain in the green direction. Be sure to include yellow lights in any transition from green in one direction to another.

18.2 *Modified traffic-light controller, II.* Using the microcoded FSM of Figure 18.2, simulate your microcode from Exercise 18.1. Be sure to instantiate a version of the FSM with enough inputs, outputs, and bits of state.

18.3 *Modified traffic-light controller, III.* How many bits in your code from Exercise 18.1 must you change in order for a light to change direction only when no cars are coming in the current direction and there is a car in the opposite direction?

18.4 *Microcoded vending machine.* Implement the control path of the vending machine of Section 16.3.1 using the microcoded controller with the instruction sequencer of Section 18.2. What are the inputs and outputs of the machine? How large a control store is required? Show the microcode for this implementation. Assume that each external input to the controller is held high until the FSM pulses a *nxt* output signal.

18.5 *Microcoded combination lock.* Implement the control of the combination lock of Section 16.3.2 using the microcoded controller and sequencer of Section 18.2. What are the inputs and outputs of the machine? How large a control store is required? Show the microcode for this implementation.

18.6 *SOS flasher, I.* Write the microcode for an SOS flasher (Exercise 17.11). While the input `flash` is high, your system should flash an SOS sequence – three short flashes (one clock each) followed by three long flashes (four clocks each) followed by three short flashes again. Spaces within each character should be one clock long. Spaces between characters within one SOS should be three clocks long. Spaces between one SOS and the next SOS should be seven clocks long. When the input is lowered, the flasher should reset back to a reset state. Use the microcoded FSM of Section 18.1. Leave your flasher completely unfactored.

18.7 *SOS flasher, II.* Modify the microcoded FSM and microcode from Section 18.1 and Exercise 18.6 to act as a control module to interface with a datapath of your design. Provide a block diagram of this datapath, the signals needed to interface between the two, and your microcode.

18.8 *SOS flasher, III.* Write and verify the VHDL for the SOS flasher (datapath + microcoded FSM) of Exercise 18.7.

18.9 *SOS flasher, IV.* Write the microcode for the SOS flasher of Exercise 18.7 using the sequenced microcoded FSM of Section 18.2 instead of the FSM of Section 18.1.

18.10 *SOS flasher, V.* Write the microcode for the SOS flasher of Exercise 18.7 using the microcoded FSM with subroutines from Section 18.5. Each character in your system should be a separate subroutine.

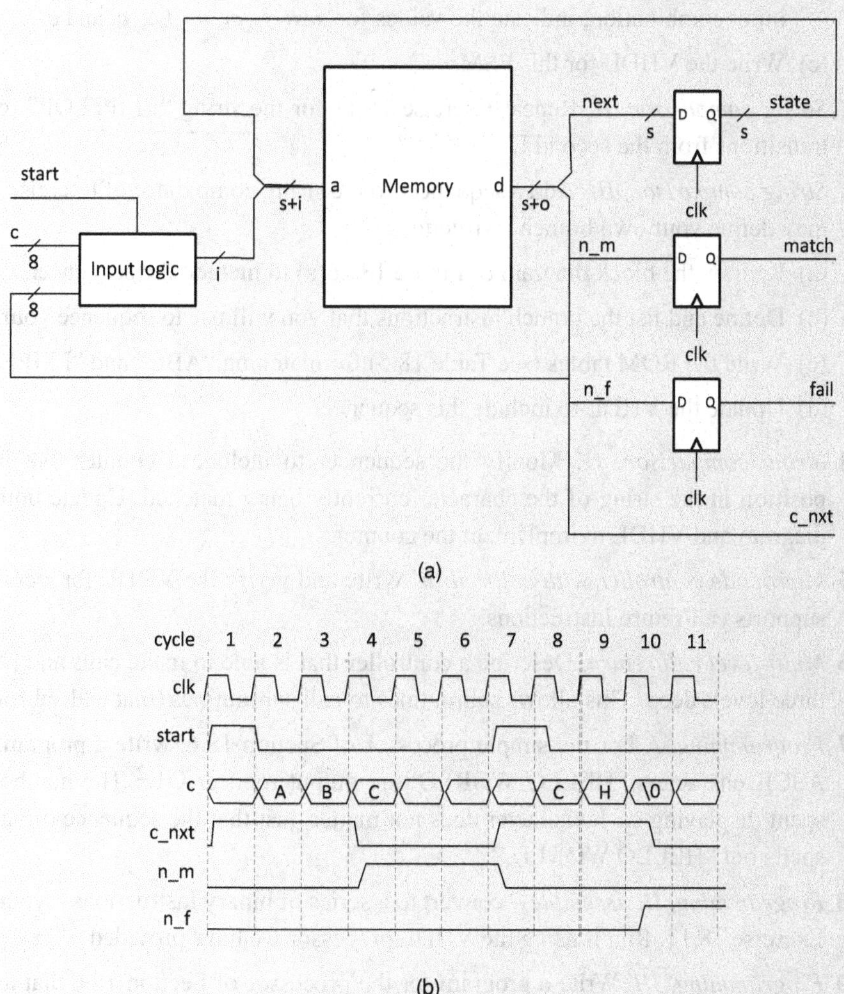

Figure 18.26. Microcode block (a) and waveform (b) of a basic string comparison engine. The input character *c* is matched against a string "ABC." If it matches, the match output will be asserted. If the string terminates '/0', a failure output is asserted.

18.11 *String comparison, I.* This exercise and Exercises 18.12–18.14 deal with building a microcoded state machine to perform an ASCII string comparison. The initial block diagram is shown in Figure 18.26(a). Timing is shown in Figure 18.26(b). A string comparison is started by asserting the input signal *start*. After comparing against the entire string, if we find a matching string, the *match* output is asserted until *start* again pulses. If the string termination character arrives (*end* = 1), the *fail* signal should be asserted until restart. The microcode machine asserts *c_nxt* to request each new character from the input. The ROM provides signal *s_c* to the input block to match against the current character. The input logic block outputs three signals: *start*, *end* (c = `8'b0`), and *match* ($c = s\_c$).

(a) If the match sequence is "11ABC," draw the state diagram.

(b) Draw the microcode table without a sequencer (see Table 18.1). For each state and input combination, indicate the values for *next*, *n_m*, *n_f*, *s_c*, and *c_nxt*.

(c) Write the VHDL for this FSM.

18.12 *String comparison, II.* Repeat Exercise 18.11 for the string "FLIPFLOP" (consider all transitions from the second L).

18.13 *String comparison, III.* Add a sequencer to the string comparator of Exercise 18.12. You may define your own branch instructions.

(a) Redraw the block diagram of Figure 18.26(a) to include the sequencer.

(b) Define and list the branch instructions that you will use to sequence your states.

(c) Write the ROM tables (see Table 18.5) for matching "ABC" and "FLIPFLOP."

(d) Update the VHDL to include this sequencer.

18.14 *String comparison, IV.* Modify the sequencer to include a counter that indicates in position in the string of the character currently being matched. Update both the block diagram and VHDL to implement the counter.

18.15 *Microcode controller with call/return.* Write and verify the VHDL for a controller that supports call/return instructions.

18.16 *Multi-level call/return.* Describe a controller that is able to make calls and returns up to three levels deep. This allows subroutines to call subroutines (that call subroutines).

18.17 *Programming, I.* For the simple processor of Section 18.6, write a program to put the ASCII characters "HELLO WORLD" on output register O1.[2] The number of cycles spent displaying each character does not matter, just that the sequence of values on O1 spells out "HELLO WORLD."

18.18 *Programming, II. Assemble* – convert to a series of binary instructions – your code from Exercise 18.17. Run it using the VHDL processor we have provided.

18.19 *Programming, III.* Write a program for the processor of Section 18.6 that averages the first 32 values stored in memory (addresses 0–31). Output the result on register O1.

[2] In ASCII, the word "HELLO" (with the trailing space) corresponds to the hex values 0x48, 0x45, 0x4C, 0x4C, 0x4F, 0x20.

19 Sequential examples

This chapter gives some additional examples of sequential circuits. We start with a simple FSM that reduces the number of 1s on its input by a factor of 3 to review how to draw a state diagram from a specification and how to implement a simple FSM in VHDL. We then implement an SOS detector to review factoring of state machines. Next, we revisit our tic-tac-toe game from Section 9.4 and build a datapath sequential circuit that plays a game against itself using the combinational move generator we previously developed. We illustrate the use of table-driven sequential circuits and composing circuits from sequential building blocks like counters and shift registers by building a Huffman encoder and decoder. The encoder uses table lookup along with a counter and shift register, while the decoder traverses a tree data structure stored in a table.

19.1 DIVIDE-BY-3 COUNTER

In this section we will design a finite-state machine that outputs a high signal on the output for one cycle for each three cycles the input has been high. More specifically, our FSM has a single input called `input` and a single output called `output`. When `input` is detected high for the third cycle (and sixth, ninth, etc.), `output` will go high for exactly one cycle. This FSM divides the *number* of pulses on the input by 3. It does not divide the binary number represented by the input by 3.

A state diagram for this machine is shown in Figure 19.1. At first it may seem that we can implement this machine with three states; however, four are required. We need states A to D to distinguish having seen the input high for zero, one, two, or three cycles so far. The machine resets to state A. It sits in this state until the input is high on a rising clock edge, at which time it advances to state B. The second high input takes the machine to C, and the third high input takes the machine to D, where the output goes high for one cycle. We can't simply have this third high input take us back to A because we need to distinguish having seen three cycles of high input – in which case the output goes high – from having seen zero cycles of high input.[1]

The FSM always exits state D after one cycle. The input during this cycle determines the next state. If the input is low, the machine advances to state A to wait for three more high inputs before the next output. If the input is high, this counts as one of the three high inputs, so the machine advances to state B to wait for two more.

A VHDL description of this divide-by-3 FSM is shown in Figure 19.2. A `case` statement is used to implement the next-state function, including reset. A single concurrent signal

[1] See Exercise 19.3 for an approach that does require only three states.

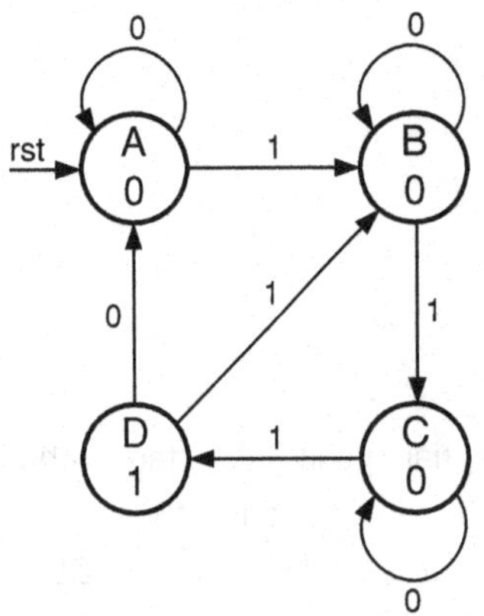

Figure 19.1. State diagram for a divide-by-3 counter FSM. The four states represent having seen the input high for zero, one, two, and three cycles so far.

assignment statement implements the output function, making the output high in state D. The constant declarations used to define the states and their width are not shown. Waveforms from simulating this VHDL model are shown in Figure 19.3.

19.2 SOS DETECTOR

Morse code, once widely used for telegraph and radio communication, encodes the alphabet, numbers, and a few punctuation marks into an on/off signal as patterns of dots and dashes. Spaces are used to separate symbols. A *dot* is a short period of *on*, a *dash* is a long period of *on*. Dots and dashes within a symbol are separated by short periods of *off*, and a space is a long period of *off*. The universal distress code, SOS, in Morse code is three dots (S), a space, three dashes (O), a space, and three dots again (the second S).

Consider the task of building a finite-state machine to detect when an SOS is received on the input. We assume that a *dot* is represented by the input being high for exactly one cycle, a *dash* is represented by the input being high for exactly three cycles, dots and dashes within a symbol are separated by the input being low for exactly one cycle, and that a *space* is represented by the input being low for three or more cycles. Note that the input going either high or low for exactly two cycles is an illegal condition. With this set of definitions, one legal SOS string is 101010001110111011100010101000.

We can build an SOS detector as a single, flat state machine as shown with the state diagram of Figure 19.4. The FSM resets to state R. States S11 to S18 detect the first "S" and the associated space. States O1 to O11 detect the "O," and states O12 to O14 detect the space following the "O." Finally states S21 to S28 detect the second "S" and the subsequent space. State 28 outputs a "1" to indicate that SOS has been detected.

For clarity, many transitions are omitted from Figure 19.4. The transitions along the horizontal path from state R through state S28 represent the transitions that occur when an SOS is detected. If at any point along this path a 1 is detected when a 0 is expected, the machine transitions to state E1. Similarly, if a 0 is detected when we expect a 1, the machine transitions to state E2. These transitions are shown for the first row (via boxes E and E1) and then omitted to avoid cluttering the figure. States E1 to E3 are error-handling states that wait for a space after an error condition and then restart the detection.

```
------------------------------------------------------------------------
--Divide by 3 FSM
--   in - increments state when high
--   out - goes high one cycle for every three cycles in is high
--     it goes high for the first time on the cycle after the third cycle
--     in is high.
------------------------------------------------------------------------
library ieee;
use ieee.std_logic_1164.all;
use work.ff.all;

entity Div3FSM is
  port( clk, rst, input: in std_logic;
        output: out std_logic );
end Div3FSM;

architecture impl of Div3FSM is
  constant AWIDTH: integer := 2;
  constant A: std_logic_vector(AWIDTH-1 downto 0)  := 2d"0";
  constant B: std_logic_vector(AWIDTH-1 downto 0)  := 2d"1";
  constant C: std_logic_vector(AWIDTH-1 downto 0)  := 2d"2";
  constant D: std_logic_vector(AWIDTH-1 downto 0)  := 2d"3";

  signal state, n: std_logic_vector(AWIDTH-1 downto 0); -- current, next state
begin
  -- state register
  state_reg: vDFF generic map(AWIDTH) port map(clk, n, state);

  -- next state function
  process(all) begin
    case state is
      when A => if rst then n <= A; elsif input then n <= B; else n <= A; end if;
      when B => if rst then n <= A; elsif input then n <= C; else n <= B; end if;
      when C => if rst then n <= A; elsif input then n <= D; else n <= C; end if;
      when D => if rst then n <= A; elsif input then n <= B; else n <= A; end if;
      when others => n <= A;
    end case;
  end process;

  -- output function
  output <= '1' when state = D else '0';
end impl;
```

Figure 19.2. VHDL description of the divide-by-3 counter.

The transition from O1 to S12 handles the case where the input includes the string SSOS. After detecting the first S, we are expecting an O but instead receive a second S. If we were to transition to state E2 on receiving a 0 in O1, we would miss this second S and hence the SOS. Instead we must recognize the *dot* and go to state S12.

The transition from S28 to S11 (via the box labeled D) is needed in order to allow back-to-back SOSs with minimum-sized spaces to be detected. After detecting SOS, including the

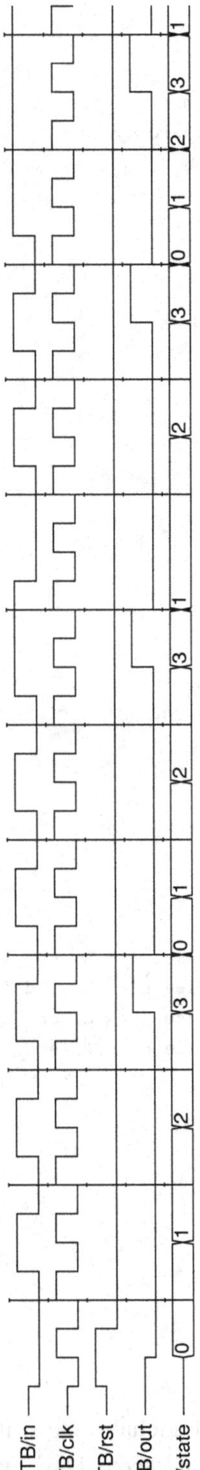

Figure 19.3. Waveforms from simulating the divide-by-3 counter.

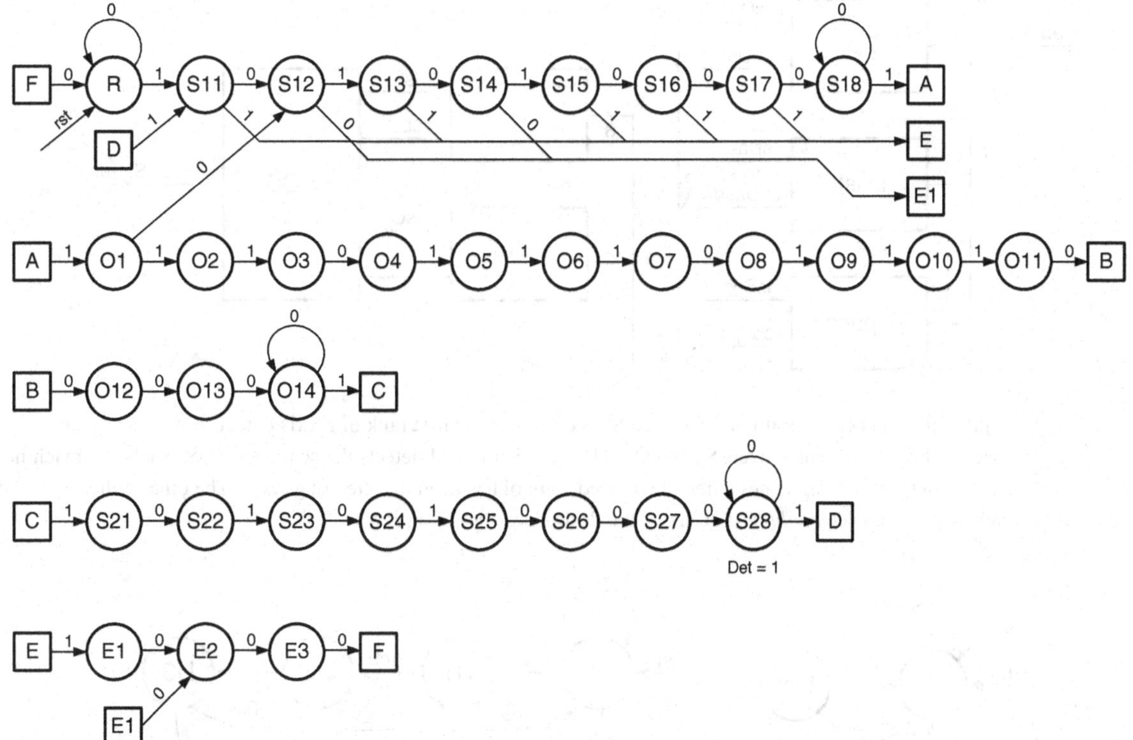

Figure 19.4. State diagram for an SOS detector realized as a single, flat FSM. The square boxes indicate connections.

subsequent space, in state S28, the next 1 may be the first dot of the next SOS, and must be recognized by going to state S11.

While the flat FSM of Figure 19.4 works, it is not a very good solution for a number of reasons. First, it is not modular. If we were to change the definition of a dot to be the input going high for one or two cycles, the flat machine would need to be changed in eight places (every place a dot is recognized). Similar global changes would be needed to accommodate a change to the definition of a dash or a space. Also, the machine would need to be completely reworked if the sequence we are detecting is different than SOS, say ABC. Second, the machine is large, 34 states, and would become even larger with more flexible definitions of dots and dashes. Finally, some aspects of the machine, like the transition from O1 to S12, are subtle.

The SOS machine is a perfect candidate for *factoring*. We can build FSMs to detect dots, dashes, and spaces, and then use the outputs of these FSMs to build FSMs that detect S and O. Finally, a simple top-level FSM detects SOS. A block diagram for a factored version of our SOS-detection FSM is shown in Figure 19.5. The input bit stream (*sequence*) is input to three element-detecting FSMs – *Dot*, *Dash*, and *Space*. Each of these FSMs has two outputs: one that indicates when the element has been detected, and one that indicates that the current input sequence *could be* part of that element. For example, the *Dot* FSM outputs *isDot* when a dot is detected and *cbDot* when the current sequence could be a dot, but needs additional input before deciding.

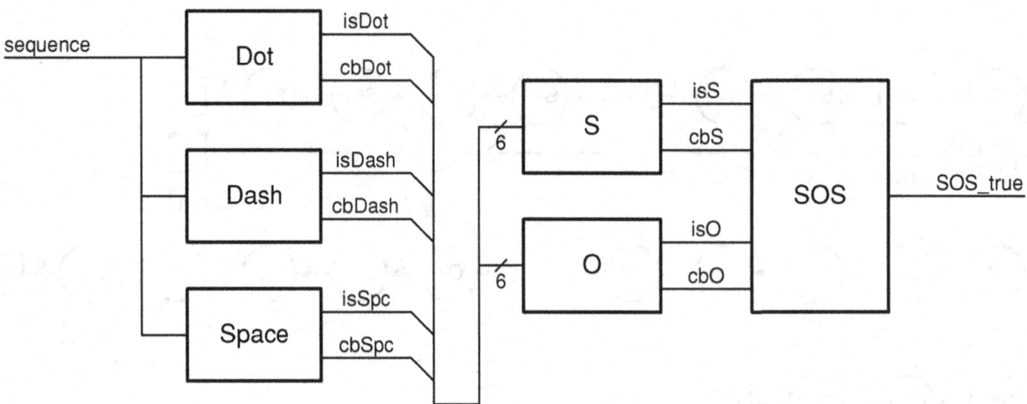

Figure 19.5. Block diagram of a factored SOS detector. The first rank of FSMs detects dots, dashes, and spaces. The second rank detects Ss and Os. The final SOS FSM detects the sequence SOS. Each sub-machine has two outputs: one indicates when the desired symbol has been detected (e.g., *isS*). The other indicates when the current sequence *could be* a prefix of the desired symbol (e.g., *cbS*).

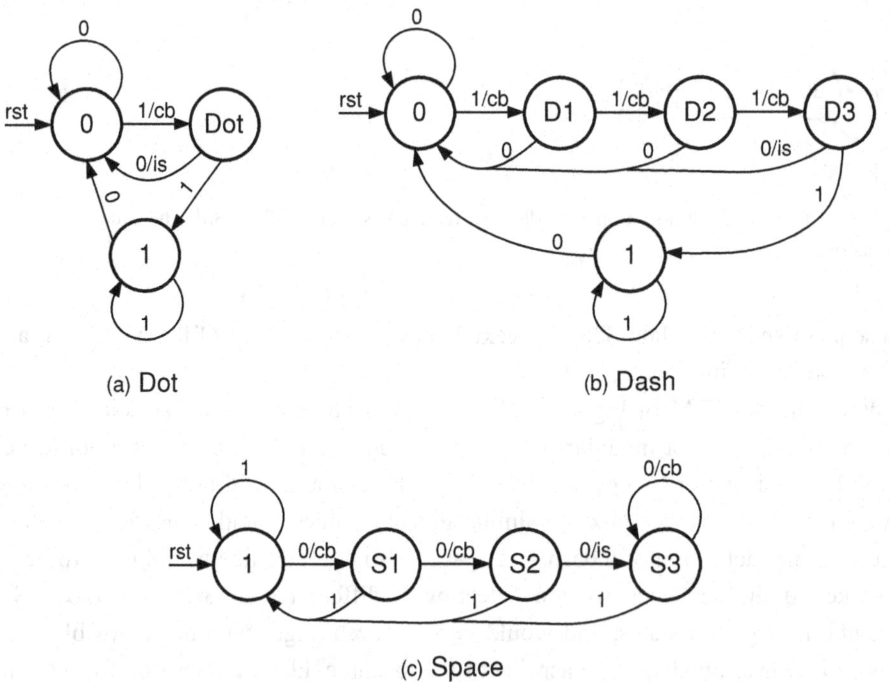

Figure 19.6. Element finite-state machines. (a) Dot, (b) Dash, and (c) Space. Each outputs when the current sequence *could be* (*cb*) the respective element and when the element has been detected (*is*).

The six signals out of the three element detectors feed a pair of character detectors, one each for S and O. Like the element detectors, each character detector also has an *is* and a *could be* output. The four signals out of the two character detectors are input to a top-level SOS FSM that indicates when SOS is detected.

Figure 19.6 shows the three FSMs that detect the elements Dot, Dash, and Space. The Dot FSM resets to state 0. Upon detecting a 1 on the input, it indicates that the current sequence *could be* a dot by asserting output *cb* and transitions to state *Dot*. In state *Dot*, a 0 on the input

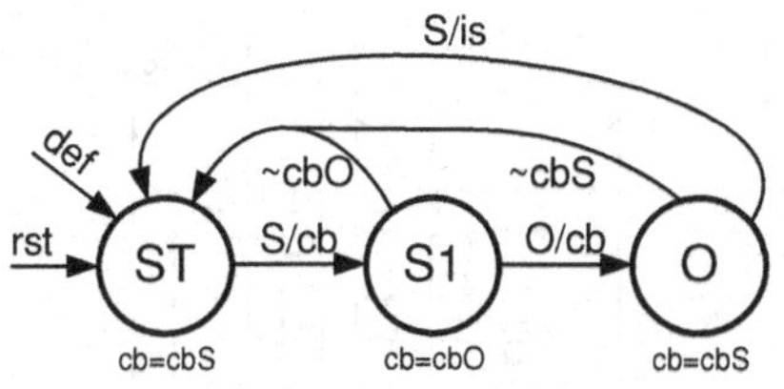

Figure 19.7. State diagram for S-detecting FSM.

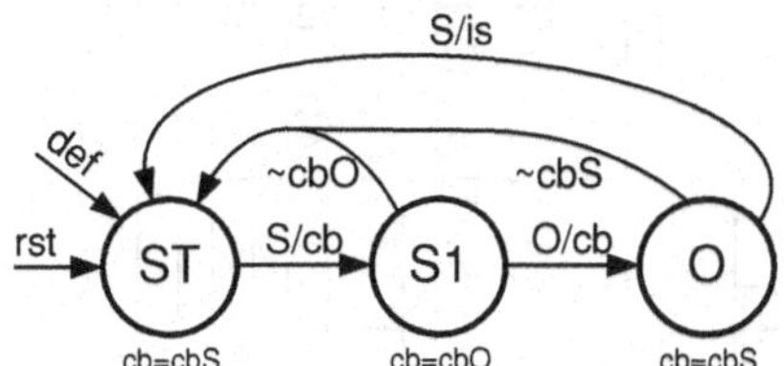

Figure 19.8. State diagram of main SOS-detecting FSM.

results in a dot being detected with the *is* output asserted and returns the machine to state 0. Note that when the *is* output is asserted, the *cb* output is also asserted. If a 1 is detected in state *Dot*, the machine goes to state 1 to wait for a zero. The Dash and Space machines operate in a similar manner.

The character FSM for the character S is shown in Figure 19.7. The machine resets to state OTH (other), which is also the target of the default (def) transition, which covers unexpected inputs. Upon detecting a space, the machine enters state SPC. Detecting the first dot moves the machine to state D1, and subsequent dots move the machine to states D2 and D3. Detecting a space in state D3 returns the machine to state SPC and asserts *is* to signal that an S has been detected.

If, at any point during the sequence from SPC through D1, D2, and D3 and back to SPC, the input could not be the element being waited for (e.g., if *cbDot* is false in state D1), then the machine returns to state OTH. This is why we need the *could-be* outputs on the element detectors. They allow us to detect illegal elements between the elements we are looking for. Consider, for example, the input sequence 00010110101000. The machine detects the space 000, and the first dot 10, but then returns to state OTH on the illegal element 110 because *cbDot* falls when the second 1 is detected. If we just wait for *isDot*, we would erroneously determine that this sequence is an S because it has three dots. Without monitoring *cbDot* we would not see that the three dots are not contiguous, and hence not an S.

The main SOS-detecting FSM, shown in Figure 19.8, contains only three states. It waits in state ST (start) until an S is detected, when it moves to state S1. From state S1, if an O is detected it moves to state O. However, if at any point in S1 input *cbO* goes false, indicating an illegal sequence between the S and the O, the machine returns to ST. If the machine detects a second S while in state O, it asserts its *is* output, detecting an SOS, and returns to ST. If input *cbS* goes false while in state O, detecting an illegal sequence between the O and the S, the machine returns to ST without detecting SOS.

Waveforms showing the operation of the factored SOS detector are shown in Figure 19.9. The waveforms show two good SOS detections with one SOT (T is a single dash) between them. Note the *cb* and *is* waveforms for each element, for the characters S and O, and for SOS itself. On the second 1 of the T's dash, *cbDot* falls, causing *cbS* and *cbSOS* to fall in turn (combinationally).

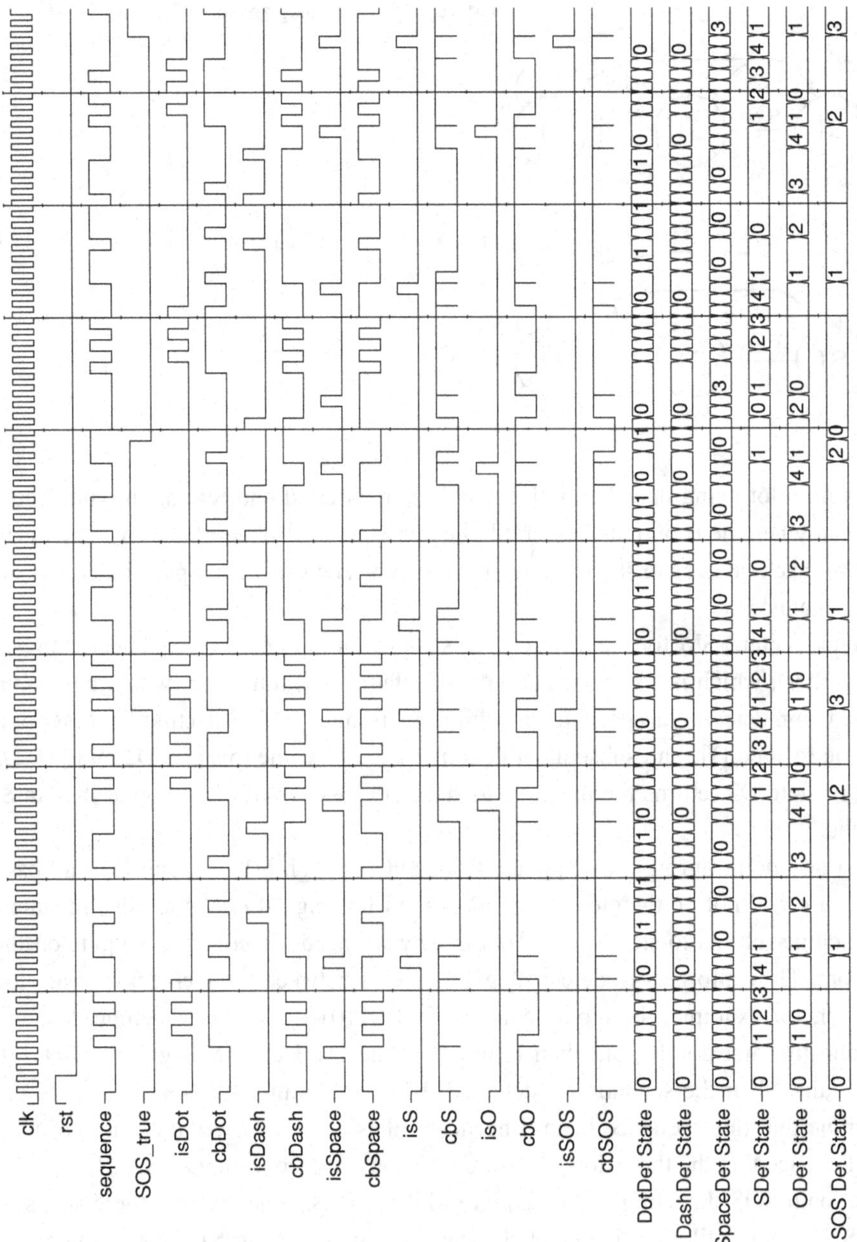

Figure 19.9. Waveforms showing operation of factored SOS detector.

Factoring the SOS detector gives us a much simpler system that is far easier to modify and maintain. Instead of 34 states in one brittle, monolithic state machine, we have a total of 20 states divided over six small, simple FSMs. The largest single FSM in the factored machine has a total of five states. Should we modify our specification to change the definition of a dot to be one or two 1s in a row, we would make one simple change to the Dot FSM.[2]

[2] For some practice modifying this factored SOS detector, see Exercises 19.4 and 19.5.

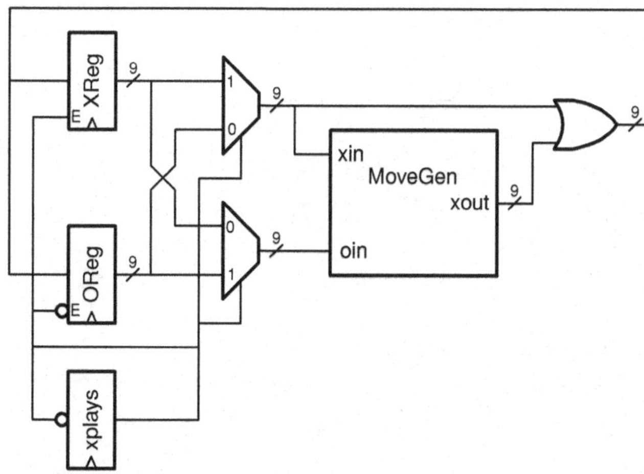

Figure 19.10. Block diagram of a tic-tac-toe playing system using the move generation module of Section 9.4.

19.3 TIC-TAC-TOE GAME

In Section 9.4 we designed a combinational module that generated moves for the game of tic-tac-toe. In this section we will use this module as a component in a sequential system that plays a game of tic-tac-toe against itself.

A block diagram of the system is shown in Figure 19.10. The state in the system resides in the three registers on the left side of the figure. The nine-bit *Xreg* and *Oreg* hold bit maps that reflect the current positions of Xs and Os, respectively. The one-bit *xplays* register is true if X plays next and false if O plays next. Reset is not shown in the figure. *Xreg* and *Oreg* reset to all zeros; *xplays* resets to true.

When it is X's turn to play ($xplays = 1$) the multiplexers direct *Xreg* to the *xin* input of the move generator and *Oreg* to the *oin* input. The move generator generates the next move on *xout* and this is ORed with the current X position to generate the new X position that is stored back in *Xreg* at the end of the cycle. The write to *Xreg* is enabled by *xplays*. When *xplays* is false, the multiplexers switch the move generator inputs to generate a move for O, and this move is written back to *Oreg* at the end of the cycle.

A VHDL description of the tic-tac-toe playing system is shown in Figure 19.11. After the declaration of the three state registers, an assignment statement toggles `xPlays` each cycle. The input multiplexers are implemented as conditional signal assignment statements. Two conditional signal assignment statements compute the next state for `Xreg` and `Oreg`. The OR of the move with the previous state is included in these statements.

19.4 HUFFMAN ENCODER/DECODER

A Huffman code is an *entropy* code that encodes each symbol of an alphabet with a bit string. Frequently used symbols are encoded with short bit strings, whereas infrequently used symbols are encoded with longer bit strings. To be able to distinguish short bit strings from the first parts of longer bit strings, each short bit string must not be used as a prefix for any longer bit string. The net result is *data compression*; that is, a typical sequence of symbols is encoded into fewer bits than would be required if all symbols were encoded with the same number of bits.

```
--------------------------------------------------------------------------
-- Sequential Tic-Tac-Toe game
--   Plays a game against itself
---------------------------------------------------------------------
library ieee;
use ieee.std_logic_1164.all;
use work.ff.all;
use work.ch9.all;

entity SeqTic is
  port( clk, rst: in std_logic;
        xreg, oreg: buffer std_logic_vector(8 downto 0);
        xplays: buffer std_logic );
end SeqTic;

architecture impl of SeqTic is
  signal nxreg, noreg, move, areg, breg: std_logic_vector(8 downto 0);
  signal nxplays: std_logic;
begin

  -- state
  X:  vDFF generic map(9) port map(clk, nxreg, xreg);
  O:  vDFF generic map(9) port map(clk, noreg, oreg);
  XP: sDFF port map(clk, nxplays, xplays);

  -- x plays first, then alternate
  nxplays <= '1' when rst else not xplays;

  -- move generator - mux inputs so current player is x
  areg <= xreg when xplays else oreg;
  breg <= oreg when xplays else xreg;
  moveGen: TicTacToe port map(areg, breg, move);

  -- update current player
  nxreg <= 9d"0" when rst else
           xreg or move when xplays else
           xreg;
  noreg <= 9d"0" when rst else
           oreg or move when not xplays else
           oreg;
end impl;
```

Figure 19.11. VHDL description of the tic-tac-toe playing system.

19.4.1 Huffman encoder

For this example we will build a Huffman encoder and decoder for the letters of the alphabet, A–Z. The input to our encoder is a five-bit code, where A = 1 and Z = 26.[3] To prevent input characters from arriving faster than our encoder can handle them, our encoder generates an input ready (*irdy*) signal that indicates when the encoder is ready for the next input character.

[3] This corresponds to the low five-bits of the ASCII code for both upper-case and lower-case letters.

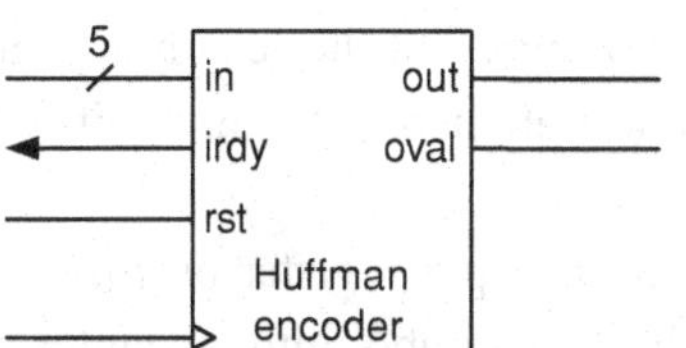

Figure 19.12. Schematic symbol for our Huffman encoder. The encoder accepts a five-bit character on *in* each time *irdy* goes high and generates a bit serial output stream on *out* when *oval* is high.

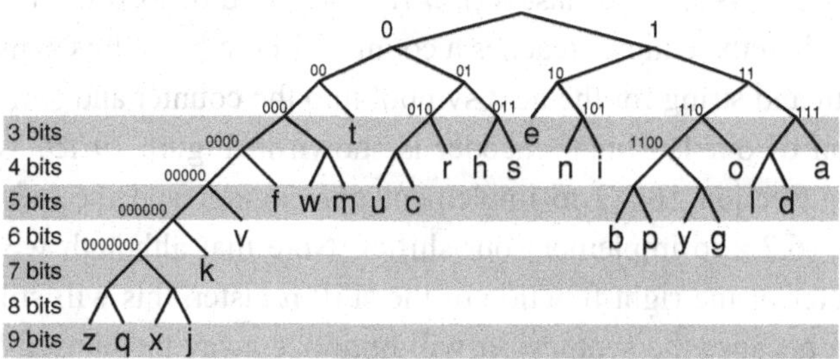

Figure 19.13. Huffman code for the alphabet shown as a binary tree. Starting at the root, each branch to the left denotes a 0 and each branch to the right denotes a 1. Hence the letter W, which is reached by a sequence of left, left, left, right, left, is encoded as 00010.

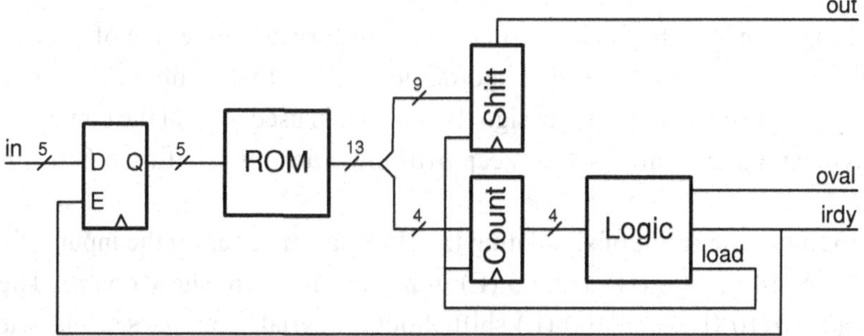

Figure 19.14. Block diagram of a Huffman encoder. A ROM stores the length and string for each character. A counter counts down the length while a shift shifts out the string.

The output is a serial bit stream of the encoded characters. To make it easy for the decoder to find the start of the bit stream, our encoder also generates an output valid (*oval*) signal that signals when the bits in the output stream are valid. A block diagram symbol for our encoder showing inputs and outputs is shown in Figure 19.12.

Figure 19.13 shows the code we will use for our example in tree form. The path from the root of the tree to a character gives the code for that character. The letter E, for example, is reached by going right, left, left, and hence is represented by the three-bit string 100. The letter J is reached by going left seven times and then right twice, so it is represented by the nine-bit string 000000011. Very frequently occurring characters like T and E are represented with just three bits. Very infrequently occurring characters like Z, Q, X, and J are represented with nine bits. Representing the code as a tree makes it clear that a short string used to represent one symbol is not a prefix of a longer string used to represent another symbol, since each leaf node of the tree terminates the path used to reach that leaf.

A block diagram of a Huffman encoder is shown in Figure 19.14. A five-bit input register holds the current symbol and is loaded with a new symbol each time *irdy* is asserted. The symbol is used to address a ROM that stores the string and string length associated with each symbol. For example, the ROM stores string 0011 001000000 for the symbol T. This indicates that the string representing T is three bits in length, and the three bits are 001. Because the maximum length string is nine bits, we use four bits to represent the length and nine bits to represent the string. Strings shorter than nine bits are left-aligned in the nine-bit field so they can be shifted out to the left.

One cycle after a new symbol has been loaded into the input register by *irdy*, signal *load* is asserted to load the length and string associated with that symbol into a counter and a shift register. The counter then counts down while the shift register shifts bits onto the output. When the counter reaches a count of 2 (second to last bit), *irdy* is asserted to load the next symbol into the input register, and when the counter reaches a count of 1 (last bit of this symbol), *load* is asserted to load the length and string for the next symbol into the counter and shift register.

A VHDL implementation of our Huffman encoder is shown in Figure 19.15. We use the up/down/load counter from Section 16.1.2 to implement our counter and the left/right/load shift register from Section 16.2.2 to implement our shifter. Note that although we don't use the up function of the counter or the right function of the shift register, this will still result in an efficient implementation because the synthesizer will optimize away the unused logic. The table is implemented in module `HuffmanEncTable` (not shown), which is coded as a large `case` statement.

The control logic for the Huffman encoder is straightforward. One line of code asserts *irdy* on a count of 2 or 0 – the latter is needed in order to load the first symbol after a reset. A DFF then delays *irdy* by one cycle to generate signal *dirdy* that is used to load the counter and shifter. One line of code and a DFF are used to keep *oval* low after reset until the first time *dirdy* is asserted.

Figure 19.16 shows the result of simulating the Huffman encoder on the input string "THE." The three symbols 14 (T), 08 (H), and 05 (E) in hexadecimal are shown on *in*. The resulting output is 001 (T), 0110 (H), and 100 (E) shifted out bit serially on *out* starting with the first cycle in which *oval* is asserted. The value in the counter can be seen on counting down from the string length (3 or 4) to 1 for each symbol, while the value in the shift register *obits* is shifting the string associated with each symbol left.

19.4.2 Huffman decoder

Now that we have encoded a character string using a Huffman code, we will look at building the corresponding decoder. To decode a Huffman-encoded bit string we simply traverse the encoding tree of Figure 19.13, traversing one edge for each bit of the input bit stream – the left branch for each 0, and the right branch for each 1. When we encounter a terminal node during this traversal, we emit the corresponding symbol on the output and restart our traversal at the root of the tree.

To facilitate storing the decoding tree in a table, we relabel the nodes of the tree as shown in Figure 19.17. Each node is assigned an integer that serves as its address in the table. Note that the root does not need to be stored in the table, so we start labeling nodes at 0 with the left child of the root. At each entry in the table we store a type and a value. The type indicates whether this node is an internal node (type = 0) or a terminal node (type = 1). For a terminal node, the value holds the symbol to emit. For an internal node, the value holds the address of the left child of this node (which will always be an even number). The address of the right child can be found by adding 1 to the value.

```
--------------------------------------------------------------------------
-- Encoder
--    in - character 'a' to 'z' - must be ready
--    irdy - when high accepts the current input character
--    out - bit serial Huffman output
--    oval - true when output holds valid bits
--
--    input character accesses a table RAM with each entry having
--    length[4], bits[9]
--------------------------------------------------------------------------

library ieee;
use ieee.std_logic_1164.all;
use work.ch16.all;
use work.ff.all;

entity HuffmanEncoder is
  port( clk, rst: in std_logic;
        input: in std_logic_vector(4 downto 0);
        irdy, output, oval: buffer std_logic );
end HuffmanEncoder;

architecture impl of HuffmanEncoder is
  component HuffmanEncTable is
    port( input: in std_logic_vector(4 downto 0);
          length: out std_logic_vector(3 downto 0);
          bits: out std_logic_vector(8 downto 0) );
  end component;

  signal length, count: std_logic_vector(3 downto 0);
  signal bits, obits: std_logic_vector(8 downto 0);
  signal char, nchar: std_logic_vector(4 downto 0);
  signal dirdy: std_logic; -- irdy delayed by one cycle - loads count and sr
  signal noval: std_logic;
begin
  -- control
  output  <= obits(8); -- MSB is output
  irdy <= '0' when rst else -- 0 count for reset
          '1' when count = 4d"2" or count = 4d"0" else
          '0';
  noval <= '0' when rst else
           dirdy or oval; -- output valid cycle after load

  -- instantiate blocks
  CNTR: UDL_Count2 generic map(4) port map(clk=>clk, rst =>rst, up => '0', down => not
          dirdy, load => dirdy, input => length, output => count);
  SHIFT: LRL_Shift_Register generic map(9) port map(clk =>clk, rst => rst, left => not
          dirdy, right => '0', load => dirdy, sin => '0', input => bits, output => obits);
  nchar <= input when irdy else char;
  IN_REG: vDFF generic map(5) port map(clk, nchar, char);
  IRDY_REG: sDFF port map(clk, irdy, dirdy);
  OV_REG: sDFF port map(clk, noval, oval);
  TAB: HuffmanEncTable port map(char, length, bits);
end impl;
```

Figure 19.15. VHDL description of the Huffman encoder.

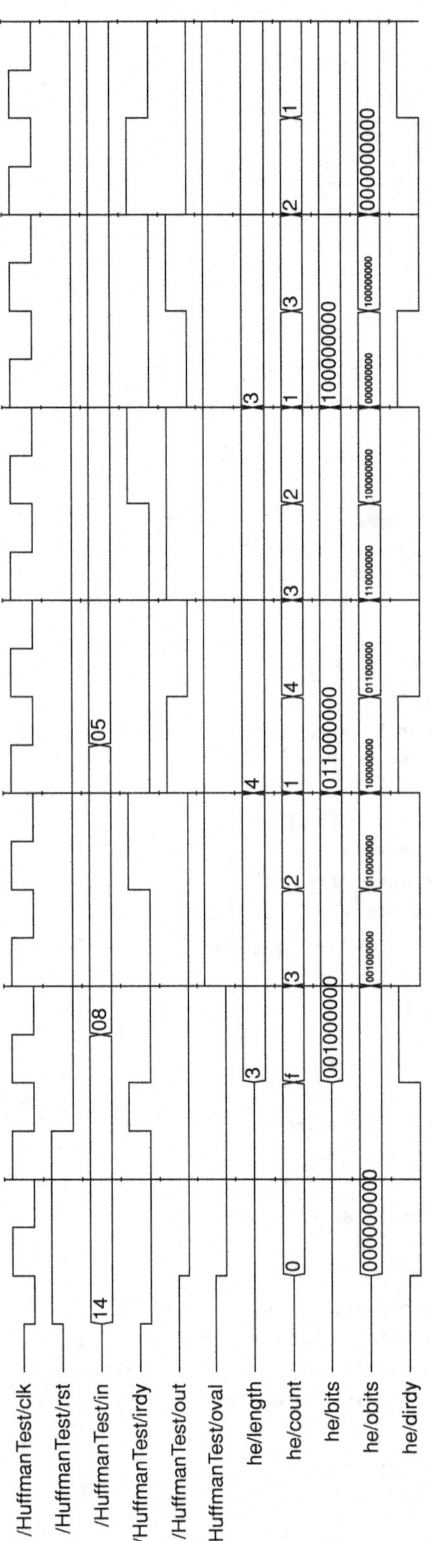

Figure 19.16. Waveforms from simulating the Huffman encoder on the string "THE."

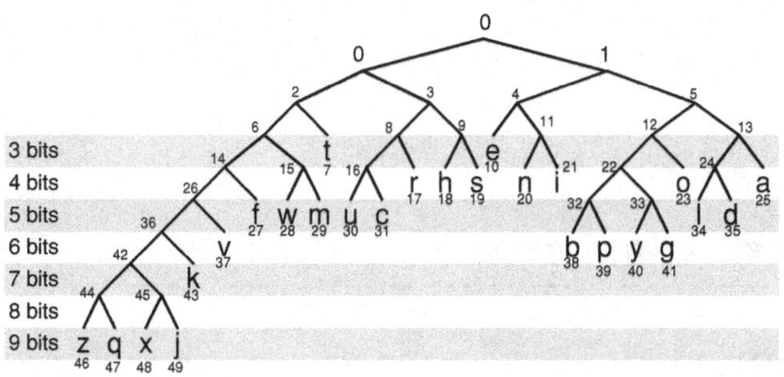

Figure 19.17. The Huffman code tree of Figure 19.13 relabeled to facilitate its storage in a decoding table. Each node of the tree is assigned a unique integer that serves as its address in the table.

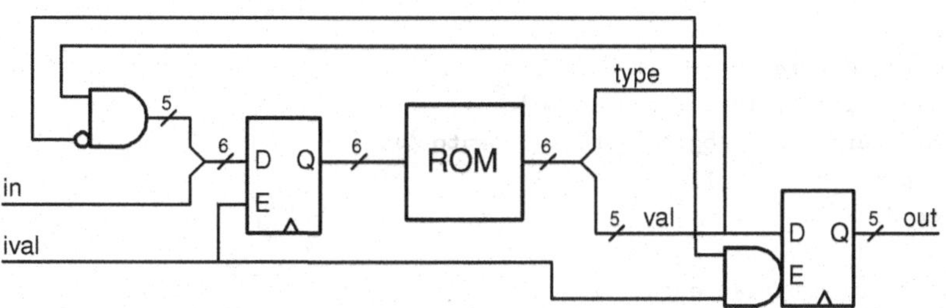

Figure 19.18. Block diagram of a Huffman decoder. The *node* register holds the address of the current tree node. The tree itself is stored in the ROM.

To see how we traverse the table representing the tree to decode a bit string, consider decoding the bit string 001. We start at the root of the tree which has a left child with address 0, and the first 0 of the string directs us to this child. We read the entry for address 0 and find that it is an internal node with a value of 2. The second bit of the string is a 0, so we proceed to address 2 (if this bit were a 1, we would have gone to address 3). We read the entry for address 2 and find that it is an internal node with a value of 6. The third bit of the string is a 1, so we proceed to one more than this value, address 7. Reading the entry for address 7, we find that it is a terminal node with a value of "T" (hex 14). We emit this value and reset our machine to start again from the root.

A block diagram of the Huffman decoder is shown in Figure 19.18, and VHDL code for the decoder is shown in Figure 19.19. The address of the current table node is held in the *node* register. When *type* is asserted – indicating a terminal node – *node* is set to the value of the next input bit (which selects one of the two children of the root to restart the search), the value field from the table is enabled into the output register, and *oval* is asserted on the following cycle. This outputs the current symbol and starts the machine at one of the children of the root depending on the first bit of the next symbol. If *type* is not asserted – indicating an internal node – the input value is combined with the value field from the table to select the left or right child of the current node – traversing the tree. The input value provides the least significant bit of the node address, and the remaining bits come from the value field of the table. This simple concatenation is possible because all left children in the table have even addresses. If *ival* goes low, the machine stalls, holding its present state until a valid input bit is available. Signal *ftype* in the VHDL model forces the machine to start from the root on the first valid input following reset.

Waveforms showing the combined operation of the Huffman encoder feeding the Huffman decoder are shown in Figure 19.20. The first 11 lines are the same as Figure 19.16 and represent the operation of the encoder encoding the symbol string "THE" into the bit string 0010110100. Signals *mid* and *mval* are output from the encoder and input (as *in* and *ival*) to the decoder.

```
------------------------------------------------------------------------
-- Huffman Decoder - decodes bit-stream generated by encoder
-- Figure 19.19
--   in - bit stream
--   ival - true when new valid bit present
--   out - output character
--   oval - true when new valid output present
------------------------------------------------------------------------
library ieee;
use ieee.std_logic_1164.all;
use work.ff.all;

entity HuffmanDecoder is
  port( clk, rst, input, ival: in std_logic;
        output: buffer std_logic_vector(4 downto 0);
        oval: out std_logic );
end HuffmanDecoder;

architecture impl of HuffmanDecoder is
  component HuffmanDecTable is
    port( input: in std_logic_vector(5 downto 0);
          output: out std_logic_vector(5 downto 0) );
  end component;
  signal node, nnode, hdeco: std_logic_vector(5 downto 0);
  signal value, tmp, nout: std_logic_vector(4 downto 0);
  signal typ: std_logic;  -- type from table
  signal ftyp: std_logic; -- fake a type on first ival cycle to prime pump
begin
  tmp <= 5d"0" when typ or ftyp else value;
  nnode <= 6d"0" when rst else
           (tmp & input) when ival else
           node;

  nout <= 5d"0" when rst else
          value when ival and typ else
          output;

  NODE_REG: vDFF generic map(6) port map(clk, nnode, node);
  TAB: HuffmanDecTable port map(node,hdeco);
  typ <= hdeco(5);
  value <= hdeco(4 downto 0);
  OUT_REG: vDFF generic map(5) port map(clk, nout , output);
  OVAL_REG: sDFF port map(clk, not rst and typ and ival, oval);
  FT_REG: sDFF port map(clk, rst or (ftyp and not ival), ftyp);
end impl;
```

Figure 19.19. VHDL description of the Huffman decoder.

The state of the decoder is shown in the *node* variable, and the *type* and *value* variables show what is read from the table at each node address. Note that each time *type* is asserted – indicating a leaf node – the search restarts on the next cycle with *node* at 0 or 1 (depending on *mid*). Also on the cycle following *type* the just decoded symbol is output on *out* (values shown are hexadecimal) and *oval* is asserted to indicate a valid output.

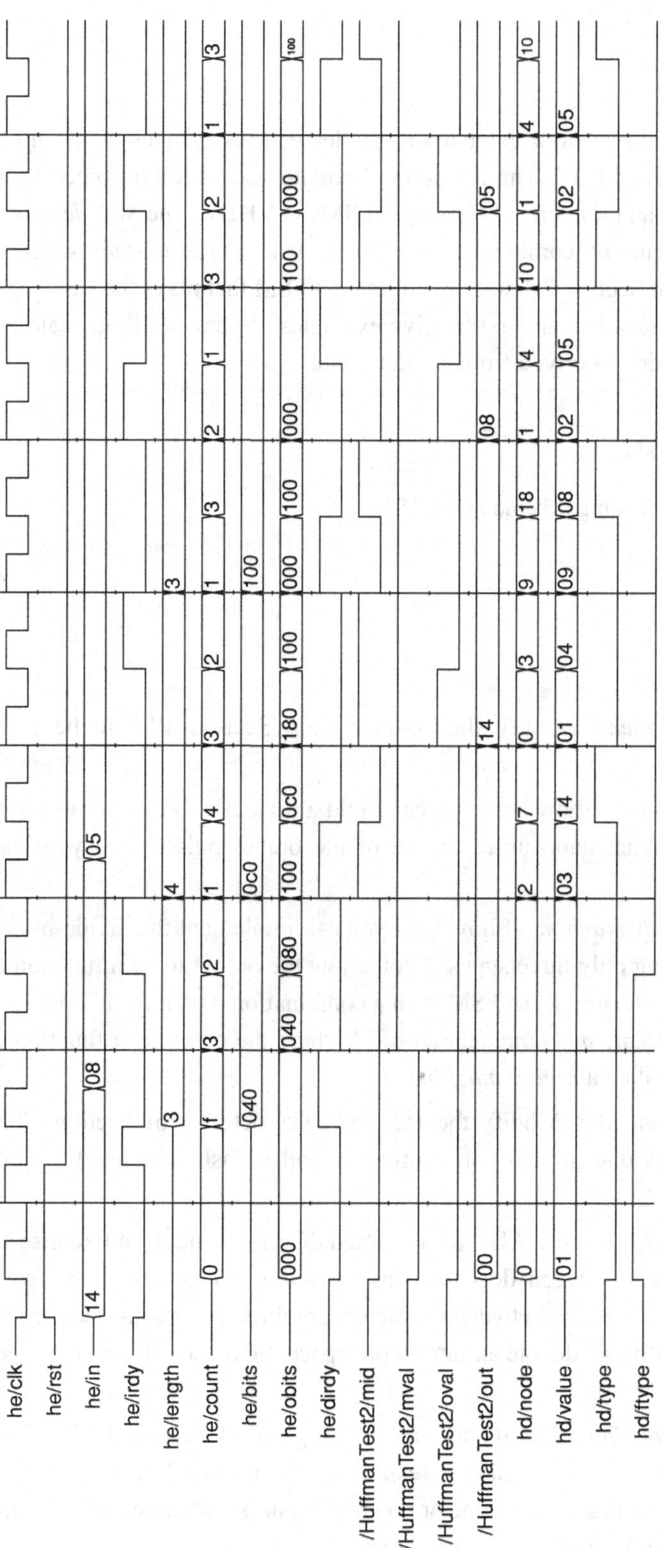

Figure 19.20. Waveforms of Huffman encoder and decoder, encoding the string "THE" into 0010110100 and then decoding this bit string back to "THE."

Summary

In this chapter you have seen four extended examples that bring together much of what you have learned in Chapters 14–18. The *divide-by-3* counter example reinforced basic skills in drawing state diagrams and implementing simple FSMs in VHDL. The *SOS detector* is another example of FSM factoring by combining common sequences. The *tic-tac-toe game* takes the combinational circuit of Section 9.4 and turns it into an FSM that plays the game against itself. Finally, the *Huffman encoder* and *decoder* give examples of datapath finite-state machines in which the control is entirely derived from the data state.

BIBLIOGRAPHIC NOTE

Huffman describes his encoding scheme in ref. [51].

Exercises

19.1 *Divide-by-4 counter.* Modify the counter from Section 19.1 to be a divide-by-4 counter.

19.2 *Divide-by-9 counter.* Show how you can build a divide-by-9 counter using two divide-by-3 counters. What happens to timing of the output pulse when you combine two counters?

19.3 *Divide-by-3 Mealy machine.* Show how you can implement the divide-by-3 counter of Section 19.1 using only three states if you allow the output to be a function of both the present state and the input. An FSM with a combinational path from input to output like this is called a *Mealy machine*, and an FSM where the output is a function only of the present state is called a *Moore machine.*

19.4 *Modified SOS machine.* Modify the factored SOS detector of Section 19.2 so that a dot is defined as one or two consecutive 1s and a dash is defined as three or four consecutive 1s.

19.5 *Further SOS modifications.* Alter the modified SOS machine from Exercise 19.4 further so that the pauses between dots and dashes within a character may be one or two consecutive 0s, the spaces between characters are three or four consecutive 0s, and five or more consecutive 0s denote an inter-word space. SOS should appear as a single word to be recognized.

19.6 *Tic-tac-toe completion.* Modify the tic-tac-toe game of Section 19.3 to include three new output signals: `gover` (game over), `xwin` (X has won), and `owin` (O has won). When either player has won a game or no empty spaces exit, assert `gover` and stop play until reset. If X (O) is the winner, also assert `xwin` (`owin`).

19.7 *Tic-tac-toe competition.* Further modify the tic-tac-toe game of Exercise 19.6 to include two different move generators capable of playing against each other instead of sharing a single generator. Improve your move generator and demonstrate it winning against the baseline of Figure 9.12.

19.8 *Tic-tac-toe tournament.* Build a module that instantiates eight different tic-tac-toe modules and plays them against each other in a single-elimination tournament. Each match should be best of two, alternating the module with the first play. If modules are tied at the end of a match, your controller can pick an arbitrary winner. The outputs of your system should be a three-bit signal `champion` indicating the tournament winner and a valid signal `cvalid` indicating tournament completion.

19.9 *Huffman encoder with flow control.* Modify the Huffman encoder of Section 19.4.1 to accept an input valid signal *ival* that is true when a valid symbol is available on the input. A new symbol will be accepted only when both *ival* and *irdy* are asserted. Note that the output valid signal *oval* may need to go low after a string has been shifted out if there is a wait for the next input signal.

19.10 *More flow control.* Take the Huffman encoder from Exercise 19.9 and extend it further to accept an output ready *ordy* signal that is true when the module connected to the output is ready to accept the next bit.

19.11 *One-bit strings.* Modify the Huffman encoder of Section 19.4.1 so it will work with a length of 1, that is, for codes where a symbol may be represented by a one-bit string.

19.12 *Pattern counter, I.* Design a sequential logic circuit with a one-bit input `input` and outputs `output` and `count(3 downto 0)`. Your FSM should generate `output` = 1 when `input` = 0101 and be 0 otherwise. Overlapping patterns need to be recognized, and the number of 0101 patterns must be shown on output `count`. An example of correct behavior is

```
in:      0010101011011101010
out:     0000010101000000001
count:   0000011223000000001
```

Draw a block diagram for your control and datapath FSM.

19.13 *Pattern counter, II.* Write and verify the VHDL to implement the pattern counter of Exercise 19.12.

Part V

Practical design

20 Verification and test

Verification and test are engineering processes that complement design. Verification is the task of ensuring that a design meets its specification. On a typical digital systems project, more effort is expended on verification than on the design itself. Because of the high cost and long delays involved in fabricating a chip, thorough verification is essential to ensure that the chip works the first time. A design error that is not caught during verification would result in costly delays and retooling.

Testing is performed to ensure that a particular instantiation of a design functions properly. When a chip is fabricated, some transistors, wires, or *contacts* may be faulty. A manufacturing test is performed to detect these faults so the device can be repaired or discarded.

20.1 DESIGN VERIFICATION

Simulation is the primary tool used to verify that a design meets its specification. The design is simulated using a number of *tests* that provide stimulus to the unit being tested and check that the design produces correct outputs. The VHDL *testbenches* we have seen throughout this book are examples of such tests.

20.1.1 Verification coverage

The verification challenge amounts to ensuring that the set of test patterns, the *test suite*, written to verify a design is complete. We measure the degree of completion of a test suite by its *coverage* of the specification and of the implementation. We typically insist on 100% coverage of both specification *features* and implementation *lines* or *edges* to consider the design verified.

The *specification coverage* of a set of tests is measured by determining the fraction of *features* in the specification that are exercised and checked by the tests. For example, suppose you have developed a digital clock chip that includes a day/date and an alarm function. Table 20.1 gives a partial list of features to be tested. Even for something as simple as a digital clock, the list of features can easily run into the hundreds. For a complex chip it is not unusual to have 10^5 or more features. Each test verifies one or more features. As tests are written, the features covered by each test are checked off.

The hierarchy of the feature set makes it easier to manage the development of tests. For example, the tests for alarm features can be developed largely independently of the tests for time keeping and the tests for display. This enables several groups to work on these tests simultaneously.

Table 20.1. **Partial list of features to be tested for a hypothetical digital clock chip**

Designation	Name	Description
I	increment	Time increments properly
I.s	inc. seconds	Seconds register increments once per second
I.sw	seconds wrap	Seconds register rolls over from 59 to 0
I.m	inc. minutes	Minutes register increments when seconds roll over from 59 to 0
...	...	(similar definitions apply for I.mw, I.h, I.hw, I.days, I.daysw, I.months, I.monthsw, I.years)
I.leap	leap year	February wraps from 29 to 0 during a leap year
A	alarm	Alarm features
A.set	alarm set	Alarm can be set as specified
A.set.s	alarm seconds set	Seconds can be set as specified
...		(similar features are used to set minutes and hours)
A.act	alarm activate	Alarm sounds when alarm time reached
A.quiet	alarm deactivate	Alarm can be turned off as specified
A.snooze	alarm snooze	Alarm can be delayed for specified snooze interval
D	display features	Clock state is properly displayed for the current mode
D.time	time display	LCD display driven to display properly hours, minutes, and seconds
...	...	(other display functions are used for day, date, and alarm)

In addition to checking for specification coverage, we also check our test suite for *implementation coverage*. Our test suite should exercise every *line* of our VHDL code. Every case of a `case` statement, for example, should be activated. For every state machine in our design, every *edge* between states should be traversed.

If after achieving 100% specification coverage we discover that some lines of our VHDL code have not been activated by our test suite, we need to take a close look at those lines and determine whether (a) they describe a feature that was initially left off our feature list, (b) they are not needed, or (c) they are assertions for error conditions that are not expected to occur during test conditions.

20.1.2 Types of tests

Ideally we would like to verify a feature using an *exhaustive* test, one that applies every possible input stimulus and checks for the correct results. Unfortunately, for all but the simplest modules, it is usually not feasible to try all possible input combinations and all possible states. Consider a 64-bit binary adder. There are $2^{128} = 3.4 \times 10^{38}$ possible input patterns. Even if you could test a million patterns per second, it would still take over 10^{25} years to test all possible combinations.

In place of exhaustive tests, we typically perform a combination of *directed tests* and *random tests*. A directed test is written to cover some interesting test cases, such as an edge condition or an extreme value. With our clock chip, for example, we would make sure to test that the clock rolls over properly from `23:59:59` to `00:00:00`. For our adder we would check an add that produces the largest positive (and negative) number and an add one more than this to check for overflow.

We write random tests to complement our directed tests. As the name implies, these tests are randomly generated. For our adder, such a test generates two random input operands and checks that the adder produces the correct result. For our clock, such a test generates a random time and alarm setting and verifies correct operation. For a processor, a random test generates a random sequence of instructions along with interrupts, faults, and other conditions, and verifies that the registers are left in the proper states.

Random tests that uniformly sample the space of all inputs will find bugs that happen with some frequency – e.g., 1 in 10^8 patterns. However, they are not likely to find a bug that occurs for only one of 10^{38} possible test patterns. To stack the deck in our favor, we often write random tests to sample the input space *non-uniformly* around areas of interest. For our clock, for example, we might sample alarm times within a few seconds of the current time – because these patterns are more likely to have interesting alarm behavior within the duration of our test. For a processor we might create many tests that include combinations of exceptions with mis-predicted branches because such conditions put stress onto a great deal of the processor's logic.

A typical test suite may involve 10^9 or more test patterns. We clearly cannot look through all of the results manually to see whether they are correct each time we rerun the test. Instead, we need to make our tests self-checking. A common approach is to compare our design with a higher-level model of the same function. These high-level models are often written in a programming language like "C" and may not match the exact cycle-by-cycle timing of the design being verified. Of course, the high-level model may have bugs. However, it is unlikely that they are the same bugs as in the VHDL design.

20.1.3 Static timing analysis

In addition to verifying that our design is functionally correct, we must also verify that it satisfies setup- and hold-time margins (Chapter 15). As discussed in Section 15.6, this is performed by a static timing analyzer – either one built into the synthesis tool or a separate program, or both.

While, in theory, timing verification could be performed via simulation, in practice it is very hard to construct a set of tests that is guaranteed to test the timing of the worst-case path. With static timing analysis, all paths are checked without the need to generate test vectors. The downside of static timing analysis is that it often reports issues on paths that are guaranteed never to be used.

20.1.4 Formal verification

For some modules, a proof technique can be used to verify (prove) functional correctness without the need for simulation. For example, the tool *Formality* uses equivalence-checking techniques to prove that two versions of a design are equivalent. Proof techniques are also often used to verify protocols, such as cache coherence protocols in a microprocessor. They are able to verify that all state transitions preserve certain properties without the need to write tests that cover all of the transitions.

20.1.5 Bug tracking

During the design verification process, a *bug-tracking* system is often used to keep track of all discrepancies (bugs) that are identified by tests. *Bugzilla* is an example of an open-source bug-tracking system.

When a test finds an error, a bug report is *opened*. When the designer corrects the bug and verifies the solution, or if the feature the bug was testing is abandoned, the bug is *closed*. One can get a good sense of the state of a design project by tracking the number of open bugs at a given point in time. Initially there are no bugs. As tests are written, the number of bugs increases, reaching a maximum when full test coverage is reached. The number of bugs then monotonically decreases as bugs are closed, without introducing too many new bugs, one hopes. Eventually the number of bugs reaches zero, and the design can be released for manufacturing.

By extrapolating the number of bugs as a function of time, one can estimate when the number of open bugs will reach zero. The distribution of closure times also gives some insight into the debugging process. The majority of bugs are closed quickly, within one day. However, a small fraction of difficult bugs remain open for a week or more.

20.2 TEST

The test process verifies that a particular instance of a design in fact implements the design. As with verification, we desire to come up with a set of tests that achieves 100% coverage. However, rather than covering features, a manufacturing test must cover potential *faults* in the design. Anything that could be manufactured incorrectly must be checked for. We use a *fault model* to reason about the set of potential faults.

Because test time is costly, and every chip manufactured must be tested, there is an emphasis on completing a test as quickly as possible. Thus, we desire to find the shortest test that achieves 100% coverage. This is in contrast to verification where, to a first approximation, verification time is not critical.

20.2.1 Fault models

A *fault model* is an abstract model of physical faults that could cause a chip not to work. On a modern integrated circuit, connections (wires and contacts) could be open or shorted. Similarly, a transistor can fail in a way that causes it to be always on or always off.

The fault model that is almost universally used to abstract all of these potential failure modes is the *stuck-at fault model*. This model abstracts all potential faults as resulting in a logical node of the circuit being either stuck at logic "0" or stuck at logic "1."

Consider the two-input NAND gate of Figure 20.1. Real failures in a chip implementing this gate could cause one or more of the four transistors M1–M4 to be shorted or opened. We model

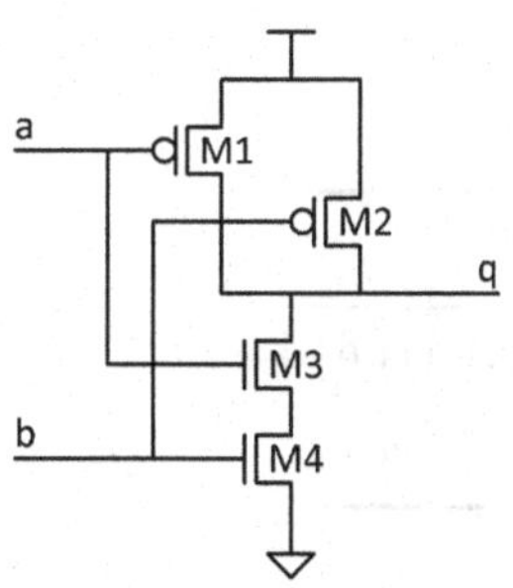

Figure 20.1. A two-input NAND gate. With the stuck-at fault model, all possible physical faults in this gate are modeled as the output q being stuck at 0 or stuck at 1.

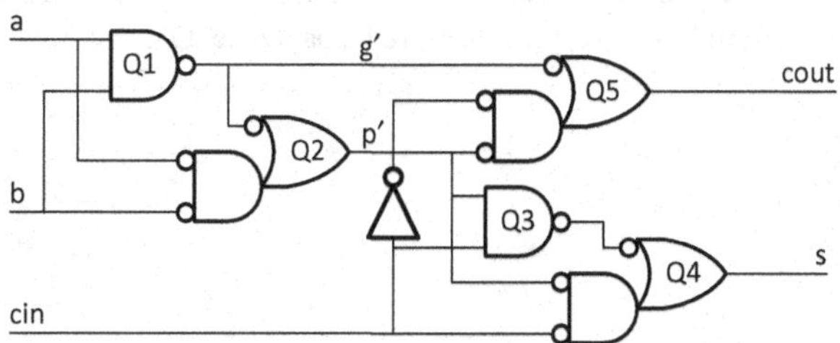

Figure 20.2. CMOS gate-level implementation of a full adder.

all of these potential failures by the output of the gate being stuck at either 0 or 1. This model is not completely accurate. For example, if PFET M1 is open, the gate does not function properly, but the output is not stuck in one state. The output will still go high when $b = 0$ and will go low when $a = b = 1$. Thus, a test that detects a stuck-at-0 fault at the output may not detect an M1-open fault. Despite this shortcoming, good coverage of a stuck-at fault model results in good coverage of actual manufacturing faults.

20.2.2 Combinational testing

To test a block of combinational logic, we need a set of *test patterns*, sometimes called *test vectors*,[1] that detect whether any node in the logic block is stuck at 0 or 1. Consider the full adder of Figure 20.2 (repeated from Figure 10.5). With the stuck-at fault model there are ten possible faults. Each node, namely g', p', s, $cout$, or the output of Q3, could be stuck at 0 or stuck at 1. To detect a particular fault, we need to drive the node in question to the opposite state while the output is sensitive to that node. As shown in Table 20.2, two test vectors – all 1s and all 0s – suffice to cover all ten stuck-at faults (labeled *signal*-0 (−1) for stuck at 0 (1)) in this circuit.

Automatic test-pattern generation (ATPG) tools automate the process of generating test vectors for combinational logic. Given a netlist, an ATPG tool will generate a minimal set of test vectors that provides 100% fault coverage – if such coverage is possible.

20.2.3 Testing redundant logic

Some logic includes redundant gates. This is common, for example, in asynchronous logic and other situations where hazards must be covered. It is not possible to achieve 100% fault coverage on such logic unless additional signals are added to disable the redundancy.

[1] So called because they are bit *vectors* used for testing.

Table 20.2. **Two test vectors are sufficient for 100% stuck-at fault coverage of the full adder of Figure 20.2**

a	b	cin	g'	p'	Q3	$cout$	s	Faults covered
0	0	0	1	1	1	0	0	g'-0, p'-0, $cout$-1, $sout$-1, Q3-0
1	1	1	0	1	0	1	1	g'-1, p'-1, $cout$-0, $sout$-0, Q3-1

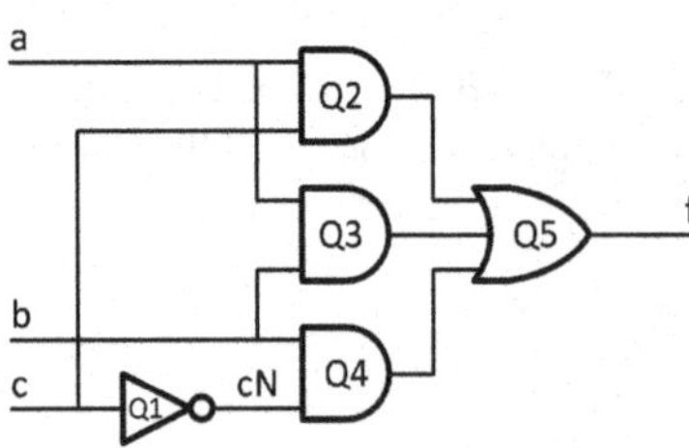

Figure 20.3. Hazard-free two-input multiplexer circuit (repeated from Figure 6.19(b)) has a redundant gate, Q3, that cannot be tested.

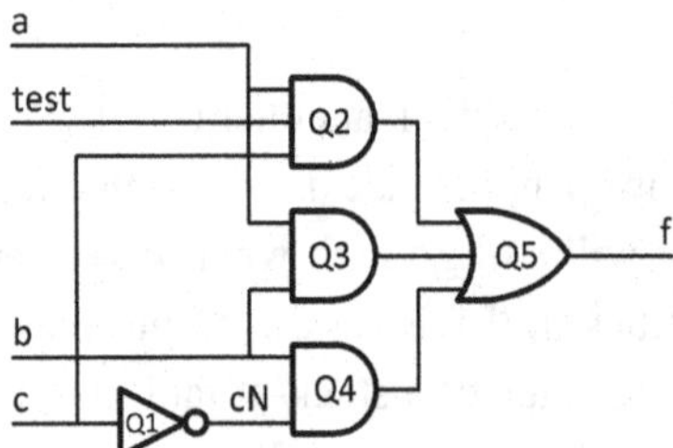

Figure 20.4. Adding a test input to gate Q2 makes the circuit of Figure 20.3 testable.

Consider, for example, the two-input multiplexer with no hazards of Figure 20.3 (repeated from Figure 6.19(b)). It is not possible to test for a stuck-at-0 fault at the output of gate Q3. Any time $a = b = 1$, the output of either Q2 or Q4 will be high, causing the output f to be high regardless of the state of Q3.

To test this circuit we have to introduce an auxiliary input to one of the other two gates. Figure 20.4 depicts adding an input *test* to the upper AND gate. During normal operation $test = 1$ and the multiplexer operates as before. We allow *test* to take on either state during testing. With this input, the vector $a = 1, b = 1, c = 1, test = 0$ detects a stuck-at-0 fault on Q3.

20.2.4 Scan

The problem of testing sequential logic can be reduced to the testing of combinational logic through the use of *scan chains*. In a design that uses scan, a multiplexer is included on the input of every flip-flop, as shown in Figure 20.5. When the select input of the multiplexer is in the *scan* state, the flip-flops are connected together into a shift register.

To apply one test vector to every combinational block on a chip, the chip is placed in scan mode and a test pattern is shifted into the resulting shift register. One clock cycle is then run with the *scan* input low to sample the output of all of the logic blocks into the flip-flops. Finally, *scan* is asserted again and the result of the test is shifted out to be checked. A new vector can be shifted in at the same time as the result is being shifted out.

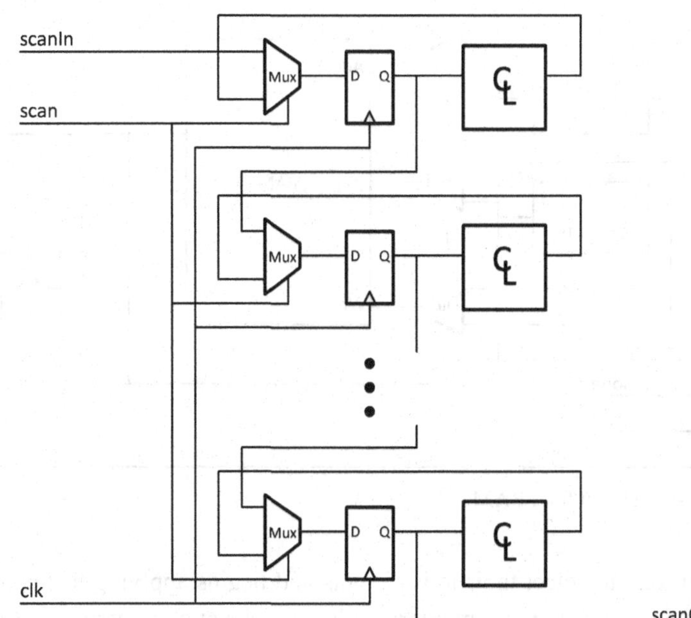

Figure 20.5. Adding a multiplexer to each flip-flop enables all the flip-flops on a chip to be connected in a *scan-chain* shift register when signal *scan* is asserted. Test vectors can be shifted in, and test results shifted out, via the scan chain.

For scan testing of sequential logic to be complete, all inputs and outputs of combinational blocks must be accessible via the scan chain. This implies that macros like I/Os and SRAMs must include scannable registers on their inputs and outputs.

The scan technique can also be applied to test the printed-circuit boards that chips are assembled on. By connecting the chip inputs and outputs together in a scan chain, one can shift in a test pattern to test the connections between chips on a board. This technique is often called *boundary scan* because the inputs and outputs are the *boundary* of a chip.

A common interface used in many integrated circuits for scan testing is the IEEE's JTAG interface [57].

20.2.5 Built-in self-test (BIST)

On-chip memories, RAMs and ROMs (Sections 8.8 and 8.9), require many test patterns for thorough testing. For a RAM, every bit must be written with both a "1" and a "0" and then read back. To check for addressing faults, locations must be written with contrasting patterns, so that a read from location *a* cannot be mistaken for a read from location *b*. While such testing could be done through the scan chains, it would take a prohibitive amount of time, driving up test cost.

To shorten the testing time for on-chip memories, most modern chips employ *built-in self-test* (BIST). A BIST circuit is a state machine associated with a memory or group of memories that generates and checks test patterns. While a scan chain requires thousands to millions of cycles to apply a single test pattern, a BIST circuit can apply a test pattern every clock cycle – accelerating the testing of RAMs by a factor equal to the length of the scan chain.

A typical BIST circuit is a simple datapath finite-state machine as illustrated in Figure 20.6. During normal operation the multiplexers select their upper input and the RAM operates normally. When the *start* input to the BIST controller is asserted, the BIST controller

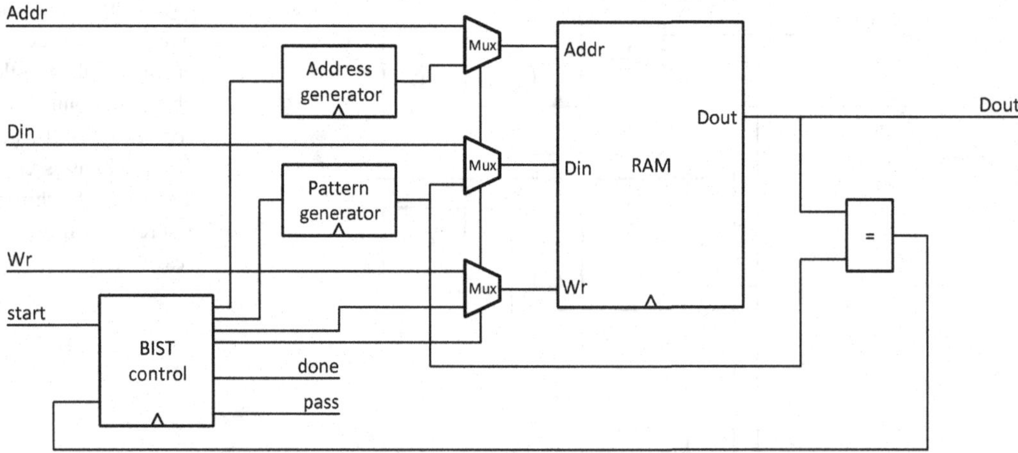

Figure 20.6. Built-in-self-test circuit for a RAM.

commands the multiplexers to select their lower inputs and begins applying test patterns. An address generator sequences through the memory addresses. A pattern generator selects from a number of predefined patterns, e.g., 01010101 and 10101010, to write to the memory, and a comparator checks data values read from the memory. When the test is complete, the *done* line is asserted, whether the test was passed or not is indicated on the *pass* output, and the multiplexers switch back to their upper inputs. The details of the operation of this FSM are examined in Exercise 20.13.

Some BIST circuits go one step further and repair errors they find during testing. A RAM can be made repairable by providing spare rows and/or spare columns. The BIST circuit writes a register with the row or column address that is to be replaced by the spare. The details of memory repair are explored in Exercise 20.14.

Another class of BIST exists for pseudo-random testing of logic. On-chip LBIST state machines generate a series of test patterns that are scanned into registers, clocked, and then scanned out into a multiple-input signature register (MISR) also located on-chip. This MISR reduces all of the state on a chip into a handful of 64-bit words using a series of shifts and XORs. This process is repeated many times, and finally the MISRs are scanned off-chip and compared with their expected values. Because the FSMs are located entirely on-chip, LBIST can run many more test cases than would be run with a traditional scan.

20.2.6 Characterization

Characterization tests are performed on samples of a design to determine typical and extreme parameters of the design, to determine the *operating envelope* of the design, and to measure the aging properties of the device. Unlike manufacturing tests, which are performed on every chip produced, characterization tests are performed on only a small sample of chips.

The parameters measured by a characterization test include the electrical parameters of the chip's inputs and outputs, and the power consumed by the chip in different operating modes. Input and output circuits may be characterized by measuring a few operating points, for example V_{OH} and V_{OL} under a particular load, or a complete V–I curve may be measured.

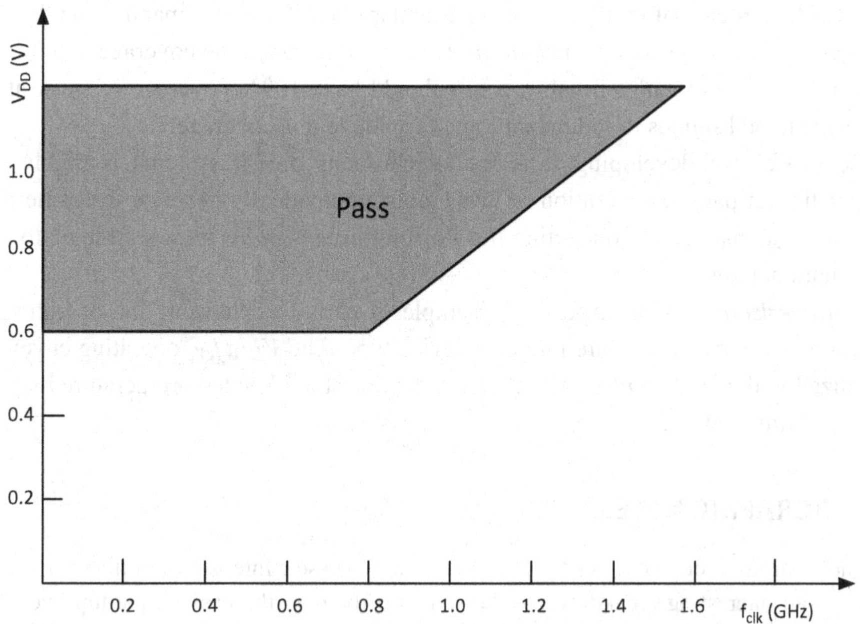

Figure 20.7. An example *shmoo* plot shows the range of V_{DD} and f_{clk} over which a part operates.

The *operating envelope* of a chip is the range of supply voltage V_{DD} and clock frequency f_{clk} over which a chip operates properly. The envelope is measured by running a functional test of the chip at all combinations of V_{DD} and f_{clk} and plotting which combinations resulted in a passing test. Such a plot, an example of which is shown in Figure 20.7, is often called a *shmoo* plot because at one point such a plot resembled a cartoon character called a shmoo.

The example shmoo plot shows that the part in question cannot operate below a minimum voltage of $V_{min} = 0.6$ V. At V_{min} it will operate up to 800 MHz. Increasing voltage enables higher-frequency operation, reaching a maximum of 1.6 GHz at 1.2 V.

A characterization test may include an *accelerated-life test*, sometimes called a *burn-in* test to measure the failure rate of the device. Such a test is typically performed at elevated temperature (100 °C or more) to accelerate aging of the part according to the Arrhenius equation. A sufficient sample of parts is operated at a sufficiently high temperature for a sufficiently long period of time to give a statistically significant measure of device failure rate, or at least to guarantee with a specified confidence factor that the rate is below a required level.

Summary

In this chapter you have learned the basics of verification and testing. Design verification is the process of verifying that a design meets its specification. We write verification tests to give complete *specification coverage* – to make sure all features of the design are correctly implemented – and *implementation coverage* – to ensure that every line of VHDL code has been exercised. To manage the verification process, we track the number of *open* bugs. An extrapolation of this number gives a good estimate of when the verification process will be complete.

Test is the process of verifying that a particular device has been manufactured correctly. We use a *fault model*, typically a *stuck-at fault model*, to measure the coverage of a manufacturing test. To detect faulty chips reliably, a test should have 100% coverage of stuck-at faults. We may have to add signals to redundant logic to achieve this coverage.

The process of developing tests for synchronous digital systems is highly automated. Automatic test-pattern generation (ATPG) tools automatically generate test patterns (vectors) for combinational logic. Connecting flip-flops into *scan chains* allows all logic to be tested as combinational logic.

Characterization involves testing a sample of parts to determine the operating envelope, critical parameters, and failure rate of a device type. The V_{DD},f_{clk} operating envelope may be visualized with a *shmoo plot*. An accelerated-life test at elevated temperature may be used to estimate failure rate.

BIBLIOGRAPHIC NOTES

For more information about verification on a real processor, Intel's Pentium 4®, read ref. [10]. It provides an interesting breakdown of the source of bugs in the project, the top two of which are "goofs" and "miscommunication."

For more information on pattern generation for test, refer to refs. [61] and [115]. Many generation algorithms rely on the idea of fault-equivalence [73].

One of the first scan latches was IBM's LSSD latch [40]; ref. [9] gives an overview of how it is used in systems.

Two papers that give an overview of memory BIST and repair are refs. [12] and [65]. BIST can also be used to test logic, as explained by McCluskey [72]. Riley *et al.* [96] detail the test strategy used in IBM's Cell® processor.

Shmoo was invented in the 1970s by Huston, and further reading about shmoo in a modern setting can be found in ref. [7]. If the reader prefers comic books to academic papers, one such book featuring the shmoo is ref. [24].

More information about the maths behind accelerated-life testing can be found in [87].

Exercises

20.1 *Feature list, I.* Write a feature list for a simple four-function calculator chip. The chip connects to a key pad and drives individual segments of a four-digit seven-segment display.

20.2 *Feature list, II.* Write a feature list for a digital watch chip.

20.3 *Feature list, III.* Write a feature list for the traffic-light controller of Figure 17.10.

20.4 *Feature List, IV.* Write a feature list for the vending machine of Figure 16.21.

20.5 *Directed tests.* Write a VHDL testbench that applies six directed test patterns to a 32-bit 2's complement adder. Explain what each of your patterns is checking for.

20.6 *Random tests.* Write a VHDL testbench that applies 100 random patterns to a 32-bit adder.

20.7 *Implementation coverage.* Write a testbench for the traffic-light controller of Figure 17.11 that exercises every edge of the state diagram.

20.8 *Combinational testing, I.* Write a minimal set of test vectors that gives 100% coverage of the decoder of Figure 8.3.

20.9 *Combinational testing, II.* Write a minimal set of test vectors that gives 100% coverage of the multiplexer of Figure 8.10(a).

20.10 *Fault models, I.* Consider a fault model where we consider the set of gate-input stuck-at faults. For example, in the full adder of Figure 10.5 the bottom input of Q5 may be stuck-at-1 independently of the other two inputs driven by signal p'. With the 13 gate inputs in the figure, there are 26 possible faults. Write a set of test vectors that covers the six faults associated with Q5.

20.11 *Fault models, II.* Repeat Exercise 20.10 but for the four faults associated with Q3.

20.12 *Fault models, III.* Repeat Exercise 20.10 but for the six faults associated with Q4.

20.13 *Built-in self-test.* Write a VHDL model for a BIST unit like that shown in Figure 20.6 that tests an 8 KB RAM. Your BIST unit should perform the following test procedure.

(a) Write every location in the RAM with the binary value 01010101.

(b) For each location i write $M[i] = 10101010$. Verify that $M[i] = 10101010$ and that $M[j] = 01010101\ \forall i \neq j$.

(c) Write $M[i] = 01010101$.

(d) Repeat steps (a)–(c) with the data values complemented.

(e) Assert a *done* signal and, if successful, a *pass* signal.

20.14 *Memory repair.* Write a VHDL model for a memory that uses an eight-bit register to act as a spare byte for an 8 KB RAM array. Use a 13-bit register to hold the address of the byte being replaced and an additional bit to indicate that replacement should occur. When the replacement bit is set, reads and writes to the address in the address register should go to the eight-bit data register rather than to the RAM array.

Part VI

System design

21 System-level design

After reading to this point in the book, you now have the skills to design complex combinational and sequential logic modules. However, if someone were to ask you to design a DVD player, a computer system, or an Internet router you would realize that each of these is not a single finite-state machine (or even a single datapath with associated finite-state controller). Rather, a typical system is a collection of modules, each of which may include several datapaths and finite-state controllers. These systems must first be decomposed into simple modules before the design and analysis skills you have learned in the previous chapters can be applied. However, the problem remains that of how to partition the system to this level where the design becomes manageable. This *system-level design* is one of the most interesting and challenging aspects of digital systems.

21.1 SYSTEM DESIGN PROCESS

The design of a system involves the following steps.

Specification The most important step in designing any system is deciding – and clearly specifying in writing – what you are going to build. We discuss specifications in more detail in Section 21.2.

Partitioning Once the system has been specified, the main task in system design is dividing the system into manageable subsystems or modules. This is a process of divide and conquer. The overall system is divided into subsystems that can then be designed (conquered) separately. At each stage, the subsystems should be specified to the same level of detail as the overall system was during our first step. As described in Section 21.3, we can partition a system by state, task, or interface.

Interface specification It is particularly important that the interfaces between subsystems be described in detail. With good interface specifications, individual modules can be developed and verified independently. When possible, interfaces should be independent of module internals – allowing modules to be modified without affecting the interface, or the design of neighboring modules.

Timing design Early in the design of a system, it is important to describe the timing and sequencing of operations. In particular, as work flows between modules, the sequencing of which module does a particular task on a particular cycle must be worked out to ensure that the right data come together at the correct place and time. This timing design also drives the performance tuning step described below.

Module design Once the system has been partitioned, modules and interfaces have been specified, and the system timing has been worked out, the individual modules can be designed and verified independently. Often, the exact performance and timing (e.g., throughput, latency, or pipeline depth) of a module is not known exactly until after the module design has been completed. As these performance parameters are finalized, they may affect the system timing and require performance tuning to meet system performance specifications. The test of a good system design is if such independently designed modules can be assembled into a working system without reworking.

Performance tuning Once the performance parameters of each module are known (or at least have been estimated), the system can be analyzed to see whether it meets its performance specification. If a system falls short of a performance goal – or if the goal is to achieve the highest possible performance at a given cost – performance can be tuned by adding parallelism. This topic is treated in more detail in Chapter 23.

21.2 SPECIFICATION

All too often people start designing a system without a clear specification, only to discover halfway (or further) through the design that they are building the wrong system. Much work is then discarded as they restart the design process. Another problem with vague specifications is that two designers may read the specification differently and design incompatible system parts.

A system design may start from an oral discussion of requirements. However, writing the specification down is a critical step to make sure that there are no misunderstandings about what is being designed. A written specification can also be used to validate that the right system is being designed by reviewing the specification with prospective customers and users of the system.

At a minimum, a good specification must include the following descriptions.

(1) The overall system: what the system is, what it does, and how it is used.
(2) All inputs and outputs: their formats, range of values, timing, and protocols.
(3) All user-visible state. This includes configuration registers, mode bits, and internal memories.
(4) All *modes* of operation.
(5) All notable *features* of the system.
(6) All interesting *edge cases*, i.e., how the system handles marginal cases.

The remainder of this section gives specifications for three example systems: a “Pong” game, a DES cracker, and a music player.

21.2.1 Pong

Overall description Pong is a video game that was designed by Atari in the early 1970s. It displays a ping-pong-like game on a VGA video screen. Users control the game using push-buttons to move the paddles and serve. Games are played to 11 points. The player

Table 21.1. **Inputs and outputs of Pong system**

Name	Direction	Width	Description
leftUp	input	1	When true, moves the left paddle up
leftDown	input	1	When true, moves the left paddle down
leftStart	input	1	When true, starts the game or serves the ball from left to right
rightUp	input	1	When true, moves the right paddle up
rightDown	input	1	When true, moves the right paddle down
rightStart	input	1	When true, starts the game or serves the ball from right to left
red	output	8	The intensity of the red color on the screen at the current pixel
green	output	8	The intensity of the green color on the screen at the current pixel
blue	output	8	The intensity of the blue color on the screen at the current pixel
hsync	output	1	Horizontal synchronization – when asserted, this starts a horizontal retrace of the screen
vsync	output	1	Vertical synchronization – when asserted, this starts a vertical retrace of the screen

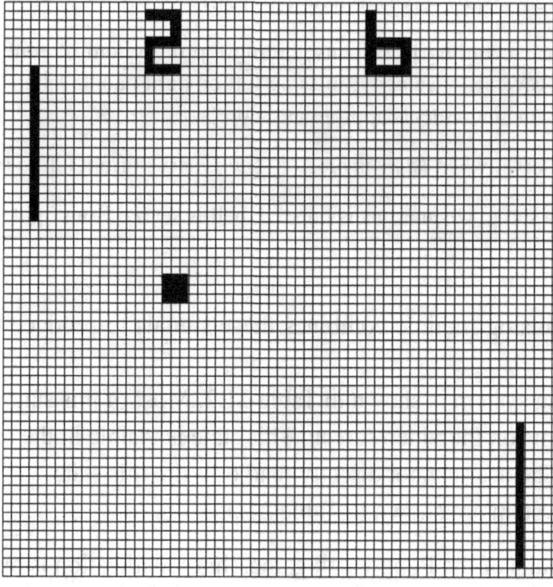

Figure 21.1. The Pong display consists of two paddles, a ball, and the score on a 64 × 64 grid.

winning the previous point serves. The screen is considered to be a 64 × 64 grid for purposes of the game – point (0, 0) is top left. An example screen-shot is shown in Figure 21.1.

Inputs and outputs The inputs and outputs of our Pong system are specified in Table 21.1. Note that our digital module produces as output digital red, green, blue, and sync outputs

Table 21.2. **User-visible state of the Pong system**

Name	Width	Description
rightPadY	6	*y*-position of the top of the right paddle
leftPadY	6	*y*-position of the top of the left paddle
ballPosX	6	*x*-position of the ball
ballPosY	6	*y*-position of the ball
ballVelX	1	*x*-velocity of the ball (0 = left, 1 = right)
ballVelY	2	*y-velocity* of the ball (00 = none, 01 = up, 10 = down)
rightScore	4	Score of right player
leftScore	4	Score of left player
mode	2	Current mode – *idle, rserve, lserve, play*

Table 21.3. **Modes of the Pong system**

Name	Description
idle	Score is zero, awaiting first serve: first start button pressed zeroes score and starts game with serve from that direction (e.g., *lstart* serves from left to right)
play	Ball in play: ball advances according to velocity; hitting top or bottom of court reverses *y*-velocity; hitting paddle reverses *x*-velocity; missing left or right paddle enters *rserve* or *lserve* mode, respectively, and increments appropriate score
lserve	Waiting for left player to serve: when *lstart* is pressed, ball is served from left to right
rserve	Waiting for right player to serve: when *rstart* is pressed, ball is served from right to left

for the display. A separate analog module combines these signals to produce the analog signals to drive the display.

State The visible state of the Pong system is shown in Table 21.2. Most of this state represents the positions of game elements on the video screen. This state is visible in the sense that it can be seen on the display. The user cannot directly read or write this state.

Modes The modes of the Pong system are shown in Table 21.3.

While we have given many details of the Pong video game, this specification is by no means complete. Things left unspecified include the value of the ball position and velocity on a serve, how the *y*-velocity of the ball changes when hitting a paddle, and the height of a paddle. A complete specification should leave nothing to the imagination. In Exercise 21.1 the reader is given the task of completing the specification of the Pong game. In practice, specifications are typically completed in an iterative manner, with additional details specified on each iteration. Specification reviews are often held, where a specification is presented to a group and reviewed critically to identify missing or incorrect items.

21.2.2 DES cracker

Overall description The DES cracker is a system that accepts a block of ciphertext encrypted using the data encryption standard and searches the space of possible keys to find the key that was used to encrypt the ciphertext. Since DES is a symmetric-key algorithm, the same key is used to encode and decode the data. Finding this key enables the user to read and send coded text. To check that the cracker has the correct key, the system checks the output to see whether it is plaintext. We assume that the original plaintext is ASCII text that uses only capital letters and numbers.

The DES standard works on eight-byte pieces of code, not a full encrypted message. Our cracker will therefore iterate through eight-byte chunks with a key, checking each for plaintext. If a key decodes all pieces into plaintext, success is declared.

Inputs and outputs The inputs and outputs of the DES cracker are shown in Table 21.4. The user places one block of ciphertext onto the *cipherText* at a time, asserting the *cipherTextValid* signal. Once all of the text has been loaded by the user, the *start* signal is asserted and the system begins decryption. For the remainder of this section, we assume that all the data have been loaded into storage RAM. We task the reader with specifying the input protocol in Exercise 21.2.

With the ciphertext loaded, the user pulses the *start* signal. This causes the cracker to run until completion. When a key successfully decrypts the ciphertext into plaintext, *found* is asserted and the key is placed on *key*.

State The different visible state in the DES cracker is shown in Table 21.5. The key is set at each decryption iteration and its value is held upon decryption success. The *blockNumber* state selects a *cipherTextBlock* for decryption; *mode* indicates whether the system is currently reading new data, doing a decryption, or is idle.

Modes The different operation modes of the DES cracker are shown in Table 21.6. We use three modes *idle*, *dataIn*, and *cracking*.

Table 21.4. **Inputs and outputs of the DES cracker system**

Name	Direction	Width	Description
cipherText	input	8	Ciphertext to be cracked: this text is input one byte at a time; a byte is accepted on each clock for which *cipherTextValid* and *cipherTextReady* are asserted
cipherTextValid	input	1	Asserted when *cipherText* has the next valid byte of ciphertext to load
cipherTextReady	output	1	Asserted when the system is able to accept a byte of ciphertext
start	input	1	Start key search: asserted when loading of ciphertext is complete to direct the system to start search of key space
found	output	1	Asserted when the key is found
key	output	56	When *found* is asserted, the recovered key is output on this signal

Table 21.5. **Visible state of the DES-cracking system**

Name	Width	Description
key	56	DES key currently being used to decrypt
cipherTextBlock	64	Encoded block of text being decrypted
blockNumber	16	Ciphertext block number currently under decryption
mode	2	Current mode – idle, reading, computing
cipherTextStore	512	Ciphertext block storage

Table 21.6. **Modes of the DES-cracking system**

Name	Description
idle	After reset or the successful decryption of a block of text: remains in the idle state until *cipherTextValid* is asserted to begin reading in new data
dataIn	Data being read into the ciphertext storage, one byte at a time
cracking	The engine continuously iterates over possible keys until a match has been found

This specification has been for a DES cracker with a limited flexibility. In the exercises, we will ask you to extend this specification to include the ability to interrupt cracking, input new data, and output plaintext blocks.

21.2.3 Music player

Overall description The music player is used to read a song out of a RAM module and synthesize an audible waveform. The song is stored in a format that indicates a series of note values, each with a fixed 100 ms duration. The output of the synthesizer serves as the input to an audio codec. The codec requires one new input, in s0.15 format, every 20.8 μs (48 kHz). Our initial design will allow only a single note to be played at once, but we will enable multiple harmonics of that note.

Inputs and outputs Table 21.7 lists the inputs and outputs of the music player. The RAM that stores the song in the music player is assumed to have a song preloaded. The user simply indicates the beginning of playback. After some number of cycles, a value for the next time step and a high *valueValid* appear at the output of the system. This value is held until the *next* signal is asserted by the codec, triggering computation of the next output value.

State The states for our music player are shown in Table 21.8. The *noteNumber* state indicates the current note in playback. This note is read from the song RAM and converted into a frequency. This frequency value is a 0.16 number and represents the interval between two 48 kHz samples in radians. A value of 1 indicates π radians between samples, or a 24 kHz note (which you cannot hear). Using this frequency and the time value, the synthesizer calculates the harmonics and output wave value.

Table 21.7. **Inputs and outputs of the music player**

Name	Direction	Width	Description
start	input	1	Begin playing of the selected song
value	output	*s*0.15	The current value in the audible waveform, this signal must be valid once every 20.8 μs
valueValid	output	1	Indicates that the output value is valid
next	input	1	The codec is ready for the next value

Table 21.8. **Visible states of the music player**

Name	Width	Description
noteNumber	16	Note of the song to be synthesized
noteFrequency	0.16	Frequency of the current note
time	12	Current time-step for the output, in units of 20.83 μs; every note has a 4800 time-step duration
mode	1	Idle or playback

Table 21.9. **Modes of the music playing system**

Name	Description
idle	No music being played
playback	Generating audible output

Modes The music player has only two modes, *playback* and *idle*, shown in Table 21.9. The user is able only to initiate playback from the *idle* state, and the song plays until completion.

21.3 PARTITIONING

Much of system design involves partitioning a system into modules. While many consider this to be an art, most systems are partitioned by state, task, or interface. Most systems employ some combination of these three partitioning techniques. This partitioning may be hierarchical, with a different type of partitioning at each level. For example, a system may employ task partitioning at one level and then have one task further partitioned by state and another task further partitioned by interface.

With state partitioning, the system is divided into modules associated with different pieces of state (user-visible or strictly internal). Each module (e.g., VHDL design entity) is responsible for maintaining its portion of system state and communicating appropriate *views* of this state to other modules in the system.

With task partitioning, the function performed by a system is divided into tasks, and separate modules are associated with each task. There are many sub-genres of task partitioning: *pipelining* (Chapter 23) divides one large task into a series of subtask modules, with each subtask passing its output to the next subtask in the series. With *master–slave* partitioning, a *master* module oversees the operation of a number of *slave* modules – doling out work to the slaves and handling their responses. With *resource partitioning*, a module is associated with, and arbitrates access to, a shared resource. For example, a memory module may be shared across many clients in a complex system. As another example, a route-computation module in a router is a resource that may be shared across many input ports in a router. With model–view–controller partitioning, the system is partitioned into a *model* module (which contains most of the system function), a *view* module (responsible for all output (views of the model)), and a *controller* module (responsible for all input (controlling the model)).

Associating modules with input and output (as with the view and controller parts of model–view–controller partitioning) is actually a form of interface partitioning. With interface partitioning, a separate module is associated with each interface (or related set of interfaces) of a system. For example, a system with a DDR3 DRAM interface will typically have a separate module that controls this interface, arbitrates among clients sharing the interface, and provides a simpler, more abstract interface to the rest of the system.

21.3.1 Pong

Figure 21.2 shows how the Pong video game system is partitioned along two axes. Horizontally, the system is partitioned by task, into model, view, and controller. The model portion of the

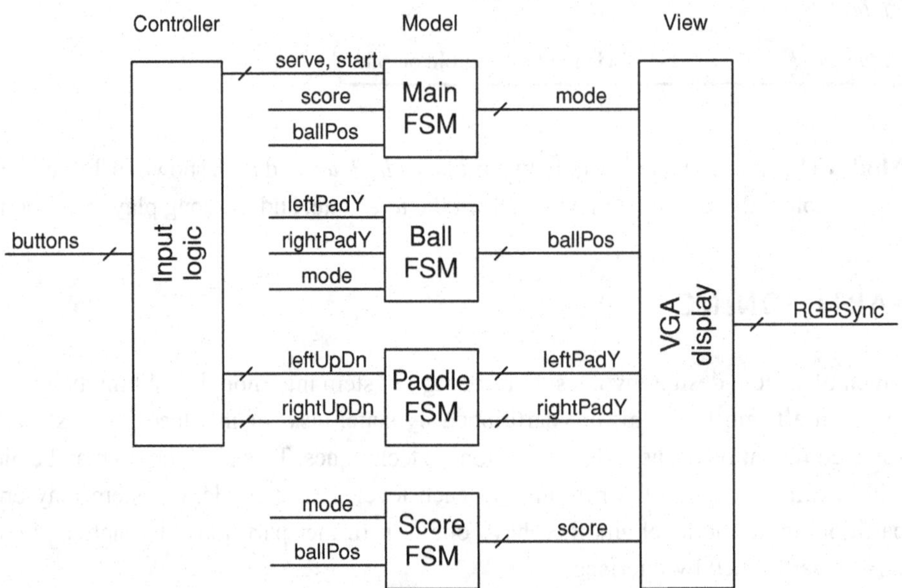

Figure 21.2. The Pong game is partitioned using the model–view–controller decomposition. The score and position of the ball and paddles constitute the *model*. This model is viewed by a VGA display module. A controller module conditions input buttons to affect the model. Note that the model is further partitioned by state into separate modules for ball, paddles, and score.

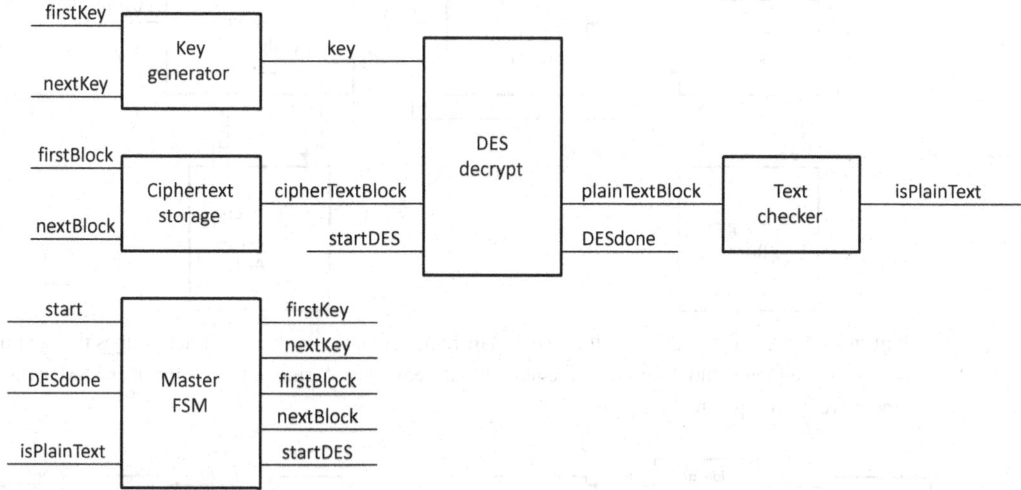

Figure 21.3. A DES cracker is partitioned by task. The modules perform the subtasks of generating keys, sequencing the ciphertext, decrypting the ciphertext to plaintext, and checking the decryption output to see whether it is plaintext. A master FSM module controls overall timing and sequencing.

system is further partitioned vertically by state with the ball state, the paddle state, the score, and the mode in separate modules. The figure also shows the interfaces defined between the modules. In most cases, a module simply exports all or part of its state (e.g., *score*, *ballPos*).

21.3.2 DES cracker

The DES cracker is partitioned into several separate modules, as shown in Figure 21.3. The master FSM acts as the system controller. The ciphertext storage block reads and stores the ciphertext. When the master FSM pulses the *firstBlock* signal (for one cycle), the storage puts the first block onto the *cipherTextBlock* bus (and holds it). The storage FSM outputs the next block in storage when the master pulses *nextBlock*. The key generator iterates through a series of DES keys in order to find a match. The *firstKey* and *nextKey* signals work in the same way as the *firstBlock* and *nextBlock* signals. The DES decrypt block takes multiple cycles to decrypt a block and asserts *DESdone* upon completing an iteration. Finally, the text checker asserts *isPlainText* when the "plain" text is indeed plaintext.

21.3.3 Music synthesizer

We have partitioned the music player to have a pipelined series of tasks, shown in Figure 21.4. The note FSM will iterate through the song, reading each note from memory. This note is converted into a frequency in the next block, and stored into the synthesizer. The synthesizer FSM will begin at time-step 0 and calculate the output waveform at each time-step for the duration of the note. This computation itself involves taking the sine of a value, and we use a RAM to look up these sine values.

A more complex synthesizer is shown in Figure 21.5. It includes two new state machines for computing harmonics and an attenuation envelope. The harmonic FSM communicates with the

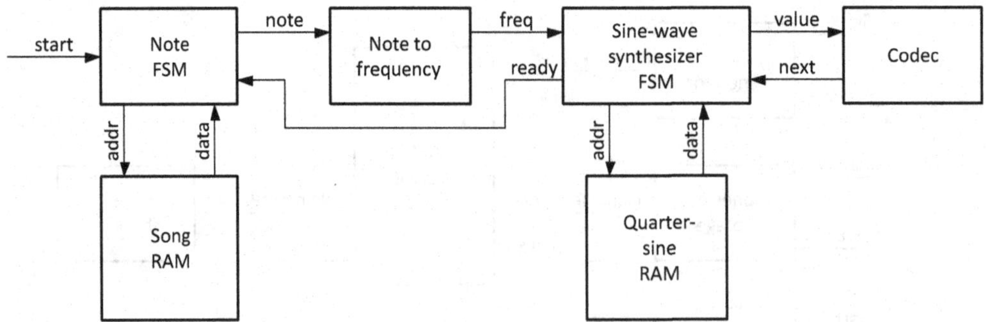

Figure 21.4. A simple music synthesizer is partitioned by task. A note FSM determines the next note to play. A note-to-frequency block converts the note into a frequency. A sine-wave synthesizer FSM synthesizes a sine wave of the specified frequency.

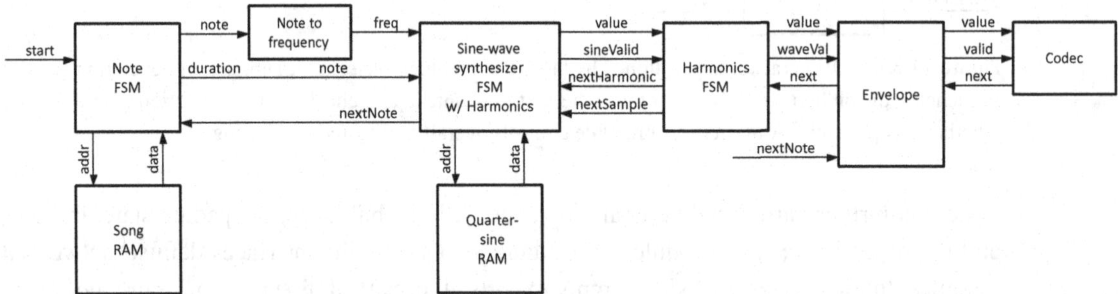

Figure 21.5. An expanded version of the music synthesizer uses a sine-wave synthesizer module that returns the value of multiple harmonics for each note, a harmonics FSM that combines these harmonics, and an envelope FSM that modulates the resulting waveform with an attack–decay envelope.

synthesizer FSM and receives a series of s0.15 waveform values. The harmonic FSM combines these waves as input to the envelope FSM. This final FSM attenuates the wave with an attack–decay envelope using the current time-step. A *nextNote* signal from the note FSM indicates when to reset the time counter.

Flow control is enforced in the music playback system through the use of ready signals. The output codec notifies the synthesizer FSM when it is ready for the next waveform value. When a note has completed playback, the sine-wave synthesizer asserts the ready signal to the note FSM.

Summary

In this chapter you have started to explore the process of system design, namely how to take the specification for a complex system and decompose it into modules that are simple enough to be designed directly. We have introduced a six-step process for system design:

(1) specification,
(2) partitioning,
(3) interface specification,

(4) timing design,
(5) module design, and
(6) performance tuning.

We then explored the first two steps via three extended examples. It is important to *specify* a system completely before starting the design process – to avoid repeated backtracking, confusion, and inconsistent interpretations. The specification should have sufficient detail that no externally visible feature is left to the designer's imagination during the remainder of the design process.

The process of *partitioning* a design is at the heart of system design. Designs are usually partitioned into modules associated with a part of system *state*, a *task*, or an *interface*. Partitioning is often done recursively. Our Pong example partitions first by tasks using the *model–view–controller* decomposition and then partitions the model by state.

Chapters 22–25 describe the remaining steps of the design process in more detail.

BIBLIOGRAPHIC NOTES

The DES standard [35] describes in detail the implementation of DES encryptors and decryptors. Olson and Belar describe one of RCA's first music synthesizers in ref. [91]. The paper, written in 1955, includes images of the machine taking up an entire office wall.

The actual block diagram from the Pong video game system is shown in ref. [17]. This paper, written by Atari's Vice President of Engineering in 1977, provides schematics for blocks such as the sound generator and score tracker. For a description of many early items of consumer electronics, including video games, consult ref. [98].

Exercises

21.1 *Pong specification.* The specification of the Pong video game in Section 21.2.1 is purposely incomplete. Some of the missing specifications are listed in the text. Identify as many unspecified issues as you can and give a specification of each.

21.2 *DES: data write.* In the DES block, we included inputs for writing data into the cipher block storage, but did not include that logic in the partitioned diagram (Figure 21.3). Update the block diagram to include the ability to read in data. Specify how this works.

21.3 *DES: interrupt.* In the DES specification and block diagram, add an input *interrupt* that stops DES computation. Be sure to specify the values of the outputs on interrupt, the state the system goes into, and what happens during the interruption of data input. How does execution resume?

21.4 *DES: plaintext output.* At the end of DES cracking, we would like to output the plaintext data. Specify this feature and include it in the partitioned block diagram.

21.5 *Music player: input song.* In the music player specification and simple block diagram, add the ability to input a song file into the RAM.

21.6 *Music player: pause, stop buttons.* Add the ability to stop and pause playback of a song. Discuss where playback should begin when the user again asserts the *start* signal.

21.7 *String searcher.* Give a specification and partitioning of a string searcher. Given three different sequences to find and a single long string to search, your block should output the number of occurrences of each search term.

21.8 *Elevator controller.* Specify and partition an elevator controller. You should take floor requests as inputs and output the door state (open or closed), destination floor, and a signal to start the elevator moving.

22 Interface and system-level timing

System-level timing is generally driven by the flow of information through the system. Because this information flows through the interfaces between modules, system timing is tightly tied to interface specification. Timing is determined by how modules sequence information over these interfaces. In this chapter we will discuss interface timing and illustrate how the operation of the overall system is sequenced via these interfaces, drawing on the examples introduced in Chapter 21.

22.1 INTERFACE TIMING

Interface timing is a convention for sequencing the transfer of data. To transfer a datum from a source module S to a destination module D, we need to know when the datum is valid (i.e., when the source module S has produced the datum and placed it on its interface pins) and when D is ready to receive the datum (i.e., when D samples the datum from its interface pins). We have already seen examples of interface timing in the interfaces between the modules of factored state machines in Chapter 17. In the remainder of this section, we look at interface timing in more depth.

22.1.1 Always valid timing

As the name implies, an *always valid* signal (Figure 22.1), is always valid. An interface with only always valid signals does not require any sequencing signals.

It is important to distinguish an *always valid* signal from a *periodically valid* signal (Section 22.1.2) with a period of one clock cycle. An always valid signal represents a value that can be dropped or duplicated. A temperature sensor that constantly outputs an eight-bit digital value representing the current temperature is an example of such a signal. We could pass this signal to a module operating at twice the clock rate (duplicating temperature signals) or a module operating at half the clock rate (dropping temperature values), and the output of the module still represents the current temperature (with perhaps a slight lag).

The state interfaces in the Pong game of Figure 21.2 are another example of always valid interfaces. The *mode*, *ballPos*, *leftPadY*, *rightPadY*, and *score* signals are all examples of always valid signals. Each of these signals always represents the current value of a state variable.

A *static* or *constant* signal is a special case of an always valid signal where the signal is guaranteed not to change values between specified events (e.g., system resets). A static signal

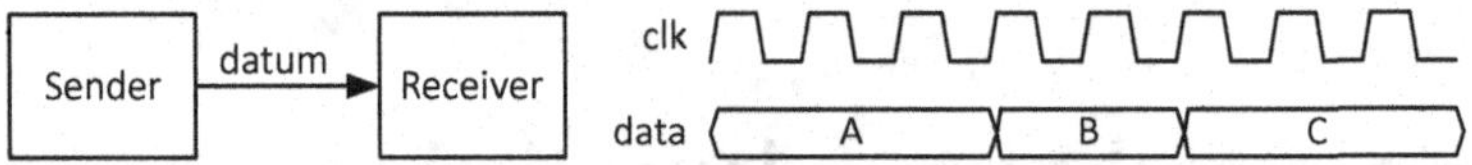

Figure 22.1. An *always valid* interface requires no sequencing signals or flow control. The datum is valid every cycle and can be sampled by the receiver at any time.

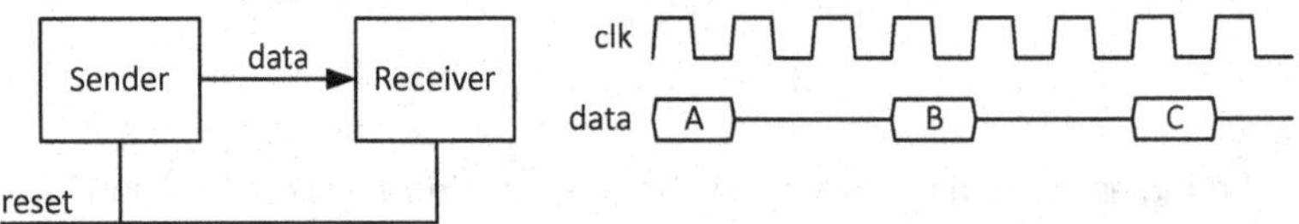

Figure 22.2. A *periodically valid* interface transfers a datum once every N cycles. There is no flow control, and values cannot be dropped or repeated. The figure shows a periodic signal with period $N = 3$.

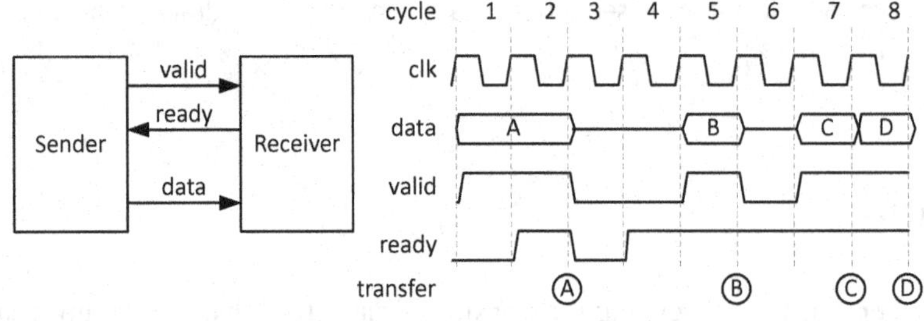

Figure 22.3. With *ready–valid* flow control, a datum or token is passed between two modules only when the sender indicates that the datum is *valid* and the receiver indicates that it is *ready* to receive the datum. If either *ready* or *valid* is not asserted, no transfer takes place.

is even easier to deal with when crossing clock domains. Because it doesn't change during the period of interest, it need not be synchronized at all.

22.1.2 Periodically valid signals

A signal with *periodically valid timing* or *periodic timing* (Figure 22.2) is valid once every N cycles. The interval N is the period of the signal. Unlike an always valid signal, each value of a periodically valid signal represents a particular event, task, or token, and cannot be dropped or duplicated. The *key* output of the key-generator module in the DES cracker (Section 21.3.2) is an example of a periodically valid interface (assuming that the *nextKey* input is a periodic signal). A new key appears on this interface once every N cycles. Each key represents a particular task (decrypt the ciphertext with this key) and cannot be dropped or duplicated.[1]

A periodically valid signal with a period of 1 is valid every cycle but it is not the same as an always valid signal because its values cannot be dropped or duplicated. For example, suppose we design the DES cracker of Figure 21.3 so that the key generator produces a new key each cycle. This is a periodically valid signal with a period of 1. We cannot drop or duplicate the key values produced. Each must be considered exactly once.

This distinction between *always valid* and *periodically valid* signals becomes apparent when we move signals between clock domains (Chapter 29). It is easy to move always valid signals between clock domains as long as we avoid synchronization failure because it is acceptable

[1] Strictly speaking, we could duplicate keys – this would simply result in extra work as we attempt to decrypt the ciphertext block multiple times with the same key.

to duplicate or drop values. On the other hand, flow control is required in order to move a periodically valid signal with a period of one clock cycle across clock domains.

Of course, with periodic signaling, the sending and receiving modules must synchronize at some point – so that they agree on which cycle to start counting to *N*. This is typically achieved by initializing counters in each module on the transition out of reset.

Periodically valid interfaces tend to be brittle and in most cases should be avoided in favor of interfaces with flow control (Section 22.1.3). Periodic signaling violates modularity. If one needs to redesign a module to change the value of *N*, all modules connected to this module need to change as well – and may have trouble accommodating the new value. Using signaling with flow control isolates changes in timing to one module.

22.1.3 Flow control

An interface with flow control uses explicit sequencing signals, usually called *valid* and *ready*, to sequence the transfer of data over the interface. Figure 22.3 shows an interface that uses *ready–valid* flow control. The sending module signals when a valid datum is present on the interface by asserting *valid*. The receiving module indicates that it is ready to accept a new datum by asserting *ready*. The datum is transferred only when both *valid* and *ready* are asserted.

With a flow-controlled interface, the module (sender or receiver) that is ready for the transfer first must wait for the other module before the transfer takes place. In the first transfer in Figure 22.3, the sender waits for the receiver. The sender puts *A* on the *data* lines and asserts *valid* in cycle 1, but has to wait until the receiver signals *ready* in cycle 2 for the transfer to take place. In the second transfer, the receiver waits for the sender. The receiver asserts *ready* in cycle 4, but has to wait until the sender puts *B* on the *data* lines and asserts *valid* in cycle 5.

With the ready–valid interface there is no need for either *ready* or *valid* to go low between transfers. In the third transfer in the figure *ready* remains high in cycle 6, but the receiver must wait for the sender to provide *data* and assert *valid* in cycle 7. When both *ready* and *valid* remain high, a new datum is transferred each cycle – as with *D* in cycle 8.

When one of the two modules is known always to be ready for a transfer, one side of the ready–valid interface can be omitted. In the music player of Figure 21.4, for example, the *valid* side of the flow control is omitted; only the *ready* side of the flow control signaling is provided. The sending module is assumed to provide valid data before *ready* (or *next* for the codec) is asserted. This one-sided flow control, where transfers are controlled by the receiver, is sometimes called *pull timing* because the receiver is *pulling* new values by asserting *ready*.

Similarly, a one-sided interface, where only the *valid* signals are provided and the receivers are assumed to be ready before the next datum, is called *push timing* because the sender *pushes* data through the system with *valid* signals.

In some cases, the *valid* signal can be encoded in the *data* signal by using an unused or invalid data code to imply *not valid*. When this convention is used, the presence of a valid data code on the *data* signal implies *valid*. A separate *ready* signal is always required because the receiver has no other way of signaling whether it is able to accept a datum.

In some cases, it is advantageous to allow the sending module to get a few data ahead of the receiving module. This can be accomplished by inserting a FIFO buffer between the sender and the receiver. The FIFO provides a few words of storage with a ready–valid interface both on its input and on its output. Its output is *valid* unless it is empty, and its input is *ready* unless it is full. We defer describing details of FIFOs until Chapter 23.

EXAMPLE 22.1 Flow-controlled registers

Design a register in which the input and output data uses flow-controlled signaling. Such a register can be used to split long, high-delay wires that are on the critical path of the design.

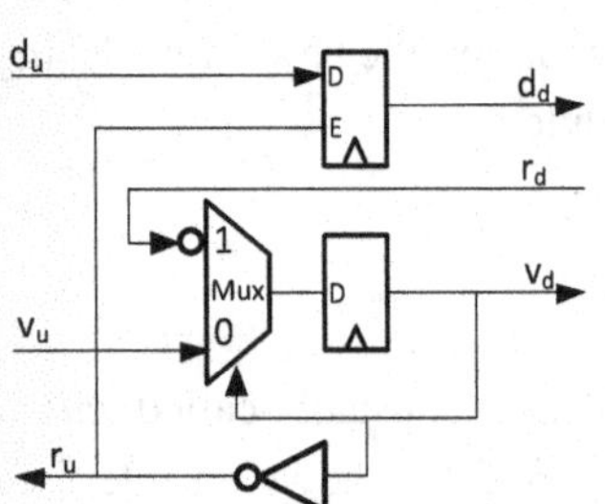

Figure 22.4. Schematic for the flow-controlled register described in Example 22.1.

Figure 22.4 shows the design of our module. The signals for communicating with the upstream module are d_u (data input), r_u (ready output), and v_u (valid input); d_d (output), r_d (input), and v_d (output) transfer data downstream. When designing this module, we make it a point *not* to include any combinational paths passing through our register. Doing so defeats the purpose of the module: eliminating long delay paths.

Every cycle, when no valid data are stored ($v_d = 0$), both the valid register and data are updated with the upstream value. Our buffer also signals upstream that it is ready to accept new data ($r_u = 1$). When valid data are stored in the register ($v_d = 1$), the data register is disabled, holding the stored value. If the downstream unit is not ready, v_d is held high and r_u is kept low. When ready, the downstream unit asserts r_d, causing the v_d to be deasserted on the next cycle. With only one register and no combinational paths between stages, this module can accept new data only every other cycle. Exercise 22.6 asks you to make a version that doubles this data rate.

The always valid and periodically valid versions of this register comprise a D-flip-flop that is updated on each clock edge with the most recent value. Designers must be aware that a periodic output will be valid one cycle later with the register than without. This may require changing downstream modules.

22.2 INTERFACE PARTITIONING AND SELECTION

The *data* portion of an interface is often partitioned into a number of fields. For example, in the Pong game of Figure 21.2, the *model* subsystem has an output that includes five fields representing distinct state variables. At the next level down, the Paddle FSM has an always valid interface with two data fields, *leftPadY* and *rightPadY*. The Ball FSM outputs a single signal, *ballPos*. However, this signal is logically divided into X and Y components, *ballPos.X* and *ballPos.Y*.

In an interface with multiple data fields, often one field determines how the remainder of the interface is interpreted. For example, in Figure 18.13 the first part of our microcode instruction is a field that determines how the rest of the instruction is interpreted.

A common interface technique provides separate fields for *control*, *address*, and *data*. The control and address fields of the interface are *selection* fields. The control field selects the operation to be performed and the address field selects which location the operation is performed on. For example, in a memory system, the *control* field may specify one of *read, write, no-operation, refresh, set-parameter*, and other operations. The address field selects the location or parameter to be read, written, or refreshed. The data field then provides (or receives) the data associated with the operation. Both the data field and the selection fields are sequenced using one of the timing conventions described in Section 22.1.

22.3 SERIAL AND PACKETIZED INTERFACES

When an interface must transfer a large datum with a low duty factor, it may be advantageous to *serialize* the datum, transferring it over many cycles, one part per cycle over a narrower interface. For example, suppose an interface transfers a 64-bit block of data once every four cycles. We can transfer this block over a 16-bit interface by sending one-quarter of the block each cycle, as shown in Figure 22.5. On the first cycle a_3 (*a(63 downto 48)*) is transferred, in the second cycle a_2 is transferred, and so on.

With a serialized signal, there must be a convention for the sending and receiving module to determine which cycle a transfer starts on. This can be done using either flow control or periodically valid timing. Figure 22.5 shows an interface with one-way flow control that uses an explicit *frame* signal to denote the first cycle of each transfer. This is an example of push timing. The *frame* signal is a *valid* signal and the receiver is assumed to be always ready. An explicit *ready* signal could be added to the interface if two-way flow control is required.

With flow control, a transfer need not start every four cycles and need not even start on a multiple of four cycles. In the example shown in Figure 22.5, the link goes idle in cycle 9, because the transmitter has no data to send. The transfer of datum *c* then starts in cycle 10. With explicit framing, the transmitter need not wait until cycle 12, as would be required with periodically valid timing.

The interface shown in Figure 22.5 is an example of *nested timing* with push timing used at the frame level and periodic timing used at the cycle level. Frame-level timing is determined by the one-way flow control using the *frame* signal. However, once the frame has been started, the

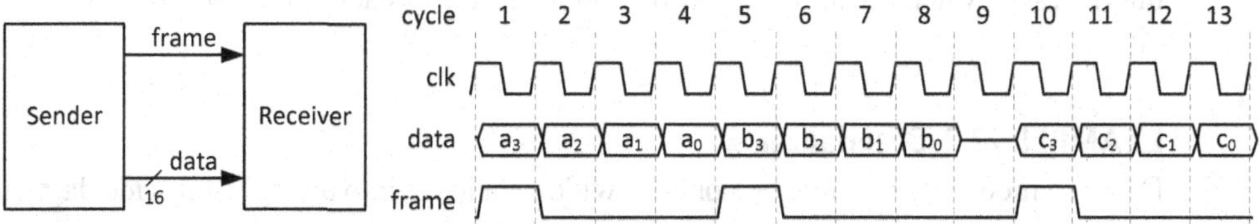

Figure 22.5. A *serialized* interface passes data over a number of cycles. In this case, a *frame* signal denotes the first cycle of each transfer, a form of *push* signaling.

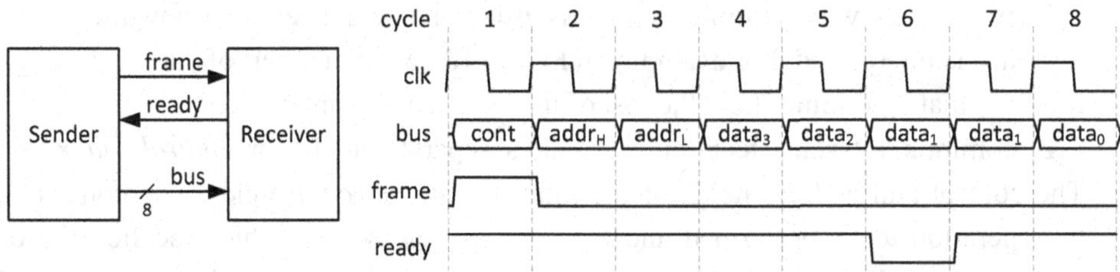

Figure 22.6. A memory or I/O interface may be serialized to send the control, address, and data over a shared bus. In this example, two-way flow control is used with *frame*, indicating a valid frame, and *ready*, indicating receiver readiness on a cycle-by-cycle basis.

remaining transfers take place each cycle (periodic timing with $N = 1$) with no flow control. It is possible to use flow control at both levels of timing by using a *valid* signal to sequence the subsequent words of a transfer as well. As long as the frames are of fixed size, a single *valid* signal can be used both at the frame level (for the first word of a transfer) and at the cycle level (for subsequent words). If two-way flow control is required, a single *ready* signal can be used for both levels.

Memory and I/O interfaces often serialize the command, address, and data fields to transmit them over a shared, narrow bus, as shown in Figure 22.6. In the figure a memory transaction is serialized over a byte-wide interface over seven cycles with the control sent on the first cycle, the address sent over two cycles, and data sent over four cycles. The example uses cycle-valid/frame-ready flow control, with the *frame* signal indicating that an entire frame of data is ready and the *ready* signal indicating receiver readiness on a cycle-by-cycle basis. The receiver signals *not ready* during cycle 6, causing $data_1$ to be retransmitted in cycle 7.

Serialized interfaces can be thought of as being *packetized*. Each item transmitted is a *packet* of information containing many fields and possibly of variable length. The packet is serialized for transmission over an interface of a given width and deserialized on the far side. The width of the packet's fields have no relation to the width of the interface. The information transmitted in a given cycle may include all or part of several fields, and a given field may span multiple cycles.

The decision to serialize an interface or leave it parallel is based on cost and performance. The advantage of a serialized interface is a reduction in the number of pins or wires required for the interface. The disadvantages are an increase in latency and the complexity of the serialization, deserialization, and framing. On-chip, where the cost of additional wires is small, it is almost always better to leave an interface wide – unless the signals being transmitted were serialized to begin with. Off-chip, where chip pins and system-level signals are expensive, interfaces are often serialized to keep the duty factor of each pin close to unity.

EXAMPLE 22.2 Deserializer

Design a module that converts a one-bit wide data input into an eight-bit wide data output. Both the input and the output use push flow control. The output is not valid until eight valid inputs (bits $0, 1, \ldots, 7$) have been received.

```
library ieee;
use ieee.std_logic_1164.all;
use ieee.numeric_std.all;
use work.ff.all;

entity deserializer is
  generic( width_in: integer := 1;
           n: integer := 8 );
  port( clk, rst, vin: in std_logic;
        din: in std_logic_vector(width_in-1 downto 0);
        dout: out std_logic_vector(width_in*n-1 downto 0);
        vout: out std_logic );
end deserializer;

architecture impl of deserializer is
  signal en_nxt, en_nxt_rst, en_out, en_out_rst: std_logic_vector(n-1 downto 0);
  signal din_rst: std_logic_vector(width_in-1 downto 0);
  signal vout_nxt: std_logic;
begin
  en_nxt <= std_logic_vector(unsigned(en_out) rol 1) when vin else en_out;
  en_nxt_rst <= (n-1 downto 1 => '0') & '1' when rst else en_nxt;
  en_out_rst <= (others => '1') when rst else en_out;
  din_rst    <= (others => '0') when rst else din;

  cnts: vDFF generic map(n) port map(clk,en_nxt_rst,en_out);

  DATA: for i in n downto 1 generate
    reg: vDFFE generic map(width_in) port map(clk, en_out_rst(i-1),
              din_rst, dout(width_in*i-1 downto width_in*(i-1)));
  end generate;

  vout_nxt <= en_out(n-1) when not rst else vin;
  vout_r: sDFF port map(clk, vout_nxt, vout);
end impl; -- deserializer
```

Figure 22.7. VHDL design entity that implements a deserializer with push flow control for Example 22.2.

Shown in Figures 22.7 and 22.8 are the VHDL and testbench output of our module. The general idea is to write each incoming bit into the appropriate one of eight D-flip-flops (LSB first). We keep a (one-hot encoded) count that keeps track of which bit is to be written next, shifting it each cycle the input is valid. In the VHDL, we have registers to store three states:

(1) `en_out`: an n-bit long one-hot signal that encodes which bit of the deserializer is to be written next;

(2) `dout`: the data itself, fed directly to the output (each individual flip-flop is enabled by the `en_out` signal above);

(3) `vout`: if the output is valid this cycle (set the cycle after the eighth bit has been captured).

The output of our deserializer is shown in Figure 22.8, inputting words of all 1s, all 0s, and then all 1s. We can verify the timing of this module by comparing it with the *timing table* we specify in Example 22.3.

```
# vin: 1 din: 1 vout: 0 dout: 00000000 en: 00000001
# vin: 1 din: 1 vout: 0 dout: 00000001 en: 00000010
# vin: 1 din: 1 vout: 0 dout: 00000011 en: 00000100
# vin: 1 din: 1 vout: 0 dout: 00000111 en: 00001000
# vin: 1 din: 1 vout: 0 dout: 00001111 en: 00010000
# vin: 1 din: 1 vout: 0 dout: 00011111 en: 00100000
# vin: 1 din: 1 vout: 0 dout: 00111111 en: 01000000
# vin: 1 din: 1 vout: 0 dout: 01111111 en: 10000000
# vin: 1 din: 0 vout: 1 dout: 11111111 en: 00000001
# vin: 1 din: 0 vout: 0 dout: 11111110 en: 00000010
# vin: 1 din: 0 vout: 0 dout: 11111100 en: 00000100
# vin: 1 din: 0 vout: 0 dout: 11111000 en: 00001000
# vin: 1 din: 0 vout: 0 dout: 11110000 en: 00010000
# vin: 1 din: 0 vout: 0 dout: 11100000 en: 00100000
# vin: 1 din: 0 vout: 0 dout: 11000000 en: 01000000
# vin: 1 din: 0 vout: 0 dout: 10000000 en: 10000000
# vin: 1 din: 1 vout: 1 dout: 00000000 en: 00000001
# vin: 1 din: 1 vout: 0 dout: 00000001 en: 00000010
# vin: 1 din: 1 vout: 0 dout: 00000011 en: 00000100
# vin: 1 din: 1 vout: 0 dout: 00000111 en: 00001000
# vin: 0 din: 1 vout: 0 dout: 00001111 en: 00010000
# vin: 1 din: 1 vout: 0 dout: 00011111 en: 00010000
# vin: 0 din: 1 vout: 0 dout: 00011111 en: 00100000
# vin: 1 din: 1 vout: 0 dout: 00111111 en: 00100000
# vin: 1 din: 1 vout: 0 dout: 00111111 en: 01000000
# vin: 1 din: 1 vout: 0 dout: 01111111 en: 10000000
# vin: 1 din: 1 vout: 1 dout: 11111111 en: 00000001
```

Figure 22.8. Set of outputs showing the verification of the VHDL of Example 22.2 and Figure 22.7.

22.4 ISOCHRONOUS TIMING

Some interfaces, such as an LCD display or an audio codec, require *isochronous timing.* These devices have hard real-time constraints on timing – requiring that each data element be delivered within a bounded window of time to avoid missing a sample. The timing constraint can be thought of as periodic timing with a margin – due to a FIFO. Sample i must be delivered between cycle $N(i - B)$ and cycle Ni, where N is the period and B is the size of the FIFO buffer. The interfaces themselves employ flow control – to allow some variation in the timing – but they require that the flow-control signals respond within the required timing interval.

The audio codec in the music player of Figure 21.4, for example, requires isochronous timing. This is satisfied by using pull timing and ensuring that the rest of the system responds more rapidly than required.

Isochronous timing can be challenging in a system that includes arbitration for resources – particularly when the arbitration is between many isochronous flows. The amount of time spent waiting for arbitration must be bounded to prevent the worst-case delay from exceeding the timing constraint.

Table 22.1. **Timing table for the Huffman encoder of Section 19.4.1 encoding the character string "THE"**

cycle	*rst*	*irdy*	*in*	*char*	*load*	*count*	*value*	*oval*	*out*
0	1			×		×	×		
1		1	14	×		×	×		
2				14	1	×	×		
3				14		3	001000000	1	0
4		1	08	14		2	01000000	1	0
5				08	1	1	1000000	1	1
6				08		4	011000000	1	0
7				08		3	11000000	1	1
8		1	05	08		2	1000000	1	1
9				05	1	1	000000	1	0
10				05		3	100000000	1	1

22.5 TIMING TABLES

Since Chapter 14, we have been using *timing diagrams* such as Figure 22.6 to illustrate timing relationships. In these diagrams time advances from left to right, and we show the evolution of each signal over time on the vertical axis, either as a waveform, for binary signals, or with a set of values. While timing diagrams are useful for visualizing binary signals for a few cycles, when we are visualizing multi-bit signals the waveforms don't help, and we often need to see many more cycles.

In situations where we want to visualize more than a few cycles, or where most signals are not binary, a *timing table* is more useful than a timing diagram. Table 22.1 shows a timing table for the Huffman encoder of Section 19.4.1 operating on the same input data as used for the simulation shown in Figure 19.16.

While the data are the same whether horizontal or vertical, it is easier to interpret them as a table. Showing successive values of *value* one on top of the other makes the shifting operation clearer. Also, the updating of state from line to line is easier to follow. In a cycle during which *irdy* is asserted, *char* takes on the value of *in* in the next row. Similarly, in a cycle during which *load* is asserted, *count* and *value* are updated in the next row.

EXAMPLE 22.3 Deserializer timing table

Show a timing table for the deserializer of Example 22.2.

In Table 22.2, we show the deserializer operating over two iterations (data *A* and *B*). Every cycle when the input is valid (*ival*), the counter (in a binary-weighted format) is incremented and the data are stored into the appropriate bit position. On the cycle after the counter is 7 and the input data is valid, the output valid signal (*oval*) is asserted.

Table 22.2. **Timing table showing example execution of the deserializer of Example 22.2**

cycle	*rst*	*ival*	*in*	*count*	*out*	*oval*
0	1	×	×	×	×	×
1	0	1	A_0	0	00000000	0
2	0	1	A_1	1	$0000000A_0$	0
3	0	1	A_2	2	$000000A_1A_0$	0
...	0	1	A			0
8	0	1	A_7	7	$0A_6A_5A_4A_3A_2A_1A_0$	0
9	0	1	B_0	0	$A_7A_6A_5A_4A_3A_2A_1A_0$	1
10	0	0	X	1	$A_7A_6A_5A_4A_3A_2A_1B_0$	0
11	0	1	B_1	1	$A_7A_6A_5A_4A_3A_2A_1B_0$	0
12	0	1	B_2	2	$A_7A_6A_5A_4A_3A_2B_1B_0$	0
...	0	1	B			0
18	0	0	X	0	$B_7B_6B_5B_4B_3B_2B_1B_0$	1

22.5.1 Event flow

A timing table makes the *event flow* of a digital system clear. The event flow is the sequence of events that through cause and effect drive the system forward. For our Huffman encoder, the event flow is driven entirely off of the counter. When the counter reaches 2, the encoder asserts *irdy* to load another input character into the *char* register. When the counter reaches 1, *load* is asserted, and the counter and shift register are loaded on the next cycle with the bit string to encode the character.

For most non-trivial digital systems the event flow is driven from a key interface and events are synchronized between modules with a ready–valid interface.

22.5.2 Pipelining and anticipatory timing

The Huffman encoder is an example of a pipeline (Chapter 23) that uses anticipatory timing. Each input to the encoder passes through two pipeline stages. If a character arrives at the input and *irdy* is asserted in cycle i, the character appears at the output of the input register in cycle $i+1$, and the length and bit string for that character are loaded into the counter and shift register at the start of cycle $i+2$.

Because of the two-cycle delay from input to output, the control logic must *anticipate* the end of the current character's bit string and assert *irdy* two cycles in advance – when the value in the counter is 2 – to load the next character into the input register. One cycle in advance – (when the count is 1) it asserts the *load* signal to load the output of the ROM into the counter and the shift register.

This anticipatory timing becomes more interesting when the pipeline is longer and when the minimum length of an output string is 1.

22.6 INTERFACE AND TIMING EXAMPLES

We will now revisit our examples from Chapter 21 and examine their system timing and interface timing.

22.6.1 Pong

The bulk of the Pong system uses *always valid* timing. Each of the modules in the center column of Figure 21.2 generates its portion of the global state. The Ball FSM, for example, generates the X and Y coordinates of the current time. These signals are always valid and can be sampled at any point in time. Each FSM samples the state generated by the other FSMs to carry out its function. For example, the Ball FSM uses the paddle positions from the Paddle FSM to check for the ball hitting a paddle. Note that the clock must be run fast enough (relative to the speed of the ball) that the ball does not move more than one pixel in X or Y per clock cycle, so that key events – such as the ball intersecting the top of the screen or hitting a paddle – won't be missed between time-steps. The VGA display module samples the always valid state of the four FSMs as needed to generate the display.

The signals out of the input logic use single-ended flow control (*push* timing). In contrast to the always valid signals that represent a continuous state, these signals represent *events*. When a button is pressed, one of the input signals *serve*, *start*, *leftUp*, *leftDn*, *rightUp*, or *rightDn* is asserted for exactly one cycle – signaling an input event. Each of these signals can be thought of as a *valid* signal with implied data. That is, these signals combine flow control – signaling the event – and data – whether the event is a left paddle up or a left paddle down – into a single signal. The Main FSM and Paddle FSM respond to these events by updating their state.

22.6.2 DES cracker

The timing for the DES cracker using a deterministic schedule is shown in Table 22.3. The cracker has eight ciphertext blocks stored, a 16-cycle DES descriptor, a one-cycle key generator, and a combinational plaintext checker. Every 16 cycles, a new ciphertext block is decrypted. On the 128th cycle, a new key is generated, and block 0 is again decrypted.

The event flow in this version of the DES cracker is driven entirely by a series of counters in the master FSM block. These counters sequence timing of the peripheral blocks to synchronize with the 16-cycle period of the DES decrypt block. The individual blocks (the key generator, the cyphertext storage, and the plaintext checker) operate with one-sided flow control (pull timing)

Table 22.3. In periodic DES timing, we decrypt all eight blocks of ciphertext before advancing to the next key

The columns labeled FK, NK, FB, and NB refer to signals *first_key*, *next_key*, *first_block*, and *next_block*, respectively.

Cycle	FK	NK	Key	FB	NB	CT block	DES	Check
−1	1			1				
0			key 0			block 0		
1							round 1	
2							round 2	
...							...	
15					1		...	
16						block 1	round 16	
17							round 1	PT block 0
18							round 2	
...							...	
31					1		...	
32						block 2	round 16	
33							round 1	PT block 1
...							...	
111					1		...	
112						block 7	round 16	
113							round 1	
...							...	
127		1		1			...	
128			key 1			block 0	round 16	
129							round 1	PT block 7

(Section 22.1.3). The signals *first_key*, *next_key*, *first_block* etc. are *ready* signals indicating when the DES cracker has accepted the last input produced by a block and that the block should advance to the next value. These first/next signals are an example of combining a data signal with a flow-control signal. They indicate both when the master FSM is *ready* for the next value and whether the next value should reset to the beginning of the sequence or advance to the next element of the sequence.

This system is another example of a pipeline with anticipatory timing. To be able to start round 1 with a new key in cycle 129, we need to assert *next_key* in cycle 127 so the key will be on the inputs of the DES unit in cycle 128. Similar timing is needed for the *first_block* and *next_block* signals.

We can greatly accelerate our DEC cracker by using a less rigid schedule in which we advance to the next key as soon as we detect a plaintext block that fails our plaintext check.

Table 22.4. **Modification of the DES cracker to move immediately to the next block when the plaintext check fails**

Cycle	FK	NK	KGen	FB	NB	CT	SD	DES	Check
−1	1			1					
0			key 0			block 0	1		
1								round 1	
2								round 2	
...								...	
15					1			round 15	
16						block 1	1	round 16	
17		1		1				round 1	not PT
18			key 1			block 0	1	round 2	
19								round 1	
20								round 2	
...								...	
33					1			round 15	
34						block 1	1	round 16	
35								round 1	OK

Table 22.4 shows the timing of such a cracker. Here the DES unit finishes decrypting ciphertext block 0 in cycle 16, with the plaintext block appearing on its output in cycle 17. The plaintext checker combinationally checks this block and signals that it is not plaintext during cycle 17. We use this indication to assert *next_key* and *first_block* before the end of cycle 17.[2] The new key and block 0 appear on the inputs of the DES unit in cycle 18. The *start_DES* signal (SD in the table) is asserted during this cycle, interrupting the DES unit's decryption of block 1, which is already on round 2. If the majority of incorrect keys give a bad plaintext block on the first ciphertext block, this optimization accelerates our cracker by 7.1×. It reduces the time between keys from 128 cycles to 18 cycles.

The operation of the DES unit in cycles 17 and 18 is an example of *speculation*. We don't know in cycle 15 whether the last block is going to pass the plaintext check. However, rather than wait around until cycle 17 to find out, we assert *next_block* in cycle 15 and *start_DES* in cycle 16 to start decrypting the next ciphertext block speculatively. We are speculating, that is guessing, that the plaintext check will pass, and then proceeding with that assumption. If

[2] The designer needs to check that using this status signal to generate control signals in the same cycle does not lengthen the critical path. If it does, these signals should be deferred to cycle 18.

Table 22.5. **Timing table for the music player**

Cycle	*next Sample*	*next Harm*	*sine Valid*	*next Note*	Comment
0	1				Next sample requested by codec
1	1				
2	1		1		Value of fundamental
3		1			Read second harmonic
4		1			
5		1	1		Value of second harmonic
6		1			Read third harmonic
7		1			
8		1	1		Value of third harmonic
...					Idle until next 48 kHz sample
2084	1				Read fundamental for next sample
2085	1				
2086	1		1		Value of fundamental
2087		1			Read second harmonic
2088		1			
2089		1	1		Value of second harmonic
2090		1			Read third harmonic
2090		1			
2091		1	1		Value of third harmonic
...					Repeat 4800 times per note
X+6		1			Read third harmonic of last sample of note
X+7		1			
X+8		1	1		Value of third harmonic
X+9				1	Request next note

the assumption is incorrect, as we discover in cycle 17, we cancel the speculative work by reasserting *start_DES* in cycle 18, and we are no worse off (time-wise[3]) than if we had waited to operate non-speculatively. If the speculation is correct, as in cycle 35, we are two cycles ahead of where we would have been if we had waited to operate non-speculatively.

[3] Speculation often has an energy penalty because the speculative work dissipates energy.

22.6.3 Music player

We conclude with the isochronous music player of Figure 21.5. The event flow for the music player is driven by the isochronous codec which requests a new sample, via signal *next* every 20.83 μs (48 kHz). This is an example of pull timing. The rest of the system has 2083 10 ns (100 MHz) clocks to provide this sample before the next request. There is no FIFO to buffer samples before the codec, so the system must compute each sample in real time.

Timing for the music player is shown in Table 22.5. Each block in the task chain pulls signals from the block before using ready–valid flow control. In cycle 0 the *next* signal from the codec is passed through the *envelope* module and triggers the *nextSample* signal out of the Harmonics FSM. This signal causes the sine-wave synthesizer to advance one time-step (20.83 μs) and output the value of the *fundamental* frequency in cycle 2. The harmonic unit responds by asserting *nextHarmonic* in cycle 3 to request the second harmonic. The synthesizer replies with a *valid* signal in cycle 5. A similar sequence returns the third harmonic in cycle 8. The hardware then idles until the codec requests the next sample in cycle 2084.

The nine cycles from cycle 2084 to 2091 repeat this sequence of requesting the next sample and then two additional harmonics. This nine-cycle sequence is repeated every 2084 cycles until the synthesizer has determined that all of the samples for the current note have been provided. After providing the third harmonic of the last sample of the current note, the synthesizer requests the Note FSM to provide the next note in cycle X + 9. The Note FSM has until cycle X + 2084 to provide the note.

Summary

In this chapter you have learned how to reason about the timing of signals between modules and how to use timing tables to analyze system timing.

The simplest interfaces, like our Pong example, use *always valid timing*, in which a signal can be sampled at any point in time. Modules that produce one valid result every *N* cycles use *periodic timing*. Such timing tends to be brittle – it breaks if any part of the system is changed or if an event that requires a module to be restarted occurs.

More robust timing can be realized using *flow control*, in which a *ready* signal signals when a receiver can accept a value and a *valid* signal indicates when a sender has a value to send. A value is passed when *ready* = *valid* = 1. Flow control can be implemented unidirectionally or bidirectionally. A unidirectional interface with a *ready* signal but no *valid* signal is said to use *pull timing*. An interface with only a *valid* signal is said to use *push timing*.

A *timing table* is a useful tool to design and visualize the timing of a system. These tables show time on the vertical axis with a separate row for each cycle. Columns are devoted to key signals.

In systems with delay between a request and a response, *anticipatory timing* – where the request is made several cycles before the response is needed – can be used to compensate for the delay.

Much insight into system timing can be generated by following the *event flow* – the cause and effect relationship between key timing signals. Event chains are typically triggered by external events arriving at interfaces or by the completion of bottleneck modules.

Exercises

22.1 *Always valid timing.* Aside from sensors and games, give three other examples of where always valid timing is used.

22.2 *Periodically valid timing.* Aside from what was mentioned in the text, give three examples of where periodically valid timing is used.

22.3 *Flow-controlled timing.* Aside from what was mentioned in the text, give three examples of where timing with flow control is used.

22.4 *Conversion to periodic timing.* Design a VHDL design entity that acts as a receiver for an interface with an eight-bit wide data signal and ready–valid flow control and as a sender for an interface with periodic timing with a period of $N = 5$. Explain how you handle the case where your module is empty when the next period comes up.

22.5 *Conversion from periodic timing.* Design a VHDL design entity for receiving periodic ($N = 10$) eight-bit signals and outputting them to a ready–valid interface. Your module should include the ability to save up to two of the periodically valid signals if the output is not ready. You may drop the third such packet.

22.6 *Fully utilized flow control.* In Example 22.1, we designed a module that can be used to buffer a ready–valid flow-controlled communication channel. The problem with this, however, was that we were able to accept new data only every other cycle. Design a module that can accept a new value every cycle, enabling full throughput. You are not allowed to have any combinational paths from the downstream interface to the upstream interface (or vice versa). You will need two registers to store incoming data.

22.7 *Credit-based flow control.* Credit-based flow control is an alternative to ready–valid signaling. The sending module starts with *n* credits. On every cycle in which the module still has at least one credit remaining, it can emit an eight-bit data signal and signal valid. The receiver is guaranteed to capture this value. Sending valid data subtracts one credit from the count of remaining credits. A periodically valid (period of 1) signal from the receiver to the sender, *creditRtn*, is used to "return" credits back to the sender. Every cycle during which *creditRtn* is asserted, the sender increments the credit count. Design and write VHDL for this credit-based sending module. The inputs are an always valid eight-bit data signal, reset, and *creditRtn*. The outputs are the *valid* signal and *data*.

22.8 *Serialization, I.* Design a VHDL design entity to convert a 64-bit data signal with periodic timing (eight-cycle period) into a series of eight-bit signals with periodic timing (one-cycle period). You must store the input data, since it can change to unknown values when it is not considered to be valid.

22.9 *Serialization, II.* Repeat Exercise 22.8, but instead assume that the 64-bit input uses ready–valid flow control. The output interface also uses a ready–valid protocol, but at a frame granularity. When the output signals ready and the input is valid, the input sends all eight eight-bit packets in eight consecutive cycles. You are not allowed to have a purely combinational path from the upstream to the downstream interface (nor from the downstream to upstream interfaces). What is the maximum utilization of the output?

22.10 *Frame and cycle-level flow control.* Design a VHDL design entity that receives a serialized signal over eight cycles using frame-level flow control and acts as a sender for a serialized signal over eight cycles using cycle-level flow control.

22.11 *Isochronous timing and predictability.* Describe three different situations that can insert unpredictability into a system with an isochronous timed output. For each reason, explain how you could bound the worst case error.

22.12 *Timing tables, I.* Draw a timing table for both of the serialized interfaces described in Exercises 22.8 and 22.9.

22.13 *Timing tables, II.* Convert the waveform of Figure 17.5 into a timing table. Be sure to include all relevant signals.

22.14 *Timing tables, III.* Draw a timing table for the Pong game described in Section 22.6.1. You should start at cycle i with the ball in the middle of the screen and moving directly left at a rate of one pixel every 20 cycles. When the ball reaches the end of the grid, it should register a score (assume the left paddle missed it). Finally, your timing table should include the service of the ball and the beginning of its movement rightward. Include all relevant state and control signals.

22.15 *DES, faster speculation.* A student observes that if a register is provided to hold the old key, the new key can be produced during cycle 1 in Table 22.4. Any subsequent start of a DES decryption (indicated by asserting *start_DES*) can select between the two keys using a multiplexer. Redraw the timing table of Table 22.4 assuming this facility is available. What is the interval between keys if the first plaintext block fails the plaintext check with this change?

22.16 *DES, speculate on failure.* The speculation shown in Table 22.4 speculates on success. That is, it assumes the plaintext check of the current block will succeed, and thus proceeds with anticipatory timing to decrypt the next block with the same key. We can speculate in the other direction as well. We can guess that the current block will fail the plaintext check and assert *new_key* and *first_block* in cycle 15 of Table 22.4. If our speculation is correct, we proceed with the new key with no delay. If our speculation is incorrect, we need to restore the appropriate state.

(a) What state needs to be saved to allow us to recover from incorrect speculation in this case?

(b) Adding a column for RS (*restore state*) to Table 22.4, redraw this table to speculate on failure. Include the same two cases where key 0/block 0 fails and then key 1/block 0 passes.

(c) Suppose the first plaintext block fails the plaintext check 95% of the time. Is it faster overall to speculate on failure or to speculate on success?

22.17 *DES with two-way flow control.* Suppose we design an improved plaintext checker that performs a two-step check of a decrypted block to see whether it is valid plaintext. Step one takes one cycle and rejects a block 95% of the time. However, if a block passes the first step, a second step taking six additional cycles is employed, which rejects 90% of the remaining blocks. Explain how to interface this improved module to the DES system of Figure 21.3 and how it affects system timing. Draw a timing table showing the timing of the revised system. In your table include examples of both a fast reject and a slow reject by the new module.

22.18 *DES, idle resources.* In Table 22.3, our key generator is idle for 127 of 128 cycles. In order to utilize this resource fully, describe how you can instantiate multiple DES decoders to run in parallel.

(a) Draw the new timing table with 128 different DES decoders.

(b) Repeat this exercise with the implementation shown in Table 22.4. How many parallel DES decoders does the system need?

23 Pipelines

A pipeline is a sequence of modules, called *stages*, that each perform part of an overall task. Each stage is like a station along an assembly line – it performs a part of the overall assembly and passes the partial result to the next stage. By passing the incomplete task down the pipeline, each stage is able to start work on a new task before waiting for the overall task to be completed. Thus, a pipeline may be able to perform more tasks per unit time (i.e., it has greater *throughput*) than a single module that performs the whole task from start to finish.

The throughput, or work done per unit time, of a pipeline is that of the worst stage. Designers must load balance pipelines to avoid idling and wasting resources. Stages that take a variable latency can stall all upstream stages to prevent data propagating down the pipeline. Queues can be used to make a pipeline elastic and tolerate latency variance better than global stall signals.

23.1 BASIC PIPELINING

Suppose you have a factory that assembles toy cars. Assembling each car takes four steps. In step 1 the body is shaped from a block of wood. In step 2 the body is painted. In step 3 the wheels are attached. Finally, in step 4, the car is placed in a box. Suppose each of the four steps takes 5 minutes. With one employee, your factory can assemble one toy car every 20 minutes. With four employees your factory can assemble one car every 5 minutes in one of two ways. You could have each employee perform all four steps – producing a toy car every 20 minutes. Alternatively, you could arrange your employees in an assembly line with each employee performing one step and passing partially completed cars down to the next employee.

In a digital system, a pipeline is like an assembly line. We take an overall task (like building a toy car) and break it into subtasks (each of the four steps). We then have a separate unit, called a *pipeline stage* (like each employee along the assembly line), performing each task. The stages are tied together in a linear manner so that the output of each unit is the input of the next unit – like the employees along the assembly line passing their output (partially assembled cars) to the next employee down the line.

The *throughput*, Θ, of a module is the number of problems a module can solve (or tasks a module can perform) per unit time. For example, if we have an adder that is able to perform one add operation every 10 ns, we say that the throughput of the adder is 100 Mops (million operations per second). The latency, T, of a module is the amount of time it takes the module to complete one task from beginning to end. For example, if our adder takes 10 ns to complete a problem from the time the inputs are applied to the time the output is stable, its latency is

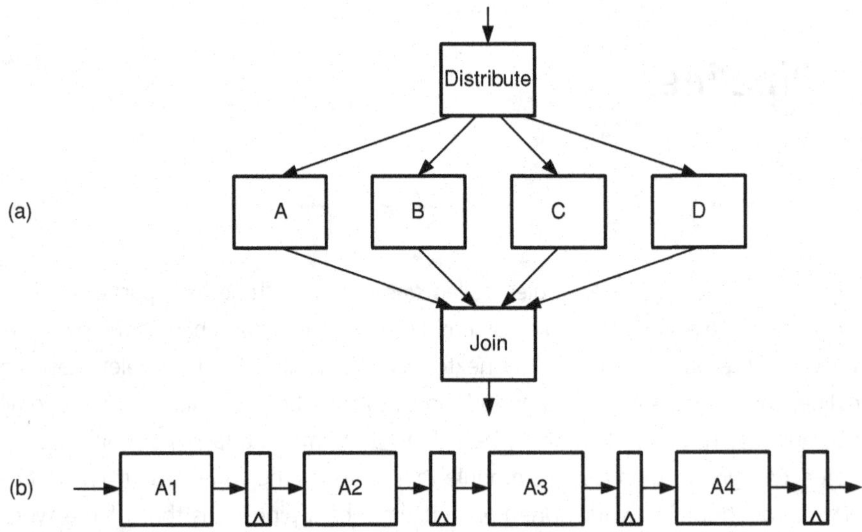

Figure 23.1. Throughput of a module can be increased by (a) using parallel copies of the module, or (b) pipelining a single copy of the module.

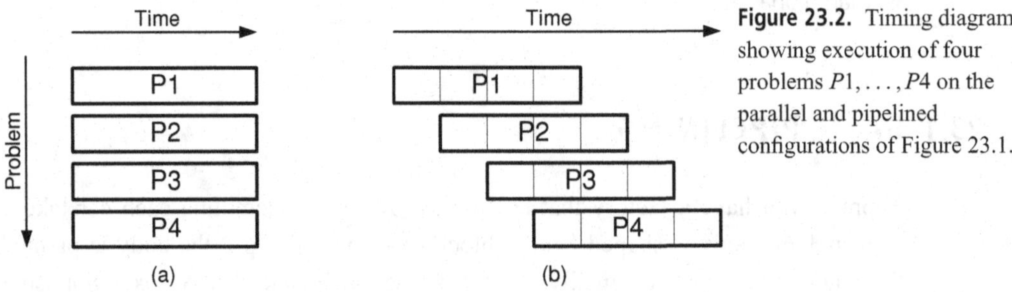

Figure 23.2. Timing diagram showing execution of four problems $P1, \ldots, P4$ on the parallel and pipelined configurations of Figure 23.1.

10 ns. Latency is just another word for delay. For a simple module, throughput and latency are reciprocals of one another: $\Theta = 1/T$. If we accelerate modules through pipelining or parallelism, however, the relation becomes more complex.

Suppose we need to increase the throughput of a module with $T = 10$ ns, $\Theta = 100$ Mops, by a factor of 4. Furthermore, suppose the module is already highly optimized so that we are unlikely to get much increase in throughput by redesigning the module. Just as with our toy car factory, we have two basic options. First, we could build four copies of our module, as shown in Figure 23.1(a). Modules *A–D* are identical copies of our original module. The distribute block distributes problems to the four modules, and the join block combines the results. Here we can start four copies of our problem in parallel as shown in Figure 23.2(a). Our latency is still $T = 10$ ns because it still takes 10 ns to complete one problem from beginning to end. Our throughput, however, has been increased to $\Theta = 400$ Mops since we are able to solve four problems every 10 ns.

An alternative method of increasing throughput is to *pipeline* a single copy of the module, as shown in Figure 23.1(b). Here we have taken a single module A and divided it into four parts $A1, \ldots, A4$. We assume that we were able to do this partition evenly so that the delay of each of the four submodules Ai is $T_{Ai} = 2.5$ ns. (Such an even division of modules is not always possible, as discussed in Section 23.6.) *Pipeline registers* between stages hold the result of the preceding submodule for one problem, freeing that submodule to begin working on the next problem. Thus, as shown in Figure 23.2(b), this pipeline can operate on four problems at once in a staggered fashion. As soon as submodule $A1$ finishes work on problem $P1$, it starts working on $P2$, while $A2$ continues work on $P1$. Each problem continues down the pipeline, advancing one stage each clock cycle, until it is completed by module $A4$. If we ignore (for now) register overhead, our latency is still $T = 10$ ns (2.5 ns $\times$ 4 stages) and our throughput has been increased to 400 Mops. The system completes a problem every 2.5 ns.

Compared with using parallel modules, pipelining has the advantage that it multiplies throughput without the cost of duplicating modules. Pipelining, however, is not without its own costs. First, pipelining requires inserting registers between pipeline stages. In some cases, these registers can be very expensive. Also, a pipelined implementation has more register delay overhead than a corresponding parallel design.

As an example of register overhead, suppose each register has a total overhead $t_{reg} = t_s + t_{dCQ} + t_k = 200$ ps. For a single module or a parallel combination (Figure 23.1(a)), we pay this overhead once only, increasing our latency to $T = 10.2$ ns and reducing the throughput of four parallel modules to $\Theta = 4/10.2 = 392$ Mops. Pipelining the module, on the other hand, incurs the register overhead once per stage. Thus the latency is increased to $T = 10.8$ ns and the throughput is reduced to $\Theta = 1/2.7 = 370$ Mops.

In practice, many systems use parallel pipelined modules. Designers often pipeline a module until the register overhead becomes expensive, obtaining further throughput increases by using multiple copies of the pipelines working in parallel. In general, if we are replicating a module with latency T_m, throughput Θ_m, and area a_m p times, this gives

$$T = T_m + t_{reg}, \tag{23.1}$$

$$\Theta = \frac{p}{T}, \tag{23.2}$$

$$a = p(a_m + a_{reg}). \tag{23.3}$$

The same module pipelined into n stages yields

$$T = T_m + nt_{reg}, \tag{23.4}$$

$$\Theta = \frac{1}{T_m/n + t_{reg}}, \tag{23.5}$$

$$a = a_m + na_{reg}. \tag{23.6}$$

EXAMPLE 23.1 Simple pipelining

Calculate the throughput, latency, and clock period of a module that has a 10 ns delay per unit of work. Assume flip-flops at the input and output with $t_{reg} = 100$ ps. Repeat this for a 10$\times$ replication and a ten-stage pipeline.

The baseline module has

$$T = T_m + t_{\text{reg}} = 10.1\,\text{ns},$$

$$\Theta = \frac{1}{T} = 99\,\text{Mops},$$

$$T_{\text{clk}} = 10.1\,\text{ns}.$$

Replicating the module 10× (Equations (23.1) and (23.2)) yields

$$T = T_m + t_{\text{reg}} = 10.1\,\text{ns},$$

$$\Theta = \frac{p}{T} = 990\,\text{Mops},$$

$$T_{\text{clk}} = 10.1\,\text{ns}.$$

Pipelining the module into ten stages (Equations (23.4) and (23.5)) yields

$$T = T_m + nt_{\text{reg}} = 11\,\text{ns},$$

$$\Theta = \frac{1}{T_m/n} + t_{\text{reg}} = 909\,\text{Mops},$$

$$T_{\text{clk}} = \frac{T_m}{n} + t_{\text{reg}} = 1.1\,\text{ns}.$$

23.2 EXAMPLE PIPELINES

Figure 23.3 shows three examples of how pipelines are applied to digital systems.

A graphics pipeline (Figure 23.3(a)) *renders* a sequence of triangles from a scene, compositing the rendered *fragments* into a frame buffer, from which they are displayed. The triangles go through a series of transformations to filter out unseen objects and light others. Triangles are then broken into fragments, shaded, and *composited* into the frame buffer – depending on their depth and transparency. Graphics pipelines must have high throughput to render complex scenes with complex shaders at 60 frames per second (16 ms per frame) or faster.

The shading stage of the pipeline uses parallelism to hide the latency of texture accesses. For each fragment, one or more accesses are made into the texture cache. Often, this access is a hit and completes quickly. A miss, however, takes a long time to come back from memory. In order to avoid paying this penalty for every fragment, GPUs launch multiple requests at the same time. The multiple requests *hide* the latency of the texture access by providing other useful work to do whilst waiting. As each texture returns from the memory subsystem, it is matched to its requesting fragment and proceeds through the pipeline. The matching must be done in a manner that maintains the original ordering of the fragments because compositing must take place in order to obtain the correct final image. By parallelizing texture retrieval, the overall throughput of the GPU is increased.

Pipelining is also applied to increase the throughput of modern processors. In a simple five-stage pipeline, shown in Figure 23.3(b), an instruction is *fetched* from memory in the first stage.

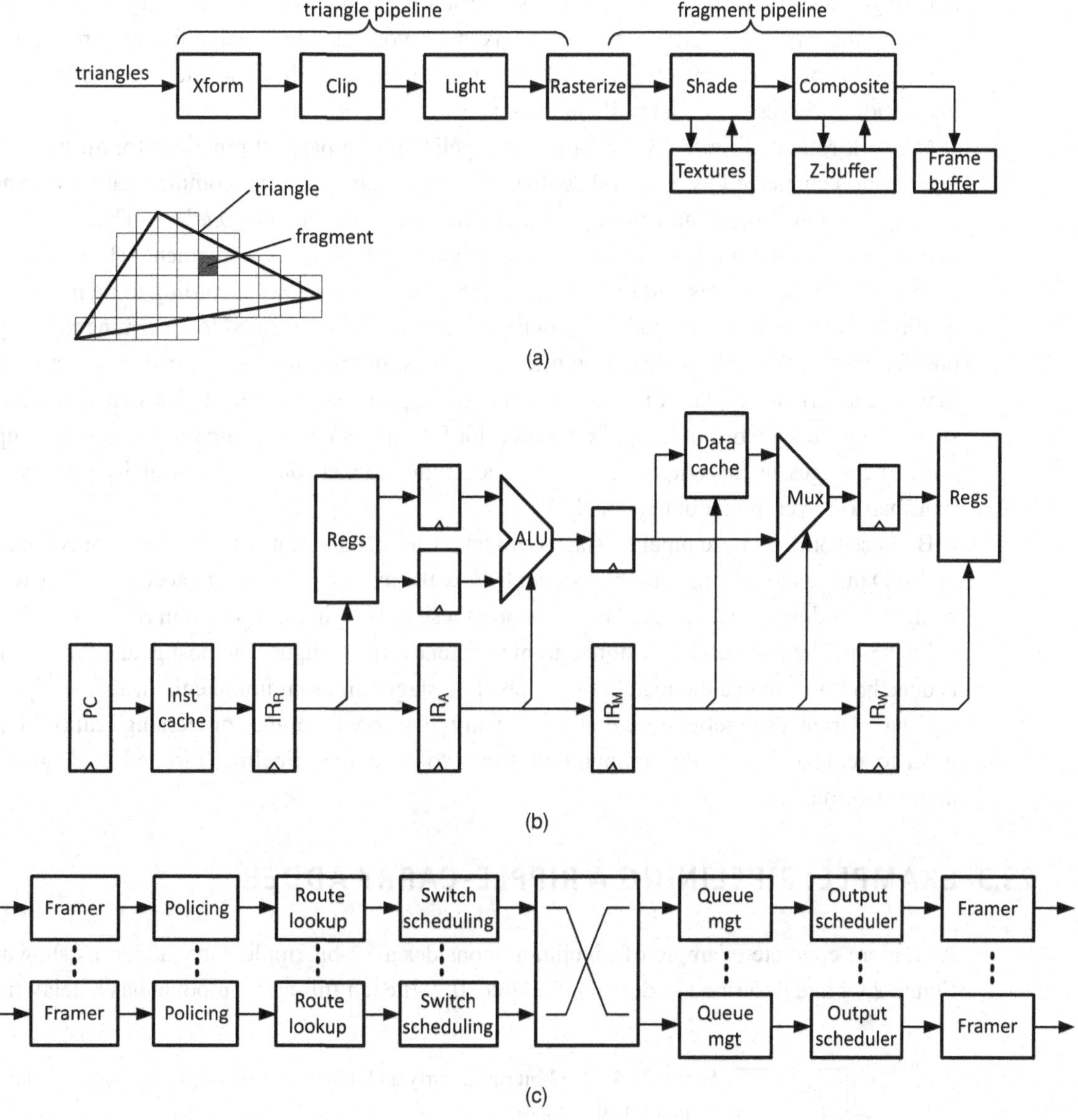

Figure 23.3. (a) Graphics pipeline; (b) processor pipeline; (c) packet processing pipeline.

Next, the registers are read in the second stage and operands are passed into the third stage, which performs an arithmetic operation. If needed, memory is accessed in the fourth stage. Finally, the result of the instruction is written back into the register file.

In a CPU pipeline, both the instruction fetch stage and memory stage can cause accesses to a shared level-two cache (Section 25.4). If such a conflict occurs, the memory access stage will typically win arbitration.[1] Giving priority to the *downstream* stage avoids a potential deadlock situation. The register file in a CPU pipeline is shared between the register read and writeback stages. However, each stage has exclusive access to one or more register *ports*. The register

[1] The instruction fetch stage needs to wait until the memory stage has been completed before proceeding.

read stage accesses the two read ports of the register file while the writeback stage has exclusive access to the write port. Care must still be taken to avoid reading a value from the register file before it has been written, since an earlier instruction reaches the writeback stage after a later instruction accesses the register file in the read stage.

Network routers, Figure 23.3(c), are also pipelined with parallel pipelines for all input and output ports connected by a central switch. Bits coming off of serial communication channels are aligned and formed into words by a *framer* and grouped into *packets*. The next block *polices* packets, making sure they are in compliance with their use agreement. This stage may drop misbehaving packets and may assign priority to other packets according to their assigned quality of service. In the next stage a routing calculation is performed to determine the output port for each packet. This calculation often involves a *prefix search* in a *trie* data structure and may take a variable amount of time. Once an output port has been selected, a *switch scheduler* stage assigns each word of the packet a timeslot for traversing the switch to the desired output port. The stages in the output pipeline schedule the packet, on the basis of its priority, for transmission over the output channel.

Because some of these pipeline stages can take variable amounts of time, they are connected by FIFO buffers to average out the load and allow the input to continue to accept packets while waiting on a downstream stage. We will study these buffers in more detail in Section 23.7.

Each pipeline is carefully designed to meet latency, throughput, and cost goals. Partitioning is done both to balance the load between pipeline stages and to minimize the *width* of pipeline registers. Graphics pipelines are replicated many times on a graphics processing unit (GPU) in order to achieve desired throughput. CPUs tend to have long pipelines (around 20 stages) for high performance.

23.3 EXAMPLE: PIPELINING A RIPPLE-CARRY ADDER

As a more concrete example of pipelining, consider a 32-bit ripple-carry adder, as shown in Figure 23.4 and described in detail in Section 10.2. If each full-adder module has a delay from

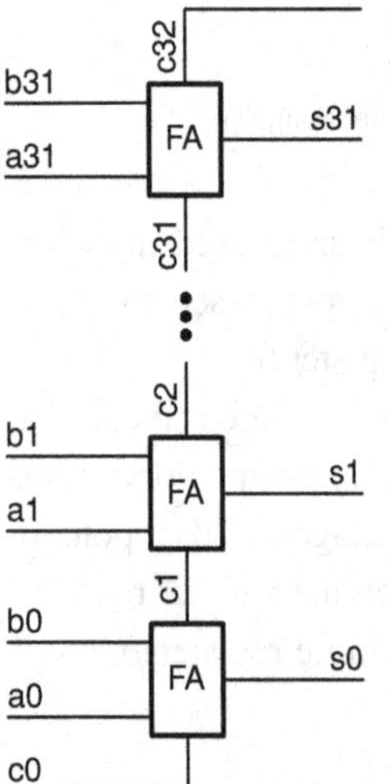

Figure 23.4. A 32-bit ripple-carry adder. If each stage requires 100 ps, an add can be performed in 3.2 ns.

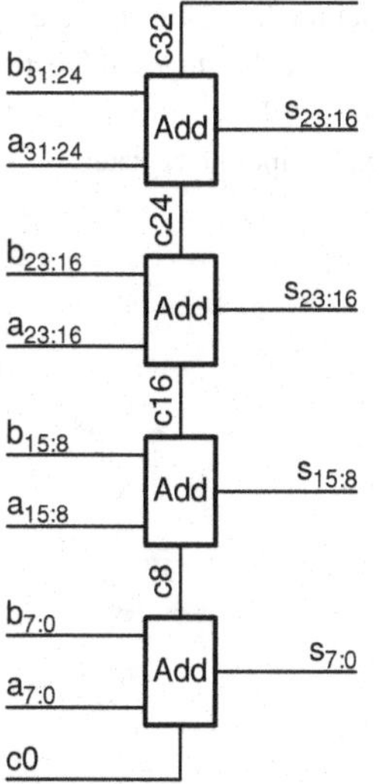

Figure 23.5. Dividing the 32-bit ripple-carry adder into four eight-bit adders.

carry-in to carry-out of $t_{dcc} = 100$ ps, then in the worst case (where the carry must propagate all the way from the LSB to the MSB), the delay of the adder is $t_{dadd} = 32t_{dcc} = 3.2$ ns. This single adder is capable of performing one add every 3.2 ns for a throughput of 312.5 million adds per second. Suppose our goal is to achieve a throughput of one billion adds per second – starting a new add every 1 ns. We can pipeline this adder to achieve this goal.

In operation, only a single bit of this adder is *busy* at any point in time. For example, 1 ns after the inputs are applied only bit 10 is active. Bits 0–9 have completed operation, and bits 11–31 are waiting on their carry inputs. (They have computed p and g, but cannot compute s until the carry input is available.) By pipelining the adder n ways, we can get n of the 32 bits working at once.

The first step in pipelining a unit is to divide it into submodules. Figure 23.5 shows our 32-bit adder module divided into four submodules, each of which performs an eight-bit add. Each eight-bit adder accepts an eight-bit slice of each input vector, $a_{i+7:i}$ and $b_{i+7:i}$, along with a carry-in ci and generates an eight-bit slice of the sum $s_{i+7:i}$ and a carry-out c_{i+8}. The delay (from carry-in to carry-out) of each of these eight-bit adders is $t_{d8} = 8t_{dcc} = 800$ ps, so it will fit in a 1 ns clock cycle, leaving 200 ps for register overhead.

To allow our partitioned adder to work on multiple problems at the same time, we insert pipeline registers between the submodules. The thicker lines in Figure 23.6 show where the registers are to be inserted. The circuit, redrawn after inserting the registers, is shown in Figure 23.7. The combination of a submodule and its register is referred to as a *pipeline stage*. Pipeline register $R1$ is inserted after the first eight-bit adder. This register captures the partial result of the add after 800 ps – this includes the low eight bits of the sum $sR0_{7:0}$, bit 8 of the carry $c8R0$, and the upper three bytes of the input vectors $aR0_{31:8}$ and $bR0_{31:8}$, for a total of 57 bits.

We label all signals before $R1$ (including the primary inputs) with the suffix $R0$ to denote that they are in the zeroth pipeline stage and to distinguish these signals from signals in other pipeline stages. In a similar manner, the outputs of $R1$, and all other signals in the first pipeline stage, are labeled with a suffix $R1$. Note that signals $sR1_{7:0}$ and $sR2_{7:0}$ are different signals. While they are both the low byte of a sum, at any point in time they are the low bytes of

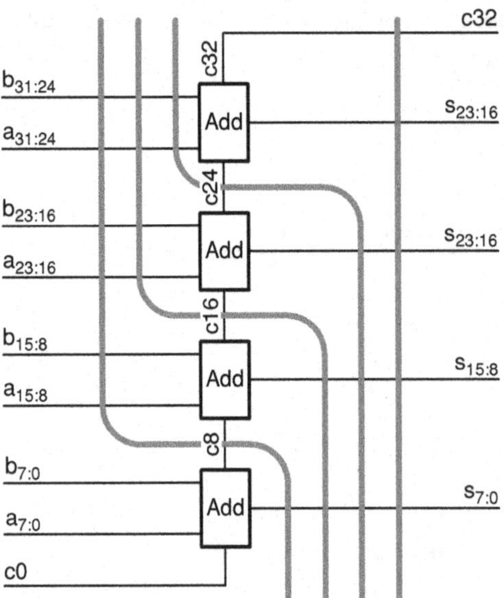

Figure 23.6. To pipeline the 32-bit ripple-carry adder, we insert registers to separate the stages at locations shown by the thick, gray lines. All paths from input to output must pass through each pipeline register.

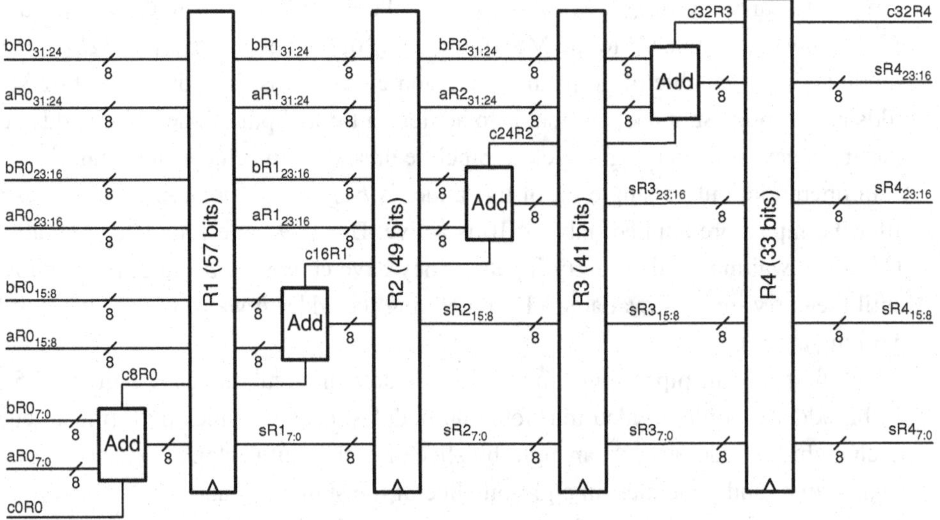

Figure 23.7. Pipelined 32-bit ripple-carry adder. Registers have been inserted and the schematic redrawn to make each pipeline stage one vertical region. Signal names are augmented with the name of the pipeline register producing them.

different sums. By labeling signals with their pipeline stage, we can easily spot the common pipeline error of combining signals from different stages. We know that an expression *fooR1 & barR2* is an error because signals from different pipeline stages should not be combined.[2]

The second stage of the pipeline adds the second byte of data $aR1_{15:8}$ and $bR1_{15:8}$, using $c8R1$ as the carry-in, giving $sR1_{15:8}$ and $c16R1$. The outputs of this second byte adder are captured by $R2$, along with $sR1_{7:0}$, $aR1_{31:16}$, and $bR1_{31:16}$, a total of 49 bits. In a similar manner, the third and fourth pipeline stages add the third and fourth bytes of data. At the output of the fourth

[2] As with all rules, there are exceptions, two examples of which are stall signals (Section 23.4) and result forwarding (Exercise 23.10).

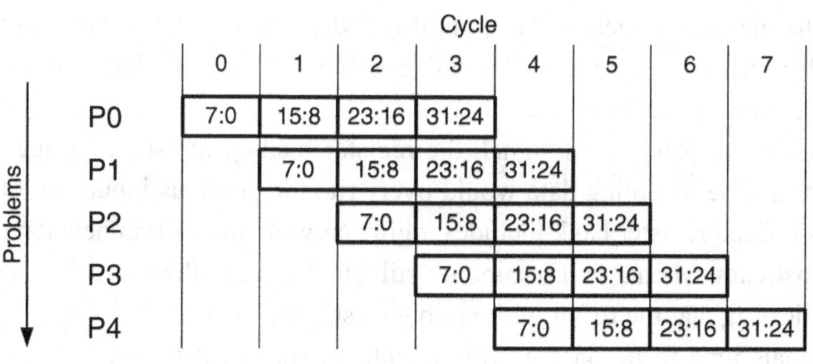

Figure 23.8. Pipeline diagram showing the timing of the pipelined 32-bit ripple-carry adder. The diagram illustrates on which cycle each sum byte of problems $P0, \ldots, P4$ is computed.

stage, register $R4$ captures the full 32-bit sum $sR3_{31:0}$ and the carry-out $c32R3$. The output of $R4$ is the result of the add $s_{31:0}$ and $c32$.

The timing of the pipeline is illustrated in the *pipeline diagram* of Figure 23.8. The figure shows five problems, $P0, \ldots, P4$, proceeding down the pipeline. Problem Pi enters the pipeline on cycle i. Byte j (bits $8j+7:8j$) of the sum of Pi is computed during cycle $i+j$. The result of the add for problem Pi appears on the output four clock cycles later, on cycle $i+4$. At the time at which the result for a problem $P0$ appears on the output during cycle 4, four subsequent problems are in process in the various pipeline stages.

Pipelining our ripple-carry adder has increased both its throughput and its latency. If we assume that the register overhead $t_{\text{reg}} = t_s + t_{\text{dCQ}} + t_k$ is 200 ps for each register, then the delay of each pipeline stage is 1 ns (800 ps for the eight-bit adder and 200 ps for register overhead). Hence we can operate our pipeline at 1 GHz, achieving our throughput goal of $\Theta = 1$ Gops. The latency of our pipeline is $T = 4$ ns, compared with the original 3.2 ns for the unpipelined adder. The difference is the overhead of the four registers.

In general, if the delay of the combinational logic in the longest pipeline stage is $t_{\max}$, then the total delay of a pipeline stage is given by

$$t_{\text{pipe}} = t_{\max} + t_{\text{reg}} = t_{\max} + t_s + t_{\text{dCQ}} + t_k. \tag{23.7}$$

From t_{pipe} we see that the latency of n stages is given by

$$T = n\left(t_{\max} + t_{\text{reg}}\right) = n\left(t_{\max} + t_s + t_{\text{dCQ}} + t_k\right), \tag{23.8}$$

and the throughput is given by

$$\Theta = \frac{1}{t_{\max} + t_{\text{reg}}} = \frac{1}{t_{\max} + t_s + t_{\text{dCQ}} + t_k}. \tag{23.9}$$

23.4 PIPELINE STALLS

In some situations, a pipeline stage may not be able to complete its work in the allotted amount of time. For example, in the toy factory, the wheel assembler at the toy plant may drop a wheel. In a processor pipeline, some of the inputs to a computation may not be ready. In either case, the offending pipeline stage must notify all upstream stages of its failure to complete the work. This is done via a big red button in a factory and a stall signal in a digital system. Asserting such a signal halts upstream pipeline stages until it is deasserted. Alternatively, we can use a *ready* signal and set it to 0 (not ready) on a stall condition. This is similar to the ready–valid flow-control mechanism of Section 22.1.3.

In a digital system without a stall signal, the register feeding the stalled stage would still clock in new inputs. The incoming data would overwrite the previous inputs, and the stalling task would be lost. Each register bank includes logic that will not clock in new data if *ready* is deasserted. Downstream, when a stall occurs, we still clock the flip-flops, but deassert a control signal (*valid*) indicating that the result of the previous stage is invalid.

These stall signals tend to be on the critical path, as shown in Figure 23.9(a). Often the assertion of a stall signal does not occur until late in the pipeline stage. Once computed, these signals must propagate up the length of the pipeline (wire delay) and combine with other stall signals (logic delay). The *valid* signals are not on the critical path, since they are fed into the register only at the end of a stage.

The logic at each pipeline register to handle a stall, Figure 23.9(b), is relatively simple. If the *ready* signal is set to 0, the register re-clocks its old value. If the downstream stages are all ready, then the next task becomes input to the register.

Figure 23.9(c) shows an example of a four-stage pipeline where a stall occurs during cycle 2 at stage 3. The *ready* signal will be set to 0 and values from computation B and C will not

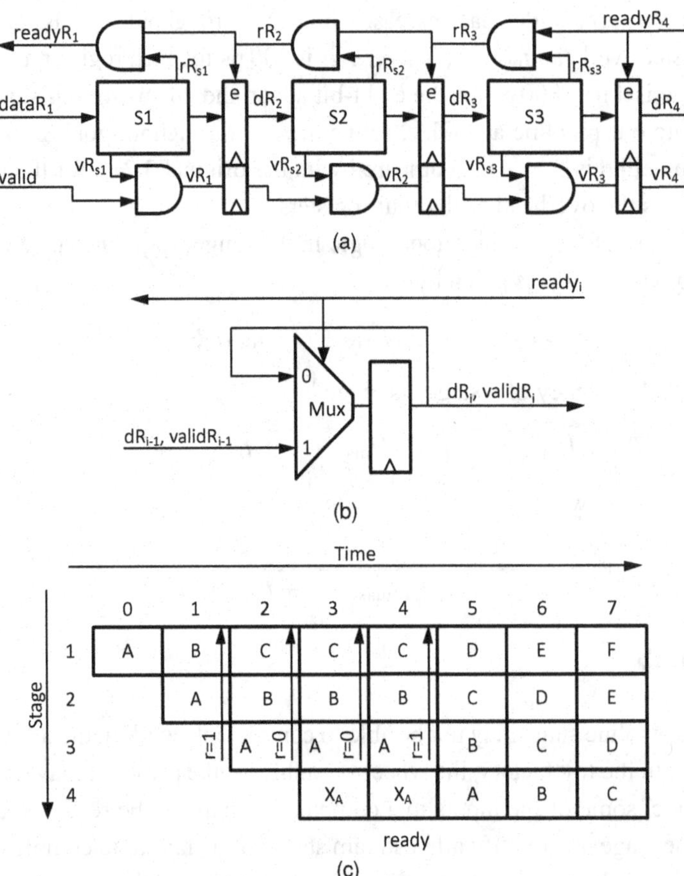

Figure 23.9. Three-stage pipeline with (a) stall signals, (b) stall logic, and (c) sample execution. If a stage has a stalling condition, it will deassert its ready signal (rR_i). This combines with all other ready signals upstream to prevent pipeline registers from clocking in new data. When stage 3 is not ready in cycle 2, all upstream pipeline stages immediately stall until stage 3 becomes ready.

```
library ieee;
use ieee.std_logic_1164.all;
use work.ff.all;

entity single_buffer is
  generic( bits: integer := 32 );
  port( rst, clk: in std_logic;
        upstream_data: in std_logic_vector(bits-1 downto 0);
        downstream_ready, upstream_valid: in std_logic;
        downstream_data: buffer std_logic_vector(bits-1 downto 0);
        upstream_ready, downstream_valid: buffer std_logic );
end single_buffer;

architecture impl of single_buffer is
  signal stall, valid_nxt: std_logic;
  signal data_nxt: std_logic_vector(bits-1 downto 0);
begin
   upstream_ready <= downstream_ready;
   stall <= not upstream_ready;

   data_nxt <= (others => '0') when rst else
               downstream_data when stall else
               upstream_data;

   valid_nxt <= '0' when rst else
                downstream_valid when stall else
                upstream_valid;

   dataR: vDFF generic map(bits) port map(clk, data_nxt, downstream_data);
   validR: sDFF port map(clk, valid_nxt, downstream_valid);
end impl; -- single_buffer
```

Figure 23.10. The VHDL for a single buffer. The downstream ready signal is propagated upstream via combinational logic. If the downstream stages are not ready, the register does not clock in a new value.

propagate. The output of stage 2 will advance to the next stage, but the *valid* bit will be set to 0. We note this in the diagram as X_A.

The VHDL for a stalling pipeline buffer is shown in Figure 23.10. Its inputs are the `upstream_data`, `upstream_valid`, and `downstream_ready` signals. The `ready` signal is immediately propagated upstream while the other two inputs are clocked into registers when no stall condition exists. If the pipeline does need to stall, the registers will hold their outputs.

23.5 DOUBLE BUFFERING

The problem with a global stall signal is one of delay. In many systems it is infeasible to broadcast the stall condition in one cycle. Instead, we must clock the ready signal into a flip-flop at every stage and propagate it upstream one cycle at a time. To do so without losing data, we must double buffer, holding in the register both the old data and the new data arriving until the upstream stage can be stopped.

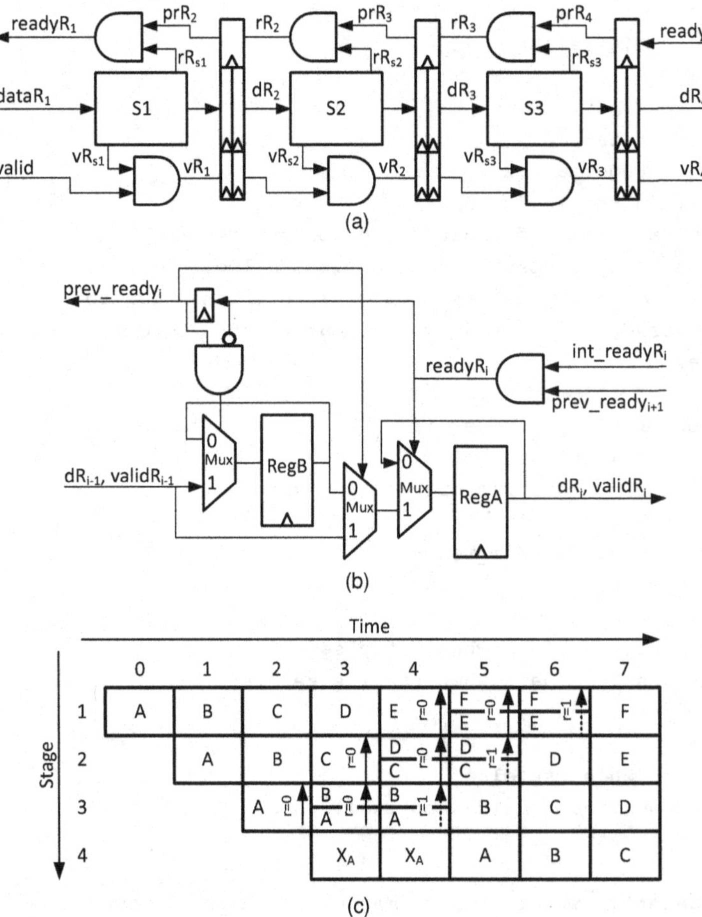

Figure 23.11. (a) Three-stage pipeline with double buffering; (b) the buffer logic; (c) sample execution. In a double-buffered pipeline, each stall signal propagates back only to the beginning of the stage. This will reduce the latency of a system by removing the stall from the critical path. Stalling causes a second register to be written, propagating the change in ready status (r) upstream one stage at a time.

Figure 23.11(a) shows an example of a double-buffered pipeline. Between every stage are two registers, Figure 23.11(b). When a pipeline stage i becomes not *ready*, it sets the internal ready signal to 0. This causes the values coming from stage $i-1$ to be buffered into the second of two registers (stage $i-1$ has not received the not ready signal). On the next cycle, stage $i-1$ input $prev\_ready_i$ is set to 0, causing $ready_{i-1}$ to be zero.

The logic for controlling the two buffers is shown in Figure 23.11. When stage i has been ready for more than two cycles, both $readyR\_i$ and $prev\_ready_i$ are 1. This causes dR_{i-1} to be written into *RegA* every cycle. In the first cycle during which either int_readyR_i or $prev\_ready_{i+1}$ is asserted, *RegA* does not clock in new data, and *RegB* buffers data from the upstream stage. In the subsequent cycles until stage i becomes ready, both *RegA* and *RegB* hold their values. In the first cycle after the ready signal has again been asserted, the contents of register B will be promoted to register A.

In the execution example of Figure 23.11(c), stage 3 causes a stall at cycle 2. The registers between stages 0–1 and stages 1–2 have not received this signal and clock in data D and C as normal. To avoid losing datum B, it is buffered into the extra register between stages 2–3. On the next cycle, the not ready signal is available at stage 2, double buffering D (since E is clocked into the register at the end of stage 1). When the ready signal is reasserted, it too will

```
library ieee;
use ieee.std_logic_1164.all;
use work.ff.all;

entity double_buffer is
  generic( bits: integer := 32 );
  port( rst, clk: in std_logic;
        upstream_data: in std_logic_vector(bits-1 downto 0);
        downstream_ready, upstream_valid: in std_logic;
        downstream_data: buffer std_logic_vector(bits-1 downto 0);
        upstream_ready, downstream_valid: buffer std_logic );
end double_buffer;

architecture impl of double_buffer is
   signal data_a: std_logic_vector(bits-1 downto 0);
   signal data_b: std_logic_vector(bits-1 downto 0);
   signal valid_a, valid_b: std_logic;
   signal data_a_nxt, data_b_nxt: std_logic_vector(bits-1 downto 0);
   signal valid_a_nxt, valid_b_nxt, upstream_ready_nxt: std_logic;
begin
   downstream_data <= data_a;
   downstream_valid <= valid_a;

   data_b_nxt <= (others => '0') when rst else
                 upstream_data when upstream_ready and downstream_ready else
                 data_b;

   data_a_nxt <= (others => '0') when rst else
                 data_a when downstream_ready else
                 upstream_data when upstream_ready else
                 data_b;

   valid_b_nxt <= '0' when rst else
                  upstream_valid when upstream_ready and downstream_ready else
                  valid_b;

   valid_a_nxt <= '0' when rst else
                 valid_a when downstream_ready else
                 upstream_valid when upstream_ready else
                 valid_b;

   upstream_ready_nxt <= '1' when rst else downstream_ready;

   dataRa:  vDFF generic map(bits) port map(clk, data_a_nxt, data_a);
   dataRb:  vDFF generic map(bits) port map(clk, data_b_nxt, data_b);
   validRa: sDFF port map(clk, valid_a_nxt, valid_a);
   validRb: sDFF port map(clk, valid_b_nxt, valid_b);
   readyR:  sDFF port map(clk, upstream_ready_nxt, upstream_ready);
end impl; -- double_buffer
```

Figure 23.12. VHDL for a double buffer. The `downstream_ready` signal is propagated upstream after one cycle. If the downstream stages are not ready, the second buffer (`data_b`) clocks in the upstream value. Once the stages have again become ready, `data_b` is shifted into `data_a`.

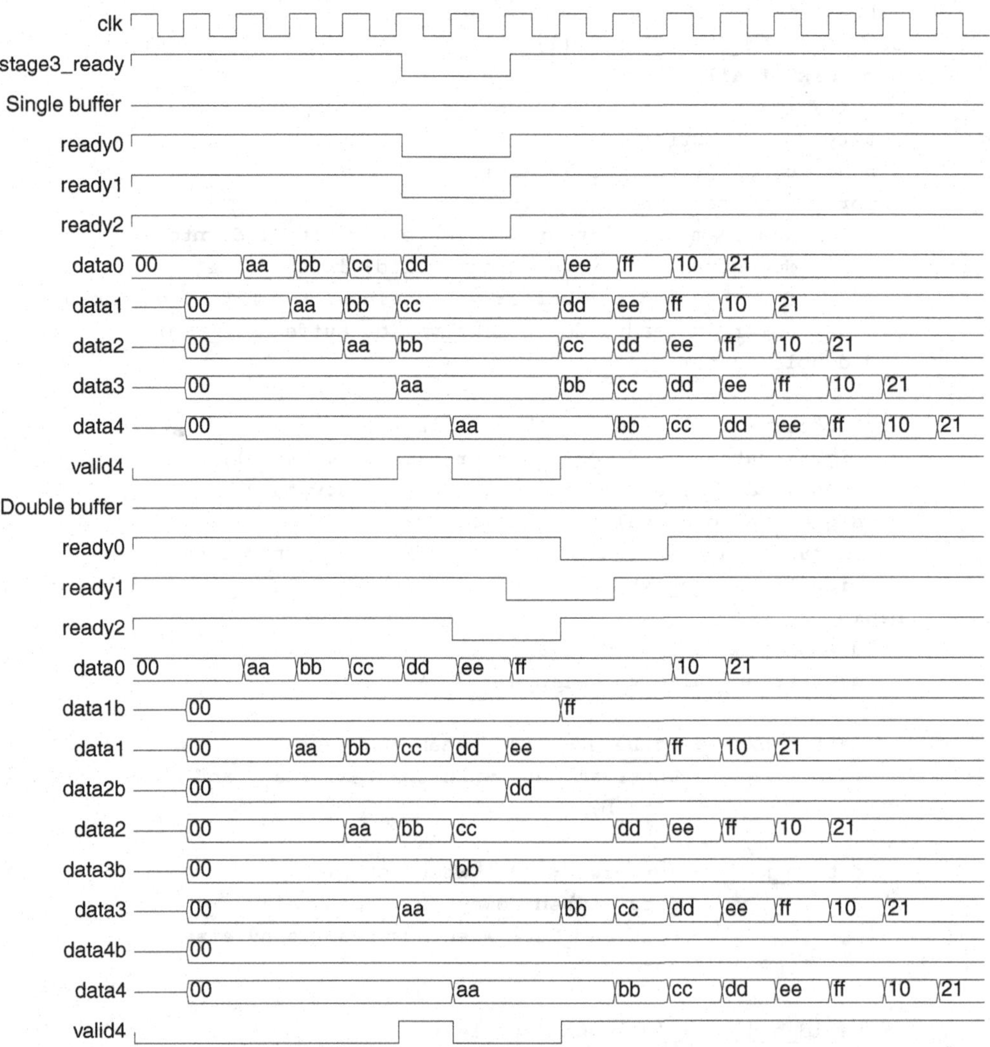

Figure 23.13. Waveform of a single- and double-buffered pipeline with stalls. We set the `stage3_ready` signal to 0 and hold it for two cycles. The single buffer causes a global stall (**ready**$_i$) and no data are clocked into the buffers. With double buffering, the ready signal propagates a cycle at a time. The signal **data**$_i$**b** represents the value stored in the second buffer.

propagate upstream. Stage 3 will execute A and B before C is clocked into its primary register. This type of buffering is also called a *skid buffer* because the data skid to a halt into each of the double buffers.

Double buffering can decrease the latency and increase the throughput of systems. On removing the stall signal from the critical paths, modules run faster. The additional registers impose an area overhead compared with a global stall signal.

We have provided the VHDL code for a synchronous double buffer in Figure 23.12. When the ready signal remains high, the first register (`data_a`) clocks in the incoming values. Unlike the single buffer, we propagate the ready signal through a flip-flop. When the system is not ready, the upstream data are stored in the second register (`data_b`).

A waveform showing the juxtaposition of the two buffer types is displayed in Figure 23.13; `data0–data4` represent the inputs to their respective stages of the pipeline. After some time, we force the stage 3 ready signal to be low. In the single-buffer case (top), this is a globally visible event. All stages stop clocking in new data. With double buffering, we see that the change in ready stages flows upstream one clock at a time. At each stage, the second buffer (`data_ib`) clocks in the upstream data. After the stage has become ready again, this buffered data is shifted into the first buffer.

EXAMPLE 23.2 Double buffering to reduce delay

Compute the cycle time for a ten-stage pipeline with single buffers and with double buffers. Assume that each stage has a 10 ns delay (including $t_{\rm reg}$) and that the stall condition is evaluated 8 ns into the cycle ($t_{\rm rdy}$). It takes 1 ns to propagate a ready signal from one end of a pipeline stage to the other (t_{sd}). This delay must be incurred even by the stage that generated the stall condition.

With single buffering, the clock period is given by

$$T_{\rm clk} = \max(T_{\rm stage}, t_{\rm rdy} + nt_{sd}) = \max(10, 8 + 10 \times 1) = 18\,\text{ns}.$$

With double buffering, ready signals traverse only a single pipeline stage:

$$T_{\rm clk} = \max(T_{\rm stage}, t_{\rm rdy} + t_{sd}) = \max(10, 8 + 1) = 10\,\text{ns}.$$

23.6 LOAD BALANCE

When designing a pipeline, it is important to ensure that each stage has equal throughput because the overall throughput of the system is determined by the slowest stage. The pipeline stage with the lowest throughput is often called the *bottleneck* stage because, like the narrow neck of a bottle, it limits the flow. Faster stages upstream from the bottleneck must idle to prevent overflowing buffers. Faster stages downstream from the bottleneck must idle because they lack input data much of the time.

Figure 23.14(a) shows an unbalanced pipeline. Each stage other than stage 3 has a throughput $\Theta = 1$ Gops. Stage 3, however, has only $\Theta_3 = 250$ Mops. The throughput of the entire system is thus limited to 250 Mops. The clock period is set to 4 ns, while the latency of the entire module is 16 ns.

To increase the throughput of stage 3 to match the rest of the pipeline (1 Gops), we can make four copies of it (Figure 23.14(c)), pipeline it four deep (Figure 23.14(b)), or some combination of the two. Pipelining stage 3 allows each stage to produce a new output every 1 ns. The latency of stage 3 as a whole remains at 4 ns (plus register overhead), but the throughput has increased fourfold. If stage 3 cannot be pipelined efficiently, we can replicate it. Doing so requires that the system clock be set to 4 ns, the latency of stage 3. To achieve full 1 Gops throughput, every other stage must be able to produce four results in a 4 ns cycle. Exercise 23.17 asks you to design a pipeline where stage 3 is replicated and the cycle time is 1 ns.

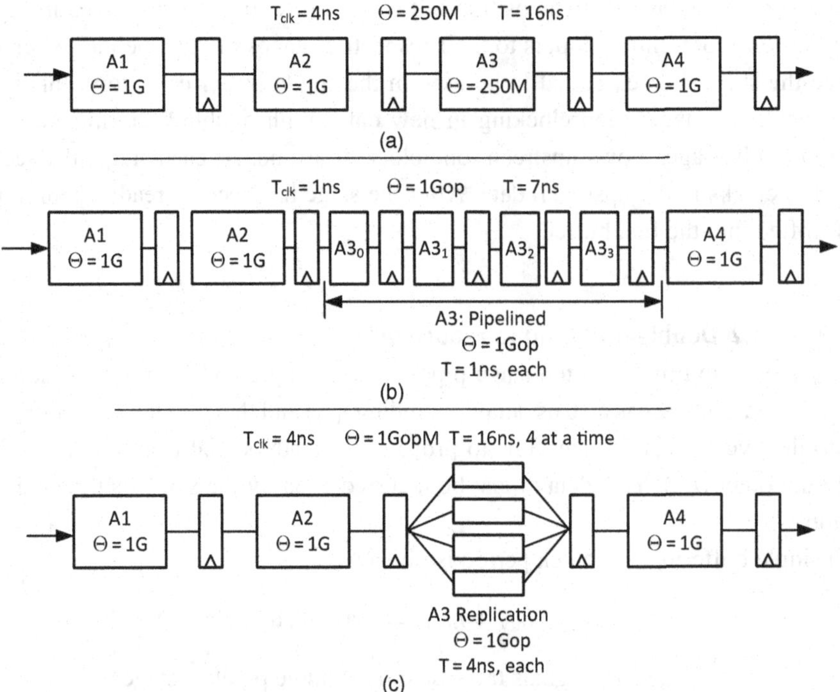

Figure 23.14. Total throughput, latency, and clock times for three implementations of the same pipeline. In (a), stage 3 limits the overall throughput. We pipeline stage 3 into four 1 ns stages in (b), achieving full throughput and low latency. By replicating stage 3 in (c), the full 1 Gops throughput is obtained, but not the fast latency. In (c), all other stages must produce four outputs per 4 ns clock cycle.

23.7 VARIABLE LOADS

The delay of a pipeline stage need not always be constant. For example, the time it takes to decode a frame of a movie depends on the difference between the current and the previous frame. With a rigid pipeline, the upstream stages will be stalled each time a downstream stage takes an excessive amount of time to process one problem instance. However, by inserting first-in-first-out (FIFO, see Section 29.4) buffers between the pipeline stages, variation in throughput can be averaged out – making the full throughput dependent only on the average throughput of each stage.

An upstream pipeline stage inserts results into the FIFO unless it is full. The downstream stage takes problem instances from the head of the queue unless it is empty. If the queue becomes full, the upstream stage stalls. On an empty queue, the downstream stage idles.

In this type of decoupled system, a global clock need not exist. Each stage of the pipeline computes until either the upstream queue is empty or the downstream queue is full. In the steady-state condition, neither of these should occur. To provide full throughput, queues must be deep enough to accommodate all uncertainty between stages. Factors in queue sizing include latency distribution and burstiness.

Figure 23.15 shows the need for queuing with variable latencies. Pipeline stage A has a latency of ten cycles, whereas stage B has a latency of either five or 15 cycles. With

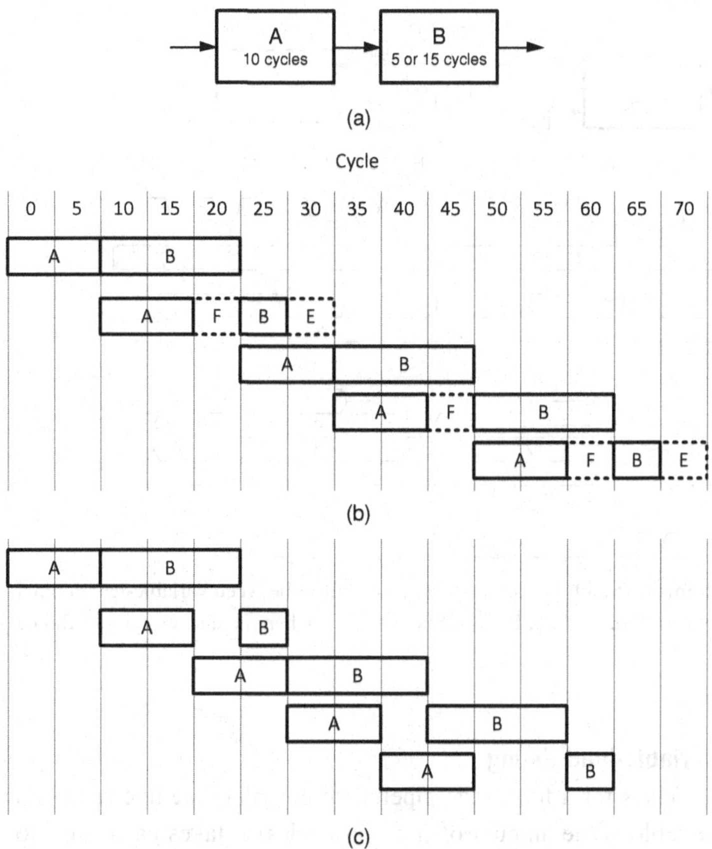

Figure 23.15. Pipelines with variable loads. Pipeline stage B can take either five or 15 cycles. In (b) only a single register sits between A and B. This causes A to stall when A has completed but B has not (F), and B to idle when B has completed but A has not (E). Replacing the register with a FIFO, as in (c), eliminates the stalls.

only a single pipeline latch between the two stages, as in (b), A must stall whenever B has not completed its work on the previous iteration (condition F). B must idle when the single buffer has not yet been written by A (condition E). With a buffer with multiple slots, shown in (c), stage A continuously executes without stalling – it simply queues up problem instances if B is still busy. As long as the queue doesn't become empty, B does not idle. The only time B would idle is if it had a burst of five cycle executions and drained the queue completely.

As a detailed example of decoupling variable pipeline stages, consider the waveform of Figure 23.16. A three-entry FIFO separates the fixed-latency stage A from the variable-latency stage B. The FIFO interfaces with both stages using ready–valid flow control (Section 22.1.3). From the FIFO, signal *A_ready* represents not-full and *B_valid* represents not-empty. When stage A asserts its valid signal and the FIFO is ready, the datum from A is written into the FIFO on the rising clock edge. The datum is written into the first available register, called the *tail*. When the FIFO is full in cycles 11, 12, and 13, stage A must stall. When stage B asserts its ready signal and the FIFO is valid, the FIFO will pop the queue and shift the rest of the data. The FIFO continuously outputs the first valid datum, called the *head* of the queue. When the FIFO is empty in cycles 1, 2, and 4, stage B idles. The FIFO should be sized to limit the number of cycles stage A spends stalled and stage B spends idling.

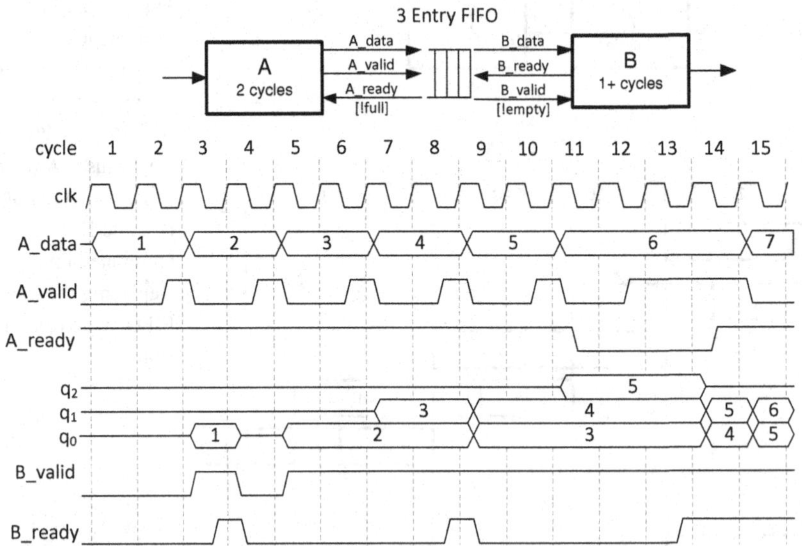

Figure 23.16. Waveforms when a FIFO queue is used as a buffer between variable-length pipeline stages. When the queue is empty (*B_valid* = 0), B idles. Stage A stalls when the queue becomes full (*A_ready* = 0).

EXAMPLE 23.3 Variable-load timing

Compute the time it takes for a four-stage pipeline to complete the five tasks shown in Table 23.1. Shown in the table is the amount of time (in cycles) it takes each stage to complete a given problem. For example, it takes stage 3 ten cycles to complete problem 1 but one cycle to complete problem 2. Complete this example once assuming a global stall signal, and again assuming FIFOs between each stage.

Table 23.1. **Timing for the problem described in Example 23.3: each row represents one unit of work, and the cells are the number of cycles each problem must take in a given pipeline stage**

	S1	S2	S3	S4
P1	1	1	10	1
P2	1	1	1	1
P3	1	1	1	10
P4	1	10	1	1
P5	10	1	1	1

We will draw timing tables for each implementation. Note that, in these tables, time is labeled vertically, not horizontally. Each entry is the problem currently being computed at each stage. First, for a global stall, we have Table 23.2. Next, with FIFOs, we have Table 23.3.

Table 23.2. **Execution timing for the series of problems in Example 23.3: with a global stall signal, execution takes 35 cycles**

Cycle	S1	S2	S3	S4
0	P1			
1	P2	P1		
2	P3	P2	P1	
...	P3	P2	P1	
11	P3	P2	P1	
12	P4	P3	P2	P1
13	P5	P4	P3	P2
...	P5	P4	P3	P2
22	P5	P4	P3	P2
23		P5	P4	P3
...		P5	P4	P3
32		P5	P4	P3
33			P5	P4
34				P5

Table 23.3. **Execution timing for the series of problems in Example 23.3: with FIFOs used to tolerate variable load, execution takes 26 cycles**

Cycle	S1	S2	S3	S4
0	P1			
1	P2	P1		
2	P3	P2	P1	
3	P4	P3	P1	
4	P5	P4	P1	
...	P5	P4	P1	
11	P5	P4	P1	
12	P5	P4	P2	P1

Table 23.3. (*cont.*)

Cycle	S1	S2	S3	S4
13	P5	P4	P3	P2
14		P5	P4	P3
15			P5	P3
16				P3
...				P3
23				P3
24				P4
25				P5

The FIFO-based implementation is better able to handle the variable latency and runs the workload in 26 cycles instead of 35.

23.8 RESOURCE SHARING

An expensive but lightly used resource may be shared between pipeline stages. For example, in Figure 23.17(a), stages A and D of the pipeline access memory. The memory is expensive and is used less than 50% of the time by each stage, making sharing attractive. Figure 23.17(b) shows two pipelines that share a module to compute the cosine of value, a very infrequent operation.

An arbiter (Section 8.5) is used to prevent more than one sharer from accessing the resource at any given time. For each resource cycle, sharers needing the resource assert a *request* line. The arbiter grants the resource to at most one requester. A sharer that loses arbitration must stall until it is granted the resource. To prevent one requester from being starved (repeatedly losing), a fair arbiter, such as a round-robin arbiter (see Exercise 8.13), should be used. Priority arbiters are used when one stage or pipeline is more critical than the other. For example, granting permission to downstream stages will cause fewer stages to stall than granting an upstream stage access and avoids a potential deadlock issue. A variable number of cycles may be spent arbitrating for a shared resource – sometimes enough to warrant placing FIFOs before sharing stages.

Resource sharing is most beneficial in cases where a resource is lightly used or costly to replicate. In a router, for example, the routing logic calculates where a packet must be sent. It is used only once per packet, at the header, but packets may take dozens of cycles to traverse the router. This routing logic may be shared across multiple or even all input ports of the router because of its low utilization.

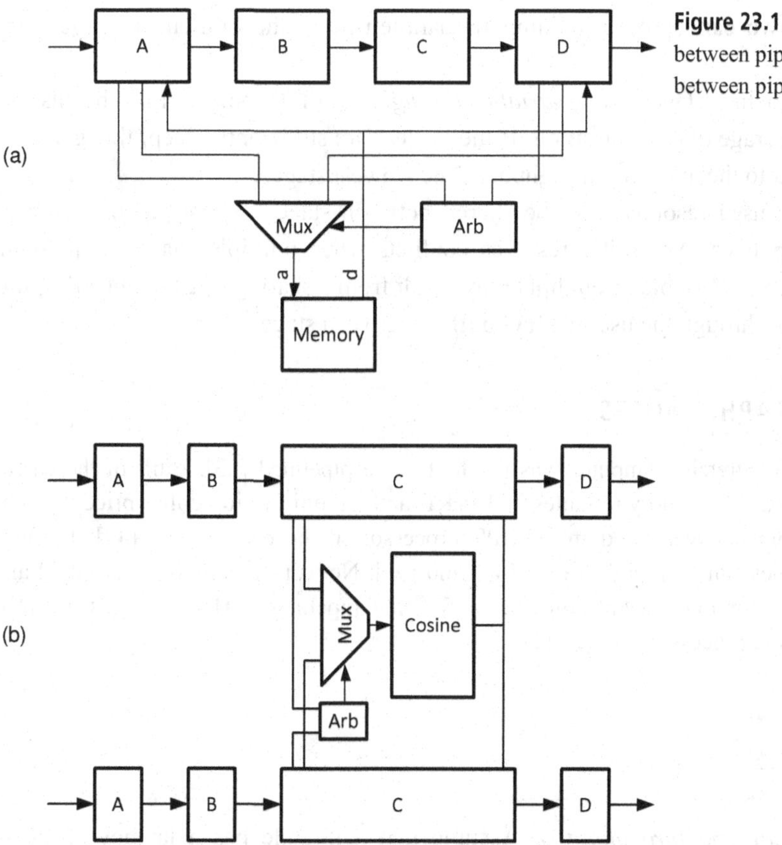

Figure 23.17. Resource sharing between pipeline stages (a) and between pipelines (b).

Summary

In this chapter you have learned how to use pipelining and parallelism to increase the performance of a module of your system. We measure the performance of a module in terms of its *latency* and its *throughput*. Pipelining involves dividing a module into a number of *stages* so that information flows in one direction from one stage to the next. Registers are inserted between the stages so that each stage can work on a separate instance of a problem at the same time.

Either parallelism or pipelining can be used to increase throughput while holding latency roughly constant. Pipelining a module n stages deep to first approximation increases its throughput by a factor of n while increasing latency only slightly (by nt_{reg}) and increasing area (cost) by only the cost of the inserted registers. In contrast, applying n parallel units also increases throughput by a factor of n, but at a cost of an $n\times$ increase in area.

A pipeline stage that cannot complete its task in the allotted time for a stage may *stall* the pipeline, halting all upstream stages until it completes its work. The timing difficulty associated with stalling an entire pipeline can be avoided by *double buffering*, in which the registers between stages are replaced by buffers capable of holding two (or more) entries.

An efficient pipeline must be *balanced*, with each stage having equal throughput, because the throughput of the overall pipeline will be set by the slowest, *bottleneck* stage. To balance a pipeline we can apply pipelining or parallelism to the bottleneck stage to increase its throughput.

When pipeline stages have *variable throughput*, FIFO buffers may be inserted between stages to average out the variation. If the buffers are sufficiently deep, this gives a throughput that is equal to the average throughput of the slowest stage.

A lightly used resource may be shared between stages of one pipeline or between independent pipelines. An arbiter resolves conflicts when multiple sharers request the resource simultaneously. Variable throughput may result from a sharing stage losing arbitration, and can be mitigated through the use of FIFO buffers between stages.

BIBLIOGRAPHIC NOTES

IBM's 7030 (Stretch) computer was the first to be pipelined [23]. Built in the early 1960s, the computer boasted memory modules of 128 KB and a multi-million dollar price tag. The canonical five-stage pipeline was used in a MIPS processor in the early 1980s [47]. Further details on CPU pipelines can be found in refs. [48] and [93]. Network pipelines are detailed and described in ref. [34]. For more details on one of the first pipelines – Henry Ford's assembly line and manufacturing process – read ref. [2].

Exercises

23.1 *Latency and throughput, I.* Assume that a module has a latency of 20 ns and that $t_{reg} = 500$ ps. Including the output register, what are the latency and throughput of this module?

23.2 *Latency and throughput, II.* Assume that a module has a latency of 20 ns and that $t_{reg} = 500$ ps. Including the output register, what are the latency and throughput of this module when it is replicated five times?

23.3 *Latency and throughput, III.* Assume that a module has a latency of 20 ns and that $t_{reg} = 500$ ps. Including the output register, what are the latency and throughput of this module when pipelined into five equal stages?

23.4 *Latency and throughput, IV.* Assume that a module has a latency of 20 ns and that $t_{reg} = 500$ ps. Including the output register, what are the latency and throughput of this module when pipelined into five equal stages and then this pipeline is replicated five times?

23.5 *Latency, throughput, and area, I.* Assume you have a module that consumes 100 units of area and has a delay of 10 ns. The pipeline registers have an area of 2 units, and $t_{reg} = 500$ ps. By varying the number of pipeline stages, make a graph with throughput on the y-axis and area on the x-axis. What can you say about the cost (area) versus benefit (throughput) for very deep pipelines?

23.6 *Latency, throughput, and area, II.* Repeat Exercise 23.5, but instead set the area per pipeline register at 20 units (keeping all other parameters the same). Compare the two plots with each other, commenting on what you think is the "best" pipeline depth for each.

23.7 *Batch latency.* We have shown that pipelining incurs a latency penalty of nt_{reg}. This problem looks at the latency of processing a batch of jobs. Given a module that has a latency of 20 ns and $t_{\text{reg}} = 500$ ps, how long does it take from start to finish to do ten units of work:

(a) with a single unpipelined instance of the module;

(b) if the module is replicated five times;

(c) if the module is divided into five pipeline stages.

(d) Repeat (a), (b), and (c) for 1000 units of work.

23.8 *Latency and throughput design.* You have been given the task of designing a new media chip for HD applications. Specifications call for the rendering of a 1920 × 1080 pixel image at 60 Hz. With a single module, it takes 10 μs to process one pixel, and t_{reg} is 500 ps.

(a) What is the throughput needed, in pixels per second, for your processor?

(b) Your coworker suggests making one long pipeline. Can you do this and still meet your throughput goal? If so, how many stages does it require? If not, why not?

(c) Why could a single pipeline be a bad idea?

(d) Another coworker suggests just replicating the processing module; how many do you need?

(e) Why would this be a bad idea?

(f) After talking to the logic designers, you decide to use replicated ten-stage pipelines. How many do you need?

23.9 *Pipelined adders.* What are the latency, throughput, clock time, and number of flip-flops used for each of the following implementations of the 32-bit adder of Section 23.3? Recall that $t_{dcc} = 100$ ps (delay of a one-bit add) and that $t_{\text{reg}} = 200$ ps.

(a) Four stages of eight bits each.

(b) Two stages of 16 bits each.

(d) Thirty-two stages of one bit each.

23.10 *Data forwarding.* The adder pipeline of Figure 23.7 prevents the use of $a + b$ until the entire add has completed (four cycles). Modify the adder so that it can begin executing an add of $a_1 + (a_0 + b_0)$ on the cycle after $a_0 + b_0$ begins. You will need to include a *dataFwd* signal to indicate a forwarding condition. The throughput and latency of the adder should stay the same, except for the minimal logic you will need to add.

23.11 *Speculation, single buffer.* In Section 22.6.2 we briefly discussed *speculation*, where a downstream pipeline stage can trigger all upstream stages to drop their current problem. Add a *fail* signal into the block diagram of Figure 23.9 and the VHDL of Figure 23.10. When asserted, all upstream data should be invalidated.

23.12 *Speculation, double buffer.* Repeat Exercise 23.11 with the double buffer of Figure 23.11 and the VHDL of Figure 23.12. The *fail* signal should propagate by only one stage per cycle (like the *ready* signal).

23.13 *Stall critical paths.* In an eight-stage pipeline, assume that each logic block has latency of 5 ns and $t_{\text{reg}} = 0$. Each logic block's stall signal becomes stable (evaluated) after 4 ns. The wire delay from the beginning to end of a stage is 500 ps.

(a) Without double buffering, what is the maximum legal clock rate?

(b) With double buffering, what is the maximum legal clock rate?

23.14 *Load balancing, bottleneck detection.* Suppose you have a system where every problem must pass through four pipeline stages with delays of 30 ns, 60 ns, 15 ns, and 20 ns. Which stage is the bottleneck? What is the utilization (time spent doing useful work divided by total time) of each stage?

23.15 *Load balancing, replication.* Suppose you have a system where every problem must pass through four pipeline stages with delays of 30 ns, 60 ns, 15 ns, and 20 ns. Each stage cannot be further pipelined, but can be replicated. How many of each module do you need in order to have full utilization of your system? (That is, no module should ever be idle, after a warm-up period.)

23.16 *Load balancing, pipelining.* Suppose you have a system where every problem must pass through four pipeline stages with delays of 30 ns, 60 ns, 15 ns, and 20 ns. Each stage cannot be replicated, but can be pipelined. What is the minimum number of stages, from beginning to end, in a pipeline that has no load imbalance?

23.17 *Bottleneck replication, control.* In Figure 23.14, we replicated the bottleneck stage four times and kept the clock rate at 4 ns. This requires the stages with 1 ns delays to produce four results every cycle. For this exercise, design control logic for both the input and the output of the bottleneck stage that uses a 1 ns clock. The control should give new work to each of the 4 ns stages once every 4 ns in a staggered fashion (cycle 1 to module 1, cycle 2 to module 2, etc.). At the output of these replicated modules, the data from the completing module should be fed into the pipeline register before the final stage.

23.18 *Resource sharing.* You have four replicated pipelines in a system. One stage in each of these pipelines accesses a shared resource 50% of the time (at random). For each value of n, the number of shared resources, compute the utilization of the shared resources and the probability that more than n requests are made in one cycle:

(a) $n = 2$;

(b) $n = 3$;

(c) $n = 4$.

23.19 *Variable loads.* Using the simple pipeline of Figure 23.15(a) and a queue depth of three buffers, how long will it take to finish computation if stage B has the following delay profiles. Assume all three buffers are initially empty. Draw diagrams indicating the stall conditions (if any) as in Figure 23.15(b).

(a) 15, 5, 15, 5, 15, 5, 15, 5, 15, 5;

(b) 15, 15, 15, 15, 15, 5, 5, 5, 5, 5;

(c) 15, 15, 5, 5, 5, 5, 5, 15, 15, 15.

24 Interconnect

The interconnect between modules is as important a component of most systems as the modules being connected. As described in Section 5.6, wires account for a large fraction of the delay and power in a typical system. A wire of just 3 μm in length has the same capacitance (and hence dissipates the same power) as a minimum-sized inverter. A wire of about 100 μm in length dissipates about the same power as one bit of a fast adder.

Whereas simple systems are connected with direct point-to-point connections between modules, larger and more complex systems are better organized with a bus or a network. Consider an analogy to a telephone or intercom system. If you need to talk to only two or three people, you might use a direct line to each person you need to talk to. However, if you need to talk to hundreds of people, you would use a switching system, allowing you to *dial* any of your correspondents over a shared interconnect.

24.1 ABSTRACT INTERCONNECT

Figure 24.1 shows a high-level view of a system using a general interconnect (e.g., a bus or a network). A number of clients are connected to the network by a pair of links to and from the interconnect. The links may be serialized (Section 22.3), and flow control is required on at least the link into the interconnect – to back-pressure the client in the event of contention.

To communicate, client S (the source client), transmits a packet over the link i_S into the interconnect. The packet includes, at minimum, a *destination address*, D, and a *payload*, P, which may be of arbitrary (or even variable) length. The interconnect, possibly with some delay due to contention, delivers P to client D over link o_D out of the interconnect. The payload P may contain a request type (e.g., read or write), a local address within D, and data or other arguments for a remote operation. Because the interconnect is addressed, any client A can communicate with any client B while requiring only a single pair of unidirectional links on each client module.

A packet (D, P) sent from S to D may result in D sending a *reply packet* (S, Q) with payload Q back to S. However, this is not required. Communications may be one-way.

The interconnect may or may not permit multiple concurrent operations. For high throughput, we would like an interconnect that permits multiple independent client pairs to communicate simultaneously. Also, if a reply packet is required, we would like S to be able to send several packets to the same or different destinations before having to wait for a reply. Lower-cost interconnects, like the bus in Section 24.2, however, do not support this degree of concurrency.

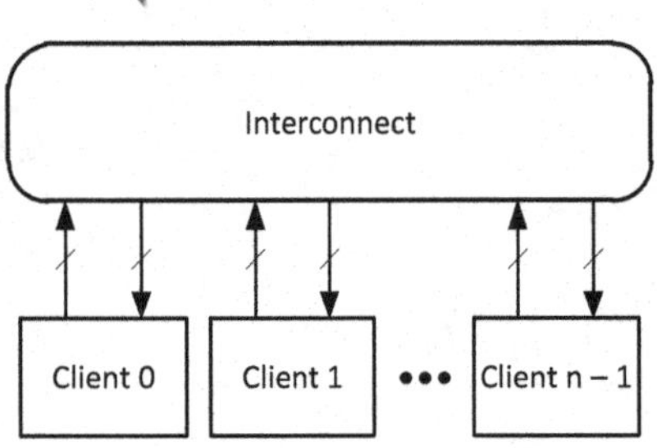

Figure 24.1. An abstract interconnect provides arbitrary connections between a number of clients. Each link into the interconnect is flow-controlled and transfers a packet containing a destination address.

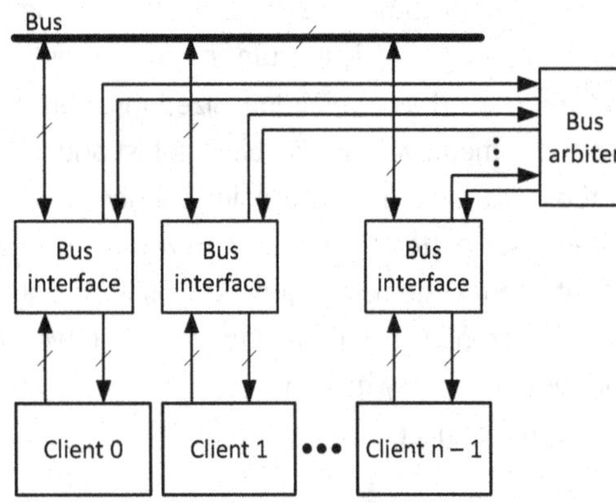

Figure 24.2. Bus interconnect. Modules connect to the bus via a bus interface. A source module arbitrates for access to the bus and then drives its packet onto the bus. All destination interfaces monitor the address field of the bus and supply the packet to their client if it has a matching address.

24.2 BUSES

The broadcast bus is one of the simplest general-purpose interconnects and is widely used in applications that have modest performance requirements. A bus has the advantages of simplicity, a broadcast facility, and serialization (ordering) of all transactions. The major disadvantage of a bus is its performance. It allows only one packet to be sent at a time.

Figure 24.2 shows a typical bus interconnect. Each module connects to the bus via a bus interface that converts the module's ready–valid flow control to bus arbitration. In addition to the ready and valid signals, each module's connection to the interface includes address and data fields.

VHDL code for a purely combinational bus interface is shown in Figure 24.3. The figure uses a signal-naming convention where the first letter of the name denotes the side of the interface: `b` for bus or `c` for client, and the second letter of the name denotes the direction: `r` for for outbound (bus receive), and `t` for inbound (bus transmit).

When a client wishes to communicate with another module on the bus, it places the address of the destination module on its address field `cr_addr`, puts the data to be communicated on its data field `cr_data`, and asserts valid `cr_valid`. The bus interface routes the valid signal from the module to the central bus arbiter (`arb_req = cr_valid`), which performs an arbitration and sends a single grant back to one requesting interface (see Section 8.5). The bus interface routes the grant signal back to the requesting module as a ready signal (`cr_ready = arb_grant`), and also uses this signal to gate the module's data and address onto the bus.

```
-- Combinational Bus Interface
-- t (transmit) and r (receive) in signal names are from the
-- perspective of the bus
library ieee;
use ieee.std_logic_1164.all;

entity BusInt is
  generic( aw: integer := 2;    -- address width
           dw: integer := 4 ); -- data width
  port( cr_valid, arb_grant, bt_valid: in std_logic;
        cr_ready, ct_valid, arb_req, br_valid: out std_logic;
        cr_addr, bt_addr, my_addr: in std_logic_vector(aw-1 downto 0);
        br_addr: out std_logic_vector(aw-1 downto 0);
        cr_data, bt_data: in std_logic_vector(dw-1 downto 0);
        br_data, ct_data: out std_logic_vector(dw-1 downto 0) );
end BusInt;

architecture impl of BusInt is
begin
  -- arbitration
  arb_req <= cr_valid;
  cr_ready <= arb_grant;

  -- bus drive
  br_valid <= arb_grant;
  br_addr <= cr_addr when arb_grant else (others => '0');
  br_data <= cr_data when arb_grant else (others => '0');

  -- bus receive
  ct_valid <= '1' when (bt_valid = '1') and (bt_addr = my_addr) else '0';
  ct_data <= bt_data ;
end impl;
```

Figure 24.3. VHDL code for a combinational bus interface.

If a requesting client loses arbitration, it simply waits with `cr_valid` asserted until it wins arbitration as signaled on `cr_ready`.

The bus drive logic assumes that the bus performs an OR of the signals driven from each bus interface; each interface drives a zero when not selected. Off-chip, a tri-state drive is sometimes used. However, on-chip tri-state buses are problematic for a number of reasons (see Section 4.3.4). Thus, on-chip buses are usually implemented by ORing together the signals from all of the bus interfaces and then distributing the result back to all bus interfaces.

On the inbound side, each bus interface monitors the address field of the bus for its address. When a matching address is detected, and the bus valid signal is asserted, the bus data are routed to the destination module along with a valid indication. Push flow control is commonly used on the receive side of buses (i.e., the modules must immediately accept data addressed to them). We explore adding full flow control to the bus receivers in Exercise 24.1.

The bus can be adapted to perform either *multicast* or *broadcast* communication. To specify a multicast, the address signals are replaced by an output-selection bit vector, where each bit corresponds to one of the clients. To send a packet to a single client, only a single bit of the bit vector is set. To multicast the packet to multiple clients, multiple bits are set – one for each destination. When all bits of the bit vector are set, the packet is broadcast to all of the clients. The transformation of the VHDL to perform multicast is straightforward for push flow control, but requires some care in handling the *ready* signals for full two-way flow control on the output side. This is explored in Exercise 24.4.

The logic in Figure 24.3 is strictly combinational. This assumes that all clients share a common clock. At the end of each clock cycle, a client with `cr_valid = cr_ready = 1` has completed an outgoing transaction and can move on to its next transaction or deassert `cr_valid`. Similarly, a client with `ct_valid` at the end of the clock cycle must accept the incoming transaction. The clock cycle must be long enough to allow both the bus arbitration and the propagation of the signals over the bus to take place in a single clock cycle. In Exercise 24.2 we explore how these two functions can be pipelined to improve bus speed.

The logic in Figure 24.3 is also completely parallel – transferring the address and all of the data in a single cycle. On-chip, where wires are plentiful, a fully parallel bus is often the correct solution. Off-chip, however, or on-chip in cases where the data are already serialized, a serialized bus that performs transactions over several cycles as shown in Figure 22.6 is preferred, in order to reduce pin count or avoid deserialization. We explore this approach in Exercise 24.3.

We have described both our abstract interconnect (Section 24.1) and our bus as performing one-way communication – delivering a packet from a sender to a receiver. If a reply is required, for example when reading a memory location located in the receiver, a separate packet is sent over the interconnect in the reverse direction. Performing an operation with two separate communications, one for request and one for reply, is sometimes called a *split transaction*. This is to contrast the operation to some historic buses that included a reply with the ready signal completing the request – eliminating the need for a separate reply communication.

In general, sending a reply as a separate communication is preferred for speed and generality. First, a combined request–reply transaction is very slow and holds the bus idle while the receiver is performing the action needed to compose the reply. In the case of a memory read, for example, it may take 100 ns (100 1 GHz clock cycles) actually to read the memory. Holding the bus idle for these cycles wastes a valuable communication resource. Second, performing a combined transaction over a multi-stage interconnect like a network is difficult (Section 24.4). Thus, a client that expects this type of interconnect is limited in what it can connect to.

24.3 CROSSBAR SWITCHES

When an interconnect with more performance than a bus can deliver is needed and the number of clients is small (typically fewer than 16), a crossbar (sometimes called *crosspoint*) switch is often a good solution. Figure 24.4 illustrates a crossbar switch that connects m sending clients to n receiving clients. In the most common case, $m = n$, and every client is both a sender and a receiver. Each of the lines in the figure includes data and valid signals in the forward direction and a ready signal in the reverse direction. The sending clients also provide an address to select the receiving client(s) to which their data should be delivered.

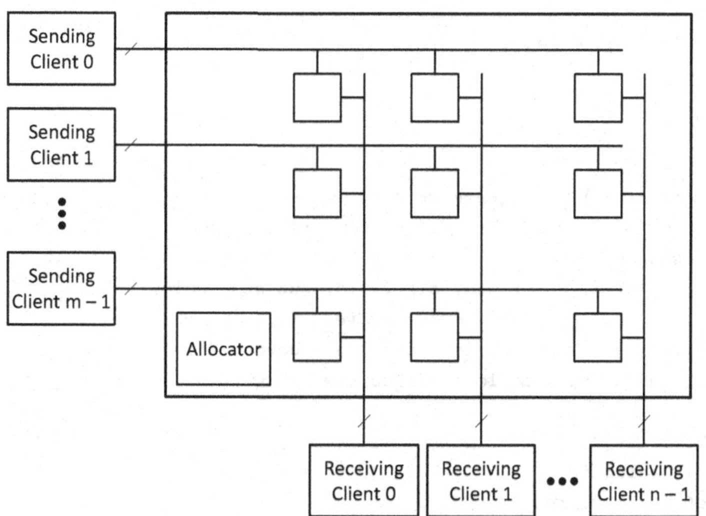

Figure 24.4. A crossbar switch can connect any idle input to any idle output, supporting many simultaneous connections.

When a sending client i wants to communicate with a receiving client j, it puts j on its address signal, puts the data to communicate on its data signal, and asserts *valid*. The allocator considers all connection requests and generates a set of non-conflicting grants (an $m \times n$ matrix of binary signals that enable the crosspoints). If client i's request to send a packet to j is granted, the allocator asserts g_{ij}, enabling the crosspoint at row i and column j. This connects the data and valid signals from row i to column j and connects the ready signal from column j to row i.

The allocator considers an $m \times n$ matrix of connection requests and generates an $m \times n$ matrix of non-conflicting grants. To be non-conflicting, each row and column in the grant matrix can have at most a single 1. In the case where each input specifies only a single output (on its address signal), there is at most a single 1 in each row of the request matrix, and the allocation reduces to n arbitrations, one for each column.

In the more general case, each sending client may buffer several packets and request multiple destinations (one for each buffered packet). In this case there may be multiple 1s per row of the request matrix, and the allocator must solve a bipartite matching problem. Such general allocators are described in detail in ref. [34].

VHDL code for a 2×2 crossbar switch with full ready–valid flow control at both inputs and outputs is shown in Figure 24.5. The arbitration in this example always gives priority to client 0. This code, which generates the grant matrix from the request matrix, can be replaced with a fairer arbiter if required. The grants, once generated, are used to enable both the forward connections (for *valid* and *data*) and the reverse connection (for *ready*). Once the addressing and arbitration have been performed, the crossbar connects the *ready*, *valid*, and *data* lines of the flow-controlled interface between a source client and a destination client.

Like the bus, the crossbar can be pipelined with arbitration performed one cycle ahead of data communication. If needed, the horizontal part of switch communication can be done in parallel with the arbitration – with a pipeline register at each crosspoint. The vertical part of communication is then done on the following cycle. This organization is considered in Exercise 24.10.

```
-- 2 x 2 Crossbar switch - full flow control
library ieee;
use ieee.std_logic_1164.all;

entity Xbar22 is
  generic( dw: integer := 4 ); -- data width
  port( c0r_valid, c0t_ready, c1r_valid, c1t_ready: in std_logic;
         -- r-v handshakes
        c0r_ready, c0t_valid, c1r_ready, c1t_valid: out std_logic;
        c0r_addr, c1r_addr: in std_logic; -- address
        c0r_data, c1r_data: in std_logic_vector(dw-1 downto 0); -- data
        c0t_data, c1t_data: out std_logic_vector(dw-1 downto 0) );
end Xbar22;

architecture impl of Xbar22 is
  signal req00, req01, req10, req11: std_logic;
  signal grant00, grant01, grant10, grant11: std_logic;
begin
  -- request matrix
  req00 <= '1' when not c0r_addr and c0r_valid else '0';
  req01 <= '1' when     c0r_addr and c0r_valid else '0';
  req10 <= '1' when not c1r_addr and c1r_valid else '0';
  req11 <= '1' when     c1r_addr and c1r_valid else '0';

  -- arbitration 0 wins
  grant00 <= req00;
  grant01 <= req01;
  grant10 <= req10 and not req00 ;
  grant11 <= req11 and not req01 ;

  -- connections
  c0t_valid <= (grant00 and c0r_valid) or (grant10 and c1r_valid);
  c0t_data <=  (c0r_data and (dw-1 downto 0 => grant00))  or
               (c1r_data and (dw-1 downto 0 => grant10));
  c1t_valid <= (grant01 and c0r_valid) or (grant11 and c1r_valid);
  c1t_data <=  (c0r_data and (dw-1 downto 0 => grant01)) or
               (c1r_data and (dw-1 downto 0 => grant11));

  -- ready
  c0r_ready <= (grant00 and c0t_ready) or (grant01 and c1t_ready);
  c1r_ready <= (grant10 and c0t_ready) or (grant11 and c1t_ready);
end impl;
```

Figure 24.5. VHDL code for a 2 × 2 crossbar switch with full flow control.

Also like the bus, the crossbar can perform multicast and handle serialized interfaces. We explore these variations in Exercises 24.8 and 24.9.

The throughput of a crossbar can be increased by providing buffering for entire packets at the crosspoints. This decouples input and output scheduling. An input can stack up packets destined for multiple outputs at different crosspoints in its row. These packets then arbitrate with other packets in their column for access to their output.

24.4 INTERCONNECTION NETWORKS

When more than 16 clients must be connected, an interconnection network is usually required in order to provide communication between the modules. An interconnection network consists of a set of *routers* connected by *channels* and is characterized by a *topology*, a *routing* algorithm, and a *flow-control* method.

The network topology specifies a set of routers and channels and how they are connected. For example, Figure 24.6 shows an interconnection network that connects 18 clients using a 3×3 two-dimensional *mesh* topology with two clients per router. The network has nine routers, with up to six bi-directional ports each, and 12 bi-directional channels. Each router is connected to its neighbors in the 3×3 mesh.

The routing algorithm specifies a *path* through the network from a source client to a destination client. One possible routing algorithm for this network is *dimension-order routing*, in which a packet is routed first in the x-dimension to the destination column, then in the y-dimension to the destination row, and finally to the destination client port.

For example, consider routing from client 0 to client 11. Client 0 injects the packet (over the same interface as introduced in Section 24.1) into router 00, and the packet is routed first

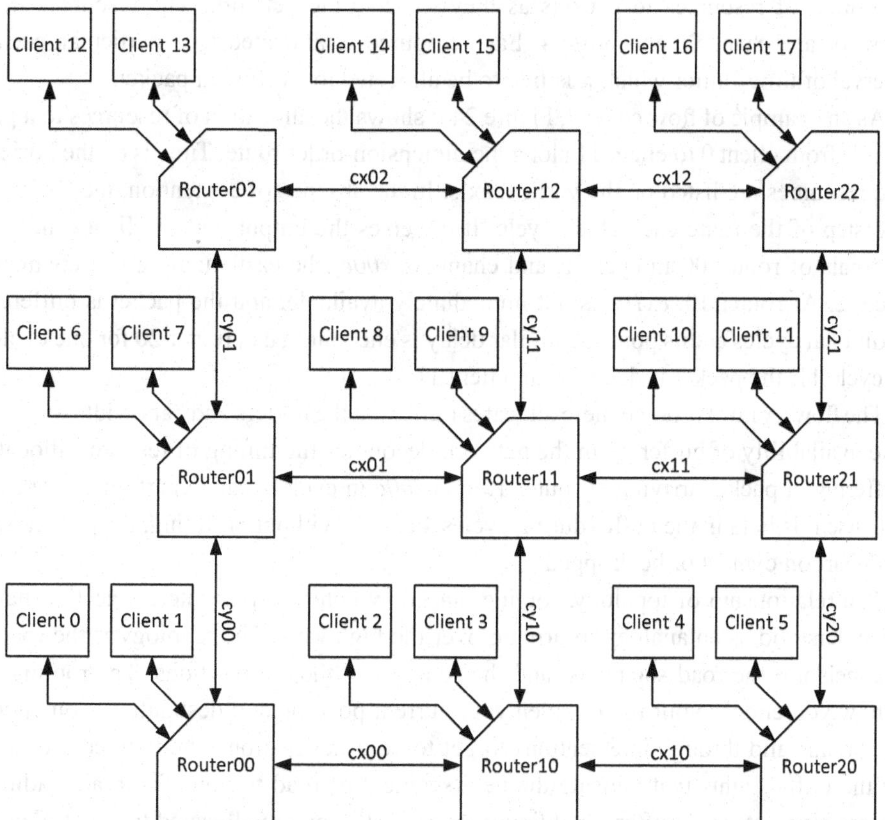

Figure 24.6. An interconnection network connects clients via a network of *channels* connected by *routers*.

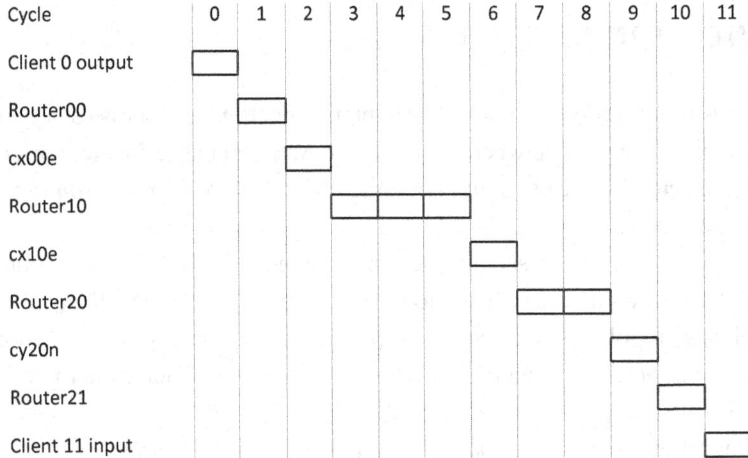

Figure 24.7. Time–space diagram illustrating the flow of one packet from client 0 to client 11. The packet is blocked for two cycles at router 10 and one cycle at router 20.

in the x-dimension to router 20. This route takes place over x-channels, *cx00* and *cx10*, and intervening router 10. Next the packet is routed in the y-dimension to router 21 via y-channel *cy20*. Finally, the packet is delivered to the client 11 port on router 21.

Interconnection-network flow control, as opposed to interface flow control, deals with the allocation of resources to packets as they traverse the network. The resources in a typical network are channels and buffers. Each resource is allocated to a particular packet for an interval of time – after which it is free to be allocated to a different packet.

As an example of flow control, Figure 24.7 shows the allocation of resources to a packet as it travels from client 0 to client 11 along the dimension-order route. Time is on the horizontal axis and resources are listed on the vertical axis. In the absence of contention, the packet advances one step of the route each clock cycle. It traverses the output port of client 0 in cycle 0, the internals of router 00 in cycle 1, and channel *cx00e* (the eastbound half of channel *cx00*) in cycle 2. At router 10, *cx10e* is not immediately available, and the packet is buffered for two additional cycles at this point. A similar delay is encountered at router 20 for one cycle. Finally, in cycle 11, the packet is delivered to client 11.

The flow-control protocol here allocates buffers in the routers and bandwidth on the channels. The availability of buffering in the network decouples the timing of resource allocation. With buffering, a packet arriving at router 10 on *cx00e* in cycle 2 can depart on *cx10e* in cycle 6 – because it is held in the buffer during cycles 4 and 5. Without buffering, the packet would need to depart on cycle 4 or be dropped.

The relationship of topology, routing, and flow control in an interconnection network can be understood as an analogy to driving over the highways. The topology is the road map, the channels are the road segments, and the routers are the intersections. The routing algorithm is how you choose your route; given your current position and destination, you choose a path over roads and through intersections to get to your destination. The flow control is governed by the traffic lights that allocate the next segment of road to you. The road leading into the intersection acts as a buffer – holding your car until you are allocated the next *channel* by the traffic light.

So far, we have assumed the entire packet is delivered in parallel in a single clock cycle. As with the other interconnects we have discussed, an interconnection network can be serialized – with a packet being delivered over several cycles over a narrower interface. With a serialized network, the flow control can be performed either at the level of a whole packet (in which case routers must have buffers large enough to hold whole packets) or at the level of flow-control digits, or *flits* (usually the amount of information delivered in one clock cycle). Packet-level flow control is similar to the frame-level flow control shown in Figure 22.6, while flit-level flow control is similar to a normal ready–valid interface.

A full treatment of interconnection networks is beyond the scope of this book. If they are not properly designed, these networks are subject to deadlocks from routing or higher-level protocol interactions. One way to eliminate deadlock, and more generally to isolate traffic, is to allocate buffers in a way that makes a single physical channel in the network appear to be multiple *virtual channels*. These concepts, and many other topics, are discussed in detail in ref. [34].

Summary

The interconnect between modules is a critical component in most digital systems. When connecting more than a few modules, a switched interconnect, such as a bus or a network, is preferable to providing dedicated point-to-point links between communicating modules. Much like a telephone exchange, a switched interconnect allows each client module to communicate with any other module using only a single pair of links – one in each direction.

A *bus* is the simplest implementation of a switched interconnect. In a bus system, clients arbitrate for access to a single shared communication channel. The winning client drives its data, along with a destination address, onto the channel. The receiving module, upon recognizing its address, reads the data from the bus. Buses have advantages of simplicity, providing ordering of all communications, and providing a broadcast facility. However, they have limited performance – because only one client can transmit at a time.

A *crossbar switch* can be used to connect a small number of clients when more performance is needed than a bus can provide. A crossbar allows multiple communication actions to be performed simultaneously as long as conflicts are avoided.

To connect larger numbers of clients, an *interconnection network* is typically used. An interconnection network consists of a number of *routers* connected by *channels*. The connection of routers and channels is called the *topology* of the network. A *routing algorithm* selects a path, through routers and over channels, to forward a packet from source to destination. Resources along this path are allocated using a *flow-control* method.

BIBLIOGRAPHIC NOTES

For further information about interconnection networks as a whole, consult refs. [32] and [34] by Dally. Reference [66] provides an overview of a bus that was in a recent commercial processor, IBM's Cell.

Exercises

24.1 *Bus receiver with flow control.* The simple combinational bus of Figure 24.3 provides full ready–valid flow control on the outgoing side (signals `cr_xxx`), but only one-way push flow control on the incoming side (signals `ct_xxx`). Modify this interface to provide full ready–valid flow control on the incoming side. (Hint: this will require adding signal `ct_ready` to the client side of the interface and signals `bt_ready` and `br_ready` to the bus side.)

24.2 *Pipelined bus arbitration.* The simple combinational bus of Figure 24.3 can be made faster by pipelining the arbitration one cycle ahead of the bus transfer. Sketch a block diagram and write (and test) the VHDL code for a bus interface that pipelines arbitration in this manner. The interface signals should be identical to those in Figure 24.3. Also, your module should be able to perform back-to-back transactions from a single bus interface. (Hint: to perform back-to-back transfers, the module must buffer one transaction (address and data) internally while accepting the next transaction for arbitration.)

24.3 *Serialized bus.* Modify the bus interface of Figure 24.3 to perform serialized transfers as illustrated in Figure 22.6. Assume that the bus itself has a single four-bit wide path to carry both address and data. Assume the address is four bits wide and the data is 20 bits wide (four bits of control and 16 bits of payload data). Send the address first. Assume frame-level, two-way flow control at the source and one-way push flow control at the destination.

24.4 *Multicast bus with full flow control.* Modify the VHDL of Figure 24.3 to handle multicast with full flow control on the receiver (as in Exercise 24.1). Assume that `cr_addr` is replaced with `cr_vector`, a bit vector that can specify multiple destination clients. (Hint: `bt_valid` must not be asserted until *all* selected outputs assert `ct_ready`.)

24.5 *Daisy-chain bus arbitration.* Design (and code) a controller and arbiter for a combinational daisy-chained bus. A daisy-chained bus does not have a centralized arbiter, but rather each controller makes a local request/grant decision. Controller 0 will always receive a grant and place its data onto the bus if it has a request. Controller 1 grants itself access to the bus only if controller 0 has not made a request, and so forth. Controller N will get the bus only if all $N-1$ downstream controllers have no requests.

24.6 *Distributed bus arbitration.* Write a controller to carry out distributed bus arbitration. During each round of arbitration (over multiple clock cycles), each controller with a request puts its priority onto a bus that ORs together all signals. On the first round, if the MSB of the bus priority is greater than that of a given controller, that controller drops out of participation. This process is then repeated using the MSB -1 bit and all remaining requesters. At the end of arbitration, only one controller will remain and become the bus master.

24.7 4×4 *crossbar.* Write the VHDL to implement a 4×4 crossbar switch with full flow control. Use the same input and output signals as in Figure 24.5, but with two more controllers.

24.8 *Multicast crossbar.* Design a 4 × 4 crossbar that supports multicast messaging. Each input can request one or more outputs, but arbitration must be done as all or nothing. That is, an input is either granted to send to all outputs or can send to none at all.

24.9 *Serialized crossbar.* Modify the 2 × 2 crossbar of Figure 24.5 to allow the serialized transport of a 20-bit payload. The crossbar wires should be only four bits wide and should perform arbitration and flow control on the head only once for each packet.

24.10 *Buffered crosspoints.* Design a crossbar to have n^2 buffered crosspoints. On each cycle, every input will write into the buffer that connects that input to the desired output (provided the buffer is not full). The output channels then arbitrate between the input crosspoints with requests, popping one of them and outputting the data.

24.11 *VHDL implementation of a simple router.* Write the VHDL code for a simple router for use in a mesh network such as that illustrated in Figure 24.6. Your router should have a single client port with ready–valid flow control in both directions and ports for channels in four directions (west, east, north, and south) also with ready–valid flow control. Each of the five inputs to your router should provide double buffering (Section 23.5) so that if the next channel is not immediately available the packet can be buffered without requiring a combinational path to the previous router. Assume that the entire route is encoded in the address field of the packet, with three bits per hop specifying the port for that hop. Each router should use the most significant three bits and then shift the address field three bits left to put the next router's routing information in this position.

25 Memory systems

Memory is widely used in digital systems for many different purposes. In a processor, SDDR DRAM chips are used for main memory and SRAM arrays are used to implement caches, translation lookaside buffers, branch prediction tables, and other internal storage. In an Internet router (Figure 23.3(b)), memory is used for packet buffers, for routing tables, to hold per-flow data, and to collect statistics. In a cellphone SoC, memory is used to buffer video and audio streams.

A memory is characterized by three key parameters: its capacity, its latency, and its throughput. Capacity is the amount of data stored, latency is the amount of time taken to access data, and throughput is the number of accesses that can be done in a fixed amount of time.

A memory in a system, e.g., the packet buffer in a router, is often composed of multiple memory *primitives*: on-chip SRAM arrays or external DRAM chips.[1] The number of primitives needed to realize a memory is governed by its capacity and its throughput. If one primitive does not have sufficient capacity to realize the memory, multiple primitives must be used – with just one primitive accessed at a time. Similarly, if one primitive does not have sufficient bandwidth to provide the required throughput, multiple primitives must be used in parallel – via duplication or interleaving.

25.1 MEMORY PRIMITIVES

The vast majority of all memories in digital systems are implemented from two basic primitives: on-chip SRAM arrays and external SDDR DRAM chips.[2] Here we will consider these memory primitives as *black boxes*, discussing their properties and how to interface to them. It is beyond the scope of this book to look inside the box and study their implementation.

25.1.1 SRAM arrays

On-chip SRAM arrays are useful for building small, fast, dedicated memories integrated near the logic that produces and consumes the data they store. While the total capacity of the SRAM that can be realized on one chip (about 400 Mb) is small compared with a single 4 Gb DRAM chip, these arrays can be accessed in a single clock cycle, compared with 25 cycles or more for a DRAM access. By operating many SRAM arrays in parallel, very high aggregate memory

[1] SRAM stands for static random-access memory; DRAM is an acronym for dynamic random-access memory.

[2] We are not considering non-volatile memories such as flash and disk that are used for persistent storage.

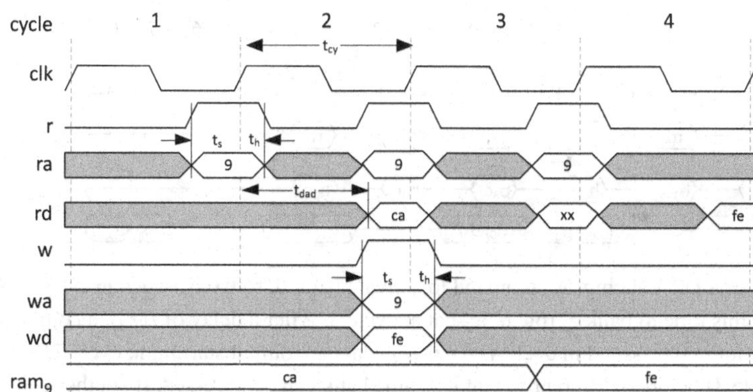

Figure 25.1. Timing of a dual-ported synchronous SRAM. When the read (write) (r and w, respectively) signal is high on a clock edge, the data at address ra (wa) is read (written). The read and write signals, addresses, and data have setup- and hold-time constraints. The time between the clock and data output on a read is t_{dad}.

bandwidths can be achieved. Consider, for example, a chip with 1024 1K × 64 SRAMs (64 Mb total) that operate at 1 GHz. The aggregate bandwidth is 8 TB/s. In contrast, a typical DRAM chip has a bandwidth of 1 GB/s or less. The location of the SRAM – on-chip, right next to the logic that uses it – is also a critical advantage. The advantage of single-cycle access is diminished if the SRAM is 20 cycles away on the other side of the chip, or, worse, on another chip.

As described in Section 8.9, an SRAM accepts an address, a data input, and a write signal, and produces a data output. An SRAM can have any number of ports P, but the vast majority of SRAMs are single-ported (because cost increases as P^2). Dual-ported SRAMs, with one read and one write port, are not uncommon. SRAMs with more than two ports are unusual – and costly.

Most SRAMs are *synchronous* – operating off of a clock – as shown in Figure 25.1. Each port can perform one read or one write (but not both) each cycle. The address and write data must be set up t_s before the rising edge of the clock and held until t_h after the clock edge. Read data are available a propagation delay (t_{dad}) after the rising edge of the clock. The clock cycle must be large enough to allow time for the write operation, internal precharge, and other internal operations to complete. For most SRAMs, reading and writing the same address in a single cycle, as illustrated in cycle 3 of the figure, results in undefined read data.

SRAMs are organized as arrays of cells with row decoders and column multiplexers, like the ROM of Figure 8.46, but with an SRAM cell at the junction of each word line and each bit line. Electrical constraints limit the maximum size of the basic array to no more than 256 rows and no more than 256 columns (64 Kb or 8 KB). Depending on the multiplexing factor, a single maximum-size 256 × 256 SRAM array can realize a 64K × 1-bit RAM, a 256 × 256-bit RAM,[3] and many sizes between these two extremes, such as a 2K × 32-bit RAM. A single SRAM array – including decoder and multiplexer – typically operates in one clock cycle.

If we need a RAM larger than 8 KB or wider than 256 bits, we must combine multiple RAM arrays via bit-slicing or banking (Section 25.2).

[3] Some RAMs have a minimum column multiplexing requirement of 2 or 4, making the widest RAM built from a single array 128 or 64 bits wide, respectively.

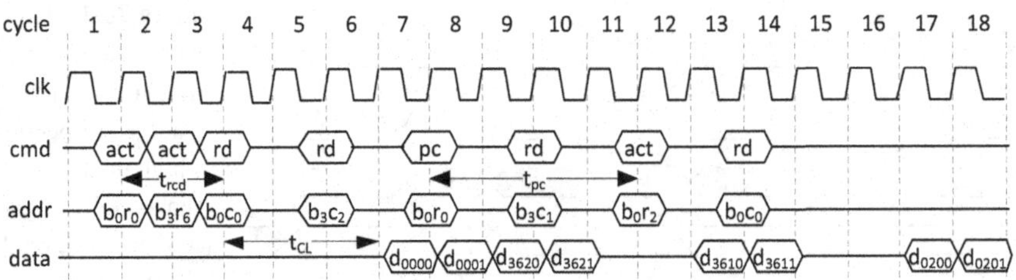

Figure 25.2. Timing of a DRAM chip. A command to activate bank 0, row 0 is issued in cycle 2. A second activation command, this time to bank 3, row 6, is sent at cycle 3. After a delay of t_{RCD}, a command to read bank 0, active row, column 0 is issued in cycle 4. After t_{CL}, data are output starting in cycle 7. A row must be precharged (as in cycle 8) before a different row in the same bank can be accessed. If another column in the same row is accessed, as at cycle 10, no precharge is necessary.

25.1.2 DRAM chips

DRAM is the *fast*[4] off-chip memory technology with the lowest cost per bit. Contemporary DRAM chips have a capacity of up to 4 Gb, significantly more storage than can be realized on a single chip using SRAM primitives. This large capacity, however, leads to a high latency. A modern SDDR3 memory chip has a latency of 20 ns compared with an on-chip SRAM array that may have a latency of 400 ps.

Dynamic memory is not inherently any slower than static memory. However, a commodity DRAM chip is much slower than an on-chip SRAM array for three reasons. First, because the DRAM chip is a discrete part, a large fraction of the latency is spent traversing the chip interface. Second, because the part has a large capacity, considerable time is required to traverse on-chip buses to access a particular sub-array within the chip. Finally, because they are part of a system with high communication latency, the sub-arrays within a DRAM chip are not optimized for speed.

While some applications use single DRAM chips, DRAM is more commonly used in modules that build a memory from several DRAM chips in a bit-sliced manner (Section 25.2). Multiple modules may be interleaved (Section 25.3) to provide a higher-bandwidth memory. The memory of high-performance CPUs, for example, is organized in this manner. For the remainder of this section, we focus on a single DRAM chip.

Reading (or writing) a DRAM module requires a three-step sequence of row activation, column access, and precharge, as illustrated in Figures 25.2 and 25.3. The interface of a DRAM module consists of *address*, *data*, and *control* buses. Because DRAM pins are limited, both address and data buses are serialized. Addresses are split into *bank*, *row*, and *column* bit fields. To read a specific address from DRAM, the first step is to *activate* a row of a particular bank on the basis of the bank (upper) and row (middle) bits of the memory address. As illustrated in Figure 25.3(a), row activation reads a row of the selected bank into the bank's *sense amplifiers*. The process of reading the data destroys the stored data, so at the end of the row activation the memory is left in an unknown state, illustrated in Figure 25.3(b). Before the column address

[4] Here, "fast" means able to make a random read or write access in 100 ns or less. Flash and disk memory is of lower cost per bit but too slow to be used for main memory in many applications.

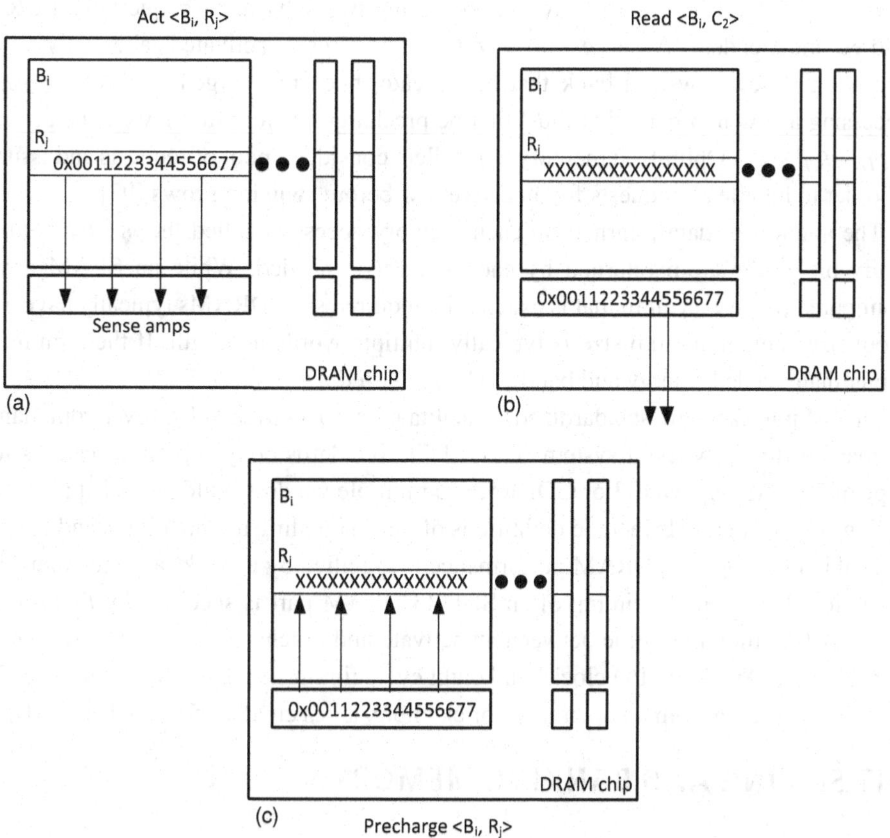

Figure 25.3. Steps taken to read a DRAM memory bank. (a) A specific row is first *activated* and read into the sense amplifiers, destroying the stored data. (b) Next, commands are used to read and write specific columns of the activated row. Finally, in (c), the *precharge* command writes the row back into the array. Multiple reads can be issued to different columns in the same row without a precharge/activate sequence. Also, rows can be activated in multiple banks at the same time.

and command are issued, the controller must wait a fixed latency, t_{RCD},[5] for the row activation to complete. Operations can be performed on other banks during this delay.

After the row activation is complete, a read (or write) command is issued along with the bank and column addresses. After t_{CL} (CAS latency or column latency), the first sub-word of data appears on the data pins. A new sub-word appears every cycle thereafter in a burst, until the read has completed.[6] Multiple read and write operations can be performed on an activated row without the need for another activation cycle.

After all read and write operations have been completed on a row, a *precharge* operation is performed to write the data in the sense amps back into the memory array (Figure 25.3(c)). From the time the precharge command is issued (cycle 8 in Figure 25.2) the controller must wait t_{pc} to allow the precharge operation to complete before performing another row activation on the same bank. Other banks can be accessed during this delay.

[5] *RCD* stands for the RAS to CAS delay, terms from the days of asynchronous DRAM. Think of it as the row to column delay.

[6] In a double-data rate (DDR) memory, separate data words are transferred on each half of the clock cycle.

Unlike SRAM, the latency between processing two subsequent requests in a DRAM is not address-independent. An access to a row that is already activated takes only t_{CL}, whereas accessing a new row in a bank that has already been precharged requires $t_{RCD} + t_{CL}$, and accessing a row in a bank that must first be precharged before the row can be activated takes $t_{RCD} + t_{CL} + t_{pc}$. Optimized memory controllers consider a pool of requests and issue them out of order to handle all requests for an active row before switching rows [97].

The amount of data returned on each memory access is called the *memory atom*. Ideally, every bit of the atom returned by each request is needed. While an SRAM array can be configured to have an atom that is exactly the required size, DRAMs typically have a minimum atom size. This minimum size is typically multiple words in length. If the minimum atom is larger than needed, energy and bandwidth are wasted.

DRAM interfaces are standardized to facilitate interoperation of DRAMs from many vendors in many different types of system. The JEDEC standards body sets specifications for a given type of DRAM, e.g., DDR3 or GDDR5, and multiple vendors build parts that meet the specification. These standards include definitions of pins, signaling methodology, and commands.

Within one standard, DRAM performance is specified by a clock rate and latencies for each operation. For example, timing of an SDDR3 DRAM part is specified by f_{clk}, t_{CL}, t_{RCD}, t_{RP}, and t_{RAS} (the minimum time between an activate and precharge command). A DDR3-1600 8-8-8-24 has an 800 MHz I/O clock[7] and eight cycle timings for everything but t_{RAS} (24 cycles). Reading a column from an open row, for example, takes eight 1.25 ns cycles, or 10 ns.

25.2 BIT-SLICING AND BANKING MEMORY

When we need a memory larger or wider than we can build from a single primitive (SRAM or DRAM), we combine several primitives using either *bit-slicing* or *banking*. With bit-slicing, we divide the primitives across the bits of our memory subsystem. With banking, the primitives are divided across the address space of our memory subsystem. The two can be combined with primitives divided across both bits and address space.

For example, suppose we need to build a 1 Mb (128 KB) SRAM array organized as 16K $\times$ 64. Figure 25.4 shows two ways to realize such a memory from 16 64 Kb memory arrays: bit-slicing (Figure 25.4(a)) and banking (Figure 25.4(b)). With bit-slicing, the address is distributed in parallel to 16 16K $\times$ 4 arrays and each array provides four bits of the output. With banking, one of 16 1K $\times$ 64 banks is selected by decoding the upper bits of the address. The low bits of the address are broadcast to all banks to select a word in each bank. The decoded high bits select between bank outputs.

The two configurations have the same capacity (1 Mb) and bandwidth (8 B per cycle). Also, in practice, both the bit-sliced and the banked memories would be laid out as a two-dimensional 4 $\times$ 4 array of memory primitives.

In a bit-sliced memory, all memory arrays must be accessed to complete an operation, since each provides a portion of the result. In a banked memory, however, only one array needs to be accessed. If the decoder output is used to activate only the selected array, the other arrays can remain idle, saving energy compared with a bit-sliced access.

[7] For a double-data rate (DDR) memory, data are transferred at twice the clock rate, i.e., 1600 MHz, hence the DDR3-1600 designation for an 800 MHz part.

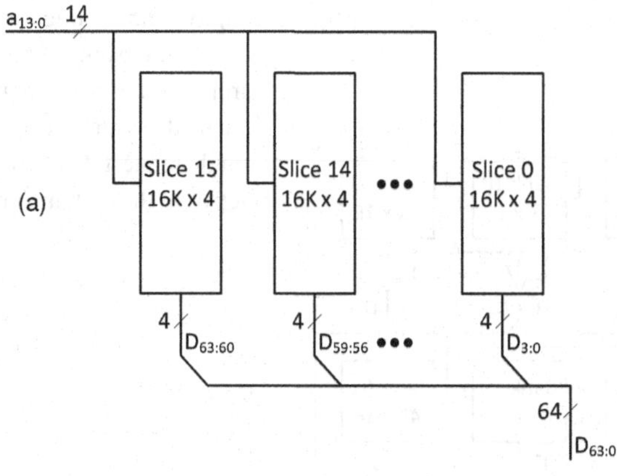

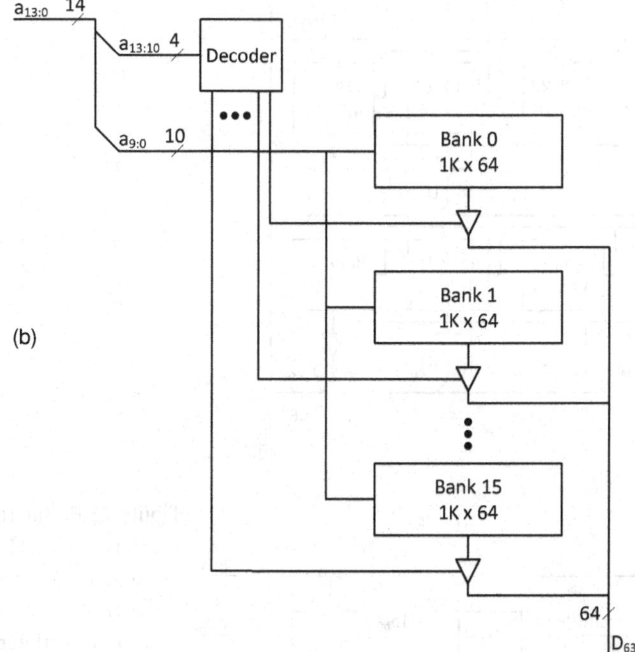

Figure 25.4. Two different multiple-primitive configurations for increasing memory capacity. Bit-slicing (a) reads a sub-word from each primitive simultaneously and combines them to form the output. Banking (b) reads the entire output word from a single primitive selected by the high bits of the address. With banking, the other primitives may remain inactive saving power.

Combining bit-slicing and banking gives a memory architecture like that shown in Figure 25.5. Each of the 16 primitives is 4K × 16 – requiring four arrays (one row) to read (or write) a full 64 bits. Four rows – addressed as banks – are needed to give the required memory capacity of 1 Mb. Every read or write request activates the four memory primitives in one row. The other 12 primitives remain idle.

While the configurations shown here use the high address bits to select the bank in a banked configuration, this is not a requirement. Any set of address bits can be used to select the bank. They need not even be contiguous. However, most memory systems use the high bits of the address for this purpose.

25.3 INTERLEAVED MEMORY

Allowing multiple requests to access multiple banks simultaneously increases our aggregate memory bandwidth. Performing simultaneous accesses requires replacing the global address

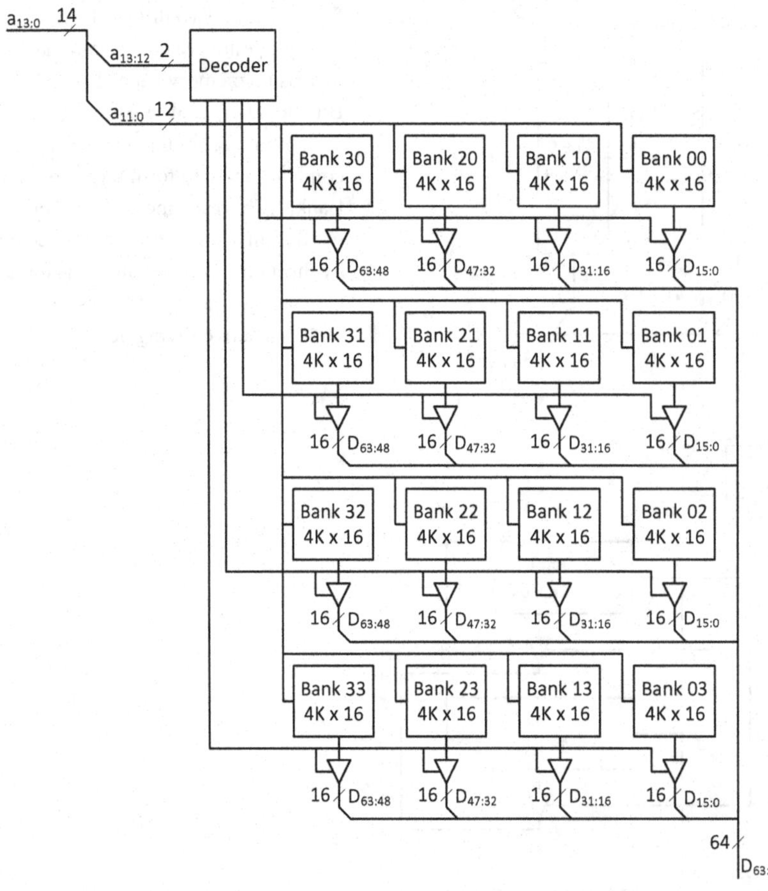

Figure 25.5. *Tiling* memory arrays into banks of bit-sliced arrays. All arrays in a given bank are read (written) to satisfy a single request. Only one request is active in the system at once.

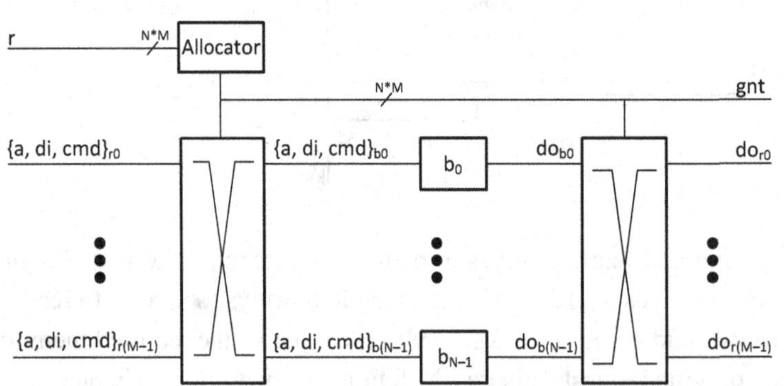

Figure 25.6. Interleaved memory, where M requesters arbitrate for N memory banks. A matrix of requests, r, is passed to an arbiter that then grants (*gnt*) requesters access to banks through a crossbar.

distribution of Figures 25.4 and 25.5 with an arbitrated crossbar (Section 24.3). We show this configuration in Figure 25.6. Multiple memory accesses arbitrate for each bank on the basis of their memory address. This enables multiple requests to be satisfied every cycle. Of course, these banks can be further bit-sliced and subdivided.

The potential bandwidth of the memory system has now increased from one word per cycle to $\min(N, M)$ words per cycle, where M is the number of requesters and N is the number of interleaved banks. This peak aggregate bandwidth is not always achieved, however. If two requests require access to the same bank, a conflict occurs and one request is postponed. One

Table 25.1. **Bit fields of an address in an interleaved memory system**

Name	Bits	Description
Byte	$b-1:0$	Byte within $B=2^b$ byte block
Bank	$n+b-1:b$	One of the $N=2^n$ banks
Block	$a+n+b-1:n+b$	One of $A=2^a$ blocks in each bank

Table 25.2. **With head-of-line blocking, the memory system does not have full throughput; requests that can be granted are stuck behind requests waiting for an over-subscribed resource**

Time	Q0	Q1	Q2	Q3	G0	G1	G2	G3
1	0, 1, 2, 3	0, 1, 2, 3	0, 1, 2, 3	0, 1, 2, 3	Q0	–	–	–
2	1, 2, 3	0, 1, 2, 3	0, 1, 2, 3	0, 1, 2, 3	Q1	Q0	–	–
3	2, 3	1, 2, 3	0, 1, 2, 3	0, 1, 2, 3	Q2	Q1	Q0	–
4	3	2, 3	1, 2, 3	0, 1, 2, 3	Q3	Q2	Q1	Q0
5	–	3	2, 3	1, 2, 3	–	Q3	Q2	Q1
6	–	–	3	2, 3	–	–	Q3	Q2
7	–	–	–	3	–	–	–	Q3

or more banks may go idle if there are no requests to their portion of the address space on a given cycle.

When the number of banks is a power of 2, bank selection is carried out by a subset of the address bits. The middle bits are typically used for this purpose, as shown in Table 25.1. The low b bits of the address are used to select a byte within a block – the amount of memory mapped to one bank before moving on to the next bank. The next $n = \log_2(N)$ bits of memory are used to select the bank. The remaining bits are used to select the block within the bank. This addressing maps successive addresses to different banks, reducing the number of bank conflicts when requests are clustered near one another.[8]

With interleaved memories, each request can be satisfied by a specific memory bank. If the switch allocator has access only to the head of each queue, the memory system may be under-utilized because of head-of-line blocking, Table 25.2. Even though there are requests for idle banks, these requests are blocked behind other requests in their respective queue. Allowing the switch allocator to consider all requests in the queue – possibly handling requests out of order – Table 25.3, gives greater memory utilization. In the example, full utilization is achieved.

Even with an allocator that avoids head-of-line blocking, memory utilization is limited by *load balance* across the banks. To maximize memory throughput, requests must be evenly distributed across the memory banks. Using the middle bits of an address for bank selection

[8] As we discuss in Section 25.4, such references are said to exhibit *spatial locality.*

Table 25.3. **When the arbiter is able to see requests beyond the head of the queue, the memory system has no head-of-line blocking; it achieves full throughput**

Time	Q0	Q1	Q2	Q3	G0	G1	G2	G3
1	0, 1, 2, 3	0, 1, 2, 3	0, 1, 2, 3	0, 1, 2, 3	Q0	Q1	Q2	Q3
2	1, 2, 3	0, 2, 3	0, 1, 3	0, 1, 2	Q3	Q0	Q1	Q2
3	2, 3	0, 3	0, 1	1, 2	Q2	Q3	Q0	Q1
4	3	0	1	2	Q1	Q2	Q3	Q0

typically provides more even distribution than using the upper bits. However, pathological cases exist where accesses are strided such that their middle bits do not change.

If a memory is interleaved with a block size of B bytes and N different banks, two memory addresses a_0 and a_1 will access the same bank if

$$\Delta a = (a_0 - a_1), \quad \mathrm{mod}(NB) = 0. \tag{25.1}$$

For example, consider a 256 × 256 array of double-precision floating-point numbers (eight bytes each). If we walk down a column of the array, $\Delta a = 2048$. These accesses will all be the same bank of any interleaved memory system when NB is a power of 2 less than 4096.

The simplest way to avoid these bank conflicts is to *pad* the array layout to give row sizes that are relatively prime with the number of banks. For example, if we store our 256 × 256 array as if it were a 257 × 256 array, Δa becomes 2056, and the references will be evenly distributed across the banks.

Hardware solutions that involve using a prime number of memory banks have been proposed in order to avoid bank conflicts. However, these solutions are costly and still result in conflicts when the stride Δa is a multiple of the number of banks. The software solution of padding array sizes – or more generally avoiding the bad strides – is preferred.

In real memory systems, not all requests have the same priority. Data load requests, for example, may have a higher priority than data stores. In these situations, we can queue each class of request in a different buffer. Higher-priority requests are then able to bypass lower-priority items. This scheme may *starve* low-priority items in the presence of constant high-priority requests by never granting low-priority requests resource access. Performance guarantees can be made by either increasing priority with wait time or allocating a static amount of resources to each packet class. For example, after four grants to high-priority requests, a lower-priority request must win arbitration.

When arbitrating for DRAM, this problem becomes even more difficult, since the time to access DRAM is address-dependent, and the controller must decide when to open a new row. Modern memory controllers must balance the goal of maximizing throughput while still providing a minimum quality of service to all requests.

25.4 CACHES

When implementing a memory system, one is faced with a tradeoff between capacity and speed. A DRAM provides gigabytes of storage, but can take over 100 cycles to access. Small SRAM

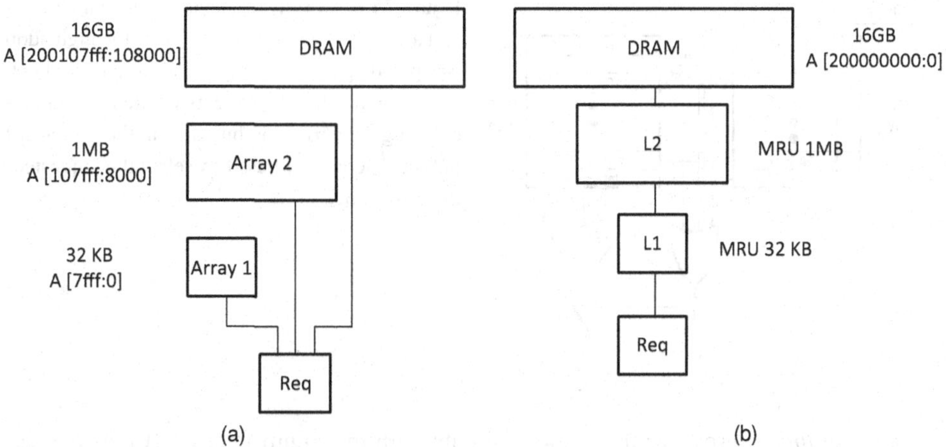

Figure 25.7. Memory hierarchies composed of memories of increasing capacity and latency. (a) Explicitly managed hierarchy, where each address maps to a specific memory. (b) Implicitly managed *cache* hierarchy, where the two smaller memories hold the most recently used (MRU) words.

arrays can be accessed in a cycle, but are limited to just 16–64 KB. Combining these memory components into a *memory hierarchy* gives the best of both worlds. Small, fast arrays store small amounts of frequently accessed data, and a large, slow memory stores everything else. If the designer or programmer knows which data fall into each category, an explicitly managed hierarchy can be used, as shown in Figure 25.7(a). A subset of the *address space* is assigned to each component. A request is handled by one of the three arrays, depending on its address.

The situation is different, however, when the specific data patterns are not known. A graphics card rendering a game accesses many textures,[9] only a small subset of which are in a given scene. The texture needed is not known ahead of time, but if it is used once it will probably be used again in the next pixel, next object, or next frame. If a player is wandering a sunny meadow, for example, only the grass texture needs to be frequently accessed. The sand, brick, snow, and moon textures can be stored in DRAM with no performance loss.

This is the principle of *temporal locality* – that data that have been referenced recently are likely to be referenced again in the near future. The grass texture is used by many nearby pixels, over several objects and several consecutive frames, before it is no longer needed. When the sand texture is finally loaded, it too will be used repeatedly.

Access patterns also exhibit *spatial locality*: data that are near (in address) the data that have just been referenced are likely to be referenced in the near future. If we reference one pixel of our grass texture, it is very likely that we will reference the adjacent pixels. We explore the use of *line size* to exploit spatial locality in Exercise 25.6.

We can exploit temporal locality by keeping the data that have been referenced most recently in a small, fast memory called a *cache*. For each data element stored in a cache, we also store a *tag*, which includes the address of the data element along with some state information. During a read operation, each cache level is checked in sequence for the requested address to see whether it has a copy of the requested data. If the L1 cache has a tag containing the requested address

[9] Images that are "painted" onto objects displayed on screen.

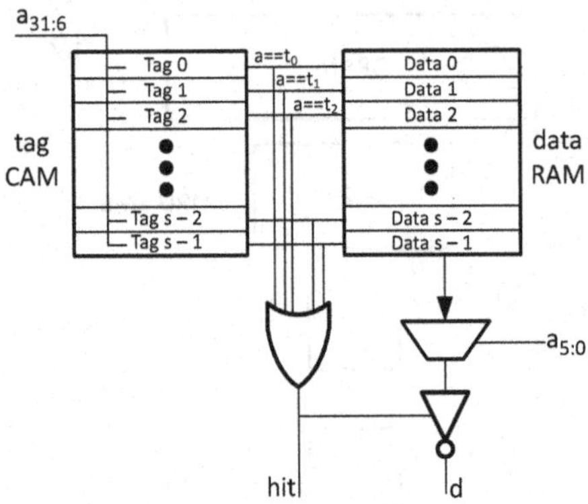

Figure 25.8. A fully associative cache includes a tag CAM and a data RAM. On a read access, the high address bits are compared against every valid address in the tag CAM. If there is a match (a hit), the associated word line goes high, enabling the corresponding data in the data RAM onto the output. The low address bits select the requested word within the cache line.

(a cache *hit*), it provides the associated data with minimum latency. If the L1 cache does not have a tag matching the requested address (a cache *miss*), the search continues with the L2 cache. If the data element is not found in any cache, it is supplied by the backing DRAM. After the data element has been returned from a cache miss, it, along with its address, is stored in the L1 cache – because it is now the most recently accessed data element. When new data are brought into the cache, some older (less recently accessed) data may need to be *evicted* to make room.

For typical access patterns, keeping the most recently accessed data in a cache results in very high hit rates. A typical microprocessor with a 32 KB L1 cache has a 98% L1 hit rate on a popular set of benchmarks. This high hit rate means that the processor almost always (98% of the time) gets the low latency and high bandwidth of the small L1 memory while enjoying the large capacity of the DRAM. With good locality, a cache hierarchy gives the behavior of a large, fast memory.

A cache consists of a tag memory, which holds addresses, and a data memory which holds lines[10] (multiple adjacent words) of data associated with each tag. In a *fully associative* cache, a content-addressable memory (CAM) is used to hold the tags, as shown in Figure 25.8. The input to the CAM is the address to be accessed. The output of the CAM is a one-hot signal indicating the location of a match. This one-hot array of *match* signals is used to enable the corresponding data line onto the output of the data memory. If no matching address is found in the tag CAM, a miss occurs and the request is propagated to the next level of the memory hierarchy. While simple, this fully associative structure is used only for very small caches (fewer than 64 entries) because CAM arrays are large and slow.

Table 25.4 shows a series of accesses to a fully associative cache. On a miss, the requested data are stored into the cache when the value returns. If the cache is full, one line must be *evicted* from the cache. There are many possible ways to select the line to be evicted. One option its to evict the *least recently accessed* line.

A *direct-mapped* cache may be used to realize a cache memory that is too large to be built using a CAM array. As shown in Figure 25.9, the tags are stored in a conventional RAM (rather than a CAM). Both the tag and data arrays are accessed using the middle bits of a memory address. After reading the tag array, the upper bits of the requested address are compared with

[10] A *cache line* – the block of data associated with one cache tag – is sometimes referred to as a *cache block.*

Table 25.4. **Example of a cache with four entries with full associativity. We show the tag stored in each set at the time of the request and whether the request hit or missed.**

Request	Tag address	H/M	S0	S1	S2	S3
1	$3ff_{16}$	M	–	–	–	–
2	400_{16}	M	$3ff_{16}$	–	–	–
3	404_{16}	M	$3ff_{16}$	400_{16}	–	–
4	400_{16}	H	$3ff_{16}$	400_{16}	404_{16}	–
5	300_{16}	M	$3ff_{16}$	400_{16}	404_{16}	–
6	200_{16}	M	$3ff_{16}$	400_{16}	404_{16}	300_{16}
7	300_{16}	H	200_16	400_{16}	404_{16}	300_{16}

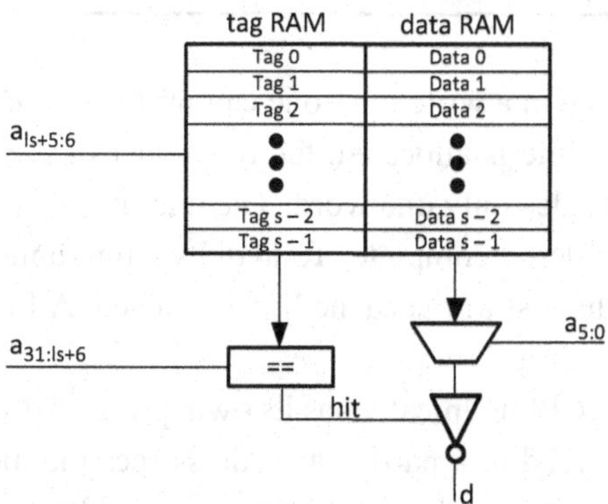

Figure 25.9. A direct-mapped cache stores tags in a RAM array. The lowest address bits are used to select the output word in a cache line. The middle address bits are used to read both the tag and data arrays. The upper address bits are compared with the output of the tag array. If the upper address bits match the tag read, a hit occurs and the data read from the data array are output.

the tag. If the upper bits of the address match the tag (a hit), the data read from the data RAM are output.

Table 25.5 shows an example access pattern and tag state in a simple four-entry cache. Since each address is directly mapped onto a single location, evictions may occur before the cache is full. If a second address comes along needing the same location, the first address is evicted – even if it was used quite recently. Exercise 25.7 asks you to explore set-associative caches where each line can be in any of w locations.

The amount of data associated with each tag in a cache, a cache line, is typically larger than a single word. Data transfers between caches and upper levels of the memory hierarchy are performed in units of cache lines. Most caches have line sizes between 32 and 128 bytes. The choice of line size is a compromise between exploiting spatial locality and avoiding the transfer of unneeded data. This tradeoff is explored in Exercise 25.6.

Memory accesses often contain more information than simply an address, operation, and data. This meta-data, such as data destination, is not needed by upper levels on an L1 cache miss and is stored locally in a miss status holding register (MSHR). Each cache miss is allocated an MSHR before being sent higher up the hierarchy, and is queued if none are available. When a miss returns, the MSHR data are retrieved and the entry deallocated.

Table 25.5. **Example of a cache with four sets, addressed with bits[7:6]. We show the tag stored in each set at the time of the request and whether the request hit or missed.**

Request	Address	Set	H/M	S0	S1	S2	S3
1	$3ff8_{16}$	3	M	–	–	–	–
2	4000_{16}	0	M	–	–	–	$3f_{16}$
3	4080_{16}	2	M	40_{16}	–	–	$4f_{16}$
4	4010_{16}	0	H	40_{16}	–	40_{16}	$4f_{16}$
5	4000_{16}	0	H	40_{16}	–	40_{16}	$4f_{16}$
6	3000_{16}	0	M	40_{16}	–	40_{16}	$4f_{16}$
7	4000_{16}	0	M	30_{16}	–	40_{16}	$4f_{16}$
8	3000_{16}	0	M	40_{16}	–	40_{16}	$4f_{16}$

Cache writes can miss just like cache reads. On a write miss one can *write around* the cache or allocate a line and write to the cache. If a line is allocated, the remainder of the line must be fetched from memory, since the write provides only one word. The missing write operation does not need this additional data to be considered complete. To avoid waiting unnecessarily, writes may be stored in a *write buffer* until the rest of the cache line is fetched. All loads must query the write buffer to read the most recent data.

In multiprocessor systems, each processor (CPU) typically has its own, private L1 cache. For correct operation, it is imperative that every read to a particular address sees the most recent write to that address by any processor. If multiple CPUs read a particular address, they each make a copy in their private caches. On a write, all outstanding versions of a line must either be invalidated or updated to keep the system *coherent* – i.e., so that only the most recently written data will be read by a subsequent read.

Summary

Memory is a key part of most digital systems. A memory subsystem is characterized by its *capacity*, *latency*, and *bandwidth*, and is composed of one or more memory *primitives* such as SDDR DRAM chips or SRAM arrays.

To build memory subsystems with larger capacity than an individual primitive, multiple primitives can be combined via *bit-slicing*, *banking*, or some combination of the two. If a subsystem with higher bandwidth than an individual primitive is required, duplication or *interleaving* can be employed. In an interleaved memory, multiple input ports are connected to multiple banks of memory via a switch. More efficient scheduling is achieved if multiple requests from each port are considered simultaneously – to avoid head-of-line blocking.

A *cache* memory exploits locality in an access pattern by using a small, fast memory to hold recently accessed data. When combined with a large, slow backing memory, the cache makes the entire subsystem appear to be a large, fast memory for most references – those that hit in the cache.

BIBLIOGRAPHIC NOTES

The design of memory array circuits can be found in refs. [26], [49], and [112].

Further reading about memory systems as a whole can be found in Jacob *et al.* [60]. More information on memory interleaving and scheduling is in Rau's classic paper [95] and Bailey's study of bank contention [6]. One example of a computer that used prime-number interleaving was the Burroughs scientific processor [69]. Memory access scheduling is detailed in ref. [97].

The impacts of caches on CPU performance are throughly explained in ref. [48], and coherency is detailed in refs. [101] and [30].

Exercises

25.1 *Memory addressing.* For each of the following memories, state how many bits are needed in order to address the full capacity. Also explain which bits are used for byte selection, bank selection, and word selection. Assume byte addressing, and that the bank selection is done with the bits just after the byte select bits.

(a) One array with 2000 32-bit words.

(b) Eight bit-sliced arrays, each with 1000 16-bit words.

(c) Sixteen banked arrays, each with 512 128-bit words.

(d) Eight banks of 16 bit-sliced arrays, each array has 1000 64-bit words.

25.2 *VHDL for SRAMs.* Using the RAM primitive of Figure 8.54, write the VHDL to implement the following:

(a) a memory of eight bit-sliced arrays, each with 1024 16-bit words.

(b) a memory of 16 banked arrays, each with 512 128-bit words. Only the needed bank should be activated.

25.3 *DRAM timings, I.* Assume a DRAM has 5–5–5–12 timings. Addresses are eight bits, with the upper four bits being row-select and the lower four column-select. Answer (a) and (b) for the following address stream: 01, 02, 03, 10, 20, a3, b3, 04, b1, b2.

(a) What is the total delay? (You must start and end with all rows precharged.)

(b) If you can rearrange the requests at will, what is the new delay?

25.4 *DRAM timings, II.* Compare a DRAM with an 800 MHz I/O clock and 8–8–8 timings with one with a 1 GHz I/O clock and 12–8–8 timings (we are ignoring t_{RAS} in this exercise). Which is faster for the following access patterns?

(a) A series of completely random addresses that are always to different rows.

(b) A series of addresses that are to an open row 99% of the time.

What percentage of accesses need to be to an open row in order to get equal performance?

25.5 *Strided accesses.* A matrix is accessed using a row index r and column index c. If the matrix is stored in row-major format, the address offset from the base is $rn_c + c$, where n_c is the number of columns. Column major offsets are accessed with an offset of $r+n_r c$. With an interleaved memory, which layout do you want to have for the following C code (assume accesses can be done in parallel):

```
for(int i=0; i<nr; i++){
   for(int j = 0; j<nc; j++){
      sum += m[i][j]; //i is row idx, j is col idx
```

25.6 *Spatial locality and line size.* Consider a cache with a one-word, four-byte line size. With a particular workload, if a word at address a is referenced, the next seven words at addresses $a+4, a+8, \ldots, a+28$ will all be referenced with a probability of $P = 0.95$. That is, if we reference a word 19 times out of 20, we also reference the next seven words. One time in 20 we don't reference the next seven words. What happens to the hit rate of this cache on this workload as we increase the line size to two words, four words, and eight words? What happens to the demand on memory bandwidth at each of these line sizes?

25.7 *Set-associativity.* Caches need not be only fully associative or direct-mapped. Designers can build caches that are two, four, ..., w-way set associative. Each address can reside in any one of w locations. Design and draw the block diagram of a four-way associative cache. Each address is 32 bits and indexes each byte. A cache line is 64 bytes and there are in total 1024 sets (each of which has four ways). You can use only SRAM arrays and must perform your access in one (potentially long) cycle. Be sure to give all array sizes and explain which address bits are used to index the arrays and which are stored as tags.

25.8 *Worst-case access patterns.* Using the minimum number of addresses, describe an access pattern that never hits in each of the following caches:

(a) a direct-mapped cache with n different sets;

(b) a fully associative cache with n entries;

(c) a w-way set-associative cache.

25.9 *Balls and bins.* Assume that every request has an equal probability of requesting one of n banks. The formula for the average number of requests ($E(r)$) needed to be examined in order to have at least one request to each bank is as follows:

$$E(r) = n\sum_{i=1}^{n}\frac{1}{i}. \tag{25.2}$$

(a) Plot this expectation as a function n.

(b) Being limited by design constraints, you can build only a 16-input arbiter. How many banks can you fully utilize on average? Your answer need not be a power of 2.

(c) Given a 256-input arbiter, how many banks, on average, would you expect to get full throughput with?

25.10 *Fair arbitration.* Write the VHDL for an arbiter that takes four high-priority requests and four low-priority requests and outputs the eight grant signals.

(a) Write the baseline module where low-priority requests can be starved.

(b) Write a module that after four cycles of granting high-priority requests will grant a low-priority request. Assume a static tie-breaking scheme for requests of equal priority.

(c) Modify the above module to implement a round-robin way of breaking ties within a class. That is, input 0 of 4 has the highest priority until granted, then input 1 of 4 has highest priority, etc.

Part VII

Asynchronous logic

26 Asynchronous sequential circuits

Asynchronous sequential circuits have state that is not synchronized with a clock. Like the synchronous sequential circuits we have studied up to this point, they are realized by adding state feedback to combinational logic that implements a next-state function. Unlike synchronous circuits, the state variables of an asynchronous sequential circuit may change at any point in time. This asynchronous state update – from next state to current state – complicates the design process. We must be concerned with hazards in the next-state function, since a momentary glitch may result in an incorrect final state. We must also be concerned with *races* between state variables on transitions between states whose encodings differ in more than one variable.

In this chapter we look at the fundamentals of asynchronous sequential circuits. We start by showing how to analyze combinational logic with feedback by drawing a flow table. The flow table shows us which states are stable, which are transient, and which are oscillatory. We then show how to synthesize an asynchronous circuit from a specification by first writing a flow table and then reducing the flow table to logic equations. We see that state assignment is quite critical for asynchronous sequential machines, since it determines when a potential race may occur. We show that some races can be eliminated by introducing transient states.

After the introduction given in this chapter, we continue our discussion of asynchronous circuits in Chapter 27 by looking at latches and flip-flops as examples of asynchronous circuits.

26.1 FLOW-TABLE ANALYSIS

Recall from Section 14.1 that an asynchronous sequential circuit is formed when a feedback path is placed around combinational logic, as shown in Figure 26.1(a). To analyze such circuits, we break the feedback path as shown in Figure 26.1(b) and write the equations for the *next*-state variables as a function of the *current*-state variables and the inputs. We can then reason about the dynamics of the circuit by exploring what happens when the current-state variables are updated (in arbitrary order if multiple bits change) with their new values.

At first, this may look just like the synchronous sequential circuits we discussed in Section 14.2. In both cases we compute a next state on the basis of the current state and input. What's different is the dynamics of how the current state is updated with the next state. Without a clocked state register, the state of an asynchronous sequential circuit may change at any time (asynchronously). When multiple bits of state are changing at the same time, it can cause a condition called a *race*. The bits may change at different rates, resulting in different end states. Also, a synchronous circuit will eventually reach a steady state in which the next state and

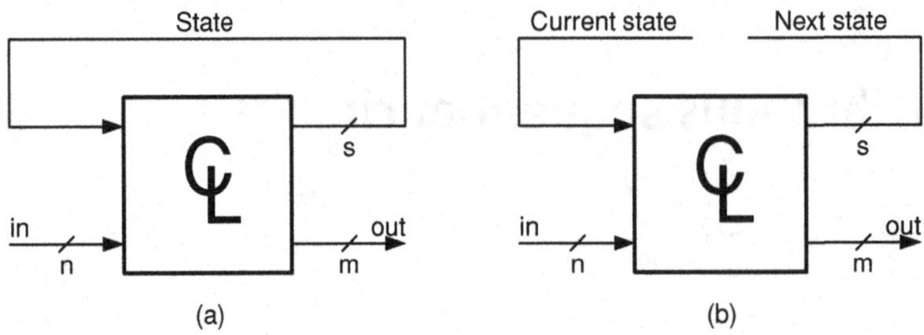

Figure 26.1. Asynchronous sequential circuit. (a) A sequential circuit is formed when a feedback path carrying state information is added to combinational logic. (b) To analyze an asynchronous sequential circuit, we break the feedback path and look at how the next state depends on the current state.

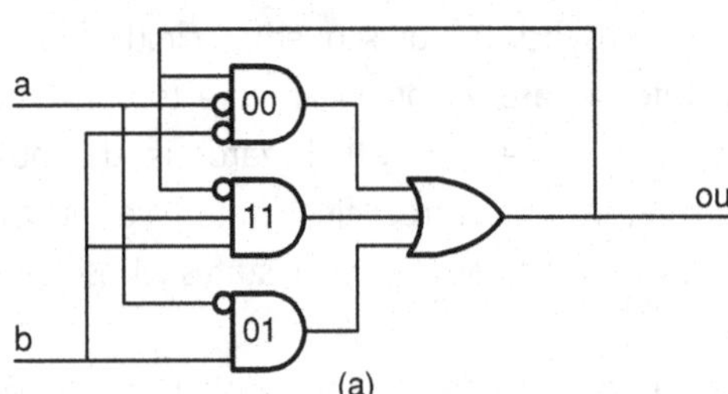

Figure 26.2. Example asynchronous sequential circuit. (a) Original circuit. (b) With feedback loop broken. (c) Flow table showing next-state function. Circled entries in the flow table are *stable* states.

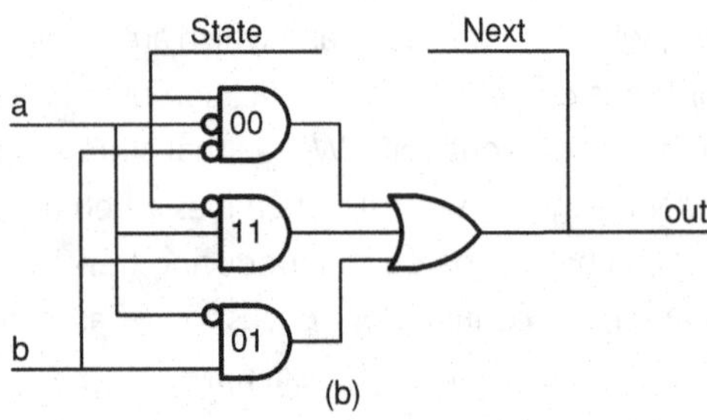

State	Next			
	00	01	11	10
0	(0)	1	1	(0)
1	(1)	(1)	0	0

(c)

outputs will not change until the next clock cycle. An asynchronous circuit, on the other hand, may never reach a steady state. It is possible for it to oscillate indefinitely in the absence of input changes.

We have already seen one example of analyzing an asynchronous circuit in this manner – the RS flip-flop of Section 14.1. In this section we look at some additional examples and introduce the *flow table* as a tool for the analysis and synthesis of asynchronous circuits.

Consider the circuit shown in Figure 26.2(a). Each of the AND gates in the figure is labeled with the input state *ab* during which it is enabled. For example, the top gate, labeled 00, is enabled when *a* and *b* are low. To analyze the circuit, we break the feedback loop, as shown in Figure 26.2(b). At this point, we can write down the next-state function in terms of the inputs, *a* and *b*, and the current state. This function is shown in tabular form in the *flow table* of Figure 26.2(c).

Figure 26.2(c) shows the next state for each of the eight combinations of inputs and current state. Input states are shown horizontally in Gray-code order. Current states are shown vertically. If the next state is the same as the current state, this state is *stable*, since updating the current state with the next state doesn't change anything. If the next state is different than the current state, this state is *transient*, since, as soon as the current state is updated with the next state, the circuit will change state.

For example, suppose the circuit has inputs $ab = 00$ and the current state is 0. The next state is also 0, so this is a stable state – as shown by the circled 0 in the leftmost position of the top row of the table. If from this state input b goes high, making the input state $ab = 01$, we move one square to the right in the table. In this case, the 01 AND gate is enabled and the next state is 1. This is an unstable or transient situation since the current state and next state are different. After some amount of time (for the change to propagate), the current state will become 1 and we move to the bottom row of the table. At this point, we have reached a stable state since the current and next state are now both 1.

If there is a cycle of transient states with no stable states we have an *oscillation*. For example, if the inputs to the circuit of Figure 26.2 are $ab = 11$, the next state is always the complement of the current state. With this input state, the circuit is never stable, but instead will oscillate indefinitely between the 0 and 1 states. This is hardly ever a desired behavior. An oscillation in an asynchronous circuit is almost always an error.

So, what does the circuit of Figure 26.2 do? By this point, the astute reader will have realized that it is an RS flip-flop with an added oscillation feature. Input a is the reset input. When a is high and b is low, the state is made 0; when a is lowered, the state remains 0. Similarly, b is the set input. Making b high while a is low sets the state to 1 and it remains at 1 when b is lowered. The only difference between this flip-flop and the one of Figure 14.2 is the addition of the middle AND gate, labeled 11. When both inputs are high, this gate causes the circuit of Figure 26.2 to oscillate, whereas the circuit of Figure 14.2 resets. If gate 11 and the $\overline{b}$ input to gate 00 are removed, the circuit becomes identical to that of Figure 14.2.

To simplify our analysis of asynchronous circuits, we typically insist that the environment in which the circuits operate obey the *fundamental-mode* restriction:

> **Fundamental-mode restriction** Only one input bit may be changed at a time and the circuit must reach a stable state before another input bit is changed.

A circuit operated in fundamental mode need only worry about one input bit changing at a time. Multiple-bit input changes are not allowed. Our setup- and hold-time restrictions on flip-flops are examples of a fundamental-mode restriction. The clock and data inputs of the flip-flop are not allowed to change at the same time. After the data input has changed, the circuit must be allowed to reach a steady state (setup time) before the clock input can change. Similarly, after the clock input has changed, the circuit must be allowed to reach a steady state (hold time) before the data input can change. We will look at the relation of setup and hold time to the design of the asynchronous circuits that realize flip-flops in more detail in Chapter 27.

In looking at a flow table, like the one in Figure 26.2, operating in the fundamental mode means that we need only consider input transitions to adjacent squares (including wrapping from leftmost to rightmost). Thus, we don't have to worry about what happens when the input

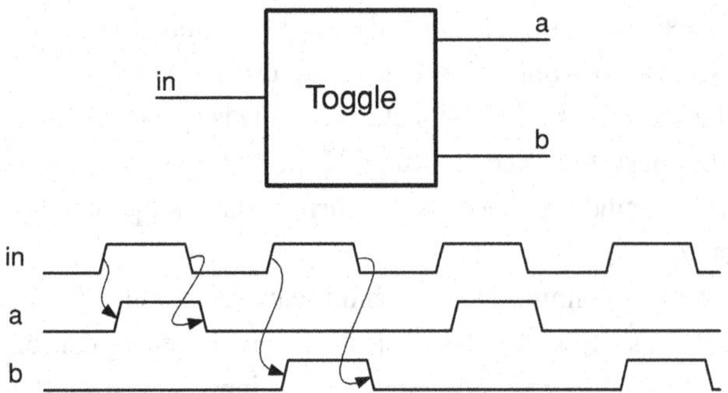

Figure 26.3. A toggle circuit alternates pulses on its input *in* between its two outputs *a* and *b*.

changes from 11 (oscillating) to 00 (storing). This can't happen. Since only one input can change at a time, we must first visit state 10 (reset) or 01 (set) before getting to 00.

In some real-world situations, it is not possible to restrict the inputs to operating in fundamental mode. In these cases we need to consider multiple input changes. This topic is beyond the scope of this book, and the interested reader is referred to some of the texts listed in the bibliographic notes at the end of this chapter.

26.2 FLOW-TABLE SYNTHESIS: THE TOGGLE CIRCUIT

We now understand how to use a flow table to analyze the behavior of an asynchronous circuit. That is, given a schematic, we can draw a flow table and understand the function of the circuit. In this section we will use a flow table in the other direction. We will see how to create a flow table from the specification of a circuit and then use that flow table to synthesize a schematic for a circuit that realizes the specification.

Consider the specification of a toggle circuit – shown graphically in Figure 26.3. The toggle circuit has a single input *in* and two outputs *a* and *b*.[1] Whenever *in* is low, both outputs are low. The first time *in* goes high, output *a* goes high. On the next rising transition of *in*, output *b* goes high. On the third rising input, *a* goes high again. The circuit continues steering pulses on *in* alternately between *a* and *b*.

The first step in synthesizing a toggle circuit is to write down its flow table. We can do this directly from the waveforms of Figure 26.3. Each transition of the input potentially takes us to a new state. Thus, we can partition the waveform into potential states, as shown in Figure 26.4. We start in state A. When *in* rises we go to state B, where output *a* is high. When *in* falls again we go to state C. Even though C has the same output as A, we know it is a different state because the next transition on *in* will cause a different output. The second rising edge on *in* takes us to state D with output *b* high. When *in* falls for the second time, we go back to state A. We know that this state is the same as state A since the behavior of the circuit at this point under all possible inputs is indistinguishable from where we started.

Once we have a flow table for the toggle circuit, the next step is to assign binary codes to each of the states. This state assignment is more critical than with synchronous machines. If two states X and Y differ in more than one state bit, a transition from X to Y requires first

[1] In practice, a reset input *rst* is also required, in order to initialize the state of the circuit.

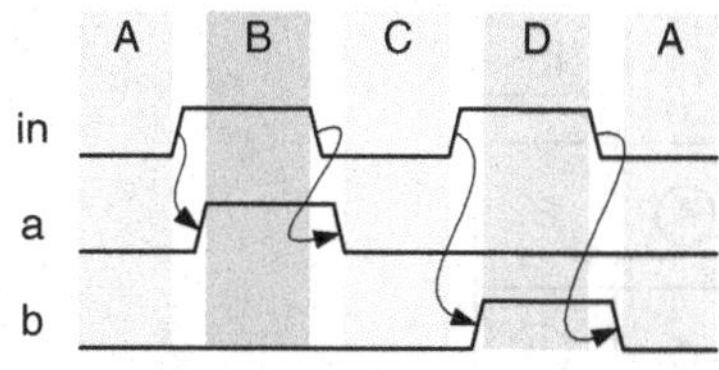

Figure 26.4. A flow table is created from the specification of the toggle circuit by creating a new state for every input transition until the circuit is obviously back to the same state.

State	Next (in) 0	Next (in) 1	Out (a,b)
A	(A)	B	00
B	C	(B)	10
C	(C)	D	00
D	A	(D)	01

visiting a *transient* state with one state bit changed before arriving at Y. In some cases, a *race* between the two state bits may result. We discuss races in more detail in Section 26.3. For now, we pick a state assignment (shown in Figure 26.5(a)) where each state transition switches only a single bit.

With the state assignment, realizing the logic for the toggle circuit is a simple matter of combinational logic synthesis. We redraw the flow table as a Karnaugh map in Figure 26.5(b). The Karnaugh map shows the symbolic next-state function; i.e., each square shows the next-state name (A through D) for that input and current state. The arrows show the path through the states followed during operation of the circuit. Understanding this path is important for avoiding races and hazards. We refer to this Karnaugh map showing the state transitions as a *trajectory map*, since it shows the trajectory of the state variables.

We redraw the Karnaugh map with state names replaced by their binary codes in Figure 26.5(c), and separate maps for the two state variables s_0 and s_1 are shown in Figures 26.5(d) and (e), respectively. From these Karnaugh maps we write down the equations for s_0 and s_1:

$$s_0 = (\overline{s_1} \wedge in) \vee (s_0 \wedge \overline{in}) \vee (s_0 \wedge \overline{s_1}), \tag{26.1}$$

$$s_1 = (s_1 \wedge in) \vee (s_0 \wedge \overline{in}) \vee (s_0 \wedge s_1). \tag{26.2}$$

The last implicant in each expression is required in order to avoid a hazard that would otherwise occur. Asynchronous circuits must be hazard-free along their path through the input/state space. Because the current state is being constantly fed back, a glitch during a state transition can result in the circuit switching to a different state – and hence not implementing the desired function. For example, suppose we left the $s_0 \wedge \overline{s_1}$ term out of Equation (26.2). When *in* goes low in state B, s_0 might go momentarily low before s_1 comes high. At this point, the middle term of both equations becomes false and s_1 never goes high – the circuit goes to state A rather than C.

All that remains to complete our synthesis is to write the output equations. Output a is true in state 01 and output b is true in state 10. The equations are thus given by

$$a = \overline{s_1} \wedge s_0, \tag{26.3}$$

$$b = s_1 \wedge \overline{s_0}. \tag{26.4}$$

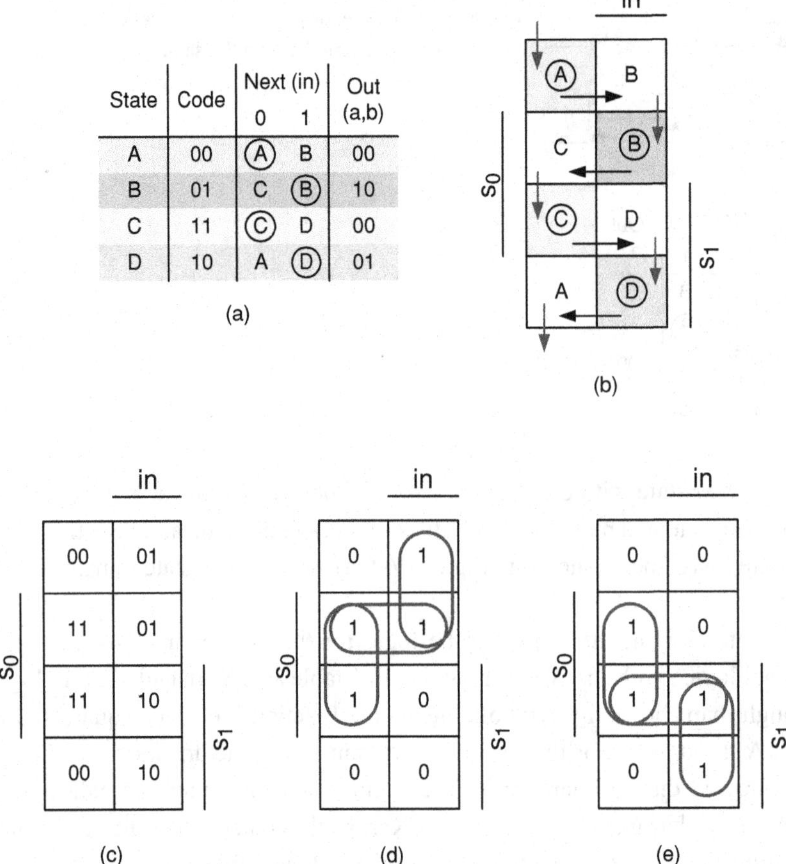

Figure 26.5. Implementing the toggle circuit from its flow table. (a) Flow table with state assignment. (b) Flow table mapped to Karnaugh map. (c) Next-state codes mapped to Karnaugh map. (d) Karnaugh map for s_0. (e) Karnaugh map for s_1.

EXAMPLE 26.1 Divide-by-3 circuit

Design an asynchronous sequential circuit with input *a* and output *q*, where output *q* transitions once for every three transitions on input *a*.

The waveform in Figure 26.6 shows operation of this circuit. Names for each asynchronous state are labeled. It is straightforward to draw the flow table from the waveform (see Figure 26.7) The second column of the flow table shows our state assignment *qrs*. We choose output *q* to be one state variable and add two additional variables *r* and *s* to distinguish the six states. We choose our assignment so that only a single state variable changes between adjacent states, in order to avoid races.

Using this state assignment we write the trajectory map and the K-maps for each state variable, as shown in Figure 26.8. From the K-maps we identify the prime implicants and write the logic equations for each state variable. We are careful to cover all of the transitions along

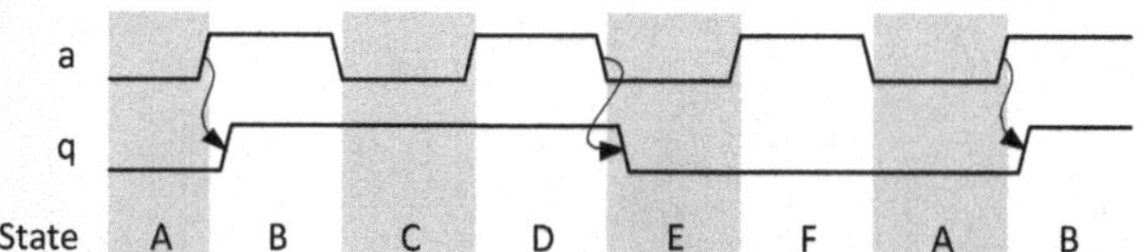

Figure 26.6. Waveform showing the operation of a divide-by-3 asynchronous circuit, designed in Example 26.1.

State	Code	Next (a) 0	Next (a) 1	q
A	000	(A)	B	0
B	100	C	(B)	1
C	101	(C)	D	1
D	111	E	(D)	1
E	011	(E)	F	0
F	010	A	(F)	0

Figure 26.7. Flow table showing state transitions for the divide-by-3 asynchronous circuit of Example 26.1.

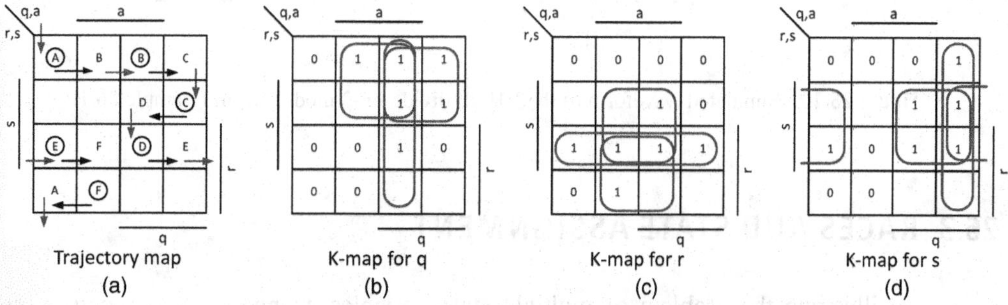

Figure 26.8. Trajectory map (a) and Karnaugh maps (b)–(d) for each state variable in the divide-by-3 circuit of Example 26.1.

the trajectory to avoid critical hazards. It is interesting that each of the three state variables is realized by a single *majority* gate. The equations are as follows:

$$q = (q \wedge \bar{r}) \vee (q \wedge a) \vee (a \wedge \bar{r}),$$
$$r = (r \wedge s) \vee (r \wedge a) \vee (s \wedge a),$$
$$s = (s \wedge q) \vee (s \wedge \bar{a}) \vee (q \wedge \bar{a}).$$

A VHDL design entity implementing these equations is shown in Figure 26.9. We add a reset signal `rst` to initialize the state variables.

Simulating this VHDL model gives the waveforms shown in Figure 26.10. In addition to generating the divide-by-3 waveform on *q* state variables, *r* and *s* also provide divide-by-3 waveforms with a phase shift of a half cycle. This makes the circuit a *three-phase* generator.

```
library ieee;
use ieee.std_logic_1164.all;

entity Div3 is
  port( rst, a: in std_logic;
        q: out std_logic );
end Div3;

architecture impl of Div3 is
  signal r, s: std_logic;
begin
  q <= not rst and ((not r and q) or (not r and a) or (q and a));
  r <= not rst and ((s and a) or ( s and r) or (a and r));
  s <= not rst and ((s and not a) or (s and q) or (q and not a));
end impl; -- Div3
```

Figure 26.9. VHDL implementation for the divide-by-3 circuit of Example 26.1.

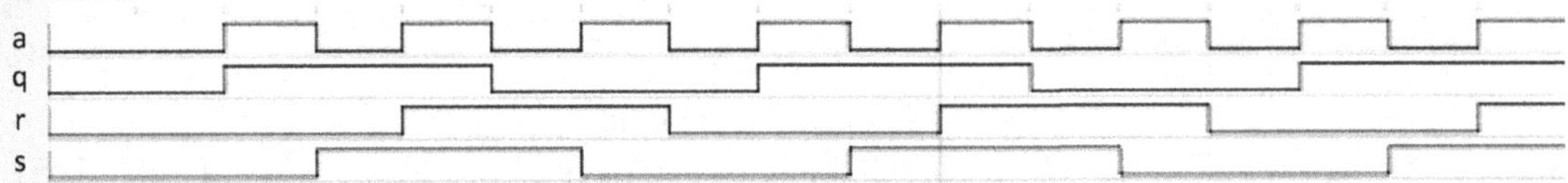

Figure 26.10. Simulated waveform of the VHDL divide-by-3 module from Example 26.1.

26.3 RACES AND STATE ASSIGNMENT

To illustrate the problem of multiple state variables changing simultaneously, consider an alternative state assignment for the toggle circuit shown in Figure 26.11(a). Here we observe that the two outputs, *a* and *b*, can also serve as state variables, so we can add to the outputs just one additional state variable *c* to distinguish between states A and C, giving the codes shown in the figure.[2]

With this state assignment, the transition from state A (*cab* = 000) to state B (110) changes both *c* and *a*. If the logic is designed so that *in* going high in state A makes both *c* and *a* go high, they could change in an arbitrary order. Variable *a* could change first, variable *c* could change first, or they could change simultaneously. If they change simultaneously, we go from state A directly to state B with no intermediate stops. If *a* changes first, we go first to state 010, which is not assigned, and then, if the logic in state 010 does the right thing, to state 110. If *c* changes first, the machine will go to state C (100), where the high input will then drive it to state D. Clearly, we cannot allow *c* to change first. This situation where multiple state variables can change at the same time is called a *race*. The state variables are *racing* to see which one can change first. When the outcome of the race affects the end state – as in this case – we call the race a *critical race*.

[2] Note that the bit ordering of the codes is *c*, *a*, *b*.

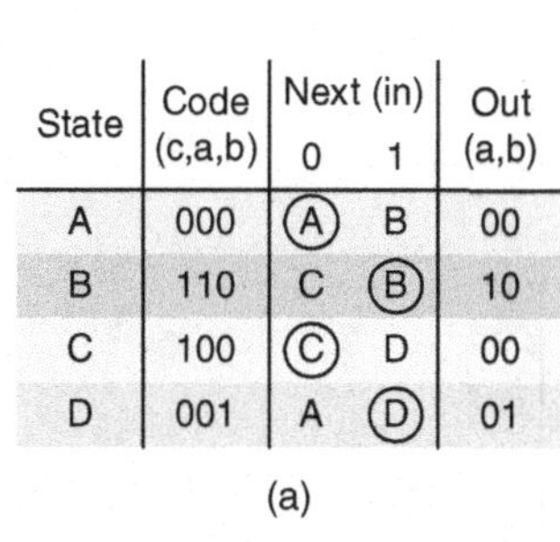

State	Code (c,a,b)	Next (in) 0	Next (in) 1	Out (a,b)
A	000	(A)	B	00
B	110	C	(B)	10
C	100	(C)	D	00
D	001	A	(D)	01

(a)

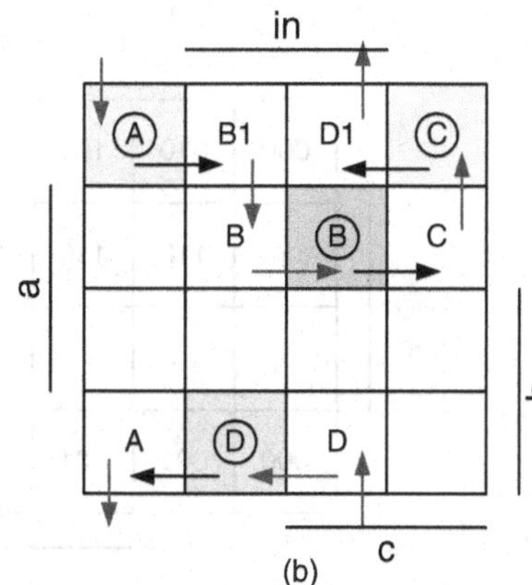

Figure 26.11. An alternative state assignment for the toggle circuit requires multiple state variables to change on a single transition. (a) Flow table with revised state assignment. (b) Trajectory map showing the introduction of transient states B1 = 010 and D1 = 101.

To avoid the critical race that could occur if we allow both *a* and *c* to change at the same time, we specify the next-state function so that only *a* can change in state A. This takes us to a *transient state* 010, which we will call B1. When the machine reaches state B1, the next-state logic then enables *c* to change.

The introduction of this transient state is illustrated in the trajectory map of Figure 26.11(b). When the input goes high in state A, the next-state function specifies B1 rather than B. This enables only a single transition, downward as shown by the arrow, which corresponds to *a* rising, to state B1 (corresponding to the square marked B). A change in *c* is not enabled in state A, in order to avoid a horizontal transition into the square marked D1. Once the machine reaches state B1, the next-state function becomes B, which enables the change in *c*, a horizontal transition, to stable state B.

A transient state is also required for the transition from state C 100 to state D 001. Both variables *c* and *b* change between these two states. An uncontrolled race in which variable *c* changes first could wind up in state B 110, which is not correct. To prevent this race, we enable only *b* to change when *in* rises in state C. This takes us to a transient state D1 (101). Once in state D1, *c* is allowed to fall, taking us to state D (001).

Figure 26.12 illustrates the process of implementing the revised toggle circuit. Figure 26.12(a) shows a Karnaugh map of the next-state function. Each square of the Karnaugh map shows the code for the *next* state for that present state and input. Note that where the next state equals the present state, the state is stable. Transient state B1 (at 010 – the second square along the diagonal) is not stable since it has a next state of 110.

From the next-state Karnaugh map, we can write the individual Karnaugh maps for each state variable. Figures 26.12(b)–(d) show the Karnaugh maps for the individual variables *a*, *b*, and *c*, respectively. Note that the states that are not visited along the state trajectory (the blank squares in Figure 26.12(a)) are don't cares. The machine will never be in these states, thus we don't care what the next-state function is in an unvisited state.

From these Karnaugh maps, we write the state variable equations as follows:

$$a = (in \wedge \overline{b} \wedge \overline{c}) \vee (in \wedge a), \quad (26.5)$$

$$b = (in \wedge \overline{a} \wedge c) \vee (in \wedge b), \quad (26.6)$$

$$c = a \vee (\overline{b} \wedge c). \quad (26.7)$$

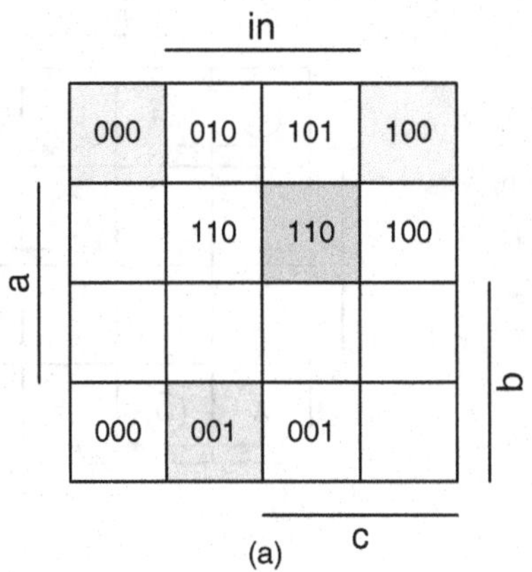

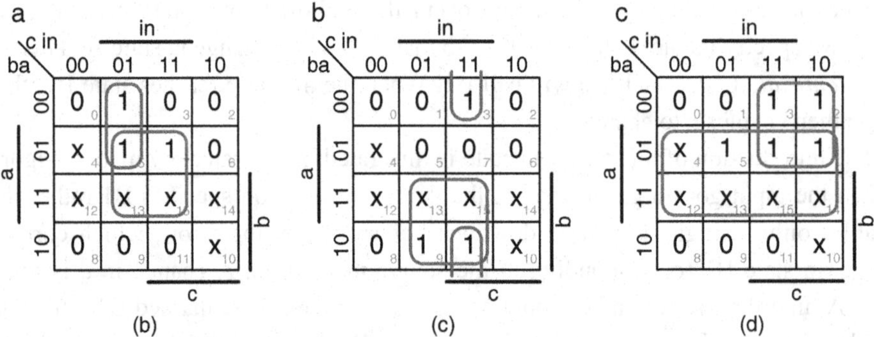

Figure 26.12. Implementation of the toggle circuit with the alternative state assignment of Figure 26.11(a). (a) Karnaugh map showing three-bit next-state function *c*, *a*, *b*. (b) Karnaugh map for *a*. (c) Karnaugh map for *b*. (d) Karnaugh map for *c*.

Note that we do not require separate equations for output variables since *a* and *b* are both state variables and output variables.

EXAMPLE 26.2 One-hot toggle circuit

Design a toggle circuit using a one-hot state assignment. Describe how you resolve any potential races.

We use our outputs *a* and *b* as the one-hot variables representing states B and D, respectively, and add two new state variables *r* and *s* representing states A and C, respectively. So, writing the state as *bsar*, our state assignment is A = 0001, B = 0010, C = 0100, and D = 1000. Two state variables need to change on every state transition, leading to four potential races.

A five-variable transition map for the one-hot toggle circuit is shown in Figure 26.13. When *in* goes high in state A = 0001, we direct the machine to transient state B1 = 0011, which then switches to state B = 0010. The flow of total state here (*in* plus state) is $\{\overline{in}, \mathrm{A}\}$ (00001), $\{in, \mathrm{A}\}$ (10001), $\{in, \mathrm{B1}\}$ (10011), $\{in, \mathrm{B}\}$ (10001). Note that we could allow a race here, by allowing a transition through either B1 (0011) or B2 (0000), but have chosen not to. We revisit this decision in Exercises 26.12 to 26.15.

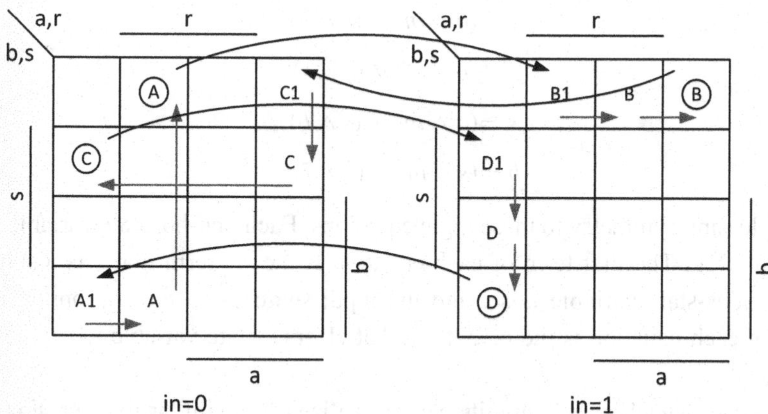

Figure 26.13. Five-variable state transition map for the one-hot toggle circuit of Example 26.2.

When *in* goes low in B, we transfer first to C1 = 0110 and then to C = 0100. Again, we could choose to allow a race through either C1 or C2 = 0000 to C, but have chosen not to. Note that C2 = 0000 does not conflict with B2 = 0000 because C2 is a transient state when *in* is low and B2 is a transient state when *in* is high. In other words, the *total state* – input and internal state – of C2 is 00000, while the total state of B2 is 10000.

Similarly, on the transfer from C to D we force a transition through D1 = 1100. We could allow a race here through either D1 or B2, but only if we choose not to allow a race to B – since B2 can transfer to D or B but not both.

Finally, we force the transition from D to A to pass through A1 = 1001. We could allow a race to go through either A1 or C2, but only if we choose not to allow a race to C.

The K-maps for the four state variables are shown in Figure 26.14.

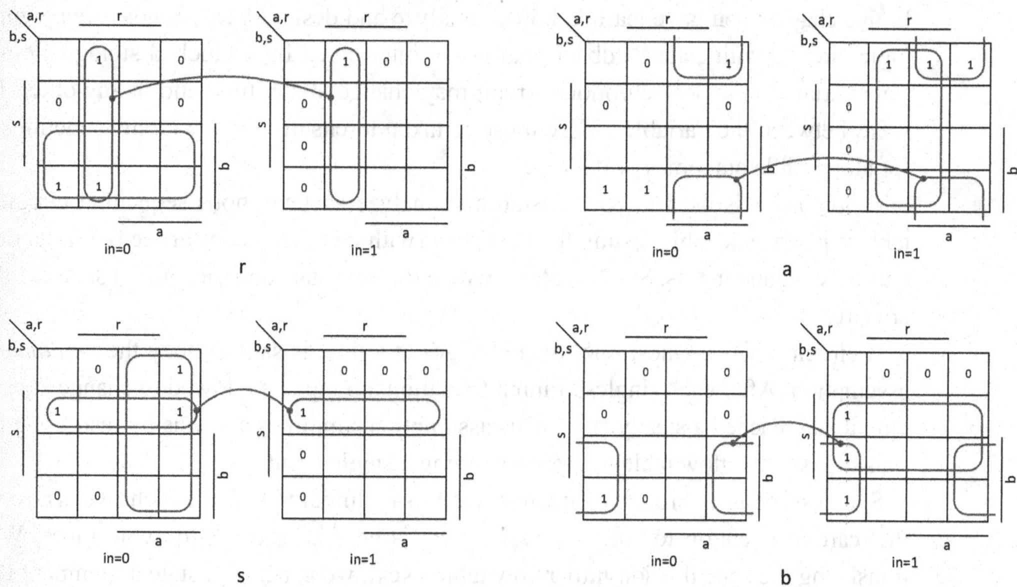

Figure 26.14. Karnaugh maps used to derive the logic equations for the four state variables in Example 26.2.

Writing the logic equations for the state variables from these K-maps, we have

$$r = (b \wedge \overline{in}) \vee (r \wedge \overline{a}),$$

$$a = (r \wedge in) \vee (a \wedge \overline{s}),$$

$$s = (a \wedge \overline{in}) \vee (s \wedge \overline{b}),$$

$$b = (s \wedge in) \vee (b \wedge \overline{r}).$$

There is a pleasing symmetry to these logic equations. Each one-hot state variable is realized with an RS flip-flop. The first term of each equation is the *set* term that sets a state variable when the previous-state variable is set and the input switches to the appropriate state. The second term of each equation is the *reset* term that clears a state variable when the next-state variable is set.

General one-hot asynchronous circuits can be realized in a similar manner. For each state variable x, write a *set equation* S with a term for every transition into that state and a *reset set* R of states that can follow that state. The equation for x can then be written as

$$x = S \vee \left(x \wedge \overline{\sum R} \right). \tag{26.8}$$

That is, any transition into state X corresponding to variable x sets x and any state that follows x clears x. These rules can be used only for asynchronous state machines, where the minimum cycle includes at least three states.

Summary

In this chapter you have learned how to analyze and design *asynchronous sequential circuits*, logic circuits with state feedback that is not interrupted by a clocked storage element. Each state variable in an asynchronous circuit may change at any time and in any order, leading to *races* between the variables. Care must be taken to ensure that the circuit functions correctly for all possible outcomes of the race.

A *flow table* is used to both design and analyze asynchronous sequential circuits. A flow table is just a state table, listing the next state (with feedback loops broken) as a function of the current state and inputs. Stable states – where the next state and the current state are the same – are circled.

A circuit where at most one input changes at a time is said to obey the *fundamental-mode* assumption. After each single-bit input transition, no input is allowed to change again until the circuit has reached a steady state. This assumption simplifies analysis by restricting horizontal transitions in the flow table to those involving a single input.

Synthesis of asynchronous sequential circuits is similar to that of synchronous circuits except that careful attention to the state assignment is needed in order to avoid races. We start by translating the specification into a flow table. Next, we generate a state assignment that avoids multiple state bits changing simultaneously. This may require inserting new *transient* states. Finally, we generate the logic for the resulting next-state function. It is important that the logic be *hazard*-free along trajectories that may be followed in the flow table.

For the rare cases when you need an asynchronous circuit, flow-table analysis and synthesis is an invaluable skill. Too often engineers without this skill attempt to design a circuit like the toggle module in an ad hoc manner. They cobble together flip-flops and latches clocked by both edges of the clock along with a random assortment of gates. The results are at best disappointing and often disastrous. With an understanding of flow tables and races, you can build efficient asynchronous circuits that operate reliably.

BIBLIOGRAPHIC NOTES

Many early computer designs were completely asynchronous, such as the ORDVAC [78] and ILLIAC II [16]. Over time, however, the simplicity of synchronous design won out. Almost all modern digital systems are largely synchronous, with asynchronous design reserved for a few special cases. Ivan Sutherland's Turing Award Lecture [103] provides an interesting perspective on asynchronous logic. Research on asynchronous design continues and is reported in the IEEE International Symposium on Asynchronous Circuits and Systems each year. Texts on the topic include refs. [68], [86], and [108].

Exercises

26.1 *Schematic analysis.* Write the flow table for the schematic in Figure 26.15. Explain the function of the circuit.

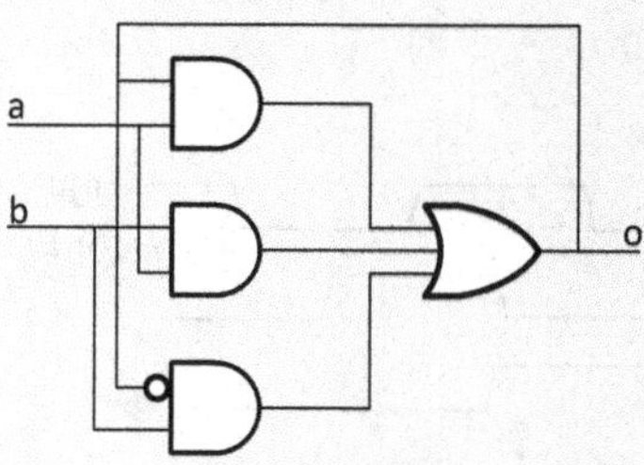

Figure 26.15. Schematic for Exercise 26.1.

26.2 *Synthesis.* A phase comparator takes two input signals with the same frequency and transforms them into a voltage, indicating the degree to which they are in phase (aligned). This problem explores the digital portion of the design. Write a flow table for a basic phase comparator that has two inputs a and b and two outputs A and B. The initial state is to have $A = B = 0$. On a rising edge of a, the output should either transition from (AB) 00 to 10 or from 01 to 00, and should be stable in any other state. On a rising edge of b the transition should be from 10 to 00 or from 00 to 01. Synthesize a gate-level circuit from your flow table. The analog portion of the design translates the amount of time either A or B is high into a voltage magnitude.

26.3 *Toggle synthesis.* Synthesize a version of the toggle circuit where the state assignments are as in Figure 26.11(a), except that state B is encoded as $cab = 010$.

26.4 *Edge toggle.* The toggle circuit of Section 26.2 is a *pulse toggle* – a circuit in which alternate pulses on the input alternate between the two outputs. For this exercise, design an *edge toggle* – a circuit in which edges on the input cause edges that alternate between the two outputs. The waveform of this circuit is shown in Figure 26.16.

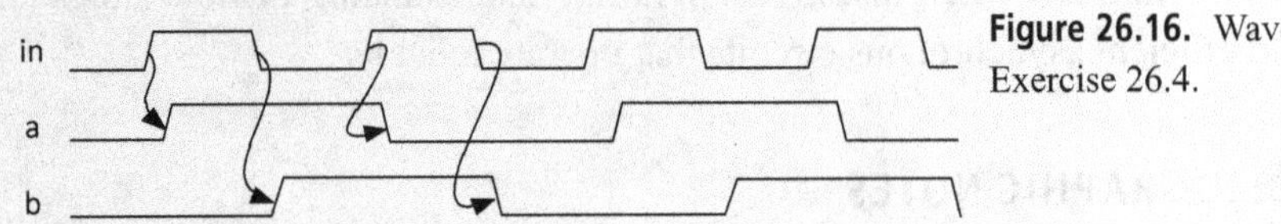

Figure 26.16. Waveform for Exercise 26.4.

26.5 *Three-way toggle.* Design a toggle circuit like the one in Section 26.2 except that pulses on the input alternate over three outputs.

26.6 *Three-way edge toggle.* Modify the edge circuit of Exercise 26.4 to alternate edges between three outputs.

26.7 *Divide-by-3 circuit.* Re-implement the divide-by-3 circuit of Example 26.1 using the following state assignment: $\mathrm{A} = 000, \mathrm{B} = 100, \mathrm{C} = 110, \mathrm{D} = 111, \mathrm{E} = 011, \mathrm{F} = 001$.

26.8 *Divide-by-5 circuit.* Design a divide-by-5 asynchronous circuit, like the divide-by-3 circuit of Example 26.1, except that it produces one transition on the output for every five transitions on the input.

26.9 *Divide-by-3 halves circuit.* Design a circuit that accepts quadrature inputs i and q (inputs that have a 50% duty factor and are 90 degrees out of phase) and generates an output x that has one transition for every three transitions on i and q combined. The desired behavior is illustrated in Figure 26.17.

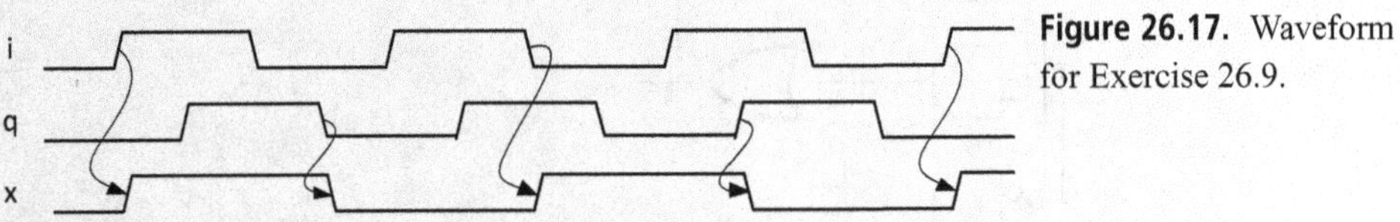

Figure 26.17. Waveform for Exercise 26.9.

26.10 *State reduction.* Write a flow table for the toggle circuit, but this time creating a new state for each of the first eight transitions on *in*. Then determine which states are equivalent, in order to reduce your flow table to a four-state table.

26.11 *Races.* Design a circuit that when input c is high counts from 0 to 4 using standard binary notation (and wraps). If c is low, the circuit will simply hold its state. In your first implementation (part (a)), ignore all data races. For part (b), enumerate all the possible races and potential incorrect sequences of states.

26.12 *One-hot toggle, I.* Write the logic equations for the one-hot toggle circuit of Example 26.2, allowing a race in the transfer from A to B.

26.13 *One-hot toggle, II.* Write the logic equations for the one-hot toggle circuit of Example 26.2, allowing a race in the transfer from B to C.

26.14 *One-hot toggle, III.* Write the logic equations for the one-hot toggle circuit of Example 26.2, allowing races in the transfers from A to B and from D to A.

26.15 *One-hot toggle, IV.* In the one-hot toggle circuit of Example 26.2 is it possible to allow races both from A to B and from C to D? If it is, write the logic equations for a circuit that does so; if not, explain why this is not possible.

27 Flip-flops

Flip-flops are among the most critical circuits in a modern digital system. As we have seen in previous chapters, flip-flops are central to all synchronous sequential logic. Registers (built from flip-flops) hold the state (both control and data state) of all of our finite-state machines. In addition to this central role in logic design, flip-flops also consume a large fraction of the die area, power, and cycle time of a typical digital system.

Until now, we have considered a flip-flop as a black box.[1] In this chapter, we study the internal workings of the flip-flop. We derive the logic design of a typical D flip-flop and show how the timing properties introduced in Chapter 15 follow from this design.

We first develop the flip-flop design informally – following an intuitive argument. We start by developing the latch. The implementation of a latch follows directly from its specification. From the implementation we can then derive the setup, hold, and delay times of the latch. We then see how to build a flip-flop by combining two latches in a master–slave arrangement. The timing properties of the flip-flop can then be derived from its implementation.

Following this informal development, we then derive the design of a latch and flip-flop using flow-table synthesis. This serves both to reinforce the properties of these storage elements and to give a good example of flow-table synthesis. We introduce the concept of *state equivalence* during this derivation. This formal derivation can be skipped by a casual reader.

27.1 INSIDE A LATCH

A schematic symbol for a latch is shown in Figure 27.1(a), and waveforms illustrating its behavior and timing are shown in Figure 27.1(b). A latch has two inputs, data d and enable g, and one output, q. When the enable input is high, the output follows the input. When the enable input is low, the output holds its current state.

As shown in Figure 27.1(b), a latch, like a flip-flop, has a setup time t_s and a hold time t_h. An input must be setup t_s before the enable *falls* and held for t_h after the enable has fallen in order for the input value to be correctly stored. Latch delay is characterized by both delay from the enable rising to the output changing, t_{dGQ}, and delay from the data input changing to the output changing, t_{dDQ}. For the enable to dominate the delay, the input must be set up at least t_{s1} before the enable rises. Usually, this is just a question of which signal (d or g) is on the critical path, and $t_{s1} = t_{dDQ} - t_{dGQ}$. As we shall see below, these times can be derived from the logic design

[1] A *black box* is a system for which we understand the external specifications, but not the internal implementation – as if the system were inside an opaque (black) box that keeps us from seeing how it works.

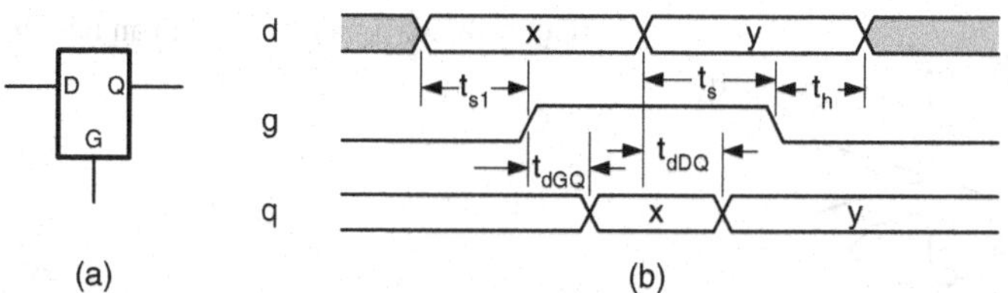

Figure 27.1. A latch: (a) schematic symbol; (b) waveforms showing timing properties.

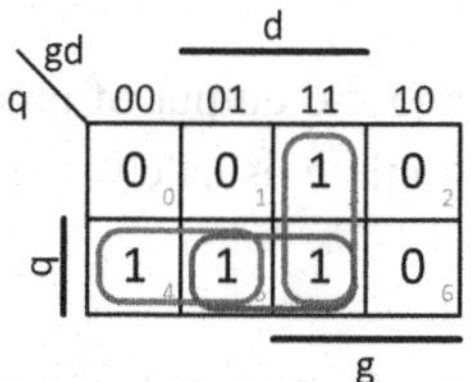

Figure 27.2. Karnaugh map for a latch.

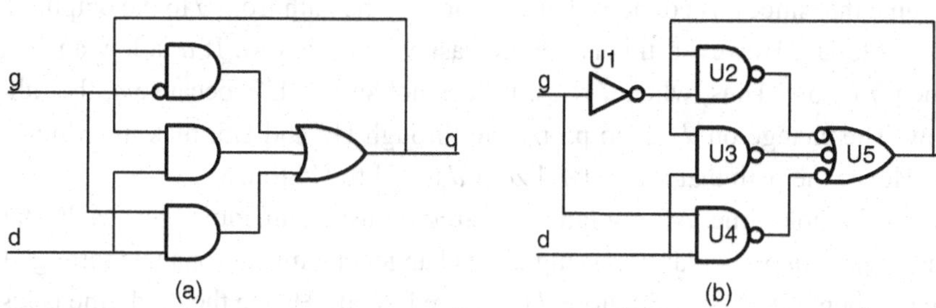

Figure 27.3. Gate-level schematic of an Earle latch. (a) Using abstract AND and OR gates. (b) CMOS implementation using only inverting gates.

of the latch and the need to meet the fundamental-mode restriction (Section 26.1) during latch operation.

From the description of a latch, we can write down its logic equation:

$$q = (g \wedge d) \vee (\overline{g} \wedge q). \tag{27.1}$$

That is, when g is true, $q = d$, and when g is false, q holds its state ($q = q$). This is almost correct. As can be seen from the Karnaugh map of Figure 27.2, there is a hazard that may occur when g changes state and both d and q are high. To cover this hazard, we must add an additional implicant to the equation, as follows:

$$q = (g \wedge d) \vee (\overline{g} \wedge q) \vee (d \wedge q). \tag{27.2}$$

From Equation (27.2) we can draw a gate-level schematic for a latch, as shown in Figure 27.3(a). This implementation of a latch is often called an *Earle* latch after its original developer. For a CMOS implementation, we can redraw the Earle latch using only inverting gates (an inverter and four NAND gates), as shown in Figure 27.3(b).

From the schematic of Figure 27.3, we can now derive the timing properties of the latch. Let the delay of gate *Ui* in the figure be t_i – in practice, we would calculate the delay of these gates as described in Section 5.4 – and the delays may be different for rising and falling edges and for different states. First consider the setup time, t_s. To meet the fundamental-mode restriction, after changing input d, the circuit must be allowed to reach a stable state before input g falls. For the circuit to reach a stable state, the change in d (rising or falling) must propagate through

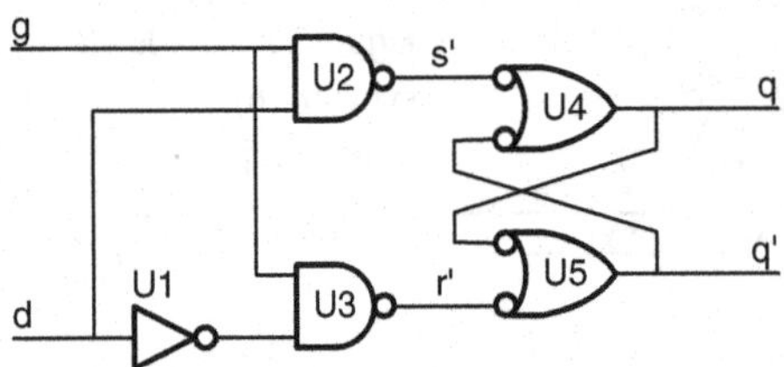

Figure 27.4. Latch built from an RS flip-flop.

gates U4, U5, and U3 in that order, and both state variable q and the output of gate U3 must reach a stable state before g falls. This delay, and hence the setup time, is calculated as

$$t_s = t_4 + t_5 + t_3. \tag{27.3}$$

This setup calculation illustrates some of the subtlety of asynchronous circuit analysis. One might assume that since d is connected directly to U3, the path from d to the output of U3 would be direct – d to U3. However, this is not the case when d is low. If d is low and the circuit is stable, then q is low. Thus, when d rises, it does not switch U3, because q, the other input of U3, is low. The change on d has to propagate through U4 and U5, making q high, before U3 switches. Hence the path that must stabilize is d to U4 to U5 to U3.

Now consider hold time. After g falls, the circuit must again come to a steady state before d can change again. In particular, if d is high, the change on g must propagate through U1 and U2 to enable the loop of U2 and U5 before d is allowed to fall. Hence the hold time is just given by

$$t_h = t_1 + t_2. \tag{27.4}$$

To complete our analysis, we see that each propagation delay is just the delay from an input g or d to the output q. For t_{dDQ} this path includes gates U3, U4, and U5, and for t_{dGQ} the path is through U4 and U5 (recall that we are worried only about g rising here). Thus we have

$$t_{dDQ} = \max(t_3, t_4) + t_5, \tag{27.5}$$

$$t_{dGQ} = t_4 + t_5. \tag{27.6}$$

An alternative gate implementation of a latch is shown in Figure 27.4. Here we construct a latch by appending a gating circuit to an RS flip-flop (Section 14.1). The RS flip-flop is formed by NAND gates U4 and U5. When the upper input of U4, $\bar{s}$, is asserted (low), the flip-flop is set ($q = 1, \bar{q} = 0$). When the lower input of U5, $\bar{r}$, is asserted (low), the flip-flop is reset ($q = 0$, $\bar{q} = 1$).

The gating circuit, formed by U1, U2, and U3, sets the flip-flop when $g = 1$ and $d = 1$, and resets the flip-flop when $g = 1$ and $d = 0$. Thus, when g is high, the output q follows input d. When g is low, the flip-flop is neither set nor reset, and holds its previous state.

While the latch of Figure 27.4 has identical logical behavior to the Earle latch of Figure 27.3, it has very different timing properties. We leave derivation of these timing properties as Exercises 27.1–27.4.

27.2 INSIDE A FLIP-FLOP

An edge-triggered D-type flip-flop updates its output with the current state of its input on the rising edge of the clock. At all other times the output holds its current state. The schematic

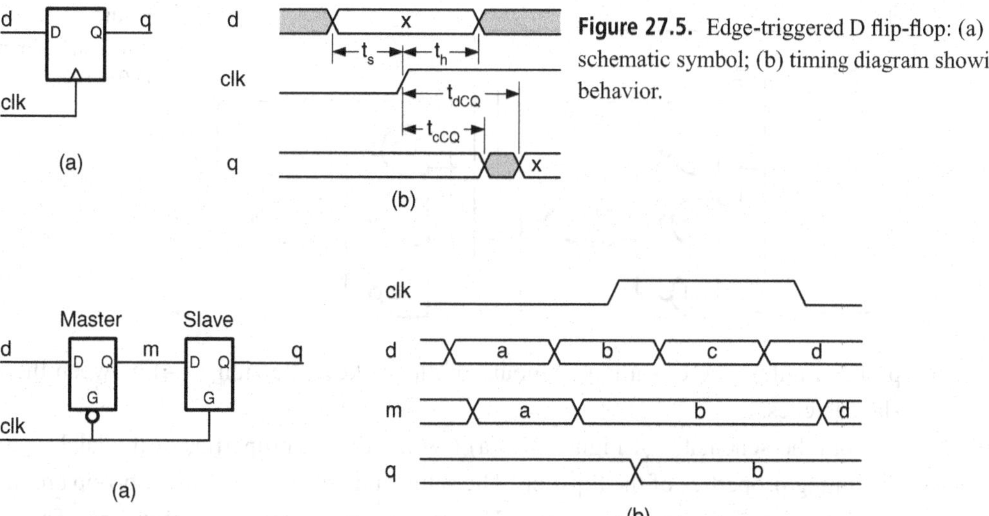

Figure 27.5. Edge-triggered D flip-flop: (a) schematic symbol; (b) timing diagram showing behavior.

Figure 27.6. A master–slave D flip-flop is constructed from two latches.

symbol and timing diagram of a D flip-flop are shown in Figure 27.5 (repeated from Figure 15.5). In Section 14.2 we saw how D flip-flops are used for the state registers of synchronous sequential circuits, and in Chapter 15 we looked at the detailed timing properties of the D flip-flop. In this section, we derive a logic design for the D flip-flop and see how this logic design gives rise to the timing properties of the flip-flop.

A latch with a negated enable does half of what we need to make a flip-flop. It samples its input when the enable rises, and holds the output steady while the negated enable is high. The problem is that its output follows its input when the negated enable is low. We can use a second latch, in series with the first, to correct this behavior. This latch, with a normal enable, prevents the output from changing when the enable is low.

A D flip-flop implemented with two latches in series with complemented enables is shown in Figure 27.6(a). Waveforms illustrating the operation of the flip-flop are shown in Figure 27.6(b). The first latch, called the *master*, samples the input on the rising edge of the clock onto intermediate signal m. In the waveforms, value b is sampled and held steady on m while the clock is high. Signal m, however, follows the input when the clock is low. The second latch, called the *slave* (because it follows the master), enables the data captured on signal m onto the output when the clock is high. When the clock falls, it samples this value – still held steady by the master – and holds this value on the output when the clock is low. The net result of the two latches is a device that acts as a D flip-flop. It samples the data on the rising edge of the clock, and holds it steady until the next rising edge. The master holds the value when the clock is high – while the slave is transparent, and the slave holds the value when the clock is low – while the master is transparent.

For correct operation of the master–slave flip-flop, it is critical that the output of the master does not change until t_h after the slave clock has fallen. That is, the hold-time constraint of the slave must be met. If the output of the master latch were to change too quickly after the clock has fallen, the new value on intermediate signal m (value of d in the waveform) could race through to the output before the slave latch blocks its flow. In practice, this is rarely a

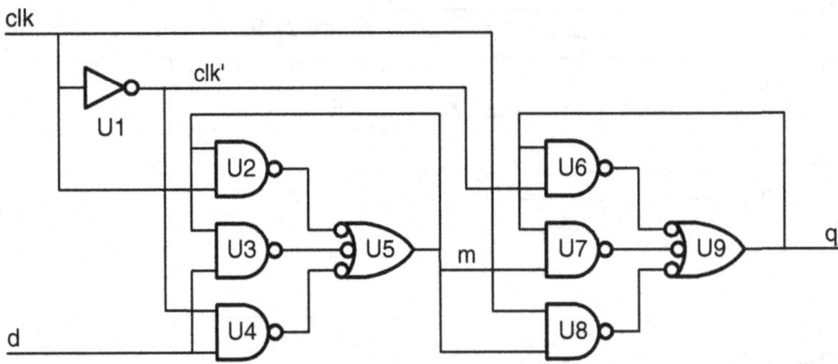

Figure 27.7. Gate-level schematic of a master–slave D flip-flop.

problem unless there is a large amount of clock skew (Section 15.4) between the master and slave latches.

From the schematic of Figure 27.6(a) and the timing properties of the latch, we can derive the timing properties of the flip-flop. The setup and hold times of the flip-flop are just the setup and hold times of the master latch. This is the device doing the sampling. The delay of the flip-flop, t_{dCQ}, is the delay of the slave latch, t_{dGQ}.[2] The output q changes to the new value, t_{dGQ}, after the clock rises.

To illustrate the timing parameters of the D flip-flop in a more concrete way, consider the gate-level schematic of Figure 27.7. This figure shows the master–slave flip-flop of Figure 27.6 expanded out to the gate level. The inverter on the enable input of each latch has been factored out to a single inverter U1 that generates $\overline{clk}$. The polarities of the clock connection to the master and slave are opposite.

The setup time, t_s, of the flip-flop shown in Figure 27.7 is the amount of time required for the circuit to reach a steady state after d changes when clk is low.

Strictly speaking, the path here is through U4, U5, and U3 or U7 (if q is high). However, we don't really care whether the output of U7 reaches steady state or not since the clock is about to go high, enabling U8. So we ignore U7 and focus on the path that ends at the output of U3, giving

$$t_s = t_4 + t_5 + t_3 = t_{sm}, \tag{27.7}$$

where t_{sm} is the setup time of the master latch.

The hold time of this flip-flop is the time for the master part of the circuit to settle after the clock has risen. This is just the maximum delay of gates U1 and U2. If the data changes before U1 has brought $\overline{clk}$ low, state m might be affected. Similarly, the data cannot change until the output of U2 has stabilized or we may lose the value stored in the master latch. Thus we have

$$t_h = \max(t_1, t_2). \tag{27.8}$$

Compared with Equation (27.4), we don't add t_1 to t_2 here because, with the low-true enable of the master latch, U1 is not in the path from clk to the output of U2.

For hold-time calculation, we are concerned only with the master latch reaching steady state. When the clock rises, it may take longer for the slave to reach steady state. However, we will capture the data reliably as long as the master reaches steady state.

[2] This assumes that the setup time is large enough that signal m is stable t_{s1} before the clock rises. See Exercise 27.11 for an example where this is not the case.

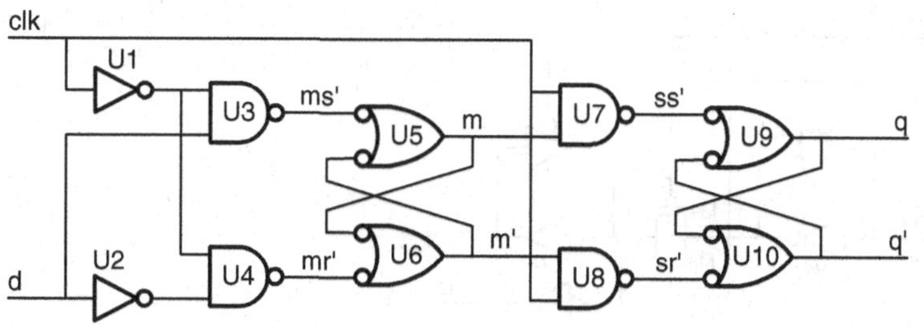

Figure 27.8. Alternative gate-level schematic of a master–slave D flip-flop.

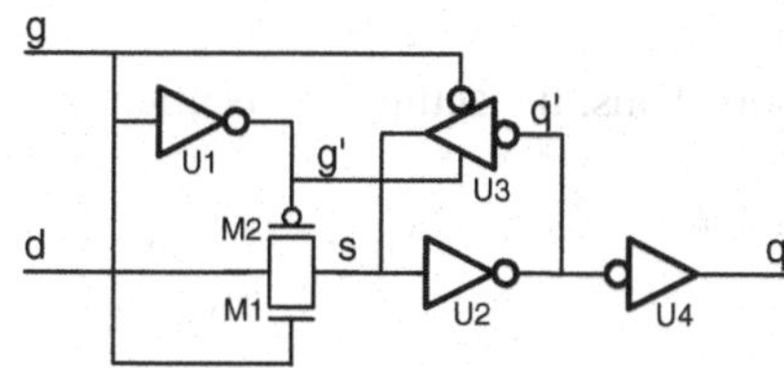

Figure 27.9. CMOS latch circuit using a transmission gate and a tri-state inverter.

Finally, t_{dCQ} is the delay from the clock rising to the output of the flip-flop. The path here is through U8 and U9, giving

$$t_{dCQ} = t_8 + t_9 = t_{dGQs}, \tag{27.9}$$

where t_{dGQs} is t_{dGQ} for the slave latch.

An alternative gate-level schematic for a master–slave D flip-flop is shown in Figure 27.8. This circuit uses the RS-flip-flop-based latch of Figure 27.4 for the master and slave latches of the flip-flop. We leave the derivation of t_s, t_h, and t_{dCQ} for this flip-flop as Exercises 27.5–27.7.

27.3 CMOS LATCHES AND FLIP-FLOPS

The latch circuits of either Figure 27.3 or Figure 27.4 use static CMOS gates. CMOS technology, however, also permits us to construct a latch with a transmission gate and a tri-state inverter, as shown in Figure 27.9. Most CMOS latches use transmission gates in this style because this results in a latch that is both smaller and faster than the alternative gate circuits.

When enable g is high (and $\bar{g}$ is low), the transmission gate formed by NFET M1 and PFET M2 is *on*, allowing the value on input d to pass to storage node s. If $d = 1$, PFET M2 passes the 1 from d to s, and, if $d = 0$, NFET M1 passes the 0 from d to s. Output q follows storage node s buffered by inverters U2 and U4. Thus, when g is high, the output q follows input d.

When enable g goes low, the transmission gate formed by M1 and M2 turns off isolating storage node s from the input. At this time, the input is sampled onto the storage node. At the same time, tri-state inverter U3 turns on, closing a storage loop from s back to s through two inverters. This feedback loop reinforces the stored value, allowing it to be retained indefinitely. The tri-state inverter is equivalent to an inverter followed by a transmission gate.

We can calculate the setup, hold, and delay times of the CMOS latch in the same manner as we did for the gate-based latches. When the input changes while g is high, the effect of the change on the storage loop must settle out before g is allowed to fall. The input change must propagate through the transmission gate and inverter U2. We need not wait for U4 to drive

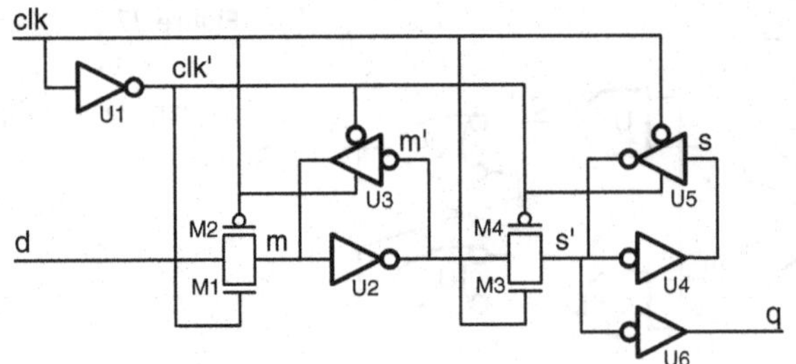

Figure 27.10. A CMOS flip-flop circuit is constructed from two CMOS latches.

output q since that does not affect the storage loop. Thus, the setup time is given by

$$t_s = t_g + t_2, \tag{27.10}$$

where t_g is the delay of the transmission gate.

After g goes low, we need to hold the value on input d until the transmission gate is completely shut off. This is just the delay of inverter U1:

$$t_h = t_1. \tag{27.11}$$

We do not need to wait for feedback gate U3 to turn on. Its output, storage node s, is already in the correct state, and there is sufficient capacitance on node s to hold its value steady until U3 turns on.

The delay times are calculated by tracing the path from input to output:

$$t_{dGQ} = t_1 + t_g + t_2 + t_4, \tag{27.12}$$

$$t_{dDQ} = t_g + t_2 + t_4. \tag{27.13}$$

A CMOS flip-flop can be constructed from two CMOS latches, as shown in Figure 27.10. The master latch, formed by NFET M1, PFET M2, inverter U2, and tri-state inverter U3, connects input d to storage node m when enable g is low and holds m when g is high. The slave latch, formed by NFET M3, PFET M4, inverter U4, and tri-state inverter U5, connects master latch state $\overline{m}$ to storage node $\overline{s}$ when enable g is high. An additional inverter, U6, generates output q. Output q is logically identical to node s at the output of U4. Isolating the output from the storage loop in this manner, however, is critical for the synchronization properties of the flip-flop, as discussed in Chapter 29. We leave the analysis of this flip-flop as Exercises 27.8–27.10.

27.4 FLOW-TABLE DERIVATION OF THE LATCH

The latch and flip-flop are themselves asynchronous circuits and can be synthesized using the flow-table technique we developed in Chapter 26. Given an English-language description of a latch, we can write a flow table for the latch, as shown in Figure 27.11. To enumerate the states, we start with one state – all inputs and outputs low (state A) – and from this state we toggle each input. We then repeat this process from each new state until all possible states have been explored.

In state A, toggling g high takes us to state B, and toggling d high takes us to state F. This gives us the first line of the flow table of Figure 27.11(b). In state B, toggling g low takes us

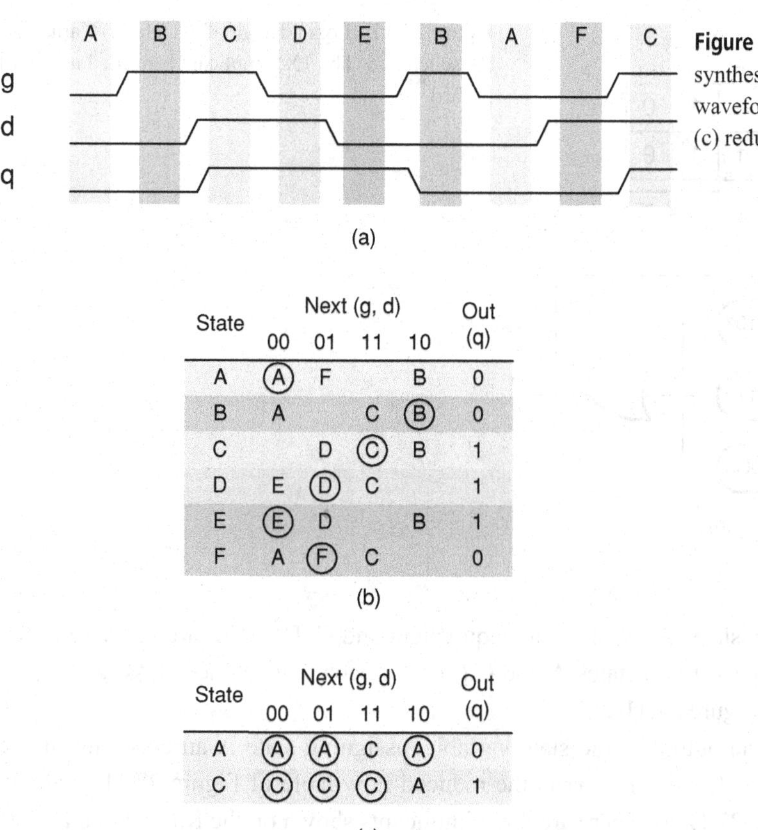

State	Next (g, d) 00	01	11	10	Out (q)
A	(A)	F		B	0
B	A		C	(B)	0
C		D	(C)	B	1
D	E	(D)	C		1
E	(E)	D		B	1
F	A	(F)	C		0

(b)

State	Next (g, d) 00	01	11	10	Out (q)
A	(A)	(A)	C	(A)	0
C	(C)	(C)	(C)	A	1

(c)

Figure 27.11. Flow-table synthesis of a latch: (a) waveforms; (b) flow table; (c) reduced flow table.

back to state A, and toggling d high takes us to state C, with q high. This gives us the second line of the flow table. We continue in this way, constructing the flow table line by line until all states have been explored. The end result is the table of Figure 27.11(b).

In constructing this flow table we have created a new state for each input combination that didn't obviously match a previous state. This gives us six states, A–F, which it would take three state variables to represent. However, many of these states are *equivalent* and can be combined. Two states are equivalent if they are indistinguishable from the inputs and outputs of the circuit.

We define equivalence recursively. Two states are *0-equivalent* if they have the same output for all input combinations. Here, states A, B, and F are 0-equivalent, as are states C, D, and E. Two states are *k-equivalent* if they have the same output for all input combinations and their next states for each input combination are $(k-1)$-equivalent. Here we see that the next states for A, B, and F aren't just equivalent, they are the same, so A, B, and F are also 1-equivalent. Similarly for C, D, and E.

In practice, we find equivalent states by forming sets of states that are 0-equivalent (e.g., $\{A, B, F\}$ and $\{C, D, E\}$), then sets of states that are 1-equivalent, and so on until the sets don't change. That is, when we find a set of states that is both k-equivalent and $(k+1)$-equivalent we're done – these sets are equivalent.

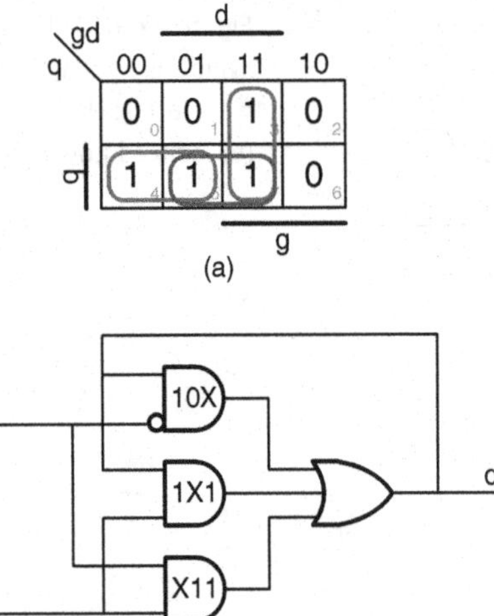

Figure 27.12. Logic design of a latch: (a) Karnaugh map; (b) schematic. The 1X1 implicant is needed in order to prevent a hazard.

In this case, states A, B, and F are equivalent and C, D, and E are equivalent. Rewriting the flow table with just two states A and C (one for each equivalence class) gives us the reduced flow table of Figure 27.11(c).

If we use the output as the state variable, assigning state A an encoding of 0 and state C an encoding of 1, we can rewrite the reduced flow table of Figure 27.11(c) as the Karnaugh map of Figure 27.12(a). There are three implicants shown on the Karnaugh map. While we can cover the function with just two implicants (qgd = X11 $\vee$ 10X), we need the third, 1X1, to avoid a hazard (see Section 6.10). If the output were to dip low momentarily when the enable g input falls (going from qgd = 111 to 101), the end state could be 001 – with the latch losing the stored 1. The added implicant (1X1) avoids this hazard. A schematic for the latch circuit is shown in Figure 27.12(b).

27.5 FLOW-TABLE SYNTHESIS OF A D FLIP-FLOP

Figure 27.13 shows the derivation of a flow table for a D-type flip-flop. We start by showing a waveform that visits all eight states shown in the flow table. As with the latch, we construct the flow table by toggling each input (*clk* and *d*) in each state. We create a new state for each input combination unless it is obviously equivalent to a state that we have already visited. The resulting flow table has eight states and is shown in Figure 27.13(b).

The next step is to find state equivalence classes. We start by observing that the 0-equivalent sets are {A, B, C, D} and {E, F, G, H}. For 1-equivalence, we observe that the next state of D with an input of 11 is in a different class than that of the rest of its set. Similarly for H with an input of 10. Thus, the 1-equivalent sets become {A, B, C}, {D}, {E, F, G}, break and {H}.

A reduced flow table with just these four states is shown in Figure 27.13(c). Our output q will be one of our state variables. We assign a second state variable as shown in the "Code"

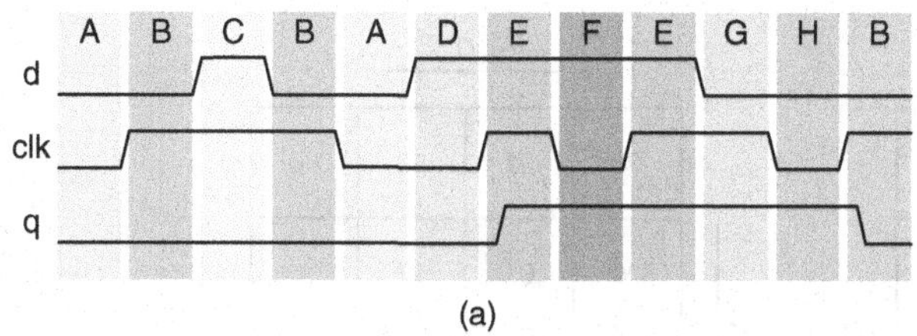

(a)

State	Next (clk d) 00	01	11	10	Out (q)
A	(A)	D		B	0
B	A		C	(B)	0
C		D	(C)	B	0
D	A	(D)	E		0
E		F	(E)	G	1
F	H	(F)	E		1
G	H		E	(G)	1
H	(H)	F		B	1

(b)

State	Code	Next (clk d) 00	01	11	10	Out (q)
ABC	00	(A)	D	(C)	(B)	0
D	01	A	(D)	E		0
EFG	11	H	(F)	(E)	(G)	1
H	10	(H)	F		B	1

(c)

Figure 27.13. Flow-table synthesis of an edge-triggered D flip-flop: (a) waveforms; (b) flow table; (c) reduced flow table.

column of the table. Here q is the MSB of the two-bit state code, and our new variable is the LSB.

Figure 27.14 shows how the flow table of Figure 27.13(c) is reduced to logic. We start by redrawing the flow table as a Karnaugh map with symbolic entries in Figure 27.14(a). Variable q is the output, and variable x is our additional state variable. The next step is to replace the symbolic next states with their state codes qx, as shown in Figure 27.14(b). We next write down the Karnaugh maps for the two state variables in Figure 27.14(c) – q is on the left, x is on the right.

The two next-state variable functions can each be covered with two implicants. As with the latch, we must add a third to each to avoid a hazard. For variable q, we have implicants $qxcd =$ X11X, 1X0X, and, to eliminate the hazard, 11XX. For x we have implicants X11X, XX01, and, for the hazard, X1X1. Drawing out these implicants as the logic diagram of Figure 27.14(d), we see that we have synthesized the master–slave D flip-flop using Earle latches of Figure 27.7. The astute reader will notice that the top AND gate of the master latch and the bottom AND gate of the slave latch have identical inputs and can be replaced by a single shared AND gate.

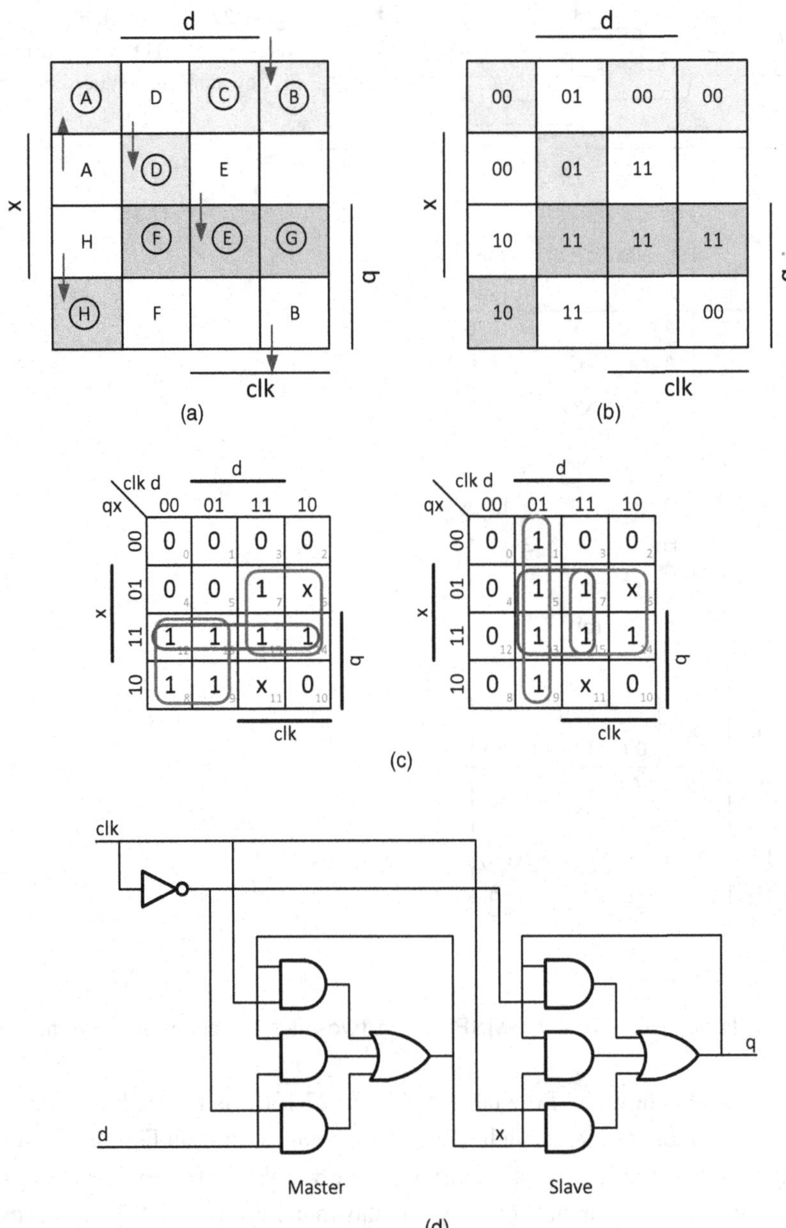

Figure 27.14. Deriving the logic design of a D flip-flop from the flow table of Figure 27.13(c). (a) Flow table drawn on a Karnaugh map. (b) Karnaugh map showing next-state function. (c) Karnaugh maps for each bit of next-state function. (d) Logic diagram derived from Karnaugh maps.

Summary

In this chapter you have looked inside a flip-flop to understand better its behavior and timing constraints. We first derived a latch circuit, the *Earle* latch, and, using this circuit we derived the setup, hold, and delay times of a latch, in terms of internal gate delays within the latch.

Combining two latches forms a *master–slave* flip-flop. The *master* latch, enabled by clock low, follows the input while the clock is low and samples the input on the rising edge of the clock. The *slave* latch, enabled by clock high, holds the output steady while the clock is low. Analyzing this circuit, we derived the setup, hold, and delay times for a flip-flop.

In CMOS logic, latches and flip-flops are often implemented using *transmission gates*. Their behavior and analysis, however, are identical to those of the Earle latch.

Finally, we performed flow-table analysis of both the latch and the flip-flop. This analysis gives more insight into the behavior of the flip-flop and provides a good example of flow-table analysis.

BIBLIOGRAPHIC NOTES

The Earle latch was developed at IBM in the mid 1960s. It was disclosed in 1965 [38] and granted a patent in 1967 [39]. For a more thorough survey of flip-flops, consult a circuit text [112] or comparison-based research papers such as refs. [28] and [67].

Exercises

27.1 *Latch timing properties, I.* Compute t_s for the latch of Figure 27.4 in terms of the delays of the individual gates. Assume that the delay of gate *Ui* is t_i.

27.2 *Latch timing properties, II.* Compute t_h for the latch of Figure 27.4 in terms of the delays of the individual gates. Assume that the delay of gate *Ui* is t_i.

27.3 *Latch timing properties, III.* Compute t_{dDQ} for the latch of Figure 27.4 in terms of the delays of the individual gates. Assume that the delay of gate *Ui* is t_i.

27.4 *Latch timing properties, IV.* Compute t_{dGQ} for the latch of Figure 27.4 in terms of the delays of the individual gates. Assume that the delay of gate *Ui* is t_i.

27.5 *Flip-flop timing properties, I.* Compute t_s for the D flip-flop of Figure 27.8 in terms of the delays of the individual gates. Assume that the delay of gate *Ui* is t_i.

27.6 *Flip-flop timing properties, II.* Compute t_h for the D flip-flop of Figure 27.8 in terms of the delays of the individual gates. Assume that the delay of gate *Ui* is t_i.

27.7 *Flip-flop timing properties, III.* Compute t_{dCQ} for the D flip-flop of Figure 27.8 in terms of the delays of the individual gates. Assume that the delay of gate *Ui* is t_i.

27.8 *CMOS flip-flop timing properties, I.* Compute t_s for the D flip-flop of Figure 27.10 in terms of the delays of the individual gates (and transmission gates). Assume that the delay of gate *Ui* is t_i and that the delay of each transmission gate is t_g.

27.9 *CMOS flip-flop timing properties, II.* Compute t_h for the D flip-flop of Figure 27.10 in terms of the delays of the individual gates (and transmission gates). Assume that the delay of gate *Ui* is t_i and that the delay of each transmission gate is t_g.

27.10 *CMOS flip-flop timing properties, III.* Compute t_{dCQ} for the D flip-flop of Figure 27.10 in terms of the delays of the individual gates (and transmission gates). Assume that the delay of gate *Ui* is t_i and that the delay of each transmission gate is t_g.

27.11 *Flip-flop contamination delay, I.* Consider a flip-flop constructed by placing a delay line with delay t_d between the master and slave latches. When the flip-flop input d changes exactly t_s before the clock rises, signal m at the input of the delay line becomes valid exactly at the clock edge, and signal $m1$ at the output of the delay line and input to the slave latch becomes valid t_d after the rising edge of the clock. What is the contamination delay t_{cCQ} and propagation delay t_{dCQ} of this modified flip-flop?

27.12 *Flip-flop contamination delay, II.* What is the maximum legal value of t_d for the flip-flop of Exercise 27.11?

27.13 *Pulsed-latch flip-flop, I.* Another D flip-flop design is shown in Figure 27.15. This flip-flop consists of a single latch that is gated by a pulse generator. When *clk* rises, a narrow pulse is generated on enable $\overline{g}$ to sample *d*. The latch then holds the sampled value until the next rising edge of the clock. Answer the following questions about the circuit of Figure 27.15(c). Assume that the delay of gate *Ui* is t_i.

(a) What is the minimum pulse width generated by U2 for which this circuit will work properly?

(b) What is the maximum pulse width generated by U2 (if any) for which this circuit will work properly?

27.14 *Pulsed-latch flip-flop, II.* Assuming that the pulse width out of U2 is the value you computed in Exercise 27.13(a), compute t_s, t_h, and t_{dCQ} for this pulsed-latch flip-flop and compare these values with the timing parameters of the master–slave flip-flop analyzed in the text. In particular, compare the *overhead*, $t_s + t_{dCQ}$, of the two flip-flops.

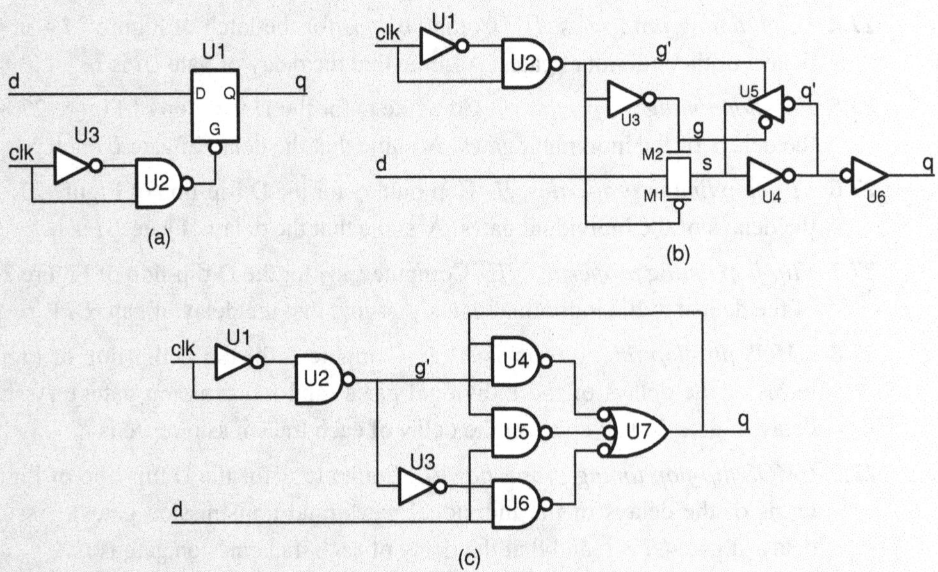

Figure 27.15. Pulsed-latch D-type flip-flop: (a) schematic using latch symbol; (b) schematic showing internals using a CMOS latch; (c) schematic showing internals using an Earle latch. When *clk* rises, NAND gate U2 generates a narrow low-going pulse on $\overline{g}$ that enables the latch to sample *d*.

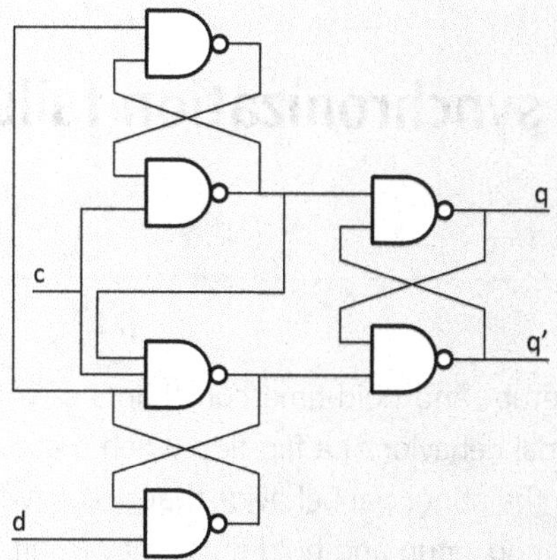

Figure 27.16. The flip-flop design of the 7474, used in Exercises 27.17 and 27.18. It consists of a four-state asynchronous circuit feeding an RS latch with inverted inputs.

27.15 *CMOS pulsed-latch flip-flop, I.* Repeat Exercise 27.13 for the CMOS pulsed latch of Figure 27.15(b).

27.16 *CMOS pulsed-latch flip-flop, II.* Repeat Exercise 27.14 for the CMOS pulsed latch of Figure 27.15(b).

27.17 *The 7474 flip-flop, I.* Figure 27.16 shows the schematic of another type of D flip-flop: the 7474. Like other D flip-flops, it has a clock (c) and data (d) inputs. The circuit itself can be broken into an RS latch (rightmost two NAND gates) and a four-state asynchronous circuit (leftmost four NAND gates). Draw the flow diagram for this four-state asynchronous circuit, circling the stable states. Explain how this circuit operates. Include the state transitions, showing that the output can change only on the rising clock edge.

27.18 *The 7474 flip-flop, II.* What is the setup and hold time of the 7474 D flip-flop of Figure 27.16? Assume that the delay of each NAND gate is a constant t.

27.19 *Combining logic and storage.* This question explores building a latch that takes in two inputs (a and b) and stores the AND of their values.

(a) Using the Earle latch of Figure 27.3, derive the timings of a system that places a two-input AND gate before input d. Assume that the delay of gate Ui is t_i.

(b) Modify the Earle latch by adding another input to both U3 and U4, such that the AND is performed inside the feedback loop. What are the new timings with respect to g?

(c) What are the setup and hold times of a with respect to b?

28 Metastability and synchronization failure

What happens when we violate the setup- and hold-time constraints of a flip-flop? Until now, we have considered only the normal behavior of a flip-flop when these constraints are satisfied. In this chapter we investigate the abnormal behavior that occurs when we violate these constraints. We will see that violating setup and hold times may result in the flip-flop entering a *metastable* state in which its state variable is neither a 1 nor a 0. It may stay in this metastable state for an indefinite amount of time before arriving at one of the two stable states (0 or 1). This *synchronization failure* can lead to serious problems in digital systems.

To stretch an analogy, flip-flops are a lot like people. If you treat them well, they will behave well. If you mistreat them, they behave poorly. In the case of flip-flops, you treat them well by observing their setup and hold constraints. As long as they are well treated, flip-flops will function properly, never missing a bit. If, however, you mistreat your flip-flop by violating the setup and hold constraints, it may react by misbehaving – staying indefinitely in a metastable state. This chapter explores what happens when these good flip-flops go bad.

28.1 SYNCHRONIZATION FAILURE

When we violate the setup- or hold-time constraints of a D flip-flop, we can put the internal state of the flip-flop into an *illegal* state. That is, the internal nodes of the flip-flop can be left at a voltage that is neither a 0 nor a 1. If the output of the flip-flop is sampled while it is in this state, the result is indeterminate and possibly inconsistent. Some gates may see the flip-flop output as a 0, while others may see it as a 1, and still others may propagate the indeterminate state.

Consider the following experiment with a D flip-flop. Initially both *d* and *clk* are low. During our experiment, they both rise. If signal *d* rises t_s before *clk*, the output *q* will be 1 at the end of the experiment. If signal *clk* rises t_h before *d*, the output *q* will be 0 at the end of the experiment. Now consider what happens as we sweep the rise time of *d* relative to *clk* from t_s before *clk* to t_h after clock. Somewhere during this interval, the output *q* at the end of our experiment changes from 1 to 0.

To see what happens when we change the input during this *forbidden interval*, consider the master latch of a CMOS D flip-flop, as shown in Figure 28.1(a). Here we assume that the supply voltage is 1 V, so a logic 1 is 1 V and a logic 0 is 0 V. Figure 28.1(b) shows the initial-state voltage of the flip-flop (the voltage across the inverter $\Delta V = V_1 - V_2$ the instant after the clock has risen) as a function of data transition time. If *d* rises at least t_s before *clk*, the node labeled V_1 is fully charged to 1 V and the node labeled V_2 is fully discharged to 0 V before the clock falls. Thus the state voltage $\Delta V = V_1 - V_2$ is 1 V. As *d* changes later, the state voltage is

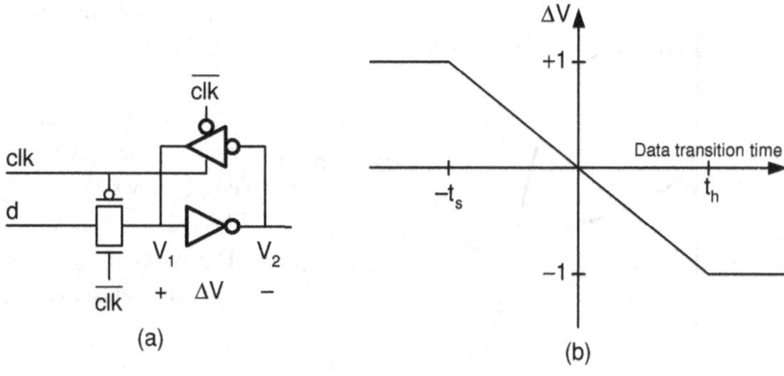

Figure 28.1. Abnormal operation of the master latch of a flip-flop. (a) Schematic of the latch. (b) State voltage vs. data transition time. As the data transition time sweeps from t_s before the clock to t_h after the clock, the state voltage – i.e., the voltage across the storage inverters ($\Delta V = V_1 - V_2$) – changes from +1 to −1.

reduced as shown in Figure 28.1(b). At first, V_1 is still fully charged, but V_2 doesn't have time to discharge fully. As d rises still later, V_1 doesn't have time to charge fully. Finally, when d rises t_h after the clock, V_1 doesn't have time to charge at all, and we have $V_1 = 0$ and $V_2 = 1$ so $\Delta V = -1$.

The initial-state voltage ΔV is a continuous function of the data transition time as it sweeps from $-t_s$ to t_h. It may not be an exact linear function as shown in the figure, but it is continuous, and it does cross zero at some point during this interval. Over this entire interval, the flip-flop has an initial state voltage that is not +1 or −1, hence its output is not a fully restored digital signal and may be misinterpreted by following stages of logic.[1]

28.2 METASTABILITY

The good news about synchronization failure is that most of the illegal states that flip-flops can be left in after violating timing constraints decay quickly to a legal 0 or 1 state. Unfortunately, it is possible for the circuit to be left in an illegal *metastable* state where it may remain for an arbitrary amount of time before decaying to a legal state.

After the clock has risen, the latch of Figure 28.1(a) becomes a regenerative feedback loop (Figure 28.2(a)). The input transmission gate of the latch is off and the feedback tri-state inverter is enabled so that the equivalent circuit is that of two back-to-back inverters.

The DC transfer curve of these back-to-back inverters is shown in Figure 28.2(b). This figure shows the transfer curve of the forward inverter, V_2, as a function of V_1 ($V_2 = f(V_1)$, solid line), and the transfer curve of the feedback tri-state inverter, V_1, as a function of V_2 ($V_1 = f(V_2)$, dashed line). There are three points on the figure where the two lines cross. These points are *stable* in the sense that, in the absence of perturbations, the voltages V_1 and V_2 do not change. Since $V_1 = f(V_2) = f(f(V_1))$, the circuit can sit at any of these three points indefinitely.

At any point other than these three stable points, the circuit quickly converges to one of the outer two stable points. For example, suppose that V_2 is just slightly below the center point, as shown in Figure 28.3. This drives V_1 to a point found by going horizontally from this point to the dashed line. This in turn drives V_2 to a point found by going vertically to the solid line, and so on. The circuit quickly converges to $V_1 = 1$, $V_2 = 0$.

[1] Technically, any output voltage above V_{IH} and below V_{IL} will not be misinterpreted. For the rest of the chapter, however, we classify initial-state voltages between −1 and +1 as illegal.

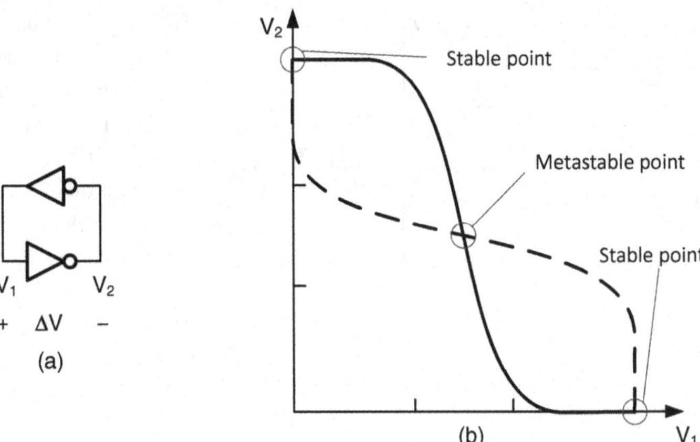

Figure 28.2. (a) The input latch of a flip-flop with the clock high acts as two back-to-back inverters. (b) DC transfer characteristics of the back-to-back inverters. The solid curve shows $V_2 = f(V_1)$ and the dashed curve shows $V_1 = f(V_2)$. The system has two stable points and one metastable point.

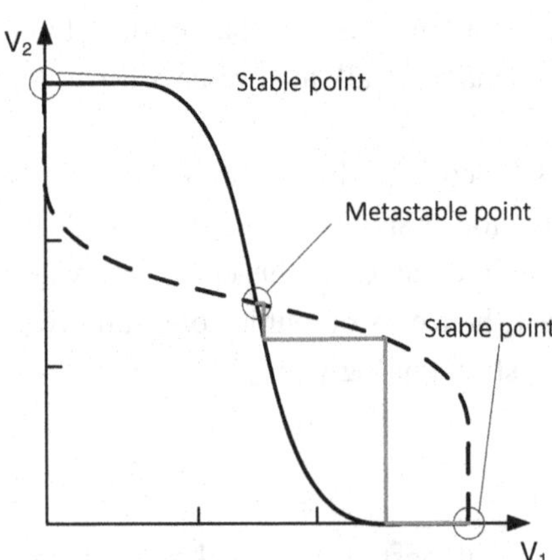

Figure 28.3. The dynamics of the back-to-back inverter circuit can be approximated by repeatedly applying the DC transfer characteristics of the two inverters, as shown by the gray line on this figure.

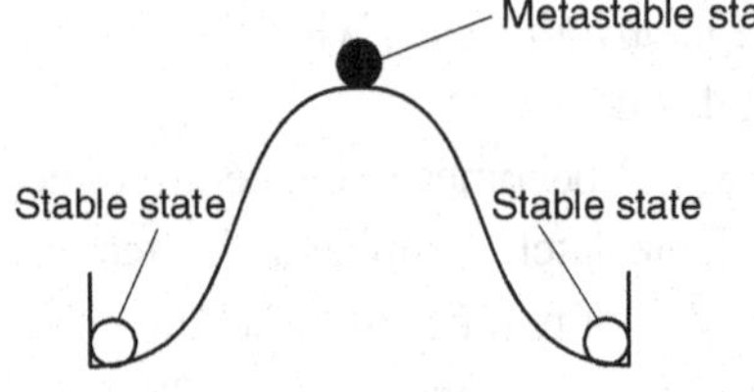

Figure 28.4. A ball at the top of a hill is in a *metastable* state. At rest, there is no force acting to push the ball left or right. However, a slight perturbation to the left or right will cause the ball to leave this state and fall to the left or right stable states at the bottom of the hill.

This iteration of the DC transfer curves is an oversimplification, but it gets the main point across. The circuit is stable at any of the three stable points. However, if we perturb the state slightly from either of the two end-points, the state returns to that end-point. If we disturb the state slightly from the midpoint, the state will quickly converge to the nearest end-point. A state, like the central stable state, where a small perturbation causes a system to leave that state, is called *metastable*.

There are many physical examples of metastable states. Consider a ball at the top of a curved hill, as shown in Figure 28.4. This ball is in a stable state. That is, it will stay in this state since, if it's exactly centered, there is no force acting to pull it left or right. However, if we give the ball a slight push left or right (a perturbation), it will leave this state and wind up in one of the two stable states at the bottom of the hill. In these states, a small push will result in the ball returning to the same state.

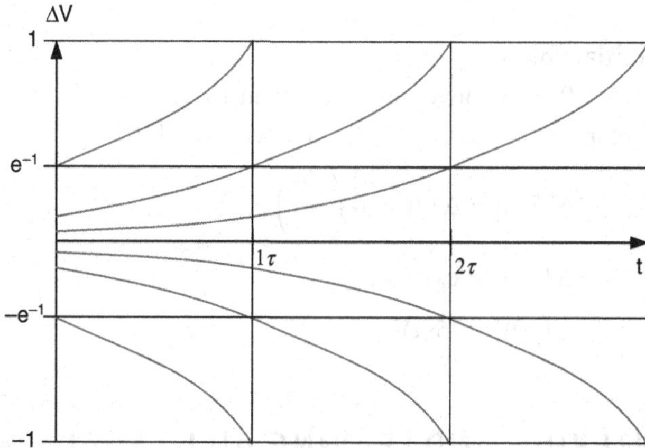

Figure 28.5. Plot of $\Delta V(t)$ for different values of $\Delta V(0)$. The magnitude of ΔV increases by e each time constant.

The ball on the hill is exactly analogous to a flip-flop. The flip-flop in the metastable state (center point of Figure 28.2) is exactly like the ball on top of the hill. All it takes is one little push (or not being exactly centered to begin with) and the circuit *rolls* downhill to one of the two stable states.

The dynamics of the back-to-back inverter circuit are, in fact, governed by the following differential equation:

$$\frac{d\Delta V}{dt} = \frac{\Delta V}{\tau_s}, \tag{28.1}$$

where τ_s is the time constant of the back-to-back inverters. In simple terms, the rate of change of the flip-flop state ΔV is directly proportional to its magnitude. The further it is from zero, the faster it moves away – until it is limited by the power supply.

The solution of this differential equation is given by

$$\Delta V(t) = \Delta V(0)\exp\left(\frac{t}{\tau_s}\right). \tag{28.2}$$

This solution is plotted in Figure 28.5 for several initial values of ΔV. With the exponential regeneration of the back-to-back inverters, the magnitude of ΔV increases e-fold each τ_s. Thus, if the circuit starts with $\Delta V(0) = e^{-1}$, it takes time τ_s for the circuit to converge to $\Delta V(t) = 1$. Similarly, if the circuit starts at $\Delta V(0) = e^{-2}$, it takes $2\tau_s$ to converge, and so on. Generalizing, we see that the amount of time that it takes to converge to $\Delta V = +1$ or -1 is given by

$$t_s = -\tau_s \log(\Delta V(0)). \tag{28.3}$$

EXAMPLE 28.1 Time to converge

What is the time taken to converge to $\Delta V = 1$ when $\Delta V(0) = 0.25$? By plugging the values into Equation (28.3), we find t_s:

$$t_s = -\tau_s \log(0.25) = 1.4\tau_s.$$

EXAMPLE 28.2 Minimum initial voltage

What is the smallest value of $\Delta V(0)$ that converges in less than $10\tau_s$?

To solve this problem, we rearrange Equation (28.2) to solve for $\Delta V(0)$:

$$|\Delta V(0)| < \Delta V(t)\exp\left(\frac{-t}{\tau_s}\right),$$

$$|\Delta V(0)| < \exp(-10),$$

$$|\Delta V(0)| < 45\ \mu\text{V}.$$

28.3 PROBABILITY OF ENTERING AND LEAVING AN ILLEGAL STATE

If we sample an asynchronous signal (one that can change at any instant), with a clock (Figure 28.6), what is the probability that our flip-flop will enter a metastable state? Suppose our clock period is t_{cy}, and that our flip-flop has setup time t_s and hold time t_h. Each transition of the asynchronous signal is equally likely to occur at any point during the cycle. Thus, there is a probability of

$$P_E = \frac{t_s + t_h}{t_{cy}} = f_{cy}(t_s + t_h) \tag{28.4}$$

that a given transition will violate the setup or hold time of the flip-flop and hence cause it to enter an illegal state.

If the asynchronous signal has transitions with frequency f_a s^{-1}, then P_E of the asynchronous edges will fall during the *forbidden* setup and hold region of the cycle. Thus, the frequency of errors (violating setup or hold time) is given by

$$f_E = f_a P_E = f_a f_{cy}(t_s + t_h). \tag{28.5}$$

For example, suppose we have a flip-flop with $t_s = t_h = 100$ ps and our cycle time is $t_{cy} = 2$ ns. Then the probability of error is

$$P_E = \frac{t_s + t_h}{t_{cy}} = \frac{200\text{ ps}}{2\text{ ns}} = 0.1. \tag{28.6}$$

Now consider the case where the asynchronous signal has a transition frequency of 1 MHz. Then the frequency of errors is given by

$$f_E = f_a P_E = 1\text{ MHz}(0.1) = 100\text{ kHz}. \tag{28.7}$$

We see that, for realistic numbers, sampling an asynchronous signal can lead to an unacceptably high error rate.

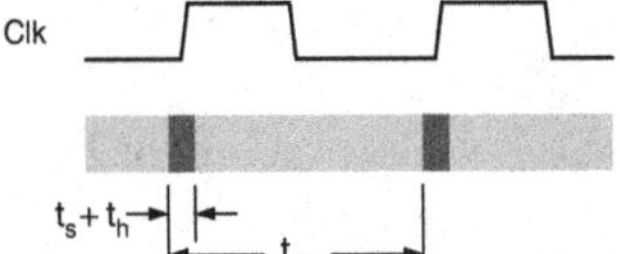

Figure 28.6. Sampling an asynchronous signal with a clock. If a transition of the asynchronous signal happens during the $t_s + t_h$ interval around the clock edge, our flip-flop may enter an illegal state.

As we shall see in Chapter 29, a partial solution to the problem of frequently entering illegal states is to *hide* the condition for a period of time, t_w, to allow the illegal state to decay to one of the two stable states. We can calculate the probability of a flip-flop still being in an illegal state after waiting time t_w by considering the probability of certain initial states and the time required for them to decay.

Given that a flip-flop enters an illegal state, it lands at a particular state voltage ΔV with some probability. We can (conservatively) estimate this probability distribution to be uniform over the interval $(-1, 1)$. Then, from Equation (28.3), the probability of taking longer than t_w to exit an illegal state is the same as the probability that

$$|\Delta V(0)| < \exp\left(\frac{-t_w}{\tau_s}\right). \tag{28.8}$$

With $|\Delta V(0)|$ uniformly distributed between 0 and 1, the probability of taking longer than t_w to reach a stable state is simply:

$$P_S = \exp\left(\frac{-t_w}{\tau_s}\right). \tag{28.9}$$

For example, suppose we have a flip-flop with $\tau_s = 100\,\text{ps}$ and we wait $t_w = 2\,\text{ns} = 20\tau_s$ for an illegal state to decay. To still be in an illegal state after $20\tau_s$, the flip-flop must have started with $|\Delta V(0)|$ smaller than $\exp(-20)$. With $|\Delta V(0)|$ uniformly distributed over $[0, 1)$, the probability of this is $P_S = \exp(-20)$.

28.4 DEMONSTRATION OF METASTABILITY

To many students, metastability is just an abstract concept until they see it for themselves. At that point, something clicks and they realize that metastability is real and it can happen to their flip-flops. This is best done via a laboratory exercise or a classroom demonstration. Ideally, you will have a chance to experience a live demonstration of metastability. If not, then this section will at least show pictures of what it really looks like.

Figure 28.7 shows a schematic of our metastability demonstration circuit. The circuit consists of six 4000-series CMOS integrated circuits, five resistors, and three capacitors. The small numbers next to the gate and transistor terminals are the pin numbers of the integrated circuits. You can easily build your own copy of our metastability demonstration circuit from this schematic and repeat the demonstration for yourself.

A photograph of our implementation of the demonstration circuit is shown in Figure 28.8. U1, U2, and U3 are along the bottom of the board from left to right, and U4, U5, and U6 are along the top of the board. The double-pole, double-throw switch that selects between the two RS flip-flops is out of the image to the right.

The circuit has three main sections: an oscillator, a pair of voltage-controlled delay lines, and two RS flip-flops. At the far left, one gate of IC U1 – a CD4093 quad Schmitt-trigger NAND gate – is wired as a relaxation oscillator. The 1 kΩ resistor and 3.3 nF capacitor set the time constant of the oscillator at 3.3 μs, giving a period of about 6 μs. Integrated circuits U3 and U2 are wired as voltage-controlled delay lines. The circuit is an inverter with its supply *starved* by a PFET, whose gate is connected to a control voltage V_C. The higher V_C, the lower the current flowing through the PFET and the larger the delay of a rising edge on the output of the inverter.

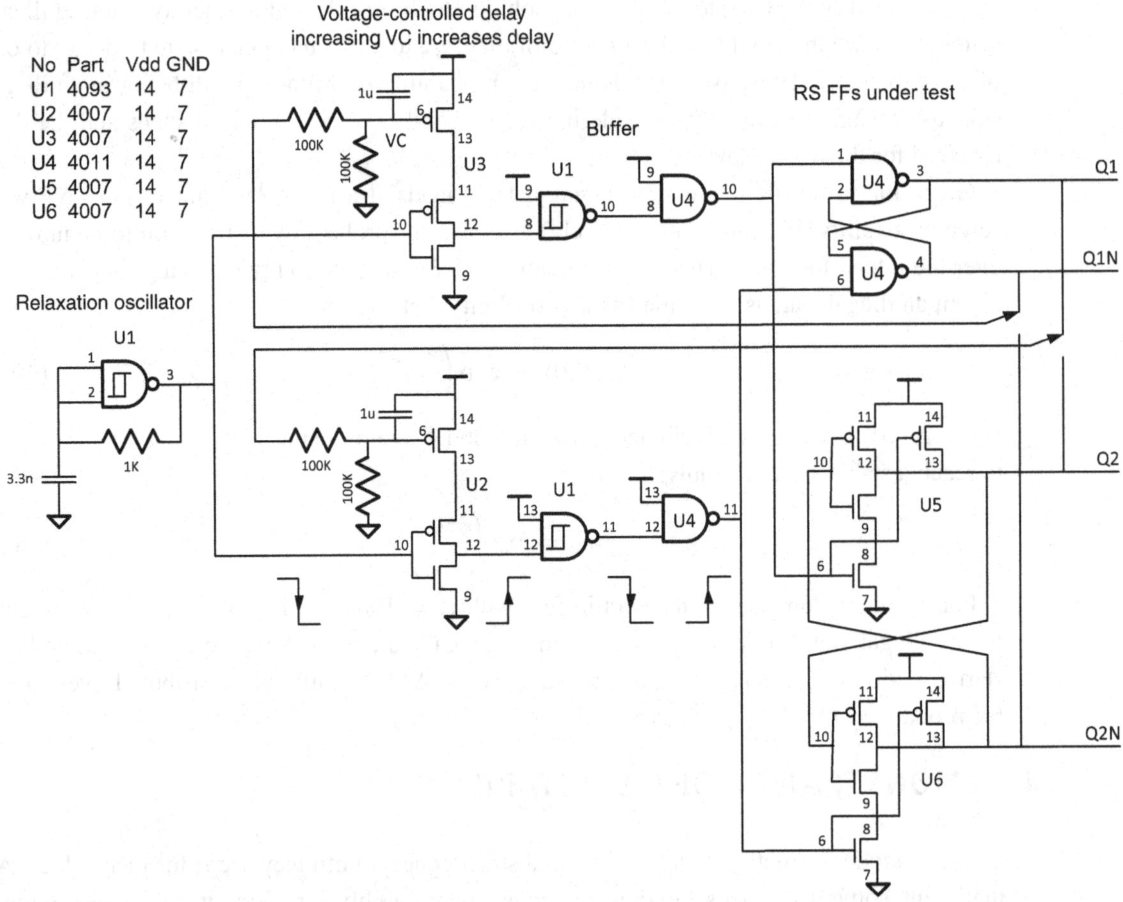

Figure 28.7. Schematic of our metastability demonstration circuit. The circuit consists of an oscillator, voltage-controlled delay lines, and two RS flip-flops. Feedback acts to push the flip-flop under test into the metastable state.

Extra gates of U1 and U4 are used to buffer the output of the delay lines – to give sharp rise and fall times into the RS flip-flops being tested. Finally, the right side of the schematic shows two RS flip-flops. The top flip-flop is constructed from two gates of U4 – a CD4011 quad NAND gate. The bottom flip-flop is constructed from individual transistors of U5 and U6 – a CD4007.

Feedback from the output of the flip-flops controls the delay lines to drive the flip-flops into a metastable state. A switch selects one of the two flip-flops under test to control the voltage-variable delay line via an *RC* filter. The *Q* output of the selected flip-flop is connected to the control voltage of the lower delay line – which drives the NAND gate driving the *QN* output. Similarly, the *QN* output of the selected flip-flop drives the upper delay line, which drives the NAND gate connected to *Q*.

This feedback connection drives the flip-flop into a metastable state. When the inputs to the flip-flop go high, the faster input will *win* and cause its output to go low. This, in turn, lowers the control voltage of the delay line driving the other input – speeding it up and causing the slow input to catch up with the fast input. Once the flip-flop is in the metastable state, its two outputs *Q* and *QN* will go low equally often – holding the circuit in equilibrium.

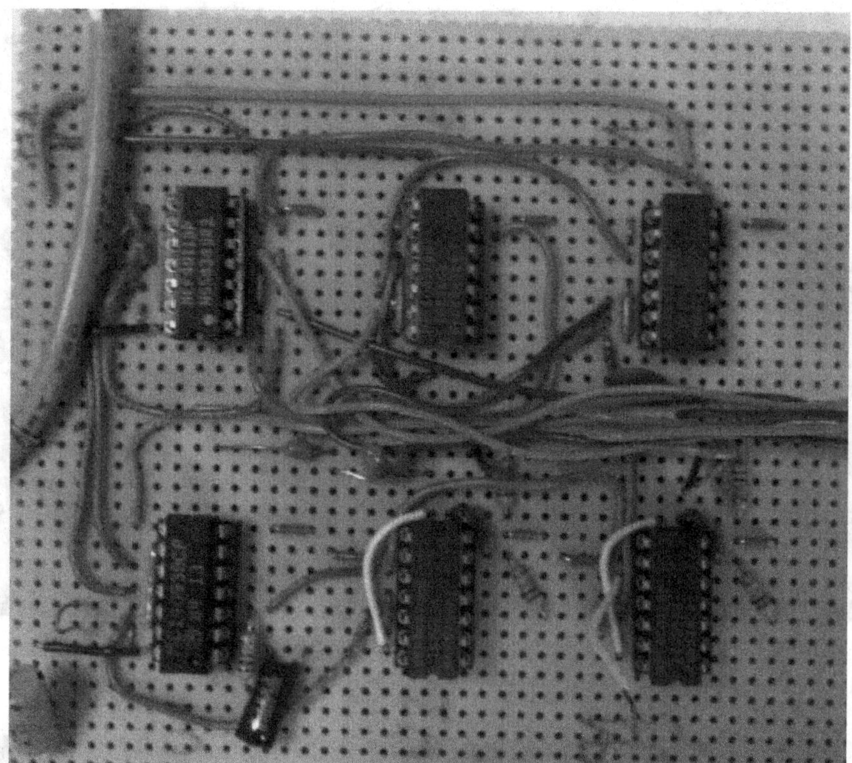

Figure 28.8. Metastability demonstration circuit implemented on a prototyping board.

Figure 28.9 shows the RS flip-flop entering and leaving a metastable state. This figure is a screen capture from an oscilloscope in *infinite-persistance* mode. In this mode the scope leaves the old waveforms on the screen as it writes new ones – accumulating all waveforms over a period of time. The horizontal scale is 200 ns per division and the vertical scale for both signals is 2 V per division. The scope was recording for 1s when this screen was captured. With the oscillator running at about 400 kHz, this recording is the superposition of about 400 000 traces.

The top trace shows one input to the RS flip flop and the bottom trace shows one output. The output enters a metastable state, a voltage midway between GND and V_{DD} about 250 ns after the input has fallen. It leaves the metastable state to become either a 1 or a 0 with an exponential decay, like that sketched in Figure 28.5. In most cases, the metastable state decays in about 250 ns. However, on a few occasions it takes about 400 ns for the metastable state to decay. This reflects the fact that, with decreasing probability, the initial state can be arbitrarily close to the center of the metastable region – and hence can take arbitrarily long to decay. In this figure, the output decays to a high state slightly more often than it decays to a low state. This is due to an offset in the circuit, causing the initial probability distribution to be off center.

Figure 28.10 shows a screen capture of the oscilloscope after 1 h – an accumulation of more than one billion traces. On a few of these billion, the flip-flop has landed very close to the center of the metastable region, as indicated by the few traces that have taken up to 800 ns to decay to a stable state. The left side of the waveform has filled in – indicating a large number of trials that decay in 250 ns to 600 ns, with the recording becoming progressively sparser from 600 ns to 800 ns, reflecting the exponential drop in probability with time.

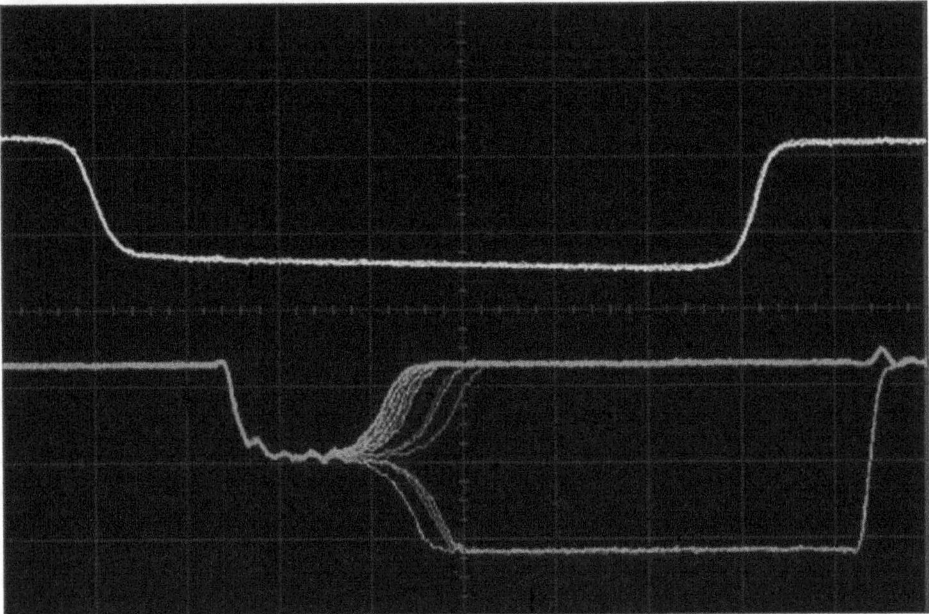

Figure 28.9. Input and output of the lower RS flip-flop in Figure 28.7. Waveforms accumulated for one second.

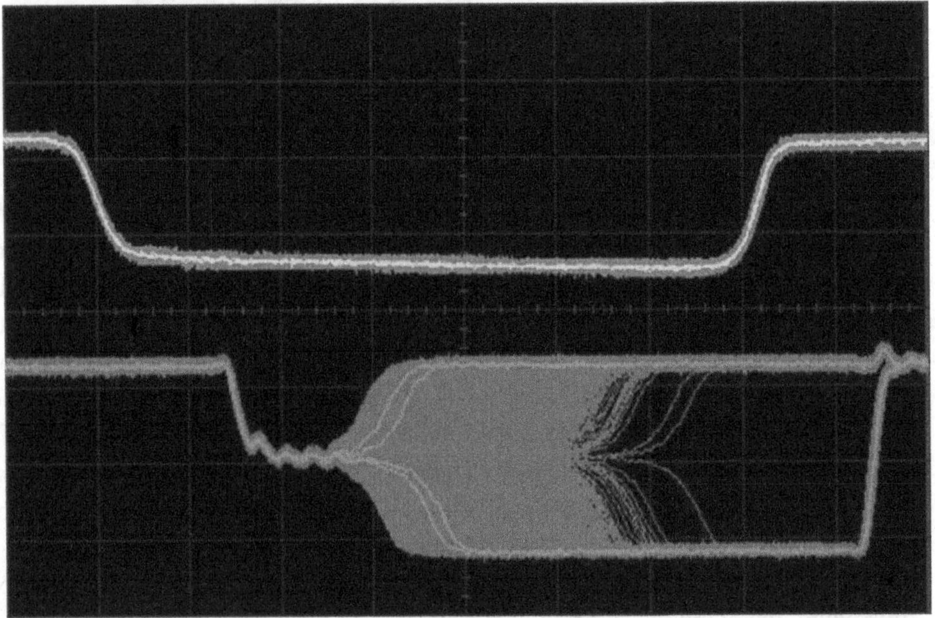

Figure 28.10. Input and output of the lower RS flip-flop in Figure 28.7. Waveforms accumulated for one hour.

With just one billion trials, we have metastable states lasting up to 800 ns. A modern high-end GPU chip has millions of flip-flops[2] running at more than 1 GHz. A large supercomputer may have 20 000 such chips. In such a machine, over 10^{16} flip-flop clock events happen per

[2] However, only a small fraction of these flip-flops must deal with asynchronous signals.

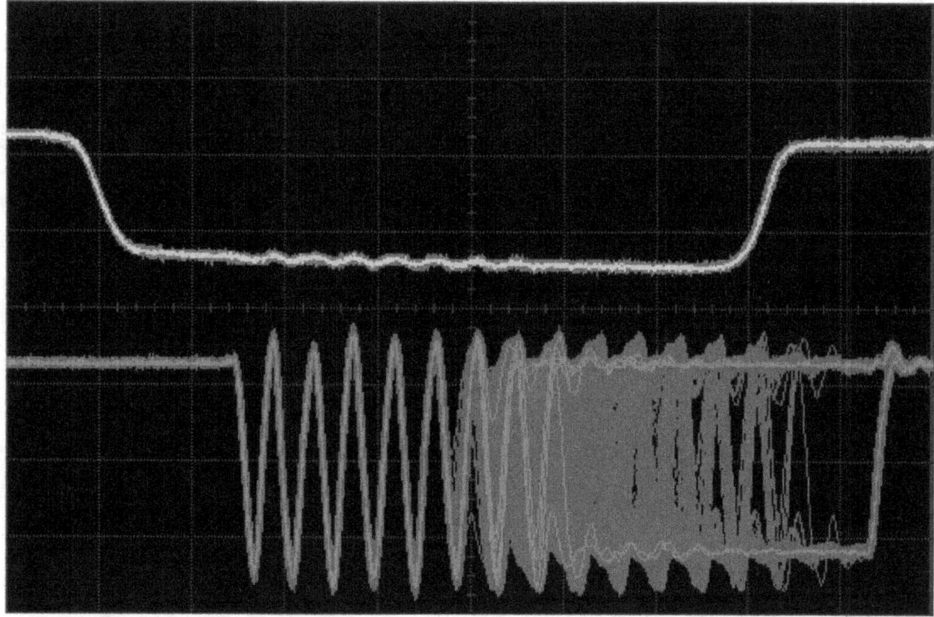

Figure 28.11. Input and output of the upper RS flip-flop in Figure 28.7, using CD4011 quad NAND gate integrated circuit, with this flip-flop selected.

second, and over 10^{20} flip-flop clock events occur per year. Synchronization and metastable states must be handled carefully, because even very unlikely events can be a problem when multiplied by such a high frequency.

When the other RS flip-flop, constructed from two gates of a CD4011 quad NAND gate, is selected, the waveform in Figure 28.11 results. We leave the explanation of this waveform as Exercise 28.13.

Summary

In this chapter you have developed an understanding of metastability and the failures that can occur when flip-flops sample asynchronous signals. We modeled a flip-flop as a regenerative circuit with positive feedback and saw how this model has two *stable* states and one *metastable* state. The two stable states correspond to the two normal states of the flip-flop. In these normal states, one storage node of the flip-flop is high and the other is low. In the metastable state, both internal nodes of the flip-flop are at an intermediate voltage – neither high nor low. Like a stable state, the metastable state can persist indefinitely. Unlike a stable state, a small perturbation from the metastable state will cause it to diverge to one of the two stable states. Like a ball perched on top of a hill, a flip-flop in a metastable state can stay in that position forever. However, if the ball, or the flip-flop, is displaced slightly to one side or the other, it will fall down the hill, or into a stable state.

We developed formulae to estimate the rate at which flip-flops used to sample asynchronous signals enter and leave the metastable state. The probability of entering the metastable state on an asynchronous event is just the probability that the event occurs during the setup and hold window of the flip-flop:

$$P_E = \frac{t_s + t_h}{t_{cy}}.$$

Once a flip-flop is in a metastable state, any small deviation from the metastable state grows exponentially, growing by a factor of e every τ_s. The probability that the flip-flop is still in the metastable state after waiting time t_w is given by

$$P_S = \exp\left(\frac{-t_w}{\tau_s}\right).$$

Finally, we saw a demonstration of metastability with oscilloscope waveforms showing a flip-flop entering and leaving the metastable state.

BIBLIOGRAPHIC NOTES

One of the earliest papers on the problems caused by metastability is ref. [27]. A more detailed treatment of this problem is given in ref. [33].

Exercises

28.1 *Time to settle, I.* What is the time to settle when the initial voltage difference is 16 mV?

28.2 *Time to settle, II.* What is the time to settle when the initial voltage difference is 0.16 μV?

28.3 *Maximum voltage difference, I.* What is the maximum initial voltage difference for a metastable system to settle to $\Delta V = 1$ V within $7\tau_s$?

28.4 *Maximum voltage difference, II.* What is the maximum initial voltage difference for a metastable system to settle to $\Delta V = 1$ V in $3.5\tau_s$?

28.5 *Time to settle vs. transition time.* Assume a flip-flop with $t_s = t_h = 500$ ps and $\tau_s = 100$ ps. Plot a graph of data transition time vs. time to settle on the y-axis for the flip-flop. Assume that the initial ΔV of the system is linear between 1 V and -1 V from -500 ps to 500 ps (as shown in Figure 28.1(b)). The time to settle should reflect the total time to get to $|\Delta V| = 1$.

28.6 *Probability of error, I.* What is the probability of an asynchronous signal transitioning and violating setup and hold constraints of a flip-flop where $t_s = t_h = 100$ ps and $f_{cy} = 2$ GHz?

28.7 *Probability of error, II.* What is the probability of an asynchronous signal transitioning and violating setup and hold constraints of a flip-flop where $t_s = t_h = 20$ ps and $f_{cy} = 4$ GHz?

28.8 *Probability of error, III.* What is the probability of an asynchronous signal transitioning and violating setup and hold constraints of a flip-flop where $t_s = t_h = 300$ ps and $f_{cy} = 1$ MHz?

28.9 *Frequency of error, I.* With a target of having only one illegal asynchronous transition every 0.01 s, compute the maximum f_a in a system where $t_s = t_h = 100$ ps and $f_{cy} = 2$ GHz.

28.10 *Frequency of error, II.* With a target of having only one illegal asynchronous transition every 0.01 s, compute the maximum f_a in a system where $t_s = t_h = 20$ ps and $f_{cy} = 4$ GHz.

28.11 *Frequency of error, III.* With a target of having only one illegal asynchronous transition every 0.01 s, compute the maximum f_a in a system where $t_s = t_h = 300$ ps and $f_{cy} = 1$ MHz.

28.12 *Probability of staying in an illegal state.* You are designing a circuit and are given two different flip-flop options. Flip-flop 1 has $t_s = t_h = 50$ ps and $\tau_s = 20$ ps while flip-flop 2 has $t_s = t_h = 250$ ps and $\tau_s = 10$ ps. Explain which flip-flop you use to minimize errors in each of the following scenarios:

(a) $t_w = 0$ ps;

(b) $t_w = 50$ ps.

Plot the error probability as a function of t_w for both flip-flop 1 and flip-flop 2. Assume $f_c = 1$ GHz and $f_a = 10$ MHz.

28.13 *Oscillatory metastability.* Explain the waveform of Figure 28.11. (Hint: look at a schematic showing the internal circuit of the CD4011 NAND gate.)

28.14 *False synchronizer.* An engineer in your company has suggested building a fast synchronizer by using a flip-flop, followed by a comparator, followed by a second flip-flop. The comparator has a threshold set at 0.2 V, low enough that a metastable output of the flip-flop is sensed as a logic 1. Does this reliably synchronize one-bit signals? Explain your answer, and give the probability of synchronization failure, if it is non-zero, in terms of circuit parameters.

29 Synchronizer design

In a synchronous system, we can avoid putting our flip-flops in illegal or metastable states by always obeying the setup- and hold-time constraints. When sampling asynchronous signals or crossing between different clock domains, however, we cannot guarantee that these constraints will be met. In these cases, we design a *synchronizer* that, through a combination of waiting for metastable states to decay and isolation, reduces the probability of synchronization failure.

A brute-force synchronizer consisting of two back-to-back flip-flops is commonly used to synchronize single-bit signals. The first flip-flop samples the asynchronous signal and the second flip-flop isolates the possibly bad output of the first flip-flop until any illegal states are likely to have decayed. Such a brute-force synchronizer cannot be used on multi-bit signals unless they are encoded with a Gray code. If multiple bits are in transition when sampled by the synchronizer, they are independently resolved, possibly resulting in incorrect codes, with some bits sampled before the transition and some after the transition. We can safely synchronize multi-bit signals with a FIFO (first-in first-out) synchronizer. A FIFO serves both to synchronize the signals and to provide flow control, ensuring that each datum produced by a transmitter in one clock domain is sampled exactly once by a receiver in another clock domain – even when the clocks have different frequencies.

29.1 WHERE ARE SYNCHRONIZERS USED?

Synchronizers are used in two distinct applications, as shown in Figure 29.1. First, when signals are coming from a truly asynchronous source, they must be synchronized before being input to a synchronous digital system. For example, a push-button switch pressed by a human produces an asynchronous signal. This signal can transition at any time, and so must be synchronized before it can be input to a synchronous circuit. Numerous physical detectors also generate truly asynchronous inputs. Photodetectors, temperature sensors, pressure sensors, etc. all produce outputs with transitions that are gated by physical processes, not a clock.

The other use of synchronizers is to move a synchronous signal from one *clock domain* to another. A clock domain is simply a set of signals that are all synchronous with respect to a single clock. For example, in a computer system it is not unusual to have the processor operate from one clock, *pclk*, and the memory system operate from a different clock, *mclk*. These two clocks may have very different frequencies. For example, that of *pclk* may be 2 GHz while that of *mclk* is 800 MHz. Signals that are synchronous to *pclk* – i.e., in the *pclk* clock domain – cannot be directly used in the memory system. They must first be synchronized with *mclk*.

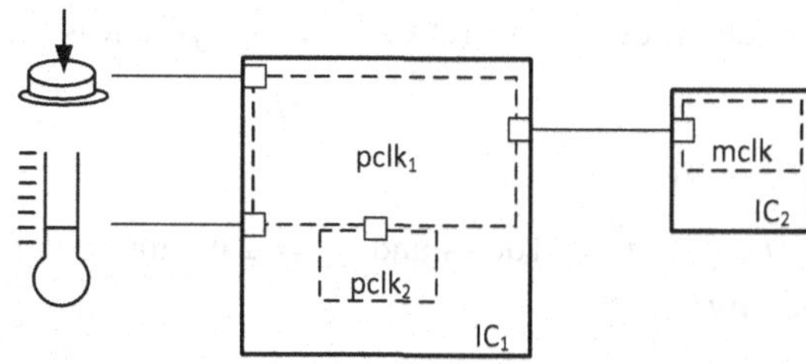

Figure 29.1. Synchronizers are needed between asynchronous and synchronous systems (left). They are also needed between clock domains both on- and off-chip.

Similarly, signals in the memory system must first be synchronized with *pclk* before they can be used in the processor.

In moving signals between clock domains, there are two distinct synchronization tasks. If we wish to move a sequence of data where each datum in the sequence must be preserved, we use a *sequence synchronizer*. For example, to send eight words in sequence (one word at a time over a data bus) from the processor to the memory system, we need a sequence synchronizer. On the other hand, if we wish to monitor the state of a signal we need a *state synchronizer*. A state synchronizer outputs a recent sample of the signal in question synchronized to its output clock domain. To allow the processor to monitor the depth of a queue in the memory system, we use a state synchronizer – we do not need every sample of queue depth (one each clock), just one recent sample. On the other hand, we cannot use a state synchronizer to pass data between the two subsystems; it may drop some elements and repeat others.

29.2 BRUTE-FORCE SYNCHRONIZER

Synchronization of single-bit signals is often achieved using a *brute-force synchronizer*, as shown in Figure 29.2. Flip-flop FF1 samples an asynchronous signal a, producing output a_w. Signal a_w is unsafe because of the high frequency with which FF1 will enter an illegal state. To guard the rest of the system from this unsafe signal, we wait one (or more) clock periods for any illegal states of FF1 to decay before resampling it with FF2, to produce output a_s.

How well does the synchronizer of Figure 29.2 work? In other words, what is the probability of a_s being in an illegal state after a transition on a? This will happen only if (1) FF1 enters an illegal state and (2) this state has not decayed before a_w is resampled by FF2. FF1 enters an illegal state with probability P_E (see Equation (28.4)), and it will remain in this state after a waiting time t_w with probability P_S (see Equation (28.9)). Thus, the probability of FF2 entering an illegal state is given by

$$P_{ES} = P_E P_S = \left(\frac{t_s + t_h}{t_{cy}}\right) \exp\left(\frac{-t_w}{\tau_s}\right). \tag{29.1}$$

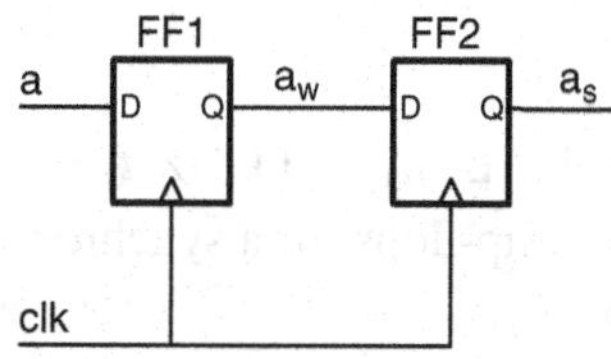

Figure 29.2. A brute-force synchronizer consists of two back-to-back flip-flops. The first flip-flop samples asynchronous signal a, producing signal a_w. The second flip-flop waits one (or more) clock cycles for any metastable states of the first flip-flop to decay before resampling a_w, to produce synchronized output a_s.

The waiting time t_w here is not a full clock cycle, but rather a clock cycle less the required overhead:

$$t_w = t_{cy} - t_s - t_{dCQ}. \tag{29.2}$$

For example, if we have $t_s = t_h = t_{dCQ} = \tau_s = 100$ ps and $t_{cy} = 2$ ns, then the probability of FF2 entering an illegal state is given by

$$\begin{aligned} P_{ES} &= \left(\frac{t_s + t_h}{t_{cy}}\right) \exp\left(\frac{-t_w}{\tau_s}\right) \\ &= \left(\frac{100\,\text{ps} + 100\,\text{ps}}{2\,\text{ns}}\right) \exp\left(\frac{-1.8\,\text{ns}}{100\,\text{ps}}\right) \\ &= 0.1 \exp(-18) = 1.5 \times 10^{-9}. \end{aligned}$$

If signal a has a transition frequency of 100 MHz, then the frequency of failure is given by

$$f_{ES} = f_a P_{ES} = (100\,\text{MHz})(1.5 \times 10^{-9}) = 0.15\,\text{Hz}. \tag{29.3}$$

If this synchronizer failure probability is not low enough, we can make it lower by waiting longer. This is best accomplished by adding clock enables to the two flip-flops and enabling them once every N clock cycles. This lengthens the wait time to

$$t_w = N t_{cy} - t_s - t_{dCQ}. \tag{29.4}$$

In our example above, waiting two clock cycles reduces our failure probability and frequency to

$$P_{ES} = 0.1 \exp(-38) = 3.1 \times 10^{-17}, \tag{29.5}$$

$$f_{ES} = (100\,\text{MHz})(3.1 \times 10^{-17}) = 3.1 \times 10^{-9}\,\text{Hz}. \tag{29.6}$$

Using a clock enable to wait longer is more efficient than using multiple flip-flops in series because with the clock enables we pay the flip-flop overhead $t_s + t_{dCQ}$ just once, rather than once per flip-flop. Each clock cycle after the first adds a full t_{cy} to our waiting time. With flip-flops in series, each additional flip-flop adds $t_{cy} - t_s - t_{dCQ}$ to our waiting time.

How low a failure probability is low enough? This depends on your system and what it is used for. Generally we want to make the probability of synchronization failure significantly smaller than that of some other system failure mode. For example, in a telecommunication system where the bit-error rate of a line is 10^{-20}, it would suffice to make a synchronizer with a failure probability P_{ES} of 10^{-30}. For some systems used for life-critical functions, the mean time to failure (MTTF = $1/f_{ES}$) must be made long compared with the lifetime of the system times the number of systems produced. So, if the system is expected to last 10 years (3.1×10^8 s), and we build 10^5 systems, we would like to make our MTTF much larger than 3.1×10^{13} (i.e., f_{ES} should be much less than 3×10^{-14}). Here we might set a goal of $f_{ES} = 10^{-20}$ (fewer than one failure every 10^{11} years – per system).

EXAMPLE 29.1 Brute-force synchronizer

Calculate the mean time to failure for a system with $f_a = 1$ kHz, $f_{cy} = 1$ GHz, $t_s = t_h = 50$ ps, $\tau_s = 100$ ps, and $t_{dCQ} = 80$ ps that uses three back-to-back flip-flops for a synchronizer.

The probability of entering the metastable state is given by Equation (28.4):

$$P_E = \left(\frac{t_s + t_h}{t_{cy}}\right) = \left(\frac{50 + 50}{1000}\right) = 0.1.$$

Our three back-to-back flip-flops wait two clock cycles less flip-flop insertion overhead:

$$\begin{aligned} t_w &= 2(t_{cy} - t_s - t_{dCQ}) \\ &= 2(1000 - 50 - 80) \\ &= 1740\,\text{ps} \\ &= 17.4\tau_s. \end{aligned}$$

Thus, the probability of failure on each event is given by

$$\begin{aligned} P_{ES} &= P_E P_S \\ &= (0.1)\exp(-t_w/\tau_s) \\ &= (0.1)\exp(-17.4) \\ &= (0.1)(2.78 \times 10^{-8}) \\ &= 2.78 \times 10^{-9}. \end{aligned}$$

Thus our failure frequency is given by

$$f_{ES} = f_a P_{ES} = (10^5)(2.78 \times 10^{-9}) = 2.78 \times 10^{-4}\,\text{Hz}.$$

The MTBF is therefore

$$\text{MTBF} = 1/f_f = 3.60 \times 10^3\,\text{s}.$$

29.3 THE PROBLEM WITH MULTI-BIT SIGNALS

While a brute-force synchronizer does a wonderful job synchronizing a single-bit signal, it *cannot* be used to synchronize a multi-bit signal unless that signal is Gray-coded (i.e., unless only one bit of the signal changes at a time). Consider, for example, the situation shown in Figure 29.3. The output, *cnt*, of a four-bit counter clocked by one clock, *clk*1, is synchronized with a second clock, *clk*2, operating at a different frequency. Suppose on the count from 7 (0111) to 8 (1000) all of the bits of *cnt* are changing when *clk*2 rises – violating setup and hold times on register R1. The four flip-flops of R1 all enter illegal states. During the next cycle of *clk*2 these states all decay to 0 or 1 with high probability, so that when *clk*2 rises again, a legal four-bit digital signal is sampled and output on cnt_s.

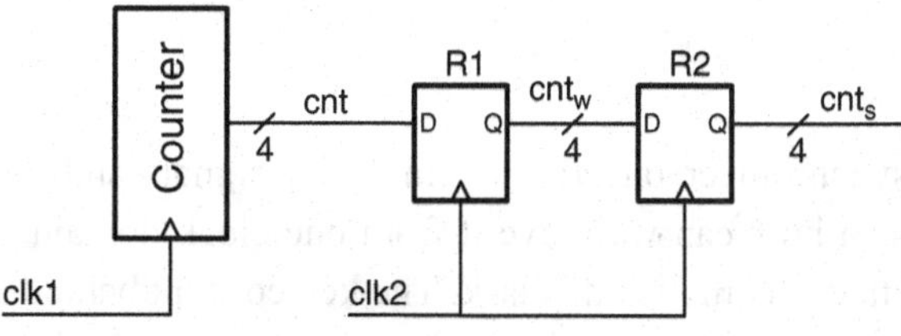

Figure 29.3. Incorrect method of synchronizing a multi-bit signal. Counter clocked by *clk*1 is sampled by synchronizer clocked by *clk*2. On a transition where multiple bits of signal, *cnt*, change, the synchronized output, cnt_s, may see some of the bits change, but not others, giving an incorrect result.

```
library ieee;
use ieee.std_logic_1164.all;
use work.ff.all;

entity GrayCount4 is
  port(clk, rst: in std_logic;
       output: buffer std_logic_vector(3 downto 0) );
end GrayCount4;

architecture impl of GrayCount4 is
  signal nxt: std_logic_vector(3 downto 0);
begin
  COUNT: vDFF generic map(4) port map(clk, nxt, output);

  nxt(0) <= not rst and not (output(1) xor output(2) xor output(3)) ;
  nxt(1) <= '0' when rst else
            not (output(2) xor output(3)) when output(0) else
            output(1);
  nxt(2) <= '0' when rst else
            not output(3) when output(1) and output(0) else
            output(2);
  nxt(3) <= '0' when rst else
            output(2) when not (output(1) or output(0)) else
            output(3);
end impl; -- GrayCount4
```

Figure 29.4. VHDL description of a four-bit Gray-code counter.

The problem is that, with high probability, the output on cnt_S is wrong. For each of the changing bits of *cnt*, the synchronizer can settle to either a 0 or a 1 state. In this case, when all four bits are changing, the output of the synchronizer could be any number between 0 and 15.

The only time a brute-force synchronizer can be used with multi-bit signals is when these signals are guaranteed to change at most one bit between synchronizer clock transitions. For example, if we wish to synchronize a counter in this manner, we must use a Gray-code counter, which changes by exactly one bit on each count. A four-bit Gray-code counter uses the sequence 0, 1, 3, 2, 6, 7, 5, 4, 12, 13, 15, 14, 10, 11, 9, 8. Only one bit changes between adjacent elements of this sequence. For example, the eighth transition is from 4 (0100) to 12 (1100). Only the MSB changes during this transition. If we replace the counter of Figure 29.3 with a four-bit Gray-code counter, and the 4 to 12 transition takes place during a rising edge of *clk*2, then only the MSB of R1 will enter an illegal state. The low three bits will remain steady at 100. Either way the MSB settles, 0 or 1, the output is a legitimate value, 4 or 12.

VHDL code for a four-bit Gray-code counter that generates this sequence is shown in Figure 29.4.

29.4 FIFO SYNCHRONIZER

If we can't use a brute-force synchronizer on arbitrary multi-bit signals and our signal is not amenable to Gray coding, then how can we move it from one clock domain to another? There are several synchronizers that accomplish this task. The key concept behind all of them

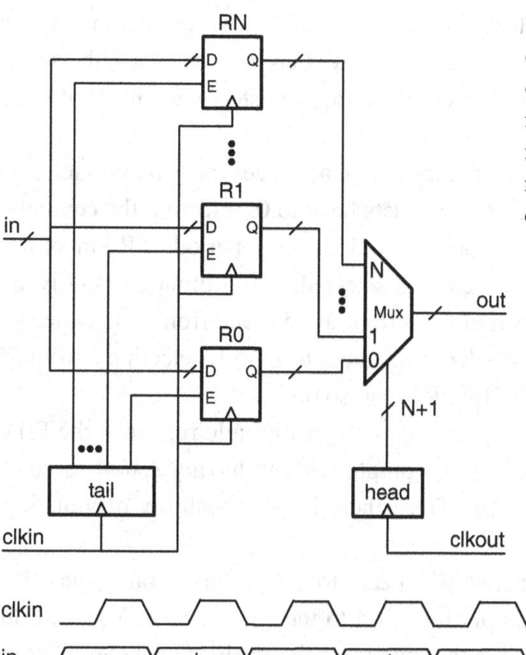

Figure 29.5. Datapath of a FIFO synchronizer. Input data are placed in registers *R*0, . . . , *RN* by the input clock *clkin*. Output data are selected from among the registers by an output clock *clkout*. A control path, not shown, ensures that (1) data are placed in a register before being selected for output and (2) that data are not overwritten before being read.

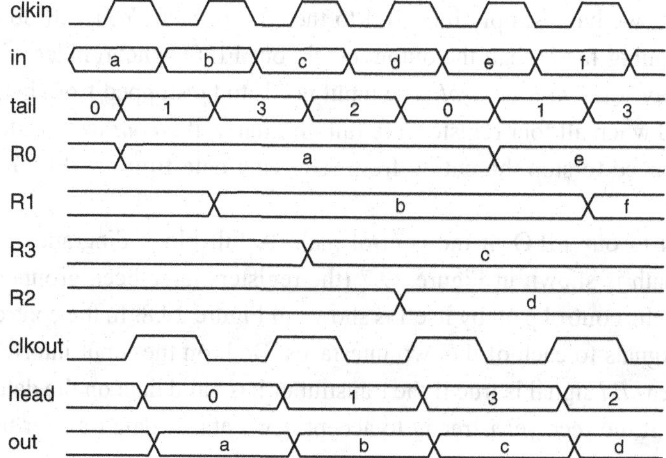

Figure 29.6. Timing diagram showing operation of a FIFO synchronizer with four registers.

is removing synchronization from the multi-bit datapath. The synchronization is moved to a control path that is either a single bit or a Gray-coded signal.

Perhaps the most common multi-bit synchronizer is the FIFO synchronizer. The datapath of a FIFO synchronizer is shown in Figure 29.5. The FIFO synchronizer works by using a set of registers R0 to R*N*. Data are stored into the registers under control of the input clock and read from the registers under control of the output clock. A *tail* pointer selects which register is to be written next, and a *head* pointer selects which register is to be read next. Data are added at the tail of the queue, and removed from the head of the queue. The head and tail pointers are Gray-coded counters that are decoded to one-hot to drive the register enables and multiplexer select lines. Using Gray-code encoding for the counters enables them to be synchronized using brute-force synchronizers in the control path.

A timing diagram showing operation of a FIFO synchronizer with four registers (R0 to R3) is shown in Figure 29.6. The input clock, *clkin*, in this example is faster than the output clock, *clkout*. On each rising edge of the input clock, a new datum on line *in* is written to one of the registers. The register written is selected by the *tail* pointer, which increments with a Gray-code pattern (0, 1, 3, 2, 0, ...). The first datum *a* is written to register R0, the second datum *b* is written to R1, *c* is written to R3, and so on. With four registers, the output of each register is valid for four input clocks.

On the output side, each rising edge of clock *clkout* advances the *head* pointer, selecting each register in turn. The first rising edge of *clkout* sets head to 0, selecting the contents of register R0 *a* to be driven onto the output. The four-input-clock valid period of R0 more than overlaps the one-output-clock period during which it is selected, so no input-clock-driven transitions are visible on the output. The only output transitions are derived from (and hence synchronous with) *clkout*. The second rising edge of *clkout* advances *head* to 1, selecting *b* from R1 to appear on the output, the third edge selects *c* from R3, and so on.

By extending the valid period of the input data using multiple registers, the FIFO synchronizer enables this data to be selected on the output without having *clkout* sample any signal with transitions synchronized with *clkin*. Thus, there is no probability of violating setup and hold times in this datapath.

Of course, we have not eliminated the need to synchronize (and the probability of synchronization failure), we have simply moved it to the control path. You will observe that, with the input clock running faster than the output clock, our FIFO synchronizer will quickly overflow unless we apply some *flow control*. The input needs to be stopped from inserting any more data into the FIFO when all four registers are full. Similarly, if *clkout* were faster than the input clock, we would need to stop the output from removing data from the FIFO when it is empty.

We add flow control to our FIFO in the control path. A full block diagram of the FIFO including the control path is shown in Figure 29.7 (the registers have been grouped together into a RAM array), and the control path by itself is shown in Figure 29.8. In the control path we add two flow-control signals to each of the two interfaces. Both on the input interface and on the output interface, the *valid* signal is true if the transmitter has valid data on the data line, and the *ready* signal is true if the receiver is ready to accept new data. A data transfer takes place

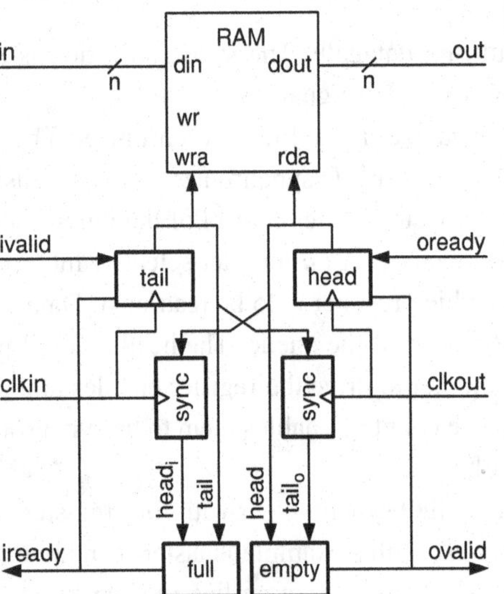

Figure 29.7. FIFO synchronizer showing control path.

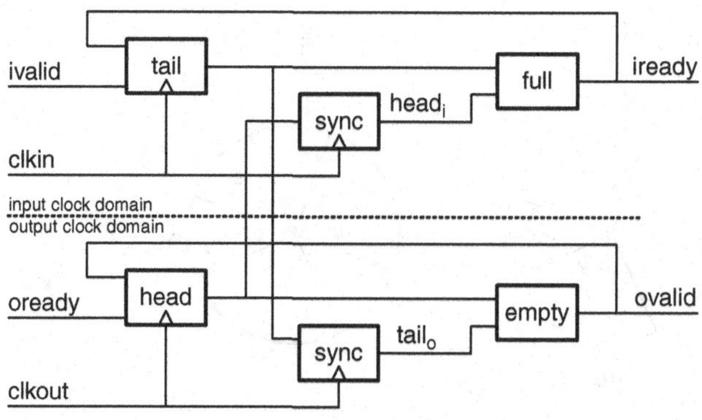

Figure 29.8. Control path of FIFO synchronizer showing clock domains.

only if *valid* and *ready* are both true. On the input side, *ivalid* is an input signal, and *iready* is asserted if the FIFO is not full. On the output side, *oready* is an input signal, and *ovalid* is asserted if the FIFO is not empty.

The *iready* (not full) and *ovalid* (not empty) signals are generated by comparing the head and tail pointers. This comparison is complicated by the fact that *head* and *tail* are in different clock domains. Signal *head* is synchronous with *clkout*, whereas *tail* is synchronous with *clkin*. We solve this problem by using a pair of multi-bit brute-force synchronizers to produce a version of *head* in the input clock domain, $head_i$, and a version of *tail* in the output clock domain, $tail_o$. This synchronization is allowed only because *head* and *tail* are Gray-coded and thus will have only one bit in transition at any given point in time. Also, note that the synchronization delays the signals, so that $head_i$ and $tail_o$ are up to two clock cycles behind *head* and *tail*.

Once we have versions of *head* and *tail* in the same clock domain, we compare them to determine the full and empty conditions. When the FIFO is empty, *head* and *tail* are the same, so we can write

```
empty <= '1' when (head = tail_o) else '0';
ovalid <= not empty ;
```

When the FIFO is full, *head* and *tail* are also the same. Thus, if we were to allow all registers to be used, we would need to add additional state to discriminate between these two conditions. Rather than add this complexity, we simply declare the FIFO to be full when it has just one location empty, and write

```
full <= '1' when (head_i = inc_tail) else '0';
iready <= not full;
```

where `inc_tail` is the result of incrementing the tail pointer along the Gray-code sequence. This approach always leaves one register empty. For example, with four registers, only three would be allowed to contain valid data at any time. Despite this disadvantage, this approach is usually preferred because of the high complexity of maintaining and synchronizing the extra state needed to discriminate between full and empty when `head == tail`. We leave to the reader, as Exercise 29.10, the design of a FIFO where all locations can be filled.

FIFO states are illustrated in Figure 29.9. Part (a) shows the state of the head and tail pointers when the FIFO is empty. In parts (b) and (c) data are added to the FIFO, incrementing the tail. In Figure 29.9(d), the FIFO has only one location open, and (in our implementation) will assert *full* and not accept new data. In part (e) the FIFO is completely full, with *head* = *tail*. The head

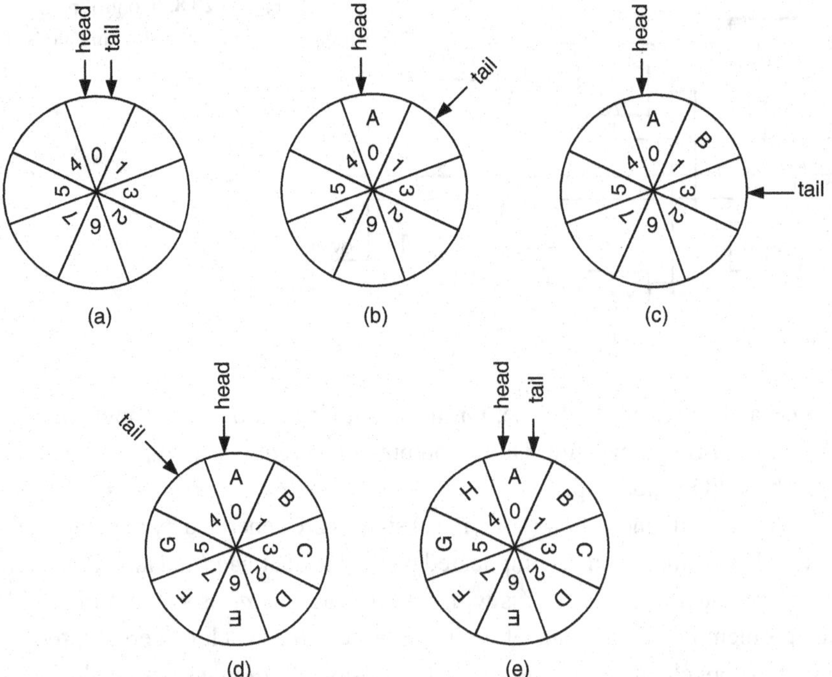

Figure 29.9. FIFO states. (a) FIFO is empty, *head* = *tail*. (b) After inserting one datum. (c) After inserting two data. (d) Almost full, *head* = *inc(tail)*. (e) Completely full, *head* = *tail*.

and tail pointers both equal zero in Figures 29.9(a) and (e); the former represents an empty FIFO and the latter shows a full one. Additional state must be maintained to distinguish these two conditions.

The VHDL code for the FIFO synchronizer is shown in Figure 29.10. The width and depth of the FIFO are parameterized – defaulting to eight bits wide with eight registers. A synchronous RAM module, not shown, contains the eight registers. A register clocked by `clkout` holds `head` and one clocked by `clkin` holds `tail`. A pair of brute-force synchronizers create signals `head_i` and `tail_o`. A pair of three-bit Gray-code incrementers (VHDL code in Figure 29.11) increments `head` and `tail` – along the Gray-code sequence – producing `inc_head` and `inc_tail`. Note that if this code were to be used for a FIFO deeper than eight registers, wider Gray-code incrementers would be required – `GrayInc3` is not parameterized. Next in the code, flow-control signals `iready` and `ovalid` are computed as described above. Finally, the next states for `head` and `tail` are computed. The head and tail are incremented only if both the `valid` signal and the `ready` signal for their respective interfaces are true.

Operation of the FIFO synchronizer is shown in the waveforms of Figure 29.12. On the input side, after reset, `iready` becomes true, signaling that the FIFO is not full and hence able to accept data. Two cycles later, `ivalid` is asserted and seven data elements are inserted into the FIFO. At this point, the FIFO is full and `iready` goes low. On each data word inserted, `tail` is incremented in a three-bit Gray-code sequence. Note that `o_tail` is delayed by two output clocks from `tail`. Later, as words are removed from the FIFO on the output side, `iready` goes high again and additional words are inserted. Signal `ivalid` goes low for three cycles after 09 has been inserted. During this period, data are not input, even though `iready` is true.

```
library ieee;
use ieee.std_logic_1164.all;
use work.ff.all;
use work.ch29.all;
entity AsyncFIFO is
  generic( n: integer := 8; -- width of FIFO
           m: integer := 8; -- depth of FIFO
           lgm: integer := 3 ); -- width of pointer field
  port( clkin, rstin, clkout, rstout, ivalid, oready: in std_logic;
        iready, ovalid: buffer std_logic;
        input: in std_logic_vector(n-1 downto 0);
        output: out std_logic_vector(n-1 downto 0) );
end AsyncFIFO;
architecture impl of AsyncFIFO is
  signal head, next_head, head_i: std_logic_vector(lgm-1 downto 0);
  signal tail, next_tail, tail_o: std_logic_vector(lgm-1 downto 0);
  signal inc_head, inc_tail : std_logic_vector(lgm-1 downto 0);
begin
  -- words are inserted at tail and removed at head
  -- sync_x is head/tail synchronized to other clock domain
  -- inc_x is head/tail incremented by Gray code

  -- Dual-Port RAM to hold data
  mem: DP_RAM generic map(n,m,lgm)
        port map(clk => clkin, input => input, inaddr => tail(lgm-1 downto 0),
                 wr => iready and ivalid, output => output,
                 outaddr => head(lgm-1 downto 0));

  -- head clocked by output, tail by input
  hp: vDFF generic map(lgm) port map(clkout, next_head, head);
  tp: vDFF generic map(lgm) port map(clkin, next_tail, tail);

  -- synchronizers
  hs: BFSync generic map(lgm) port map(clkin, head, head_i); -- head in tail domain
  ts: BFSync generic map(lgm) port map(clkout, tail, tail_o); -- tail in head domain

  -- Gray-code incrementers
  hg: GrayInc3 port map(head, inc_head);
  tg: GrayInc3 port map(tail, inc_tail);

  -- iready if not full, oready if not empty
  -- full when head points one beyond tail
  iready <= '0' when head_i = inc_tail else '1'; -- input clock for full
  ovalid <= '0' when head = tail_o else '1'; -- output clock for empty

  -- tail increments on successful insert
  next_tail <= (others=>'0') when rstin else
               inc_tail when ivalid and iready else tail;

  -- head increments on successful remove
  next_head <= (others=>'0') when rstout else
               inc_head when ovalid and oready else head;
end impl;
```

Figure 29.10. VHDL description of a FIFO synchronizer.

```
library ieee;
use ieee.std_logic_1164.all;
entity GrayInc3 is
  port(input: in std_logic_vector(2 downto 0);
       output: out std_logic_vector(2 downto 0) );
end GrayInc3;
architecture impl of GrayInc3 is
begin
  output(0) <= not (input(1) xor input(2));
  output(1) <= not input(2) when input(0) else input(1);
  output(2) <= input(1) when not input(0) else input(2);
end impl;
```

Figure 29.11. VHDL description of a three-bit Gray-code incrementer.

Just before the end of the simulation, the input (which is running on a faster clock) gets seven words ahead of the output and `iready` goes low.

On the output side, `ovalid` does not go high, signaling that the FIFO is not empty, until two output cycles after the first datum has been inserted on the input side. This is due to the synchronizer delay of signal `tail_o`, from which `ovalid` is derived. In the simulation we wait until the FIFO is full and then remove five words, one every other cycle. Note that, from removing the first word (01), it takes two input cycles for `iready` to go high. This is due to the synchronizer delay of signal `head_i`, from which `iready` is derived. After removing word 05, `iready` remains high for the remainder of the simulation and one word per cycle is removed.

EXAMPLE 29.2 FIFO depth

Consider a FIFO synchronizer that uses brute-force synchronizers composed of two back-to-back flip-flops to synchronize the head and tail pointers. Assuming the input and output clocks are running at approximately the same frequency ($\pm 10\%$), what is the minimum FIFO depth that will support data transport at full rate?

The FIFO must be deep enough that the input side does not see the FIFO as full before the head pointer is synchronized back into the input domain when the clocks are at a worst-case relative phase. Thus, the FIFO depth must be 5 – deep enough to cover a round-trip delay through (a) the tail synchronizer to the output domain (two cycles), (b) one cycle for the output logic to react to the `ovalid` signal and increment `head_o`, and (c) the head synchronizer back to the input domain (two cycles). The delay of the synchronizers is two cycles each – one for each flip-flop in the synchronizer. This assumes that the input and output clocks are nearly aligned – which is worst-case timing. With the best-case relative phase, the total synchronizer delay will be three cycles rather than four.

The total delay, and hence the required FIFO depth, can be seen in the timing table (Table 29.1), which assumes that the clocks are nearly aligned. The FIFO is originally empty – with both pointers at 0. In cycle 0, `ivalid` is asserted, inserting a word into the FIFO and causing `tail_i` to increment to 1 in cycle 1. Two cycles later, in cycle 3, `tail_o` increments to 1, causing `ovalid` to be asserted – showing the output that the FIFO is not empty. The output

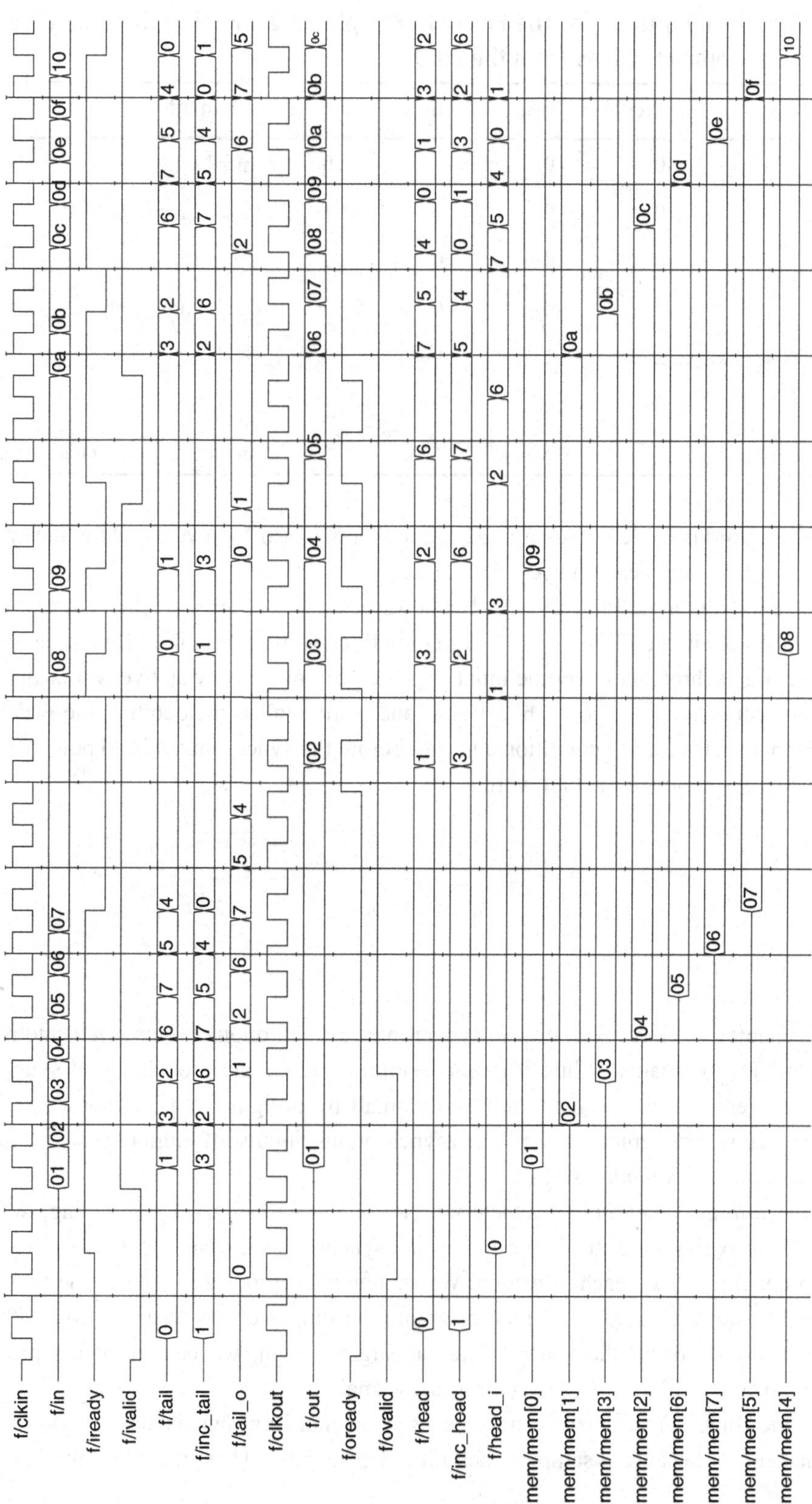

Figure 29.12. Waveforms from simulating the FIFO synchronizer of Figure 29.10.

Table 29.1. **Timing table for the FIFO of Example 29.2 showing the time taken to insert and remove values from the FIFO**

Cycle	tail_i	tail_o	head_o	head_i	T–H	Comment
0	0	0	0	0	0	Initial state
1	1	0	0	0	1	First element inserted
2	2	0	0	0	2	
3	3	1	0	0	3	Output not empty
4	4	2	1	0	4	
5	5	3	2	0	5	
6	6	4	3	1	5	Increment of head reaches input

removes a word in cycle 3, causing `head_o` to increment to 1 in cycle 4. After the synchronizer delay, `head_i` increments in cycle 6.

The sixth column of Table 29.1, T–H, shows `tail_i – head_i`, the number of words the input thinks are in the FIFO. Even though only three words are in the FIFO in the steady state, because of synchronizer delay the input sees the peak occupancy as five words and the output sees the occupancy `tail_o = head_o` as one word. Unless the depth of the FIFO is at least 5, the input will see a full condition and stall before the synchronized head pointer reflects that the output has removed the first word.

Summary

In this chapter you have seen how to synchronize asynchronous events *safely* and pass signals between clock domains. While it is impossible to reduce the probability of synchronization failure to zero,[1] it can be made arbitrarily small by designing an appropriate *synchronizer*. Synchronizers are typically used on asynchronous inputs of digital systems and on the boundaries of *clock domains*.

A *brute-force synchronizer* synchronizes a single-bit signal by sampling, waiting, and resampling. A sampling flip-flop samples an asynchronous signal, entering a metastable state with probability P_E on each transition. We wait an appropriate amount of time t_w to allow any metastable states to decay and then resample the output of the first flip-flop with a second flip-flop. By choosing the waiting interval large enough, we can make the probability of synchronization failure $P_F = P_E P_S$ arbitrarily small.

A brute-force synchronizer cannot be used on multi-bit signals where many bits change simultaneously because it samples each bit independently. Thus, its output may include some

[1] Unless the synchronization is avoided entirely.

bits from the old word and other bits from the new word. Brute-force synchronizers can be used on multi-bit signals only when they are Gray-coded so that only one bit changes at a time.

Arbitrary multi-bit signals can be synchronized with a *FIFO synchronizer*. Words are inserted into the FIFO in the input clock domain and removed from the FIFO in the output clock domain. Synchronization is required only to pass the *head* and *tail* pointers between the two domains to check for *empty* and *full* conditions. This can be safely accomplished by Gray coding the head and tail pointers and synchronizing them with brute-force synchronizers.

BIBLIOGRAPHIC NOTES

There are many more types of synchronizers than we have had time to discuss here. See ref. [33] for a survey. Even though the synchronization problem is decades old, new work is constantly being carried out on the topic [31].

Exercises

29.1 *Brute-force synchronizer, I.* Calculate the mean time to failure of a system with $f_a = 200\,\text{MHz}$, $f_{cy} = 2\,\text{GHz}$, waiting one cycle for synchronization. Use the flip-flop parameters of Table 29.2.

Table 29.2. Flip-flop parameters to use throughout the exercises

t_s	50 ps
t_h	20 ps
τ_s	40 ps
t_{dCQ}	20 ps

29.2 *Brute-force synchronizer, II.* Calculate the mean time to failure of a system with $f_a = 200\,\text{MHz}$, $f_{cy} = 2\,\text{GHz}$, waiting five cycles for synchronization. Use the flip-flop parameters of Table 29.2.

29.3 *Brute-force synchronizer, III.* Calculate the mean time to failure of a system with $f_a = 200\,\text{MHz}$, $f_{cy} = 2\,\text{GHz}$, using five back-to-back flip-flops for synchronization. Use the flip-flop parameters of Table 29.2.

29.4 *Brute-force synchronizer, IV.* For a system with $f_a = 200\,\text{MHz}$, $f_{cy} = 2\,\text{GHz}$, plot the mean time to failure on a semi-log axis as a function of the number of cycles spent synchronizing.

29.5 *Brute-force synchronizer, V.* You are designing a pacemaker that takes an asynchronous signal that pulses up to 200 times per minute. You synchronize using a flip-flop with the parameters given in Table 29.2 and a 1 GHz clock.[2] There are to be over 10^7 pacemakers in the field, with a lifetime of 30 years. How many cycles must your synchronizers wait

[2] Real pacemakers run nowhere near this clock rate; we are using this value for the sake of example.

in order to have a mean time to failure greater than that of the combined lifetime of all the pacemakers?

29.6 *Once-and-only-once synchronizer.* Design a once-and-only-once synchronizer. This circuit accepts an asynchronous input a and a clock clk and outputs a signal that goes high for exactly one clock cycle in response to each rising edge on the input a.

29.7 *Multi-bit synchronization, I.* When using a two-bit brute-force synchronizer to transfer a two-bit Gray-coded signal across clock domains, what is the minimum amount of time that needs to elapse between bit toggles? That is, what is the maximum clock rate at which the Gray codes can advance?

29.8 *Multi-bit synchronization, II.* Tired of constantly encoding your data to have only one bit transition at a time, you decide to build a different system to send data from one clock domain to another. You will send eight bits (with different propagation delays, or data skew, t_{ds}) and one valid signal from the input clock domain (cycle time t_a) to a brute-force synchronizer into the output clock domain (cycle time t_b). The module connected to the output will sample the data bits on the first rising output clock where the synchronized valid is high. Draw a waveform, labeling all relevant timing constraints, of sending eight-bit values in this manner. Make sure to allow enough time for the data to synchronize into the output domain before asserting the valid signal.

29.9 *Thermometer-coded synchronization.* Design a module that can be used to synchronize an eight-bit thermometer-coded signal (Figure 1.7(c)). The input should be the two clocks, *increment* and *decrement* signals in the input clock domain. The outputs should be the eight-bit signal. Make sure you do not use any state from the output domain with logic on the input domain.

29.10 *FIFO synchronizer.* Modify the logic of the FIFO synchronizer to allow the last location to be filled. You will need to add state to encode whether the FIFO is full or empty when $head = tail$.

29.11 *Full/empty bits.* Design and code a FIFO synchronizer that uses a presence bit for each register instead of head and tail pointers. When the input writes a value into a register, the corresponding presence bit is marked as full. When the output reads a value, the corresponding presence bit is marked empty. You will need a bank of presence bits in each clock domain. Assume that the RAM can be safely read asynchronously, but is written synchronously in clock domain 1. You should have room for four entries in the shared RAM.

29.12 *FIFO depth, I.* Consider a FIFO synchronizer that uses brute-force synchronizers composed of three back-to-back flip-flops to synchronize the head and tail pointers. Assuming the input and output clocks are running at approximately the same frequency ($\pm 10\%$), what is the minimum FIFO depth that will support data transport at full rate?

29.13 *FIFO depth, II.* Consider a FIFO synchronizer that uses brute-force synchronizers composed of four back-to-back flip-flops to synchronize the head and tail pointers. Assuming the input and output clocks are running at approximately the same frequency ($\pm 10\%$), what is the minimum FIFO depth that will support data transport at full rate?

29.14 *Clock stopper, I.* Consider a pipeline of modules each running on their own clock. An alternative to using FIFO synchronizers between the stages is to stop the clock of each stage to ensure that transfers at the input and output are *safe* from synchronization failure. Design a *clock stopper* module, an asynchronous circuit that, together with a delay line, generates the clock for one such pipeline stage. Your module should have inputs *ivalid* from the preceding pipeline stage, *oready* from the next pipeline stage, $delay_o$ from the delay line, and *done* from the logic of this pipeline stage. It should produce outputs *iready* to the preceding pipeline stage, *ovalid* to the next pipeline stage, and $delay_i$ to the delay line. You may assume that the delay of the delay line is more than half of the minimum cycle time of this pipeline stage.

Because there is no common clock to reference for the *ready* and *valid* signals, your module should obey *four-phase* flow control. In phase 1, a module signals that it has a valid datum on its output by asserting its *valid* signal, which must remain high until phase 2, when the receiving module accepts the datum by asserting its *ready* signal. The *ready* signal must remain high until phase 3, when the transmitting module lowers its *valid* signal. Finally, *valid* must remain low until phase 4, when the transmitting module lowers its *ready* signal.[3]

29.15 *Clock stopper, II.* Repeat Exercise 29.14 with *two-phase* flow control between modules. With two-phase flow control, a transmitting module signals that it has a new datum ready on its output by *toggling* its *valid* signal, and a receiving module accepts the datum by *toggling* its *ready* signal. Because signaling is done on both rising and falling edges of the signals, there is no need to wait for the signals to return to zero.

29.16 *Putting it all together.* Go design and implement a digital system that is useful, profitable, interesting, and/or really cool.

[3] Because the meaning of these flow-control signals is slightly different in the asynchronous context, they are often called *request* and *acknowledge* (or *req* and *ack* for those with repetitive stress injuries) in asynchronous systems. To avoid confusion, we will stick with the *valid* and *ready* convention here.

Part VIII

Appendix: VHDL coding style and syntax guide

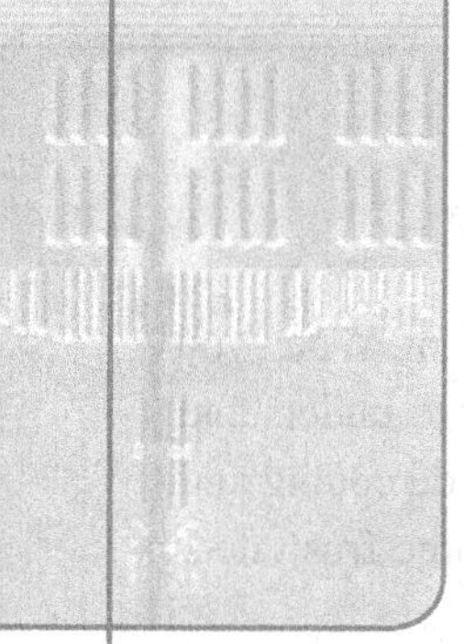

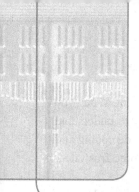

APPENDIX A
VHDL coding style

In this appendix we present a few guidelines for VHDL coding style. These guidelines grew out of experience with both VHDL and Verilog description languages. These guidelines developed over many years of teaching students digital design and managing design projects in industry, where the guidelines have been proven to reduce effort, to produce better final designs, and to produce designs that are more readable and maintainable. The many examples of VHDL throughout this book serve as examples of this style. The style presented here is intended for synthesizable designs – VHDL design entities that ultimately map to real hardware. A very different style is used in testbenches. This section is not intended to be a reference manual for VHDL. The following appendix provides a brief summary of the VHDL syntax used in this book and many other references are available online. Rather, this section gives a set of principles and style rules that help designers write correct, maintainable code. A reference manual explains what legal VHDL code is. This appendix explains what good VHDL code is. We give examples of good code and bad code, all of which are legal.

A.1 BASIC PRINCIPLES

We start with a few basic principles on which our VHDL style is based. The style presented is essentially a VHDL-2008 equivalent to a set of Verilog style guidelines based upon our experience over many years of teaching students digital design and managing design projects in industry combined with nearly a decade of experience teaching earlier versions of VHDL.

Know where your state is Every bit of state in your design should be explicitly declared. In our style, all state is in explicit flip-flop or register components, and all other portions are purely combinational. This approach avoids a host of problems that arise when writing sequential statements directly within an "**if** `rising\_edge(clk)` **then**" statement inside a process. It also makes it much easier to detect inferred latches that occur when not all signals are assigned in all branches of a conditional statement.

Understand what your design entities will synthesize to When you write a design entity, you should have a good idea what logic will be generated. If your design entity is described structurally, wiring together other components, the outcome is very predictable. Small behavioral design entities and arithmetic blocks are also very predictable. Large behavioral design entities, where the outcome of the synthesis can't be predicted, should be avoided.

Make your code readable Over the course of a project, a design entity may need to be revised. Whether this is easy or nearly impossible has a lot to do with the readability of the design entity. The functionality should be clear – e.g., use a `case` or `case?` statement for a truth table. Use descriptive signal and constant names. Include comments that capture the intent of the code – don't just duplicate what the code says with a comment. When you go to edit a design entity that has been well written, you will find it easy to understand the functionality of the design and make changes. If the design entity uses odd coding and cryptic names, you will have difficulty understanding it and you will become frustrated, particularly when you realize that you wrote it in the distant past.

Be defensive Think about what could go wrong with your design entity. What are the edge cases for inputs and outputs? Make sure each of these can be handled. Include *assertions* for invariants that must be maintained and for error conditions that can be checked.

Let the synthesizer do its job Modern logic synthesis tools (such as Synopsys Design Compiler®, Altera Quartus II®, and Xilinx Vivado®) do a great job of optimizing *small* combinational design entities (up to about ten inputs). They also do a great job of optimizing arithmetic circuitry. Except under very rare circumstances, there is no point in manually optimizing such designs. The design entity description should be written in a straightforward behavioral manner, enabling the tools to do what they are good at.

Understand what the synthesizer can't do Modern synthesis tools are not good at making high-level optimizations. They don't understand how to factor or partition logic or when a function unit can be shared. They can't make a tradeoff about when a function should be executed in one cycle or when it should be pipelined over several cycles. This is where the human designer adds value. Don't expect the tools to do the designer's job – and vice versa.

A.2 ALL STATE SHOULD BE IN EXPLICITLY DECLARED REGISTERS

All state in your design should be in explicitly instantiated register or flip-flop components. An example of correct style is shown in Figure A.1 (adapted from Figure 16.3). Here, a four-bit register is declared with input `nxt` and output `output`. The next-state function is described with a single assignment statement.

Avoid writing design entities like that shown in Figure A.2. This design entity mixes the creation of state (an assignment within an `if rising_edge(clk) then` statement) with the definition of the next-state function.

We insist on this coding style because it helps designers, and particularly students, follow our first two design principles. First, it helps the designer know where the state is. When you explicitly declare registers – and regard inferred latches as errors – you know where your state is. When coding in the style of Figure A.2, it is very easy to create state without knowing it. There is no strong distinction between a state variable and an intermediate variable that is used in computing the next-state function.

This coding style also helps a designer understand what their VHDL will synthesize to. It will generate the state register and the combinational logic that produces the next-state function. When coding in the incorrect style that mixes state and function (Figure A.2), it is very easy for designers, and particularly students, to start writing VHDL with a style similar to C code without understanding what is implemented in time and what is implemented in space. The results are often disastrous.

When teaching VHDL, our style avoids confusion about what is the current state and what is the next state. The statement

```
library ieee;
use ieee.std_logic_1164.all;
use ieee.std_logic_unsigned.all;
use work.ff.all;

entity Good_Counter is
  generic( n: integer :=4 );
  port( clk, rst: in std_logic;
        output: buffer std_logic_vector(n-1 downto 0) );
end Good_Counter;

architecture impl of Good_Counter is
  signal nxt: std_logic_vector(n-1 downto 0);
begin
  nxt <= (others=>'0') when rst else output+1;
  count: vDFF generic map(n) port map(clk, nxt, output);
end impl;
```

Figure A.1. Correct style for a counter design entity. The state of the counter is held in an explicitly declared register, component `count` of type `vDFF`.

```
library ieee;
use ieee.std_logic_1164.all;
use ieee.std_logic_unsigned.all;
use work.ff.all;

entity Bad_Counter is
  generic( n: integer := 4 );
  port( clk, rst: in std_logic;
        output: buffer std_logic_vector(n-1 downto 0) );
end Bad_Counter;

architecture impl of Bad_Counter is
  signal nxt: std_logic_vector(n-1 downto 0);
begin
  process(clk) begin
    if rising_edge(clk) then
      if rst then
        output <= (others => '0');
      else
        output <= output + 1;
      end if;
    end if;
  end process;
end impl;
```

Figure A.2. Incorrect style for a sequential design entity. State is mixed with next-state computation in a single process.

```
nxt <= (others=>'0') when rst else output+1;
```

makes it clear that nxt is the next state and that we are computing this state from the current state output. In contrast, the statement

```
if rst then
  output <= (others => '0');
else
  output <= output + 1;
end if;
```

confuses the issue by using the same signal output on both sides of an assignment. The next state is hidden.

Our style also sidesteps a tremendous amount of confusion and frustration that results from incorrect use of VHDL signals and variables *within* a process. Moreover, the bad style has no clear advantages. It isn't any shorter or easier to read than the good style.

Don't mix state and function in a single process block. There is no upside to the style of Figure A.2, and considerable downside.

A.3 DEFINE COMBINATIONAL DESIGN ENTITIES SO THAT THEY ARE EASY TO READ

Now that we have banished all state to explicitly declared registers, we can focus the remainder of this style guide on how to write good combinational logic designs.

Modern logic synthesis tools are very good at optimizing small combinational logic designs. Thus, these components should be written in the manner that is easiest to read and understand. For logic that is naturally described with a truth table, we code a design using a case or case? statement inside a process. For logic that is naturally described with an equation (datapath logic, Chapter 16) a concurrent assignment statement is in order.

Figure A.3 (extracted from Figure 9.6) illustrates how to use a case statement to implement a combinational function best described with a table, in this case the days-in-month function. A case statement on the input signal month assigns the appropriate values to the output signal days for each value of input. The most common output case is handled in the when others => case. Input cases with the same output assignment are grouped together for clarity and brevity.

```
process(all) begin
  case month is
    -- thirty days have September...
    -- all the rest have 31
    -- except for February which has 28
    when 4d"4" | 4d"6" | 4d"9" | 4d"11" => days <= 5d"30";
    when 4d"2" => days <= 5d"28";
    when others => days <= 5d"31";
  end case;
end process;
```

Figure A.3. Implement combinational functions that are best described with a table using a case statement.

```
number_of_ones <= ("00" & x(0)) + ("00" & x(1)) + ("00" & x(2)) + ("00" & x(3));
```

Figure A.4. Combinational design whose function is best described by an equation should use a concurrent signal assignment statement.

```
process(all) begin
  case opcode is
    when OP_ADD => nxt <= acc + data;
    when OP_XOR => nxt <= acc xor data;
    when OP_AND => nxt <= acc and data;
    when OP_OR  => nxt <= acc or data;
    when others => nxt <= 8d"0";
  end case;
end process;
```

Figure A.5. A logic description where a `case` statement is used to decode a control signal (in this case `opcode`) to select an equation to compute a data signal.

```
process(all) begin
  case? std_logic_vector'(rst & load & shl & inc) is
    when "1---" => nxt <= 8b"0";
    when "01--" => nxt <= input;
    when "001-" => nxt <= state(6 downto 0) & '0';
    when "0001" => nxt <= state + 1;
    when others => nxt <= state;
  end case?;
end process;
```

Figure A.6. Using a conditional `case?` statement to assign a value to the `nxt` signal on the basis of four prioritized input signals.

A combinational function that is naturally described with an equation should use a concurrent assignment statement (i.e., a statement that is NOT in a process), as illustrated in Figure A.4. Here we compute the number of 1s[1] in a four-bit signal `x`. This is very easily described and understood as an equation. Describing this function as a `case` statement would be long and confusing.

Often it is natural to describe a component as a truth table of equations. A truth table of control signal values determines how the next states of data signals are computed using equations. An example of this style is shown in Figure A.5.

When a number of binary control signals with priority are decoded to determine a function, a variant of this hybrid style using a `case?` statement can be employed as shown in Figure A.6. This statement computes a next-state variable `nxt` for a sequential circuit that can reset, shift left, shift right, load, or hold the current value depending on a set of binary control signals.

A.4 ASSIGN ALL SIGNALS UNDER ALL CONDITIONS

To avoid inferring latches – state that is not intended – it is imperative that every variable that is assigned on *any* branch of a conditional statement be assigned on *all* branches of that statement. See also Rule C2 in Section B.10.

[1] The number of 1s in a binary word is often referred to as the *population count* of the word. Some computers have a `POPCOUNT` instruction that performs this function.

```
process(all) begin
  if rst = '1' then
    nxt <= 8d"0";
  end if;
end process;
```

Figure A.7. Bad code: a latch is inferred because `nxt` is not assigned when `rst` is false.

```
process(all) begin
  if rst then
    nxt <= 8d"0";
  else
    nxt <= output+1;
  end if;
end process;
```

Figure A.8. Correct code: `nxt` is assigned on all branches.

```
signal state, next_state: std_logic_vector(n-1 downto 0);
signal next_state_rst: std_logic_vector(n-1 downto 0);
...
process(all) begin
  case? input & state is
    when "--" & FETCH_STATE =>
      output <= FETCH_OUT;
      next_state <= DECODE_STATE;
    when "1-" & DECODE_STATE =>
      next_state <= REG_READ_STATE;
    when "0-" & DECODE_STATE =>
      output <= DECODE_ZERO;
      next_state <= REG_READ_STATE;
    ...
    -- remaining cases omitted for brevity
    when others =>
      output <= (others => '-');
      next_state <= (others => '-');
  end case?;
end process;

next_state_rst <= FETCH_STATE when rst else next_state;
```

Figure A.9. Bad code: not all signals are assigned in each case.

Figure A.7 shows an example of a common error. Signal `next` is assigned when `rst` is true (`'1'`), but it is not assigned if `rst` is false (`'0'`). This is *not* a combinational circuit. It is a latch. When `rst` is false, `nxt` holds its value.

Correct use of the `if` statement is shown in Figure A.8. Signal `nxt` is assigned both when `rst` is true and when `rst` is false. At this point, it is worth contrasting the use of an `if` statement with a `case` statement. For a `case` or `case?` VHDL requires that all input combinations have a corresponding `when` statement, which is why we typically include a default condition (`when others =>`). The `when others =>` case nicely catches all branches of a `case` or `case?` that have not been handled explicitly. For this reason, a `case` or `case?` is preferred to an `if` in situations when either could be used.

```
signal state, next_state_rst: std_logic_vector(n-1 downto 0);

type fsm_t is record
  outp: std_logic_vector(m-1 downto 0);
  nxts: std_logic_vector(n-1 downto 0);
end record;

signal fsmo: fsm_t;
...
process(all) begin
  case? input & state is
    when "--" & FETCH_STATE => fsmo <= (FETCH_OUT,     DECODE_STATE);
    when "1-" & DECODE_STATE => fsmo <= (RR_OUT,       REG_READ_STATE);
    when "0-" & DECODE_STATE => fsmo <= (DECODE_ZERO, REG_READ_STATE);
    ...
    -- remaining cases omitted for brevity
    when others => fsmo <= ((others => '-'),(others => '-'));
  end case?;
end process;

output <= fsmo.outp;
next_state <= FETCH_STATE when rst else fsmo.nxts;
```

Figure A.10. Good code: assigning to a record makes it hard to omit a signal assignment from a branch. It also results in code that is more readable.

Another common error occurs when not all the signals are assigned on all of the branches. An example of this error is shown in Figure A.9. Here signal `output` is not assigned in the second case. A latch will be inferred and it will retain its previous state.

One way to avoid this problem is to make each case a single assignment to a signal declared as a VHDL record containing one field for all outputs being assigned, as shown in Figure A.10. This style of coding makes it hard to forget to assign a signal on a particular branch – it stands out if you do so. It also makes the code much more readable. It looks just like a state table.

A.5 KEEP DESIGN ENTITIES SMALL

For readability and to make sure you know what your design entity will synthesize to, behavioral design entities should be kept small – no more than 50 lines of text. This will fit on one *screen* of a text editor, making it possible to see the whole design entity at once and understand its function. Also, 50 lines of text are sufficient to describe most combinational logic functions; anything larger than this is typically several separate functions and should be broken up. If you do have a single function that can't be described in 50 lines of code, you should give serious thought as to whether it can be factored to reduce its complexity.

A.6 LARGE DESIGN ENTITIES SHOULD BE STRUCTURAL

Another way of stating the last guideline is that large design entities should be structural – instantiations of other components connected by wires. There is no doubt what a structural design entity synthesizes to. Also, while modern logic synthesis tools do well on small combinational logic designs and arithmetic circuits, they do poorly on large designs. These tools excel at making small-scale optimizations, but don't see the big picture and don't understand how to make large-scale optimizations – like sharing an adder in Figure 16.6.

A.7 USE DESCRIPTIVE SIGNAL NAMES

Your code will be much more readable if your signal names describe their functionality. Consider, for example, the statement

```
aligned_mantissa <= mantissa sll exponent_difference ;
```

The meaning of each signal and the function of the statement are easy to understand just by reading the statement. In contrast, writing the same statement as

```
i <= j sll k;
```

does not convey the meaning or intent of the statement.

Long names, however, cut both ways. While they are easily readable, they can make code appear cluttered. Statements are easier to understand if they fit on one (or a few) lines. Long names tend to make statements spill over multiple lines, making them harder to read. With an appropriate signal dictionary – and supporting comments – short or abbreviated names can be very readable, as in the following:

```
a_m <= m sll e_d;
```

When in doubt as to whether full names or abbreviated names are more readable, we find it useful to write the code both ways and pick the version that is most readable.

A.8 USE SYMBOLIC NAMES FOR SUBFIELDS OF SIGNALS

Often a long signal is broken into a number of subfields. For example, a 32-bit instruction may be split into an eight-bit opcode, three five-bit register specifiers, and a nine-bit constant. Code becomes much more readable when symbolic names are used for these subfields. Consider the statement

```
case instruction(31 downto 24) is ...
```

This conveys little meaning, and there is a danger of getting the bit-field specifier wrong – particularly if it changes. Writing the same statement as

```
case opcode is ...
```

is much more readable.

A.9 DEFINE CONSTANTS

Numbers should rarely appear in VHDL code, and any given number should never appear multiple times. Readability and maintainability are improved by defining a constant or parameter and using the symbolic name in place of the number.

```
case? opcode & fun is
  when 6x"0" & 6x"20" => aluop <= 3x"5";
  when 6x"0" & 6x"21" => aluop <= 3x"6";
  ...
  when 6x"23" & "------" => aluop <= 3x"5";
  ...
end case?;
```

Figure A.11. Bad code: constants should be defined with symbolic names.

```
case? opcode & fun is
  when RTYPE_OPC & ADD_FUN  => aluop <= ADD_OP;
  when RTYPE_OPC & SUB_FUN  => aluop <= SUB_OP;
...
  when LW_OPC    & "------" => aluop <= ADD_OP;
...
end case?;
```

Figure A.12. Good code: symbolic names are used for constants. The numerical value of a constant appearing multiple times is defined exactly once.

Consider the code fragment from an instruction decoder for a MIPS processor shown in Figure A.11. Reading this code gives no clue as to its function. Moreover, the opcode for a MIPS R-type instruction `6'h0` and the alu operation code for + (`3'h5`) both appear twice. The code is much more readable if constants are defined.

On replacing the numbers with symbolic constants, as in Figure A.12, the function of the code becomes clear. Moreover, using a symbolic constant for a value that is used repeatedly – like the opcode for an R-type instruction `RTYPE_OPC` or the alu operation code for + `ADD_OP` – the actual number is used exactly once, in the constant definition. This makes it easy to update should it ever change. For example, if we were to redesign our ALU so that the operation code for + changed from 5 to 7, only the define statement would need to be changed. The change is propagated to all uses of the constant.

Symbolic names can be bound to constant values either as *constants* (using VHDL **`constant`**) or as *generic constants* (using a **`generic`** clause in the entity declaration). Unless a symbolic name takes on different values in different instantiations of a component, it should be defined as a constant. Generic constants should be reserved for, well, generic constants, values that may differ in different instantiations of a component. The bit width of a register or a counter is a good example of a generic constant. A function code for an ALU or an opcode for a processor is a good example of a constant – it doesn't change from component instance to component instance. Constants should be defined with an architecture if they are used only within the architecture body or in a package if they are used across multiple design entities.

A.10 COMMENTS SHOULD DESCRIBE INTENTION AND GIVE RATIONALE, NOT STATE THE OBVIOUS

Well-written code should have lots of *high-quality* comments. Good comments give the big picture of what the code is doing, capture the designer's intent, and give a rationale for design choices. Unfortunately, many designers confuse quantity with quality and write lots of comments that don't add any information.

Consider the following code fragment:

```
case aluop is
  when ADD_OP => c <= a + b ; -- add a and b
  ...
end case;
```

Duh! Of course this statement adds a and b. This comment doesn't add any value. It just clutters the code and should be deleted.

Now consider the following comment, which should have been added to Figure 16.7:

```
-- factor the adder/subtractor out of the case statement
-- because the synthesizer won't combine two adders
-- increments when down=0, decrements when down=1
outpm1 <= output + ((n-2 downto 0 => down) & '1');
```

While a bit verbose, this comment adds value. It gives the big-picture view of the code – that this adder has been factored out of a case statement. It gives the rationale – because the synthesizer would have generated two adders if we didn't do the factoring manually. It also explains some functionality that isn't completely obvious on looking at the code.

A.11 NEVER FORGET YOU ARE DEFINING HARDWARE

When we are writing VHDL code, it is very easy to fall into the trap of thinking we are writing a computer program – in which one statement happens at a time. After all, VHDL looks a lot like C code and they share many constructs. It is particularly easy to fall into this trap when you switch back and forth regularly between writing VHDL and writing C, Python, or some other programming language.

This is a very dangerous trap. Unlike with a computer programming language, such as C, where one statement happens at a time, in a VHDL program *all* statements happen at once. When you write a VHDL program you are defining hardware. All of this hardware operates in parallel. Every concurrent assignment statement of every component happens at the same time.

Never forget that it is hardware – and that everything happens at the same time.

A.12 READ AND BE A CRITIC OF VHDL CODE

One becomes a better writer of English by reading the works of others, being critical of their style, and emulating styles that work. One can become a better VHDL designer in the same way. Read VHDL code whenever you can. Be critical of code – point out what is both good and bad in your own code and in the code of others. Do this constructively – judge the code, not the person. Borrow unashamedly things that work, that make code better, more readable, or easier to maintain. When your own code is being criticized, don't be defensive. Be open to criticism, listen, and learn.

The better technology companies foster an engineering culture that encourages designers to read and critique each other's code. Many companies formalize this via design reviews – where a team of top engineers reviews a completed piece of code. Others use *pair programming*, where all coding is done by pairs of people – with one programming and the other critiquing. However it is done, what is important is that a culture is created where designers review the code of others, offer helpful criticism, and are open to criticism of their own code.

APPENDIX B
VHDL syntax guide

In this appendix we provide a summary of the VHDL syntax employed in this book. Before using this appendix you should first have read the introductions to VHDL in Section 1.5 and Section 3.6. In addition to this appendix you may find the subject index at the end of this book helpful when looking for information about specific VHDL syntax features. Excellent references documenting the *complete* VHDL language can be found elsewhere [3, 55]. However, due to the complexity of the VHDL language such references tend to lack detailed discussion of hardware design topics. An abridged summary of the key aspects of VHDL syntax, such as that found in this appendix, can be very helpful when learning hardware design.

This book uses VHDL syntax features from the most recent standard, VHDL-2008, that enable greater designer productivity and are supported by the FPGA CAD tools typically used in introductory courses on digital design. Many CAD tools by default still assume an earlier version of VHDL even though they have support for VHDL-2008. Hence, you should consult your CAD tool's documentation to learn how to enable support for VHDL-2008 before trying the examples in this book.[1]

To keep descriptions brief yet precise Extended Backus-Naur Form (EBNF) is employed in this appendix. Non-terminals are surrounded by angle brackets ("<" and ">") and definitions of non-terminals are denoted by the symbol "::=". A list of choices is separated by a pipe symbol ("|") and the interpretation is that only one of the items should appear. Zero or more repetitions of a construct are indicated by surrounding the construct with curly braces ("{" and "}"). An optional construct (i.e., zero or one instances) is indicated by surrounding it with square brackets ("[" and "]"). The EBNF descriptions in this appendix are simplified versions of those found in the VHDL language standard [55]. The simplified EBNF descriptions here correspond to the VHDL syntax subset commonly used for synthesis.

The use of hardware description languages (HDLs) for hardware design differs from the use of programming languages for software development. Software is implemented by converting a program written in a programming language into computer instructions that then appear to execute one at a time. In contrast, a hardware description is synthesized into logic gates

[1] For ModelSim enable VHDL-2008 support by selecting "Compiler Options..." then "Use 1076-2008". For Altera Quartus II adding `--synthesis VHDL_INPUT_VERSION VHDL_2008`" at the top of a VHDL file enables support for VHDL-2008. Xilinx Vivado 2014.3 has beta support for VHDL-2008 for synthesis enabled by typing "`set_property vhdl_version vhdl_2008 [current_fileset]`" in the Vivado TCL Console.

connected by wires. While different instructions from a software program execute at varying times, the logic gates and wires corresponding to a hardware description operate continuously.

This appendix focuses on VHDL syntax relevant to synthesizable design descriptions. Such a VHDL description must be written using a *syntax subset* that is *synthesizable*. The requirement to restrict language usage to a subset of the complete syntax during hardware design differs from style concerns encountered during software development. The necessity of such syntax restrictions results from the general-purpose nature of HDL languages, including VHDL, that have come to dominate hardware design, combined with the difficulty of *automatically* generating an efficient hardware design from an arbitrary algorithmic specification. Indeed, there are even formal standards that specify the minimum language subset that should be considered to be synthesizable [56]. The last part of this appendix (Section B.10) provides simplified rules that you can use to ensure your VHDL is synthesizable. If these rules are violated the hardware generated from your VHDL during synthesis will very likely behave differently than your VHDL does during simulation.

Below we describe the various features of VHDL syntax beginning with basic elements (comments, identifiers, keywords), then moving on to types before considering design entities which provide the basic framework for describing logic circuits in VHDL. The remainder of this appendix considers the various forms of statements found inside design entities.

B.1 COMMENTS, IDENTIFIERS, AND KEYWORDS

In VHDL single-line comments are preceded by two dashes (`--`) and extend to the end of the line. VHDL-2008 introduces syntax for multi-line comments. A multi-line comment starts with `/*` and extends until `*/`.

Identifiers and keywords (reserved words) in VHDL are case-insensitive. An identifier must begin with an alphabetical character and can include any number of alphabetical characters, digits 0 through 9, and underscores. Examples of valid identifiers are `Mux3` and `state_next`.

VHDL-2008 defines 115 reserved words. This is a large number when compared with contemporary programming languages such as C, which defines around 30 keywords, Java, which defines around 50 keywords, and C++, which defines around 70 keywords. The large number of keywords can make learning VHDL intimidating. However, a designer can be fully productive after learning a core subset of 58 keywords that are listed in Table B.1. The keywords listed in Table B.1 are discussed throughout this appendix. Under the heading *VHDL keywords* the subject index at the end of this book also lists page references to each of these keywords where they are discussed throughout the book.

B.2 TYPES

VHDL is a strongly typed language. This means the designer is required to invest time specifying the semantics (meaning) of binary values while coding by declaring the type of signals used in the design. The "reward" for this additional effort is that the root cause of many design errors is identified via error messages when a VHDL description is compiled. Below we consider some basic types used throughout the book and discuss how to declare new types.

Table B.1. **VHDL-2008 keywords used in this book**

all	downto	in	of	rol	to
and	else	inout	or	ror	type
architecture	elsif	is	others	select	use
array	end	library	out	signal	variable
begin	entity	loop	package	sla	wait
buffer	file	map	port	sll	when
case	for	mod	process	srl	with
component	function	nand	record	subtype	xor
configuration	generic	nor	report	then	xnor
constant	if	not	return		

B.2.1 `Std_logic`

A type used in most VHDL designs is `std_logic`. A signal or variable of type `std_logic` represents a single bit value. In an electronic circuit logic values of 0 and 1 are represented using voltages (as noted in Chapter 1). In a VHDL specification these values can be represented as a `std_logic` value of `'0'` or `'1'`, respectively (the single quotes are required). When specifying a digital circuit it can also be helpful to indicate that the designer does not care whether a value is logic value `'0'` or `'1'`. This can be indicated using the `std_logic` value `'-'` (e.g., a hyphen). The `std_logic` type also includes two values `'U'` and `'X'` that are helpful for catching design errors while simulating a digital circuit specification during verification testing (Section 20.1). The value `'U'` stands for *undefined* and indicates that no value has been assigned. The value `'X'` stands for *unknown* and is usually used to indicate that a single wire is being driven simultaneously to both logic value 1 and logic value 0 – a situation which would cause a short circuit in a real circuit. In some situations, such as distributed multiplexers and buses connecting many components, it is desirable to employ tri-state buffers in a digital circuit (tri-state circuits are discussed in Chapter 4). To help tri-state circuits, `std_logic` defines the value `'Z'` which represents a *high-impedance* state corresponding to a tri-state buffer that has its enable input set to logic value 0.

B.2.2 `Boolean`

The result of comparison operations is the VHDL predefined type `boolean` which is distinct from `std_logic` and has values `true` and `false`. VHDL-2008 introduces support for automatically converting a `std_logic` to a `boolean` value where a `boolean` is expected.

B.2.3 `Integer`

VHDL includes a predefined type `integer` which is distinct from `std_logic_vector` (described below). The range of values of `integer` is guaranteed to be at least $-(2^{31}-1)$

to $2^{31} - 1$. Integers are typically used only for specifying parameters such as the width of a multi-bit std_logic_vector signal. They are *not* typically used to store values computed by a circuit.

Integer literals are written in base-10 (decimal) using the digits 0–9. For example, the value one hundred is written 100 (without quotes).

B.2.4 Std_logic_vector

While a software developer uses types that have predefined bit widths (e.g., eight-bit characters or 64-bit integers), a digital designer typically has to specify the precise bit width that suits the needs of the design. To do this they employ multi-bit bus signals (Chapter 8). VHDL represents multi-bit values as an array of std_logic values using the type std_logic_vector. To fully specify a std_logic_vector type it is necessary to include an *index range*. The index range accomplishes two goals: first, it indicates how many bits wide the std_logic_vector is, which helps check uses of std_logic_vector for errors. Second, it associates with each bit a numeric index value that can be used to read or write individual bits. A std_logic_vector is fully specified by including the index range in parentheses using the syntax std_logic_vector(<index_range>) where, using EBNF,

```
<index_range> ::= <numeric_constant> downto <numeric_constant> |
                  <numeric_constant> to <numeric_constant>
```

For example, std_logic_vector(2 **downto** 0) is a three-bit std_logic_vector with indices 2 through 0.

Multi-bit literals of type std_logic_vector are written as bit string literals using the notation <size><base>"<value>", where <size> is the bit width of the constant (in decimal); <base> specifies the base, where <base> is b for binary, o for octal, d for decimal, and x for hexadecimal; and <value> gives the value using digits of the corresponding base. Optionally, the base specifiers b, o, and x can be prefaced with s to indicate that <value> should be sign extended to <size> bits (sign extension is described in Chapter 10). For example, 6d"13" is a six-bit constant with value 001101_2, 5x"f" is a five-bit constant with value 01111_2, 4so"4" is a four-bit constant with value 1100_2 (the leftmost 1 is due to sign extension), and 7sb"-" is a seven-bit constant with each bit equal to "-" ("don't care"). Versions of VHDL prior to VHDL-2008 support neither <size> nor "d" for <base>. If <size> is omitted, the width is inferred from the number of digits in <value>. For example, x"00" is eight bits wide since a single hexadecimal digit is four bits (hexadecimal is described in Section 10.1). If <size> and <base> are both omitted then in addition the base is assumed to be binary. For example, "0101" is a four-bit binary number having value of five.

B.2.5 Subtypes

Sometimes it is convenient to give an alternative name for fully specified std_logic_vector type if it occurs in multiple places. A new type name can be introduced for this purpose using the following syntax:

```
subtype <identifier> is <subtype_indication> ;
```

where

```
<subtype_indication> ::= <type_mark> [ ( <index_range> ) ]
```

and <type_mark> is the name of a type or subtype. For example,

```
subtype state_type is std_logic_vector( 2 downto 0 );
```

declares a new subtype called state_type that is a three-bit std_logic_vector.

B.2.6 Enumeration

VHDL provides the capability of specifying enumeration types using the syntax

```
type <identifier> is ( <enum_literal> {, <enum_literal> }  );
```

where <identifer> is a user-defined name for the type that conforms to the rules for identifiers (described in Section B.1) and <enum_literal> is typically a meaningful name given by the designer. For example,

```
type state_type is ( SA, SB, SC, SD );
```

declares a new type called state_type that can take on the values SA, SB, SC, or SD. Some VHDL designers prefer to specify the names of states in a state machine (Chapter 14) using such enumerated types. In this textbook we employ VHDL constants (described in Section B.4) for this purpose. The reason is that constants enable the use of explicitly instantiated state which has advantages for reducing designer errors.[2]

B.2.7 Arrays and records

VHDL provides two forms of composite types: arrays and records. Array declarations are helpful for defining memory structures (Sections 8.8 and 8.9). The syntax for declaring a new array type is

```
type <identifier> is array (<index_range>) of <subtype_indication>;
```

For example,

```
type mem_type is array (0 to 255) of std_logic_vector( 15 downto 0 );
```

declares a new type called mem_type that is an array with 256 entries, each of which is a 16-bit value of type std_logic_vector. To refer to a single element of an array, use parentheses using the syntax

```
<array_identifier>( <index> )
```

For example, if memory has type mem_type then memory(0) refers to the first word of memory which has type std_logic_vector(15 **downto** 0). As noted earlier, std_logic_vector is actually an array of std_logic. So, if input is the identifier of a signal with subtype std_logic_vector(7 **downto** 0) then input(3) refers to the bit with index 3 and it has type std_logic.

A VHDL record is somewhat similar to a struct in C and can be used for grouping related signals into a single bus connecting different hardware components. Another use is to enable

[2] VHDL-2008 specifies support for generic types which would enable enumerated types to be succinctly combined with explicitly instantiated state. However, few CAD tools support this feature.

a single assignment statement for all outputs of a combinational logic block described with a process (see Section A.4). The syntax for declaring a new record type is

```
type <identifier> is record
                        <element_declaration>
                      { <element_declaration> }
                     end record;
```

where

```
<element_declaration> ::= <identifer> {, <identifer> } : <sub_type> ;
```

For example,

```
type inst_type is
   record
     opcode  : std_logic_vector(6 downto 0);
     dst     : std_logic_vector(2 downto 0);
     src1    : std_logic_vector(2 downto 0);
     src2    : std_logic_vector(2 downto 0);
   end record;
```

declares a record type named `inst_type` containing four fields named `opcode`, `dst`, `src1`, and `src2`.

The elements of a record are accessed using a period (`.`). Continuing the example above, if `inst` is a signal with type `inst_type`, then one can access the `opcode` element using `inst.opcode`.

B.2.8 Qualified expressions

The type of a VHDL expression may be ambiguous, in which case it can be specified by using a qualified expression. The syntax for a qualified expression is

```
<qualified_expression> ::= <type_mark>'(<expression>)
```

An example of using a qualified expression is illustrated in Figure 14.13.

B.3 LIBRARIES, PACKAGES, AND USING MULTIPLE FILES

The types `std_logic` and `std_logic_vector` are not part of the VHDL language but rather part of the `std_logic_1164` *package* within the `ieee` *library* defined as part of the VHDL language standard. As illustrated in Figure B.4 later, to use `std_logic` in your designs you must add the lines "**`library`** `ieee;`" and "**`use`** `ieee.std_logic_1164.`**`all`**`;`". The first line tells the VHDL compiler that we wish to use the `ieee` library. The second line tells the VHDL compiler that we wish to use the `std_logic_1164` package within the `ieee` library. You can define your own packages using the syntax illustrated in Figure 7.23.

For larger designs it is desirable to split your VHDL into multiple files. Each file should refer to declarations from the same file or an earlier file. To ensure this, all VHDL compilers provide a means for specifying the compilation order of multiple files.

B.4 DESIGN ENTITIES

A VHDL description of a hardware design is composed of a hierarchy of VHDL design entities. A single design entity can be viewed as a blueprint or specification for the interface and operation of an individual hardware block. A summary of the syntax for a design entity is shown in Figure B.1. Each design entity is composed of two parts: an entity declaration and an architecture body.

The entity declaration begins with the keyword **entity** and it defines the interface between the internals of the hardware design and the outside world via a list of *port declarations*. The entity declaration may also provide a set of parameters that can be used to specialize a design entity via a list of *generic constants*. To do this an optional generic clause is declared within an entity declaration using the keyword **generic**. It declares a list of generic constants using the syntax

```
generic( <generic_constants> );
```

where

```
<generic_constants> ::= <generic_constant> { ; <generic_constant> }
<generic_constant> ::= <identifier> : <sub_type> := <expression>
```

A common use for generic constants is to parameterize the bit width of inputs and outputs to a design entity. A port clause declares the inputs and outputs to the design entity and has the syntax

```
port( <port_declarations> );
```

where

```
<port_declarations> ::= <port_declaration> { ; <port_declaration> }
<port_declaration> ::= <identifer> {, <identifier> } : <mode> <sub_type>
<mode> ::= in | out | buffer | inout
```

```
entity <entity_name> is
  [ generic( <generic_constants> ); ]
  port( <port_declarations> );
end <entity_name>;

architecture <implementation_name> of <entity_name> is
  <type_declarations>
  <constant_declarations>
  <component_declarations>
  <internal_signal_declarations>
begin
  <concurrent_statements>
end <implementation_name>;
```

Figure B.1. A VHDL design entity consists of an entity declaration containing a list of input and output signal port declarations followed by an architecture body composed of an architecture declarative part containing various declarations and an architecture statement part containing a set of concurrent statements. The logic of the design entity is implemented by the concurrent statements.

The *mode* of a port indicates whether it is an input (**in**), output (**out**), output which can also be read (**buffer**), or bi-directional signal (**inout**).

The architecture body begins with the keyword **architecture** and it is divided into an architecture declarative part and an architecture statement part. The declarative part contains declarations used inside the architecture body and the statement part contains concurrent statements that implement the logic of the hardware. The two portions are separated by the keyword **begin**.

The architecture declarative part may contain type, constant, component, and internal signal declarations. Type declarations follow the syntax from Section B.2. Constants can be declared to help ensure that code is readable and maintainable using the following syntax:

```
constant <identifier> : <sub_type> := <const_expression>;
```

where the value of the constant is defined by the expression after the := operator. Examples of the syntax for defining a constant can be found in Figures 7.23 and 14.14. A component declaration is used to enable one design entity to be used within an architecture body of another using component instantiation (Section B.6.2). The syntax for a component declaration is similar to that of an entity declaration:

```
component <component_name> is
  [ generic( <generic_constants> ); ]
  port( <port_declarations> );
end component;
```

Adding a component declaration matching an existing entity declaration enables you to use that design entity in the architecture statement part. The concurrent statements within the architecture statement part communicate with each other through *signals*. A signal can be thought of as a set of wires carrying logical values from one piece of hardware to another. The syntax for declaring a signal is

```
signal <identifier> : <sub_type>;
```

The architecture statement part of the architecture body follows the keyword **begin** and is composed of a set of concurrent statements. Each concurrent statement operates in parallel with the others. This parallel operation is fundamental to hardware description languages and has no corresponding syntax in common programming languages such as C and C++. We elaborate more on the syntax of concurrent statements in Section B.6. A simple example of a design entity is shown in Figure 1.11.

B.5 SLICES, CONCATENATION, AGGREGATES, OPERATORS, AND EXPRESSIONS

In digital designs it is often necessary to operate upon individual bits of a multi-bit value. While we can operate on an individual bit using array index notation, it is sometimes more helpful to access a range of bits of a multi-bit value using *slice* notation:

```
<slice> ::= <array_identifier>( <index_range> )
```

For example, `input(7 downto 4)` refers to only the first four bits of `input`.

One can concatenate array elements and slices to create a larger array value out of individual elements and smaller array values. The concatenation operator in VHDL is the ampersand, `&`. For example,

```
'1' & '0' & '1' & '0'
```

is equivalent to `"1010"`. Another example is

```
input(3 downto 0) & input(7 downto 4)
```

which results in the eight-bit value formed by swapping the rightmost and leftmost four-bits of `input`. One can also concatenate elements with slices. For example,

```
'1' & "01" & '0'
```

is also equivalent to `"1010"`.

Similarly to concatenation, VHDL provides a means for forming arrays out of array elements that is known as an aggregate. The syntax for an aggregate is

```
<aggregate> ::= ( <element> {, <element> } ) |
                ( <index_range> => <element> {, <index_range> => <element> } )
```

and the result is an array of the element types. For example, if `a`, `b`, and `c` are signals of type `std_logic`, then `(a,b,c)` is an array aggregate of type `std_logic_vector` which has three elements. One use of aggregates is to replicate a single bit. For example, if `foo` is a signal of type `std_logic` then `(7 downto 0 => foo)` is an eight-bit signal in which each bit has the same value as `foo`.

There are two main difference between aggregates and concatenation. First, the result of a concatenation can be used only on the right-hand side of an assignment statement (assignment statements are discussed below), whereas an aggregate formed by combining signals or variables can be assigned to. Second, while the VHDL-2008 standard specifies that aggregates can be formed by combining elements with slices much like with concatenation, some CAD tools support only array aggregates formed from element types or slices that are of equal size.

VHDL expressions are composed of terminals (signals, variables, or constants, or slices thereof) and operators potentially grouped by parentheses. VHDL-2008 includes the operators listed in Table B.2. The rows contain operators of equal precedence with highest precedence at the top.

The ** operator, in the first row, is the exponentiation operator. The **`not`** operator operates bit-wise so that **`not`** `4b"1001"` is `4b"0110"`. The next two rows contain the common arithmetic operators: multiplication (*), division (/), modulo (**`mod`**), addition (+), and subtraction (-).

The fourth row in Table B.2 contains the shift and rotate operands. The shift operators (**`sll`**, **`srl`**, **`sla`**, **`sra`**) begin with "s", followed by the direction ("l" for left, "r" for right) followed by "l" if the shift is "logical" and "r" if the shift is arithmetic. Figure 8.3 provides an example of using the **`sll`** operator. A rotate operation is like a shift except that bits shifted out one side are shifted in the other side. For example, rotate left by one position applied to `"1000"` gives `"0001"`. The keyword **`ror`** means rotate to the right and **`rol`** means rotate to the left.

The next row contains comparison operators which take two operands of equivalent type and produce a result of type `boolean`. The `/=` operator checks for inequality. Some of the comparison operators and all of the arithmetic operators for `std_logic_vector` are defined

Table B.2. **VHDL-2008 operators used in this book, grouped by precedence (highest to lowest)**

**	not				
*	/	mod			
+	-	&			
sll	srl	sla	sra	rol	ror
=	/=	<	<=	>	>=
and	or	nand	nor	xor	xnor
??					

in the package `ieee.std_logic_unsigned.all` and `ieee.std_logic_signed.all`. Adding a use clause for either will cause all `std_logic_vector` values within the design entity to be interpreted as unsigned or signed, respectively (see Chapter 10 for a discussion of signed versus unsigned numbers).

The next row contains the logical operators which combine two subexpressions. The logical operators are bit-wise. This means that when they are applied to two `std_logic_vector` operands each bit of the resulting `std_logic_vector` is computed by combining the corresponding bits of the operands. For example, the result of

```
"1010" and "1100"
```

is `"1000"`.

The VHDL-2008 condition operator "`??`" converts a `std_logic` value to type boolean. In contexts where the conversion can be inferred it is not necessary to include the condition operator. For examples of implicit conversion in an *if* statement see Figures 17.3 and 17.3. An example of explicit conversion can be found in Figure 9.4.

B.6 CONCURRENT STATEMENTS

The syntax for concurrent statements is given by

```
<concurrent_statements> ::= <concurrent_statement> ; { <concurrent_statement> ; }
```

where

```
<concurrent_statement>  ::= <concurrent_signal_assignment_statement> |
                            <component_instantiation_statement> |
                            <process_statement>
```

Each concurrent statement operates in parallel. Hence, the order of individual concurrent statements within an architecture body does *not* matter. The rest of this section examines the first two forms of concurrent statement: concurrent signal assignment statements and component instantiation statements. We leave discussion of the process statement to Section B.9.

B.6.1 Concurrent signal assignment

There are three forms of concurrent signal assignment:

```
<concurrent_signal_assignment_statement> ::= <concurrent_simple_signal_assignment> |
                                             <concurrent_conditional_signal_assignment> |
                                             <concurrent_selected_signal_assignment>
```

Concurrent simple signal assignment

The *concurrent simple signal assignment* is the simplest form and has the syntax

```
<name> <= <expression>;
```

where `<name>` is either the identifier for a signal or an array slice (described below). This statement re-evaluates `<expression>` and assigns the result to `<name>` any time any of the signals inside `<expression>` changes. The "<=" compound delimiter indicates an assignment to a signal. For example, assuming signals `output`, `a`, `b`, and `c` have type `std_logic`, the concurrent simple signal assignment

```
output <= (a and b) or (b and c));
```

defines the signal `output` to be `'1'` whenever `a` and `b` are both `'1'` or both `b` and `c` are `'1'`. In this example, whenever one of the input signals `a`, `b`, or `c` changes `output` will be be updated immediately. For example, the sequence of two concurrent simple signal assignments

```
t <= (a and b);
output <= t or (b and c);
```

has identical effect when the two statements are reversed as follows,

```
output <= t or (b and c);
t <= (a and b);
```

and both produce the same values for `output` as the single line

```
output <= (a and b) or (b and c);
```

Perhaps the best way to understand how concurrent signal assignments work is to visualize them as a circuit schematic. Figure B.2 illustrates the circuit corresponding to the above concurrent simple signal assignments. Chapter 3 introduces logic gates such as **and**, **or** and **not** along with schematic diagrams.

Concurrent conditional signal assignment

The concurrent conditional signal assignment statement has the syntax

```
<name> <= <expression> when <condition> else
        { <expression> when <condition> else }
          <expression>;
```

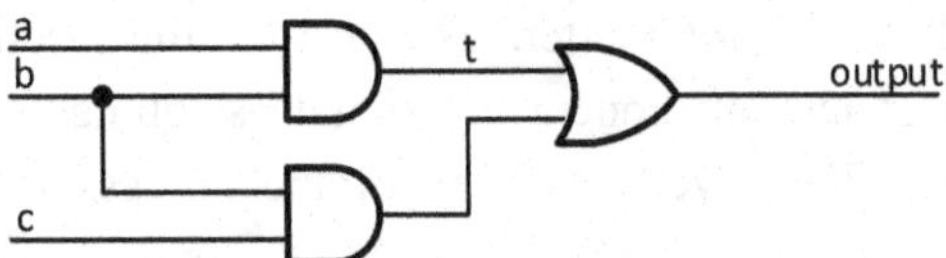

Figure B.2. Schematic diagram corresponding to concurrent simple signal assignment example.

where <condition> is an expression that evaluates to a boolean type. The value assigned to <name> is given by the first <expression> whose corresponding <condition> evaluates to true. The assignment is continuous, so whenever any signal read on the right-hand side of the signal assignment operator changes the value assigned will change. For example a 4 : 2 priority encoder (Section 8.5) can be specified using the following concurrent conditional signal assignment statement:

```
Enc <= "11" when input(3) = '1' else
       "10" when input(2) = '1' else
       "01" when input(1) = '1' else
       "00";
```

Here Enc is assigned the value "11" whenever input(3) equals '1'. If input(3) is not equal to '1' but input(2) equals '1' then Enc is assigned the value "10". If input(3) is not equal to '1' and input(2) is not equal to '1' but input(1) equals '1' then Enc is assigned the value "01". In all other cases Enc is assigned the value "00".

Concurrent selected signal assignment

The concurrent selected signal assignment statement has the syntax

```
with <name1> select
     <name2> <= { <expression> when <choices>, }
                  <expression> when others;
```

where the non-terminal <choices> is given by

```
<choices> ::= <const_expression> { "|" <const_expression> }
```

and <const_expression> is an expression with a value that does not change. Note that the pipe symbol (|) in the pattern for <choices> is part of VHDL. The double quotes should not appear – they are used here to differentiate it from the EBNF pipe symbol. The operation of the selected signal assignment is to assign to <name2> the value of the <expression> with a corresponding <choices> that contains a <const_expression> equal to <name1>. The values of each <const_expression> in the selected assignment must be distinct. The VHDL keyword **others** will match any value of <name1> that does not have a corresponding <const_expression>. For example,

```
with sel select
  output <= in_a when 2d"3",
            in_b when 2d"1" | 2d"2",
            in_c when others;
```

describes a combinational logic circuit built from a 4 → 1 binary select multiplexer (multiplexers are described in Section 8.3). The circuit has inputs in_a, in_b, in_c, and sel. If sel is equal to 2d"3" then the value assigned to output is whatever the current value of in_a is. If sel is equal to either 2d"1" or 2d"2" then the value assigned to output is whatever the current value of in_b is. As indicated by the keyword **others**, in all other cases the value assigned to output is the current value of in_c.

You may be wondering what the difference is between conditional signal assignment and selected signal assignment. One key difference is that in selected signal assignment the choices must be non-overlapping, whereas in conditional signal assignment overlap is permitted. Conditional signal assignment handles overlapping choices by allowing you to express priority in what value is assigned. Another key difference is that the choices for selected signal assignment must be a constant, whereas the choices for conditional signal assignment can be an arbitrary expression.

B.6.2 Component instantiation

The second form of concurrent statement is component instantiation. Component instantiation enables you to use one design entity as a component within another design entity. By doing so it enables you to provide a *structural description* showing how to build a complex circuit by connecting a collection of smaller circuits together using signals. A structural description is essentially equivalent to drawing a schematic diagram. Conceptually, each *instance* of a design entity inside the larger design can be thought of as a discrete chip, much like those shown in Figure 28.8, which illustrates six chips wired together on a prototyping board. In practice the component instances in a VHDL structural description are combined into a single circuit by the CAD tools. Owing to some similarities between VHDL component instantiation syntax and the syntax for a function call in popular programming languages such as C, C++, and Java, students new to hardware design tend to initially confuse component instantiation with the action of calling a function as encountered in software design. Do *not* make this mistake!

The syntax for a component instantiation statement is

```
<instance_label>: <unit> [ generic map ( <assoc_list> ) ] port map ( <assoc_list> );
```

where <instance_label> is an identifier that uniquely identifies this particular instance of the component and <unit> identifies the component being instantiated. A simple example demonstrating component instantiation is given in Figure B.4 later. This figure will be described after we have looked at the details of the component instantiation syntax above.

For the component instantiation syntax description above there are three formats for specifying <unit>:

```
<unit> ::= <component_identifier> |
           entity <entity_identifer> [ ( <architecture_identifier> ) ]  |
           configuration <configuration_identifier>
```

Here <component_identifier> is the identifier of a component from a component declaration (Section B.4), <entity_identifier> is the identifier from an entity declaration (Section B.4), and <configuration_identifier> is the identifier from a configuration declaration (Section 7.1.8). For all three formats the identifier must be visible in the scope of the component instantiation statement. There are two ways to make a component identifier visible. The first is to include a component declaration in the architecture declarative part of the architecture body containing the component instantiation statement. This is the approach taken in Figure B.4) – notice the component declarations for NOT_GATE and AND_GATE inside the architecture body for NOR_GATE. The second approach is to include a component declaration inside a package declaration and include a use clause for that package before the entity declaration associated with the architecture body containing the component instantiation (e.g., see Figures 7.23 and 7.26).

A single VHDL entity declaration can be associated with multiple architecture bodies, each with its own architecture identifier. Hence, VHDL enables you to specify which architecture body should be used with a component instantiation either by specifying it explicitly using `<architecture_identifier>` using the second format above or by using a configuration statement and the third format above.

The second format for specifying `<unit>` is sometimes called *direct instantiation*. An example of direct instantiation is shown in Figure 7.18. In this figure the entity identifier is prefaced by "`work.`" to indicate that the VHDL compiler should look for the entity declaration in the working library containing all design entities that have been compiled so far. We recommend using direct instantiation only for testbenches, since direct instantiation prevents you from using a configuration statement to later configure which architecture is used for a given component instantiation statement (see the discussion of configuration statements in Section 7.1.8). An example of the third format for specifying `<unit>`, when using a **`configuration`**, is shown in Figure 7.21.

Returning to the syntax for a component instantiation statement, we note that the **`generic map`** portion is optional. If it is present, the **`generic map`** enables you to specify the constants declared within the generic clause of the entity declaration. To use an analogy, the **`generic map`** allows the designer to customize the circuit much like one might ask a car dealer to customize a car by tinting the windows. The **`port map`** portion is required because it is used to specify how to connect external signals to the inputs and outputs of the component being instantiated.

Both **`generic map`** and **`port map`** are followed by an association list (`<assoc_list>`). There are two forms of association list that may be used in a component instantiation – positional and named association:

```
<assoc_list> ::= <named_association_list> |
                 <positional_association_list>
```

where

```
<named_association_list> ::= <formal> => <actual> { , <formal> => <actual> }
```

and

```
<positional_association_list> ::=  <actual> { , <actual> }
```

In the named association list format `<formal>` refers to the identifier name in the **`generic`** or **`port`** clause of the corresponding entity or configuration declaration. The arrow compound delimiter "`=>`" indicates that the formal identifier is connected to the signal identified by `<actual>`. Here `<actual>` is the name of a signal or a slice (Section B.5) of a signal declared in the architecture declarative part of the architecture body containing the component instantiation. In Figure B.4 the component instances with labels `U1` and `U2` use named association. The named association format is preferred when instantiating a component with many inputs and outputs or generic parameters because it reduces the chances of introducing an error when making changes. For components with only a few inputs and outputs or generic parameters, and especially for those that do not change (e.g., fundamental building blocks like those described in Chapter 8), the positional association list is preferred. In the positional association syntax each `<actual>` is connected to a formal in the order in which the formals

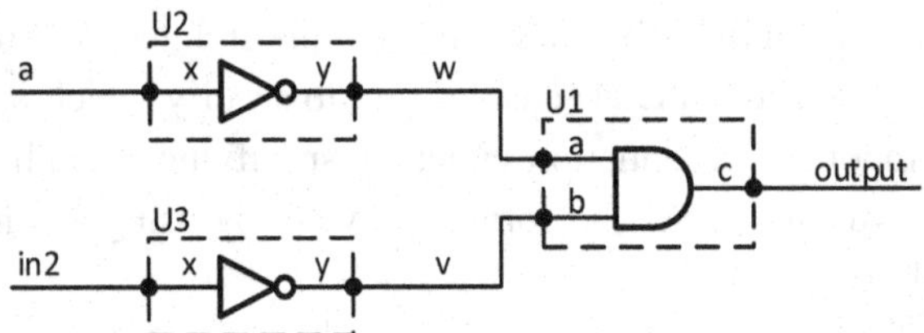

Figure B.3. Schematic diagram corresponding to VHDL component instantiation example in Figure B.4.

are declared. In Figure B.4 the component instance with label `U3` uses positional association. VHDL-2008 introduces the ability to use an expression for `<actual>` for ports with mode **`in`**. We make use of this feature in this book because it can make VHDL more readable. Below we discuss two examples of component instantiation. Many more examples can be found throughout the book.

Figure B.3 is a schematic diagram showing one way to construct a NOR gate by combining two separate instances of a NOT gate with a single instance of an AND gate. Figure B.4 illustrates a corresponding VHDL structural description of this circuit using component instantiation. Note that since they are concurrent statements the order of the component instantiations in the architecture body for `NOR_GATE` does *not* matter. Also notice that we can use the same name for signals and ports in different design entities. For example, the identifier `a` is used for the first input of `NOR_GATE` as well as the first input of `AND_GATE`. The component instantiations with label `U1` and `U2` use named association, whereas the component instantiation with label `U3` uses positional association.

A more complex example demonstrating the use of generic constants and **`generic map`** in a component instantiation is provided in Figures B.5 and B.6. The SORTER design entity in Figure B.6 instantiates a single copy of the magnitude comparator design entity. Note that the generic clause of `MagComp` sets `k` equal to 8, which would be the default value of `k` if there were no **`generic map`** clause in the component instantiation with label `CMP`. However, the component instantiation *does* include a generic map clause. The generic clause uses named notation, though we could also have used positional notation.

B.7 MULTIPLE SIGNAL DRIVERS AND RESOLUTION FUNCTIONS

What happens if multiple concurrent statements try to drive the *same* signal? In an actual circuit the signal corresponds to a wire and the values correspond to voltages. If one circuit, corresponding to a first concurrent statement, drives the signal to logic 1 while another circuit, corresponding to a second concurrent statement, drives the signal to logic 0, then a large current would flow and the circuit would likely be damaged. When used for simulation, VHDL models such cases through the use of a *resolution function*. A resolution function takes as input all the different values being driven on a signal and outputs the value that will be given to the signal. In practice you will not likely need to define your own resolution functions, since they are predefined for important types such as `std_logic` and `std_logic_vector`. For example, for a signal of type `std_logic` with two drivers with one driving a value of `'1'` and the other driving a value of `'0'` the resolved value during simulation will be `'X'`. Hence, a value of `'X'` observed while simulating a testbench typically indicates a serious bug in the design that should be identified. If instead the second driver drives `'Z'` (high impedance), the resolved

```
library ieee;
use ieee.std_logic_1164.all;

entity AND_GATE is
  port ( a, b : in std_logic; c : out std_logic );
end AND_GATE;

architecture impl of AND_GATE is begin
  c <= a and b;
end impl;
---------------------------------------------------------------------------
library ieee;
use ieee.std_logic_1164.all;

entity NOT_GATE is
  port ( x : in std_logic; y : out std_logic );
end NOT_GATE;

architecture impl of NOT_GATE is begin
  y <= not x;
end impl;
---------------------------------------------------------------------------
library ieee;
use ieee.std_logic_1164.all;

entity NOR_GATE is
  port ( a, in2: in std_logic; output : out std_logic );
end NOR_GATE;

architecture impl of NOR_GATE is
  component AND_GATE is
    port ( a, b : in std_logic; c : out std_logic );
  end component;
  component NOT_GATE is
    port ( x : in std_logic; y : out std_logic );
  end component;
  signal w, v: std_logic;
begin
  U1: AND_GATE port map( c => w, a => v, b => output );
  U2: NOT_GATE port map( x => a, y => w );
  U3: NOT_GATE port map( in2, v );
end impl;
```

Figure B.4. Example of component instantiation. Here a NOR gate is created by combining two NOT gates with a single AND gate. A corresponding schematic diagram is shown in Figure B.3.

value will be `'0'`, corresponding to the value of the first driver. The high impedance state of the second driver is essentially its way of saying it has "no opinion" as to what the value (i.e., voltage) of the signal should be.

An example of how to specify a 2 $\rightarrow$ 1 binary-select multiplexer using tri-state buffers illustrates how resolution functions operate in practice. The VHDL is shown in Figure B.7 and the corresponding circuit diagram is shown in Figure B.8. In this example the

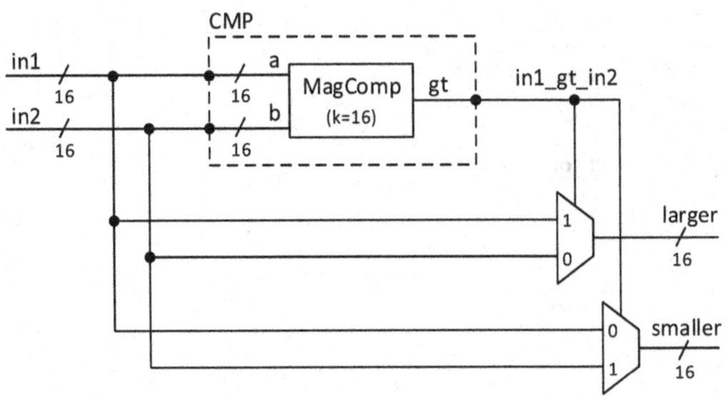

Figure B.5. Schematic diagram corresponding to the VHDL component instantiation example with generic map in Figure B.6.

line "`b <= a1 when s = '1' else 'Z';`" uses a conditional signal assignment statement (Section B.6.1) that assigns `a1` to `b` when signal `s` is equal to `'1'` and `'Z'` otherwise. Similarly, the line "`b <= a0 when s = '0' else 'Z';`" drives `a0` onto `b` when `s` equals `'0'` and `'Z'` otherwise. Hence, if `s` equals `'1'` then `b` is driven with the value of `a1` and `'Z'`. The resolved value will be equal to whatever value `a1` has. If `a1` is currently `'0'` then `b` will be driven to `'0'` and if `a1` is `'1'` then `b` will be driven to `'1'`. As noted in Section 4.3.4, because of the potential for a short circuit, tri-state circuits should be avoided.

B.8 ATTRIBUTES

VHDL provides a syntax feature called an *attribute* to obtain or specify additional information about elements of a design. To use an attribute one specifies the *attribute name* using the syntax

```
<prefix> ' <attribute_name> [ ( <expression> ) ]
```

where `<prefix>` can be any syntactic element with a name (e.g., a signal name, package name, entity name, etc.), `<attribute_name>` is the name of the attribute, and `<expression>` is an optional expression passed as an argument to a function for those attributes that return a function. Depending upon the attribute name, the result of applying the attribute name to a prefix can be a type, value, signal, index range, or function.

One reason for a user to specify an attribute is to convey additional information about the design to a CAD tool. An example of specifying an attribute for this purpose can be found in Figure 8.51, where two attributes are declared with the keyword **`attribute`** and used to help specify the contents of a read-only memory. Since such use is CAD-tool-specific, in this appendix we will not describe the syntax for declaring new attributes. CAD-tool-specific attributes are naturally described in the associated CAD tool's documentation. Unless you are developing CAD tools yourself, you will probably never have a specific need to define your own VHDL attributes. However, if you want to learn about the general syntax for declaring and defining new attributes consult the VHDL language reference manual [55] or Ashenden [3].

The VHDL standard specifies numerous *predefined attributes*, and some of the more widely supported ones are listed in Table B.3. The `T'left`, `T'right`, `T'length`, and `T'range` attributes can be helpful when declaring signals or accessing an element or slice of a signal. The `T'event` and `T'stable` attributes are particularly relevant to the following discussion on the process statement. Both of these attributes are related to the notion of an *event* on a signal,

```
library ieee;
use ieee.std_logic_1164.all;

entity MagComp is
  generic( k: integer := 8 );
  port( a, b: in std_logic_vector(k-1 downto 0); gt: out std_logic );
end MagComp;

architecture impl of MagComp is
begin
  gt <= '1' when a > b else '0';
end impl;
---------------------------------------------------------------------------
library ieee;
use ieee.std_logic_1164.all;

entity SORTER is
  port ( in1, in2: in std_logic_vector(15 downto 0);
         larger, smaller : out std_logic_vector(15 downto 0) );
end SORTER;

architecture impl of SORTER is
  component MagComp is
    generic( k: integer := 8 );
    port( a, b: in std_logic_vector(k-1 downto 0); gt: out std_logic );
  end component;

  signal in1_gt_in2: std_logic;
begin
  CMP: MagComp generic map( k => 16) port map( a => in1, b => in2, gt => in1_gt_in2 );

  with in1_gt_in2 select
    larger <= in1 when '1',
              in2 when others;

  with in1_gt_in2 select
    smaller <= in1 when '0',
               in2 when others;
end impl;
```

Figure B.6. Example of component instantiation with generic parameters. A corresponding schematic diagram is shown in Figure B.5. Note that while `MagComp` has entity and component declarations with `k` equal to 8 this value of `k` is overridden by `generic map` when it is instantiated to create `CMP`.

which in turn is related to the notion of a *transaction*. A transaction is said to occur at time instants when an assignment is made to a signal. An event is said to occur when an assignment results in a signal changing its value. Thus all events are transactions, but not all transactions are events. The attribute `T'active` is true when an assignment occurs to a signal or variable irrespective of whether or not the value changes. We list it here to emphasize the distinction between events and transactions. The `T'active` attribute should not be used in synthesizable VHDL. The `T'delayed(N)` attribute is similarly not synthesizable.

Table B.3. **Selected predefined attributes in VHDL**

T'left	Integer value of leftmost index for the signal or variable T
T'right	Integer value of rightmost index for the signal or variable T
T'length	Integer value indicating number of elements in multi-bit signal or variable T
T'range	Index range of signal or variable T
T'event	Boolean signal equal to TRUE at instants when an event occurs on T, otherwise FALSE
T'stable	Boolean signal equal to TRUE at times when an event did not occur on T and FALSE at instants when an event does occur
T'active	Boolean signal equal to TRUE at times when a transaction occurs on T, otherwise FALSE
T'delayed(N)	Signal equal to T delayed by N time units

```
library ieee;
use ieee.std_logic_1164.all;

entity Mux2b is
  port( a1, a0 : in std_logic;
        s : in std_logic;
        b : out std_logic );
end Mux2b;

architecture tri_impl of Mux2b is
begin
  b <= a1 when s = '1' else 'Z';
  b <= a0 when s = '0' else 'Z';
end tri_impl;
```

Figure B.7. Tri-state buffer in VHDL.

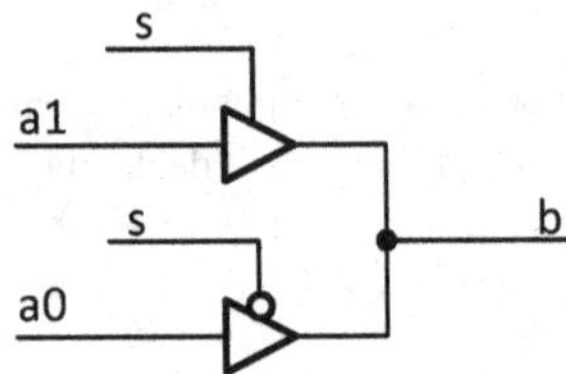

Figure B.8. Schematic of 2 → 1 multiplexer corresponding to that VHDL in Figure B.7.

B.9 PROCESS STATEMENTS

The final form of concurrent statement is the process statement. A process statement provides a way to describe a logic circuit's behavior at a higher level of abstraction using sequentially ordered statements. Since a process statement is a concurrent statement, each process may "execute" the sequential statements it contains in parallel with sequential statements contained inside other process statements. Every concurrent statement in an architecture body can be viewed conceptually as a separate process statement.

The main obstacle to correctly and effectively using process statements is that the process syntax *looks* so similar to software that it can be hard not to *think* it operates exactly like a program running on a computer. There are important differences with respect to timing discussed below. In addition, you must learn to use a restricted syntax subset to ensure the hardware synthesized from your VHDL descriptions behaves as you expect. Indeed, the sequential statements inside a VHDL process describe only the *behavior* of the logic circuit. The actual implementation of the circuit is determined by the CAD tool used to synthesize a digital circuit that *matches* the behavior of your VHDL description.[3] As an example, note the vast difference in appearance between the VHDL process statement in Figure 7.2 and the corresponding gate-level representation generated by a CAD synthesis tool shown in Figures 7.3 and 7.4.

This book uses process statements both for specifying the behavior of digital logic circuits (e.g., Figure 7.2) and when specifying test scripts (e.g., Figure 7.18). This section describes the general syntax relevant to both, while Section B.10 focuses on the syntax restrictions required for hardware synthesis.

The process statement syntax is summarized in Figure B.9 and a simple example describing a 2 → 1 binary-select multiplexer (Section 8.3) is given in Figure B.10. In Figure B.10, the process writes the value of `A` to `C` when `sel` is `'1'` and writes the value of `B` to `C` otherwise. Below we first discuss the sensitivity list and the conceptual model of a process, including the important question of *when* sequential statements execute, then look at the various forms of sequential statement syntax.

B.9.1 The process sensitivity list and execution timing

In the terminology of the VHDL language specification [55] a VHDL process is said to "execute" the sequential statements it contains whenever any signal in the *sensitivity list* of that process has an event – i.e., whenever a signal in the sensitivity list has a change in value. Conceptually, a process executes the sequential statements it contains one after the other until it reaches a **`wait`** statement or the last sequential statement inside the process. At that point the process becomes suspended. The time elapsed between executing different sequential statements inside the same process is infinitesimally small. If the process reaches the last statement then it is suspended until a new event occurs on one of the signals in the sensitivity list of that process.

For example, in Figure B.10 a change of value on any of the signals `sel`, `A`, and/or `B` listed in the sensitivity list will cause the sequential statements inside the process to execute completely until reaching the end of the process, possibly changing the value of `C`. The process in Figure B.10 will completely execute the sequential statements it contains again, *starting from the beginning of the process*, whenever there is a *new* change on one of the three signals `sel`, `A`, and `B` listed in the sensitivity list.

[3] Indeed this is true for all VHDL syntax.

```
process [ ( <sensitivity_list> ) ]
  { <process_declarative_item> }
begin
  <sequential_statments>
end process;
```

where:

```
<sensitivity_list>          ::= all |
                                 <signal_name> { , <signal_name> }

<process_declaritive_item> ::= <type_declaration> |
                                 <variable_declaration>

<sequential_statments>      ::= <sequential_statement>
                               { <sequential_statement> }

<sequential_statement>      ::= <wait_statement> |
                                 <report_statement> |
                                 <if_statement> |
                                 <case_statement> |
                                 <signal_assignment_statement> |
                                 <variable_assignment_statement>
```

Figure B.9. Syntax summary for a process in VHDL-2008.

```
process (sel,A,B)
begin
  if sel = '1' then
    C <= A;
  else
    C <= B;
  end if;
end process;
```

Figure B.10. Example VHDL process that describes a 2 → 1 binary-select multiplexer.

Figure B.11 illustrates the timing of these steps by plotting the inputs `sel`, `A`, and `B` along with the output `C`. Initially, all inputs are at logic level zero and `C` is also zero. At time t_1 input `B` changes from logic level zero to logic level one. Since `B` is listed in the sensitivity list, the process is triggered to execute at time t_1. The process causes the output `C` to change to logic level one. The transition of `C` occurs an infinitesimal time after the change on `B`. In a simulation waveform the change in `C` appears to occur at the same time as the change in `B`. In essence, you could say that the if statement executed "infinitely fast." In a real multiplexer circuit there *would* be some delay between the change on the inputs and the output. Indeed, CAD tools can be configured to simulate these delays by taking into account the results of synthesizing our VHDL description to a particular target technology. However, as far as the VHDL language standard is concerned, there is essentially no delay between the input and output changing for the VHDL in Figure B.10. Similar changes are shown at times t_2, t_4, and t_5. At time t_3 input `B` changes from logic one to logic zero and the process executes, but the value assigned to output `C` is zero so no change occurs on `C`.

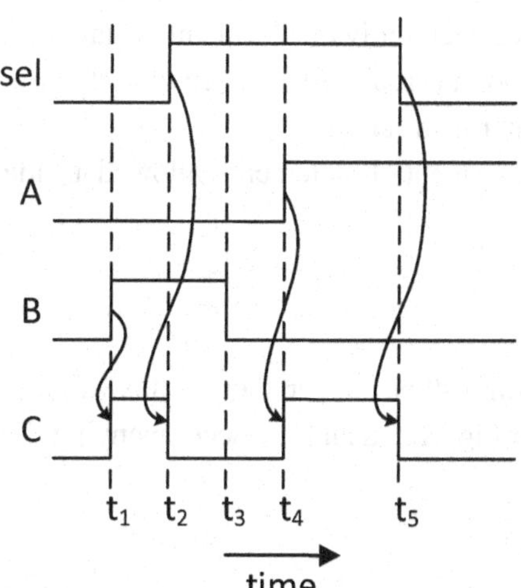

Figure B.11. Simulation waveform for VHDL example in Figure B.10. The values of inputs `sel`, `A`, and `B` were chosen to illustrate the effect of the VHDL in Figure B.10 on output `C`. Arrows indicate causality.

In this book the only uses of a process *without* a sensitivity list are for test scripts (e.g., see Figure 3.8) and for defining a clock signal to input to a synthesizable design entity being tested (see Figure 14.17). Both of these forms of process are not intended to be synthesized to hardware.[4] If there is no sensitivity list, the process begins execution immediately and will execute again from the beginning immediately after executing the last sequential statement it contains (i.e., forming an infinite loop).

If the sensitivity list contains the VHDL-2008 keyword **`all`** then it is as if all signals read by any sequential statement inside the process are listed in the sensitivity list. For example, in Figure B.10 every signal that is read inside the process, namely `sel`, `A`, and `B`, is also listed in the sensitivity list. Hence, we could have written "**`process`** `(all)`" instead of "**`process`**`(sel,A,B)`". However, note that the sensitivity list has a very different meaning than the arguments of a function in a typical programming language. In particular, as you will see in Section B.10, sometimes you *must not* include some signals that are read inside a process in the sensitivity list of that process. Do *not* make the mistake of thinking a process is like a function!

The actual hardware synthesized from a VHDL process description will be implemented using logic gates that have small delays. As illustrated in Figure 7.4, these gates will not correspond to the lines of VHDL in an easy-to-identify way. However, if the VHDL used is restricted to a synthesizable subset of the language, such as that summarized below in Section B.10, then the hardware circuit will have the same *behavior* as that described by the VHDL process statement except for minor differences due to gate and wiring delays. Such differences do not matter for synchronous circuits when proper allowances are made for timing (Chapter 15).

The contrast between a VHDL process and a software program is instructive. When debugging a compiled software program it is possible to stop at a computer instruction corresponding to a single line and to inspect the state of the running program using a debugger program (e.g., gdb – the GNU Debugger). In contrast, after synthesizing a VHDL process to

[4] It is possible to write a synthesizable process without a sensitivity list by either beginning the process with a single **`wait until`** statement or ending it with a single **`wait on`** statement. Since they are less commonly used, we do not discuss these forms in this book.

hardware it is *not* possible to peek inside to see each individual sequential statement executing inside a process because the sequential description is replaced by a network of wires and digital logic gates (which are not the same as computer instructions).

Below we consider the various forms of sequential statements allowed inside a process statement.

B.9.2 Wait and report statements

The wait and report statements are used only for testbenches, so their syntax will not be covered in detail in this appendix. See the example in Figure 3.8 and the accompanying description in Section 3.6.

B.9.3 If statements

The syntax for the if statement is illustrated in Figure B.12, where `<condition>` is an expression of type `BOOLEAN`. Only the sequential statements following the first `<condition>` part that evaluates to `TRUE` are executed and the rest are skipped. Note that, since it is a sequential statement, an if statement may appear only inside a process statement. A simple example of an if statement can be found in Figure B.10, which was discussed above. In the general form there can be any number of **`elsif`** parts. Regardless of how many **`elsif`** parts there are, as far as the VHDL language standard is concerned, the entire if statement takes an infinitesimal time to execute. An example of an if statement using **`elsif`** can be found in Figure 7.8.

B.9.4 Case and matching case statements

The syntax for the case statement is summarized in Figure B.13. When the case statement is executed the value of `<expression>` is evaluated and compared with `<choices>` following the keyword **`when`** in each case alternative. There should be exactly one case alternative

```
  if <condition> then
    <sequential_statements>
{ elsif <condition> then
    <sequential_statements> }
[ else
    <sequential_statements> ]
  end if;
```

Figure B.12. If statement syntax. Can only be used inside a process.

```
case [?] <expression> is
  <case_alternative>
{ <case_alternative> }
end case [?];
```

where:

```
<case_alternative> ::= when <choices> => <sequential_statements>
```

Figure B.13. Case statement syntax. Can only be used inside a process.

```
library ieee;
use ieee.std_logic_1164.all;

entity case_example is
 port( A: in std_logic_vector(2 downto 0); B: out std_logic );
end case_example;

architecture impl of case_example is
begin
  process(all)
  begin
    case? A is
      when "000"         => B <= '1';
      when "001"         => B <= '0';
      when "01-"         => B <= '1';
      when "101" | "111" => B <= '1';
      when others        => B <= '0';
    end case?;
  end process;
end impl;
```

Figure B.14. Example of a matching case statement. Output `B` is set to `'1'` when input `A` is either `"10"` or `"11"`. A **`when others`** case alternative should be included to ensure all possible input conditions are covered including `std_logic` meta values such as `'U'`.

that matches for all possible values of `<expression>`. The sequential statements following the arrow compound delimiter (=>) for this case alternative are executed. For the last case alternative `<choices>` can be the VHDL keyword **`others`** which will match any value of `<expression>` that does not match an earlier case alternative.[5] If the question mark (`?`) is included the case statement is said to be a *matching* case statement. In that case the `<choices>` may include don't care values ('`-`' in `std_logic`) that will match any value in the corresponding bit position of `<expression>`.

An example of a matching case statement is shown in Figure B.14, and Figure B.15 illustrates an example waveform, showing how it behaves as input `A` changes. Initially, `A` equals `"000"` and the output `B` equals `'1'`. At time t_1 input `A(0)` changes from `'0'` to `'1'`. Since `A` is in the sensitivity list the process is triggered and the case statement executes. The value of `A`, which is now `"001"` is compared against each case alternative. The second case alternative, "**`when`** `"001"=> B <= '0';`", matches, so the statement `B <= '0'` is executed and the value of `B` changes from `'1'` to `'0'`. A key point to observe is that, as far as the VHDL language standard is concerned, the change in `B` occurs an infinitesimal amount of time after the change in `A(0)`. At time t_2 input `A(1)` changes from `'0'` to `'1'` and the process is executed again from the beginning. Because `A` is now `"011"`, the case alternative "**`when`** `"01-"=> B <= '1';`" matches, since `"01-"` matches both `"010"` and `"011"`. Consequently the output `B` changes to `'1'`. At time t_3 input `A(0)` changes back to `'0'` and the process executes from the beginning again. This time `A` is `"010"`, which matches the same case alternative, and the output does not change. At time t_4 input `A(2)` changes and now `A` is `"110"`, which matches the case alternative "**`when others`** `=> B <= '0';`", so `B`

[5] Some synthesis tools impose only the requirement that all combinations of logic one and zero are covered.

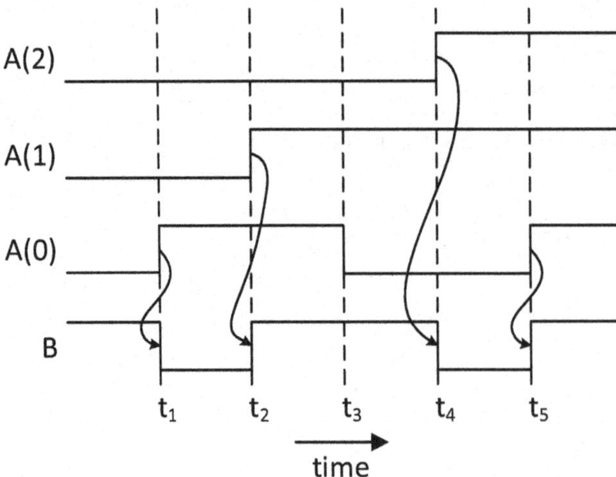

Figure B.15. Simulation waveform for the VHDL example in Figure B.14. The value of input A was chosen to illustrate the effect of the VHDL in Figure B.14 on output B. Arrows indicate causality.

changes to `'0'`. At time t_5 input `A(0)` changes and when the process executes `A` equals `"111"`, which matches "`when "101"| "111"=> B <= '1';`", so `B` changes back to `'1'`.

B.9.5 Signal and variable assignment statements

In this section we discuss an important nuance in the timing of assignments to signals after first introducing VHDL variables. A variable declaration has the following syntax:

```
variable <identifier> {, <identifier> } : <subtype_indication> [ := <expression> ];
```

Variable declarations are placed in the declarative portion of the process in which the variable is to be used, before the keyword **begin** (see Figure B.9). A variable is used to communicate a value *within* a single process, whereas a signal is typically used to communicate values *between* sequential statements inside a process and other concurrent statements outside the process. Assignments to a variable always use a colon followed by an equal sign (:=), whereas assignments to a signal always use a less-than symbol followed by an equal sign (<=). However, a more important difference is that a variable assignment takes effect before the next sequential statement is executed, whereas a signal assignment takes effect only after the process has been suspended – i.e., has finished executing all sequential statements or has reached a wait statement.

Figure B.16 provides an example that helps illustrate the difference between assignments to variables and signals. It also provides another example of a sensitivity list. In this example the process executes when input signal `clk` changes – i.e., goes from logic value `'0'` to `'1'`, which is called a *rising edge*, or goes from `'1'` to `'0'`, which is called a *falling edge*. The *if* statement condition evaluates to true if clk is `'1'`. Now, since the *if* statement is evaluated only on the rising or falling edge of `clk`, and since the *if* condition is evaluated an infinitesimal amount of time *after* the change in the signal value, it can be inferred[6] that the sequential statements inside the body of the *if* statement are executed only on the *rising* edge of `clk`. An

[6] Note that for purposes of synthesis CAD tools essentially ignore the possibility of transitions to or from meta values, such as the `std_logic` value for unknown (`'U'`), which are meaningful only in simulations.

```
library ieee;
use ieee.std_logic_1164.all;

entity example is
  port( clk, w: in std_logic;
        x: buffer std_logic;
        y, z: out std_logic );
end example;

architecture impl of example is
begin
  process(clk)
    variable tmp: std_logic;
  begin
    if clk = '1' then
      tmp := w;
      x <= tmp;
      y <= x;
      z <= not w;
    end if;
  end process;
end impl;
```

Figure B.16. Example of a process containing both variable and signal assignment statements.

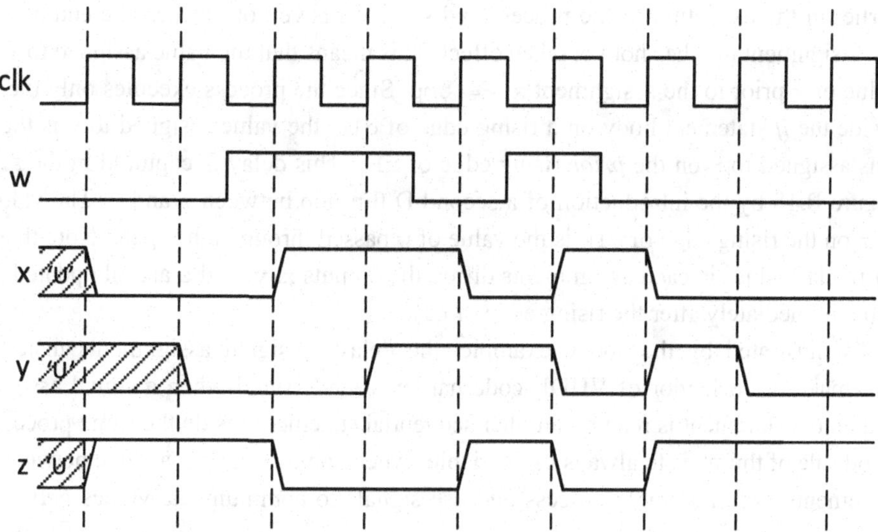

Figure B.17. Simulation waveform for VHDL example in Figure B.16. Initially x, y, and z have value 'U', which means that in a real circuit they can be either logic level zero or logic level one.

equivalent but more readable way to write the if statement using the VHDL built-in function `rising_edge` is "**if** `rising_edge(clk)` **then**".

Figure B.17 illustrates how outputs x, y, and z change for the given inputs clk and w. Transitions on the outputs are drawn slanted to emphasize that they change in response to changes on the inputs. Figure B.18 illustrates the hardware synthesized from the VHDL description in Figure B.16.

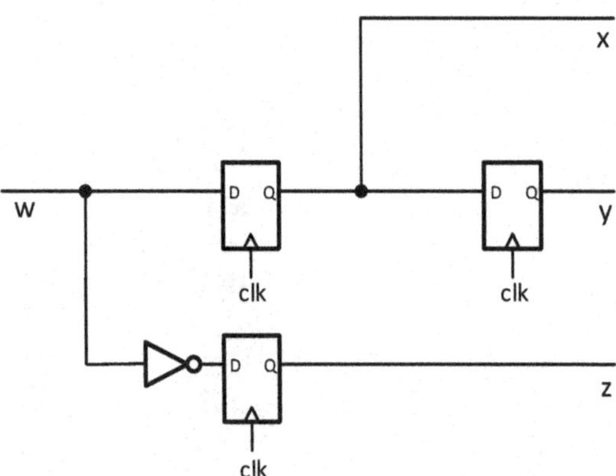

Figure B.18. Schematic diagram corresponding to the VHDL example in Figure B.16.

To better understand the waveforms in Figure B.17 and the schematic in Figure B.18, let us consider the four assignment statements inside the *if* statement body in Figure B.16. The first assignment statement, `tmp := w`, assigns the value of `w` at the time of the rising edge of `clk` to the variable `tmp`. Since `tmp` is a variable, the assignment becomes effective before the next sequential statement, `x <= tmp`, is executed. This means that when `x <= tmp` is executed the value read for `tmp` and assigned to `x` is *also* the value that `w` has at the time of the rising edge of `clk`. Now, consider the next assignment statement, `y <= x`. Since `x` is a signal assigned earlier in the execution of the process and since we have not reached the end of the process, the assignment to `x` has not yet taken effect. This means that the value assigned to `y` will be the value of `x` prior to the assignment `x <= tmp`. Since the process executes only the statements inside the *if* statement body on a rising edge of `clk`, the value assigned to `y` is the value that was assigned to `x` on the *prior* rising edge of `clk`. This delay is captured in the schematic in Figure B.18 by the introduction of a second D flip-flop between `w` and `y`. The value assigned to `z` on the rising edge of `clk` is the value of `w` passed through a not-gate. Note that, while the particular values in each assignments differ, the outputs `x`, `y`, and `z` are all updated at the same time, immediately after the rising edge of `clk`.

As illustrated by the above example, the delay of signal assignments inside a process can make the behavior of VHDL code harder to understand when a signal assigned by one sequential statement is read by another sequential statement inside the same process. Hence, a good rule of thumb is to always use variables when communicating a value between sequential statements *within* a single process and use signals to communicate values *between* different processes.

B.10 SYNTHESIZABLE PROCESS STATEMENTS

The prior section discussed the syntax and semantics of the process statement as defined in the VHDL language standard [55]. As noted earlier, CAD tools used for synthesis cannot generate hardware for an arbitrary VHDL description that is syntactically valid but instead guarantee only matching behavior between simulation and synthesis for a restricted subset of the language. This section discusses a set of rules that ensure the process statements you write are synthesizable. These rules are broader than the style guidelines. The style guidelines provide

```
process(sel)
begin
  if sel = '1' then
    c <= a;
  else
    c <= b;
  end if;
end process;
```

Figure B.19. Example of a process violating rule C1. The signals a and b are missing from the sensitivity list.

rules for ensuring that your VHDL is readable and maintainable as well as synthesizable. However, in cases where one finds it necessary to break the style guidelines it is still *essential* that one follows the rules described here.

We describe several rules that ensure a process is synthesizable. Note that which styles are synthesizable is CAD-tool-specific, but these rules are simple enough that almost every tool should support them. Also, any given CAD tool may be able to synthesize VHDL that does not follow these rules, but restricting your VHDL to these rules will ensure your VHDL is more portable in case you need to synthesize your VHDL using a different CAD tool at some later time.

Probably the most important rule to remember is that each and every process in your design must be one of three[7] types: type 1, purely combinational; type 2, edge-sensitive; and type 3, edge-sensitive with an asynchronous reset.

B.10.1 Type 1: purely combinational

To describe a combinational logic circuit using a process two rules *must* be observed.

- **Rule C1**. All inputs to the combinational logic block must appear in the sensitivity list or the sensitivity list must use the VHDL-2008 keyword **all**.
- **Rule C2**. All outputs of the combinational logic circuit *must* be assigned for any pattern of inputs.

Figures B.19, B.20, and B.21 illustrate examples that violate these rules. In Figure B.19 the signals a and b are missing from the sensitivity list, thus violating rule **C1**. In Figure B.20 the signal y is not assigned when input a is equal to '1' and input b is equal to '0', thus violating rule **C2**. Figure B.21 shows a similar example involving a matching case statement. Section A.4 of the VHDL style guide shows additional examples.

B.10.2 Type 2: edge-sensitive

For a process describing a circuit that updates its output on the rising or falling edge of a clock, also known as *edge-sensitive* logic, to be synthesizable, the following rules must be observed.

[7] In Figure 8.54 we employed a fourth type that describes the level-sensitive behavior of read–write memories.

```
process(all)
begin
  if a = '0' then
    y <= b;
  elsif b = '1' then
    y <= c;
  end if;
end process;
```

Figure B.20. Example of a process violating rule C2. The output `y` is not assigned if input `a` is `'1'` and input `b` is `'0'`.

```
process(all)
begin
  case? state & x is
    when "000"   => state_next <= "01"; y <= '0';
    when "01-"   => state_next <= "10"; y <= '1';
    when "100"   => state_next <= "10";
    when others => state_next <= "00"; y <= '0';
  end case?;
end process;
```

Figure B.21. Second example of a process violating rule C2. The output `y` is not assigned if input `state` is `"10"` and input `x` is `'0'`.

```
process (CLK, A, B) begin
  if (CLK = '1') then
    if ( A = B ) then
      Y <= '1';
    else
      Y <= '0';
    end if;
  end if;
end process;
```

Figure B.22. Example of a process violating rule E1. The sensitivity list should not contain `A` or `B`.

- **Rule E1**. The sensitivity list should contain only the clock.
- **Rule E2**. The process should describe behavior occurring on just a single edge, either rising or falling, of a single clock.
- **Rule E3**. In the sequential statements contained in the process the clock signal should appear only in the condition of the outermost *if* statement, which should not contain an **`else`** or **`elsif`** part. This *if* statement should contain all other sequential statements within the process.

We have already seen an example of a process following the above rules in Figure B.16. Figure B.22 illustrates an example violating rule **E1**.

B.10.3 Type 3: edge-sensitive with asynchronous reset

In some designs it is desirable to employ flip-flops with an asynchronous reset. An asynchronous reset sets the output of the flip-flop immediately when the reset signal is

```
process (clk, reset) begin
  if reset = '1' then
    Q <= '0';
  elsif clk'event and clk = '1' then
    Q <= D;
  end if;
end process;
```

Figure B.23. Example of a synthesizable process describing asychronous reset behavior using the `X'event` attribute.

```
process (clk, reset) begin
  if reset = '1' then
    Q <= '0';
  elsif rising_edge(clk) then
    Q <= D;
  end if;
end process;
```

Figure B.24. Example of a synthesizable process describing asychronous reset behavior using the `rising_edge` function.

asserted – i.e., before the next rising edge of the clock. Although it is not employed in any of the examples in this book, we describe the synthesis rules applicable for this type of hardware to help provide contrast with the Type 1 and Type 2 processes above. This contrast may help you to better understand the syntax restrictions required in order to ensure your VHDL is synthesizable. This final type is interesting because it combines level-sensitive and edge-sensitive behavior. A synthesizable process describing a circuit with asynchronous reset should follow the pattern in Figure B.23 or Figure B.24. Both patterns conform to the following rules.

- **Rule A1**. The sensitivity list contains only the clock and reset signals.
- **Rule A2**. All outputs should be set to a constant value by the first branch of the outermost *if* statement, which must test whether the reset signal is asserted.
- **Rule A3**. The single **`elsif`** part of the outermost *if* statement must explicitly test for the desired clock edge.
- **Rule A4**. All other *if* and *case* statements within the process must be nested within the **`elsif`** part of the outermost *if* statement.

To follow rule **A3** the **`elsif`** condition must check for the desired edge of the clock. The reason is that a change in the reset input from asserted to deasserted will cause the process to execute even though this event should not cause the output of the circuit to change. VHDL provides two ways to test for a clock edge.

The first approach is to use the event attribute, `X'event`. This is illustrated in Figure B.23, where the **`elsif`** condition tests whether `clk'event` is true. Since `clk'event` is true on both rising and falling edges of `clk`, the **`elsif`** condition must also check that the event which caused the process to run and which causes `clk'event` to be `TRUE` was a rising edge. It does this by checking the current value of `clk`. If `clk` is `'1'` then, as in Figure B.16, it can be inferred that the event was a rising edge of `clk`.

The second approach VHDL provides to test for a clock edge is to use the built-in function `rising_edge(X)` which evaluates to true only at times when `X` has a rising edge. This approach is illustrated for the same example in Figure B.24. VHDL also defines a function `falling_edge(X)`. The `rising_edge(X)` and `falling_edge(X)` functions are recommended both because they are more readable and because events involving `std_logic` metavalues such as `'U'` are handled in such a way that the results of simulation more closely correspond to the synthesized circuit.

REFERENCES

[1] Altera Corporation, *The LPM Quick Reference Guide*. (San Jose, CA: Altera Corporation, 1996.)

[2] Arnold, H. L. and Faurote, F. L., *Ford Methods and the Ford Shops*. (New York: The Engineering Magazine Company, 1915.)

[3] Ashenden, P. J., *The Designer's Guide to VHDL*, 3rd edn. (New York: Morgan Kaufmann, 2008.)

[4] Atanasoff, J. V., Advent of electronic digital computing. *Annals of the History of Computing* **6**: 3 (July–September 1984), 229–282.

[5] Babbage, H. P., ed., *Babbage's Calculating Engines: A Collection of Papers*. (Los Angeles, CA: Tomash, 1982.)

[6] Bailey, D., Vector computer memory bank contention. *IEEE Transactions on Computers* **C-36**: 3 (March 1987), 293–298.

[7] Baker, K. and Van Beers, J., Shmoo plotting: the black art of IC testing. *IEEE Design & Test of Computers* **14**: 3 (July–September 1997), 90–97.

[8] Bakoglu, H. and Meindl, J., Optimal interconnection circuits for VLSI. *IEEE Transactions on Electron Devices* **32**: 5 (May 1985), 903–909.

[9] Bassett, R. W., Turner, M. E., Panner, J. H. *et al.*, Boundary-scan design principles for efficient LSSD ASIC testing. *IBM Journal of Research and Development* **34**: 2.3 (March 1990), 339–354.

[10] Bentley, B., Validating the Intel® Pentium® 4 microprocessor. In *International Conference on Dependable Systems and Networks, 2001. DSN 2001* (Göteborg, 1–4 July, 2001), pp. 493–498.

[11] Berlin, L., *The Man Behind the Microchip: Robert Noyce and the Invention of Silicon Valley*. (New York: Oxford University Press, 2005.)

[12] Bhavsar, D., An algorithm for row–column self-repair of RAMs and its implementation in the Alpha 21264. In *Proceedings of International Test Conference*, Atlantic City, NJ, 27–30 September, 1999 (Washington, D.C.: IEEE ITC), pp. 311–318.

[13] Boole, G., *The Mathematical Analysis of Logic*. (Cambridge: Macmillan, Barclay, and Macmillan, 1847.)

[14] Boole, G., *An Investigation of the Laws of Thought on which Are Founded the Mathematical Theories of Logic and Probabilities*. (Cambridge: Macmillan, 1854.)

[15] Booth, A. D., A signed binary multiplication technique. *The Quarterly Journal of Mechanics and Applied Mathematics* **4**: 2 (1951), 236–240.

[16] Brearley, H. C., ILLIAC II – a short description and annotated bibliography. *IEEE Transactions on Electronic Computers* **EC-14**: 3 (June 1965), 399–403.

[17] Bristow, S., The history of video games. *IEEE Transactions on Consumer Electronics* **CE-23**: 1 (February 1977), 58–68.

[18] Bromley, A. G., Charles Babbage's analytical engine, 1838. *Annals of the History of Computing* **4**: 3 (July–September 1982), 196–217.

[19] Brooks, F., *The Mythical Man-Month: Essays on Software Engineering*. (Reading, MA: Addison-Wesley, 1975.)

[20] Brown, S. and Vranesic, Z., *Fundamentals of Digital Logic with VHDL Design*, 3rd edn. (New York: McGraw-Hill, 2008.)

[21] Brunvand, E., *Digital VLSI Chip Design with Cadence and Synopsys CAD Tools*. (Boston, MA: Addison-Wesley, 2010.)

[22] Bryant, R. E., MOSSIM: a switch-level simulator for MOS LSI. In Smith, R.-J. II (ed.), *DAC '81, Proceedings of the 18th Design Automation Conference*, Nashville, TN, 29 June–1 July, 1981 (New York: ACM/IEEE, 1981), pp. 786–790.

[23] Buchholz, W., ed., *Planning a Computer System – Project Stretch*. (New York: McGraw-Hill, Inc., 1962.)

[24] Capp, A., *The Life and Times of the Shmoo*. (New York: Simon and Schuster, 1948.)

[25] Cavanagh, J., *Computer Arithmetic and Verilog HDL Fundamentals*. (Boca Raton, FL: CRC Press, 2009.)

[26] Chandrakasan, A., Bowhill, W., and Fox, F., eds., *Design of High-Performance Microprocessor Circuits*. (New York: IEEE Press, 2001.)

[27] Chaney, T. and Molnar, C., Anomalous behavior of synchronizer and arbiter circuits. *IEEE Transactions on Computers* **C-22**: 4 (April 1973), 421–422.

[28] Chao, H. and Johnston, C., Behavior analysis of CMOS D flip-flops. *IEEE Journal of Solid-State Circuits* **24**: 5 (October 1989), 1454–1458.

[29] Chen, T.-C., Pan, S.-R., and Chang, Y.-W., Timing modeling and optimization under the transmission line model. *IEEE Transactions on Very Large Scale Integration (VLSI) Systems* **VLSI-12**: 1 (January 2004), 28–41.

[30] Culler, D., Singh, J., and Gupta, A., *Parallel Computer Architecture: A Hardware/Software Approach*. (San Francisco, CA: Morgan Kaufmann, 1999.)

[31] Dally, W. and Tell, S. The even/odd synchronizer: a fast, all-digital, periodic synchronizer. In *2010 IEEE Symposium on Asynchronous Circuits and Systems (ASYNC)*, Grenoble, 3–6 May, 2010 (New York: IEEE), pp. 75–84.

[32] Dally, W. and Towles, B., Route packets, not wires: on-chip interconnection networks. In *Proceedings of the 38th Design Automation Conference, DAC 2001*, Las Vegas, NV, June 18–22, 2001 (New York: ACM, 2001).

[33] Dally, W. J. and Poulton, J. W., *Digital Systems Engineering*. (Cambridge: Cambridge University Press, 1998.)

[34] Dally, W. J. and Towles, B., *Principles and Practices of Interconnection Networks*. (New York: Morgan Kaufmann, 2004.)

[35] *Data Encryption Standard (DES)*, Federal Information Processing Standards Publication 46, 3 (1999).

[36] DeMorgan, A., *Syllabus of a Proposed System of Logic*. (London: Walton and Maberly, 1860.)

[37] Dennard, R., Gaensslen, F., Rideout, V., Bassous, E., and LeBlanc, A., Design of ion-implanted MOSFETs with very small physical dimensions. *IEEE Journal of Solid-State Circuits* **9**: 5 (October 1974), 256–268.

[38] Earle, J., Latched carry-save adder. *IBM Technical Disclosure Bulletin* **7** (March 1985).

[39] Earle, J. G., Latched carry save adder circuit for multipliers, US Patent 3 340 388 1967.

[40] Eichelberger, E. B. and Williams, T. W., A logic design structure for LSI testability. In Brinsfield, J. G., Szygenda, S. A., and Hightower, D. W. (eds.), *Proceedings of the 14th Design Automation Conference, DAC '77*, New Orleans, LO, June 20–22, 1977 (New York: ACM, 1977), pp. 462–468.

[41] Elmore, W. C., The transient response of damped linear networks with particular regard to wideband amplifiers. *Journal of Applied Physics* **19**: 1 (January 1948), 55–63.

[42] Ercegovac, M. D. and Lang, T., *Digital Arithmetic*. (New York: Morgan Kaufmann, 2003.)

[43] Flynn, M. J. and Oberman, S. F., *Advanced Computer Arithmetic Design*. (New York: Wiley-Interscience, 2001.)

[44] Golden, M., Hesley, S., Scherer, A. *et al.*, A seventh-generation x86 microprocessor. *IEEE Journal of Solid-State Circuits* **34**: 11 (November 1999), 1466–1477.

[45] Hardy, G. H. and Wright, E. M., *An Introduction to the Theory of Numbers*, 5th edn. (Oxford: Clarendon Press, 1979.)

[46] Harris, D., A taxonomy of parallel prefix networks. In Matthews, M. B. (ed.), *Conference Record of the Thirty-Seventh Asilomar Conference on Signals, Systems and Computers, 2003*, Pacific Grove, CA, 9–12 November, 2003 (New York: IEEE), vol. 2, pp. 2213–2217.

[47] Hennessy, J., Jouppi, N., Przybylski, S. *et al.*, MIPS: a microprocessor architecture. In *Proceedings of the 15th Annual Workshop on Microprogramming, MICRO 15*, Palo Alto, CA, 5–7 October, 1982 (New York: IEEE Press), pp. 17–22.

[48] Hennessy, J. L. and Patterson, D. A., *Computer Architecture: A Quantitative Approach*, 5th edn. (New York: Morgan Kaufmann, 2011.)

[49] Hodges, D., Jackson, H., and Saleh, R., *Analysis and Design of Digital Integrated Circuits*, 3rd edn. (New York: McGraw Hill, 2004.)

[50] Horowitz, M. A., *Timing Models for MOS Circuits*. Unpublished Ph.D. thesis, Stanford University (1984).

[51] Huffman, D., A method for the construction of minimum-redundancy codes. *Proceedings of the IRE* **40**: 9 (September 1952), 1098–1101.

[52] Huffman, D., The synthesis of sequential switching circuits. *Journal of the Franklin Institute* **257**: 3 (1954), 161–190.

[53] Hwang, K., *Computer Arithmetic: Principles, Architecture and Design*. (New York: John Wiley and Sons Inc, 1979.)

[54] *IEEE Standard for Radix-Independent Floating-point Arithmetic*, ANSI/IEEE Standard 854-1987 (1987).

[55] *IEEE Standard VHDL Language Reference Manual*, IEEE Standard 1076-2008 (2008).

[56] *IEEE Standard for VHDL Register Transfer Level (RTL) Synthesis*, IEEE Standard 1076.6-2004 (2004).

[57] *IEEE Standard Test Access Port and Boundary-scan Architecture*, IEEE Std 1149.1-2001 (2001), i–200.

[58] *IEEE Standard for Floating-point Arithmetic*, IEEE Standard 754-2008 (2008), 1–58.

[59] *ITU-T Recommendation G.711.*, Telecommunication Standardization Sector of ITU (1993).

[60] Jacob, B., Ng, S., and Wang, D., *Memory Systems: Cache, DRAM, Disk.* (New York: Morgan Kaufmann, 1998.)

[61] Jain, S. and Agrawal, V., Test generation for MOS circuits using d-algorithm. In Radke, C. E. (ed.), *Proceedings of the 20th Design Automation Conference, DAS '83*, Miami Beach, FL, June 27–29, 1983 (New York: ACM/IEEE, 1983), pp. 64–70.

[62] JEDEC, 2.5 V $\pm$ 0.2 V (normal range) and 1.8 V–2.7 V (wide range) power supply voltage and interface standard for nonterminated digital integrated circuits. *JESD8-5A.01* (June, 2006).

[63] Karnaugh, M., The map method for synthesis of combinational logic circuits. *Transactions of the American Institute of Electrical Engineers* **72**: 1 (1953), 593–599.

[64] Kidder, T., *The Soul of a New Machine.* (New York: Little, Brown, and Co., 1981.)

[65] Kinoshita, K. and Saluja, K., Built-in testing of memory using an on-chip compact testing scheme. *IEEE Transactions on Computers* **C-35**: 10 (October 1986), 862–870.

[66] Kistler, M., Perrone, M., and Petrini, F., Cell multiprocessor communication network: built for speed. *IEEE Micro* **26**: 3 (May–June 2006), 10–23.

[67] Ko, U. and Balsara, P., High-performance energy-efficient D-flip-flop circuits. *IEEE Transactions on Very Large Scale Integration Systems* **VLSI-8**: 1 (February 2000), 94–98.

[68] Kohavi, Z. and Jha, N. K., *Switching and Finite Automata Theory*, 3rd edn. (Cambridge: Cambridge University Press, 2009.)

[69] Kuck, D. and Stokes, R., The Burroughs scientific processor (BSP). *IEEE Transactions on Computers* **C-31**: 5 (May 1982), 363–376.

[70] Ling, H., High-speed binary adder. *IBM Journal of Research and Development* **25**: 3 (March 1981), 156–166.

[71] Lloyd, M. G., Uniform traffic signs, signals, and markings. *Annals of the American Academy of Political and Social Science* **133** (1927), 121–127.

[72] McCluskey, E., Built-in self-test techniques. *IEEE Transaction on Design and Test of Computers* **2**: 2 (April 1985), 21–28.

[73] McCluskey, E. and Clegg, F., Fault equivalence in combinational logic networks. *IEEE Transactions on Computers* **C-20**: 11 (November 1971), 1286–1293.

[74] McCluskey, E. J., Minimization of boolean functions. *The Bell System Technical Journal* **35**: 6 (November 1956), 1417–1444.

[75] MacSorley, O., High-speed arithmetic in binary computers. *Proceedings of the IRE* **49**: 1 (January 1961), 67–91.

[76] Marsh, B. W., Traffic control. *Annals of the American Academy of Political and Social Science* **133** (1927), 90–113.

[77] Mead, C. and Rem, M., Minimum propagation delays in VLSI. *IEEE Journal of Solid-State Circuits* **17**: 4 (August 1982), 773–775.

[78] Meagher, R. E. and Nash, J. P., The ORDVAC. In *AIEE–IRE '51, Papers and Discussions Presented at the Joint AIEE–IRE Computer Conference: Review of Electronic Digital Computers*, New York, December 10–12, 1951. (New York: ACM, 1951), pp. 37–43.

[79] Mealy, G. H., A method for synthesizing sequential circuits. *The Bell System Technical Journal* **34**, 5 (September 1955), 1045–1079.

[80] Micheli, G. D., *Synthesis and Optimization of Digital Circuits*. (New York: McGraw-Hill, Inc., 1994.)

[81] Mick, J. and Brick, J., *Bit-Slice Microprocessor Design*. (New York: McGraw-Hill, 1980.)

[82] Montgomerie, G., Sketch for an algebra of relay and contactor circuits. *Journal of the Institution of Electrical Engineers – Part III: Radio and Communication Engineering* **95**: 36 (July 1948), 303–312.

[83] Moore, E. F., *Gedanken Experiments on Sequential Machines*. (Princeton, NJ: Princeton University Press, 1956), pp. 129–153.

[84] Moore, G. E., Cramming more components onto integrated circuits. *Electronics* **38**: 8 (April 1965), 114–117.

[85] Muller, R. S., Kamins, T. I., and Chan, M., *Device Electronics for Integrated Circuits*. (New York: John Wiley and Sons, Inc., 2003.)

[86] Myers, C. J., *Asynchronous Circuit Design*. (New York: Wiley-Interscience, 2001.)

[87] Nelson, W., Analysis of accelerated life test data – part i. The Arrhenius model and graphical methods. *IEEE Transactions on Electrical Insulation* **EI-6**: 4 (December 1971), 165–181.

[88] Nickolls, J. and Dally, W., The GPU computing era. *IEEE Micro* **30**: 2 (March–April 2010), 56–69.

[89] Noyce, R. N., Semiconductor device-and-lead structure, US Patent 2981877, 1961.

[90] O'Brien, F., *The Apollo Guidance Computer: Architecture and Operation*. (Chichester, UK: Praxis, 2010.)

[91] Olson, H. F. and Belar, H., Electronic music synthesizer. *Journal of the Acoustical Society of America* **27**: 3 (1955), 595–612.

[92] Palnitkar, S., *Verilog HDL*, 2nd edn. (Mountain View, CA: Prentice Hall, 2003.)

[93] Patterson, D. A. and Hennessy, J. L., *Computer Organization and Design: The Hardware/Software Interface*, 4th edn. (New York: Morgan Kaufmann, 2008.)

[94] Rabaey, J. M., Chandrakasan, A., and Nikolic, B., *Digital Integrated Circuits – A Design Perspective*, 2nd edn. (Upper Saddle River, NJ: Prentice Hall, 2004.)

[95] Rau, B. R., Pseudo-randomly interleaved memory. In Vranesic, Z. G. (ed.), *Proceedings of the 18th Annual International Symposium on Computer Architecture*, Toronto, May 27–30, 1991. (New York: ACM Press, 1991), pp. 74–83.

[96] Riley, M., Bushard, L., Chelstrom, N., Kiryu, N., and Ferguson, S., Testability features of the first-generation CELL processor. In *Proceedings of 2005 IEEE International Test Conference, ITC 2005*, Austin, TX, November 8–10, 2005. (New York: IEEE, 2005), pp. 111–119.

[97] Rixner, S., Dally, W., Kapasi, U., Mattson, P., and Owens, J., Memory access scheduling. In *Proceedings of the 27th International Symposium on Computer Architecture, 2000* (June 2000), pp. 128–138.

[98] Russo, P., Wang, C.-C., Baltzer, P., and Weisbecker, J., Microprocessors in consumer products. *Proceedings of the IEEE* **66**: 2 (February 1978), 131–141.

[99] Segal, R., *BDSYN: Logic Description Translator BDSIM; Switch-level Simulator*, Technical Report UCB/ERL M87/33. EECS Department, University of California, Berkeley (1987).

[100] Shannon, C. E., A symbolic analysis of relay and switching circuits. *Transactions of the American Institute of Electrical Engineers* **57**: 12 (December 1938), 713–723.

[101] Sorin, D. J, Hill, M. D., and Wood, D. A, *A Primer on Memory Consistency and Cache Coherence*. (San Rafeal, CA: Morgan & Claypool Publishers, 2011.)

[102] Sutherland, I., Sproull, R. F., and Harris, D., *Logical Effort: Designing Fast CMOS Circuits*. (New York: Morgan Kaufmann, 1999.)

[103] Sutherland, I. E., Micropipelines. *Communications of the ACM* **32** (June 1989), 720–738.

[104] Sutherland, I. E. and Sproull, R. F., Logical effort: designing for speed on the back of an envelope. In Séquin, C. H. (ed.), *Proceedings of the 1991 University of California/Santa Cruz Conference on Advanced Research in VLSI*, University of California, Berkeley (Cambridge, MA: MIT Press, 1991), pp. 1–16.

[105] Swade, D., The construction of Charles Babbage's Difference Engine no. 2. *IEEE Annals of the History of Computing* **27**: 3 (July–September 2005), 70–88.

[106] Texas Instruments, *The TTL Data Book for Design Engineers*, 1st edn. (Dallas, TX: Texas Instruments, 1973.)

[107] Tredennick, N., *Microprocessor Logic Design: The Flowchart Method*. (Bedford, MA: Digital Press, 1987.)

[108] Unger, S. H., *The Essence of Logic Circuits*, 2nd edn. (New York: Wiley–IEEE Press, 1996.)

[109] Veitch, E. W., A chart method for simplifying truth functions. In *Proceedings of the 1952 ACM National Meeting*, Pittsburgh, May 2–3, 1952. (New York: ACM, 1952), pp. 127–133.

[110] Wallace, C. S., A suggestion for a fast multiplier. *IEEE Transactions on Electronic Computers* **EC-13**: 1 (February 1964), 14–17.

[111] Weinberger, A. and Smith, J. L., A one-microsecond adder using one-megacycle circuitry. *IRE Transactions on Electronic Computers* **EC-5**: 2 (June 1956), 65–73.

[112] Weste, N. and Harris, D., *CMOS VLSI Design: A Circuits and Systems Perspective*, 4th edn. (Boston, MA: Addison Wesley, 2010.)

[113] Wilkes, M. V. and Stringer, J. B., Micro-programming and the design of the control circuits in an electronic digital computer. *Mathematical Proceedings of the Cambridge Philosophical Society* **49**: 02 (1953), 230–238.

[114] Wittenbrink, C., Kilgariff, E., and Prabhu, A., Fermi GF100, a graphics processing unit (GPU) architecture for compute, tessellation, physics, and computational graphics. In *Proceedings of Hot Chips 22*, August 22–24, 2010 (http://hotchips.org/archives/hot-chips-22).

[115] Yau, S. and Tang, Y.-S., An efficient algorithm for generating complete test sets for combinational logic circuits. *IEEE Transactions on Computers* **C-20**: 11 (November 1971), 1245–1251.

INDEX OF VHDL DESIGN ENTITIES

SUBJECT INDEX

Printed in the United States
By Bookmasters